PASSAGE THROUGH
ARMAGEDDON

ALSO BY W. BRUCE LINCOLN

Nikolai Miliutin: An Enlightened Russian Bureaucrat

Nicholas I: Emperor and Autocrat of All the Russias

Petr Semenov-Tian-Shanskii:
The Life of a Russian Geographer

The Romanovs: Autocrats of All the Russias

In the Vanguard of Reform: Russia's Enlightened Bureaucrats

In War's Dark Shadow: The Russians Before the Great War

Passage through Armageddon: The Russians in War and Revolution,
1914–1918

Red Victory: A History of the Russian Civil War

The Great Reforms: Autocracy, Bureaucracy, and the Politics
of Change in Imperial Russia

The Conquest of A Continent:
Siberia and the Russians

PASSAGE THROUGH
ARMAGEDDON

The Russians in War and Revolution
1914–1918

W. Bruce Lincoln

New York   Oxford
OXFORD UNIVERSITY PRESS

Oxford University Press

Oxford New York Toronto
Delhi Bombay Calcutta Madras Karachi
Kuala Lumpur Singapore Hong Kong Tokyo
Nairobi Dar es Salaam Cape Town
Melbourne Auckland Madrid

and associated companies in
Berlin Ibadan

Copyright © 1986 by Brumar Associates, Inc.
Copyright © 1994 by W. Bruce Lincoln

First published in 1986 by Simon and Schuster
A Division of Simon & Schuster, Inc.
Simon & Schuster Building, Rockefeller Center
1230 Avenue of the Americas
New York, New York 10020

First issued as an Oxford University Press paperback, 1994

Oxford is a registered trademark of Oxford University Press

Library of Congress Cataloging in Publication Data
Lincoln, W. Bruce.
Passage through Armageddon.
Bibliography: p.
Includes index.
1. Soviet Union — History — Nicholas II, 1894 – 1917
2. World War, 1914 – 1918 — Soviet Union. 3. Soviet
Union — History — Revolution, 1917 – 1921 — Causes.
4. Soviet Union — History — Revolution. 1917 – 1921.
I. Title.
DK262.L53 1986   947.08′3   86-3696
ISBN: 0-671-55709-2
ISBN 0-19-508954-5 (pbk.)

The author is grateful for permission to reprint the following material:
Excerpts from Robert Paul Browder and Alexander F. Kerensky, eds., *The Russian Provisional
Government, 1917*, 3 vols. (Stanford, Calif.: Stanford University Press, 1961), reprinted by permission of the publisher.
Excerpts from *Untimely Thoughts* by Maxim Gorky, translated by Herman Ermolaev (1968),
reprinted by permission of the publisher, Paul S. Eriksson.
Excerpts from Tsuyoshi Hasegawa, *The February Revolution* (Seattle: University of Washington Press, 1981), reprinted by permission of the publisher.
Excerpts from Leon Trotsky, *History of the Russian Revolution*, translated by Max Eastman
(Ann Arbor: University of Michigan Press, 1960), reprinted by permission of the publisher.

For Mary, with Love

CONTENTS

PART FOUR: 1917

PART FIVE: 1918

PREFACE

In the pages that follow, I have recounted the tale of the Russians' passage through the shattering experiences of the First World War and the revolutions of 1917. Invaded by foreign armies and threatened by the terrors of civil strife, Russia's leaders mobilized more than fifteen million fighting men between 1914 and 1918 only to find that at least a quarter of them had no boots, rifles, or ammunition. Outgunned and outmaneuvered, desperate generals sent legions of these ragged peasant soldiers across the wastes of no-man's-land to be "churned into gruel," as one observer reported, by the heavy guns of the Germans. More quickly than any other nation's, Russia's casualties soared into the millions. Then, the scourges of starvation and disease joined the enemy's guns to double and treble Russia's human losses. Never in modern history had war so devastated a nation.

Bled white by thirty months of struggle against the armies of Germany, Austria, Bulgaria, and Turkey, Russia's desperate people drove the Romanovs from their throne in February 1917. At that moment, Russians embarked upon a new and daring path as they tried to replace centuries of government by a tsar and tyrannical bureaucrats with statesmen sworn to govern by and for the people. More resilient than any could have imagined, Russians hurried to rebuild their shattered hopes and dreams around an ill-fated experiment with democracy during those late winter days. Yet the Great War refused to release them from its grip, and they had not the strength to break free. Through spring, summer, and fall, the war ground on, and none could count the deaths as hunger, disease, and the enemy's guns continued their fatal sweep across the land. Reduced to living only for today, and not daring to look beyond tomorrow, men and women

11

lost hope, their February visions shrunk to September's struggle to stay alive. Then, in October, Russia's course veered sharply once again. Determined to rule where others had failed, Lenin and the Bolsheviks seized power, only to find that triumph in the long-awaited proletarian revolution brought more crushing burdens than rewards and that immense effort would be required before their oft-repeated promises of peace, land, and bread could come true in a country where soldiers walked away from the war, workers refused to work, and peasants yielded up only at gunpoint the long-hoarded stores of grain that stood between Russia's urban folk and starvation. The Bolsheviks' first efforts therefore brought only more death, more hunger, and more tragedy.

The sources for a book such as this are endlessly varied and incredibly rich, and many of the fascinating tales that I have been obliged to leave untold for want of space deserve more attention. The struggle of Russia's Jews against persecutions more vicious than any before Nazi Germany must someday be recounted in more detail than I have done in these pages, and by someone able to explore the wealth of sources available only in Yiddish. Readers looking for accounts of battles on the Rumanian and Caucasian fronts will find that I have been obliged to confine my attention to the main areas of the conflict on Europe's eastern front, while those hoping to read about the struggles that the national minorities in the empire of Nicholas II waged against their Russian masters during these years will be disappointed not to find them recounted here. Important though they are, the dramas of the Armenian, Georgian, Ukrainian, Polish, Lithuanian, Latvian, Estonian, and Finnish struggles for independence cannot be a part of my story—in large part because they fit better into a history of the civil war and Allied intervention that followed directly after the events set forth here—while a full-fledged account of the complex infighting that drained the energies of so many revolutionaries must remain confined to the many scholarly monographs that have already examined that confusing phenomenon in minute detail.

Still, this abundance of material ultimately has proved beneficial, for in many cases I have been able to let the actors on history's stage tell their tale in their own words. Letters, diaries, memoirs, government reports, military dispatches, testimony given to the revolution's first Supreme Commission of Inquiry, and the Bolsheviks' unceasing efforts to preserve every shred of testimony left by those who supported their victorious revolutionary effort in October, provide a wealth of firsthand accounts that permit the historian to step directly into army headquarters, state council chambers, boudoirs, trenches, and underground revolutionary hideaways of the men and women who shaped the events of this crucial era.

A word of caution should be offered about the matter of dates, where, on occasion, I fear that I may have sacrificed clarity for the sake of historical precision. From 1699 until 1918, Russians reckoned dates according to the Julian calendar which, in the twentieth century, was thirteen days behind the

Gregorian calendar used in the West. Readers therefore need to remember that the dates cited in this book, until February 1, 1918, when Russia's new Bolshevik government adopted the Western calendar, are given according to the Julian calendar. This, of course, means that the Great War began on July 19, not August 1, and that the Bolshevik Revolution occurred on October 25, not on November 7, when its anniversary now is celebrated. In cases where confusion seemed all but certain, I have given both dates—as in July 19/August 1, 1914 —for the reader's convenience. Russian names and place-names pose certain difficulties as well, but I shall spare readers further explanations and simply note that I have kept to the rules I set down in previous books, using Russian transliterations for all but the names of emperors and a handful of people and places especially well known to Westerners.

In times such as these, few historians have the good fortune to be able to pursue the study of Russia's past in archives and libraries halfway around the globe, and I could not have done so had it not been for financial and logistical support so generously rendered by research foundations, academic institutions, libraries, and archives. Among those to whom I am especially grateful are:

—The Academy of Sciences of the U. S. S. R., Leningrad
—The American Council of Learned Societies, New York
—The American Educational Foundation, Helsinki
—Archives des Affaires Étrangères, Quai d'Orsay, Paris
—Archives de la Guerre, Château de Vincennes, Vincennes
—The Bodelian, Oxford
—The British Museum, London
—The Central State Historical Archive, Leningrad
—The Fulbright-Hays Faculty Research Abroad Program, U. S. Department of Education, Washington, D. C.
—The Imperial War Museum, London
—The International Research and Exchanges Board, New York
—The John Simon Guggenheim Memorial Foundation, New York
—The Kennan Institute, Woodrow Wilson Center, Washington, D. C.
—The Lenin Library, Moscow
—The National Endowment for the Humanities, Washington, D. C.
—Northern Illinois University, DeKalb, Illinois
—The Public Records Office, London and Kew, England
—The Russian and East European Center, University of Illinois at Urbana-Champaign, Urbana, Illinois
—St. Antony's College, Oxford
—The Saltykov-Shchedrin Public Library, Leningrad
—The Slavic Library, University of Helsinki, Helsinki

Beyond these, the University of Illinois Library deserves an additional note of thanks. Without access to its outstanding Slavic collection and without the generous help given by Marianna Choldin, June Pachuta Farris, Harold Leisch, Laurence Miller, and Helen Sullivan, my task would have been far more difficult.

I am also especially grateful to Robert Gottlieb, who has played a part in my work on this book from the very beginning as one of those rare literary agents whose virtuosity at balancing enthusiasm, reassurance, and encouragement, and applying them all at the right moment and in proper measure never ceases to amaze me. Simon and Schuster's Herman Gollob has applied himself to this volume in a manner for which I cannot express my full appreciation here. Jenny and Mary Margaret Livingston deserve no small amount of thanks for bearing with the oddity of having someone who works seven days a week suddenly appear in their midst. Most of all, I owe gratitude and very much more to their mother, my wife Mary. Dedicating this book to her can be but a small payment against what is destined always to remain a very large unpaid debt.

W. BRUCE LINCOLN

DeKalb, Illinois
Summer 1985

PART ONE

1914

PROLOGUE

On January 1, 1900, Americans and Europeans greeted the twentieth century in the proud and certain belief that the next hundred years would make all things possible. All facets of their experience supported that conviction, for never had a century changed the lives of men and women more dramatically than the one just past. Steam power, railroads, the telegraph, telephones, electric lighting, mass production, the manufacture of modern steel, and countless other discoveries all had come upon the Western world at dizzying speed to expand the visions of those who stood on the threshold of the twentieth century far beyond their grandfathers' wildest imaginings. More than unfounded optimism thus lay behind the *New York World*'s New Year prediction that the twentieth century would "meet and overcome all perils and prove to be the best that this steadily improving planet has ever seen."[1] Men now confidently believed that control of man and nature soon would lie within their grasp and bestow upon them undreamed-of powers to alter the destinies of peoples and nations. With a sure and certain conviction unknown in centuries past, they proclaimed imperialism as the new century's crusade and prepared to shape the world in the image of the society that had given them birth. Theirs was the White Man's burden, the self-righteous mission to bestow the blessings of Western science, technology, culture, and religion upon the other peoples of the world.

During the brief span that had separated the end of the Franco-Prussian War in 1871 from the death of England's Queen Victoria in 1901, daring young men had set out to bear the fruits of Western civilization to those distant uncharted regions of the world in which their nations' rulers and statesmen had begun to maneuver for influence, position, and power. Cloaked in mantles of humanitarian virtue, and sternly resolved to enlighten mankind and bring glory to their homelands, missionaries, explorers, soldiers, and traders from Europe and North America broke new paths across vast continents. Yet more than dedication to what the French called *la mission civilisatrice* drove them to probe the dark mysteries of Africa and the fabled wonders of the East, for forces far more subtle than man's persistent urge to know the unknown had begun to shape the destiny of nations as the nineteenth century neared its end. From the blast furnaces of Pittsburgh, to the mill towns of England's Midlands, to the ring of heavy industries that surrounded Russia's St. Petersburg, throbbing factories cried out unceasingly for raw materials to fashion into the machines, cloth, and weapons that poured from their production lines. New sources of such materials must be found, and new markets for the finished products secured, if the process were to continue. By 1900, ruin from within and destruction from without faced any nation whose leaders failed in those quests.

In the Western world's competition for vital natural resources, imperial envoys and governors held sway over lands far more vast than those they represented. Tiny Belgium controlled Africa's great Congo, France held huge tracts of mineral-rich lands in Africa and Asia, and England ruled domains that stretched from Hong Kong and Gibraltar to the Yukon and the Cape of Good Hope. As Germany, Russia, Italy, and the United States joined the race for empires, international tension heightened in North Africa, Central Asia, and the Near and Far East. Inevitably, it seemed, the nations of the West must come to blows. Yet a certain balance and discretion still marked their maneuverings. "All were fitted and fastened—it seemed securely—into an immense cantilever," Winston Churchill once explained. "A sentence in a despatch, an observation by an ambassador, a cryptic phrase in a Parliament seemed sufficient to adjust from day to day the balance of the prodigious structure. Words counted, and even whispers," Churchill recalled as he cast a fond, longing glance back from the turmoil of less certain times. "A nod could be made to tell."[2] Time and again the monarchs and statesmen of Europe approached the brink of conflict. Always, they drew back. As it had for thirty years, peace still reigned across the length and breadth of Europe in 1900.

Europe's great nations had forged their awesome power to alter the world's destiny from a fortuitous marriage of science and industry that made it possible to send messages beneath oceans in minutes, move troops across entire continents in a matter of weeks, and destroy foes on a scale unprecedented in the history of mankind. Yet those who reaped the harvest of modern technology found its fruits unexpectedly bittersweet, for a world in which new machines made it possible to manufacture goods and weapons faster, communicate more quickly, and kill more efficiently posed disturbing dichotomies. This brave new world of science harnessed in the service of industry offered broader opportunities for greater success to more people. At the same time, it scarred those who could not reap its benefits with a deeper sense of bitterness because they suffered poverty in an age that had produced more prosperity. Modern science began to teach man how to preserve life through the wonders of sterile surgery and immunology, but, at the same time, increased ten-thousand-fold his ability to kill. Men now seemed on the verge of ruling nature to a degree never thought possible before, but that very fact weakened their faith in God and eroded their belief in a higher, divinely-sanctioned order. To men and women of lessened faith, not God's will but man's desires appeared destined to shape the world in which they lived. With their attention firmly fixed upon the here-and-now, they gave little thought to the eternal concerns of the hereafter.

As Europe entered the twentieth century, almost all of its nations hurried to produce automatic weapons, high-powered repeating rifles, and long-range artillery that could hurl projectiles far beyond the horizon. Anxious to create still more awesome weapons, England and Germany became locked in a race to build fighting ships that could engage foes from beneath the sea's surface, move at speeds undreamed of a generation before, and carry weapons capable of piercing the heaviest armor. Men still spoke of peace but girded more sternly for war, as tranquility and tension, toleration and hatred, certainty and doubt, all stood suddenly juxtaposed in the life of Europe. A passion for military preparedness seized Europe's statesmen as peace hung by the slenderest of threads. "Military technique—how competent in peace to gain war," Britain's foremost military historian once lamented. "And how impotent in war to gain victory!"[3]

Among the conflicting tensions that Europe's tenuous political equilibrium struggled to balance at the beginning of the twentieth century, none proved more potentially explosive than overinflated national pride held in check by the false certainty that mankind now possessed weapons so awesome that a great war simply could not occur. "With the weapons now adopted, the effectiveness of fire presents the possibility of total mutual annihilation," one self-proclaimed prophet of Armageddon warned.[4] At the Hague Conferences of 1899 and 1907, diplomats spoke of limiting the world's armaments, while their nations' industries labored as never before to produce newer, more deadly, more terrifying weapons. Monarchs and statesmen now began to speak of a glorious rendezvous with destiny, yet all remained for the moment secure in their misplaced certainty that others would not fail to draw back from the brink. War, the Anglo-American writer Norman Angell explained loftily in his best-selling book *The Great Illusion,* had been rendered unprofitable (and thus impossible) because the economic destinies of modern nations had become so intertwined that armed conflict threatened victor and vanquished equally. Yet the possibility of war remained far more real, and its dangers infinitely more threatening, than Europe's rulers and statesmen realized. Especially in Russia, poets sensed those dangers more clearly and tried to warn men and women unwilling to heed their preachings of the threat that loomed above them all. In Moscow, the Decadent poet Valerii Briusov clearly heard the "iron tread" of "marching Huns."[5] Four hundred miles to the north in St. Petersburg, his young and passionate rival Aleksandr Blok sensed "the smell of burning, blood, and iron in the air" and thought that Armageddon could not be far away.[6]

In a contradictory world, in which the future seemed equally likely to be painted in brilliance or drenched in blood, men and women no longer cherished the values that had sustained the stability of Europe for so long. Lured by visions of better lives, millions had flocked to new industrial centers during the nineteenth century only to languish in vast urban slums where hunger and disease wasted their bodies and drained them of hope. At times, the anger of these new proletarians had burst forth in violence, but raging mobs rarely wrested economic advantages from stubborn factory owners and resolute rulers. Only when the aristocratic institutions of absolute monarchy gave way to more democratic instruments of parliamentary government across Western and Central Europe could poor men and women challenge their rulers' policies, and only then did Europe's reluctant statesmen respond to the calls for social justice that well-meaning reformers and irate revolutionaries had hurled forth for so long. Like the bittersweet harvest of modern technology, their response proved a mixed blessing, and the concessions that working folk wrenched from these representatives of the old order soon stirred new problems. Without appropriate new spiritual edifices in which to house the new political and social order of late-nineteenth century Europe, many men and women slipped into a deepening moral and spiritual vacuum that made their response to the threat posed by the bellicose policies of their rulers all the more uncertain.

Upon no nation's countenance did these tensions etch themselves more sharply than Russia's. Spurred by the policies of several astute statesmen who rose to prominence under the stern regime of Alexander III, Russia had telescoped an industrial revolution into a single generation and drawn down upon herself all the blessings and curses that the process had long since bestowed upon her Western neighbors. Unable to join the race for empire in Africa, and barred from reaching further into the Balkans by the prohibitions that the Great Powers had set down in 1878 at the Congress of Berlin, Russia in the 1880s and 1890s had turned eastward to consolidate her vast holdings in Asia. No longer satisfied to probe her southeastern frontier and maintain a token presence in her Pacific lands, she had seized the legendary Central Asian Khanates of Bukhara, Khiva, and Samarkand. Even more rapidly than the United States, she had pushed her administrative outposts eastward, moving more than a thousand miles each decade until, with her Asian lands tied to her European domains by the newly-built Trans-Siberian Railroad, she had thrust resolutely into the Far East at the beginning of the twentieth century. There, she provoked the anger of Japan, a nation that had itself just leaped from the medieval to the modern world in a scant three decades. A disastrous war ensued at the beginning of 1904, heightening tensions that soon threatened to rend the fabric of Russian society. In cities and industrial centers all across the empire, working men and women turned against their employers, their government, and their tsar. Especially in St. Petersburg, the threat of discontent grew more ominous as Father Georgii Gapon, a slightly-built young Ukrainian peasant priest whose piercing black eyes seemed able to penetrate the deepest recesses of men's souls, spoke words that stirred the anger in the hearts of the city's proletarians and began to weld them into a formidable instrument for political action.

As defeat in the war against Japan bred discontent in Russia, discontent spawned revolution. On Sunday, January 9, 1905, soldiers guarding the Winter Palace shot down hundreds of peaceful workers and peasants who had come to plead with their young sovereign to lighten their suffering. "Bloody Sunday" ushered in what the Dowager Empress Maria Feodorovna called the "year of nightmares,"[7] a full eleven months of revolutionary turmoil that saw mutinies in Russia's land and sea forces, a nationwide general strike, the opening of the first revolutionary soviets of workers' deputies,* and an armed uprising in Moscow. Many hundreds died in street violence. Thousands more perished when punitive expeditions burned entire villages in a brutal effort to restore order. Only the cruel fact that Russia's peasant army remained willing to shoot their breathren in town and country enabled Nicholas II to survive the turbulence of 1905. Of all the lessons that Russia's ill-fated emperor failed to learn from his nation's revolutionary experiences that year, the extent to which his power depended upon the army's loyalty proved the most dangerous to ignore.

The Revolution of 1905 forced Nicholas II to concede a parliamentary regime to subjects ill-prepared for any experiment with constitutional politics. Inexperienced and naive, Russia's first elected politicians assembled in the Duma, their nation's newly-formed National Assembly, to test the paths of parliamentary government with all the timidity of men unused to political action. Slowly they made their way, but their halting steps, and the clumsy workings of untried parliamentary institutions, offered too little solace to workers and peasants too long abused. In April 1912, conflict again flared between tsar and people when police shot down more than two hundred striking workers in Siberia's Lena goldfields. One after another, workers joined the growing outcry against the policies of the tsar and the great lords of Russia's industry. As scattered strikes grew into large-scale protests, discontent deepened to the point where every second factory worker in the empire took part in some form of labor protest during the year before the Great War broke out.[8] Russia thus approached the Great War in turmoil and uncertainty, with her workers pitted

---

*A unique organization of laboring men and women, the first soviet of workers' deputies in Russia evolved spontaneously from a mass meeting at the great Ivanovo-Voznesensk textile center on May 15, 1905, during which striking workers elected a group of their fellows to carry their demands to the factory owners of the entire region. Later in 1905, striking workers organized soviets in a number of other cities. The one that appeared in St. Petersburg during the revolutionary days of October became the most famous and exercised the most authority because it stood so close to Russia's centers of administrative and political power.

against factory owners, poor peasants arrayed against well-to-do landowners, and her people aroused against their tsar. None among them had a clear vision of how, or in what form, their nation might emerge from these conflicts. "Whither are you rushing in such headlong flight?" Nikolai Gogol, the most tormented of Russia's realist writers, had asked of his homeland more than seventy years before.[9] As the clouds of war darkened Russia's western horizon, no one yet could know the answer.

Uncertain of their course, tsarist statesmen nonetheless remained convinced that the wealth of Russia's resources had given her an important role to play in world politics. Set firmly astride the huge Eurasian continent, Russia embraced fully a sixth of the earth's surface. From Germany's eastern frontier, her domains stretched ever eastward, seemingly endless in their expanse, until, nearly seven thousand miles and eleven time zones away, they reached their limit at the Bering Strait on the other side of the world. The sun never set upon the tsar's colossal empire. As night's first shadows lengthened across its western lands, dawn had long since lit the sky over its Pacific coast. A land of vivid contrasts, of startling and contradictory natural phenomena, Russia embraced Arctic wastes and burning deserts. In the grain fields of the Ukraine, it seemed that a man might never reach the horizon, while it seemed equally certain that he could never emerge from the virgin forests of birch and pine in the far north. The Russian Empire held more than 150,000 rivers, some more than two thousand miles in length. Among its 180,000,000 inhabitants, scores of nationalities spoke more than a hundred different languages and dialects, some of them still unrecorded and without written form. Europeans faced this vast, enigmatic domain with deep forboding. For the Russians themselves, this Russian land held a mystical significance. In it they worshipped God—"the Russian God," they often said—in their particular Orthodox fashion.[10] Their land boasted a thousand-year history, and a dynasty that had reigned for three centuries. For her people, an enthusiastic publicist once wrote, Russia had become "a whole world, self-sufficient, independent, and absolute."[11]

In terms of her natural resources, and particularly of those raw materials needed to fight a modern war, Russia was indeed comfortably self-sufficient, independent, and absolute. While European nations searched the far corners of the earth for the minerals and raw materials their industries demanded, Russia's own lands yielded great quantities of gold, silver, platinum, and precious stones as well as every other mineral then known to man. A quarter of the world's timberlands lay within her borders, as did more than a third of its iron ore, a full half of its coal, and most of its manganese. Since the emancipation of the serfs in 1861, a soaring birthrate had endowed Russia with an unusually large proportion of males under the age of fifty, so that her lands promised to yield up men to bear arms in the tens of millions.[12] Although other European states no longer could feed their people at the beginning of the twentieth century, Russia in 1913 exported well over a hundred million tons of grain, or about five hundred pounds for every man, woman, and child in Europe.[13] Beyond that, her sheer enormity offered a defense against invaders that no European nation enjoyed. In centuries past, those who had dared penetrate her borders had found only death and defeat in the Russian land. As they watched their men perish in the tens of thousands from exposure and starvation, Sweden's Charles XII (in 1709) and Napoleon (in 1812) had learned that Nature herself would destroy any armies reckless enough to be drawn too deeply into the Russian hinterland.

Nonetheless, serious economic, political, and military deficiencies offset Russia's great natural wealth and huge manpower reserves. First of all, her economy had become almost fatally linked with Germany, the very nation that seemed certain to be her major opponent in any European war. Fully a third of Russia's exports went to Germany in 1913, and she relied upon her for nearly half her imports.[14] Yet military planners were so insensitive to this unhealthy connection that, as late as the spring of 1914, a bitter dispute raged in St. Petersburg's War Ministry over whether the French firm of Creusot or the Krupp armaments works in Germany should receive the army's order for new field guns. "We had not even laid down the sound principle that orders like these should be placed only with countries allied to us," the High Command's one-time chief of staff recalled some years later.[15] Even after almost a year of fighting, Russia would still trade with Germany to the tune of almost a million rubles a month.[16]

Although foreign and native experts agreed that the rifles and light field guns of Russia's infantry ranked among the world's best, the antiquated strategies of her conservative planners repeatedly failed to exploit that advantage. Even after most Western strategists had condemned fixed defenses as obsolete, senior Russian artillery officers continued to allocate the lion's share of their nation's military budget to new purchases of gigantic fixed guns for the great frontier fortresses Osowiec, Grodno, Kovno, Ivangorod, and Novogeorgievsk that they had long cherished as a first line of defense against Austro-German attack. Designed to serve a crucial role in siege warfare, such guns were of little use against mobile modern armies that could sweep all too easily around the installations they defended and continue their march into Russian territory.

Not long after he took office in 1908, Russia's war minister, Vladimir Sukhomlinov, marshaled a number of sensible military planners to support his warnings that the heavy field artillery of the Central Powers would render these vaunted fortress defenses useless, but he could neither inspire the confidence nor command the trust that a successful frontal attack against the sacred cows of Russia's defense establishment required. Pudgy and balding, his lidded almond eyes and upturned mustache giving him the look of a mildly startled cat, Sukhomlinov seemed too much the fawning courtier and too little the field commander, and his widely rumored misuse of army funds to support his young wife's passion for extravagant balls, expensive restaurants, and gowns from the best Parisian couturiers left him too vulnerable to resist the demands of his nation's conservative aristocratic artillerymen. He therefore lost his battle for more versatile defense systems, and his defeat left his nation's field armies with almost no mobile heavy artillery as the storm clouds of 1914 drew near. Told not to expect any new heavy guns before 1920, Russia's commanders knew that the weapons of their German and Austrian enemies would outnumber theirs by more than five to one if war came earlier. At the beginning of 1914, Russia's comparative handful of mobile heavy guns had a reserve of only a thousand shells each, while the Germans had stockpiled over three times that number. Worst of all, Sukhomlinov had failed to push through any plan to guarantee accelerated ammunition production if a war broke out.[17]

Such gross failings were not entirely of Sukhomlinov's making. Although notorious for his lack of energy, he faced grievous supply problems in the War Ministry that would have challenged any number of far more able men. Even divisions and corps stationed as a first line of defense along Russia's western frontier suffered acute shortages of clothing, transport, weapons, and ammunition. General Aleksei Brusilov found "only one pair of boots per man, and those in disrepair," when he took command of the XIV Army Corps in 1909. "We had machine guns," he added, "but only eight per regiment, and without carriages, so that, in the event of war, it became necessary to mount them atop carts requisitioned from local farmers. . . . [Even] rifle cartridges and shells for the light artillery," he remembered, "were in very short supply." Other corps commanders faced the same difficulties. Few had mortars; all had machine guns that had to be taken into combat atop farm wagons. None had enough weapons and supplies to send their men into battle fully equipped.[18]

Russia's commanders were plagued not only by shortages of heavy guns, shells, machine guns, bullets, and boots, but also by a primitive and sparsely-laid railroad system which threatened to cripple any forces sent into combat. In 1914, European Russia had only one *versta* * of railroad for each one hundred square *versty* of territory, while Germany had more than ten times that amount of track. Beyond that, Russia had far fewer locomotives and boxcars than did her opponents, her military trains moved at less than half the speed, and her rail system could accommodate only about a third as many trains on any given section of track on any particular day.[19] Russia's future enemies supplemented their far denser rail networks by water transport and, especially in Germany, by well developed systems of paved roads. By contrast, all but a handful of Russia's roads remained unpaved tracks that fall rains and spring thaws turned into knee-deep mud, and her rivers flowed only north and south, while any war against Austria and Germany would require men, weapons, and supplies to be moved east and west.

To complicate matters further, most of Russia's sealanes passed through waters easily controlled by others, and once the Great War began, the German

*A *versta* (plural, *versty*) is equal to two-thirds of a mile.

blockade of the Baltic and the Turks' closure of the Straits left only two seaports open to receive shipments of war materiel from Russia's allies. Neither proved capable of handling the vast quantities of Allied weapons and supplies that Russia's war effort required. More than fifteen hundred miles of narrow-gauge railroad stood between the White Sea port of Arkhangelsk and the front, while Vladivostok, the other open port, lay seven thousand miles further to the east on the northern Pacific. According to the calculations of General Sir Alfred Knox, the dapper and caustic British military attaché who never ceased being appalled by his ally's failings, these two ports together could accommodate just over twenty ships a week, while England's ports dealt with at least a hundred times that number.[20] To transport troops, armaments, and supplies over this meager network of land and water routes posed immense problems for Russia's war effort from the very outset. Worst of all, very few steps could be taken to resolve the problems that shortsightedness and neglect had created. Russia did not even finish laying a modern rail line from the White Sea to the front until the last year of the war.

To add to her problems in any major war, Russia had to mobilize soldiers from a population scattered across eight million square miles of territory. In 1914, each newly-mobilized Russian recruit had to travel almost three times as far as his counterpart in Germany, Austria-Hungary, or France before he could be sent into battle,[21] and it took considerably longer for him to begin the journey because news of the war's declaration reached Russia's citizen soldiers far more slowly. Because only one conscript in three could read and write, newspaper announcements and printed proclamations could not call them to war in Russia as they did in Europe. When the time came, Russia's General Staff had to assemble reservists and draftees by sending local officials to post large red cards on the main streets of tens of thousands of hamlets throughout the empire to announce mobilization to the millions of men who could be told of it in no other way.[22]

Since the time of Peter the Great, illiterate peasants had wielded musket and bayonet for the glory of tsar and country, and had always done so with valor. From such simple but stalwart men, Catherine the Great's legendary Marshal Aleksandr Suvorov had forged the rock-hard infantry that had defeated the armies of Republican France in Northern Italy. Sturdy peasant soldiers had driven Napoleon from Moscow, and their sons and grandsons had defended Russia's Black Sea bastion at Sevastopol during the Crimean War. But it was one thing to train such men to use a flintlock musket, and a very different matter to teach them to fire the complex modern rifles, machine guns, and howitzers that had poured from Europe's great armaments complexes during the half-century since breech-loading weapons had first come into use.

In the comparatively brief span between the Austro-Prussian War of 1866 and the eve of the Great War, Europe's General Staffs had witnessed the greatest revolution in tactical weaponry that the world would see before the nuclear age. By 1914, infantrymen of every modern nation could fire twenty times more ammunition twenty times faster and at ten times greater range than had been possible just fifty years before. Modern artillery now measured its range in tens of kilometers, not hundreds of meters. And there were new weapons so terrible that men had not even contemplated their existence until the technological wizardry of modern industry made their production possible. Almost every European power at the beginning of the twentieth century could mobilize more than a million men, and they had the means to move them rapidly to the front. Never had mankind assembled weapons capable of wreaking such devastation as those which then filled Europe's arsenals. Nations so girded for war grew increasingly restless with peace. A fatalistic acceptance of war's inevitability began to consume many of Europe's monarchs and statesmen.

Nowhere was this more evident than in the writings and public pronouncements of Kaiser Wilhelm II and the men who served him, although few among them spoke as frankly as the German General Staff's hard-eyed chief of military history General Friedrich von Bernhardi. The first officer to ride through the Arc de Triomphe when the Germans had marched into Paris in 1870,[23] Bernhardi had studied history's lessons well and drew from them conclusions about the social, political, and physical value of war that led him to insist upon its "biological necessity." War, Bernhardi explained in his bluntly-phrased *Germany and the Next War*, was "the greatest factor in the furtherance of culture and power," and "an indispensable factor of culture, in which a truly civilized nation finds the highest expression of strength and vitality." Unabashedly, he spoke of "the blessing of war, as an indispensable and stimulating law of development," and claimed that "war . . . evokes the noblest activities of the human nature." Germans had an "urgent duty towards civilization to perform," he insisted, and could "only fulfil it by the sword." As always, France remained the chief enemy. As the "first and foremost condition of a sound German policy," Bernhardi urged, *"we must square our account with France. . . .* France must be so completely crushed that she can never again come across our path." To succeed in her mission, Germany "must not . . . wait until our opponents have completed their arming and decide that the hour of attack has come." In order to "stamp a great part of humanity with the impress of the German spirit," Germany must strike first. "World power or downfall!" Bernhardi proclaimed in 1911, must become Germany's "rallying cry" in the months and years ahead.[24]

Certainly, the kaiser shared Bernhardi's belief in Germany's might and Germany's mission. From the moment he mounted his nation's throne in 1888, Wilhelm II had excelled in arrogance and bellicosity, and his immoderate rantings were widely-known and too often remarked upon. Nearly every European ruler and nation had been the object of his crude disdain at one time or another, and his comments about European politics had earned him the enmity of more statesmen than he could ever know. Wilhelm had publicly offered to rescue England from "the Socialist gang that is ruling the country" by sending several German army corps across the Channel,[25] and he once had urged his soldiers to model their conduct on that of Attila's brutal Huns by insisting that "there will be no quarter given and no prisoners taken!"[26] Neurotically sensitive about his standing among Europe's monarchs, he reacted furiously to imagined slights, and railed against what he thought to be his fellow sovereigns' reluctance to accord the proper measure of deference to Germany and her ruler.[27] Ill at ease with the all too readily imagined insults he encountered in the international arena, Wilhelm took refuge in the more predictable world of the precisely executed maneuvers, splendid dress uniforms, and rousing military parades of his army and navy. At their head, he imagined himself a worthy successor to Frederick the Great, and felt himself in command of a truly great and powerful Germany. That was the calling he thought most fitting for a monarch as great as he considered himself to be, and the one best suited to reminding any people of their duty to king and fatherland. "My advice," he wrote to Nicholas II during the Russian Revolution of 1905, "is more speeches and more parades, more speeches, more parades."[28]

Quick to offer advice to his Russian cousin, Wilhelm stood equally ready to condemn his failings in the bluntest terms. Nicholas, he remarked nastily to Britain's foreign secretary at Queen Victoria's funeral, was "only fit to live in a country house and grow turnips," and the "only way to deal with him was to be the last to leave the room."[29] Such cruel and vulgar remarks cut all the more deeply because they held more than a small grain of truth, for Nicholas II had very little to recommend himself as a monarch and statesman. Usually timid and indecisive, he sometimes could be hopelessly stubborn and petty in defending obviously bad decisions into which he had been coaxed by ill-informed or unfit advisers. Bigoted in his judgments about nations and people, he called the Japanese "little short-tailed monkeys," despised Jews, whom he called *zhidy* (kikes) and told one of his chief advisers that, at heart, the English were no better than *zhidy* themselves.[30] Early in 1906, he drew great satisfaction from the brutal manner in which his army's punitive expeditions suppressed the last vestiges of revolutionary dissent in the Russian countryside. "This really tickles me," he remarked to an aide who brought reports that specially-chosen army units had burned the farms of thousands of peasants suspected of revolutionary sympathies and shot any who tried to flee.[31]

Unlike his brash German cousin, who always gloried in his role as kaiser, the unassertive and inept Nicholas II preferred a simple life with simple pleasures readily enjoyed. His readiness to shun unpleasant encounters and direct personal confrontations may have stemmed in part from memories of that terrible moment in 1881 when, as a shy lad of twelve, he had looked on in horror while Russia's Tsar-Liberator, his grandfather Alexander II, died in agony, his legs shattered by a terrorist's bomb. That day, Nicholas had seen his father, the tallest Romanov sovereign since Peter the Great, speed off "accompanied by a whole regiment of Don Cossacks galloping in attack formation, their red lances shining brightly in the last rays of a crimson March sunset"[32] to begin his reign as Alexander III. Awed by his father's great physical and moral strength, and acutely conscious of his own comparative weakness, Nicholas idolized the stern autocratic *bogatyr* who, like the folk heroes in ancient legends, stood larger than life in his role as Tsar and Autocrat of All the Russias. At times, he vowed to be like him, to rule "as firmly and as steadfastly" as he,[33] but always he knew in his heart that he lacked the strength and the will to do so.

Fearful of failure, Nicholas had stood aside from state affairs while his father had imposed his stern and unyielding will upon Russia. Certain that he had many years to wait for the throne, Russia's heir apparent had preferred tennis, croquet, and mock battles in which he and his cousin Prince George of Greece pelted each other with pine cones to serious thought about the nation he must one day rule.[34] Then, without warning, on October 20, 1894, Alexander III died of nephritis. Fate suddenly had placed Nicholas, the carefree young prince whose neatly-trimmed Vandyke beard made him the mirror image of his English cousin George V, upon Russia's throne. "What is going to become of me?" Nicholas asked mournfully as he prepared to take his father's place. "I never wanted to become [a tsar] . . . I know nothing of the business of ruling. I have no idea of even how to talk to the ministers!"[35]

Less than a month later, Nicholas II married the beautifully regal Alix, a princess from the small German duchy of Hesse-Darmstadt, whose soulful blue eyes, flaming red-gold hair, and finely chiseled features had captured his heart. From the moment she agreed to become his wife in the spring of 1894, Alix had grasped her "Nicky" with a firm, unrelenting love, whose iron grip she cushioned with a flood of romantic musings and sentimental reassurances, while her obsessive sexual passion enthralled and consumed them both. Almost immediately, she invaded his diary, that well-ordered small refuge in which Nicholas had stored his thoughts and recorded his experiences since childhood, all neatly compressed and set down in concise, precise phrases. Firmly, she set herself into its very center, adding to what he had written, and even projecting herself into the future by jotting reminders of her love upon pages he would come to later on. "Your Guardian Angel is keeping watch over you," she promised a few months after their betrothal. Although it was not then clear whether she referred to the celestial or terrestrial variety, Alix had firmly assumed the latter role two decades later. "God will help me being your guardian angel," she assured Nicholas on the eve of his abdication. In words that were all too prophetic of the sad future that awaited them, she wrote: "The past is past, and will never return—the future we know not—and only the present can be called our own."[36]

In Nicholas, Alix found her life's passion and her life's mission. For the next twenty-four years, she dedicated herself as mother, lover, and devoted friend to the prince whose "lonely, pale face with big sad eyes" made her "heart cry out."

In her new homeland, Alix took the Russian name Aleksandra Feodorovna, and, from their very first days as man and wife, this intensely introspective and deeply neurotic woman urged, cajoled, and pleaded with her "darling boysy" to be the autocrat he could never be. Far more resolute than he, and with the firm mind and stern spirit needed to follow difficult courses, she urged Nicholas always to be strong. "Being firm is the only saving," she once assured him. "You are *Autocrat* & they *dare not* forget it," she insisted in one of her letters during the Great War. To him, she said, all men must submit or else he must "crush them all," for he was "the *Autocrat* without wh[ich] Russia cannot exist." "Russia loves to feel the whip," she explained. "Its their nature—tender love & then the iron hand to punish and guide."* Tenderly, yet firmly, she counseled this "softhearted child—wh[o] needs guiding," that he must take comfort from the certainty that "those who fear & cannot understand your actions will be brought by events to realise your great wisdom." She begged him "to heed to what I say, [for] its not my wisdom, but a certain instinct given by God *beyond myself* so as to be your help." At one point desperate for Nicholas to end "the time of great indulgence & gentleness," she begged him to "be Peter the Great, John [Ivan] the Terrible, [and] the Emperor Paul." "How I wish I could pour my will into your veins!" she once cried out. "I am fully convinced that great & beautiful times are coming for yr. reign & Russia," she insisted a scant three months before the revolution.[37]

It would be Russia's tragedy that Nicholas and Aleksandra, who adored their children, treasured the solitude of their family circle, and delighted in the comfortable small joys of bourgeois family life, were as inept at ruling their empire as they were devoted to each other. Acutely flawed in his ability to judge men, Nicholas rarely stood on good terms with the handful of able statesmen who struggled to serve him. More at ease with courtiers of limited intellect and mild disposition, whose willingness to flatter made them untroublesome companions, he allowed himself to be drawn into the circle of fawning mediocrities that his insecure and superstitious empress assembled around them: seers claiming to speak with the dead, quacks promising miraculous cures, and unscrupulous men and women hurrying to advance themselves by preying upon the deep insecurity and gnawing sense of personal inadequacy that lay beneath the icy mantle of disdainful arrogance that Russia's unhappy empress usually wore in public. Such were not the sort of men and women to whom rulers of great and complex empires usually turned for advice or even daily conversation. But as Russia passed the Revolution of 1905 and turned toward 1914, these were the very people from whom Nicholas and Aleksandra drew much of their limited understanding about the land and people they ruled.

---

*Aleksandra's letters to Nicholas, always written in English, are littered with spelling errors, incorrect grammar, and faulty usage. They are quoted, here and throughout this entire volume, precisely as she wrote them.

Among the leading figures in Aleksandra's entourage of shocking nonentities was Philippe Vachot, the unlicensed French charlatan known as "Dr. Philippe," who claimed to communicate with the world beyond the grave. For several years this pseudo-healer, whom the chief of Russia's secret police office in Paris had denounced as an out-and-out fraud, stood in regular attendance upon Russia's rulers to help Nicholas contact his father's departed spirit and to aid Aleksandra in conceiving the heir whose birth she and Nicholas had awaited for so long. Although he failed to find the proper prescriptions, Dr. Philippe at one point so thoroughly convinced Aleksandra of his ability that she hurried to announce the impending birth of an heir only to have to admit her mistake some six months later. Driven from Russia by outraged court opinion, Dr. Philippe nonetheless retained his imperial patrons' friendship and returned to France with their blessings and a recommendation, addressed by Russia's emperor directly to France's president, that he be admitted to the French Academy.[38]

Not long after Dr. Philippe's ignominious departure, Anna Vyrubova, the daughter of the chief of Nicholas's personal chancery, became Aleksandra's closest friend and the central figure in her innermost circle. Vyrubova was easily one of the most obvious mediocrities ever to enter the imperial entourage where her simple-minded and whole-hearted adoration quickly endeared her to the empress. "An ideal grammophone disc" in the words of one who encountered her frequently during the war years, Vyrubova stood ready to receive messages from diplomats and statesmen and repeat them to her imperial patrons, never faltering, but clearly conveying the impression that she understood none of what she reported.[39] One imperial tutor thought she had the "mind of a child," whose "limited and puerile understanding" left her hopelessly adrift in the treacherous political currents that swirled around her. Gullible beyond belief, and even more willing to believe in the pronouncements of charlatans who claimed to speak with God, Vyrubova proved a willing accomplice in her empress's search for spiritual and physical cures. Beginning in 1905, she dedicated herself to her empress and her children with a passion that one of the more sensible members of the imperial household termed "a positive danger because it was uncritical and divorced from all sense of reality." Aleksandra fiercely rejected that opinion of Vyrubova's friendship. Passionately certain that this woman of dull wit and limited intellect had not entered her life by chance, Russia's empress concluded that Vyrubova was a friend sent by God and kept her near at hand until Petrograd's outraged workers and soldiers drove the Romanovs from their throne in 1917.[40]

Perhaps more than any other, Vyrubova knew her empress's and emperor's daily thoughts and shared the greatest of their joys and sorrows that centered around the birth and childhood of the Tsarevich Aleksei. Born on July 30, 1904, Aleksei suffered from hemophilia, an agonizing disease that claimed the lives of many of its innocent victims long before they reached adulthood. Because his blood could not clot properly, the little tsarevich had to be protected at all times from those scrapes and bruises that normal children suffer as a matter of course, else minor bumps turn into great bruises and small cuts into life-threatening hemorrhages. Inevitably, even the best efforts of the imperial household failed from time to time, and at such terrible moments even the best doctors could only stand by in helpless frustration, powerless to stem the flow as the tsarevich's blood seeped away. No medication offered relief, and no physician knew a cure. Frantically, Aleksandra searched the corners of her empire for the man who could accomplish what St. Petersburg's best doctors could not, her frustration and guilt all the more bitter because she knew that it was she alone whose genetic heritage had inflicted such suffering upon her son. Finally, Vyrubova brought to the tsarevich's sickbed a Siberian peasant whose sinister magnetism simultaneously repelled and attracted nearly all who crossed his path. Joyfully, Aleksandra looked on as his strange pronouncements and muttered assurances slowed the bleeding and relieved the pain. "A man of God, Grigorii, from Tobolsk province," Nicholas called him. Soon shocked and dismayed Russians came to know him as Rasputin, a name that, in Russian, conjured up all too vivid images of the man's outrageous public behavior.*

Anxious to conceal the fact that Russia's heir suffered this dread genetic ailment, Nicholas and Aleksandra shrouded his illness in the strictest secrecy, only to have the very nature of the disease work against their best efforts. Emotional stress, doctors learned more than a half-century later, aggravates bleeding in hemophilia's victims.[41] Some hemorrhaging thus was almost certain to occur just before any occasion on which the shy Aleksei had to appear in public. "The emperor and empress are in despair," the tsarevich's tutor Pierre Gilliard confided to his diary on one such occasion. "It is almost always this way when . . . [Aleksei] is supposed to appear in public," he continued. "Fate seems to pursue him."[42]

Such had been said more than once of Nicholas himself. Born on the feast day of Job the Sufferer, May 6, 1868, many thought him to be unlucky. Certainly he would never be fortunate in politics, diplomacy, and war, the fields in which rulers must excel. Alone among Europe's monarchs, he had endured a modern war before 1914, and his armies had been tested in battle against modern weapons. Yet neither Nicholas nor his armies had stood the test well, and the lessons that Europe's statesmen and generals had learned by observing Russia's flawed performance remained lost upon the Russians themselves. The ordeal had begun in late 1903 and early 1904, when Nicholas had sunk into a mood of fatalistic resignation that had caused him to lose all control over events in the Far East. "War is war, and peace is peace," he had told one of his generals plaintively, as he abandoned any effort to control his nation's course, "but this business of not knowing either way is agonizing."[43] Convinced that she had no

---

*In Russian, the noun *rasputstvo* means "dissipation," "profligacy," or "libertinism," while the verb *rasputnichat'* means "to lead a life of debauchery."

other recourse against Russia's expansionist policies, Japan had attacked the naval bastion at Port Arthur to begin the Russo-Japanese War, and Russia endured a series of humiliating defeats at the hands of Japan's army and navy.

Nicholas's failure to perceive the constraints that European and Japanese imperialism placed upon Russian expansion in the Far East thus had brought his empire to war in 1904. The next year, his inability to comprehend the structure of European politics threatened the alliance with France that formed the cornerstone of Russia's security in the West. Ever since General Nikolai Obruchev and France's General Boisdeffre had signed the first military convention between Russia and the Third Republic in August 1892, the two nations had obligated themselves to come to each other's aid if attacked by Germany. Since there seemed little likelihood that Russia and Germany would come to blows given their long history of cooperation, this agreement seemed far more likely to benefit the French, whose long-standing vow to regain their lost provinces of Alsace and Lorraine made war between them and the Germans far more probable. At the same time, Germany had sworn to come to the aid of Austria if that nation's efforts to combat Russia's growing influence among the Slavic peoples of the Balkans, especially in newly independent Serbia, should lead to war. Thus, long-standing Franco-German and Austro-Russian rivalries threatened to bring Russia and Germany, who had been closely allied throughout much of the nineteenth century and remained bound by ties of tradition and trade, into a war that was neither of their choice nor of their making.

Yet, while Russia's alliance with France raised the specter of an unwanted conflict with Germany, it also offered a vital measure of security. If her conflicts with Austria in the Balkans led to war, Russia's only protection against facing the full weight of Germany's awesome power lay in France's guarantee that the Germans must fight her armies in the west if they attacked the Russians in the east. Only too aware of the dangers posed by such a two-front war, Wilhelm had searched for ways to lift that awesome specter from Germany's eastern horizon, but Alexander III and his ministers had proved too wise to relinquish their advantage. Far less clever than his father, Nicholas II might be more easily convinced, or so Wilhelm thought. For more than a decade, however, efforts to entice the tsar into abandoning Russia's French connection came to naught as Russia's ever-wary ministers kept their hapless sovereign from falling into the fatal trap that his German cousin continued to bait in various tempting ways. Finally, in mid-summer 1905, Wilhelm coaxed Nicholas to a meeting at Björkö, a small island off the coast of Finland, at which neither monarch's chief ministers would be present. There, with passionate references to their historic common interests, he hoped to convince Russia's unwise tsar to abandon the alliance with France that kept Russia from falling prey to Germany's tender mercies.

The events surrounding the abortive Treaty of Björkö that tsar and kaiser signed in mid-1905 are amazingly straightforward, and all the more appalling for that reason. At their private meeting the kaiser convinced Nicholas, without prior consultation with France, to sign an alliance obligating Russia to come to Germany's aid if she were attacked by England in Europe. "The morning of July 24, 1905, at Björkö became a turning point in the history of Europe, thanks to the Grace of God," Wilhelm wrote ecstatically.[44] For a moment, he told his stiff and stern chancellor Prince Bernhard von Bülow, it made him think that "[Kaiser] Wilhelm I and Tsar Nicholas I had clasped their hands in Heaven and were looking down with satisfaction upon their grandsons."[45] His mustache slightly upswept to emulate his kaiser's, his winged collar precisely in place as always, Bülow did not entirely share his master's delight, for he knew that Germany could reap little advantage from any treaty confined to Europe. "Russia could not be of service to us against England in Europe," he explained later. "Only if India were threatened would the English be hit in a sensitive spot."[46]

The kaiser's delight only added to the appalled disbelief with which Russia's statesmen received news of the Björkö treaty. Just returned from his brilliantly successful effort to negotiate an end to the Russo-Japanese War at Portsmouth, New Hampshire, Count Sergei Witte read the treaty in dumbfounded amazement. "This really is a dirty trick!" he exclaimed to Russia's foreign minister Count Vladimir Lamsdorf as he read the text of the alliance that Russia's tsar had signed with the bitterest enemy of his closest ally. "Can it be that His Majesty does not know that we have a treaty with France?" Unfailingly business-like, his monochromatic personality diluting the harshness of his words, Lamsdorf insisted that ignorance alone could not explain their emperor's irresponsible action. "His Majesty knows that very well," he told Witte. "Perhaps he forgot about it, or, what is more likely, he didn't grasp the essence of the matter in the midst of all the verbiage with which Wilhelm surrounded the matter."[47] At the combined urging of his angry ministers, Nicholas repudiated the treaty he had signed so thoughtlessly. Russia's alliance with France remained intact, but rumors about the Björkö treaty must have disturbed the men destined to become her allies in the Great War. Almost from the moment they received word of the Great War's first reverses, the French lived in fear that Russia would betray the Allies and conclude a separate treaty with Germany.[48]

While sober statesmen in Russia and Central Europe struggled to keep their sovereigns' clumsy diplomatic efforts from upsetting the carefully balanced structure they had erected over more than three decades, new tensions strained the peace in Europe. As the twentieth century entered its second decade, these tensions shifted away from Europe's traditional centers of power—now sharply polarized between the Triple Entente of France, Russia, and England, and the Triple Alliance of Germany, Austria, and Italy*—to Southeastern Europe and

*In 1904, the British and the French had formed an *entente cordiale* which had enabled them to stand more closely together in diplomatic matters than they had for a quarter of a century. Three years later, England and Russia finally had set aside their century-long antagonism and had signed the Anglo-Russian Convention of 1907. Although England continued to resist being drawn into the Balkans, where someday, Germany's Iron Chancellor Bismarck once had warned, "some damned foolish thing" would set off Europe's next great war.[49]

During the three decades before 1914, the Turks had released their centuries-long grip upon the lands and peoples that had once made up the European portion of their empire, there by creating a political vacuum that had drawn a number of small emerging Balkan nations toward its center. Dedicated to realizing the fullest dimensions of their nationalist dreams, and anxious to avenge long centuries of Ottoman oppression, these nations now turned to try their strength against their former masters. In 1912, Bulgaria, Serbia, and Greece joined Montenegro to wrest the last of the Turks' Balkan territories from their grasp, only to find themselves at odds about how to divide the spoils. Apprehensively, Europe's diplomats looked on as squabbles among the victors burst into a second Balkan war, in which Rumania helped Serbia and Greece to stave off the surprise attack that Bulgaria launched against them at the end of June 1913.

What diplomatic historians have called "one of the shortest and fiercest conflicts in modern history"[50] ended in less than six weeks with Bulgaria being stripped of all the lands she had gained in the previous year's war against the Turks, but the peace that the belligerents signed at Bucharest on August 10, 1913, offered little solace to Europe's nervous diplomats. None expected Bulgaria to acquiesce for very long in a settlement that deprived her of access to the sea and denied her territories she had long thought hers by right. Nor did the reactions of the war's victors promise the men who spoke for Europe's Great Powers any greater comfort. "The first round is won," Serbia's volatile prime minister Nikola Pašić announced as he and his allies triumphantly divided the spoils of the Second Balkan War. "Now we must prepare for the second against Austria."[51]

During the last quarter of the nineteenth century, Austria's Habsburg rulers had moved greedily to add new lands to their empire as the Ottomans had retreated from Southwest Europe and the Balkans. Beginning in 1878 at the Congress of Berlin, at which Europe's Great Powers had obliged the Russians to relinquish some of the spoils they had seized in their victory over the Turks during the previous year, the Austrians had gained the right to "occupy and administer" Bosnia, even though it remained nominally under Ottoman suzerainty. Exactly thirty years later, the Austrians had annexed it directly, despite the outcries of outraged Serb nationalists who insisted that Bosnia must inevitably become theirs as part of Serbia's effort to gather the South Slavs into one Iugoslav nation.

Serbia had vowed to seek vengeance. Their irredentist dreams inflamed

---

any formal military alliance such as Russia and France had signed in 1894, nonetheless, the three powers had begun to act together in European affairs from 1907 onward, thereby opposing to the older Triple Alliance (based upon the 1879 military alliance between Germany and Austria to which Italy had been admitted in 1882) a newer, although somewhat looser, Triple Entente.

further by the Austrians' refusal to relinquish the South Slav lands of Croatia and Slovenia that lay immediately to Bosnia's north, Serb patriots swore that Austria must be driven from the Slavic lands of Southeastern Europe just as the Turks had been driven from the Balkans. None hated the Austrians more passionately than the handful of Serb officers who formed the secret society *Ujedinjenje ili Smrt* (Unification or Death) in 1911. Bound by stern oaths sworn upon dagger, pistol, and crucifix, these men promised death to any who betrayed their cause and war against all who stood with Austria against their dream of a greater Serbia. Assassination became their main weapon, secrecy their chief protection. Dedicated to fighting the Austrians at every turn, they offered training, weapons, and a safe haven in Serbia to any who wished to fight in Austrian-held lands for their dreams of South Slav unity. None proved more eager to accept their help than the young terrorists of *Mlada Bosna* (Young Bosnia), a revolutionary group of students and intellectuals dedicated to driving the Austrians from Bosnia. Given weapons and training by the men of *Ujedinjenje ili Smrt, Mlada Bosna* planned the crime that drove Europeans from their chancery tables to the battlefields of Belgium, France, East Prussia, and Galicia.[52]

On Sunday, June 15/28, 1914,* Austria's heir, the Archduke Franz Ferdinand, and his Czech wife, the Duchess of Hohenburg, celebrated their fourteenth wedding anniversary in Sarajevo, the ancient capital of Bosnia and a volatile hotbed of *Mlada Bosna* loyalties. That same day, South Slavs celebrated *Vidov Dan,* the Festival of St. Vitus, to commemorate the moment when, just hours after the Turks had defeated the South Slav armies at the battle of Kossovo in 1389, a young hero had stridden into the sultan's tent and assassinated him. Patriotic passions seethed in the city's coffee houses and cafés, for more than a few of those young Bosnians who had ample reason to hate their Austrian rulers could be expected to recall how one of their forebears had won eternal glory by striking down a foreign oppressor. As if to stir the fires of anti-Austrian passions further, Franz Ferdinand had spent the previous day observing the nearby maneuvers of the Austrian army, an ever-present reminder to the Bosnians of the Austrian yoke that still weighed upon them. Almost inevitably, the event for which so many Bosnians had longed, and which Europe's sober statesmen had so feared, occurred. Shortly after noon, as the archduke's motorcade passed through the city, Gavrilo Princip, a Bosnian university student who had pledged his life to his countrymen's long struggle against the Austrians, stepped from the crowd that stood near the Lateiner Bridge and fired two shots pointblank into Franz Ferdinand and his wife, killing them both within minutes.[53]

All across Europe, people reacted sharply to the Sarajevo crimes. Convinced that Princip had enjoyed the backing of the Serbian authorities, enraged Austri-

---

*The first of these dates is according to Russia's Julian calendar, while the second reflects the Gregorian calendar used in Europe.

ans demanded revenge. "There is ground to regard almost all sections of the population as being just now blindly incensed against the Serbians," Britain's ambassador warned his superiors from Vienna, as he reported that even "many persons holding quite moderate and sensible views on foreign affairs" were demanding that their armies strike "such a blow [against Serbia] as will reduce the country to impotence for the future."[54] As in earlier times of crisis, sensible statesmen urged their governments to pursue a sensible course. Deeply apprehensive, Germany's ambassador urged his Austrian hosts to be prudent, while, from London, Britain's unimaginative foreign secretary, Sir Edward Grey, his closely-clipped mustache as stiff as his words, spoke the pious hope that Austria would handle the crisis "in such a way as not to involve Europe in its consequences."[55] This time, cooler heads could not prevail. Insisting that "Austria-Hungary cannot let the challenge pass," the Habsburgs' chief of staff, General Franz Conrad von Hötzendorff, proclaimed that his nation must "choose between letting itself be strangled and making a last effort to defend itself."[56] Ready for war with Serbia, Austria would accept no other course.

Whatever her feelings, Austria could not retaliate against Serbia without the full support of Germany, the dominant partner in the Triple Alliance. To the misfortune of Europe, the kaiser stormed ahead, tossing all counsels of moderation aside. "It is high time a clean sweep was made of the Serbs," he raged, his better judgment now drowning in a flood of bellicose self-righteousness. "Serbia must be disposed of, and that right *soon!*"[57] "Now we'll settle scores with Serbia,"[58] Austria's foreign minister gloated the moment he held Germany's "blank check" in hand, while in St. Petersburg, Russia's foreign minister, Sergei Sazonov, hastened to warn his government that the kaiser's thoughtless belligerence, when combined with Austria's outrage, promised nothing less than "a European war."[59]

Quickly, events moved toward their terrible denouement, as the abyss of the Great War opened before the nations of Europe. Carefully choosing a moment on July 10/23 when France's president Raymond Poincaré was en route from St. Petersburg to Paris and briefly out of touch with the Russians, the Austrians presented an ultimatum to the Serbs that left them to choose between war or the humiliation of allowing Austrians onto their soil to assist in investigating Serbians' role in the Sarajevo assassinations. "Well, there is nothing to do but die fighting," Serbia's minister of education Ljuba Jovanović remarked to his colleagues when the Serbian cabinet discussed the document that evening.[60] All that could be done now, they concluded, was to give way on a number of lesser points so as to win European public opinion to their side. For that purpose, they spent nearly two days drafting a reply. Then, at 5:45 p.m. on July 12/25, just fifteen minutes before the Austrian deadline, Serbia's prime minister Nikola Pašić delivered his nation's reply. With scarcely a glance at its contents, Austria's ambassador Baron Giesl pronounced it unacceptable. Both fearful that the other might gain even an hour's advantage, Austria and Serbia raced to mobilize their armies. Within hours, the sound of heavy gunfire crashed across the Balkans.

Could the fires of war that seared the Balkans in the predawn hours of July 13/26 be kept from spreading? That remained the last hope of Europe's statesmen as they hurried to quench the flames. For a moment, war and peace appeared to hinge upon whether Nicholas II and his generals would order mobilization or dare to stand aside while Austria's armies assembled along Russia's southwestern frontier. It seemed almost certain that they must take the former course, for Russia had already entered what her *General Staff Gazette* called a "premobilization period" on July 12/25. This, Germany's High Command insisted, could be counted as nothing less than mobilization, as they demanded that Russia cease all further preparations or face a full mobilization by the kaiser's armies. Yet to do so Russia had to risk allowing Austria's mobilized armies to stand unchallenged, for her General Staff had prepared no plan for ordering troops to her southwestern borders without sending them to Germany's frontiers as well. "Any partial mobilization," General Iurii Danilov later confessed about this incredible gap in his nation's war plan, "would have been an improvisation at best," and would have seriously threatened Russia's ability to launch a general mobilization successfully if that were needed later.[61]

After four days of torturous indecision, Nicholas ordered general mobilization on July 16/29, only to withdraw his order moments later. "Everything possible must be done to save the peace," he insisted to one of his aides. "I will not be responsible for a monstrous slaughter."[62] Irate, aghast, and, above all, fearful that quickly mobilized German armies might overwhelm Russia's frontier defenses, Nicholas's generals stormed against his hesitation. "Mobilization is not a mechanical process that one can arrest at will, as one can a wagon, and then set it in motion again," Sukhomlinov warned, as he insisted that Russia's armies needed every precious moment to prepare for Germany's first assault.[63] Still Nicholas delayed, absurdly hopeful that the blustering kaiser might still mediate the crisis, while the now-desperate Sukhomlinov hurried to enlist the support of the General Staff's chief Nikolai Ianushkevich and the archconservative Duma president Mikhail Rodzianko. A favorite with the emperor, with far greater talents as a courtier than a commander, Ianushkevich warned that "war has already become inevitable and we are in danger of losing it even before we have time to draw our sword from its sheath," while the pessimistic Rodzianko self-importantly threatened Nicholas with the wrath of the Russian people, who would "never forgive any delay that might throw the nation into fatal complications."[64] Together, these men prevailed upon Foreign Minister Sergei Sazonov to speak with the tsar at the Romanovs' summer palace at Peterhof the next afternoon. Using every bit of diplomatic finesse he could muster, Sazonov insisted that they must act. Either they must "unsheath the sword to defend our vital interests and await the enemy's attack with weapons in hand . . . or cover ourselves with everlasting shame, turn away from battle, and throw ourselves upon [the Germans' and Austrians'] mercy."[65] Sadly, Nicholas agreed. His last words to Sazonov that day were: "Give my order for mobilization to the chief of the General Staff."[66]

Awed by what lay ahead, his first elation already tinged by fear, Sazonov telephoned Ianushkevich the moment he left his emperor's room. Now the General Staff could give the orders for mobilization, he told him, adding that "then you can smash your telephone."[67] Without a moment's hesitation, Ianushkevich ordered the hard-working General Sergei Dobrorolskii, who headed the General Staff's Mobilization Section, to have the tsar's order countersigned by the ministers of Internal Affairs, War, and the Admiralty, as the law required. At six o'clock that evening, July 18/31, 1914, the machines in St. Petersburg's Central Telegraph Office flashed a brief message to all parts of the empire:

HIS IMPERIAL MAJESTY ORDERS: THE ARMY AND NAVY TO BE PLACED ON WAR FOOTING. TO THIS END RESERVISTS AND HORSES TO BE CALLED UP ACCORDING TO THE MOBILIZATION PLAN OF THE YEAR 1910.[68]

Within hours, the never-before-seen red cards summoning men to mobilization points appeared on signposts in cities, towns, and tiny villages all across Russia. The Great War now stood but a scant hair's breadth away.

Only if Germany dared leave her armies unassembled in the face of Russia's general mobilization along her eastern frontier could war still be avoided. Yet such a course threatened perils greater than any German ruler dared contemplate. Ever since General Count Alfred von Schlieffen had drafted his famous plan for a two-front war in the 1890s, Germany had based her defense in the east upon the premise that Russia's great distances, undeveloped rail system, and ponderous bureaucracy would slow her mobilization to a comparative snail's pace. If Germany fought only a holding action in the east during the early weeks of a general war, Schlieffen reasoned, she then could throw the weight of her armies against France. Belgium formed a vital link in Schlieffen's strategy, for he planned to attack France through Belgium while tempting the French to invade Alsace and Lorraine, the provinces they had lost to Germany in 1870. Such a plan required armies to advance at great speed with machine-like precision. Brussels must be taken no later than the nineteenth day of the war, the Franco-Belgian border crossed not more than three days later, and the occupation of Paris and France's final defeat accomplished by the war's thirty-ninth day. In less than six weeks, Schlieffen intended to turn the armies of Germany back toward the east, to carry the war to the Russians before the tsar's armies had mobilized and crossed the frontier.[69]

In Schlieffen's plan, every day counted, for every hour that Germany's armies fell behind schedule gave the Russians that much more time to assemble their forces for a massive assault into East Prussia. Even a brief Russian headstart at mobilization therefore posed risks to Schlieffen's timetable that were too grave for Germany's statesmen and generals to accept. War would be a virtual certainty if Russia persisted in mobilizing. Acting on very precise orders from his kaiser, Count Friedrich Pourtalès, Germany's rigidly dutiful ambassador to St. Petersburg, delivered an ultimatum to Foreign Minister Sazonov at midnight on July 18/31: Russia must halt her general mobilization by noon on the following day. For decades, the cleverness of great statesmen had kept European affairs in balance, but men now seemed suddenly to lose all control over the events upon which the fate of Europe's nations hinged. "We could foresee the course of events," Sazonov remembered with sadness, "[but] we were powerless to change them."[70]

Pourtalès was even more helpless than Sazonov. A diplomat who believed unerringly in his kaiser and his reich, Pourtalès had slept very little during those quiet nighttime hours when Russia's telegraph began to summon her reserves to arms. Now, as the last moments of peace that Europe would know for 2,383 days slipped away, he made ready to return to Russia's Foreign Office to receive Sazonov's response.

From the front windows of his nation's newly-built embassy on St. Isaac's Square, Pourtalès could see the beginning of Morskaia Street, its slight arc bending to the left as it crossed the Nevskii Prospekt to disappear beneath the great arch that bisected the building housing Imperial Russia's General Staff Headquarters, her Ministry of Finance, and her Foreign Office. From this one vast edifice flowed what Peter the Great once had called the "sinews of war" —strategy, diplomacy, men, and treasure—that had enabled Russians to emerge victorious from all but two of her wars in the two hundred years that had passed since Peter's time. As evening approached on July 19/August 1, Pourtalès knew that the brief journey he was to make down the Morskaia would be a mission of war, for his embassy's telegraph room had just received two carefully encoded declarations and accompanying instructions from Berlin. If the Russians refused to accept the kaiser's ultimatum, Pourtalès must declare war; if they proposed further negotiations in a last effort to preserve the peace, he must declare war still. Only an abject admission of Russia's "pitiful dependence upon the arbitrary will of the Central Powers," Sazonov later wrote, could have saved the peace, and Pourtalès, an ardent collector of Renaissance bronzes and marbles who had spent his life thinking that a great war could never come to Europe, now had to face its certainty when he returned that evening for Sazonov's reply.

"From his very first words," Sazonov remembered, Pourtalès "demanded to know if the Russian government was prepared to give a satisfactory answer to the ultimatum he had presented the evening before." Sazonov replied that Russia hoped for a peaceful solution, but that his government could not accept a German ultimatum. Visibly shaken, Pourtalès repeated his question twice more. Each time, Sazonov gave the same reply. Pourtalès then took a folded sheet from his pocket and thrust it forward, his hand shaking with the emotion he could not restrain. "In that case," he concluded, "I am instructed by my government to give you this note." Now overwhelmed by the frightful consequences of the duty he had just performed, Pourtalès leaned against the window that looked out on the towering Alexander Column and the rococo splendor of the Winter Palace beyond, and wept. "Who would have thought," he asked over and over, "that I would ever have to leave St. Petersburg under such conditions?"[71]

History had veered onto a new and terrible course. "Russia is ready!" Sukhomlinov had written not long before. "If you want peace, be ready for war. Russia . . . wants peace, but she is ready."[72] Now the war minister's claim would be put to the supreme test. Except for Italy, who quickly declared neutrality, none of the great nations of Europe remained at peace by midnight, July 22/August 4, 1914. "So the die is cast!" France's ambassador to St. Petersburg, Maurice Paléologue, confided to his diary. "The part played by reason in the government of peoples is so small that it has taken merely a week to let loose universal madness!"[73]

CHAPTER I

"It Is a Wide Road That Leads to the War"

What France's deeply saddened ambassador Maurice Paléologue called "world madness" began in Russia, as wars so often begin everywhere, with an outburst of patriotic fervor and outraged national feeling as Petersburgers greeted Germany's declaration with hysterical outpourings of support for their tsar and country. The next day, July 20, several thousand senior army officers, courtiers, ladies-in-waiting, and high officials assembled in the great Nicholas Hall, the largest in the Winter Palace, to hear their emperor announce Russia's declaration of war. Scene of the Romanovs' great balls and celebrations for almost a century, the hall center now held an altar upon which stood the ancient icon of the Virgin of Kazan, taken from Kazan almost four centuries before by the conquering armies of Ivan the Terrible. Hidden away during the Time of Troubles, brought back to Moscow to celebrate its liberation from the Poles in 1612, and moved to St. Petersburg by order of Peter the Great to commemorate his victory over the armies of Charles XII at Narva, this image of the Holy Mother had become the symbol of national unity and hope to which Russians turned in the direst hours of their nation's history. Before its age-darkened image, the great Marshal Kutuzov had prayed for victory as he made ready to face Napoleon's Grand Army in 1812.

Now, as Russians once more confronted the terrible specter of war, the imperial court, led by Nicholas and Aleksandra, prayed to the Virgin of Kazan for help and comfort. "Faces were strained and grave," a young grand duchess recalled as she described how "hands in long white gloves nervously crumpled handkerchiefs, and [how], under the large hats fashionable at that time, many eyes were red with crying. Men frowned thoughtfully," she continued, "shifting from foot to foot, readjusting their swords, or running their fingers over the brilliant decorations pinned on their chests."[1] Was it the start of a new beginning, the birth of that longed-for national regeneration that might unite tsar and people as Nicholas dreamed? Or was it, as others feared, the beginning of the end?[2]

As the ambassador of Russia's only full-fledged ally among Europe's Great Powers, and, therefore, the only foreign representative present, Paléologue witnessed the entire scene. By far the best schooled in Russian history, language, and culture of any ambassador ever to serve at the Romanovs' court, Paléologue watched with deep fascination, recording the details in the same sensitive manner as he would report events in Russia's capital throughout the Great War. "Nicholas II," he remembered, "prayed with a deep fervor, that gave his pale face an intensely mystical expression, [while] Aleksandra Feodorovna stood near him, her breast thrust forward, head held high, lips crimson, gaze fixed, and eyes glazed. Every now and then," he added later that day, "she shut her eyes and then her livid face made one think of a death mask."[3] As the mass neared its end, Nicholas approached the altar. In a low voice, but with great feeling, he vowed before God and his people never to accept defeat. "I hearby solemnly swear that I shall never make peace so long as a single enemy soldier remains on our soil," he promised. "Through you, the representatives of my dear Guards and Petersburg garrison here assembled, I greet my entire army, united as it is in body and spirit, standing firm as a wall of granite, and give it my blessing."[4] Reminiscent of the great oath sworn by Alexander I in 1812, these proud words stirred an outburst of true patriotism and pride among his listeners. For almost ten minutes they cheered. As their voices soared from the hall's sixteen great windows that had been thrown open to relieve the summer heat, the crowd on the Palace Embankment took up their cries.

From the Nicholas Hall, Russia's tsar moved to the balcony that overlooked the great Palace Square where, on January 9, 1905, his soldiers had shot down hundreds of unarmed workers when they had marched, icons held aloft, their voices raised in hymns, to beg their *batiushka*—their "Little Father," as they called the tsar in those days—to take pity on their plight. This cruel tragedy now seemed forgotten as almost a quarter-million Russians assembled around the famous column that Nicholas's great grandfather had erected in the 1830s to the memory of Alexander I, the conqueror of Napoleon.[5] When Russia's emperor and empress appeared on the palace balcony, the crowd knelt as one. A few moments later, the well-known words of Russia's national anthem burst from a quarter-million throats: "God save the Tsar, / Sovereign and all powerful! / Reigning in glory, / For the glory of us all!" As one, their voices rose in a mighty crescendo of supplication, calling once again upon their "Russian God" to stand with them in their hour of need. Gazing down from one of the palace windows, the Grand Duchess Maria Pavlovna thought the scene expressed "almost sublime patriotism."[6] To Paléologue, it seemed as if "the tsar really was the autocrat appointed by God, the military, political, and religious leader of his people, the absolute sovereign of their bodies and souls,"⁷ while Sir Bernard Pares, a young British expert on Russia who stood on the Square itself that day, described the feeling as one of *"sobornost,* [a Russian word] typifying the religious unity which may be felt in a great cathedral or at any moment when the community is everything and the individual only has significance as a member of it."⁸ In the space of a few hours, Russia's tsar and people seemed almost miraculously reunited for the first time since their bitter parting during the revolutionary days of 1905. They now seemed to share a common will. "It was the only time that I ever remember when Russians . . . set themselves to their tasks with precision and zest," Maria Pavlovna recalled. "The vagueness of purpose, which is one of our national characteristics, had, for the moment, disappeared."⁹

Two days later, the crowd turned its passion upon the Germans, its hatred every bit as intense as its patriotism. Completed only two years before, the German embassy in St. Petersburg stood as a monument to all that was tasteless in Wilhelm II's Second Reich. A man of carefully-chosen preferences and refined taste, Paléologue thought it "abominable as a work of art" and found the "vulgar and gaudy eloquence" of its massive red granite walls and pillars offensive. Less sophisticated in their judgment, Petersburgers reacted far more violently to this "powerfully symbolic" building that their nation's enemy had erected on the southwestern side of their city's most important square. On the evening of July 22 a mob stormed the embassy while indulgent police allowed them to smash its windows, destroy its furniture and paintings, and hurl Ambassador Pourtalès's priceless collection of Renaissance bronzes and marbles onto the square below. As a final gesture of contempt, the crowd dragged down the colossal bronze horses and naked giants that adorned the embassy's roof and added them to the wreckage that littered the square, while Russia's minister of internal affairs and the newly appointed mayor of St. Petersburg stood only a few feet away.¹⁰

So often critical of the government, many of the intelligentsia now rallied to their nation's cause during the days of euphoria that Russia enjoyed in the late summer of 1914. War caught Andrei Belyi, Russia's renowned prophet of the apocalypse, in Switzerland where he grandly announced that "the catastrophe of Europe and the explosion of my personality are one and the same event." Belyi still looked much as he had during his student days in Moscow, when he had always seemed "not to walk but to fly," his "azure-enamel" eyes set beneath what one friend described as "extremely thick, marvelous eyelashes that shaded them like fans." Now, his inner torments having driven him near the brink of madness, he saw himself as a bomb destined to destroy Russia's enemies. "I am a bomb," he insisted, as he elaborated upon the theme he had developed two years earlier in his strange and brilliant novel *Petersburg,* a bomb "hurtling to burst into pieces."¹¹

While Belyi immediately tied the war and Russia's salvation to his inner torments, young Vladimir Maiakovskii, the politically suspect Futurist poet who became an ardent artistic spokesman for Lenin's regime half a decade later, proclaimed that "the Vandal fiends have robbed Russia," and cursed the Germans for having trod upon centuries of Europe's cultural achievements. "The rings from the treasures of Liège are on the fat, beer-sodden fingers of Prussian Uhlans, and candy-filled bakers' wives sweep the streets of Berlin with petticoats of Brussels lace," he wrote in disgust in the days before his outrage at the failings of Russia's generals and statesmen overwhelmed his patriotism.[12] Nor were Maiakovskii and Belyi exceptions. Most Russian writers and artists turned to support the war in August 1914, and only a few dared confess their doubts. "Everyone has gone out of their minds," lamented Zinaida Gippius, the poet and high priestess of one of Petersburg's most famous salons, whom Belyi once had called "a wasp in human attire" during the days when he had found her cold beauty captivating and repelling at the same moment. "Why is it," Gippius asked her diary, as she filled its pages with fearful doubtings, "that, in general, war is evil but this war alone is somehow good?" This middle-aged woman, who still set off her green eyes and red-dyed blond hair with dresses cut to fit her body like a second skin, stood firmly against her countrymen's passion to go to war. "War is war," she insisted, "and to war I say: it is always wrong, and it is never necessary, not now and not ever."[13]

Like their counterparts all across Europe, politicians and statesmen in Russia chose not to think in Gippius's terms, and readily put from their minds the obvious fact that, as Paléologue wrote in his diary, "every great war has stirred a profound crisis of conscience among the Russian people."[14] Instead, they took to heart the emperor's plea that "disputes about domestic issues be forgotten, that the union of tsar and people be strengthened, and that all Russia stand united to repel the criminal attack of the enemy."[15] Resolved to unite with all political foes for Russia's defense, Vladimir Purishkevich, leader of the extreme right wing in the Duma, publicly embraced Pavel Miliukov, chief of the powerful Constitutional Democrats, or Kadet Party, whose liberal defense of parliamentary government and constitutional monarchy had stirred the hatred of Russia's ultraconservatives ever since the Revolution of 1905. A renowned historian turned politician, whose political views had been shaped by years spent in studying Russia's past and in observing constitutional politics as a lecturer at the University of Chicago and in Britain, Miliukov replied effusively in kind.[16] "Despite our attitude toward the government's policy, our first duty is to preserve the integrity and unity of our country, and to defend her position as a world power," he insisted in the Kadets' influential newspaper, Rech. "Let us remember well," he added, "that, at this moment, our first and only task is to support our soldiers, inspiring them with faith in the rightness of our cause, with calm courage, and with hope for the triumph of our arms."[17]

Soldiers, not politicians, intelligentsia, or even factory workers, held the first key to victory or defeat. But what did Russia's soldiers think about the war? And what were the feelings and thoughts of the peasant masses whence most soldiers came? None of the men who spoke so ardently in support of the war knew for certain. "In the depths of rural Russia, eternal silence reigned," Miliukov remembered when he recalled the awesome events of mid-1914,[18] while Russia's imperturbable prewar minister of finance Vladimir Kokovtsev once warned a foreign journalist that as soon as one traveled more than a hundred *versty* from any of his country's larger cities, all political issues simply vanished into utter silence. Throughout July and August, while Russia's city folk raucously swore allegiance to tsar and country, this silence continued to blanket the countryside to form an impenetrable invisible barrier between statesmen and the men and women they governed. None knew its meaning, nor could they. The peasantry, the peasant-centered Socialist Revolutionary Party's mercurial leader Viktor Chernov once insisted, had always been "the sphinx in the political history of Russia."[19]

The riddle of this peasant sphinx thus remained the most dreadful of the terrible imponderables that Russia's leaders faced in 1914. Nicholas felt certain, as he told his children's tutor Pierre Gilliard, that "there will now take place a [national] movement in Russia like that which occurred during the great war of 1812,"[20] and that Russia's peasants would stand firmly behind their tsar in defense of the empire's greater glory just as peasant partisans had united behind Alexander I to drive Napoleon from their land. This was typical of the acutely flawed political judgments that Nicholas made throughout his reign. The simple truth remained that no one knew how Russia's peasants might react to the stresses of a modern war, and the tsar's ministers could do little more than utter hopeful and dangerously meaningless platitudes about the peasantry's "age-old devotion to its homeland," and praise for its "limitless devotion to the Sovereign Emperor."[21]

In reality, the brutal riflefire of Bloody Sunday had shattered the peasantry's belief in their "Little Father," and none but the most insensitive could imagine that their devotion remained unchanged. Beyond that, Russia's peasants identified themselves not with their nation but with the province or district in which they lived. "We are *kalutskie*"—"we are natives of Kaluga province"—Miliukov remembered some of them saying when they heard that war had been declared, which meant that the war was of no real concern to them so long as it did not come to Kaluga.[22] Peasants from Tambov, Orel, or any other of Russia's provinces, all responded to outside events in similar fashion. They cared nothing for the politics of France, or Germany, or Austria, or Serbia, and had no interest in the German kaiser or the assassinated Austrian Archduke Franz Ferdinand. Nor could they have been much impressed when Nicholas and his ministers changed the name of St. Petersburg (thought by many to be of German origin but, in fact, derived from the Dutch) to the more Slavonic-sounding Petrograd at the end of August.[23] This gesture catered to the anti-German feelings of the city's educated residents, and appealed to foreign diplomats as "a political demonstration and as a protest of Slav nationalism against Germanic intrusion."[24] For Russia's rural masses, however, the change had no meaning. For them, their capital continued to be "Piter"—a name comparable to "Frisco" in the American jargon—as it had since the time of Peter the Great.

For centuries, Russia's peasants had called all foreigners *nemtsy,* a noun derived from the verb that means "to become dumb," and they perceived no difference between Germans, Austrians, Frenchmen, Englishmen, or anyone else who was not Russian. "Almost no one knew what a Serb was, and they were almost as much in the dark about what a Slav was," General Aleksei Brusilov wrote about his soldiers. "They did not even know that such a country [as Germany] existed. . . . Soldiers knew . . . even less about Austria and they hardly even knew what their own homeland was like."[25] Foreign people, territories, and politics thus had no place in peasants' narrowly-defined, uncomplex world. "A Tambov peasant is willing to defend the province of Tambov," Russia's first wartime chief of staff General Nikolai Ianushkevich once explained, "but a war for Poland [where much of the fighting took place in 1914 and 1915], in his opinion, is foreign and useless."[26] In this, the Russian common soldier stood a world apart from his British counterpart who, although nearly as ignorant of the war's causes and the enemy to be fought, had no hesitation about leaving his country in 1914 to fight for his king. "I'm a'goin' to fight the bloody Belgiums, that's where I'm a'goin'," was the way one Tommy reservist explained the purpose of his mobilization.[27]

Whether it was to be of long standing or brief duration, Nicholas was naturally and irresistibly drawn to the new union of tsar and people that seemed to emerge in his empire's few large cities in August 1914. Before the Duma assembled to offer more promises of national unity at a special one-day session on July 26, Nicholas invited its members to the Winter Palace, and patriotic words again echoed through the great Nicholas Hall as the tsar asked for its members' support. "Your people are ready to fight for the honor and glory of their homeland," its unceasingly pompous president Rodzianko replied. "The Russian people are with You. Firmly placing their trust in the mercy of God, they will find no sacrifice too great until the enemy has been overcome and the honor of our Motherland defended."

From the Winter Palace, the Duma deputies drove along the Neva embankment to the Taurida Palace, where they held a formal session to announce their support for the war. In Paléologue's words "a skeptical old man, laden with labors, honors, and experience," President of the Council of Ministers Ivan Goremykin insisted that Russia must not draw back from Germany's challenge. "The enemy's troops have entered the Russian land," Foreign Minister Sazonov announced portentously, his pale face accentuating his nervousness. "We are fighting for our Motherland, for her dignity and her position as a Great Power. We cannot tolerate the dominion of Germany and her ally in Europe." Politicians of all persuasions vied to find immoderate phrases with which to add their approval. "We set no conditions and we ask for nothing," announced Miliukov, whose Kadet Party usually headed the opposition against the government in the Duma. "We simply are placing on the scales of war our firm will to achieve victory." Only five Bolshevik deputies openly opposed the war, while the rest of the socialists followed the lead of the fiery young left-wing lawyer Aleksandr Kerenskii in supporting the government in a defensive war.[28] "The Duma was completely worthy of the occasion," Nicholas told Pierre Gilliard the next day. "It truly expressed the will of the nation, for all Russians resent the insults the Germans have heaped upon them."[29]

Russia's urban masses seemed not far behind the Duma in standing with their tsar as they hastened to abandon their protests against factory owners and government and declare their support for the war. During the first seven months of 1914, disgruntled Russian workers had launched wave after wave of strikes and demonstrations in the empire's industrial centers to demand higher pay, shorter hours, and better working conditions. On the ninth anniversary of Bloody Sunday, a quarter-million workers had marched in widespread demonstrations and, on May Day, an international socialist holiday, that number had doubled. More than a million and a quarter workers had been on strike at some point during the first half of 1914, and their protests had been especially energetic in St. Petersburg, center of the nation's armaments and shipbuilding industries. Early in July, barricades had appeared in the city's Vyborg District, and there had been other serious strikes before the middle of the month in Moscow, Kolomna, Kharkov, Kiev, and Tver. On several occasions, police had fired upon strikers.[30]

In the space of a few days, all this appeared to change. At city mustering stations, a full 96 percent of the draftees summoned to arms answered their government's first call.[31] Grieving women accompanied their husbands for a last farewell, little knowing that most would be dead before winter's end. One young couple who crossed his path on the road between St. Petersburg and Tsarskoe Selo especially touched Ambassador Paléologue. "She was very young, with delicate features and a fine neck, a red and white scarf knotted around her fair hair, a blue cotton sarafan drawn in at the waist by a leather belt, and she held an infant to her breast," he remembered. In admiration, France's ambassador looked on as the couple "spoke not a word, but gazed fixedly at each other with mournful, loving eyes," as they shared their last precious moments together while an "endless file" of ammunition wagons, baggage carts, ambulances, and field kitchens crawled past them.[32]

Hundreds of thousands of times during the next few weeks Russia's peasant men and women repeated similar scenes in countless variants. Too ready to discern the romantic pathos in these partings, Paléologue observed only a nation in arms prepared to sacrifice life and happiness for tsar and country. Far less romantic, his British colleague, Sir George Buchanan, thought of the difficult and bloody future ahead. "What would be the feelings of these people for their 'Little Father,' " Britain's hard-headed ambassador asked, "were the war to be unduly prolonged?"[33] Would Russia's masses support a war then? As always, the war's burden fell especially heavily upon the peasants, for only one out of every twelve draftees in August 1914 came from a city, even though a seventh of all Russians had moved to towns and cities by that time.[34] For the moment, peasants seemed ready to answer their government's call, although there were instances in which regular army troops had to use gunfire to put the mobilization order into effect.[35] How long the peasants would continue to march unresistingly to war no one dared predict. If ever they refused to do so, Russia's war effort must inevitably collapse.

The war was not yet two weeks old when Nicholas, Aleksandra, and their children went to Moscow where, Ambassador Buchanan said, "the heart of Russia voiced the feelings of the whole nation."[36] At the tsar's invitation, Buchanan and Paléologue joined a host of temporal and spiritual Russian lords assembled in the Kremlin Palace's Hall of St. George, resplendent in white and gold furnishings and dedicated to the knights who wore Russia's highest order for bravery. Beneath the hall's six great chandeliers, each with more than five hundred brightly-lit electric bulbs, Nicholas told the representatives of Russia's ancient capital that he had come among them to renew his strength through prayer. "From this place in the very heart of the Russian land," he announced, "I send my warmest greetings to my valiant troops and my brave allies. God is with us!"[37]

With the tsar and his family and relatives at its head, the procession moved majestically through the palace's Vladimir Room into the Sacred Vestibule and onto the famous Red Staircase that descended to the courtyard below. Built in the 1470s, and roofless since the fire of 1696, the Red Staircase had witnessed many great events in Russia's past. There, Ivan the Terrible had received a messenger from the traitorous Prince Kurbskii, betrayer of Russia to the Poles in the Livonian War. From those same stairs a century later, rebellious palace guards had thrown the Naryshkins, relatives of Peter the Great's mother, onto the upraised pikes of their fellows in the courtyard below. All Romanov sovereigns had passed down the Red Staircase to be crowned in the Assumption Cathedral, and Nicholas and Aleksandra had done so themselves in 1896, when life had looked promising and Russia's future had seemed so bright.

As they descended the Red Staircase once again, sadness unforeseen at their coronation blighted even the joy of the reception their Muscovite subjects bestowed upon them. This time no grand dukes marched in the imperial train as they had eighteen years before, because they had gone with the army to the front. Still "as elegant and seductive as in the old days before her widowhood, when she still had inspired profane passions,"[38] Aleksandra's sister, the Grand Duchess Elisaveta Feodorovna, now wore the garb of a nun; nine years before, after a terrorist had thrown a bomb made of metal fragments and nitroglycerine into her husband's carriage, she had abandoned worldly pleasures to devote herself to good works. None could forget that Russia was about to embark upon the test of arms that men both longed for and feared. This was, "in essence, [to be] a war for peace," a Russian commentator explained, a war in which "total victory" by the Allies would bring "the establishment of eternal peace" for all the world.[39]

From the foot of the Red Staircase, the procession moved to the Cathedral of the Assumption. Paléologue remembered that "Holy Moscow . . . with her sky-blue domes, copper spires, and gilded bulbs, glittered in the sun like a fanciful mirage." Here, the Kremlin—"a curious conglomeration of palaces, towers, churches, monasteries, chapels, barracks, arsenals, and bastions," Paléologue said—personified Russia in all her complexity. "This incoherent juxtaposition of sacred and secular edifices, this complex combination of fortress, sanctuary, seraglio, harem, necropolis, and prison, this blend of sophisticated civilization and archaic barbarity; this violent contrasting of the crudest materialism and the most exalted spirituality—are they not the whole history of Russia," he asked, "the entire epic of the Russian nation?"

Inside the cathedral, three metropolitans, twelve archbishops, and over a hundred bishops waited to pray with their sovereigns. "A fabulous wealth of diamonds, sapphires, rubies, and amethysts in unheard of profusion gleamed on the brocade of their mitres and chasubles," Paléologue wrote. "At some moments," it seemed, "the church was lit by a supernatural light."[40] The mass began, chanted by one of those profound bass voices for which Russian choirs are so justly famous. Nicholas and Aleksandra paid homage to the cathedral's famous relics, including a crucifix reputedly inset with a fragment of the true cross. Then the great doors swung open and the imperial family, accompanied by Buchanan, Paléologue, and the grandees of court and city faced the deafening cheers of crowds that thronged the Cathedral Square. "Stand closer to me, *Messieurs les Ambassadeurs,*" Nicholas said generously. "These cheers are as much for you as they are for me." Together the two ambassadors returned to their hotel, conscious of the historic moment they had witnessed and the historic ground upon which they had seen it.[41]

It was, of course, no great accomplishment to win popular acclaim for a war when its battles had yet to be fought and the dying had not yet begun. In retrospect, legions of memoirists and commentators "remember" their fears that many who marched away seemed marked for death, but there is more hindsight than present-mindedness in their accounts. Countless tearful leave-takings occurred all across Russia and Europe during those days, and in peasant villages, where survival depended upon the labor of able-bodied men, families parted from their young men with a particularly deep sense of loss,[42] but it was those who stayed behind who bore the regret, not those who marched away. As in other countries, the flower of the nation's youth marched off to war expecting adventure and glory, not suffering and death. "Honour has come back. . . . And

Nobleness walks in our ways again," the romantic young English poet Rupert Brooke wrote as he greeted the war's coming,[43] while Aleksandra insisted that the war had "lifted up spirits, cleansed the many stagnant minds, brought unity in feelings, and is a 'healthy' war in the moral sense."[44] In 1914, every nation in Europe hailed the Great War's beginning as a means to the greater end of lasting peace.

In August 1914, Russia's soldiers set forth to face horrors that almost none could imagine. A few particularly thoughtful officers on the General Staff worried that the combined armies of Germany and Austria-Hungary might prove more than Russia's match, but even they quieted their fears with the conviction that the forces of their French and English allies could right the balance. Yet France and England could hope to withstand the Central Powers' assault only if the Russians disrupted the precisely arranged timetable of the Schlieffen Plan and forced the German High Command to divert men and weapons from their lightning advance in the west at precisely the moment when the British and French brought the maximum weight of their forces to bear. Russia's commanders therefore must launch an offensive into East Prussia before the Germans struck their decisive blow in the west, and that meant that their armies must be mobilized with much greater speed than the German High Command had thought possible.

The need for speed in the east became all the more pressing as the kaiser's armies smashed through Belgium into northern France. With what people in those days thought appalling brutality, Germany's soldiers applied Clausewitz's dictum that terror, judiciously and consistently directed against noncombatants, could effectively shorten a war. Well before the end of August, Andenne, Seilles, Tamines, Visé, and Louvain, home of a priceless library that held hundreds of medieval manuscripts and more than a thousand incunabula, lay in ruins. At Andenne, Germans massacred 110 civilian hostages, at Tamines 384, and at Dinant 612.[45] Desperately, France and England appealed to Russia to open the offensive that would draw extra German divisions to the east and relieve the threat against Paris. "I beg Your Majesty to order your troops to take the offensive immediately," Paléologue said at an audience with Nicholas on July 23. "Otherwise, the French army faces disaster . . . [and] the entire weight of Germany's forces will turn upon Russia."[46] As the Germans drew nearer the Marne, the French called upon Russia to make good the promise, given two years before in defiance of all common sense and caution at a joint meeting of their General Staffs, that she would launch an offensive into East Prussia with at least 800,000 men no later than M-15, the fifteenth day of her mobilization.[47] Although such an early offensive invited disaster, the Russians nonetheless felt honor-bound to keep their word. "It is our duty to support France in view of the great stroke prepared by Germany against her," Chief of Staff Ianushkevich insisted a few days later. "This support must take the form of the quickest possible advance against Germany . . . in East Prussia."[48] The first task that faced the supreme commander of Russia's armed forces thus was to launch an offensive that might well fail in order to save the crumbling armies of France.

Although the Fundamental Laws of the Russian Empire made Nicholas the supreme commander of Russia's armed forces, his chief ministers, and Foreign Minister Sazonov especially, urged him not to exercise that aspect of his imperial authority at the war's beginning. "It's to be expected that we may be obliged to retreat during the war's first weeks," Sazonov reportedly warned then. "Your Majesty ought not to expose yourself to the criticism that such a retreat would be bound to cause."[49] Many expected War Minister Sukhomlinov to assume supreme command in the tsar's stead, but on July 20, after some hesitation, and perhaps even after offering the position to Sukhomlinov, Nicholas bestowed command of Russia's armies upon his uncle, Grand Duke Nikolai Nikolaevich.[50]

A dashing cavalryman with a precisely-trimmed graying beard, six-foot-six-inch Grand Duke Nikolai Nikolaevich cut an imposing figure. Still straight as a ramrod at fifty-seven, he radiated all the aura of command that his nephew so sadly lacked. For this, and because he had no use for her treasured faith healer Rasputin, Aleksandra despised him. "I know him to be far from clever," she warned her husband. "Having gone against a man of God [Nikolai Nikolaevich had threatened to hang Rasputin if he ever dared set foot in his headquarters], his work cannot be blessed or his advice good."[51] Almost no one else shared her view. One acquaintance of many years thought the grand duke "somewhat short-tempered," and a number of senior officers found him brusque, even harsh,[52] but none besides Aleksandra saw the hand of God in his failings. Among the most charming of men when he chose to be so, Nikolai Nikolaevich readily exercised that gift upon ambassadors, foreign correspondents, and military observers who came to his headquarters. England's Bernard Pares reported that "his carriage and speech communicate confidence," and admitted to being charmed when the grand duke assured an assembly of war correspondents that "the Press, in competent and worthy hands, can do an enormous amount of good."[53]

Although charm and tact could win the hearts of men, strategic brilliance and logistical genius were needed to win battles, and these were the very elements of military command in which Nikolai Nikolaevich had the least expertise. A junior officer in the imperial *Chevaliers Gardes* during the Russo-Turkish War of 1877–1878, he had become inspector-general of the cavalry during the Russo-Japanese War, a post that had kept him well away from any serious fighting. Utterly inexperienced in wartime command, the grand duke now took control of armies whose numbers soon rose beyond five million. Urgently in need of battle-tested army commanders and experienced senior staff officers to advise him, he committed his first great blunder as supreme commander when he allowed his nephew to interfere in vital appointments rather than demand the men he needed at his Supreme Headquarters, or *Stavka* as the Russians called it.

Nikolai Nikolaevich first considered for chief of staff at *Stavka* General Fëdor Palitsyn, chief of Russia's General Staff from 1905 to 1908 and a long-time student of German military institutions. Skilled in administration, Palitsyn had impressed the grand duke by his efforts to slash administrative fat from Russia's bloated military budget in the wake of the Russo-Japanese War, even though his stubborn refusal to favor the aristocratic cavalry and fortress artillery in budget disputes had earned him the undying enmity of the army's aristocratic Old Guard.[54] Unlike many senior officers, Palitsyn had not failed to draw the proper conclusions from the army's failures in the Far East and he understood how his country's land forces must be reshaped if they hoped to fight a war in Europe. His weakness was that he sometimes spoke in unfocused generalities,[55] but he nonetheless had much of the organizational experience and administrative skill that Nikolai Nikolaevich required so desperately at *Stavka.* But strong forces stood in Palitsyn's way. None other than the tsar himself asked—in fact, ordered—that General Nikolai Ianushkevich be appointed as *Stavka's* chief of staff. Even though Ianushkevich had scarcely three months' experience as chief of Russia's General Staff to recommend him, Nikolai Nikolaevich acceded to his nephew's ill-informed choice. "I did not think that I had the right to oppose the Sovereign," he later confessed to a friend. "Let history condemn me for that."[56]

The tsar's first ill-advised meddling with his supreme commander's appointments thus brought Russia into the Great War with an inexperienced chief of staff who had served in the army for less than twenty years and had yet to hear a shot fired in anger. Rumor had it that Ianushkevich owed his success to winning the tsar's favor during a brief tour as captain of a palace guard detail when he had exercised all his arts as a courtier. Forty-four years old, General of the Infantry Ianushkevich had not commanded so much as a battalion in the field. "An extremely nice man, but of limited intelligence and a poor strategist," General Brusilov thought, while General Sir Alfred Knox, the British military attaché who worked with him almost every day, thought that he "gave the impression rather of a courtier than of a soldier." The appointment of Ianush-kevich, whom any number of observers considered "still a child" in practicing the art of war, was "of fatal consequence" for Russia's early war effort, Brusilov concluded.[57]

While Ianushkevich's elevation had, as one observer wryly remarked, "excited general surprise," most experts agreed that General Iurii Danilov, *Stavka's* quartermaster-general, was "the hardest worker and the strongest brain on the [grand duke's] staff."[58] Sometimes called the "von Moltke of the Russian army,"[59] Danilov had drafted a new war plan—Plan No. 19—in 1910 that shifted the main thrust of Russia's attack from Austria-Hungary to Germany, by far the more formidable of the Central Powers. Certain that the more destructive mobile weapons perfected during the first years of the twentieth century had made the strategies of even the previous decade obsolete, Danilov had demanded speed and mobility from Russia's armies and insisted that his nation's land forces abandon the great fortresses of Novogeorgievsk, Ivangorod, Osowiec, Grodno, and Kovno upon which tsarist engineers had lavished so much imprudent attention. Although these had guarded the river barriers of northern and central Poland for a century, such defenses now did little more than tie down tens of thousands of men and consume tens of millions of rubles, Danilov insisted, now that Germany had the means to level them with long-range artillery or to bypass them altogether. Warning that "the fate and future of Russia will be decided on her western frontier,"[60] Danilov had urged that Russia prepare to throw all of her resources into battle in East Prussia, Poland, and Galicia in the event of an European conflagration, even if it meant leaving her Asiatic frontiers momentarily undefended.

Like Palitsyn, Danilov had collided head-on with hide-bound artillerists who, in 1909, earmarked nearly a third of Russia's five-year army budget to purchase hundreds of immense new fortress guns. Certain that Russia's military expenditures should be devoted to modernizing the weapons and equipment of her land forces at a time when the first motorized transport vehicle had yet to be purchased, Danilov had rejected the arguments of his traditionalist foes and had won Sukhomlinov's agreement to raze the fortresses in which Russia's senior strategists so unwisely had placed their confidence.[61] Yet neither Danilov nor Sukhomlinov were destined to have the last word. Headed by the chiefs of the Kiev and Warsaw Military Districts, conservative opinion on the General Staff and in the Imperial Duma had urged "revisions" of Danilov's historic Plan No. 19 in 1912 that once again concentrated the bulk of Russia's forces against Austria-Hungary instead of East Prussia and strengthened the very fortresses Danilov had hoped to destroy. In a dangerous compromise that had left Russia's attacking forces on her northwestern and southwestern fronts too weak to score decisive victories, Revised Plan No. 19 stipulated that two of Russia's armies remain positioned to invade East Prussia, while another five were assigned to launch her main assault against Austria-Hungary.[62]

Bitter disputes over Danilov's Plan No. 19 had reflected a much broader split among the army's senior officer corps that would cripple Russia's military efforts in 1914. Even in the early twentieth century, Russia's army remained an instrument of social mobility because its officers received noble status, and this made military careers especially attractive to men of modest origins. After the abolition of serfdom in 1861, peasants had used this avenue to higher position so successfully that more than a third of the Russian officers below the rank of colonel in 1900 came from plebian backgrounds. The old aristocracy still dominated the cavalry and artillery. But, on the eve of the Russo-Japanese War, one out of every four new infantry officers came from a peasant family, and one out of five came from some other lower-class background.[63] Sharp social divisions between patrician and plebian therefore sharpened the historical antagonisms that had flourished between branches of Russia's armed forces. Cavalry and artillery thus remained strongholds of traditionalism, while parvenue infantry officers stood on the side of innovation.

To create a constituency for the reforms that Russia's army so urgently required, War Minister Sukhomlinov had pitted himself against the staunchly pro-aristocratic cavalryman Nikolai Nikolaevich. As both won the loyalty of dedicated followers, Russia's senior officer corps split into "traditionalists" and "Sukhomlinovites," whose bitter conflicts fatally weakened the army's fighting abilities.[64] Thus, Russia went to war in 1914 with her war minister and supreme commander still bitterly at odds, and their followers in such hopeless turmoil that army commanders sometimes refused to communicate with each other or even their own chiefs of staff if they stood in different camps! Northwest Army Group commander Iakov Zhilinskii and his Second Army commander General Aleksandr Samsonov both supported Sukhomlinov, while First Army commander General Pavel Rennenkampf resolutely favored the grand duke. Rennenkampf refused to deal personally with his chief of staff, a Sukhomlinov partisan with whom he dealt only in writing, and he communicated with Zhilinskii in such cryptic terms that his chief sometimes found it difficult to discover where First Army was or what it was doing. Relations between Samsonov and his chief of staff were nearly as bad for the opposite reason, while Ninth Army commander General Lechitskii, a staunch Sukhomlinovite, stubbornly refused even to speak to his chief of staff because the man had supported Nikolai Nikolaevich.[65] Hopelessly torn by such divisive personal loyalties, these men launched Russia's August offensive into East Prussia.

Some twenty years later, General Nikolai Golovin confessed that he found it "difficult to understand the rash of haste in the opening of the campaign [of 1914]," for it seemed absurd in retrospect that men should have rushed to the first battles of a war that soon immobilized millions of soldiers in static trench warfare.[66] Yet, unable to see even dimly into the future in 1914, the military chiefs of Europe had envisioned a brief war, where victory would go to the nation whose generals mounted the most powerful offensive most quickly. Haste therefore became an important element in the war strategies of every Great Power in the West, where frontiers lay only a day's journey from mustering points that stood only a few hundred kilometers from reservists' homes. Confident of their mobilization plans, commanders expected to launch their first offensives within days, and to storm across enemy frontiers no later than a fortnight after the war began.

While Germany, France, England, and Austria all expected to have their great armies mobilized within a fortnight (Germany, in fact, invaded Luxembourg on the third day of the war), Russia's generals realized that they must begin their offensive with only part of their army in the field. With active and reserve forces scattered over more than eight million square miles of territory serviced by the sparsest of rail networks, the authors of Revised Plan No. 19 had thought in terms of weeks, not days, and anticipated that Russia's army would reach its full mobilized strength of 5,300,000 only after about three months. At best, senior staff officers hoped to have most of their cavalry and about a third of their infantry and artillery units in the field by M-15, the fifteenth day of mobilization, with another third ready by M-30. Crack army corps trained to peak efficiency in the rigorous conditions of the Caucasus, Turkestan, and Western Siberia needed sixty days to reach Russia's northwestern and southwestern fronts, while those from the remote regions of Eastern Siberia required still a month more.

One in ten of these soldiers would come from the *opolchenie*—a Home Guard that General Golovin once said must never "be considered a real [military] force"—and these would be armed only with obsolete single-shot 10.67-mm Model 1877 Berdan Rifles that their fathers and grandfathers had carried into the last Russo-Turkish War.[67] Nonetheless, Grand Duke Nikolai Nikolaevich ordered an advance against Germany on M-15 with the expectation that the troops of Russia's First and Second armies would cross the frontier on M-18 and M-20, even though they would not reach their full strength until M-36 and M-40. Unacquainted with the specifics of Revised Plan No. 19, because he had originally been assigned only to defend St. Petersburg as the commander of Sixth Army, Nikolai Nikolaevich left the mobilization's details to the diligent and dedicated General Dobrorolskii, chief of the General Staff's Mobilization Section, who had fifteen days in which to assemble all the men, horses, fodder, supplies, weapons, and ammunition needed for the 744 infantry battalions and 621 squadrons of cavalry that made up Russia's First and Second armies.[68] A man who worked far better than the system he directed, the indefatigable Dobrorolskii nearly accomplished the impossible; 60 percent of First and Second armies' cavalry and more than three-quarters of their infantry had reached full combat readiness before they crossed the frontier into East Prussia.[69]

For the moment, Russia had men to spare, but problems related to weapons, communications, supplies, and training posed crushing difficulties for Dobrorolskii and his fellow generals from the first day of the war. Russia's first-line infantry carried the 7.62-mm Model 1891 Mosin rifle, still regarded by experts as one of the best infantry weapons then in existence, but her arsenals turned out to hold over a million fewer of these rifles than the number of men called to the colors that fall. Although army organization tables assigned eight machine guns to each infantry regiment, Russia's arsenals could supply only six, and some of those, as Brusilov had complained five years earlier, had no mounts. Not even taking into account the losses that would occur once the fighting started, War Ministry experts estimated that Russia's factories needed at least a year to produce the missing rifles and machine guns.[70]

Based on their experience in the Russo-Japanese War, Russian ordnance officers had estimated that a reserve of one thousand cartridges per rifle and seventy-five thousand cartridges per machine gun would supply their troops for several months of fighting.[71] They therefore calculated that Russia's arsenals needed to have about three and a half billion rifle and machine-gun cartridges in reserve before she embarked upon any major war. But cartridge production required time and money, both of which were in short supply, and ordnance officers never managed to obtain anything close to the quantities of ammunition needed to establish even the minimal reserves they thought necessary. Russia therefore entered the Great War with ammunition reserves that stood more than a billion cartridges below the minimum set by her own ordnance experts.

Nor was the supply of artillery shells any better. Because all European general staffs had underestimated the number of shells that fighting would consume in 1914, every nation endured a "shell crisis" early in the war, but Russia suffered most because her General Staff experts had planned far more meager reserves than had her enemies. Desperately short of heavy field guns, Russia's field commanders found that their ordnance depots held only 75 percent of the minimal shell reserves that had been promised. That Russian arsenals actually stored a few more shells than the number expected for the army's 76-mm light cannon offered little comfort, especially when beleaguered commanders tried to substitute that inadequate weapon for heavy artillery against attacking German divisions. Such failings meant that German and Austrian heavy guns outnumbered those of the Russians by a crushing margin of four to one, and the disparity grew even greater as the shell crisis worsened. Soon, Russia's armies would meet German artillery barrages with silence because they had no ammunition left to fire.[72] Nor could battered units coordinate their escape from the firestorms unleashed by the German artillery, for Russia's army had entered the Great War with almost no modern communications systems.

Nineteenth-century armies had clashed upon well-defined battlefields on which combatants fought within sight or sound of each other. Generals surveyed the battle and issued their orders from vantage points not far from where their troops struggled and died, and they did so in relative safety, for the effective range of most weapons at that time rarely exceeded two thousand meters. Not until the Russo-Japanese War had armies engaged along fronts that extended far beyond the horizon. Fought in February and March 1905, and often called history's first great modern battle, the battle of Mukden in Manchuria had seen the front extended to more than 150 kilometers, and had called into use for the first time such modern instruments of communication as the telegraph, wireless, and telephone because mounted adjutants no longer could transmit their generals' orders as they had in days gone by.

The general staffs of the world's industrialized nations had wasted little time in taking the lesson of the Russo-Japanese War into account. Only the Russians failed to profit from their own experience and entered the Great War with almost no modern communications systems. That fact alone promised disaster during the war's early days. The East Prussian offensive required close coordination between First and Second armies, yet General Samsonov had only twenty-five telephones, a handful of telegraphic devices, and a primitive Hughes teleprinter with which to communicate with all the units of his Second Army, the headquarters of Zhilinskii's Northwest Army Group, and Rennenkampf's First Army. Supply depots held no substantial supplies of communications wire, and Samsonov's men found it next to impossible to string what little wire they had in the field because only a fourth of the men who were supposed to carry wire cutters actually had received them.[73] Unable to establish reliable telephone and telegraph communications, Russia's commanders relied upon the wireless, which proved to be a catastrophic substitute. Inexperienced operators often solved the problem of lost or jumbled code books by broadcasting their messages "in clear," while amazed German operators hurriedly noted down the details of troop movements and battle plans that saturated the air waves between Russian regimental and corps headquarters.[74]

While the Germans derived much advantage from such failings, the Russians had no opportunity to turn the tables. Given the great distances involved in the battles of 1914, cavalry could not easily perform its former reconnaissance function, and the armies of Europe began to use airborne observers to report the details of enemy troop and supply movements. "Only the airborne observer, provided with a camera and a wireless, was able to become a reliable and effective scout," General Danilov once explained, "because he alone could penetrate deeply into the enemy's territory and supply rapid and accurate reports about what he had seen." In all, the Russian army had more than 250 airplanes in 1914, but a shortage of spare parts kept most of them on the ground. Far too often, those few that flew reconnaissance missions returned from the enemy's lines only to be shot down by their own infantry. Most soldiers "seriously thought that such a cunning idea as an aeroplane could only emanate from, and be used by, a German," Rennenkampf's cavalry commander General Vasilii Gurko wrote to explain why Russian soldiers actually shot down planes landing at their own hangars.[75] Such crude Russian reconnaissance left German armies free to maneuver unobserved. "As a consequence of our lack of sufficient airplanes," Danilov complained in his account of the war, "the enemy could maneuver as he wished by using his highly-developed railroad network and be certain that his plans could not be discovered by us very quickly. At the same time, with considerably better aviation facilities, the enemy was able to observe each of our steps with complete impunity."[76]

If such problems as trying to move poorly trained troops who were short of rifles, artillery, cartridges, shells, and communications equipment to the front over meager railroad networks that had too little rolling stock seemed overwhelming, Nikolai Nikolaevich, his staff, and General Dobrorolskii faced still others. Every rifle, bullet, and shell needed to replace those that Russian soldiers expended during the war's first days had to travel an average of fifteen hundred kilometers, while those used by her allies had to be shipped less than a third of that distance. Western nations could move their troops and supplies to the front with remarkable speed because of their highly developed railroad systems and numerous motorized transports, but Russia's planners had not even begun to comprehend how the internal combustion engine would reshape modern warfare and therefore had not followed the example of their Western colleagues.

While her allies had made a rapid transition from horse-drawn wagons to motorized transports, the Russians still relied upon small horse- and ox-drawn carts to move weapons and ammunition from railheads to the front. Vital supplies thus reached the tsar's fighting men at a snail's pace that could not be accelerated because all of Russia's five-and-a-quarter-million-man army in 1914 had only 420 motorized transport vehicles (including two ambulances) and 259 passenger cars. Her armies therefore lumbered into East Prussia and Galicia still trying to support twentieth-century strategies and weapons with an eighteenth-century supply and transport system.[77] For Samsonov's Second Army, the short-sightedness of General Staff planners meant crossing the East Prussian frontier with only four decrepit motorcycles and a motley assortment of ten automobiles and motor transport vehicles.[78]

Russia's General Staff had yet to reach a clear understanding of the awesome difficulties involved in supplying their nation's fighting men with weapons, munitions, and provisions when, on M-14, Northwest Army Group's commander General Zhilinskii ordered the First and Second armies into East Prussia to "attack the enemy with energy and unflagging persistence along the entire front and at every opportunity."[79] For a few days more, the euphoria of late July lingered, sustained by those expectations of success that inevitably crowd upon the beginning of any offensive. Only after they moved into German territory did the first doubts began to cloud the Russians' confidence. "Our first move into East Prussia convinced us how thoroughly the Germans had prepared for war," General Gurko remembered. "They had thought everything out, had foreseen everything, and had made a large expenditure on the preparations."[80] Like the retreating Russian armies of Field Marshal Kutuzov in 1812, the Germans burned buildings, food, and forage as they withdrew into their hinterland, but did so with more speed and greater precision. "In conjunction with the approach of any considerable body of Russian troops, thick columns of smoke started from some store filled with forage or straw, usually from the farms nearest the head of the column," Gurko reported as his cavalry advanced into East Prussia. "This appearance was so general," he concluded, "that it could only have been due to arrangement by the German authorities."[81]

Burning, destruction, and death meant war, and war, Tolstoi once explained, was "an event absolutely contrary to all human nature and reason," in which "millions of people set out to inflict upon each other evil acts so numerous . . . as could not be recorded in all the annals of all the courts of all the world, but which those who committed them did not think of as crimes at the time."[82] Now, a century after the events about which Tolstoi had written in *War and Peace*, the first guns of the Great War had begun to sound on Europe's eastern front, with fighting and dying about to begin on a scale beyond men's most awful imaginings. Soon, the Russians witnessed again the truth of the sad peasant saying that warned all men who sought glory on the field of battle that such moments were indeed fleeting and that: "It is a wide road that leads to the war, and only a narrow path that leads home again."[83]

CHAPTER II

The Fall Campaigns

Alfred, Lord Tennyson once described the grandfathers of the millions of Russians who marched along the wide road to war in August 1914 as "a great gray slope of men," stolid and unyielding, as they stood to receive the charge of Scarlett's Heavy Brigade during the first weeks of the Crimean War. Sixty years later, in 1914, the heavy gray-brown overcoat, reaching midway between calf and ankle, that had stirred Tennyson's image still remained the Russian infantryman's outermost garb. Beneath it he wore a brown-green blouse belted over matching trousers which he tucked into sturdy knee-high boots. Unlike the infantrymen of other nations, the Russian soldier wore well-greased linen or cotton footcloths in place of socks. His diet of four thousand calories a day consisted mainly of black bread, soup, meat, groats, tea, and immense quantities of sugar, and he carried a 7.62-mm five-shot repeating rifle, described by Britain's hard-eyed General Knox as "foolproof." With its quadrangular bayonet (always carried fixed in the field), this weapon weighed 9 1/2 pounds and measured a full six feet. According to regulations, each infantryman was supposed to carry 120 rounds of ammunition into battle, which, together with his rifle, kit, and extra rations, weighed a fraction over 58 pounds.[1]

Thus burdened, the tsar's infantry marched into East Prussia and Galicia in August 1914. True to the old peasant saying that "it is pleasanter to die in company, and old Mother Russia has sons enough," these men preferred close combat to long-range rifle and artillery fire. "The bullet is a fool, the bayonet is a hero,"[2] Marshal Suvorov often had told his men when they fought the eighteenth-century armies of Turkey and revolutionary France, and the bayonet, brought to Russia by Peter the Great at the end of the seventeenth century, still remained the Russian soldier's favorite fighting instrument in 1914.

Although other weapons were used to support the Russian infantryman fighting with bayonet and rifle, they were never as plentiful as required. Russia went to war in July 1914 with fewer than one machine gun for every thousand soldiers mobilized. When the war changed to one of fixed positions, automatic weapons began to figure so prominently in the defensive and offensive strategies of all combatants that Russia's General Staff called for 100,000 automatic rifles in September 1915, and over 30,000 new machine guns before the end of the following year.[3] Such demands exceeded the capacity of Russia's armaments industry many times over. During the first five months of the war, weapons factories produced an average of 165 machine guns each month, and then increased production to a peak of 1,200 machine guns by December 1916. Factories on native soil thus produced only about a third of the automatic weapons Russia's soldiers needed, and the rest had to come from France, England, and the United States in such large quantities that, before the end of 1917, foreign sources had supplied Russia with almost 32,000 machine guns, most of them made by the Colt, Marlin, and Savage arms companies in New England. Not all of these machine guns fired the same cartridge. Nor did the ten or more different types of rifles (including Japanese Arisakas, American Winchesters and Krag-Jorgensens, British Lee-Enfields, outdated French Gras-Kropatcheks, and ancient Russian Berdans) that Russian soldiers carried into battle, so that matching ammunition to weapons presented ordnance and quartermaster officers with impossibly complex logistical problems.[4]

If shortages of machine guns, rifles, and ammunition plagued Russia's soldiers from the Great War's beginning, they suffered even more from shortages of artillery. As its heavy shells smashed massive fortifications, destroyed entrenchments, and obliterated troop concentrations when enemy infantry massed for attacks, artillery assumed incalculable importance in all phases of the Great War. Again, Russia had to order great quantities of artillery from her allies, for her own factories could never produce enough to meet the war's demands. Nor could her factories even produce the shells these new guns required, any more than they could produce enough rifle and machine-gun bullets. More than thirty-seven million, or about two out of every three, artillery shells fired by Russian guns before the end of 1916 came from Japan, the United States, Canada, France, or England, along with almost a billion (nearly one out of every three) rifle and machine-gun bullets. This lengthened Russia's supply lines beyond the breaking point.[5] By the end of the war's second year, each artillery shell had to travel an average of 6,500 kilometers, and a rifle or machine-gun bullet more than 4,000, before a Russian soldier could fire it against the enemy. When the war began, Russia's railroad system could scarcely support supply lines a third that length. By 1916, the strain became intolerable.[6]

The worst of these problems still lay beyond the horizon when Nikolai Nikolaevich established his *Stavka* near the small West Russian hamlet of

Baranovichi on M-16, the sixteenth day of the mobilization. Chosen for its location at the point where the railway line from Moscow to Brest-Litovsk intersected with the north-south line from Vilna to Rovno, Baranovichi had little else to recommend it as a Supreme Headquarters. Only a primitive hostel, a crude stone barracks that at one time had housed a railroad brigade, and one larger house, formerly the residence of the brigade's commander, distinguished it from a typical Russian peasant village. Chief of Staff General Ianushkevich, about twenty-five other senior officers, and Nikolai Nikolaevich himself all lived in first-class railroad salon and sleeping cars that did double duty as offices and living quarters, while the remaining three dozen or so foreign military attachés, cryptographers, and assorted civilian and diplomatic personnel who made up *Stavka*'s regular contingent found shelter as best they could in the jumble of peasant huts and assorted odd railway cars that dotted the area.[7]

Daily life at *Stavka* reflected its pastoral setting. "We were in the midst of a charming fir-wood and everything was quiet and peaceful," Britain's General Knox, whom Sukhomlinov once called the ablest military attaché he had ever known, remembered,[8] while men less willing to do without the amenities of life in Russia's capital regretted the frugal daily fare and complained about the absence of the many creature comforts to which they had been too long accustomed. At *Stavka*, few felt war's urgent pressures. In consultation with Nikolai Nikolaevich, who met with him for an hour or two each morning and afternoon, the hard-working Quartermaster-General Iurii Danilov oversaw day-to-day operations, making certain that work kept to a fixed schedule. Drafting reports, recording the previous day's events on detailed maps, and making certain that strips of paper were hung in every doorway to remind the inordinately tall grand duke to stoop when passing through any of them consumed much of Danilov's time and made him think of *Stavka*'s routine as one of "monastic simplicity." Senior officers always had time for a good cigar and leisurely conversation after dinner so long as they observed the grand duke's preference for "diversionary themes not concerning the conduct of the war."[9] Only when Russia's ruler came to consult with his supreme commander did "a disquieting note" disturb life at *Stavka*. "Knowing the tsar's frame of mind and that of some of his ministers," one of the grand duke's staff once explained, "we always worried about the outcome of these conferences, fearing the consequences of a clash between the views of the grand duke and those of the tsar's inner circle."[10] Otherwise, life moved at a surprisingly leisurely pace for commanders responsible for military operations that involved millions of men.

Even before Nikolai Nikolaevich established his *Stavka* at Baranovichi, General Iakov Zhilinskii, commander of Russia's Northwest Army Group and a man whom Britain's General Sir Edmund Ironside thought far "more of an office soldier than a leader in the field," had launched Russia's first attack into East Prussia.[11] Zhilinskii's earlier military exploits had been strikingly unimpressive. As chief of staff to Russia's viceroy in the Far East during the Russo-Japanese

War, one observer remarked, "his influence on the war could hardly be called successful,"[12] although his talents as a courtier enabled him to reap impressive rewards for his flawed military effort. Zhilinskii had compiled a modestly better record as chief of Russia's General Staff after the fighting in the Far East ended. Like many of his peers, he had come to see *élan* as the best camouflage for the all too obvious failings that Russia's military effort had revealed in her war against Japan, and had thought it the easiest way to convince a doubting France of his nation's importance as an ally. Contrary to the counsels of any number of wiser advisers, he therefore had stupidly assured France's General Joseph Joffre on a visit to Paris in 1912 that Russia would have at least 800,000 men ready to invade Germany on M-15. Now, the French called upon him to honor that rash promise before the war's first fortnight ended. Ready to place honor above everything else, Zhilinskii swore to advance against the Germans at any cost. "History will condemn me," he told France's military attaché General Laguiche, "but I have given the order to march."[13]

From the first day of his campaign, Zhilinskii squandered manpower with prodigious abandon. He left six second-line divisions far to the rear at Kovno and Grodno where they could take no part in the war's first battles, and wasted others to reinforce several useless Russian fortresses on his flanks and rear. Zhilinskii frittered away badly needed guns on these fortresses, too, and utterly failed to establish effective communications between his headquarters and the two armies upon which the success of his offensive rested. His bitter feud with General Rennenkampf left him often unaware of First Army's whereabouts, and his communications with General Samsonov became so sporadic that he once actually dispatched General Gurko's cavalry division in an effort to find out where Second Army might be found.

Notwithstanding his reckless misuse of troops, Zhilinskii retained an impressive force to launch his first attacks. When his Northwest Army Group began its advance on M-15, it included 408 battalions of infantry, 235 squadrons of cavalry, and 1,652 guns. Across the frontier, with orders whose precision was matched only by their certainty to wound the inflated pride of all good Prussians, Germany's Eighth Army faced the oncoming Russians with 158 battalions of infantry, 68 cavalry squadrons, and 774 guns, including 156 pieces of heavy artillery that far outmatched the scant handful at Northwest Army Group's disposal. Outnumbered almost two to one, the Germans knew they could expect no reinforcements. "The East during this time had to look after itself, and only when the decisive battle had been fought in the West could it expect support from thence," Colonel Max von Hoffmann, German Eighth Army deputy operations chief later wrote. If the Eighth Army could not turn Zhilinskii's offensive, it must give up all of Prussia east of the Vistula River to the Russians, for, above all, it must take care "not to allow itself to be overpowered by superior forces." As Hoffmann confessed, such a plan "certainly contained great psychological dangers," for, in the last extremity it meant abandoning much of the

Prussian Junkers' cherished homeland to the enemy.[14] Whether such a moment would come, none could tell, but the sheer weight of the Russians' numbers made it seem probable that the Eighth Army would be forced to make a strategic retreat. Zhilinskii's crude strategy, combined with the worst sorts of communications and supply difficulties, soon came to the Germans' aid.

Accustomed to serving tsar and country from behind a desk, Zhilinskii had not the slightest notion about how to mount a vigorous offensive. While the Eighth Army's commander Colonel-General von Prittwitz und Gaffron had specific orders to keep his troops out of East Prussia's fortresses, Zhilinskii continued to order entire divisions out of the field and into fortress garrisons. Such absurdly anachronistic attempts to overprotect the flanks and rear of Russia's Northwest Army Group cost General Samsonov more than a third of Second Army's infantry and cavalry along with two-fifths of its artillery before he even entered East Prussia, at the very moment when Prittwitz was committing every available man to battle, strengthening the Eighth Army's artillery with every gun that could be stripped from the nearby fortresses of Königsberg and Posen, and advancing to meet the Russians with speed and precision that none of Zhilinskii's units could match.[15] "The German advance had been thought out and perfected in every detail for years," General Ironside explained, while on the Russian side "the final order for the advance was only issued while the troops were marching to their concentration areas, after considerable change in the [previously planned] order of battle."[16]

With unwieldy forces and primitive communications, Zhilinskii hoped to entrap Prittwitz's entire force in a gigantic two-pronged pincer movement that began to take shape only as Russia's armies approached the East Prussian frontier. According to the plan finally decided upon during a frenzied debate that still raged as M-15 dawned, General Rennenkampf's First Army would attack directly west, moving to the north and south of the kaiser's favorite hunting preserves in the Romintern Forest, then advancing through the Insterburg Gap, and thence into the Junker heartland of East Prussia. Meanwhile, Samsonov's Second Army would march northward to a point north and west of Tannenberg, where the Polish-Lithuanian armies of King Jagiello had defeated the Teutonic Knights in 1410. If the First and Second armies' badly coordinated advance came together at precisely the right moment, it would close like two giant claws around Prittwitz's force somewhere just to the west of the Masurian Lakes, destroy the only major German force in East Prussia, and open the way to Berlin. Success demanded boldness, tactical precision, and the rare ability to sense the enemy's intentions that distinguishes brilliant commanders from their numerous pedestrian colleagues. Since Zhilinskii had none of those qualities, the fate of Russia's First and Second armies rested solely in the hands of those assigned to lead them in the field.

As commander of First Army, General Pavel Rennenkampf did little to compensate for Zhilinskii's gross incompetence. Scion of an old Baltic German

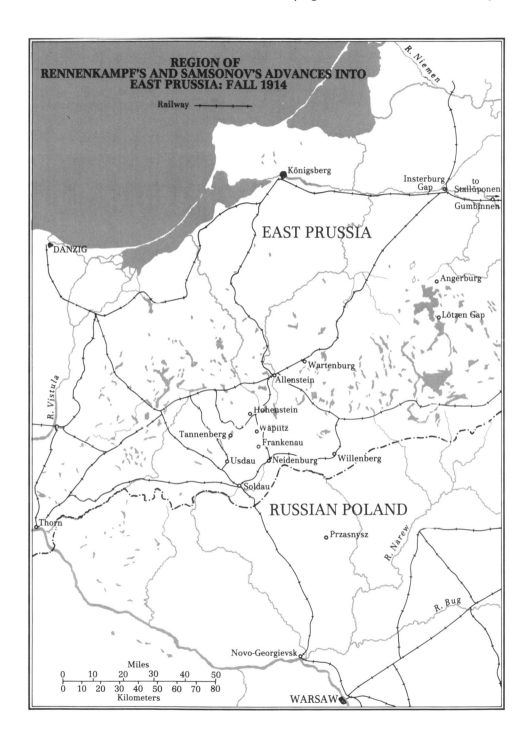

**REGION OF
RENNENKAMPF'S AND SAMSONOV'S ADVANCES INTO
EAST PRUSSIA: FALL 1914**

Railway

R. Niemen

Königsberg

Insterburg
Gap

to
Stallüponen

Gumbinnen

EAST PRUSSIA

DANZIG

Angerburg

Lötzen Gap

R. Vistula

Wartenburg

Allenstein

Hohenstein

Wäplitz

Tannenberg

Frankenau

Usdau

Neidenburg

Willenberg

Soldau

RUSSIAN POLAND

Thorn

Przasnysz

R. Narew

R. Bug

Novo-Georgievsk

Miles

0   10   20   30   40   50

0 10 20 30 40 50 60 70 80
Kilometers

WARSAW

family long known for its loyalty to the Romanovs, Rennenkampf was a dashing cavalry general with twirling mustaches and considerable combat experience. Commander of a Cossack division in the Russo-Japanese War, he had won wide respect as an able and daring officer, who, General Gurko remembered, had become known for his "great boldness, decisiveness, and resolution in working out a plan." Along with General Aleksandr Meller-Zakomelskii, Rennenkampf had crushed peasant revolutionary resistance in Siberia during the aftermath of the Revolution of 1905, had won command of the Third Corps as a reward, and finally had become chief of the Vilna Military District in 1913. Years of peacetime command seemed to have dulled Rennenkampf's talents, however, and he entered the campaign of 1914 older but certainly not wiser in the ways of modern warfare. Unable to understand that cavalry could not work effectively across the great distances that modern campaigns involved, he marched without the eyes and ears that every commander needed to guide his soldiers to victory. Perhaps not unexpected for a cavalryman, Rennenkampf's inattention to his lines of supply introduced yet another fatal element into his campaign strategy. "Rennenkampf might have been a Murat if he had lived a hundred years earlier," General Knox thought. But war in 1914 stood light years away from what it had been in 1812. "In command of an army in the twentieth century," General Knox concluded, Rennenkampf "was an anachronism and a danger."[17]

As Rennenkampf's First Army marched blindly across the frontier on August 2/15, General Hermann von François's I Corps lay in wait at the small East Prussian village of Stallüponen just five miles ahead. A dedicated disciple of Moltke's dictum that, "when the Russians arrive—no defense. Just attack, attack, attack,"[18] François had wanted to make certain that "no Russian should step on the soil of East Prussia, and that no East Prussian village should experience the horrors of war."[19] Given a choice, he would have preferred to meet the Russians on their own ground, but he had very stern orders not to do so. Momentarily held in check by Prittwitz's strategy of luring Rennenkampf beyond the limits of his primitive supply lines and defeating him before turning to meet Samsonov's slower advance from the south, François suddenly slipped his tether on August 4, and engaged the Russians at Stallüponen in direct defiance of telephone and telegraph orders from his chief. In a rage, Prittwitz sent a major-general from his staff to deliver his order to retreat. "Inform General von Prittwitz," I Corps' commander replied arrogantly, "that General von François will break off the engagement when he has defeated the Russians!"[20]

At Stallüponen, one of I Corps' brigades outflanked the Russians and took some three thousand prisoners before François finally retired to Gumbinnen, some thirty kilometers to the west. Pleased with the minor tactical success he had won by disobeying his superior's orders, François had carelessly jeopardized Prittwitz's larger plan. His advance slowed by the Germans' early attack, there was now a good chance that Rennenkampf might not reach the defensive positions the Eighth Army had prepared on the Angerapp River before Prittwitz had to contend with Samsonov's assault from the south. Although he was unaware of his adversary's strategy, Rennenkampf's tactical indolence unwittingly worked to the Russians' advantage. As he slowed his advance for no reason other than a natural reluctance to move quickly, several more precious days slipped away and Prittwitz knew that he must shift his battle plans quickly.[21] He therefore decided to engage Rennenkampf's forces east of Gumbinnen rather than wait for the Russians to continue their ponderous march to the Angerapp.

By August 6, time had become so important that Prittwitz dared not wait even long enough to bring up the troops that remained on the Angerapp before the corps commanded by generals François and Mackensen attacked the Russians. Again, François scored the first victory, and, as he had done at Stallüponen, he took advantage of the Russians' tactical blindness. Even with masses of cavalry assigned to reconnaissance, sixteen batteries of François's 1st Infantry Division artillery managed to lay down a murderous barrage at 3:30 on the morning of August 7 that caught the Russian 28th Infantry Division completely by surprise as it advanced along the northern edge of the Romintern forest. Outnumbered three to one, and with much heavier guns firing against them, six batteries of the 28th Division's light artillery tried to reply but found that the Germans had entrenched their guns to prevent the Russians' flat-shooting light weapons from firing into their positions. "Send us a battery of [high trajectory] howitzers," the 28th's desperate chief of staff begged his corps commander at 7:10 that morning.[22] Moments later, twelve more batteries from the artillery of François's 2nd Infantry Division slipped into position behind the Russians' right flank and opened fire. As more than 150 German guns rained a storm of high explosives upon their enemies' positions, the Russians for the first time learned what it meant to face heavy artillery barrages without the means to reply. "Catastrophic," one Russian general said. In less than an hour, François was ready to unleash his 1st and 2nd divisions.[23]

As the spiked helmets and gray uniforms of advancing infantry appeared in the fields that spread out before their positions, Russian riflemen returned the Germans' fire. Quickly, the crescendo of battle rose and blended into explosive waves that rolled back and forth across the terrain that stood between them.

> The noise of firing sounded more like something boiling up in a gigantic cauldron than like separate shots [wrote a young artillery officer who came under fire for the first time that day]. Our batteries began to fire in volleys [he continued] and the white expanse of the roadway [that stretched out in front of our position] suddenly turned grey from the masses of German corpses that fell there. A second wave of men in spiked helmets crossed the field. Again there was artillery fire in volleys, and again the masses of dead filled the roadway. . . . An enemy machine gun crackled briefly, then died.

Five or six hundred paces from our batteries, the German infantry began to inch forward, firing from prone positions as they came. Our batteries now only fired sporadically and then fell silent. They had no more shells left to shoot.[24]

Now unopposed by artillery fire, new waves of François's infantry stormed across the narrow gap that separated them from the Russians. By eleven o'clock that morning, 7,549 Russian officers and men—more than six out of every ten defenders—had been killed. Rennenkampf's 28th Division was no more. "Entire companies lay in rows with their officers and battalion commanders," an eyewitness reported. "It was as if they had been forzen into those poses in which they had met death."[25]

While François scored his triumph on Rennenkampf's right flank, the white-haired Mackensen moved against the Russians' left flank and center. Intending to keep pace with François's advance, he had sent the 35th and 36th Infantry divisions of his XVII Corps across the Rominte River in the late evening hours of August 6, only to become caught in a tangle of refugees, livestock, and wagons. Their advance slowed to a crawl, these units could not go into action until more than four hours after the hot-headed François's first assault; by that time, the commanders of the Russian 25th, 27th, and 40th Infantry divisions had their men well-entrenched and their artillery sighted in with enough ammunition in reserve to rain a murderous hail of metal and high explosives upon the Germans for an entire day. Unwilling to wait for his artillery, Mackensen impetuously assaulted the Russians' well-prepared infantry positions, and the effort cost him heavy losses.

Moving recklessly, the 35th German Infantry Division advanced into an artillery fire pocket upon which the guns of the 27th Russian Infantry Division fired more than 10,000 shells, while the neighboring German 36th Infantry reeled before a wall of rifle and machine-gun fire that nearly wiped out several regiments as the 108th Saratov Infantry Regiment poured more than 800,000 rifle and machine-gun bullets into its ranks before nightfall. Several German units panicked and fled, their terror infecting others near them. Abdicating his responsibility as a corps commander, the outraged Mackensen stormed from his headquarters, his square jaw jutting forward in grim determination to stem his soldiers' flight. Instead, he merely succeeded in leaving his corps without a commander at the battle's most critical moment. "How those Russians have learned to shoot since the Russo-Japanese War!" one captured German officer exclaimed.[26] Before the end of the day, Gumbinnen had become a clear Russian victory. "The Russians have shown themselves to be very dangerous opponents," a chagrined German colonel reported. "Good soldiers by nature, they were disciplined, had excellent combat training, and were well armed."[27] That day, Mackensen had learned a lesson about the importance of artillery preparation that transformed him into one of the Great War's most awesome practitioners of massive bombardment. Within a year, Russia's armies faced what contemporaries called "the Mackensen Wedge," in which huge concentrations of heavy guns battered their way forward, their heavy shells obliterating everything in their path before the infantry moved forward to occupy the ground.

Elated by the success of their first major battle against German troops, the Russians hurried to draw too many optimistic conclusions from too little evidence. "It is my general impression that, in those areas where the Germans manage to get the advantage of technology on their side, they will . . . inflict great losses upon us," Rennenkampf reported to Zhilinskii after the battle, "but in actual strategy and in moral temper, they have no real advantage over us."[28] In strategy, however, Rennenkampf continued to prove himself sadly deficient, for he failed to follow up his first success with a decisive advance. Instead, he frittered away his soldiers' hard-won victory by turning aside to lay siege to the fortress of Königsberg, which his army could have bypassed with ease. "The situation at the front is unchanged," Rennenkampf reported to Zhilinskii two days later,[29] at the very moment when changes were taking place in the German command that soon would force Rennenkampf's First Army to snatch defeat from the jaws of victory.

A vigorous Russian advance on August 8 and 9 might well have broken the German line and brought Rennenkampf's troops to Prittwitz's defenses at the Insterburg Gap, for defeat at Gumbinnen, combined with the news that General Samsonov's Second Army had crossed the frontier, had destroyed Prittwitz's will to fight. "The army shall break off the fight [and] retire beyond the Vistula River," he told Colonel von Hoffmann at that moment. With no taste left for anything but flight, Prittwitz stood ready to abandon all of East Prussia to the advancing armies of Zhilinskii's Northwest Army Group.[30] While Hoffmann begged his chief to reconsider, the unrelenting François made it clear that he was appalled by Prittwitz's timidity and demanded reinforcements to follow up his first successes. Fearful that the Vistula's low waters might not prove an effective barrier against the Russians, Prittwitz rejected the advice of both men and told German Supreme Headquarters at Coblentz that very evening of his decision to order a major retreat. Now certain that the Eighth Army's commander had lost his nerve, the German High Command removed him at once. To replace him and his chief of staff they chose General Paul von Hindenburg and Major-General Erich Ludendorff, a pair of obscure and seemingly unaccomplished officers whose names soon acquired all the trappings of legend in the annals of the Great War.[31]

Without important court and social connections, Hindenburg and Ludendorff were cut from a very different bolt of cloth than the men they replaced on Germany's eastern front. Stiff, straight, his ample jowls spreading heavily above his high-necked collar while his features seemed frozen in a perpetual scowl, Ludendorff had not received his first regimental command until the age of forty-eight, an astonishing fact which gave every indication that his prospects in the army of the Second Reich had become very dark indeed. A man whose peacetime career bore not a single mark of distinction before 1914, Ludendorff's brilliance burst forth the moment the guns began to sound, for his was a genius that could be forged only in the crucible of war. When the Belgian fortress of Liège, its fortifications defying the best-known talents of the kaiser's army, had barred the Germans' path during the first days of the war, Ludendorff shattered its defenses in a matter of hours. Suddenly the darling of the German High Command for his successful assault, he soon became the man of the hour in the east, as his superiors looked to his newly discovered genius to repair the damage that Prittwitz's rigid timidity had wrought. "You can prevent the worst from happening," Moltke wrote on August 9th. "Your task is a difficult one, but you are equal to it."[32] Convinced that Prittwitz's plan to "retreat behind the Vistula would have spelt ruin," Ludendorff left for the eastern front less than fifteen minutes after he received Moltke's message.[33] Endowed, in his chief's words, with "the almost superhuman capacity for work and untiring resolution,"[34] he became the perfect chief of staff for the man whom the kaiser chose to be Prittwitz's replacement.

Paul von Benckendorff und Hindenburg had first gone to war in Prussia's brief campaign against Austria in 1866, and had won the Iron Cross at Sedan in 1870. Sixty-seven when the Great War broke out, he had been in retirement for three years and faced the war's beginning with anxiety and uncertainty. "Would my Emperor and King need me?" he asked himself again and again. At first it seemed that he would not, for Wilhelm left him in retirement as the war's first campaigns began. Three weeks and a day later, Hindenburg received his kaiser's call. Was he prepared to return immediately to active service, the emperor wanted to know. He responded simply: "I am ready," and left his home in Hannover at three o'clock in the morning to take command of Germany's faltering defense of her Prussian lands.[35]

The specter of defeat loomed larger with each passing moment. The German Eighth Army still faced Rennenkampf's victorious divisions, while Samsonov marched from the south intent upon striking a hammer blow in its rear that would smash it against the anvil of Rennenkampf's forces. To save the Eighth Army, Hindenburg and his new chief of staff must either win two major battles in very rapid succession against armies whose combined strength was more than double their own, or retreat as Prittwitz had urged. Even for experienced generals, the outlook seemed grim and, at Germany's Supreme Headquarters at Coblentz, her chief strategists feared that all of East Prussia might fall to the advancing Russians. Convinced that Rennenkampf's victory marked the beginning of an offensive that could push the Eighth Army behind the Vistula, Moltke violated a cardinal principle of the Schlieffen Plan and reinforced the east with two corps from the western front before Germany's armies had won the battle for France. That decision tipped the balance in favor of the Allies in the west and fatally weakened the German assault against Paris. "Let us pay proper homage to our [Russian] allies," a French journalist wrote that fall. "One of the elements in our victory [on the Marne] was their debacle [at Tannenberg]."[36]

When Hindenburg and Ludendorff took command in the east on the evening of August 10, the prospects for victory thus seemed remote, and only Ludendorff's strategic genius, combined with brilliant intelligence work by Hoffmann, made it possible for the Germans to succeed in the daring maneuvers that consumed the first fortnight of Hindenburg's reign as the Eighth Army's new chief. From intercepted wireless transmissions, which the Russians, "with quite incomprehensible thoughtlessness," sent unciphered, Hoffmann learned that Rennenkampf had halted to replenish dwindling reserves of ammunition, food, and fodder and, as he watched his opponent's ponderous movements, this dedicated intelligence expert, whose far from elegant looks concealed his brilliance, concluded that Rennenkampf had little interest in following up his early successes.[37] Certain from Hoffmann's detailed reports that one claw of the Russian pincers had stopped moving, Ludendorff gambled that he could turn the Eighth Army and defeat Samsonov before Rennenkampf realized that only a screening force remained on his front. Leaving only a cavalry division and a brigade of territorials to deceive the Russians, Ludendorff ordered all of the Eighth Army onto trains that carried them toward Tannenberg and Samsonov's advancing forces.

As his forces turned to attack Samsonov, part of Ludendorff's thoughts remained fixed upon the Eighth Army's rear. "Rennenkampf's formidable host hung like a threatening thunder-cloud to the north-east," he remembered. "He need only have closed with us and we should have been beaten."[38] But Rennenkampf thought he had already accomplished his task, that the Germans had begun to withdraw behind the Vistula, and that it now lay with Samsonov to cut off their retreat and finish what the First Army had begun. As commander of the Northwest Army Group, the dully inept Zhilinskii agreed and ordered Samsonov to shift his line of march so as to intercept the "retreating" Germans. "You are to intercept the enemy as he retreats before the advancing army of General Rennenkampf with the object of cutting off his escape across the Vistula," he telegraphed to Samsonov on August 10, although he had no certainty about the positions of any of the three armies involved.[39]

Obliged to proceed by forced marches to obey his chief's new orders, Samsonov pressed forward thinking that Rennenkampf was in hot pursuit of Hindenburg's forces when, in fact, he was about to commit about three-quarters of his forces to a useless siege of Königsberg.[40] At the same time, his extreme timidity now heightened by the hope of victory, Zhilinskii withdrew still more of the First Army's front-line troops to strengthen the garrison at Grodno, even though that city fortress stood more than seventy miles to the rear of Rennenkampf's forces.[41] Clearly, this floundering staff officer had lost control of his campaign and had allowed the movements of the First and Second armies to become hopelessly muddled. Therefore, as Ludendorff and Hindenburg launched the full weight of the German Eighth Army toward Tannenberg, Zhilinskii continued to misadvise Samsonov that "the enemy evidently has left only an insignificant force at your front."[42] Now almost completely out of touch with higher headquarters, Samsonov's claw of the Russian pincers began to waver, while Rennenkampf's had already turned back upon itself toward the Königsberg fortress.

Zhilinskii had not the slightest inkling that Samsonov was about to be engulfed by a major battle, nor did he now even know the exact location of either of the two armies under his command. On the eve of the greatest battle that Russia had ever fought with Germany, Zhilinskii communicated with Samsonov by sending telegrams in care of Warsaw's Central Post Office, where an adjutant collected them once a day and carried them more than a hundred kilometers by car to Second Army headquarters.[43] "The state of communications in the Russian Army can hardly be realised by a British officer," an appalled onlooker observed some years later. "The Russian wireless gave the Germans complete information as to their proposed movements."[44] No one was more aware of that problem than Samsonov. While Ludendorff began to position his forces near Tannenberg, the Second Army commander issued a directive warning his corps commanders that the wretched state of communications made it imperative for them to make individual efforts to support each other in the event of any major battle.[45] Stated most simply, the Second Army was almost deaf. As it marched through the thick forests that separated East Prussia from Russia's Polish lands, it had also become nearly blind.

General Aleksandr Vassilevich Samsonov, to whom Zhilinskii had delegated full responsibility for defeating the German Eighth Army, was five years younger than Rennenkampf and a far more sober-minded officer. A graduate of St. Petersburg's prestigious Nikolaevskoe General Staff Academy, he had received an appointment to the Imperial General Staff at the age of twenty-five and became a major-general at forty-three. Commander of the hard-fighting Ussuri mounted brigade and the Siberian Cossack Division during the Russo-Japanese War, he had gone on to become governor-general of Turkestan, where, despite bouts with heart trouble and asthma, he compiled an unimpeachable record for valorous and efficient command. General Gurko thought him a man of "irreproachable" moral character, "possessed of a brilliant mind reinforced by a good military education." Unlike Rennenkampf and Zhilinskii, who had enjoyed considerable experience as high-ranking staff officers in commands headquartered along Russia's western frontier, Samsonov's only acquaintance with the area in which he was about to fight the battle of Tannenberg had come during a few brief months as chief of staff of the Warsaw Military District almost a decade before. Perhaps a better tactician than Rennenkampf, and certainly far more able than Zhilinskii, Samsonov had to suffer the consequences of their combined failings. Thanks to the stupidity of one, he had been deceived into thinking that the Eighth Army was retreating in disorder across his front, while the strategic blunders of the other had left Hindenburg and Ludendorff free to fight a major battle against him with no pressure against their rear.[46]

By August 12, Samsonov was far less prepared to fight a major battle than Rennenkampf had been five days before. The First Army had advanced through a region with enough rail facilities to bring up reserves of men, weapons, ammunition, food, and fodder, while Samsonov's rear lay in the frontier region of Russian Poland in which tsarist planners had neglected to build roads and railroads in the misguided hope that such a policy would help to slow any German advance into Russia. Compared to Rennenkampf, Samsonov therefore faced crushing and insoluble supply problems. From the moment they crossed the frontier, his soldiers had too little bread, and his cavalry was short of fodder. "I don't know how the troops can manage any longer," one of his adjutants concluded after a hasty inspection. "Some system of requisitioning has got to be organized."[47]

When Samsonov received Zhilinskii's orders to hasten his advance in order to cut off the "fleeing" Eighth Army after Rennenkampf's victory at Gumbinnen, his army had been without rest for six days, advancing an average of twenty-four kilometers a day over primitive roads.[48] Obliged to march in the burning August heat, their route taking them from sun-baked sandy marshes into gloomy forests of spruce and pine, his men were in no condition to fight the divisions that Ludendorff and Hindenburg were moving into position behind the carefully-planned screen formed by the delaying actions of General von Scholtz's XX Army Corps.[49] After particularly bloody fighting at Frankenau on August 10 and 11, Samsonov mistook Scholtz's withdrawal for a full-scale retreat, which made him all the more ready to believe Zhilinskii's assurances that he had only a demoralized enemy to contend with.[50] To Samsonov's misfortune, the Germans already knew of his miscalculation and made ready to turn his error to their advantage. "This voluntary retreat," Hoffmann later said of Scholtz's withdrawal, "proved to be a happy manoeuvre [for] it aroused in the Russians the belief in a general retreat of the German army."[51] Although none knew it at the time, the fighting at Frankenau marked the beginning of the battle of Tannenberg. By the morning of August 11, Hoffmann's delighted intelligence officers had extracted a full set of Samsonov's battle plans and had plotted the precise location of all his units from the flood of uncoded Russian wireless messages that had filled the air during the previous twenty-four hours.[52]

With the information Hoffmann and his staff provided, Ludendorff and Hindenburg laid their final battle plans. The most critical maneuver had been to move François's I Corps and its heavy artillery from its former position on Rennenkampf's right flank along more than 150 kilometers of rail and roads to the vicinity of Usdau, slightly to the south and west of Tannenberg, where it was to break through Samsonov's left flank and cut off his retreat. "Day and night, train after train shuttled at half-hour intervals . . . completing in 25 minutes or less unloading that normally took one or two hours in peacetime," one German officer reported proudly.[53] Still, by August 13, when Hindenburg and Ludendorff planned to begin their general attack, François had only twenty of his thirty-two batteries in position. Without artillery, he delayed his attack, insisting that it would be next to impossible to advance without heavy guns to clear the way. "If I am ordered to attack, of course my troops will do so," he announced when Hindenburg and Ludendorff pressed him that morning, "but it will mean that they will have to fight with the bayonet."[54] All this had become a moot point, for François's postponement already had won him the extra day he needed. By the morning of August 14, he was ready to begin his assault.

If it had been difficult for the Germans to move their men and weapons into position on the eve of Tannenberg, the Russians faced far more crushing difficulties. When Rennenkampf's advance patrols reached the Eighth Army's hastily abandoned positions at Insterburg Gap, they misread the Germans' hurried departure as evidence that their enemy was in headlong retreat. Rennenkampf therefore slowed his advance even more, fearful of pushing the Germans across the Vistula before Samsonov intercepted them. At the same time, Samsonov shifted his advance toward the northwest to cut the main rail line that led into the heart of Germany, and unwittingly moved further away from Rennenkampf's forces rather than narrowing the gap that separated them. Both Russian commanders now totally misunderstood the situation. Samsonov thought the Germans were falling back before him, while Rennenkampf remained utterly ignorant about their whereabouts.

Arrogantly certain that the Germans had fled their positions at Insterburg Gap, Rennenkampf slowed his advance still more and continued to issue definitive reports about the Eighth Army's "retreat," although almost a week had passed since any of his forces had made serious contact with any German units. By August 13, *Stavka* had become so concerned about his behavior that Grand Duke Nikolai Nikolaevich had hurried in person to Zhilinskii's headquarters to urge the First Army to speed its advance.[55] Rennenkampf continued as before. "Did he, or would he, not see that Samsonov's right flank was already threatened with utter ruin and that the danger to his left wing also was increasing from hour to hour?" Hindenburg wondered.[56] He did not.

August 14 dawned humid and hot. At 5:30 a.m., with more than two hundred guns massed against less than fifty on the Russian side, François opened fire against I and XXIII Corps on Samsonov's left flank. While Ludendorff and Hindenburg looked on, François's own I Corps, supported by his XX Corps on his left, stormed ahead. "There was plenty of heroism in the cause of the Tsar," Hindenburg remembered, but it was "heroism which saved the honor of arms but could no longer save the battle."[57] By afternoon, Usdau fell and the Russians were thrown back beyond Soldau, some fifteen kilometers to the south. "It was one of the most tragic sights I saw during the entire war," François confessed as he surveyed the abandoned Russian positions. "There were trenches two meters deep that were piled up with dead and seriously wounded Russians."[58]

"In my opinion," Hoffmann later wrote, the Germans' "breakthrough at Usdau . . . [was] the decisive point of the whole battle."[59] But the fighting was far from finished when François's troops smashed the Russians' defenses, for both Samsonov's center and General Blagoveshchenskii's VI Corps, which had been detached from Second Army's main body and sent toward Bischofsburg on Zhilinskii's orders, remained intact not far to the northeast. The next day, parts of Samsonov's XV and XXIII Corps won a victory to the north and east of Usdau at Waplitz, where, one German general remembered, "the Russians fought like lions,"[60] but to no avail as the ring of fire and steel that Ludendorff had begun to forge on August 13 began to close. On the morning of August 16, François's cavalry joined advance detachments of General von Mackensen's XVII Corps in Willenberg to close the trap. Before noon, the Russian Second Army was surrounded.

As Samsonov's predicament worsened, Rennenkampf remained out of the fighting and unwilling to come to his aid. Throughout the second week of August, Rennenkampf had done nothing to aid Samsonov, and the orders he received from Zhilinskii only added further confusion to his already indecisive movements. Zhilinskii still insisted that Rennenkampf's main objective must be Königsberg, and that only "the remaining forces of the army" could be used to pursue "that portion of the enemy's forces which, not having found shelter in Königsberg, may attempt to flee across the Vistula."[61] Uncertain about how, or even if, he should pursue the Germans, Rennenkampf now held his army in check, and his most serious military action on the day when François smashed Samsonov's left flank involved a staff officer, orderly, and trumpeter whom he sent to demand the surrender of the tiny German outpost that blocked the Lötzen Gap. This inglorious effort ended ignominiously when German sharpshooters wounded the staff officer and his trumpeter and made prisoners of them both.[62] Only on August 16, when Zhilinskii belatedly sensed that the Second Army might be in danger, did Rennenkampf turn to support Samsonov and then only after he received direct orders from his chief. "In view of the heavy fighting in the Second Army, the C-in-C orders you to move two corps to its support," Zhilinskii telegraphed at 7 o'clock that evening. Four hours later, he canceled the order, but still considered a proposal to send Rennenkampf's II Corps to Samsonov's aid by a 135-kilometer march followed by a rail journey of another 240 kilometers. Such a maneuver would have taken weeks to complete; by then, even if Samsonov had received help within hours it would have been too late.[63]

Their food and ammunition exhausted, and with no hope of relief from Rennenkampf, more than ninety thousand of Samsonov's soldiers surrendered. Among them was General Martòs, commander and one of the few survivors of Russia's XV Corps, who was brought before Ludendorff and Hindenburg in "a tiny, filthy hotel" in the town of Osterode. Typically, Ludendorff was rude while his chief remained the model of courtly politeness. "Seeing that I was terribly

upset, he gripped my hands for a long time and begged me to calm myself,"
Martos remembered. "As a worthy foe, I am returning to you your golden sword
[which Martos had received for bravery in the Russo-Japanese War]," Hindenburg concluded. "I wish you happier days in the future."[64]

Fortune treated Samsonov much less kindly. After he lost all contact with
Stavka and Zhilinskii during the last days of the battle, he had taken a few of
his most trusted staff officers, mounted Cossack horses, and fought with Martos's
XV Corps in the vicinity of Waplitz. When defeat seemed certain, Samsonov
tried to make his way out of Ludendorff's rapidly closing ring of steel on the
night of August 16 along with his chief of staff and a handful of officers. Their
horses gone, the men stumbled through the dark night on foot. Still recuperating
from a bout of heart trouble and asthma that he had suffered just before the
war broke out, Samsonov began to fall behind the others as they searched for
a way through the dense Tannenberg forest. About one o'clock on the morning
of August 17, Samsonov disappeared. Moments later his comrades heard a single
pistol shot, and they understood that he had killed himself, unable to bear the
shame of his defeat. In the space of five days, he had lost an entire army. Of
the XIII and XV Corps, only fifty officers and two thousand men still lived,
while something over three hundred field guns and thousands of horses had
fallen into German hands. Before Samsonov's comrades found his body, a
German machine-gun unit drove them off,[65] and it fell to the Germans to bury
his remains. They, in retrospect, pronounced the last words over his fallen body.
"One cannot deny the persistence and energy with which he commanded his
army," the German General Staff's report said. "The task given to him simply
was beyond his strength."[66]

Victory over Samsonov solved only half of the Eighth Army's problem.
Leaving two and a half divisions behind in case the Russians should draw new
forces from the fortresses of Warsaw, Augustowo, and Osowiec and attack again
from the south, Ludendorff and Hindenburg now wheeled the Eighth Army's
remaining fifteen divisions and 1,074 guns back toward the north and east. On
August 22, with men and guns in their new positions only four days after
Samsonov's defeat, they launched their new offensive against the Russian First
Army.[67] Ludendorff hoped for another Tannenberg, but neither he nor Hindenburg were rash enough to underestimate their opponent. "The Russian leader
was known to be a fine soldier," Hindenburg later said with characteristic
generosity, but both he and Ludendorff were surprised to see how readily
Rennenkampf frittered away his best opportunities. "Why did he not use the
time of our greatest weakness, when the troops were exhausted and crowded
together on the battlefield of Tannenberg, to fall upon us?" Hindenburg wondered. "Why did he give us time to disentangle our units, concentrate afresh,
rest and bring up reinforcements?"[68]

Unlike Samsonov, who had been forced to confront the enemy when his
troops were unrested and unfed, Rennenkampf had spent the middle of August

in deploying his army in defensive positions that stretched from the Baltic Sea to the northern end of the Masurian Lakes. Well-entrenched, well-rested, and well-supported by artillery, these troops could bring nearly fifteen hundred guns to bear against the enemy. But Rennenkampf had stretched his lines too thin by spreading twelve divisions over more than 130 kilometers, and concentrating four of them to await the attack he expected the Germans to launch from Königsberg. Still without adequate reconnaissance, he did not know that it would again be François's I Corps that would attack first, or that François would strike his left flank from the south where the small German-held fortress at Lötzen Gap still kept open a fatal passage into the First Army's lines of communication. With François supported by Mackensen's XVII Corps, Hindenburg and Ludendorff ordered the rest of the Eighth Army to attack directly from the west against points at which Rennenkampf could not bring the four divisions he had concentrated at Königsberg into the battle.[69]

The battle of the Masurian Lakes began at 3:30 on the morning of August 27. After making a forced march of 115 kilometers in three days, François's 1st and 2nd Infantry divisions routed the Russians in a surprise attack at Soltmahnen and seized five thousand prisoners and sixty guns. Ludendorff and Hindenburg attacked along the entire Insterburg Gap front that same day, but made little headway against the deeply entrenched Russian defenders. With his soldiers holding the main German force at bay, Rennenkampf faced the terrible prospect of "greatly superior forces"—in this case, François's victorious divisions, supported by two cavalry regiments and Mackensen's XVII Corps—breaking through his crumbling left flank, slashing into his rear, and cutting off his retreat.[70] "Another Tannenberg appeared within reach of Hindenburg's Army," one observer concluded.[71] Unwilling to take that risk, Rennenkampf ordered a general retreat.

Hindenburg thought his foe "seemed to be impatient" in his withdrawal.[72] Indeed he was. With unheard of speed, the remains of the First Army covered eighty-five kilometers in just fifty hours, with Rennenkampf and his staff in the lead every step of the way. In their haste to escape, the army corps sometimes marched along the roads three abreast. Rennenkampf moved his headquarters four times in a single day, completely broke off communications with Zhilinskii's headquarters, and left Russia's Northwest Army Group commander utterly in the dark about the First Army's whereabouts. Not until August 31 did Zhilinskii begin to hear reports that his elusive army commander might be found in the fortress of Kovno, much further to the rear than anyone had expected. At first, he thought the report "so improbable that it requires verification," only to have it prove all too accurate. Rennenkampf had indeed abandoned his army for the safety of Kovno. No longer "von Rennenkampf," but *"Rennen von Kampf"* (a German phrase meaning "flight from the battlefield"), a number of his staff remarked bitterly as they took few pains to conceal their loathing for the general whose carelessness and cowardice had cost the First Army 145,000 men, 150 guns, and more than half of its transport vehicles in less than a month.[73] In all, the Russians had lost more than a quarter of a million men during the first thirty days of fighting in East Prussia.

The first month of war in East Prussia showed that some of Russia's most vaunted commanders could not lead modern armies in the field. Zhilinskii's inadequacy, Rennenkampf's personal arrogance and tactical timidity, and Samsonov's too trusting carelessness had cost their nation the better part of five army corps, and had left their shattered First and Second armies in critical need of reinforcements, weapons, food, and rest. Within days after Rennenkampf's retreat, Grand Duke Nikolai Nikolaevich relieved Zhilinskii, although Rennenkampf's protectors at court managed to shield him from the grand duke's wrath. Still, the military situation in early September remained far from grim, for General Nikolai Ivanov's Southwest Army Group had done much better against the forces of Austria-Hungary in Galicia and southern Poland. Together with his talented chief of staff, General Mikhail Alekseev, Ivanov led Russia's armies to their first and greatest victories in the Great War, and their achievements helped to offset the terrible failures in East Prussia.

Unlike the aristocratic Rennenkampf and Zhilinskii, the grizzled, weathered Nikolai Iudovich Ivanov came from humble origins and his rise to high rank had not destroyed his homely manner. Born three years before the Crimean War, he had studied the ways of war at the Mikhailovskoe Artillery School, where most of Russia's great nineteenth-century artillerymen received their training. Sixty-three when he assumed command of Russia's Southwest Army Group, Ivanov enjoyed a widely-held reputation as an expert on modern warfare, although his first combat experience had come only during the Russo-Japanese War, where he had commanded the III Siberian Corps and twice had won the Cross of St. George for bravery under fire.

With his roughly trimmed beard spread across his chest in the style of his peasant forbears, Ivanov seemed the reincarnation of such legendary folk heroes as Ivan Susanin and Kuzma Minin, who had risen from the common people to save Russia from the invader's heavy heel in centuries past. General Knox called him "simple and unpretentious," and remembered him as a commander "beloved by his men."[74] General Brusilov, whose Eighth Army fought on the extreme left flank of Ivanov's command, thought him "utterly dedicated to his work," but "narrow in his outlook, indecisive, and overly concerned with details."[75] Certainly his memory for detail was prodigious, and his mind always quick to assemble information. For that, and a number of other reasons, Ivanov's armies proved themselves worthy heirs to the great military traditions of Peter the Great, Suvorov, Kutuzov, and Alexander I during the fall of 1914.

Although General Danilov's controversial Plan No. 19 had concentrated Russia's strength against East Prussia, conflicts within the High Command had diverted enough forces back to the Austrian front by 1914 that the Russians sent their weaker force against their strongest enemy and aimed a much more powerful thrust against Austria. By M-30, Ivanov's Southwest Army Group therefore contained fifty-three infantry divisions and eighteen divisions of cavalry totaling about one and a quarter million men spread along roughly five-hundred kilometers of front from the Vistula to the Romanian frontier. As they advanced into Galicia, Ivanov's forces took with them more than two thousand guns, seventy of them precisely that sort of heavy artillery that Samsonov had needed so desperately in East Prussia.

Against Ivanov's advance, Austria's belligerent chief of staff, Franz Conrad von Hötzendorff, assembled thirty-seven divisions of infantry, ten of cavalry, and a few *Landwehr* units sent from Germany with which he desperately hoped to duplicate the successes of Hindenburg and Ludendorff. Neurotically sensitive about Austria's honor, Conrad had always spoken as a patriot who could not "view with indifference" Austria's declining role in the affairs of Central Europe, and he bitterly resented that his nation must stand in the shadow of Germany in the councils of the Central Alliance. Time and again his had been the voice to speak loudest for war in Vienna, and he, more than anyone else, had been willing to risk drawing Europe's Great Powers to the brink of conflict, despite his emperor's reprimands that "even in politics, one should stick to the rules of decency." Appointed Austria's chief of staff at the end of 1912, he gazed covetously upon the Balkan lands that lay beyond Austria's frontiers as a means to restore his nation's greatness. On several occasions, he had urged war with Serbia "despite all qualms," and had greeted Serbia's rejection of Austria's rude ultimatum with delight.[76] As Ivanov's armies approached the domains of his Habsburg masters at the beginning of August, Conrad arrogantly looked to recoup Austria's tarnished prestige by equaling the feats of Hindenburg and Ludendorff in the north.

The first test of strength between Russians and Austrians came when the Fourth Russian Army that Baron Zalts led south from Lublin toward the Austrian fortress of Przemysl collided with General Dankl's Austro-Hungarian First Army as it marched north from the River San. Neither commander expected to encounter a major enemy force, and Zalts was so certain of that fact that he did not even throw out cavalry screens as his army entered Austrian territory. On the morning of August 10, these armies clashed near Krasnik, some forty kilometers south of Lublin where, for three days, nearly half a million men fought and died in the region between the Tanev Forest and the Wyznicza River. By late on August 12, one in ten had fallen on each side. The Austrians had taken more than six thousand Russian prisoners, seized thirty guns, and pushed Zalts's divisions back nearly ten kilometers. That very evening, while Dankl claimed the victory, an angry Grand Duke Nikolai Nikolaevich replaced the incompetent Zalts with the cautious but reasonably able General Evert.[77] "A joyful and welcome beginning," Conrad exclaimed when Dankl sent word of his success, although he knew only too well that "it was only a beginning."[78]

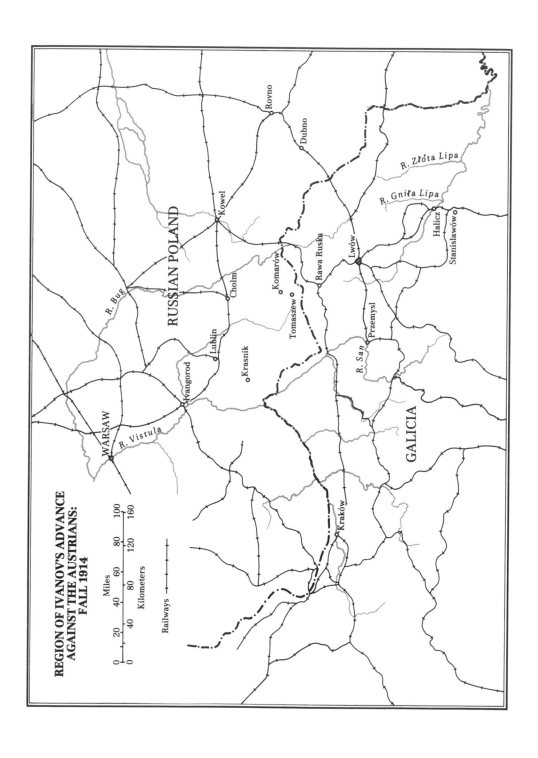

REGION OF IVANOV'S ADVANCE
AGAINST THE AUSTRIANS:
FALL 1914

Miles

| 0 | 20 | 40 | 60 | 80 | 100 |

| 0 | 40 | 80 | 120 | 160 |

Kilometers

Railways ━━┿━━┿━━

RUSSIAN POLAND

GALICIA

R. Bug

R. Vistula

WARSAW

Ivangorod

Lublin

Krasnik

Cholm

Komarów

Tomaszew

Kowel

Rawa Ruska

Lwów

Przemysl

R. San

Kraków

Rovno

Dubno

R. Złóta Lipa

R. Gniła Lipa

Halicz

Stanisławów

Like the Russians in East Prussia, the Austrians had begun their war with a victory. Also like the armies of Zhilinskii, fate decreed that their good fortune would be brief.

While *Stavka*'s senior officers reshuffled the Fourth Army's command, General Plehve's Fifth Army advanced on Evert's left. Supported by more than three hundred machine guns and five hundred pieces of artillery, the Fifth Army's 144 infantry battalions and 100 cavalry squadrons clashed with the Austrians in the battle of Tomaszew (sometimes called Komarów), which began on the very day that Dankl claimed victory over Zalts's embattled forces.[79] Although well-armed, Plehve's troops had in reserve only 135 cartridges per rifle and only 190 shells for each field gun. During the next week of fighting, both sides suffered reverses beginning on the night of August 14 when the artillery of the Russian V Corps trapped the 15th Infantry Division of the Austrians' VI Corps in a murderous crossfire in a swamp near Laszczów. After the Austrians endured almost five thousand casualties, lost four thousand prisoners, and abandoned twenty guns, the Fifth Army ran short of ammunition and their triumphs turned into reverses. For a brief moment, Conrad envisioned a decisive victory, hoping that the soldiers of Austria's Fourth Army could repeat the Germans' success at Tannenberg.

But Tomaszew was not Tannenberg, and Plehve was not Samsonov. A man who stubbornly refused to accept defeat, Plehve was equally unlikely to fall into a trap. "Small and old and bent, and weak in health," according to General Knox's account, his "logical mind and iron will" prevented him from storming ahead into the half-closed iron ring his enemies had forged.[80] After losses of almost 40 percent, Plehve therefore withdrew to the north on August 18 to regroup his forces along with Evert's army, which had lost 20,000 prisoners and close to a hundred guns. Although the Austrians claimed a great triumph, their hopes for a decisive victory to match the brilliant early successes of Hindenburg and Ludendorff had been shattered.

Desperate to match the German achievement, Conrad had drawn heavily upon his reserves to entrap Plehve and had weakened the forces that could be sent against the Third and Eighth armies that the Russian generals Nikolai Ruzskii and Aleksei Brusilov were bringing in from the east. Unlike so many of their counterparts, these Russian commanders moved resolutely, their infantry and artillery columns always deployed behind heavy cavalry screens. Willing to accept battle only at a time and place of their choosing, these were not men who could be lured into hostile terrain against their will, nor should the Austrians have dared to meet their advance with anything less than the heaviest concentrations of men and weapons.[81] Choosing the time and place, Ruzskii fought his first battle against the Third Austrian Army at Złota Lipa on August 13 where, for reasons that are not clear, the Austrians badly misjudged the Russians' strength.[82] Anxious to engage what they thought to be only "an isolated corps" of Ruzskii's command, the Austrians rushed to attack along a fifty-kilometer front at two o'clock on the afternoon of August 13 with only ninety-one infantry battalions, eighty-seven cavalry squadrons, and just over three hundred guns only to find that they had challenged Ruzskii's entire force which, although roughly equal to the Austrians in cavalry, deployed more than twice the infantry and supported them with 408 machine guns and 685 pieces of artillery.[83] As the Austrian divisions shattered against the superior Russian force, Ruzskii launched counterattacks that cost some of their units up to two-thirds of their men, so that by the morning of August 15, Conrad had no choice but to order a general retreat to the Gniła Lipa some fifteen kilometers to the rear where, from positions that could hardly have been more poorly chosen, the Austrians made a new stand.

About halfway between the Austrians' abandoned positions on the Złóta Lipa and Lwów, the easily fordable Gniła Lipa formed a slight natural barrier some ten to twenty meters wide. Swamps abounded along its banks, with the flatness of the terrain broken only by a few scattered forests. On each side of the river valley, a range of hills rose to a height of fifty to one hundred meters. It was a place, one observer remarked, in which "large masses of troops could be maneuvered easily, although it gave the Austrians no particular advantage as a defensive position."[84] Masses of troops were precisely what the Russians deployed when Ruzskii and Brusilov combined their forces on August 16 to launch a massive attack against the Austrian defenses. With 400,000 men supported by 1,304 guns, the Russians now could concentrate an average of three infantry battalions, at least one cavalry squadron, and eleven field guns along each kilometer of front. Even when reinforced, their opponents had less than 300,000 men and 828 guns. Nonetheless, Conrad ordered an attack on August 17 which proved an even greater disaster than their struggles at Złóta Lipa a few days before.[85]

The battle of Gniła Lipa stretched northward along a front of eighty kilometers, from Halicz to a point about thirty kilometers directly east of Lwów. As the Austrians began to lay down systematic artillery barrages upon the enemy positions across the river on the morning of August 17, the Russians prepared to advance. "We'll all die here, but not one step back," one Russian general told the 12th Cavalry Division as the Austrians launched their first attacks.[86] That morning the Russian line held, then pushed its way across the river to break the front of the Austrians' XII Corps and drive them back in panic. All along the line similar scenes repeated themselves with terrible brutality as Austrian troops, already exhausted and demoralized by their defeat at Złóta Lipa, faced their Russian foes. By the next evening the Austrians had been soundly beaten. "The entire field of battle, for a distance of almost a hundred *versty,* was piled with corpses, and they collected the wounded only with great difficulty," Brusilov wrote to his wife. "Even to give drink and food to all those who were suffering was impossible. This," he concluded sadly, "is the painful and seamy side of war."[87] Along roads already clogged with guns, wagons, and refugees, the Austrians fled in disorder, as the cry: *"Kosaken kommen!"*—"the Cossacks are coming!"—drove them on in deadly fear.[88]

Just as he claimed victory over Evert and Plehve at Tomaszew and Krasnik in the north, Conrad thus had to admit an even greater defeat at Gniła Lipa in the south where his Second and Third armies had lost tens of thousands killed and wounded in less than a week. Fleeing troops had left more than seventy guns in Russian hands, and twenty thousand Austrian soldiers had surrendered.[89] Nor was the disaster at Gniła Lipa the end of Conrad's troubles. During the next two weeks, Ruzskii and Brusilov captured the key city of Lwów, imposed a Russian administration upon its citizens, and continued their march westward.[90] Conrad now hoped to roll up their northern flank by turning Austria's Fourth Army back upon them, but that maneuver demanded far more than its weary men could accomplish. At the same time, Evert's and Plehve's newly reinforced armies drove back the fifteen Austrian divisions that faced them with heavy losses. Trapped between Plehve's divisions and a corps of Russian cavalry, a corps commanded by the Archduke Josef Ferdinand lost four men out of every five as Ivanov's generals stormed to retake Krasnik, Komarów, Tomaszew, and Rawa Ruska, before advancing to seize Halicz, Gródek, and Mikolajów. Leaving a large garrison to reinforce the great fortress of Przemysl, Conrad ordered the rest of the Austro-Hungarian armies back across the San River. By that time, his flawed generalship had cost the Austrians 300,000 killed and wounded, another 100,000 taken prisoner, and more than 400 guns captured. On the Russian side, the losses were less but still appalling, for 210,000 had been killed and wounded, 40,000 taken prisoner, and 100 guns seized. Exhausted by the first month of fighting, Ivanov and Conrad both stopped to take stock of the situation their armies faced.[91]

For the Austrians, the military situation at the beginning of September had become very serious, and Conrad's failure placed new burdens upon Germany's already strained resources. Fearful that Ivanov's advancing armies might all too easily overwhelm the remaining Austrian defenses and burst into Germany from the south, the German High Command concluded that not a moment could be wasted in coming to Austria's aid. "The Austro-Hungarian Army would have to be supported if it were not to be annihilated," Ludendorff concluded. "Help must be sent immediately and could not be too powerful."[92] The German High Command therefore formed a new Ninth Army out of units taken from the Eighth Army and the western front to come to the Austrians' aid. Commanded by Hindenburg and Ludendorff, the Ninth Army shouldered the heavy burden of shoring up the sagging Austrian forces and turning aside any new Russian advance, while trying to regain the initiative that Conrad's defeat had relinquished to Ivanov's forces.

The early September lull in the fighting along Ivanov's front also brought changes in command on the Russian side as Nikolai Nikolaevich named the victorious Ruzskii to replace the discredited Zhilinskii as commander of Northwest Army Group. Yet success infected Ruzskii with such a paralysis of caution that he could think only of a retreat that would move Russia's northwest front some seventy kilometers further east to the great Kovno fortress, even though any such withdrawal would expose all of Ivanov's lines of communication to German attack from the north. With one front commander urging an aggressive advance, and the other demanding immediate retreat, Grand Duke Nikolai Nikolaevich summoned both men to a conference at Cholm on September 13. When both generals refused to compromise, *Stavka* and the grand duke took the lead, as they should have done long before, and ordered a new offensive against Conrad's forces.

Delayed for some two weeks by the Ivanov-Ruzskii feud, the Russians did not begin to follow up their advantage against the Austrians until mid-September, when Nikolai Nikolaevich ordered Plehve's Fifth Army north to reinforce the left wing of Ruzskii's Northwest Group. At the same time, he sent the Fourth Army north to Ivangorod, leaving the Ninth Army to hold its position slightly to the south and west of Lublin.[93] With the First and Tenth armies detailed to guard the East Prussian frontier, and the Third and Eighth armies ready to march against Kraków, Nikolai Nikolaevich began to assemble more than half a million men and 2,400 guns upon the Warsaw-Ivangorod axis.[94] The task of this main Russian force, the grand duke's chief of staff Ianushkevich announced on September 15, was "to prepare for a deep thrust into Germany with all available force."[95]

As so often happened on Europe's eastern front, events moved far more swiftly than the Russians, and the opportunities that Ivanov's victories had presented at the end of August already had begun to slip away as the Germans once again took advantage of the Russians' slowness and indecision. While Ivanov and Ruzskii had disputed with their supreme commander, Ludendorff and Hindenburg had moved to seize the initiative, although their combined resources totaled only 311,000 men and 1,600 guns.[96] With the near-miraculous efficiency that had come to characterize all of their operations, they moved the German Ninth Army some three hundred kilometers south over what Hindenburg called "indescribable roads" to railheads to the north of Kraków. "It was as if we had entered another world," Hindenburg observed as he looked in amazement upon the "physical, moral and material squalor [in which] the Russian administration has left this part of the country."[97] All of this was concealed from the Russians until September 17, when the pocket of a German officer killed in a minor skirmish yielded papers which revealed that Ruzskii had overestimated the number of Germans facing his First and Tenth armies in East Prussia by more than 300 percent, and that four of the corps he had reported to be in East Prussia actually were marching with Hindenburg toward the Russians' Warsaw-Ivangorod line.[98]

Supported by the First Austrian Army, Hindenburg began his attack just as Russian's Fourth and Fifth armies began to move north along the east bank of the Vistula to join the grand duke's planned drive into Germany. "A curious strategic situation," Hindenburg thought as he looked on from his new head-quarters in Kielce. "While hostile [Russian] corps from Galicia were making for Warsaw on the far side of the Vistula," he later explained, "our own corps were moving in the same northerly direction but on this side of the river."[99] After a minor skirmish between advanced units of Mackensen's XVII Corps and the Russians' newly arrived Siberian Corps, the Germans in turn found documents on the body of a dead Russian officer detailing Nikolai Nikolaevich's entire plan, including the strength of the forces he planned to commit to his offensive into Germany. Fearful that the weight of the forces massing at Warsaw would overwhelm them, Ludendorff and Hindenburg hurried their attack.[100] To begin their assault, they gave the redoubtable General von Mackensen three corps. "We are on the eve of a decisive operation in the central Vistula region," Danilov warned one Russian commander. "The fate of the first period of this campaign and, perhaps, the entire war, will be decided [here]."[101]

On the Russian side, Ivanov seemed ready to surrender the initiative while he waited for reserves of men and supplies, and it required much prodding from *Stavka* to convince him to order his Fourth, Fifth, and Ninth armies across the Vistula.[102] For some days, the Germans beat back his efforts until the fierce Armenian and Georgian infantry of III Caucasus Corps forced a crossing toward the end of September. "The trails of the Russian gun-carriages were literally in the Vistula," Hoffmann reported with admiration. "The Caucasians had set foot on the left bank and clung to it fast."[103] Gradually the Russians widened their bridgehead. On September 28, they began the general battle that decided the fate of the German offensive.[104] Once again, supply and communications prob-lems plagued the Russians. "We had to fight with what we had," a young officer in the Caucasus Grenadiers later wrote, "but rifle fire simply was not enough."[105]

The Germans began the battle for Warsaw with five divisions against Iva-nov's nine, and when the sheer weight of numbers at last began to turn the tide against them at the beginning of October, Ludendorff and Hindenburg decided to withdraw rather than risk having the Russians turn their flanks. "Our far-flung battle-line was firmly held in front while superior enemy forces, reaching farther and farther west, threatened to roll up our left flank," Hindenburg remembered. When Conrad's flawed effort to begin a new offensive collapsed a few days later, Hindenburg's right flank lay open to Russian attack as well. Determined to save his army, Hindenburg chose to "break away with a view to being able to employ our army for another blow elsewhere later on."[106] To escape Ivanov's advance, Hindenburg therefore ordered Germany's Ninth Army to fall back toward Częstochowa in the southwest, while Conrad's Austrians continued their flight toward Kraków. By mid-October, with Ruzskii now in command of the Second, Fourth, and Fifth armies, Russia stood poised to invade Silesia, and Ivanov turned again to fight the Austrians.

Worn by forced marches and more than a month of continuous fighting, Hindenburg's men now faced new reserves of fresh Russian troops while their own decimated ranks remained unreinforced. Heavily committed to the bloody and exhausting struggle for the channel ports in Belgium, Moltke's successor, General Erich von Falkenhayn, could promise to release extra corps to Germany's commander in the east only after the fall's great battles ended in the west. Until then, Ludendorff and Hindenburg would have to parry Nikolai Nikolaevich's coming offensive into Silesia with the forces at hand. To do so, they must somehow seize the initiative from the Russians once again. "The problem of saving Silesia," Hindenburg decided, "could only be solved by an offensive" that would strike Ruzskii's right flank somewhere in the vicinity of Łódz.[107]

Heavily outnumbered, Hindenburg's divisions had to aim their strike at the weakest points in the Russian line with pinpoint accuracy, but even air reconnaissance, the newly-discovered eyes of modern commanders, could not provide all the information needed to gauge those points with absolute precision. Again, the Russian wireless helped the Germans. "I cannot help admitting how much the punctual knowledge of the dangers that threatened us was facilitated by the incomprehensible lack of caution, I might almost say naivete, with which the Russians used their wireless," Hindenburg confessed. "By tapping the enemy wireless, we were often enabled not only to learn what the situation was, but also the intentions of the enemy."[108] Once the Russian wireless revealed how Ruzskii planned to dispose his troops, the Germans laid their plans. Now certain of their ground, they sent the Ninth Army several hundred kilometers north from Częstochowa to Thorn on the lower reaches of the Vistula in East Prussia. As soon as Ruzskii's armies committed themselves to an attack toward the west and southwest, Hindenburg and Ludendorff planned to unleash the Ninth Army upon their flank from the northeast.

Now in command of the Ninth Army after Hindenburg's recent appointment as commander-in-chief of all German forces in the east, General von Mackensen drove his divisions into the right flank of Russia's Second Army on October 29 in a well-aimed rapier-like thrust that took Ruzskii totally by surprise. Even though some of his regional commanders had reported a large German build-up near Thorn a few days before, and Second Army's commander had even warned that the enemy was assembling large numbers of troops "for the purpose of preparing an assault against our right flank when we continue our offensive,"[109] Ruzskii had insisted that the invasion of Silesia must begin on November 1 as planned. Three days after Mackensen's first attack Ruzskii still thought that no more than two German divisions were at Thorn, and that they could be held easily by the V Siberian Corps when, in fact, the Siberians already had been overwhelmed. As Stavka urged him to delay his offensive, Ruzskii stubbornly refused. "I have set the beginning of the offensive of the Second, Fifth, and Fourth Armies for November 1," he insisted, "and I shall not postpone it."[110]

Advancing toward Silesia on Ruzskii's orders, General Plehve brought his forces within forty kilometers of the German frontier before Ruzskii ordered him to abandon his offensive and retrace his steps.[111] By that time, five corps of the German Ninth Army had burst through Ruzskii's front and had driven his Second Army back upon its supply center at Łódz. Never one to mince words, General Knox spoke openly about the possibility of another German victory "on a par with Cannae, Sedan, or Tannenberg."[112] As snow and freezing rain blanketed the battle area, the Russians struggled to keep Mackensen's divisions from closing the ring around them. On November 8, the Russians managed to halt the encircling divisions on their right and center and it seemed that they were about to close their own ring around the Germans' XXI Reserve Corps, 3rd Guards Division, and Richthofen's Cavalry Corps, all under the command of General Scheffer.[113]

"Scheffer's sixty thousand Germans were far more completely surrounded than Samsonov's army had been at Tannenberg," one commentator wrote. "Moreover, they were surrounded by vastly superior numbers."[114] But the Russians' "vastly superior numbers" proved less dangerous than they seemed for, as the appalled leader of the Octobrist Party in the Duma reported soon afterward, "reinforcements numbering 14,000 were sent to the right flank, but the men were without rifles."[115] Unaware of that particular problem, and thinking that the capture of Scheffer's force "would be a fitting revenge for the Samsonov catastrophe," General Danilov confidently ordered sixty trains to transport the fifty thousand prisoners that *Stavka* expected to fall into their grip from Scheffer's besieged force.[116] Already rumors of a great Russian victory had begun to spread through Petrograd. "Our victory at Łódz is splendid, complete, and much more important than all our successes in Galicia," Foreign Minister Sazonov promised Ambassador Paléologue, while the army's chief of staff General Mikhail Beliaev confided: "We've won a victory, a great victory. . . . I've been working the entire night to arrange transportation for 150,000 prisoners."[117] Delighted at the news, Paléologue wagered his British colleague Ambassador Buchanan that the war would be over by Christmas.[118]

"The difficult position of the Germans, surrounded as they were on all sides, would become even worse with the deepening frost and the snowless, windy weather," General Danilov thought on the night of November 9 as he lay down to await word that Scheffer had surrendered. But Scheffer was not Samsonov, and he had resolved to deny the Russians the victory they expected. Confident that his men had the will that had eluded the fatigued and hungry soldiers of Russia's Second Army at Tannenberg, Scheffer hurled his forces in a tight wedge against the walls of the Russian trap while Danilov slept. At four o'clock on the morning of November 10, an aide awakened Danilov with word that a message was being received from Ruzskii's headquarters, and the elated general hurried from his railroad sleeping car to *Stavka*'s communications center. Anxiously, he peered at the tape as it fell away from the machine. "In front of me the telegraph unconcernedly tapped out General Ruzskii's message," Danilov remembered:

ATTEMPTS OF THE 6TH SIBERIAN CORPS TO ATTACK UNSUCCESSFUL. . . .
5TH SIBERIAN CORPS COMPLETELY UNABLE TO ATTACK. . . .
IMMENSE LOSSES OF UP TO SEVENTY PERCENT ARE BEING SUSTAINED.[119]

As the crude Hughes teletype continued its unceasing clatter, Danilov learned that Scheffer's retreating troops had escaped, wiping out the VI Siberian Corps and capturing sixteen thousand prisoners as they did so.

Scheffer's escape denied the Russians their victory, but it did not end the struggle that a quarter-million of Hindenburg's Germans had waged against more than a half-million Russians at Łódz. Only when Hindenburg threw his long awaited reserves from the west into the struggle on December 6 did the Russian defenses finally collapse. Before Christmas the Polish Manchester, the rich center of the Russian Empire's huge cotton industry, lay in German hands. Hindenburg had given his kaiser the much-needed victory that had eluded Falkenhayn's efforts to seize Ypres and the channel ports in the west.[120]

The great victory in Galicia, the terrible defeats at Tannenberg and the Masurian Lakes, and the unexpected loss of Łódz left the Russians unable to claim victory or admit defeat at the end of 1914. As on the western front, the first months of the Great War had proved utterly different from the expectations of Europe's General Staffs, for great offensives by vast armies had failed to bring the war to a rapid end. In east and west, the railroad had revolutionized warfare as generals moved hundreds of thousands of men across hundreds of kilometers in the space of a few days. But, in the forests, fields, and marshes of Eastern Europe, with few roads and fewer motorized vehicles, the transportation revolution ended at the railhead, where the movement of men, weapons, and supplies reverted to horse and foot power as they had in the days of Napoleon. Even Hindenburg's victorious Eighth Army had been unable to pursue Rennenkampf's fleeing soldiers much beyond the frontiers of East Prussia because it had outrun its supply lines soon after it crossed into the undeveloped West Russian lands. Nor could the Russians take advantage of the better developed German rail system when they launched offensives into East Prussia or Silesia. Because the Russians had built wider gauge tracks than those used elsewhere as a crude defense against invasion, they had to rely upon whatever rolling stock they could capture once they crossed their frontiers, and neither Germans nor Austrians left much behind when they retreated.

During the fall and early winter of 1914, the civil and military authorities in Eastern Europe's war zone complicated Russia's war effort as the war itself took second place to their constant wrangles over prerogatives and priorities.

Special trains had to be requisitioned every week to bring fresh flowers from the far-away Crimea for the Empress Aleksandra's boudoir even when every locomotive was needed to carry men, weapons, ammunition, and rations to the front.[121] In the recently-occupied regions of Galicia, the newly-appointed Russian officials became so preoccupied with forcing the region's Uniate Christians to accept Orthodoxy that they requisitioned precious rail transport to bring priests into the region just as Hindenburg threw his new reserves into the battle for Łódz. "Here I am expecting trainloads of ammunition," Grand Duke Nikolai Nikolaevich had raged at one point. "And they send me trainloads of priests!"[122]

When the fall campaign ended, Russian losses in killed, wounded, and prisoners exceeded a million and a half men.[123] "Corps of the Line have become divisions; brigades have shrunk into regiments, and so forth," the tsar lamented.[124] Officers and noncommissioned officers made up a startlingly high percentage of these losses. "Shoot at their officers," German and Austrian riflemen were ordered before each Russian attack, and the Russian officers' distinctive uniforms made them easy targets. Many infantry companies lost two-thirds of their officers in 1914, and Nicholas even begged graduating cadets to be especially careful in battle because Russia could not afford so many casualties among her junior officers. Nor were the losses of hundreds of thousands of common soldiers as easy to replace as one might imagine despite Russia's huge population. Although peasants always had tried to avoid service with the colors by cutting off the index finger on their right hands so that they could not fire a rifle, the numbers of these so-called *"palechniki"* soared once the war began. Not only draftees continued this long tradition of self-mutilation, but many men already in uniform simply shot off their fingers in the hope of being sent back from the front lines.[125]

Russia lost hundreds of thousands of weapons as well as men before the end of 1914. During peacetime, General Danilov reported, commanders instructed every soldier that, "if he was wounded, he should give his rifle and cartridges to a noncommissioned officer or, at the very least, to the comrade nearest him, before he went to a dressing station for help." In battle, no one paid any attention to such instructions and simply left their weapons wherever they happened to fall.[126] Able-bodied men sometimes "forgot" to take their rifles when they moved to another position, and often threw them away during retreats. There were even reports that Russian soldiers used precious reserve supplies of rifles to build roofs to shield their trenches from the rain and snow.[127] Soldiers drafted after the war began therefore had no rifles to train with and, before the first snows fell, many of these new draftees actually went to the front lines unarmed. "Reinforcements are coming in well, but half of them have no rifles, as the troops are losing masses [and] there is nobody to collect them on the battlefields," Nicholas wrote to Aleksandra when the war was not yet four months old.[128] At one point, the Russian high command offered to pay bonuses of six rubles for every Russian rifle, and five rubles for each Austrian rifle collected from the battlefields, but this effort, as General Knox noted sadly, "had no useful result."[129]

Even bullets were in short supply during the 1914 fall campaign. Russia's munitions factories produced 58,561,000 rifle and machine-gun cartridges that August while a single regiment—the 108th Saratov Infantry—fired 800,000 rounds, or almost 1.5 percent of the entire month's production, in one day's fighting at Gumbinnen![130] During the first week's fighting, Russian troops were overrun when their artillery ran out of shells, and these shortages grew more frequent as the war entered its second and third months. Late in August, General Brusilov refused General Ivanov's orders to limit the Eighth Army's artillery fire even though Ivanov warned that Southwest Army Group's reserve of shells was nearly spent.[131] Because, as General Knox once remarked, the Russians suffered from a "suicidal desire to represent situations in a falsely favorable light," they at first concealed these desperate shortages from their allies.[132] Only in December, when Grand Duke Nikolai Nikolaevich told him in a sudden burst of candor during one of his visits to *Stavka* that "the artillery has used up all its shells" did Ambassador Paléologue learn the worst. Paléologue was stunned. Could it be possible, he asked himself as he returned to Petrograd that evening. During the past several months, War Minister Sukhomlinov had assured him and Ambassador Buchanan on several occasions that "there is no reason for any serious concern about the present state of munitions in the Russian army," and that "all precautions had been taken to make certain that the Russian artillery should always have an abundant supply of ammunition." Paléologue could not at first believe that the war minister's statements had been so far from the truth.

Fearful that Russia's collapsing war effort might leave Germany free to throw all her resources against the western front, France's shocked ambassador met with his British colleague, and the two hurried to see General Mikhail Beliaev, a man they thought to be "the soul of conscience and honor" at the War Ministry. "Our entire reserve [of artillery ammunition] is exhausted," Beliaev confessed. "The armies need 45,000 rounds per day, while our maximum output is 13,000 at most." But Beliaev had even worse news for the startled diplomats, who wanted to know how Russia was planning to utilize her vast reserves of manpower. "We have more than 800,000 men in our [replacement] depots," he told them. "But we haven't the rifles to arm and train these men. . . . To make up this shortage, we are going to purchase a million rifles in Japan and America." Paléologue could hardly contain himself. "I learned yesterday that the Russian artillery is short of shells!" he exclaimed. "[Now] this morning I learn that the infantry is short of rifles!"[133]

Shortages of officers, shells, rifles, bullets, even boots and uniforms, showed during the very first weeks of the Great War how difficult it was for Russia to sustain a modern war effort. Whether she could continue to fight depended upon how efficiently the home front could be mobilized. All across the Russian land that fall, women, children, and old folk struggled to bring in the grain their men had left unharvested. The fall days grew shorter, but work days in the factories grew longer as anxious factory managers tried to meet desperate army quartermasters' demands for guns, bullets, uniforms, and boots. In the cities, shortages of food, kerosene, oil, and coal appeared briefly but soon subsided. As prices began to climb, city folk worked longer for wages that bought less. As the war's first winter set in, the crucial question became: would the citizens of Russia remain steadfast in their loyalty to tsar and country? None thought war adventurous and heroic now that the terrors of battle had claimed over a million of Russia's youth.

Of those whom death did not seek out on the battlefield or in military hospitals that fall and winter, tens of thousands limped home from the front, torn, bleeding, and forever maimed. Every day that passed taught families, friends, and neighbors the bitter truth that war's horror and suffering left scars that its transient glories could never erase. "An acrid haze hung over Russia all summer," Gippius wrote at the end of September. "With the coming of fall, it began to redden and become more acrid and more terrible still. . . . This general misery does not unite, but only makes us bitter."[134] The days of optimism had already passed, and the year ended on a note of failure as General von Hindenburg drove the Russian defenders from Lódz. Fewer men thought of winning now, and many more thought only of not losing. The war had taken a new direction. Concluded Ambassador Buchanan, "the curtain had risen on the opening act of the great Russian tragedy."[135]

—————————⎯∞⎯—————————

CHAPTER III

The War's First Winter

Fall came quickly to Russia's European lands in 1914. When her young men had marched away to war in August, the grain was not yet ripe in the vast fields of the rich Black Earth region that stretched from the Ukraine eastward to the foothills of the Ural Mountains. For hundreds of miles the sea of grain flowed, broken only by those great rivers whose southern reaches had sheltered bands of Cossacks since the days of Ivan the Terrible. Samsonov's defeat and Ivanov's victories saw the grain ripen, its heavy, full heads "mistily gold as far as the eye could see," a young tramcar driver remembered.[1] All across Russia, peasants knew that even the war could not stand in the harvest's way, for the survival of tens of millions depended upon it. As always, the peasants asked "would there be grain enough?" This, Tolstoi once had said, was the most "terrible question," for its answer brought an affirmation of life or a sentence of death to Russia's villages.[2] A poor harvest would mean that Russia's peasant folk at home faced death just as surely as did their brothers, husbands, and fathers on the battlefields of East Prussia and Galicia.

While their menfolk fought Hindenburg's armies before Warsaw that October, women, children, and old men helped the men not yet called to the colors to bring in the harvest. As in the middle ages, the men wielded scythes, their backs straight and their arms bowed, while the women, bent to the ground, worked with sickles to reap what their men had left. The sharp smell of frost filled the air as anxious hands gathered the grain and carried it to the threshing floor, where hand-wielded flails broke its ears apart so that the kernels of wheat and rye could be sifted from the chaff. Each kernel was carefully collected to be put away along with every other bit of food that could be stored for the long winter months, for everyone knew that the barest of margins would stand between them and hunger at winter's end.

As Ruzskii's defeated soldiers carried their shattered dreams of victory away from Łódz in November, their mothers, wives, and sweethearts gathered the last of the fall mushrooms, dug the few remaining turnips and potatoes from the soil, and set to work pounding flax and carding wool for the yarn they would spin that winter. None knew if their men would return to help with the spring sowing, nor did they know if they would be alive to wear the shirts they would weave during the dark winter days. Usually illiterate, and isolated from regular sources of news in any case, Russia's peasant women knew little of the war's day-to-day progress, but they drew scant comfort from the rare letters their men sent home, for these spoke of suffering, hunger, cold, and fears of facing the cold without boots or winter clothing. "They still haven't given us overcoats," one soldier wrote from the 266th Infantry Battalion. "Everyone's toes are coming through his boots," a man from the 25th Mortar Division complained, while a cannoneer serving with the fortress artillery in Warsaw found himself in a city rich with food, but given daily rations made up of "the sort of [stale] bread we feed the pigs back home."

Anxious to escape richly deserved punishment for the food and clothing they sold to speculators, corrupt and incompetent officers tried to hide their under-handed dealings from their superiors. "For the tsar's inspection they prepared one company, collecting for its use all the best uniforms and equipment from each regiment, and left all the poorly clothed and ill-equipped men in the trenches," one soldier confided in a letter to a friend.[3] Some statesmen still talked of ending the fighting by Christmas. But, as winter's snows began to spread their drifts across Russia's isolated hamlets, the voice of folk wisdom spoke with greater authority. Russia's masses knew all too well that wars were never brief, that their cost in peasant blood was always high, and that their gains were never for the peasants' benefit. In some Russian villages, one peasant in five between the ages of twenty and fifty had been called into the army in 1914. Many more soon would be taken from their midst: 1,180,000 would go in January, 950,000 more in April, and another 550,000 in May. In all, 2,680,000 men would be drafted before the peasants finished their spring planting in 1915.[4]

In Russia's cities, far fewer men were conscripted, but discontent stirred among them for other reasons. Anxious to profit from the lull in labor unrest that had come with the war's outbreak, and desperate to fill the orders that had been pressed upon them by frantic military procurement officers, Russian factory owners lengthened working days and increased their work forces. Wives and children of men who had gone to war were the most obvious sources of new workers, and female and child labor in Petrograd's armament factories nearly quadrupled before the middle of 1915. Demands for more labor also drew men from Russia's villages who joined the urban work force in such numbers that, before the war was a year old, one male factory worker out of every seven in Petrograd had lived in the city for less than a year. With anxious foremen pressing them to work longer at dilapidated machines kept running without proper safety precautions in order to increase production, these workers grew dissatisfied. As newcomers to Russia's factories, they had not been part of the uneasy peace that Russian labor had made with their employers at the war's beginning, and they were even less prepared to continue it than were their more experienced proletarian breathren.

The fall's schizophrenic mixture of victories and defeats deadened the remaining shreds of enthusiasm that any of Russia's city workers might have felt for the war, and they therefore became a new source of labor unrest. The new year opened with a new wave of strikes. In all of Russia, a mere 4,200 workers had gone on strike during the last four months of 1914, but as 1915 began, their numbers soared to 8,800 in January and in February almost tripled again.[5] "Some sort of evil will, persistently striving to undermine the proper conduct of affairs," was at work in Russia's mills and accounted for factory owners' inability to produce all the weapons and ammunition that the army required, a police report concluded at the beginning of 1915.[6] Fearful of new revolutionary agitation, and certain that German agents were trying to undermine workers' morale, Russia's secret police—the Okhrana—detailed legions of new undercover agents to work in the empire's factories and report on everything they saw. Men and women everywhere came under surveillance, most of them for no good reason, and this added still another note of discontent in their lives.

Amid their growing discontent, Russia's masses drew small comfort from the war's progress that fall and winter. For most city dwellers, Ivanov's victories in mid-August seemed far away, in regions little-known and scarcely heard of. Far closer to their daily lives was the stream of wounded that flowed toward the military hospitals and rehabilitation centers that stood in their midst, and this was all the more disturbing because the flood of torn and broken men continued unabated regardless of whether the newspapers carried news of victory or defeat. More than four million men made that journey to the rear during the first three years of the war, although it seems certain that a substantial number of their wounds were self-inflicted. Studies done just after the war found that nearly one wound in five involved the wrists or fingers, and that when wounds involved fingers or toes, they were concentrated heavily on the left, the side of a right-handed soldier's body most convenient to the muzzle of his rifle.[7] Desperate to escape the war after only a few months of fighting, tens of thousands of Russia's soldiers mutilated themselves in their desperate efforts to escape the front before they suffered more serious injury or death as did so many of their comrades.

A constant reminder that war's suffering far outweighed its glories, the sad procession of ruined lives lengthened every week, as men who had not escaped the front in time fell victim to the German and Austrian bullets. Not heroic deeds, but "bloodstained bandages and filthy, crumpled overcoats," the young student Vsevolod Vishnevskii suddenly realized, was the "true essence of war." An expert observer of Russian politics who held the trust of any number of leading politicians and military leaders, England's Bernard Pares visited military hospitals all across Russia and saw again and again the sad sight of "faces bound up [and] limbs missing." For so many of them, life would never be the same again. "One will not forget the figures leaning up in bed and the young, radiant faces," Pares added some weeks later. "Many of these men are cripples who will never fight again."[8] Pares did not exaggerate. Nearly one out of every four who survived their convalescence emerged from Russia's military hospitals permanently crippled.[9]

On familiar terms with Russia's leading politicians, Pares saw more than did those of his English and French colleagues, who came away from brief visits to carefully selected hospitals full of enthusiasm about the care Russia's wounded received. Stanley Washburn, special war correspondent from *The Times*, spoke enthusiastically of "the obvious democracy" with which hospitals were managed, insisted that amputations were "comparatively rare," and felt that Russian medical skill produced very "few cripples . . . left to drag out their lives in misery." Doctors easily convinced him that men recovered from the most serious wounds with amazing speed, and Washburn solemnly reported to his readers that, "scarcely a month after the first flood of war's effects struck them, the hospitals are manifestly becoming sparsely populated" because so many of the wounded had been healed.

Washburn thought that rapid recoveries and low death rates on Europe's eastern front stemmed in part from the fortunate fact that "the Russian soldier is not highly nervous and hence . . . is little apt to die of wounds which would kill a more sensitive man merely from the nervous shock," and was certain that talented doctors and hospitals "as complete in equipment, though not perhaps so luxurious as a city hospital," also played an important part. Helpful Russian doctors assured him that the death rate among the wounded under their care hovered near 2 percent. "It comes as a relief," the readily gullible Washburn concluded, "to know that all human care, skill and kindness can do to alleviate the suffering of the afflicted is being done here in Russia during this terrible time."[10] Never did he see the true depths of "dirt, suffering, neglect, and misery" endured by Russian wounded. Most of these cases—the young Grand Duchess Maria Pavlovna, who dedicated herself to the care of Russia's wounded throughout the war, once explained—were "seldom shown in the capital and were usually sent to provincial towns."[11]

Reports such as Washburn's were nonsense, for Russians wounded in battle faced terrible torments of ill-treatment and neglect, the likes of which remained unfamiliar to the men who fought in the armies of their allies, for military medicine continued to be a crude and cruel art in the Russian army. Even Anna Vyrubova, the empress's ever-faithful confidante, whose memoirs are a paean to the noble goodness of her imperial patrons and their regime, was appalled to find that trainloads of men, whose "indescribable mutilations" were encrusted with "vermin-infected dressings," received almost no medical attention before they were shipped hundreds of miles from the front to Petrograd.[12]

What Vyrubova saw was in no way out of the ordinary. A scant decade before, America's Major Charles Lynch had been shocked at the crude surgery Russian doctors performed upon casualties in the Russo-Japanese War, and condemned some of their techniques as being "so unjustifiable as to be almost criminal." As a medical observer from the United States army's surgeon general's staff, Lynch was appalled to find that Russian surgeons carelessly sliced their way into wounds "without reference to the direction of the muscles, the fibres of which were ruthlessly cut traversely," and looked on in horror as these crude operations sent a parade of stretchers bearing men needlessly crippled to recuperate in filthy, makeshift hospitals. There, many died as much from bad sanitation and disease as from their wounds.[13]

The situation was even worse in 1914 than it had been when Lynch made his report because Russian soldiers faced even more devastating weapons and hospitals had to cope with far more casualties. Toward the end of the war's first winter, German gas shells began to empty their contents into Russian trenches in clouds that seared the lungs of men who had no gas masks, while the explosive bullets used by Austrian soldiers after the early days of the war tore terrible gaping wounds in their victims that, some experts estimated, increased the number of cripples sent from Russian hospitals by almost 40 percent during the first part of 1915.[14] Unable to care for the flood of casualties at the front, doctors hastily shipped them elsewhere. "The wounded were simply piled up on the floors of freight cars, without any medical care, where they died by the hundreds [while being shipped to the rear]," the leader of the Kadet Party in the Duma, Pavel Miliukov, reported.[15]

During long nights spent at the Moscow station to which such trainloads of wounded were shipped from Brest-Litovsk, Konstantin Paustovskii, a young streetcar operator whose sensitivity about human suffering in wartime later won him the Lenin Prize and recognition as one of the Soviet Union's most revered writers, took part in the terrible ritual that required him and his comrades "to drag living men, their bodies torn by shell fragments, into the tram cars like dead freight."[16] Duma President Mikhail Rodzianko passed through the railroad station in Warsaw that fall to find seventeen thousand wounded lying unattended "in the cold rain and mud without so much as straw litter." Furious at such neglect of helpless men, Rodzianko demanded that the authorities take action, only to find that their "heartless indifference to the fate of these suffering men" was supported by a host of bureaucratic regulations.[17]

Callous officials, unskilled, or poorly trained surgeons, and shortages of vital supplies and equipment increased the suffering of the Russian wounded. Only the compassion of Russia's women, many of whom threw themselves into the task of caring for the wounded with true dedication, brightened this gloomy picture. During the war's first winter, concern for the wounded touched all of Russian society, from the widely-publicized work of the Empress Aleksandra and the ladies of her court to the less dramatic but no less comforting efforts of poor working women. Together, these women nursed the sick and wounded and, where doctors' efforts at healing proved of no avail, they comforted untold numbers of men through the last terrible hours before an Austrian bullet or a piece of German shrapnel exacted its final toll.

For some years the Empress Aleksandra had suffered a variety of malaises that seemed to leave her in such a weakened physical condition that she appeared in public only rarely. Now her sicknesses lessened dramatically, an occurrence that lends some substance to the verdict that much of the illness that kept her apart from Russia's people was psychosomatic. Together with her daughters and a few chosen friends, she studied nursing, and, after two months' training, they all proudly pinned the red cross on their gray uniforms to announce that they were war nurses. "From that time on our days were literally devoted to toil," the faithful Vyrubova remembered. "We rose at seven in the morning and very often it was an hour or two after midnight before we sought our beds." All of them, Vyrubova claimed in her usual exaggerated fashion, bandaged wounds, assisted in operations, and prayed fervently for divine guidance and Imperial victory. On one occasion, Aleksandra thought it important to be at a particular hospital to help with an operation in which "one of the officers must have his apendicitis [*sic!*] cut off," and she later proudly reported to Nicholas that *"our apendicitis operation went off well."*[18]

Helping to nurse the sick, and even helping to "cut off" an "apendicitis" when the need arose, probably was not the best way for an empress to serve her people and nation in such terrible times. "She undertook the care of the wounded, or thought she did, forgetting that Russia had thousands of women able to do this work, while she alone, as empress, could arouse emotions and inspire loyalties which no one else could," her niece Maria Pavlovna explained.[19] Aleksandra refused to make the public appearances that might have stiffened the loyalty of Russia's common folk to their monarchs when the strains of war began to take a toll, and they, unsure of their empress, thought her an evil presence who conspired with her German relatives and plotted with dark forces represented by the charlatan peasant healer Rasputin to ruin Russia. Even those closer to the court had their suspicions. There was a "cabal around the empress," Paléologue remarked at the end of 1914, who insisted that "salvation can only come through a reconciliation with German kaiserism."[20]

Almost three-quarters of a century has now passed since the Great War's first winter, and not a shred of evidence has ever been found to implicate the Empress Aleksandra in any plot against her adopted homeland. That Russians high and low believed she had done so was a tragedy made all the more profound because it was partly of her own making. An intensely neurotic woman, Aleksandra had felt unwelcome and unloved from the day she arrived in Russia, and the psychological defenses she erected around herself had grown higher and more formidable as the years passed. At first she had been reluctant to appear in public; after the birth of her hemophiliac son and the revolutionary events of 1905, she appeared only when it became absolutely unavoidable. Even when Russians celebrated the Romanov dynasty's tricentennial in 1913, she had been unable to steel herself to sit through a special performance given at the Mariinskii Opera to commemorate the occasion. "We could see that the fan of white eagles' feathers the Empress was holding was trembling convulsively, we could see how a dull, unbecoming flush was stealing over her pallor, [and] could almost hear the laboured breathing which made the diamonds which covered the bodice of her gown rise and fall, flashing and trembling with a thousand uneasy sparks of light," the British ambassador's daughter, Meriel Buchanan, reported. "Presently, it seemed that this emotion or distress mastered her completely, and . . . she withdrew to the back of the box, to be seen no more that evening."[21]

During the war's first winter, Grand Duchess Maria Pavlovna saw her empress behave similarly on less formal occasions. Toward the end of Aleksandra's visit to military hospitals in Pskov, the young grand duchess begged in vain for her to take a few moments to greet a group of young cadets who were about to leave for the front. When Maria Pavlovna arranged their route to the railroad station so that they would pass by the cadet school where the young men were drawn up in ranks, the empress refused even to stop "just for a second." As Meriel Buchanan had noted at the Imperial Opera, Aleksandra's face again turned "a spotty red" from anger and confusion. A trivial incident, the grand duchess confessed, but she felt that it expressed "something of that increasing inner rigidity and that fading sense of proportion which in this unhappy mother, who chanced also to be the Empress of all Russia, could not but . . . send forth undercurrents of incalculable effect on history.[22]

At the root of what Maria Pavlovna called Aleksandra's "fading sense of proportion" lay her fierce dedication to nursing her hemophiliac son. Ever since his birth in 1904, she had watched the Tsarevich Aleksei's suffering in silent anguish and had grasped desperately at even the slimmest straws that might spare his life and relieve his pain. When medical science failed her, she had turned to herbalists, peasant healers, itinerant holy men, and a motley array of psychological and spiritual oddities upon whom she had long relied for counsel and comfort: Daria Osipova, an epileptic who treated women "whose babies would not stay fixed [in their wombs]," the "Holy Fool" Mitia Koliaba, a cripple incapable of speech whom she arranged to have installed in St. Petersburg's prestigious Theological Academy, and the notorious "Dr. Philippe," who, when he left Russia under a cloud, bestowed upon his benefactress a silver bell with which to warn the tsar when untrustworthy and dangerous people approached.[23] Soon, Aleksandra's quest had brought her to Rasputin, a bogus holy man who preached that salvation was most readily attained through sins of the flesh. Convinced of Rasputin's saintliness, Aleksandra invited him into the bosom of the imperial family circle where he led a bizarre life that stirred bitter resentment among Russians and further widened the chasm that separated the empress from her subjects during the war years. Rasputin's connections with German agents and German banking interests during the war could not be disputed, and Russians who knew of his strange and close connection with their empress hastened to draw conclusions that seemed obvious even though they were unfounded.[24]

Ambassador Paléologue once remarked that the Russian "revels in repentance and delights in making it impressive." He concluded that this stemmed from an "anarchy that is peculiar to the Russian nature" that combined "the highest spirituality and the lowest materialism, the exaltation of the spirit and the mutilation of the flesh . . . [with] asceticism and lust."[25] This strange combination of intense opposites had once held a central place in the writings of Dostoevskii as he had searched the dark inner recesses of the Russian mind for the means to reconcile them. "I have lived a dissipated life, but I loved goodness," Dmitrii Karamazov confessed when he stood on trial for his father's murder. "Every moment, I strived to be better, but I lived like a wild beast."[26] Sinful one moment, contrite the next, repentant Russians often sought the mediation of a *starets*, a holy man, who, in response to God's urging, had renounced the world so as to pray for the salvation of those who lived in it and could serve as a link between sinful men and a forgiving God. "A *starets*," Dostoevskii had once explained, "takes your soul, your will, into his soul and his will," so that worldly men and women might "attain complete freedom, that is, freedom from one's self."[27]

Grigorii Rasputin was a Siberian peasant who falsely claimed to be just such a *starets* and who, after some years of wandering in Siberia, the Holy Land, and European Russia, appeared in St. Petersburg to preach that sin marked the first step to salvation. "How can we repent if we have not first sinned?" he asked again and again. "Yield to [temptation] voluntarily and without resistance," he urged. "Then we may afterwards do penance in utter contrition."[28] To enter upon the path of true repentance, Rasputin recommended the sins of the flesh above all others. In Petersburg, he won a following among high-born ladies bored with lives lived only for the moment, sated with pleasures too readily enjoyed, and anxious for his guidance on their first steps to salvation. Yet there was more than simple boredom in the yielding of Rasputin's flock, for he possessed an inner power from which many found it difficult to turn away. "His eyes held mine, those shining steel-like eyes that seemed to read one's inmost thoughts," Lili Dehn, one of Aleksandra's friends, confessed after their first meeting.[29] The attraction grew stronger; for some it became compelling. Even the hard-eyed French ambassador felt its pull when he met Rasputin during the war's first winter. "His gaze was at once penetrating and caressing, artless and cunning, direct and yet distant," Paléologue remembered. "When he was excited, it seemed as if his pupils became charged with magnetic force."[30]

This "magnetic" attraction, perhaps enhanced by some form of hypnosis, seemed to calm the tsarevich and stem his hemorrhages. To Aleksandra, Rasputin thus seemed God-sent, and not even the greatest of Russia's statesmen could dislodge him from her favor. Count Kokovtsev, the sober chairman of Russia's Council of Ministers, had read the police reports on Rasputin's scandalous public behavior with disgust and had tried without success to have this "Siberian tramp" sent away from the capital.[31] Even Petr Stolypin, Kokovtsev's predecessor and easily the ablest man to serve Russia between the revolutions of 1905 and 1917, had failed to remove Rasputin from court, and the best efforts of this bearlike, barrel-chested statesman, who combined blunt words with hard work in a last-ditch effort to save Russia from revolution, had stirred Aleksandra's wrath to such a degree that she had seen his tragic assassination in 1911 as a clear expression of divine retribution. "Those who have offended God in the person of our Friend," she said of this last great servant of the Romanovs, "may no longer count on divine protection."[32]

Because no one knew the nature of the tsarevich's illness, no one knew the real reason for Aleksandra's devotion to the man all others knew only as an illiterate peasant who publicly engaged in the most shocking sexual excesses. With their emperor and empress hidden away for months on end at Tsarskoe Selo's Alexander Palace, Russians whispered all sorts of fantastic tales about orgies involving Rasputin, the empress, her children, and her friends.[33] None were true, but most enjoyed wide circulation among shocked and suspicious Russians who hurried to assign the worst motives to the empress they rarely saw and did not know. Not even Aleksandra's well-intentioned efforts to succor Russia's wounded during the war's first winter could offset the scandal. "No one liked the tsaritsa," Zinaida Gippius remembered. "Her sharp features, beautiful but bad-tempered and melancholy, with thin lips tightly pressed together, did not please."[34] Separated by a wall of suspicion and distrust that neither could breach, Aleksandra and her subjects soon became consumed by mutual hatred.

Others of the imperial family had plunged into war work even before Aleksandra, and their efforts won greater appreciation although their dedication to Russia's soldiers proved no greater than the empress's own. The Grand Duchess Maria Pavlovna graduated from one of Petrograd's city hospitals as a war nurse less than a month after the fighting began, and, by early September, was serving in field hospitals, "in a world unrelated to any other, [where] death came often," just a few miles behind the front lines at Gumbinnen. There, Maria Pavlovna came to know a touching assortment of sturdy women from all walks of life who had made the cause of the war wounded their own. Among her chief assistants, the elderly peasant woman Zandina, who had first gone to serve the wounded during the Russo-Turkish War of 1877–1878, labored long hours in the hospital, and then dedicated her free hours to laying out the dead and praying over them. "Think only of the many trials the poor soul has to go through in the other world and all alone," Zandina would lament.[35] There were others like her, some of them aristocratic, even princely, but they all shared a common commitment to tending the sick and wounded that poured in a never-ending stream from Russia's front lines in Galicia and East Prussia.

That fall and winter, the flood of wounded required a great deal more than the comforting ministrations of generous-spirited women ready to dedicate long hours to their care as the wretched state of the Russian army's medical service became painfully obvious to everyone. As in other civil and military sectors of the war effort, Nicholas and his chief advisers tried to discourage any involvement by private citizens, but this was one area in which a concerned public most readily defied the government's efforts. "This was the part of the war effort that was least defended against public opinion," Miliukov remembered, "and it was here that civic pressure first forced its way past all of the obstacles that had been placed in its way."[36] Even sensible government officials realized that such involvement had become necessary as well as useful. "The Russian public forces had so far matured by 1914 that it was impossible to restrict their role to that of mere onlookers, as had been possible during the war of 1877–1878," explained Vladimir Gurko, Russia's wartime deputy minister of internal affairs, whose lengthy commentary about the men and issues of his time offers an insider's view of the reign of Nicholas II. "Now they had to be given an active share in the common tasks. This was done," he added, "but with bad grace and with much bickering."[37]

Much of the ill-feeling to which Gurko referred came from his chief, Russia's shamelessly reactionary minister of internal affairs Nikolai Maklakov, who insisted that private citizens must never become involved in government affairs. Once described by Gurko as "a fat, rubicund, cheerful man, a typical provincial dandy, [and] a ladies' man,"[37] Maklakov was anything but cheerful when it came to dealing with men who stood for parliamentary government instead of absolutism. At one point, he spoke of Rodzianko as "stupid and bombastic," and warned that the Duma itself was "paving the way for freedom of revolution." Loyal servitors such as he must somehow stem the tide of pretentious and unwarranted public meddling in state concerns, he insisted. "I am struggling against aspirations that seem to be growing irrepressibly among everyone who, forgetting the tsar, see in public opinion alone the be-all and end-all of everything," he exclaimed[38] as he turned to block even the best-intentioned efforts by responsible citizens to help their soldiers and their government. When Rodzianko proposed late in 1914 to summon a conference of private citizens and politicians to discuss ways in which they could properly become usefully involved in the war effort, Maklakov accused him of plotting to extort political concessions from the tsar. "We know all about your conferences," he warned the astonished Duma president. "You simply want to get together and put forth various demands: ministers responsible to the Duma and, perhaps, even a revolution."[39]

Maklakov's efforts to obstruct private war relief raised a wall of bitter resentment between the government and the public-spirited citizens who searched for ways to aid Russia in her struggle against the Central Powers. Still, the stubborn minister's best efforts could not deter resolute civilians from becoming deeply involved in Russia's war effort. Urged on by Rodzianko, men who represented Russia's local rural councils (called *zemstva,* singular *zemstvo*) gathered in Moscow on July 30th to form the All-Russian Union of *Zemstva* for the Relief of Sick and Wounded Soldiers. Two weeks later, representatives of Russia's town and city councils formed the All-Russian Union of Municipalities, and together they launched their first efforts to succor their nation's war casualties. Moscow's mayor Mikhail Chelnokov headed the Union of Municipalities, but it was Prince Lvov, elected president of the All-Russian Union of *Zemstva,* who directed most of the work.[40]

Prince Georgii Evgenevich Lvov came from the province of Tula, a region famous for its samovars and metalwork, where he had been born in 1861, the year of the serfs' emancipation. Fifty-three when the Great War began, he had a quarter-century of public service to his credit, including almost a decade spent as a deputy in the Duma. Perhaps more than any other private citizen in Russia, he was experienced in dealing with precisely the sort of obstructions that Maklakov raised in the path of *Zemstvo* Union, for he had once before faced the unyielding opposition of a minister of internal affairs when he had organized a nationwide *zemstvo* program to aid sick and wounded soldiers just returned from the Russo-Japanese War. Despite the best efforts of Maklakov's far stronger predecessor, Viacheslav Plehve, these private efforts had provided valuable support services to wounded soldiers and their families that the government had proved unable or unwilling to offer, and Lvov's reputation as a public-spirited and philanthropic-minded servitor of the public good had soared. During the next decade, the prince had continued to dedicate himself to famine relief and Duma affairs, always standing aloof from the corrosive conflicts of party politics, and always seeking to widen the gap through which private citizens could participate in government affairs. A man whose name had become widely known and deeply trusted throughout Russia, he now placed his great prestige at the service of the Great War's victims at a moment when the government had neither the resources nor the organization to nurse them back to health.[41]

According to the plans of the War Ministry, wounded soldiers were to be evacuated to central clearing hospitals located hundreds of miles behind the front, after which they would be dispersed to a network of smaller hospitals spread all across European Russia to receive most of their treatment. Lvov's *Zemstvo* Union performed perhaps its greatest service in providing this last type of hospital facilities in the rear and in transporting wounded soldiers to them. The Union's effort required thousands of railroad hospital cars. When none could be found, its workers transformed freight trains into railroad ambulances overnight by suspending cots from the roofs of simple boxcars sealed with felt insulation and heated with cast-iron stoves. Although orderlies proved nearly as difficult to come by as doctors because the army had first claim on "intelligent, patient, well-disciplined, and strong men," resourceful *zemstvo* leaders staffed their trains with Mennonite conscientious objectors whom the government had required to perform alternate service.[42] Before the war was six months old, Lvov's workers had put into service nearly fifty such hospital trains capable of transporting up to sixteen thousand wounded at a time.

Concerned with results first and costs second, the *zemstvo* men worked in haste and paid considerably less attention to costs than did government agencies. Buildings were erected in as little as two months, and the price of bringing a new hospital bed into operation fluctuated wildly between three hundred and seven hundred rubles. All across Russia, the *zemstvo* men passed vital supplies on to the army: 7,500,000 suits of underwear at the end of September, a quarter of a million tents in October, and fur winter clothing for another quarter-million men in November. Some government officials were so appalled by the casual accounting procedures that Lvov and his associates employed that Vladimir Gurko later insisted that, "had the war not ended in revolution, the heads of this [*Zemstvo*] Union surely would have been called to account."[43] By the end of 1914, the Union had collected and spent more than twenty-five million rubles on medicines for the wounded and clothing for soldiers at the front. Within a year, the sum quadrupled.[44]

Certainly, waste and corruption marred the Union's efforts, but it produced results, precisely as Lvov had hoped. In two weeks, the *Zemstvo* Union provided nearly twelve thousand beds for the wounded; after six months of fighting, its agents had organized fifteen times that number. Urged on by Lvov and his allies, wealthy citizens endowed small hospitals, and groups of professional people and workers joined together to do the same. In Moscow, peasants brought in cabbages, potatoes, and other winter vegetables to help feed the wounded, and in Kaluga they collected thousands of yards of homespun linen for bandages, sheets, and underwear.[45] "From surgical instruments and materials for dressings, to evacuation trains, distribution centers, medical personnel, and hospitals —all this was provided in time and staffed by the *zemstvo* organization," wrote Miliukov in praise of the achievements of the many men and women whose help the Russian government continued to accept with such ill grace and obvious distaste.[46] An admirer of Lvov and an enemy of the government, Miliukov tended to inflate the Union's achievement. But even when the superlatives and overstatements are stripped away, the fact remains that the *Zemstvo* Union accomplished deeds in support of Russia's early war effort that the government could not match.

Aid to the wounded marked only the beginning of public involvement in the war effort during the war's first winter on Russia's home front. During the first months of fighting, Duma President Rodzianko visited *Stavka* to learn the full dimensions of the army's supply crisis from Grand Duke Nikolai Nikolaevich himself. "I have no rifles, no shells, no boots," the grand duke confessed. "Go

back to Petrograd and get my army shod. . . . My troops cannot fight barefoot."
Determined to use Nikolai Nikolaevich's revelations to enlist the support of
Russia's civic leaders, Rodzianko left *Stavka* appalled to think of men being
ordered to fight without weapons and fearful about the impact of such terrible
failings upon soldiers' morale, public opinion, and the war's outcome. He therefore returned to Petrograd certain that "you can't fight a war with nothing but
sticks" and convinced that, "without a general upsurge of everything that is
productive [in society], nothing would be accomplished." In one way or another,
Rodzianko felt certain, all Russians must be drawn into the war effort.[47]

That some of Russia's soldiers faced the winter's first snows with only canvas
wrappings for their feet, stirred public outrage nearly as deeply as the thought
of wounded men lying unattended on railway station platforms and in boxcars.
Rodzianko decided that "if one could manage to drag public opinion into the
matter of boots [for the army] then half the task would already be accomplished," and he turned to the Unions of *Zemstva* and Municipalities for that
purpose. Again, Maklakov barred the way and reportedly hurried to warn the
Council of Ministers that Rodzianko, "a man of agitated temper, [has begun]
poking around in matters that are none of his business." Uncertain about how
to draw public opinion to the side of the government, and unwilling to learn
how that might be done, Maklakov preferred coercion to voluntary public
action, as one noblewoman learned at the beginning of 1915 when she saw a
group of sullen artisans being herded through her native Kaluga by a squad of
police.

"Where are you going," she asked. "And why?"
"We're going to sew boots. There's been an order to round up all the
shoemakers in Kaluga to sew boots," one of the men replied.

It was "a characteristic stroke," Rodzianko concluded. "From that moment, at
the beginning of January, arose an obstinate, politically conscious, union of
public opinion," that brought Russia's citizens directly into opposition to their
government.[48] Once aroused, Russian public opinion urged the Duma to take
an active part in public affairs and demanded that the *Zemstvo* Union and the
Union of Municipalities be allowed to provide the army with those vital supplies
that the government seemed unable to obtain.[49] Nikolai Nikolaevich's plea to
"get my army shod" gave the *zemstvo* men the wedge they needed to edge their
way into those complex problems of military supply that had baffled Russia's
quartermasters since the war's beginning. Why were there no boots to be had
in a country where almost all men wore boots, *Zemstvo* Union officials asked.
Late in 1914, they set out to learn the answer.

Their search revealed another of those gross oversights that proved all too
sadly typical of the blindness and gross incompetence of Russia's military establishment at the time. In 1913, Russians had produced somewhere between forty

and fifty million pair of boots of which only about two million had been bought by army quartermasters. This did little more than keep the peacetime army shod, for Sukhomlinov's War Ministry laid in very few reserves for the millions of men certain to be inducted if Russia went to war. The shortage of boots therefore began just days after the Great War's first battles, and became acute within a matter of weeks. Evidently no one in the War Ministry had ever asked how millions of pairs of boots might be obtained in wartime, nor had they thought about how they could be manufactured in a country where artisans working in thousands of small workshops still produced most of the boots. Nor had they even made any plans for producing enough leather to meet the army's needs. Only after the war began did amazed investigators discover that Russian tanners had been almost completely dependent upon Germany for the chemicals they used, and that only one factory in the entire empire could produce tanning extract.[50] Russian bootmakers therefore could not find enough leather even to make as many boots in 1914 as they had the year before. Yet the demand rose so rapidly that the army soon called for a quarter of a million pair of boots a week.[51] As the spring of 1915 approached, Nikolai Nikolaevich's soldiers needed still more boots, and neither the army nor the *Zemstvo* Union had yet succeeded in producing them. In desperation, the Union turned to factories in the United States, where they bought almost five million pair of American boots for Russian soldiers before the end of the year.[52]

The boot shortage was only one crisis among many that struck Russia's industrial sector during the war's first winter. Anyone who produced any sort of military supplies, weapons, or ammunition found it difficult to assemble vital raw materials, in part because of the strange fashion in which Russian industry had developed. European and American industrialists had taken great pains during their industrial revolutions to build factories near sources of raw materials and fuel in order to minimize costs, but very different considerations had guided their Russian counterparts ever since the beginnings of Russian industry during the long-ago days of Peter the Great. For more than a hundred and fifty years, most of Russia's iron ore had come from the Ural Mountains, where peasants had smelted it in primitive back-yard charcoal furnaces, and shipped it over a complex network of rivers and canals to factories in St. Petersburg. When such crudely smelted pig iron could no longer meet the growing demands of modern industries, Russian industrialists began to import pig iron and steel over fragile Baltic sea lanes from Europe rather than develop a modern smelting industry in the south, where vast stores of iron ore in Krivoi Rog stood some eight hundred kilometers away from equally large deposits of coking coal in the Donets coalfields. Not until the very end of the century did railroads connect these two vital concentrations of natural resources, but the iron and steel they produced could be more easily exported through the Dardanelles and sold abroad than shipped nearly two thousand kilometers overland to Russia's industrial centers around St. Petersburg. On the eve of the Great War, Russia therefore found herself in the absurd situation of exporting iron and steel from the south and importing it in the north because she had not developed any efficient way to ship the products of her coalfields and blast furnaces to her factories.[53]

Other considerations that compare strangely with the experience of Europe and the United States also had figured prominently in the development of Russia's heavy industries, and these made her wartime difficulties even worse. Because the favor of the tsar or one of his ministers always promised rich profits, the first lords of Russia's heavy industry had been far more concerned to establish their factories close to court rather than to locate them nearer to distant sources of raw materials. Russia's modern armaments industry therefore grew up in Petrograd, far from any iron and coal deposits, even though armaments factories consumed huge quantities of both. Because it was cheaper to ship coal from England through the Baltic Sea to Petrograd than to transport it overland from the Donets coalfields, most of the several hundred factories that produced military goods in Russia's capital depended upon imported fuel. In 1913, the city's six largest armaments complexes alone burned almost half of the three million tons of coal that the city consumed.[54] These fuel supplies could be easily cut off by enemy ships, but it was not until June 1914 that Russian planners began to think about moving their nation's war industries away from Petrograd so that they could have more ready access to domestic resources.[55]

While Russia's industrialists looked on helplessly, coal imports from England plummeted by nearly 90 percent during the first weeks of the war. Suddenly caught with less than three days' reserves, Petrograd's war industries called for more Russian coal, only to find that the first mobilization levies had taken nearly a third of the miners from the Donets coalfields. Mine managers marshaled women, children, and even prisoners of war to mine coal,[56] but even these efforts could not prevent the empire's coal production from falling precipitously just as desperate generals called for more shells, rifles, and field guns than ever before.[57] Before the war, these generals had estimated that half a million artillery shells would supply the army for at least a month. Now they tripled their estimate and quickly doubled it again. Beyond that, battle losses of almost a quarter-million rifles a month had to be replaced, and two and a half million draftees had to be armed. Russia's factories could not hope to supply weapons in such quantities. Sukhomlinov had reduced weapons orders so drastically before the war that the entire Russian armaments industry had been called upon to produce just *forty-one* military rifles during the entire first seven months of 1914! Struggling to recover from their state of supreme unpreparedness, and short of fuel throughout the winter, Russia's armaments factories were not able to produce more than forty thousand rifles a month until the middle of 1915.[58]

To make matters worse, major coal producers had consistently underproduced during the decade before the Great War in an attempt to keep fuel prices high, and they therefore had few reserves to draw upon when bitter frosts and unusually heavy storms struck European Russia at the end of 1914. Loss of the

Dombrowa coal mines in Poland made the shortage still worse, as harried Transportation Ministry officials frantically tried to move whatever fuel could be found in the Donets coalfields to the capital. In turn, their effort disrupted vital food shipments so thoroughly that tons of grain accumulated on isolated sidings all across Russia that winter to await freight cars and locomotives that never came. After a year of war, more than two million food shipments consigned for Petrograd had been shunted aside in that fashion.[59] Every month, the problem worsened. "There is enough grain in Russia," one frustrated official wrote, but people will go hungry and even die from hunger simply because of improper distribution."[60] Prices rose faster than wages; the first of the war's many shortages began.

During the war's first winter, rising prices began to cut away those slim supports that kept many factory workers' marginal living standards from collapsing into destitution. The average price of poor quality beef almost doubled in Russia's cities during the first year of the Great War, and rye flour sold for almost half again as much as before the war. In some industrial regions, the transportation crisis drove prices even higher, so that workers in Vladimir's cotton mills saw the cost of rye flour, dried peas, oil, salt, and soap more than double in less than a year, while their wages only rose by less than an eighth. For more than a quarter-million Petrograd factory hands, who lived on a diet of coarse rye bread and a kind of watery pickled cabbage soup called *shchi*, rye flour and salt almost doubled in price and the cost of cabbage nearly tripled. Still almost the universal source of heat in lower class dwellings, wood in the spring of 1915 cost half again as much in Petrograd as it had in the fall of 1914.[61]

As the army's pleas for weapons, ammunition, supplies, and clothing grew more urgent during the war's first winter, factory managers and mine owners begged the government to set aside some of the regulations that forbade women and children from working underground, in dangerous jobs, or on night shifts. Anxious to meet the army's demands and, in any case, unsympathetic with workers' complaints, the Imperial Council of Ministers readily gave its consent.[62] Child and female labor in Petrograd's mills rose by nearly a third during the war's first year, and almost four times as many women and children worked in weapons factories by 1915 as before the war. In the clothing factories and chemical plants of Moscow, in the cotton mills of Ivanovo-Voznesensk, in the Donets coalfields, the same pattern prevailed. Women and children hired at lower pay filled the places of men who had been called to the front. Soon, millowners turned to refugees, prisoners of war, and, even, laborers imported from China and Korea, all of whom could be hired at lower wages. By the end of 1916, women and children working in Russia's factories and mills actually outnumbered the men. Overtime became compulsory, and factory managers began to add hours to the normal ten-hour work day.[63]

As they worked longer hours, and watched soaring prices erode their minimal living standard, Russian workers regretted the uneasy truce they had made at the war's beginning. "The rising cost of living is a well-known phenomenon about which we all have heard," a workers' newspaper reported. "We have not heard anything about raising the workers' wages," the reporter went on, "nor have we heard about improving working conditions." In reply, Petrograd's millowners resolved that "under no circumstances should wages be increased and, if possible, they should be reduced" because it would be "extremely difficult to lower them again when the war ends."[64] This was short-sighted and danger-ous, as the government understood all too well. "Because prices for the most basic necessities have risen sharply from the moment war was declared," an Okhrana report warned, "wage reductions make the workers particularly bit-ter."[65] Nervous authorities tried to quell the inevitable protests even before they began, and immediately sent workers who complained about such treatment to the front. Still apprehensive about labor unrest, and fearful of enemy spies in their midst, the Okhrana sent out scores of secret agents in search of "some sort of evil influence persistently seeking to destroy the proper flow of [factory] work," while foremen imposed rigid standards of labor discipline at every oppor-tunity. At the moment, there seemed to be no danger of a return to those massive labor protests that had won for workers at least part of their demands in seven strikes out of every ten in the year before the war.[66]

Despite the best repressive efforts of factory managers, police, and soldiers, however, Russian workers announced their discontent during the war's first winter by a new wave of strikes. At the beginning of November 1914, five deputies who represented the Bolsheviks in the Duma and a handful of repre-sentatives from other industrial centers met secretly in the suburbs of Petrograd to plan new protests against the conditions under which they were forced to live and work. On November 4, the third day of their meeting, Minister of Internal Affairs Maklakov's army of spies and informers learned their whereabouts and the police struck that evening as a matter of course. "At about 5 p.m., we heard a deafening knock on the door [and] . . . our room was invaded by a crowd of police and gendarmes," remembered Aleksei Badaev, a metalworker who had risen through the ranks of Lenin's Bolsheviks to become a Duma deputy. Although Badaev and the other deputies protested that any search or arrest violated their parliamentary immunity as deputies to the Duma, the police combed the apartment, confiscated their papers, and returned two evenings later to arrest them all.[67]

News about the Bolshevik deputies' arrest caused a sensation. All but a modest handful of radicals cheered the police and congratulated themselves that men who, in the words of *Pravitelstvennyi vestnik, The Government Herald,* had "devoted their efforts to shaking the military strength of Russia" no longer remained free to continue their "seditious socialist tasks."[68] Usually supremely sensitive to any infringement of its prerogatives, the Duma remained silent, and Rodzianko filed only the mildest *pro forma* protest against the police's actions with his enemy Maklakov. Nonetheless, the incident sparked the very wave of protests among Russia's workers that the government had gone to such lengths to prevent. "Comrades!" a hectographed proclamation from the Petrograd Bolsheviks proclaimed on November 11. "The autocratic government has treated the Duma delegates of thirty million members of the Russian working class with insolence and cynicism! . . . The government is trying to smother the revolutionary movement of the working class beneath the thunder of guns and rifles. It hopes to drown their aspirations for freedom in rivers of the blood that has flowed from the millions of workers and peasants that it has driven into battle."[69]

The Bolsheviks' proclamation called for meetings and one-day strikes to show the workers' disapproval of the government's policies. The next morning, strikes began at Petrograd's New Lessner Iron Foundry and spread to the Parviainen Works, where four out of every five workers hired to make bullets and desperately needed artillery shells on the day shift (more than a thousand in all) joined the protest. Workers in the Donets coalfields followed, and so did some printers in Kharkov and factory hands at the Bromley Works in Moscow. By comparison with the massive labor protests of early 1914, in which more than a million and a quarter workers had gone on strike before the end of July, these few strikes seemed unimpressive. Only two thousand workers went on strike in November, but their protest indicated that the government's shaky labor truce had run out.[70] "Even these strikes showed that the working-class movement had not been altogether stifled," Badaev wrote. "Sooner or later it would rise again in all its strength."[71] In reply, the government moved quickly to send Badaev and his comrades to exile in Turkestan, thousands of kilometers to the southeast, where they were destined to remain until the Revolution of 1917 made it possible for them to return. Other prominent labor leaders received sentences to be drafted into the army and sent immediately to the front. "There is no need to stand on ceremony with such enemies," the right-wing *Russkoe znamia, The Russian Standard,* exclaimed. "The gallows is the only means for restoring peace in our country."[72] "All workers ought to be mobilized, placed under martial law, and assigned those enterprises where they are needed," the more moderate *Utro Rossii, Russia's Morning,* added a few weeks later. "In times such as these working at a bench or a machine is just as necessary and crucial as being a sentry at an outpost."[73]

The scattered strikes of late 1914 continued into the new year and rose in a crescendo of labor protest that spring and summer. Twenty-three thousand workers struck in February. The number grew to almost thirty-six thousand in April, nearly sixty thousand in May, and eighty thousand in June.[74] June saw a major strike in the Kostroma linen mills, where the authorities ordered police and militia to use gunfire to disperse the strikers. "Soldiers!", one group of women pleaded. "We need your help. Our fathers, sons, and husbands have been taken from us and sent to the Front. . . . We are defenseless! You must defend us! We are hungry . . . but no one listens." Certainly, the authorities were not prepared to listen at that moment. "Seeing that they were not going to stop throwing stones," the deputy chief of Kostroma's police reported, "we opened fire." As their officers commanded, the soldiers aimed into the crowd, not above it, and bodies crumpled to the pavement as the shots rang out. Among the dead and gravely wounded were Katia Ivanova, Mitia Osipov, and Vania Smyslov, all not yet sixteen, and Misha Korolev, who had just turned ten. Only three of the casualties that day were over twenty, old enough to have been drafted into the army.[75]

Although he traveled more than fifteen thousand kilometers to inspect Russia's army, weapons factories, and military hospitals between mid-November and late April, the tsar neither felt the new wave of bitterness that drove his nation's working men and women into massive strikes, nor perceived the frustration that countless patriotic Russians experienced when high officials crudely interfered with their efforts to help the war effort on the home front. Nicholas made several journeys to *Stavka* during those months, inspected Russian positions at Przemysl, Lwów, and Sambor on the Austrian front, toured the empire's Crimean naval bastion at Sevastopol, and visited the Army of the Caucasus at Kars and Sarakamysh. Everywhere, he prayed at the local church, spoke comforting words to the wounded, thanked soldiers for their bravery, and decorated many with the Cross of St. George for valor. In December, he visited Moscow for a week, met with Moscow's mayor Nikolai Chelnokov, and discussed the medical work of the *Zemstvo* Union with Prince Lvov. As he journeyed through his empire, Nicholas dutifully recorded in his diary the weather, the time he arose, the hour he went to bed, the names of the cities and fortresses he visited, and recalled fond memories of previous visits. He took brisk walks after dinner, dedicated most of his evenings to reading for pleasure, and sometimes played dominoes with his staff before he went to bed. Everywhere, he drew positive conclusions about Russia's war effort and seemed happy not to question the optimistic facade that anxious generals and officials spread before him. Never had a tsar traveled so widely in so brief a time, and rarely had a tsar perceived so little of the truth about his land and people.[76]

Pleased by the results of his inspections, Nicholas paraded his optimism when he returned to celebrate the New Year at Tsarskoe Selo. On the afternoon of New Year's Day, he received Petrograd's diplomatic corps in a ceremony which, Paléologue reported, was marked by "the full range of pageantry, opulence of setting, and that unequaled pomp and grandeur in which the Russian court is without rival." It was one of those days that come so rarely in January to relieve the gloomy half-light of winter, when the air's thick frost diffuses the sun's weak rays to cast an ethereal glow upon people and their surroundings in Russia's northern capital. Nicholas met the world's diplomats in the Catherine Palace's Hall of Mirrors that the brilliant eighteenth-century architect Bartolomeo Rastrelli had built at the command of Peter the Great's flamboyant youngest daughter the Empress Elisabeth to rival the grandeur of Versailles.

With Paléologue, the tsar spoke warmly and at length. "I have just been visiting my army, in which I found the most splendid ardor and *élan,*" he began, as he tried to share his optimism with his ally's ambassador. "As soon as possible," he promised, "my army will resume the offensive to continue the struggle until our enemies sue for peace." To Paléologue, who only a fortnight before had learned that the Russian army had no reserves of bullets, shells, and weapons, the emperor's enthusiasm seemed clearly divorced from reality. To his diary he confided his admiration for Nicholas's "splendid moral resolve," but thought him "manifestly unequal to his task," a failing made worse, he explained, because "autocratic tsarism" had become a "geographical anachronism" in the twentieth century.[77]

Other events augured ill at the New Year because they weakened the counsels of good sense around Russia's weak sovereigns. The day after Nicholas spoke with Paléologue, Aleksandra's closest friend and confidante Anna Vyrubova nearly died in a train wreck. For several hours the injured woman lay trapped in the wreckage before rescuers dragged her from beneath the twisted steel and took her to a hospital where surgeons told the empress that her friend could not possibly live out the night. At her wit's end, Aleksandra summoned Rasputin from Petrograd. "Annushka," he said to the unconscious Vyrubova, "wake up and look at me." The unconscious woman stirred, opened her eyes, and whispered: "Grigorii, is it you? Thank God." Rasputin took her hand. "She will live," he told the amazed onlookers. "But she will always be a cripple."[78]

From that moment, Vyrubova's slow and painful recovery began, and Rasputin's influence with Nicholas and Aleksandra soared, even though he behaved more scandalously in public than ever. To the patrons of one of Moscow's leading restaurants, he once gave a grotesque account about how he had ravished some of Petrograd's high-born ladies, and described their nude bodies in obscene detail. Now genuinely fearful of the impact that Rasputin's evil influence might have upon Russia's war effort, former governor general of Moscow Vladimir Dzhunkovskii, who as deputy minister of internal affairs knew only too well the *starets*'s many sins, assigned a detachment of secret police agents to report on his unsavory activities.[79] His effort enraged Aleksandra and led to his disgrace. "It's so vile—always liars, enemies—I long knew Dzhunkovskii hates Grigorii," Aleksandra babbled in a letter to Nicholas at the beginning of the summer. "Oh my Boy, make one tremble before you," she begged her husband. "When at last will you thump with your hand upon the table & scream at Dzhunkovskii and others when they act wrongly?"[80] "Well, your Dzhunkovskii's had it," Rasputin told one of the police agents he knew. Less than a week later, Nicholas ordered him to resign.[81] Without much thought for the consequences, the tsar had driven away yet another of the few remaining men of good sense who tried to steer him away from the mediocre and treacherous friends whom his empress pressed upon him.

Perhaps worst of all, Aleksandra's blind faith in Rasputin's preachings made her arrogantly suspicious of everyone outside her inner circle. Combined with Nicholas's natural tendency to reject even well-intentioned public comments, this drove thoughtful Russians further from their sovereigns and made them more hostile to the men who represented them in government. "The antagonism between the imperial authority and civil society is the greatest scourge of our political life," Russia's thoughtful minister of agriculture Aleksandr Krivoshein lamented not long after the new year began. "The future of Russia will remain precarious so long as government and society insist upon regarding each other as two hostile camps."[82] A man of subtle political instincts who saw the value of broader public participation in government and had won a loyal following in the Duma, the urbane and well-connected Krivoshein feared the tension he sensed growing in Russia in late 1914. At least in public, this seemed somewhat lessened when news that Turkey had joined the Central Powers stirred a brief surge of patriotic fervor against Russia's centuries-long enemy at a three-day Duma session late in January. Private meetings between Duma members and the imperial ministers about the budget, however, made it clear that deep antagonisms lay just below the surface, ready to burst forth at any moment. "After the January session," Miliukov wrote, "the relations between the government, the Duma, and civil society in the broadest sense began to deteriorate rapidly. . . . The bureaucracy continued to suspect society of revolutionary tendencies and entered into an open struggle against it."[83]

Before the war's first winter ended, the crude paranoia of Nicholas, Aleksandra, and their favorites had all but destroyed the great sense of national unity and patriotic purpose that had swept Russia in July and August. Nicholas took to bestowing the coveted Cross of St. George *en masse* upon whole units at the front in an absurd attempt to improve morale.[84] In Russia's countryside, tension mounted at the beginning of 1915 as peasant villages made ready for the spring sowing without the men whom politicians had promised would return before the New Year. As food prices soared in the cities, politicians cursed fumbling, self-interested, and vicious ministers in whom they had lost confidence. A dark shroud of whispers about betrayal and treason settled upon the land.

That spring, angry Russians learned that the dapper Colonel Sergei Miasoedov had been accused of selling military secrets to the Germans for the third time. The former commander of a gendarme post in Russia's western frontier region whose preference for companions of questionable character was well known, Miasoedov had associated too frequently with Germans and was known to have visited the kaiser at his hunting lodge in the Romintern Forest on several occasions before the war. He had been dismissed from the army in 1907 and again in 1912 because his superiors strongly suspected him of being in the pay of Germany. On both occasions, Sukhomlinov had intervened, returned him to duty, and assigned him to more responsible and sensitive posts where his conspicuously high standard of living and notorious acquaintances quickly stirred new suspicions. By early 1915, Sukhomlinov could no longer shield his favorite, and Miasoedov was court-martialed and executed (unjustly, an inquiry by Soviet historians later proved) for selling Russian military secrets to the Germans.[85]

Known to have supported Sukhomlinov's defense of Miasoedov on one occasion, Nicholas became the object of rumors that he planned to compromise Russia's honor by a separate peace with Germany, and these persisted despite any number of vehement public denials.[86] "The days of tsarism are numbered," the great armaments manufacturer Aleksei Putilov muttered one evening in late spring, as the white nights began to return to Petrograd. "Revolution has become inevitable."[87] The time was not yet ripe for the revolution Putilov foresaw. First there would be more suffering to bear and new crises to sustain. All across Russia, the air filled with tension as men and women hoped for the best but expected the worst. "The entire country had fallen silent," Konstantin Paustovskii remembered, "as if it were wondering how to ward off the impending blow."[88]

PART TWO

1915

CHAPTER IV

The Road to Disaster

Throughout the centuries, Russia's masses had endured war, disease, and material privation with fortitude, and only rarely had sought redress for their suffering by violent confrontations with their tsar and his government. Time and again in the half-century before the Great War, peasant fatalism had thwarted the hopes of Russian radicals, whose misplaced certainty that the peasantry *must* revolt rather than endure their wretched condition condemned several generations among the nation's young revolutionaries to long decades of failed dreams. Peasant fatalism seemed without limit. *"Eto nichego!"*—"There's no point in worrying about it!" "It's of no consequence!"—Russia's common folk remarked with a shrug as they turned to face the new misfortunes that the industrial age brought to their lives. At the dawn of the twentieth century, revolutionaries despaired of ever turning these stoic peasant millions against the tsar. "Scientific socialism," the revolutionary Sergei Kravchinskii confessed in a moment of frustration, "bounces off the Russian masses like a pea off a wall."[1] Even when a half-million of their village breathren died from starvation and disease during the great famine of 1891, Russia's peasants remained passive.[2] "Fatalism and piety are the very essence of all Russian souls," Paléologue decided. "For the vast majority among them, God is nothing more than the theological synonym for Fate."[3]

Russians high and low shared the certainty that their land held a unique place in God's esteem, and if God and Fate were one, then it followed that the workings of Fate must be in keeping with God's greater plan. Ever since the sixteenth century, Russians had thought of Moscow as the Third Rome—the last bastion of Christian purity in an impious world—and had believed that that single fact had raised Russia above all other nations. Nicholas I, ruler of Russia when her army was thought to be the strongest in Europe, often spoke of "the Russian God," and assumed that he and his subjects enjoyed God's special affection.[4] For him, and for the Romanovs who followed him, Orthodoxy, Autocracy, and Nationality—all inextricably joined in the tsar's person—thus held the key to Russia's greatness. Even in the twentieth century, Russian Orthodoxy remained a central part of Russian autocracy and Russian nationality. It was not mere rhetoric that had impelled Nicholas II in 1906 to ask his ministers if he had "the right to change the form of that authority which my ancestors bequeathed to me" as he struggled against the awful certainty that he must keep his promise to grant Russia a constitution.[5] For many Russians, and perhaps for Nicholas II most of all, Fate thus expressed God's will. It was not resignation, but an ardent statement of faith, when these men said, as Nicholas I had written to one of his field marshals a few months before his sudden death: "His will be done!"[6]

New German victories all along the eastern front tested the limits of Russian faith and fatalism at the beginning of 1915. "It doesn't matter!" one member of the imperial family told Paléologue in February. "If it becomes necessary to retreat further, we shall do so."[7] At that point, a major retreat seemed very likely as Hindenburg and Ludendorff made ready to press the advantage they had seized at Łódz in December by preparing another massive two-pronged assault into Russia. "The experience of Tannenberg and the Battle of the Masurian Lakes had shown us," Ludendorff later explained, "that a great and rapid success in battle was only to be obtained when the enemy was attacked on two sides."[8] As winter struck in full force at the beginning of 1915, Germany's two victorious strategists therefore deployed the Tenth Army that they had just formed from four corps wrested from Falkenhayn in the west for an attack southward from staging areas in the vicinity of Tilsit, Insterburg, and Gumbinnen toward the right flank of Russia's Tenth Army. At the same time, they ordered the Eighth Army to make ready to strike eastward from Lötzen, Ortelsburg, and Thorn toward Lyck and Augustowo. To prevent the left flank of Russia's Tenth Army from wheeling northward and striking their Eighth Army in the rear while it moved into its new positions, the German Ninth Army attacked at Bolimów, where, on January 18, its guns opened a terrible new era in modern warfare with a barrage of eighteen thousand shells filled with poison gas.

There is a peculiar kind of blizzard in Eastern Europe—Russians call it a metel—in which a vicious gale from the east lashes the snow in blinding sheets across the vast Eurasian plain for days on end. Just a few days after the Ninth Army's attack at Bolimów, a metel roared out of Russia, across western Poland, and into East Prussia to bury the rest of the Germans' assault columns beneath its drifts.[9] Trapped in snow that the winds had piled "as high as a man," the German artillery needed teams of ten or more horses to move its guns into position, while its frozen infantry crawled forward at a snail's pace.[10] "Are not those marches in the winter nights, that camp in the icy snowstorm [nothing] . . . but the creations of an inspired human fancy?" Hindenburg later asked as he recalled the snow-clogged beginnings of the battle whose "name charms like an icy wind or the stillness of death."[11] While German men and animals struggled into position, the Russians relaxed under the snow's protective cover, certain that the Germans could not attack under such conditions. Undeterred by drifts, winds, and cold, Hindenburg and Ludendorff launched the Eighth and Tenth armies' assault at precisely that moment.

Preceding the Eighth and Tenth armies' main assault by twenty-four hours, the Germans' 2nd Division and XL Reserve Corps fell upon the Russian positions at Lyck on the morning of January 25. Caught by surprise, and with most of his staff some seventy miles to the rear in Grodno, General Rudolf Sievers, the Russian Tenth Army's commander, hastily formed a stubborn defense around the riflemen of the III Siberian Corps who, even Ludendorff conceded, defended Lyck "splendidly."[12] With most of their reserves of ammunition and supplies snowbound at rail depots, Sievers's men finally fell back on January 28, after the German XXI Corps had cut the rail line to the Russian fortress at Kovno, while its companion, the XXIX Reserve Corps, had seized ten thousand Russian prisoners near Wirballen. Four days later, German infantry entered Lyck and seized its vital rail junction. "Between the Niemen and the Vistula," a communiqué from *Stavka* reported, "our troops are gradually withdrawing from the battle zone."[13]

Stavka's cryptic message implied a mere change in strategy when, in fact, Russia's armies stood at the brink of defeat. Fighting with bayonets fixed on empty rifles, Sievers's valiant Siberians accompanied by a much-disordered XXVI Corps forced their way through the Augustowo Forest to safety the next day, but several others proved less fortunate. On February 8, advanced units of the German Tenth and Eighth armies met south of the Augustowo Forest at Lipsk and closed another ring of steel such as Ludendorff had forged at Tannenberg, "a movement recalling in some respects our celebrated [victory] at Sedan [in 1870]," Hindenburg later said.[14] This time, 110,000 soldiers and three hundred precious Russian guns fell into German hands, while another 100,000 Russians died from German bullets or from the cold. "Perhaps the first event of the war which made the public fear seriously for the successful outcome of the struggle," Vladimir Gurko remembered, "was our retreat in the Augustowo woods,"[15] after which, the court favorite General Aleksandr Bezobrazov remarked with a nasty arrogance completely worthy of his oft-demonstrated incompetence, the contradictory tripartite formula of "order, counter-order, and disorder" came to reign in the army's high command.[16]

Although a brilliant tactical success, the German victory at Augustowo Forest proved of little strategic importance. Much of the Eighth Army became bogged down after being pulled out of the line for an ill-fated attempt to seize the Russian fortress of Osowiec, and eventually had to withdraw. At the same time, the army group that soon became known as the Twelfth German Army pressed an attack against Russia's First and Twelfth armies near Przasnysz, "a very strongly fortified town," in Ludendorff's words,[17] some 150 kilometers to the west of Augustowo. Unlike the battle at Augustowo, the struggle at Przasnysz proved a draw at best, for General Sievers's stubborn defense at Lyck and Augustowo had given the Russian First and Twelfth armies time to prepare strong defenses. These held until a furious German artillery bombardment killed half the Russian defenders and breached the town's defenses on the morning of February 11, forcing the garrison's several thousand survivors to surrender. Confident of victory, the Germans' new Twelfth Army then pushed on to cut the important Przasnysz-Maków highway.

Once beyond Przasnysz, however, the Germans faced the Twelfth Russian Army of General Pavel Plehve, the one-time commander of the Don Cossacks whose iron will had continued to stand in sharp contrast to his fragile body throughout the fall and winter. Certainly not prepared to retreat, as had so many Russian commanders that the Germans had faced during the war's first winter, Plehve immediately answered the Germans' assault with an attack of his own. "The object of your attack," he told two corps of Siberians as they moved into position, "must not be [merely the relief of] the town of Przasnysz. You must attack the forces of the enemy in both flank and rear . . . [and] seize the enemy's lines of communications."[18] When the Siberians launched their attack on February 12, it became the Germans' turn to go on the defensive and, eventually, to retreat "with very heavy losses," Ludendorff confessed,[19] as Plehve's forces swelled to overwhelming numbers. By the evening of February 14, Plehve had returned Przasnysz to Russian hands, along with six thousand prisoners and fifty-eight German guns.[20]

Given the Russians' dismal performance all along the northwest front that winter, any triumph over the seemingly invincible forces of Hindenburg and Ludendorff proved welcome, but Plehve's Przasnysz success could not offset the far more serious losses at Augustowo and Lyck the week before. Now certain that he had neither the troops, the weapons, nor the ammunition to launch a counteroffensive, Nikolai Nikolaevich asked Russia's allies to mount an offensive in the west, just as his armies had done for the hard-pressed French at the beginning of the war, in order to ease the German pressure along his front. "The Grand Duke Nicholas discreetly let it be known some days ago that he would be happy to see the Army of France take the offensive with a view to preventing any transfer of German forces to the eastern front," Paléologue wrote in his diary. "The reply was just what I expected," he concluded. "General Joffre has just ordered a vigorous attack in Champagne."[21]

Even more than by news of the new French offensive, Russian public opinion was buoyed by news of victory from General Ivanov's southwestern front. Hindenburg's and Ludendorff's original plan had called for the Austrians to strike simultaneously against the Russians' southern flank in the Carpathians in order to relieve the beleaguered fortress of Przemysl. With all too predictable results, Austria's Conrad von Hötzendorff had opened his offensive on January 10, when the mountain passes were clogged with snow and their approaches slick with ice. With low-hanging clouds making accurate artillery fire impossible, and the ice clogging the complex mechanisms on their machine guns and making it impossible even to fire rifles unless soldiers first held them over campfires, the Austrians faced certain defeat. Their Second Army lost over forty thousand men to frostbite during those terrible days, and one regiment awoke to find that twenty-eight officers and eighteen hundred men had frozen to death in a single night. In all, Conrad lost over three-quarters of a million men in the mountain passes during the first six weeks of his campaign, as his effort to relieve Przemysl stumbled and then collapsed with a sigh of hopeless resignation.[22] Concluding that "there is nothing more that the troops can do,"[23] the Austrians had to face the hard fact that their greatest fortress east of the Carpathians, a vast system of defense upon which they had spent the equivalent of more than a hundred million rubles, was doomed.

Cut off from all sources of supply, the Przemysl garrison capitulated after a six-month siege that had cost them almost fifty thousand casualties. On March 9, the Russians accepted the surrender of 120,000 Austrians, including 9 generals and 2,500 officers, along with more than a thousand pieces of artillery that included eight twelve-inch mortars, the Austrian army's largest field weapon.[24] This was the victory for which the Russians had prayed throughout the war's first winter, but it had not come until the shortages of men, weapons, and ammunition had grown so desperate that neither Ivanov nor the strategists at *Stavka* could hope to follow it up with further attacks. Not far to the south of Przemysl, General Brusilov's Eighth Army now starved and froze in the Carpathian mountain passes despite their recent victories against the Austrians. By spring, Brusilov's men had so few weapons and so little ammunition that he "no longer aimed at further successes" but merely tried to hold his ground "with as few losses as possible."[25] "My regiments are from fifty to seventy-five percent understrength, replacements are arriving unarmed and there are no rifles to give them," he told Ivanov with characteristic directness a few weeks later. "Under the present conditions," he warned his chief sternly, "even one or two days of not very intensive fighting would leave my army without bullets."[26] "Given the condition of our armies," *Stavka*'s coldly realistic General Danilov concluded, Russia's generals "had no grounds for calculating that we could expect to attain any sort of decisive victory over our opponents in the immediate future."[27]

At winter's end, Russia's depots had no rifles to send Brusilov or anyone else at the front. Even at the beginning of the war, recruits had trained with antiquated Berdan rifles left over from the last Russo-Turkish War, but now even these had become so scarce that only one could be issued to every tenth man.[28] In some cases, entire battalions trained without rifles. "Could anything be more distressing?" Minister of Foreign Affairs Sazonov exclaimed one day in February as he and Paléologue crossed the great drill field on Petrograd's Champs-de-Mars. "There are perhaps a thousand men there . . . and—you can see for yourself—there's not a rifle anywhere!"[29] Nicholas himself had assured Paléologue only a few weeks before that the "general plan of campaign . . . [would] not in any way be changed" despite such shortages. Any lull in Russia's effort, he had promised, would prove no more than "a temporary suspension." At the "earliest possible moment," he had promised, Russia's armies would "return to the offensive and . . . [would] continue the struggle until our enemies sue for peace."[30] As spring approached, no one at *Stavka* could seriously contemplate a new offensive against the victorious armies of Hindenburg and Ludendorff. The Austrians, however, were another matter.

However bad things stood for the Russians at the beginning of 1915, they were worse for the Austrians. Soldiers in the Austro-Hungarian army spoke at least fifteen different languages, and that alone created monumental problems in the chain of command once the war began. Before the war, the Habsburgs' officers had taught peasant recruits the "language of service"—a basic vocabulary of about eighty German words for military commands and for various weapons and their most important components—while they, in turn, tried to become fluent in the language of their men. This meant that Habsburg officers, most of whom spoke German or Hungarian natively, always were bilingual, and often spoke three or more languages (Conrad himself spoke seven), depending upon their command experience.

The first bloody battles of 1914 had decimated this officer corps and had left the Austro-Hungarian army in the hands of poorly-trained reserve officers whose ignorance of the languages spoken by their soldiers drove them to absurd lengths to establish even the most rudimentary communication with their men. In one case, English actually became the language of command in a Slovak regiment whose officers spoke no Slovak but had studied English as schoolboys, while most of their men had learned a smattering of the language in the hope that it could help them to emigrate at some future time! Very often, Russian proved to be the language that peasant draftees understood best after their own, and they showed their resentment at being ordered to fight their Slavonic breathren by deserting to the Russians.

Austria's difficulties did not end with the loss of her officer corps. Seven months of fighting had cost her two out of every three of the men she had mobilized in 1914. Clumsily organized support services in the rear tied down so many of her remaining troops that by the end of the year she had only a quarter-million men left to hold the front against the Russians, and there was every indication that her new drafts of conscripts could not even double that number by spring. The victims of shortages that were even worse than those facing the Russians, Habsburg ordnance officers struggled to provide no less than forty-five different types of shells needed just to supply the artillery that supported their infantry.[31] By early 1915, it seemed that only rapid and massive aid from its German allies could save the Austro-Hungarian army from collapse.

Well-informed about the Austrians' weaknesses, and cheered by the success of his armies at Przemysl, General Ivanov urged *Stavka* to concentrate Russia's limited resources against the Austrians. The fall of Przemysl, he insisted, had freed the entire Eleventh Army to join with Brusilov's Third and Eighth armies in an assault that could carry them through the yet unbreached Carpathian passes into the Hungarian plain and open the way to Budapest and Vienna. In the northwest, General Ruzskii insisted that any new offensive must begin on his front, but no one at *Stavka* seemed willing to try the Germans again so soon after the disaster at Augustowo. In any case, the prospect of hurling several armies against Austria's crumbling Carpathian defenses seemed too enticing to be ignored. "The supreme commander now sees his main purpose as being to shift the entire northwestern front over to a purely defensive operation, and intends for the southwestern front to become the major focus of the campaign in the days ahead," Grand Duke Nikolai Nikolaevich announced in grand style just three days before Przemysl fell.[32] Unwilling to command a defensive operation against the Germans, and fearful of being overwhelmed if *Stavka* sent his reserves to support Ivanov's offensive against Austria, Ruzskii angrily resigned "for reasons of personal health." Anxious to end the squabbling that had enmeshed the commanders of Russia's northwest and southwest fronts in an ongoing feud for men and weapons ever since Zhilinskii's dismissal, *Stavka* quickly named Ivanov's chief of staff, the unpretentious but immensely able administrator Mikhail Alekseev, to take over Ruzskii's command.[33]

Although now seemingly committed to a new offensive against the Austro-Hungarian armies, *Stavka* denied Ivanov the unqualified support he needed. Like Ruzskii, fearful that even a defensive operation against the Germans might require more divisions than he had under his command, Alekseev quickly took up his predecessor's call for massive reinforcements. Too easily swayed by his urging, *Stavka* left hordes of troops in the northwest to guard against a possible German attack, and forced Ivanov to weaken his line all across Galicia in order to assemble a scant thirty divisions—hardly more than the number of Austrians he faced—for his Carpathian offensive. Nonetheless, for three weeks he pressed forward, shattering several Austrian divisions in the mountains and taking thousands of prisoners. By mid-March, Ivanov's forces had seized Tarnów and occupied the key Carpathian passes. Yet *Stavka*'s change of heart left him neither the men, the artillery, nor the shells to assault the main Austrian defenses and, on March 28, Ivanov halted his advance to replenish his reserves of ammunition and bring up reinforcements.[34] This gave Conrad's beleaguered Austrian divisions time to recover, regroup, and retrench, while the German High Command rushed to bolster their sagging defenses with troops drawn from the western front. The Russians, Germany's commander in chief General von Falkenhayn later explained, "threatened the Austro-Hungarian front in a way which could not be borne for any length of time on account of the decreasing *morale* of certain sections of the allied [Austro-Hungarian] troops."[35]

Every day that Ivanov delayed his advance brought German reinforcements nearer, but he could not move forward until he had enough men and shells to support an assault against the now-strengthening Austrian defenses in the mountains. At that point, Alekseev's unexpected stubborn refusal to send reinforcements shattered Ivanov's hopes. Unable to force Alekseev to recognize the danger that was building all along the southwest front, Nikolai Nikolaevich issued a sad admission that he could not control the generals under his command. "Any further strengthening of the southwest front at the expense of the northwest," *Stavka* telegraphed to Ivanov's chief of staff, "absolutely cannot be done [at the present time]." Having failed to send the necessary reinforcements, *Stavka* launched a new cycle of General Bezobrazov's cynically-named formula of "order, counter-order, disorder," and simply told Ivanov to "break through into the central Hungarian plain" with the resources he had at hand.[36] Ivanov now had forty-one divisions with which to renew his offensive and defend the entire southwestern front, while Alekseev jealously clutched sixty-six divisions in the northwest and demanded 300,000 additional reinforcements to guard against the German attack he feared might be launched from East Prussia at any moment.[37]

Coming so soon after the fall of Przemysl, and just when Italy and Rumania were about to enter the war on the side of the Entente, Ivanov's sharp thrusts into the high Carpathian passes had stirred consternation among the Central Powers' general staffs. Fearful that Germany's strategy of giving first priority to her western front might make possible a major Russian victory against Austria, Falkenhayn hurried to reshape his priorities. Because the long-hoped-for victory against France and Great Britain had not come in the west, he now decided to force a decision in the east. This time, he would not employ the pincer tactics against the Russian flanks that Hindenburg and Ludendorff had used with such marked effect during the first months of the war, for these restricted the potential gains to those that could be immediately realized from encircling enemy troops.

Falkenhayn's object in the spring of 1915 could only be obtained, he explained, "if the intended blow was so dealt that it had in view the permanent crippling of Russia's offensive powers as its ultimate aim." This could not be achieved, he insisted, by "operations against the Russian wings, [but] could only be expected from a break-through,"—what his French adversaries called *"une percée"*—in the Russian center. This was what Germans, French, and British all had strived for in the west since the war's beginning. *OHL—Oberste Heeres-Leitung,* the German Supreme Headquarters—now decided to attempt it in the narrow space of the thirty-five kilometers that separated Gorlice and Tarnów in the east, some fifty kilometers east of Krakow and a hundred kilometers west of Przemysl.[38] "The Russians had just withdrawn such strong forces from Western Galicia for their Carpathian offensive that they were no longer able to replenish this front in time," Falkenhayn explained. "Even if the break-through was only conditionally successful . . . it would render the northern portion of the Russian Carpathian Front untenable . . . [and] specially seasoned troops therefore were selected for the undertaking." Made up of four combat-hardened corps drawn from the western front, and supported by Conrad's Third and Fourth Austrian armies, General August von Mackensen's Eleventh Army formed the spearhead (later called the "Mackensen wedge") of the German assault.[39]

Unlike Ivanov's superiors at *Stavka,* Falkenhayn placed overwhelming re-sources at Mackensen's disposal, and ordered well-supported feints all along the eastern front to mask his assembly of men and heavy weapons in the Tarnów-Gorlice sector. By late April, Mackensen had more than a third of a million troops in position, and had brought up 1,272 light guns, 660 machine guns, 96 trench mortars, and 334 pieces of heavy artillery to support them. This meant that Ivanov's divisions faced ten enemy soldiers for every meter that separated Tarnów from Gorlice. If all of Mackensen's artillery had been assembled in a wheel-to-wheel formation, there would have been one cannon for every twenty-two meters of front, and every fifth one would have been a heavy gun. Mack-ensen's was the heaviest artillery concentration yet seen in the Great War, and he was about to use it with brutal effect.[40]

Facing Mackensen were 219,000 men of the Third Russian Army with 675 pieces of light artillery, 600 machine guns, no trench mortars, and 4 heavy guns that turned out to be defective. Commanded by General Radko Dimitriev, a valiant Bulgarian officer whose heroism was matched only by his carelessness and lack of foresight, the Russians were ill-prepared to turn back any major assault. Even though many of their divisions had been in position for more than four months, none had yet built reserve entrenchments that could have served as second- and third-line defensive positions against any Austro-German attack. Beginning on March 31, air observers and spies brought back reports of the German build-up, and Ivanov's intelligence officers reported that spies behind the Austro-Hungarian lines had sent word that "they are expecting German troops from the French front," but this information stirred no urgency in Dimitriev or his front-line commanders. A week later, Ivanov's southwest front headquarters again reported "increased activity of German advanced units" and "a notable strengthening of German forces,"[41] but no one thought even to order the Third Army to begin building second-line defenses until the afternoon of April 16.[42]

If Dimitriev had been careless about preparing reserve defensive positions for his infantry, he had left his artillery batteries fatally exposed. Most of the Third Army's artillery had spent the entire winter in the same positions, for, as General Knox acidly remarked, "the Austrians shot so badly that there seemed no reason why they should trouble to move."[43] The Austrian artillery may have shot badly, but its forward observers could see well enough to chart the precise

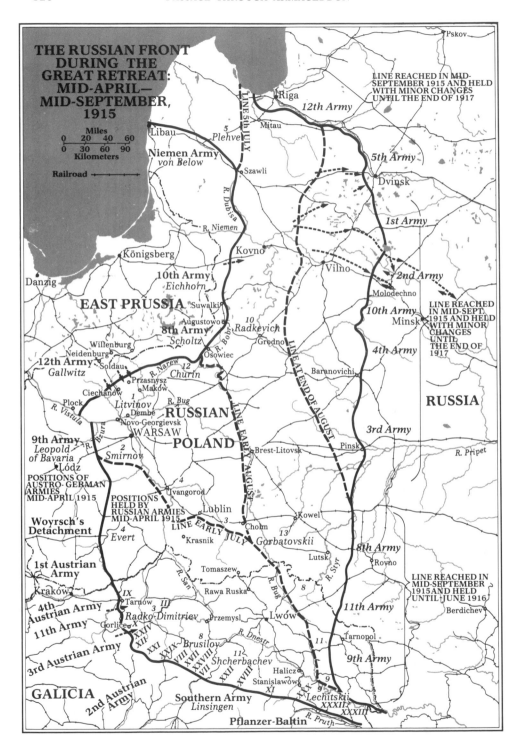

THE RUSSIAN FRONT
DURING THE
GREAT RETREAT:
MID-APRIL—
MID-SEPTEMBER,
1915

Miles
0   20   40   60
0   30   60   90
Kilometers

Railroad ←——→

LINE REACHED IN MID-
SEPTEMBER 1915 AND HELD
WITH MINOR CHANGES
UNTIL THE END OF 1917

Pskov

Riga
12th Army
Mitau
LINE 5th JULY
Plehve
5

Libau
Niemen Army
von Below
Szawli
5th Army
Dvinsk

R. Dubisa
R. Niemen
1st Army

Königsberg
Kovno
Vilno
2nd Army

Danzig
10th Army
Eichhorn
Molodechno
LINE REACHED
IN MID-SEPT.
1915 AND HELD
WITH MINOR
CHANGES
UNTIL
THE END OF
1917

EAST PRUSSIA   Suwalki
10th Army
Minsk

8th Army
Scholtz
Radkevich
10
R. Bobr
Grodno
Osowiec
4th Army

Willenburg
Neidenburg
Soldau
R. Narew
Churin
12
Baranovichi

12th Army
Gallwitz
Przasnysz
Makow
1

RUSSIA

Ciechanów
Plock
R. Vistula
Litvinov
Dembe
Novo-Georgievsk
R. Bug
R. Wkra
WARSAW
RUSSIAN
POLAND
3rd Army

9th Army
Leopold
of Bavaria
Lódz
Smirnov
2
Brest-Litovsk
Pinsk
R. Pripet

POSITIONS OF
AUSTRO-GERMAN
ARMIES
MID-APRIL 1915

POSITIONS
HELD BY
RUSSIAN ARMIES
MID-APRIL 1915
Ivangorod
4
Lublin
3
Kowel

Woyrsch's
Detachment
Evert
4
LINE EARLY JULY
Cholm
Gorbatovskii
13
8th Army

1st Austrian
Army
Krasnik
Lutsk
Rovno

Kraków
Tomaszew
Rawa Ruska
R. Bug
8
R. Styr
LINE REACHED IN
MID-SEPTEMBER
1915 AND HELD
UNTIL JUNE 1916

4th
Austrian Army
Tarnów
III
3
Przemysl
Lwów
11th Army
Berdichev

11th Army
Gorlice
Radko-Dimitriev
IX
X
XXIV
XII
XXI
R. San
R. Dnestr
11
Tarnopol
9th Army

3rd Austrian Army
XIX
VIII
XVII
VII
XXII
Brusilov
XXVIII
XVIII
Shcherbachev
11
Halicz
9
Stanisławów
XI

GALICIA
2nd Austrian Army
Southern Army
Linsingen
Lechitskii
XXXII
XXXIII

Pflanzer-Baltin
R. Pruth

LINE EARLY AUGUST
LINE AT END OF AUGUST

position of every Russian gun that faced them, so that when Mackensen moved his guns into position, his gunners knew the precise location and range of every Russian battery they faced. Unlike their Austrian allies, Mackensen's gunners put the information to deadly use, all the more so because the Russian field guns had less than forty shells apiece and their howitzers were rationed to two a day.[44]

With Teutonic punctuality, Mackensen's artillerymen began to register their guns at nine o'clock on the evening of April 18, slowly firing phosphorous shells to chart ranges and to determine precise points of impact. Within an hour, the air grew thick with the acrid smell of cordite as their firing swelled into a full-scale bombardment. Throughout the night, the massed guns spewed forth high explosives at the rate of more than a thousand shells a minute. Tongues of flame licked upward from the German side of the horizon and reached out toward Russia as Mackensen's gunners rained nearly three-quarters of a million rounds upon the Third Army's badly-prepared positions in less than twelve hours. "There were shell craters on top of shell craters," one Russian grenadier remembered.[45] General Knox calculated that the Germans "used ten medium-calibre shells for every pace and a half of front," in an effort to obliterate the crudely-constructed Russian trenches and slaughter their defenders before Mackensen launched his infantry attacks.[46] During the four hours before the assault began, the artillery fire became so intense that German infantrymen actually stood upright without danger to watch its deadly effect upon the Russians.

When the German infantry assaults began at ten o'clock on the morning of April 19, they at first met little resistance for, as one observer reported, "all the Russians in the danger zone who were not killed or wounded were stunned or contusioned" by the German artillery. "Here and there," one German recalled, "loam-grey figures jumped up and ran back weaponless, in grey fur caps and fluttering, unbuttoned great coats . . . like a flock of sheep in wild confusion."[47] Yet, where woods or hilly terrain had broken the force of the German bombardment, living Russians remained to meet the German attackers and, especially in the center, these sturdy defenders used their machine guns to good effect. But even well-aimed machine guns could stem the German advance only briefly, because the supplies of ammunition failed all too quickly. "The shortage of all sorts of ammunition has affected the activities of the entire army," Ivanov reported to General Danilov after the third day of battle. In any case, the Russians had no integrated defensive strategy, and their overall situation was made worse because, despite warnings of a German build-up, Dimitriev had left his headquarters to attend the annual celebration of the Order of the Knights of St. George. By the evening of April 21, Mackensen had driven his wedge eight kilometers into the Russian lines to accomplish in the east the breakthrough that Falkenhayn had dreamed of in the west.[48]

Russian valor counted for little during the next fortnight, as Mackensen hammered his wedge into the Third Army's lines with unrelenting brutality.[49]

During the first days of fighting, the III and V Caucasus Corps lost seventy-five men out of every hundred, while the X Corps lost eighty-five, and the IX Corps was destroyed. To fill the gap, the Russians threw reserve divisions against the Germans without artillery support, proper weapons, or even maps, only to see them melt away in a matter of hours. "Our strategic situation is completely hopeless,"[50] Ivanov's distraught chief of staff telegraphed, as he begged Stavka for permission to withdraw after Ivanov had refused his pleas to order a general retreat toward Moscow and Kiev all along the southwest front.[51] "You are hereby categorically ordered not to undertake any retreat," Nikolai Nikolaevich replied in a stern effort to stem the hysteria before it spread. But the crumbling remnants of the Third Army simply could not stand against Mackensen's assault. Within ten days, its tattered remnants—a mere forty thousand out of a quarter-million men—had fallen back to the San River, the last natural barrier that stood between the Germans and Przemysl. As they prepared to make another stand on its banks, they found they did not even have the wherewithall to build entrenchments because corrupt staff officers had sold the spades, barbed-wire, and entrenching timbers they had seized from the Austrians a scant two months before.[52] "[For] eleven days [we heard] the dreadful boom of German heavy artillery, literally razing whole rows of trenches along with their defenders," General Anton Denikin later wrote of this last effort to halt Mackensen's advance. For eleven days, the guns of Denikin's Iron Brigade stood silent, their shell racks empty and with no reserves within reach. Then, Denikin remembered, "all the riflemen breathed more easily . . . when our batteries received fifty shells [and] it was reported to all the regiments by telephone."[53]

Against the Germans' hundreds of thousands, fifty shells were a mere pinch of chaff cast upon the winds. As the Russian guns stammered halting responses to the never-ending roar of the German artillery, Ivanov begged Stavka for more shells. "Unless you have been throwing away your shell-boxes," came the curt reply, "there must be enough [at your positions] already."[54] When ammunition shortages reduced some units in the Third Army to fighting with bayonets fixed on empty rifles, Dimitriev called for ten million cartridges only to receive a severe reprimand for his request. "It is essential that measures be taken to limit the vast expenditure of rifle cartridges that has occurred in recent days," the men at Stavka informed him sternly. "It is clear that this . . . stems in large measure from the troops' negligence." Although some seventy-five million rifle cartridges remained locked away in the fortress arsenals of Kovno, Grodno, Osowiec and Brest-Litovsk, neither Dimitriev nor Ivanov could pry any of them away from the outdated champions of fortress defense who hovered around the grand duke at Stavka.[55]

Without reinforced entrenchments, artillery, bullets, or even enough men to fill the great gaps that German artillery and machine-gun fire tore in its front lines, the Third Army could not hope to stem Mackensen's onrushing divisions. On May 19, these swept across the San River, forcing the Russians to abandon

Przemysl, whose recent capture had given them brief new hope. "From all our armies comes the cry: 'Give us bullets!' " Ianushkevich reported to Sukhomlinov from *Stavka* on the day the fortress fell. In reply, Sukhomlinov wrote vaguely that "the world has never before experienced the likes of this bloodiest of struggles" and lamented the "famine of able men" that made the shortages of weapons and ammunition all the more desperate.[56] Neither complaint could alter the hard fact that Ivanov now had no reserves, and that West Russia's few clogged railways could not bring in the men he needed, even though Alekseev at long last announced himself ready to offer them.[57]

Before the end of May, Grand Duke Nikolai Nikolaevich reported to the tsar that unreplaced losses in Russia's armies were approaching 400,000 men, and that reinforcements sent directly from assembly points in Russia were worthless. Many of these arrived without weapons and many more did not even know how to fire the rifles they carried. "There are almost no rifles for the replacements that are arriving," Nikolai Nikolaevich warned, "and the few that are available are but a drop in the ocean." With only untrained levies to replace the seasoned troops he lost in battle, Ivanov continued his retreat. "There is no strategy that applies to such conditions," Nikolai Nikolaevich confessed. "It is now impossible for us to seize the initiative."[58] Six weeks after his first assault against the Tarnów-Gorlice line, Mackensen drove the Russians from Lwów. By mid-June, he had thrown Ivanov's armies across Poland's eastern frontier into Russia. Now, Falkenhayn wrote with grim satisfaction, "the threat to Hungary has been completely removed."[59]

Broken and bleeding, without enough weapons or ammunition, Russia's soldiers made ready to defend their native soil for the first time since their forefathers had driven the Grand Army of Napoleon from the Russian land at the end of 1812. During the war's first eleven months, their losses had soared beyond a million and a half men killed, wounded, or taken prisoner, a number equal to about three-quarters of Russia's entire peacetime army.[60] Nor were Ivanov's losses on the southwestern front the only disasters Russia's soldiers faced that spring and summer. Perhaps in an attempt to establish in advance a justification for any defeat he might suffer, Alekseev had grossly inflated his estimates of the men and weapons his front required, but he had been right to expect attacks from those armies of Hindenburg that had remained apart from Mackensen's Galician offensive. In mid-April, as part of the feints designed to cover Mackensen's preparations for his assault against the Tarnów-Gorlice line, Hindenburg had ordered three cavalry and three infantry divisions to probe Russia's Baltic provinces of Lithuania and Courland, where they found rich stores of food, forage, and desperately-needed leather guarded only by raw provincial militia. Convinced that this German threat to the Baltic coast was of little importance (even his superiors at Petrograd's General Staff Headquarters dismissed it as a mere foraging expedition), Alekseev reluctantly yielded to *Stavka*'s repeated proddings and sent a single cavalry division and six small field guns as reinforcements. In response, Hindenburg expanded his own divisions in the region into the Army of the Niemen, which readily seized the ports of Memel and Libau, as well as the important rail junction at Szawli before the middle of May.

From its new positions, the Niemen Army threatened the Lithuanian fortress of Kovno on its right and the key Baltic port and steel manufacturing center of Riga on its left. In panic, banks abandoned their offices in Riga, while the city's factory owners evacuated their plants to the east with so little preparation or forethought that, as General Knox later wrote, "many of them were unable to re-start work satisfactorily during the war." The mere fact of a German threat to Riga, Knox reported sadly, "deprived the great manufacturing town of all usefulness for national defence." Now fearful that a German thrust against Riga might turn all too easily into a full-scale offensive against Petrograd, *Stavka* hastily poured nine divisions of infantry and another nine of cavalry into the region, among them units that Ivanov desperately needed to shore up his crumbling defenses in the southwest.[61]

Although successful in driving the enemy from Szawli for a brief time, Russia's tardily begun effort to defend her Baltic lands could not dislodge the Germans from the positions they had seized so easily a few weeks before. With the Tenth and Niemen armies firmly in place, Hindenburg and Ludendorff turned to strike at Kovno, "the cornerstone of the Russian defense on the Niemen," whose fall, Ludendorff insisted, would open "the road to Vilno and the rear of the Russian forces" and make it possible to cut the main rail line that connected Warsaw and Vilno with Petrograd. If conducted according to his plan, Ludendorff insisted, "the summer campaign of 1915 would end in a decisive defeat of the Russian armies."[62]

Hindenburg and Ludendorff were not the only ones who conceived new plans in the late spring of 1915, nor could they be certain of their acceptance. Now certain of their enemy's weakness, Germany's High Command began to argue about how best to strike the blow that would bring Russia to her knees. During June, their debate echoed throughout *OHL* and along the telephone wires that tied it to Berlin. Mackensen insisted that his forces now must turn aside from Ivanov's collapsing front and strike northward toward Lublin and the fortress of Ivangorod on the middle Vistula. The Twelfth Army's commander saw in that plan an opportunity for his forces to strike south toward the Narew River and Mackensen's advancing Eleventh, while Prince Leopold of Bavaria hoped to strike directly eastward toward Warsaw with his Ninth Army.[63]

The debate between *OHL* and its commanders on the eastern front centered upon the large projection of Russian-held Polish territory that protruded like a hook-shaped nose westward from the advanced positions of the Niemen, Tenth, and Eighth armies in the north, and those of Mackensen's Eleventh Army and its Austro-Hungarian allies in the south. With Warsaw and the great fortress of Novo-Georgievsk near its westernmost tip, the strong Russian forces positioned in this territorial protuberance threatened the flanks and rear of the German armies advancing into Galicia and the Baltic provinces. At the same time, the rapid retreat of the Russian forces to the north and south of this area offered the Germans an immensely tempting prospect for a gigantic pincer movement which, if successful, could cut off hundreds of thousands of Russia's troops, fatally weaken her military prospects, and, perhaps, force Nicholas to sue for peace. It was to this plan that the kaiser gave his personal approval at a conference held in Poznań on June 19/July 2 as he committed his eastern armies to seizing the fortresses upon which Russia had based her western defenses ever since the 1830s. Along with the fortresses, the Germans hoped to seize huge quantities of weapons and ammunition desperately needed by the Russian war effort.

Ivangorod and Novo-Georgievsk—Russia's sprawling fortresses on the Vistula—supported by Kovno, Grodno, and Osowiec to the north and east, were all that stood between the Germans and the gateway to Russia's hinterland as the summer of 1915 approached. With remarkable lack of strategic foresight, Russia's conservative strategists had filled these huge stone and concrete bastions with over nine thousand pieces of artillery, including almost nine hundred heavy guns, and nearly a hundred million rounds of rifle and machine-gun ammunition. Novo-Georgievsk alone held almost a million of the precious artillery shells that Ivanov had begged for in April and May, and which he still needed if he hoped even to slow any new German assault. But any chance to aid Ivanov had long since been lost, for the Russians had only a small fraction of the locomotives and freight cars needed to move these too long-hoarded weapons and ammunition to his front. Now, as *OHL* readied its new assault, the Russians could not even think of moving these precious stores and weapons to safer ground. According to the best estimates, more than two thousand trains would be needed to evacuate the war materiel and troops from Warsaw, plus at least a thousand more to remove men, weapons, and ammunition from Novo-Georgievsk.[64] Because *Stavka* never could hope to assemble so much rolling stock at one time, the fortresses and their contents faced certain doom unless their defenders held them at all costs and against heavy odds.

Ivangorod, Novo-Georgievsk, Kovno, Grodno, and Osowiec all had immobilized vast quantities of men and precious armaments in the one major theater of the Great War where mobility remained a key to success. Many observers thought that keeping the armaments and garrisons of these fortresses intact had been a costly luxury when it had meant that Ivanov's armies had had to fight with empty rifles and without artillery support. The question now remained whether the luxury could prove itself worth its cost in stemming the Germans' gathering offensive. In the west, great fortresses had proved to be useless obstacles in which German heavy artillery readily destroyed the men, weapons, and supplies trapped inside, and Ludendorff had become the darling of the High Command precisely because of the ease with which he had accomplished one such mission at Liège. In mid-June 1915, Russians and Germans remained uncertain if that would prove true in the east, but many suspected that the fortresses of Poland and Western Russia would fare no better than those of Belgium and northeastern France.

Although most observers blamed the German successes in April, May, and early June upon their opponents' lack of shells and artillery, more far-ranging problems underlay the Russian failures that spring. Much heavier Allied artillery concentrations against far fewer German guns never produced such dramatic collapses in the west and, even in Flanders, where, in July 1917, England's General Sir Douglas Haig's light artillery surpassed the Germans' by almost four to one, while his heaviest guns outnumbered theirs by 128 to 14, the Allied successes would prove far less complete.[65] The sudden collapse of Ivanov's front in mid-May, and the ease with which Hindenburg and Ludendorff moved the Niemen Army into the Baltic coast, reflected much more complex Russian problems than German artillery superiority. Short of weapons, ammunition, and supplies, Russia found it difficult in mid-1915 even to assemble the manpower needed to replace the battle casualties that were running at almost 150,000 a month.[66]

Numerous and capricious exemptions from military service, bureaucratic bungling, and the government's inability to marshal its citizens in defense of their nation, meant that Russia's military authorities were not able to tap those vast reserves of manpower that had loomed so large in the expectations of Allied planners before the war. Russia's military administration entangled everyone so thoroughly in trivial bureaucratic routines that no one had time to think about what really needed to be done or to deal with the responsibilities of real command. Even company and battery commanders on the distant Central Asian frontier had to prepare between fifteen and twenty letters and reports each weekday, while their counterparts, who commanded nearer the center, had to send on as many as eight thousand such documents in a single year.[67] Military administrators became so trapped in this maze of rules and office routines that they lost all sense of direction and and common purpose. So completely did such men lose sight of the forest as they tended each individual tree that one officer in the Artillery Department actually returned a letter in which a factory owner offered to donate his factory for war work because it did not bear the proper government stamp![68]

The paperwork malaise that left officers deadened by the morphia of trivia —"sucked oranges," in one observer's words—was especially prevalent in Petrograd's General Staff Headquarters, where the average officer, the army's leading newspaper *Russkii invalid* once reported, "never decides anything and never expresses an opinion of his own, but spends his time in collating the opinions of others." Men with intelligence and initiative turned bitter in their frustration at not being able to accomplish anything they thought worthwhile, while their mediocre associates advanced quickly thanks to their mastery of the army's bureaucratic routines. "The Russian army drowns every promising reform in a sea of ink," a disgruntled officer once complained, and a great number shared his opinion.[69]

Although awash in ink and enmeshed in red tape, none of Russia's staff officers ever thought to compile accurate lists of all men liable for military service or, even, of those enrolled in the territorial militia (the *opolchenie*), for they, like their counterparts in the West, had convinced themselves that any major war would be brief and that the empire's reserves of manpower would never be needed. Certainly, the manpower for the much-vaunted "Russian steamroller" was available in the countryside, but without even lists of the men supposed to serve in the *opolchenie*, General Staff planners could mobilize only a comparative handful. If Russia had been able to assemble her reserves of manpower in the same fashion as had France, she could have mobilized upward of sixty million men between August 1914 and the end of 1917, although it is another question altogether whether the Allies could have supplied them with the weapons, ammunition, and supplies that Russian industries could not produce. As it was, Russia never managed to mobilize even a quarter of that number.[70]

Men were just one of the resources that Russia's war effort required, and the mobilization of the other resources during the first year of the Great War proved equally chaotic. This was especially true of the nation's industry, which suffered violent spasms of uneven economic development as it struggled to reach the levels of production that a modern total war demanded.[71] Although Russia's factory owners hurried to manufacture military goods at profitable prices guaranteed by war contracts with rich advances, they could not begin to produce quickly the weapons, ammunition, and supplies that their country's war effort required. In fact, they could not even manufacture enough explosives to produce an average of two artillery shells each day for each gun in service, a quantity that a visiting French officer dismissed as "practically nil."[72]

With weapons, the situation was far worse than with ammunition. It required 812 measurements, 540 templates, and 1,424 separate operations, all done at close tolerances, to manufacture the 156 moving parts for a rifle, considerably more to produce the 282 parts needed for a machine gun, and many more still to produce all the complex components that made up a field gun. Unaccustomed to this sort of precision, many factories made costly mistakes that slowed the production of new weapons to a trickle. By the summer of 1915, the Artillery Department had placed orders with Russian factories for almost nine thousand field guns, but had received only eighty-eight, a quantity many times less than the number lost in combat during the war's first year. At the same time, Russia's factories continued to produce less than a thousand rifles a day, although her soldiers consistently lost between three and six times that many at the front.

That spring, Japan's emergency sale of 300,000 Arisaka rifles relieved the dire shortage only slightly.[73]

Before the middle of 1915, Russia's War Ministry had ordered hundreds of thousands of rifles, tens of millions of bullets, and two and a half million artillery shells from Japan, England, France and, especially, the United States of America. But orders placed in countries whose factories were straining to produce the ammunition and weapons needed for their own war efforts could not be filled quickly and certainly not in time to aid the Russians as the Germans made ready to renew their assault in the middle of June.

In any case, even if the Allies had been able to ship greater quantities of ammunition and weapons that summer, only a fraction could have been moved from Russia's only wartime ports—Arkhangelsk on the White Sea and Vladivostok on the Pacific—to war zones that were thousands of kilometers away, for the port facilities were too limited, and the rail facilities too meager. Like the production problems in her war industry, Russia dealt with this obstacle very slowly. At the beginning of 1917, after the government had invested tens of millions of rubles in improving the rail connections between these ports and the front, a British attaché still would be obliged to complain to the War Office that the Russians had left nearly a hundred tons of nitrates (an essential ingredient in gunpowder) lying fully exposed to the ravages of the Siberian winter at Vladivostok, "that [the] chemical reaction [that would weaken its potency] must be occurring," and that "some of it had been lying [unattended] for well over two months."[74]

Weapons plants that could not produce rifles and field guns, shell and cartridge factories that had no gunpowder, and ports without sufficient wharves, warehouses, and facilities to forward war shipments quickly into the interior all caused the editor of Petrograd's conservative newspaper *Novoe vremia, New Times,* to lament that "from now on, we're doomed to disaster." To the age-old fatalism of the peasantry combined with the belief that Fate somehow expressed God's will, Russians now added a third, equally nebulous, element in their efforts to explain the failings of their war effort. There seemed to be something inherently flawed in the Russian character, thoughtful Russians now insisted, that made such things possible, perhaps even inevitable. "The Russian displays an extraordinary virtuosity in making all his enterprises fail," one of Paléologue's Petrograd acquaintances confided as word of Przemysl's loss reached the capital. "We set out to climb to the sky, but no sooner do we begin than we find out that the sky is very far away. Then, we can think of nothing else but to tumble down as quickly as we can, hurting ourselves as much as possible in the process."[75] Retreat, and the great suffering it would bring, thus all could be accepted as part of Russia's fate and explained by the failings of the Russian character. Russia's master storyteller Anton Chekhov had posed the question a few years earlier in somewhat different terms. "Why do we grow tired so soon?"

Chekhov had asked. "And when we fall down, why is it that we never try to get up again?"[76]

As the kaiser and his generals left their conference at Poznań on June 20/July 3, the disaster of April and May was about to continue on a much vaster scale. "The tale," Winston Churchill later wrote, would be "one of hideous tragedy and measureless and largely unrecorded suffering."[77]

CHAPTER V

Russia's Great Retreat

The summer of 1915, Minister of Agriculture Aleksandr Krivoshein said at a secret meeting of the Council of Ministers toward the end of July, was a sad time of "unending retreats and incomprehensible defeats." More attuned than ever to the tension that his finely-honed political instincts had sensed at the end of 1914, Krivoshein now felt the danger of Russia's course far more acutely. There seemed no way to avert his nation's rendezvous with disaster, as Hindenburg and Ludendorff moved Germany's armies with machine-like precision back and forth across the great distances of Europe's eastern front to inflict defeat after defeat upon any who barred their path. Russia's final hour, it seemed, approached with inexorable certainty.

Men such as Krivoshein, who knew all too well Russia's military and economic failings, could point to inept commanders, shortages of weapons and ammunition, and transportation breakdowns to explain the Germans' victories, but others turned to more nebulous explanations involving fate, failings of the Russian character—and outright betrayal, for which they found it all too easy to assemble circumstantial evidence. Among the hundred thousand losses in killed, wounded, and prisoners that Russia's armies suffered every week, four out of every ten now were men who simply walked away from the fighting and surrendered, happy to exchange life in the Russian trenches for the security of a German or Austrian prison camp. Dembe, a small fortress that the General Staff had described as "exceptionally strong," surrendered after the Germans fired only six shells into it from their heavy guns, and great fortresses once thought to be impregnable fell almost without resistance, as Russia abandoned all of the Polish lands that had formed her geographical bulwark against the

West for more than a century. Even men who knew the real reasons for Russia's defeats found it hard to accept such a stark portrait of unrelieved failure. "At the front, everything is falling apart, and the enemy is approaching the very heart of Russia," Krivoshein exclaimed at one point. "Why is poor Russia fated to live through such a tragedy?"[1]

In the minds of many, the real and imagined causes for Russia's defeats quickly mingled into a potpourri of terrible fears during the summer of 1915. "At first it rustles in a whisper, then it begins to rumble more loudly, until it ends with the erratic cry: Treason! Treason! Treason!" one shocked contemporary wrote. Now, people no longer whispered about betrayals as they had at the time of Colonel Miasoedov's execution in the spring, but spoke their fears openly. "An unseen hand"—a treasonous betrayal in high places—some said, had caused Russia's sudden loss of Przemysl so soon after Ivanov's armies had celebrated the Austrians' surrender. Others explained the shortages of weapons and ammunition in the same fashion, for it seemed incomprehensible that crises of such magnitude could be the consequence of mere stupidity and lack of foresight. Even General Anton Denikin, a sober and able officer with a long and distinguished career behind him, could never forget the ominous silence of his division artillery, "treacherously deprived of shells," he insisted, by someone in the High Command.[2]

Most of all, the shell shortage of 1915, the frightful consequence of inept planning and Russia's underdeveloped industry, became the product of treason in the minds of Russians who saw it as the underhanded work of "traitor-generals" and "traitor-ministers." Soldiers now spoke of "many traitors and spies in the High Command." Aleksandra warned Nicholas of "a spy being at Headquarters," insisted that it was none other than General Danilov, and urged that "it would only be right, tho' the man may seem perfectly charming & honest . . . [to] watch his telegrams & the people he sees."[3] Finally relieved of his duties as war minister, Sukhomlinov had to face a High Commission's inquiry that summer about his many failures. As his accusers laid out the shocking tale of Russia's unpreparedness for war, many concluded that only treason could account for such gross failings. Duma politicians and high officials brought Sukhomlinov to trial on those charges the next spring, and he spent part of 1916 locked away in the Peter and Paul Fortress until, at Rasputin's urging (there is some evidence that Sukhomlinov's beautiful, young wife paid the *starets* a special visit), Aleksandra convinced Nicholas to release him "privately, without much ado."[4]

Russians came to fear traitors not only in the High Command but at court and even in the bosom of the imperial family during the terrible summer of 1915. They had long despised their "German" empress, who had come from Hesse-Darmstadt to reign over them. Now they also turned their hatred upon the Grand Duchess Elizaveta Feodorovna, the empress's older sister, who had all the public-spiritedness and generosity that Aleksandra did not. Married to

Nicholas's uncle Grand Duke Sergei Aleksandrovich, whose "many peculiarities" led family members to describe him as "obstinate, arrogant, and disagreeable," Elizaveta Feodorovna had long been respected in Russia for her grace and forebearance. When her husband had been assassinated in 1905, she had retired from the world to found the Convent of Saint Martha and Saint Mary, where her good works among Moscow's poor had won her well-deserved accolades from a grateful city. Yet good works proved poor protection against the anti-German hysteria of 1915, and Russians turned against the pious grand duchess in an instant. "The Empress and her sister . . . must be shut up in some convent in the Urals," one great lord insisted. "Then . . . the clique of Baltic [German] barons and the entire camarilla of Vyrubova and Rasputin must be banished to Siberia [as well]." The Baltic Germans—men and women from the Baltic provinces that Russia had occupied during the time of Peter the Great—had long been among the Romanovs' most loyal servitors in the government, the army, and at court, but they all now fell under suspicion. "What kind of Tsar is it who surrounds himself with [such] thieves and chiselers?" an army cook asked that summer. Ominously, he observed: "A fish begins to stink from the head."[5]

Nowhere did the popular wrath against "Germans" burst forth more intensely in 1915 than in Moscow, the heart of Old Russia, where the pulse of Russian patriotism beat with special fervor. In June, violent outbreaks of anti-German outrage destroyed some forty million rubles' worth of private property in the city as free-wheeling mobs looted some five hundred commercial and industrial enterprises. Russian citizens with German, Swedish, or Finnish surnames suffered most of the losses, but so did some nationals from neutral and allied countries in addition to ninety Russians with *Russian* surnames.[6] Clearly, Russians' bitterness at the sacrifices their government demanded while denying them much voice in determining their own destiny had reached dangerous depths. Someone or something must bear responsibility for the artillery that went into battle without shells, the "impregnable" fortresses that surrendered even before the enemy closed his siege around them, and the men who died without rifles in their hands.

Russians began their search for a scapegoat upon whom to heap blame for their nation's crumbling war effort with the "Germans," whose success in business and government was envied by many who had to live in the shadow of their accomplishments.[7] But the "Germans" were too few and too necessary for the war effort, and Russians therefore quickly turned once again against the Jews upon whom they traditionally had vented their hatred in times of national crisis. Condemned to live in Poland and the southwestern provinces of the empire since the time of Catherine the Great, Russia's Jews had suffered centuries of persecution only slightly less vicious than that which some of their unfortunate descendants would face in Hitler's Germany. For a Jew even to have a Christian first name was a crime, and almost all had to live in the Pale of

Settlement, an infamous kidney-shaped region that lay along the empire's western and southwestern borders, in which incredible poverty limited economic opportunity for all but a scant handful. Time and again during the decades before the Great War, vicious anti-Jewish riots—pogroms—had swept the towns and cities of this region, as Russia's anti-Semitic lower classes beat, robbed, raped, and murdered defenseless Jews, all the while justifying their acts as proper retribution visited by Providence upon unrepentant "killers of Christ." *"Bei zhidov!"*—"Smash the kikes!"—was a cry known and feared by every Russian Jew, and many had fled as far as the Lower East Side tenements of New York City to escape its echoes. There could be no betterment of the Jews' life, Russian statesmen had long insisted, but some had made it a point, in the decades before the Great War, to proclaim that "the western border is open for Jews."[8] At any time, Jews were welcome to tear up roots put down centuries before and take up their ancestors' age-old quest for a homeland beyond Russia's frontiers.

Nicholas and Aleksandra shared the anti-Semitism of the most bigoted of their subjects. Despite his long dedication to public service, Aleksandra once remarked that it was "disgusting" for the Moscow businessman G. E. Weinstein, whom she called "a real Jew for sure," to be named to the Council of the Empire, and Nicholas openly supported the uncompromisingly anti-Semitic Union of the Russian People. When more than a hundred pogroms encouraged by the Union had inflicted upwards of ten thousand casualties upon the Jews living in the towns and cities of the Pale in the wake of the Revolution of 1905, Nicholas had shown no hesitation about accepting badges of honorary membership from the Union's representatives and had invited its leaders to Tsarskoe Selo to thank them for their general political support. Quick to excuse their part in the pogroms as an expression of antirevolutionary sentiment on the part of men loyal to their tsar, he placed the blame for the violence upon the Jews themselves. "Because nine-tenths of the troublemakers are Jews," he explained to his mother then, "the people's whole anger turned against them [and] that's how the programs happened."[9]

If Russia's archreactionaries used the upheavals of 1905 as a pretext for doing violence to the Jews, the revolution's call for freedom spurred others to defend them all the more strongly. Nowhere did the government act more ignominiously, nor the Jews triumph more dramatically during the next decade than in the infamous case of Mendel Beilis, a hapless Jewish clerk in an Ukrainian brickyard, who awakened on the night of July 21, 1911, to find his room filled with police and himself charged with the ritual murder of a thirteen-year-old schoolboy. Although the authorities had soon learned that the killing had been done by a band of thieves after the victim had overheard them discussing some of their plans, they had kept Beilis in prison for more than two years while they tampered with evidence and assembled the false testimony needed to convict him. Beilis's trial, before a judge who had been promised a promotion if he secured a conviction, lasted more than a month in the fall of 1913, and only as the result of an impressive array of legal counsel assembled on his behalf by men appalled by the government's cynical attempt to subvert justice, did he go free. "A legal *Tsushima!*" one police agent exclaimed, as he equated the government's defeat in the courtroom with Japan's triumph over Russia's fleet in the last sea battle of the Russo-Japanese War. Still, one victory in court, no matter how brilliant, could not bring freedom to the millions of Jews in the Russo-Polish Pale, although it helped to encourage progressive opinion to criticize the government's policies. On the eve of the Great War, Russia's Jews therefore still remained spectral shadows, second-class citizens with fewer rights and more disabilities than any other group in the empire.[10]

Ready to defend their nation despite long years of abuse, Russia's Jews had stepped forward willingly to take their place in battle alongside their Christian tormenters in 1914. Calling themselves "inseparably allied with our mother country . . . from which no power—neither persecution nor oppression—can separate us," they entered an army where regulations denied them officer rank, proclaimed them unfit to receive the Cross of St. George, and noncommissioned officers felt free even to shoot Jews in uniform for no good reason. As censors dedicated themselves to obliterating references to their valor and wartime accomplishment in the press, Jews continued to fight and die bravely for the homeland that persecuted them. Russians repaid their sacrifices by driving their families from their homes and refusing to allow wounded Jewish soldiers to be visited by their loved ones. For Russia's authorities, Jews remained enemies whose peculiar clothing and suspicious, German-sounding language made them the object of scorn, fear, and hatred. Most of all, Russia's Jews remained guilty of being Jews, a stain that not even conversion to Christianity could eradicate. The "real enemy," Russia's vocal and influential anti-Semites insisted, therefore continued to be the Jews. "No pardon for the Jew!" the ultra right-wing newspaper *Volga* urged at the beginning of July 1915. "The blood of the sons of Holy Russia, which they betray each day, cries out for vengeance!"[11]

Cursed as traitors and cowards anxious to profit from their nation's misfortunes, Russia's Jews found themselves the object of vicious rumors that gave further substance to the long-standing prejudices of those among whom they lived. "It's no secret that a *zhid* will use every underhanded means possible to get out of military service," one officer warned the 31st Infantry Battalion in July 1915. "A *zhid* will use every trick to avoid standing in the ranks of the Russian army to defend his homeland," he continued, "but after the end of the war, he'll again begin to demand equal rights along with those who really helped to defend our motherland." Other senior officers warned their quartermasters not to buy food from Jewish merchants because products of Jewish manufacture "have things put in them that can make people very sick," and the archconservative minister of roads and communications Sergei Rukhlov, once described by a colleague as "somewhat shifty," insisted that, despite all the evidence to the contrary, "it's no secret to anyone that most spies come from among the Chosen People." Jewish bankers everywhere, Rukhlov told his fellow ministers—his long years of experience in government finance and budget regulations giving added weight to his pronouncements—were making every effort "to use Russia's misfortunes as a means to further exploit the Russian people who have already been bled white" by the war. It was the fault of the Jews that Russia's armies continued to be outmaneuvered and outgunned by the Germans, that prices had soared so steeply, and that food had begun to grow scarce in the empire's towns and cities. "It is always necessary to have a scapegoat in reserve," Paléologue remarked sadly. "So infamous a calumny could only have been given birth in a despotic country."[12]

Among Russia's leaders, few entertained more pathological suspicions about the Jews than *Stavka*'s General Ianushkevich. "The experience in the present war has brought to light the patently dangerous attitudes of the Jewish population of Poland, Galicia, and Bukovina," he warned Russian commanders, as he ordered them to be on the alert for Jewish spies and provocateurs. "In every case where our troops have retreated and the enemy have occupied the area," Ianushkevich claimed, his hatred spurring his imagination to greater flights of fancy, "there have poured forth upon those of the population who were friendly to us countless retributions, mainly on the basis of denunciations made by Jews." Unrelenting in his hatred, he let no opportunity to persecute Jews pass. Taken hostage and obliged to swear that they would "answer personally, and in accordance with all severities of martial law" for any anti-Russian acts that might occur in the Jewish community, many Jews soon found themselves facing death for crimes they had not committed, while others, who had the misfortune to be standing at the roadside when Russian troops marched by, were summarily executed as spies by embittered anti-Semitic Russian commanders anxious to find easy explanations for new defeats.[13]

Jewish elders protested that "indiscriminate accusations of Jewish betrayals . . . almost always prove unfounded when [properly] investigated," only to have their pleas ignored by authorities who had no intention of looking too closely into what one appalled attorney described as "a jumble of naive assumptions and absurd analogies." Yet there were men and women in Russia who had finally begun to speak out on the Jews' behalf, and perhaps none more loudly than the left-wing Duma deputy Nikolai Chkheidze, who took the Russian government severely to task for its persecutions of men and women whose sons had given their lives so readily in Russia's defense. "What humanity is this which forbids the offering of food to hungry Jewish fugitives kept in sealed wagons at the [railroad] stations, as our authorities have done?" Chkheidze asked in a dramatic speech before the Duma. "What brotherhood is this when a part of the army is set against the Jewish soldiers who are risking their lives in the same trenches with the others?" Declaring that Ianushkevich's persecutions had "no precedent in history," Chkheidze went so far as to ask from the Duma's rostrum if "[ever] there was a government so cynical as to take hostages from its own subjects? "[14]

A handful of others, among them the dean of Russia's proletarian writers, Maksim Gorkii, agreed. "Hatred for the Jew is a brutal phenomenon," Gorkii wrote. Jews, he insisted, "must be free." Taking up Chkheidze's lead, Gorkii denounced the mistreatment of Jews as "a disgrace to Russian culture," and proclaimed that "the Jews, as the old, strong leaven of humanity, have always exalted its spirit, bringing restless and noble ideals into the world." Along with him, the noted economist Mikhail Bernatskii warned that "anti-Semitism is the worse foe of economic prosperity," and Pavel Miliukov, the Kadets' leading spokesman, insisted that "anti-Semitism serves . . . the old order from which we still are struggling to free ourselves." Still, Russian statesmen continued to move cautiously in dealing with the Jewish question, and even such ardent defenders of human rights and political freedom as Miliukov's Kadets insisted that, as a matter of political expediency, "slow embarkation on the road to Jewish equality" would more effectively enable them to win support for a broader program of far-reaching political change.[15]

Despite these few denunciations of anti-Semitism, most Russian politicians and statesmen expected to postpone any serious effort to improve the Jews' position until after the war, but Ianushkevich's vicious policies in Poland and Russia's western provinces provoked such outrage abroad that they had to take up the question of granting civil rights to Russia's Jews much sooner than they had expected. Without some serious effort to mitigate the effects of Ianushkevich's policy of mass deportations of Jews from the provinces in his rear, Russia's ministers knew for certain that they could not continue to borrow abroad the credits needed to finance their nation's war effort. While the anti-Semitic Rukhlov complained that "all this is taking place under the pressure of the Jewish purse," Finance Minister Petr Bark, his three decades of experience in banking and finance investing his words with indisputable authority, urged his colleagues to "be guided not by feelings, but by the imperatives of this extremely critical moment," and hastened to point out that these demanded some moderation in the Jews' condition. "Money for the war is being demanded from our government," Bark explained, "but we cannot obtain it because it is in the hands of those very people upon whose co-religionists General Ianushkevich has visited violence and injuries such as are intolerable in any civilized state." Russia now had no choice, he concluded, but to "give concessions to the Jews and reestablish our credit so as to obtain the means for continuing the war."

Perhaps hoping to soften the blow, the ever-politic Minister of Agriculture Krivoshein suggested that "the transfer of Jewish enterprises from the western provinces will perhaps give some impetus to the development of industry and will ensure the improvement of local conditions," but few of his colleagues were inclined to look for benefits in the decision they were being forced to make. "What can we do when the knife is at our throat," one of them asked, when they returned to the discussion a few days later. "If the evil influence of the Jews is undeniable, the need for money to carry on the war is equally undebatable, and that money is in Jewish hands." In August 1915, Russia's Council of Ministers therefore abolished the Pale, and decreed that, at least until the war ended, Jews could live in any town or city they chose. "It is painful for me," the newly-appointed director general of Russia's Holy Synod, Aleksandr Samarin, confessed as he spoke in favor of the measure so many of his colleagues found distasteful. "But there come those times in the life of any nation when a certain combination of circumstances makes it necessary to make great sacrifices for the needs of the moment."[16]

More than the sacrifices Samarin had in mind were needed, for neither Ianushkevich's persecution of the Jews nor the Council of Minister's reluctant agreement to lessen their disabilities could alter the indisputable fact that the Austro-German armies stood nearer to Russia's borders at summer's end than at the war's beginning. As the war's first year approached its end, the sense of hopelessness first stirred by Mackensen's breakthrough in May deepened all across the empire. "Blood—one sees it all the time, everywhere," Zinaida Gippius wrote in her diary as she watched the war erode her friends' long-standing dedication to art and poetry. "They say that peace is not possible—but is war?" Greater sacrifices of life, limb, and treasure had been demanded and given than at any time in Russia's history, but Russians now had begun to wonder if millions of their countrymen had spilled their blood to no purpose. "We need a leader," a high-ranking aristocrat told Paléologue. The army, he complained, felt itself "sacrificed beforehand, like a herd driven to the slaughterhouse."[17]

Overwhelmed by the scope of the tragedy that loomed above Russia's western frontier as the Germans launched their new advance, Gippius asked, "Where's the way out?" A few weeks later she concluded that "everyone is all tangled up," and that "no one understands anything at all." Then, she decided that "the dark masses go to war on orders from above, by the inertia of blind obedience," and insisted that "the *narod* knows absolutely nothing about the war." But the *narod*, those masses in whom Russia's rulers and generals always had put their faith in times of great crisis, understood all too well, and their fatalism grew tinged with the sort of cynical black humor that Russians reserve for events that hurt their lives the most. "We will retreat to the Urals [on the western border of Siberia]," soldiers told General Knox. "When we get there the enemy's pursuing army will have dwindled to a single German and a single Austrian. The Austrian will, according to custom, give himself up as a prisoner, and then we will kill the German."[18]

Highly placed Russians began to wonder if the grimly fatalistic *narod* would remain steadfast. "When a *muzhik* [a peasant] thinks himself victim of some injustice, he usually submits without a word, because he is a fatalist and meek by nature, but he continues to ponder his injury, and tells himself that *a price will have to be paid* some day," the grand master of the Russian court once explained. "At some point . . . [the peasants] will demand an accounting," he added. "And, when the *muzhik* ceases to be meek, he becomes terrifying." Did Russia's *muzhiki* now think themselves the unjustly chosen victims of a war that had spilled so much of their blood? Would they remain meek or, as some had begun to fear, turn ferocious as they had in the long-ago days when the Cossack chief Emelian Pugachëv had led the *narod* in their last great peasant war against the authority of Catherine the Great? The *narod* had almost no voice in their destiny. Would they remain satisfied to have none? Many expected the answers to these terrible questions to come before the summer's end; some feared to think what they would be. "Could there be a revolution in the midst of war?" Gippius asked. "Do we even dare wish for it?"[19]

Still resisting a flood of public pleas for the government to make common cause with its people, Nicholas and his advisers tried to impose still greater demands upon their disillusioned and unwilling subjects that summer in a desperate effort to sustain Russia's crumbling war effort. Sternly but unwisely, Aleksandra urged her husband to pay no attention to opposition or criticism, and to accept no excuses. "You must simply order things to be done," she wrote, as she insisted that even the shortages of rifles and ammunition could be overcome if only Nicholas would take to heart the admonitions of the bogus Dr. Philippe and Rasputin that all in Russia "must learn to tremble" before him. Russians were as able as any other people, the empress reminded her husband, but they were "lazy & without initiative." Certain that "there is some way in wh[ich] a woman can be of help," she warned him against his ministers' "treachery," and begged him to "speak about all to me, talk it out, cry even." While Nicholas grappled with problems grown to dimensions that no mortal could hope to solve alone, Aleksandra urged him to "hearken unto our Friend [Rasputin] . . . [for] it is not for nothing [that] God sent Him to us—only we must pay more attention to what He says—His words are not lightly spoken—& the gravity of having not only His prayers but His advise—is great."[20]

Despite Aleksandra's urgings, Rasputin could not solve Russia's problems during the summer of 1915. "The government has gone on strike," one of the editors of the Kadet newspaper *Rech* lamented. "One used to say that, for war, one needed money, money, and more money. Now it has become clear that nerves, nerves, and more nerves are more important."[21] Certainly nerves were needed as Russia's war effort hemorrhaged and stumbled. Mackensen's brutal artillery assaults in April and May had reduced many units of the Russian army to meager skeletons. Whole regiments now numbered less than companies. Divisions had shrunk to the size of regiments, and there were corps facing the Austro-German armies in the middle of June that were smaller than divisions. Even with reinforcements, Russia's Third Army had to face Mackensen's renewed attacks with only two-fifths of its usual 232,000-man strength.[22] More men had died under Mackensen's merciless artillery barrages in April and May than could be found in Russia's reserve depots and there was no time to train replacements. In mid-June, *Stavka* estimated that unreplaced Russian casualties along the entire front had climbed well past a half-million.

Desperate for bodies to fill the gaping holes that Mackensen's artillery had torn in their western defenses, War Ministry officials combed Russia's villages for more cannon fodder. Peasant lads who had tended their fields in the eternal calm of Russia's countryside at the beginning of June found themselves in the trenches a scant fortnight later. Unarmed and untrained, these unfortunates entered the front's infernal world of bursting shells, screaming wounded, and desperate, dying men untrained and unarmed. "Rifles are now more precious than gold," General Aleksei Polivanov confided to his diary soon after he replaced Sukhomlinov as war minister.[23] An able military engineer, whose genius for organization and administration would eventually produce the men and weapons to support Brusilov's victorious offensive the following summer, Polivanov could do nothing to shield Russia's soldiers from the carnage that awaited them once the German guns renewed their attack. "Our army is drowning in its own blood!" General Beliaev reported from the War Ministry. "In recent battles, more than a third of the men had no rifles . . . [and] had to wait patiently under a hail of shrapnel until their comrades fell before their very eyes and they could pick up their weapons.[24] Now there would be even more men hoping against hope that some sort of weapon would fall into their hands before they had to charge against German machine guns. In the middle of June, the usually optimistic General Staff estimated that even if the half-million replacements needed to fill the line could be assembled, there were no more than forty thousand rifles available to share among them.[25]

Front-line officers had neither the time nor the facilities to give such men even the simplest training before they marched them into the grinding jaws of the German heavy artillery.[26] Nor were there enough men to lead these frightened raw replacements when they reached the front, for the number of officers killed, wounded, or captured during the first year of fighting exceeded the size of Russia's entire 1914 officer corps by more than 50 percent.[27] "I need your lives," Nicholas once pleaded with a group of newly-commissioned cadets. "I am sure that every one of you will give his life willingly when it becomes necessary, but do it only in case of exceptional urgency." Because Russian officers still led their men to battle, the losses continued. "The Russian soldier is a dogged fighter and goes wherever he is ordered so long as there is an officer leading him," a field commander once explained. "But take away the officer . . . and our soldiers for the most part lose their heads. This means that officers must always be in front of their men and for that reason they suffer immense casualties."[28]

Because many Russians with the education and talent to become officers continued to evade military service in 1915, it proved impossible to replace the officers killed during the war's first year with men of similar ability. "There was no strong sense of duty among the intelligentsia," General Danilov explained as he recalled "the great numbers seeking the means or opportunity to escape service," and one expert has estimated that more than half a million young men of middle or upper class origin continued to evade military service at this point in the war.[29] Infantry companies met the Germans' renewed assaults under the command of ensigns, while surprised young second lieutenants found themselves commanding battalions. In mid-1915, Russia's General Staff became so desperate for junior officers that it rounded up draftees with high school educations and commissioned them as ensigns after only a few weeks of basic training.

As officers continued to fall in battle faster than replacements could be trained, War Minister Polivanov set up special schools to train subalterns and ensigns from men with four years' of formal schooling and four months' of active service. Most of the volunteers who entered these schools were literate or semiliterate peasants—often the poorest peasants—as was Vasilii Chapaev, who graduated as a subaltern from one of Polivanov's schools and rose to command a cavalry division in the Red Army a scant three years later. Together with those who came from families of petty traders, artisans, and shopfolk, men of origins similar to Chapaev's accounted for more than eight out of every ten who graduated in classes where only one in twenty was an aristocrat. Among officers below the rank of captain in the fall of 1915, almost none had been in the army for more than a year. For all practical purposes, the first year of the Great War had annihilated Russia's junior officer corps.[30]

Noncommissioned officers were in even shorter supply and even more difficult to replace. In Western nations, life in a modern industrialized society had given great numbers of men experience in directing the work of others, but life in prewar Russia had offered few such opportunities. Drafts of men consequently yielded few who could be trained to replace noncommissioned officers who fell in battle, and this made it difficult to connect the masses of untrained peasant conscripts pouring into depleted front-line units with the army's chain of command. Companies and battalions now went into battle with only scattered fragments of that combat-hardened corps of NCO's whose command abilities had held peasant recruits in the trenches under the crushing artillery barrages of earlier battles.

Fearful that without enough capable leadership crumbling reserve units might expose their flanks to surprise attacks at any moment, seasoned front-line commanders distrusted their fellow officers and became disillusioned with their men. Perhaps General Ianushkevich was the bitterest of all. "Heroes such as one reads about in folk-tales, idealistic fighters, [and] altruists are encountered only in the scantest numbers [nowadays]," he wrote in letters to War Minister Polivanov and the Council of Ministers that summer. "It is beautiful to fight for Russia," he added, "but the masses do not understand it." It was a jolting thought. Suddenly, perhaps for the first time, one of the top men in the Russian army had realized that loyalty to the Romanovs' regime was not a mysterious elixir destined to flow forever in the veins of Russia's fighting men, and that peasant soldiers had hopes and dreams that could not go forever unnourished.

For centuries, Russia's peasants had been summoned to lay down their lives for a tsar who gave them little, and a social and economic order that offered them even less. None in the High Command or in the tsar's inner councils had thought to reward the loyal peasant soldiers who had driven Napoleon from their land in 1812 or those who had stood firm against the Allies' superior weapons in the Crimean War. Against the Turks in 1877, again against the Japanese in 1904, and, yet again, in July 1914, the tsar had summoned his people to arms, and the peasants had continued to respond because, as General Danilov once explained, "they were accustomed in general to doing everything that the Supreme Power asked of them." Only under the iron hail of German shrapnel did the loyalty of Russia's peasant soldiers begin to falter during the summer of 1915. Once recognized, the implications of that fact were staggering. Only the fidelity of Russia's peasant army had enabled Nicholas and Aleksandra to weather the revolutionary storm of 1905. Without it, they were bound to be overwhelmed if revolution struck again.

To prevent Russia's peasant soldiers from abandoning tsar and country, Ianushkevich set out to buy the loyalty that Nicholas and Aleksandra had failed to earn from their people. "We are hanging by a thread," he warned. "The Russian soldier must be made materially interested in resisting the enemy and it is necessary to entice him with allotments of land and the threat of confiscating his land if he surrenders." *Stavka*'s chief of staff therefore urged the tsar to promise a reward of at least sixteen acres to every peasant soldier who served his country loyally during the war. "I beg you to forgive my importunity," Ianushkevich concluded gloomily, "but, like a drowning man, clutching at a straw, I am trying to find a way out of this wretched position by any means possible."[31]

Foreign Minister Sazonov had once called Ianushkevich "a low, filthy-minded man" whose "main pleasure comes from rooting around in information gathered by the lowest kind of spying on peoples' personal lives," and all of Russia's senior statesmen were outraged to think that one of the closest advisers to their commander in chief could approach the problem of mass loyalty so cynically. Firmly, they rejected Ianushkevich's proposal, but could not shut their eyes to the fact that, for the first time in Russia's history, her peasant soldiers had become reluctant to fight for a tsar and country that gave them nothing in return. Life at the front was all death and no glory now, and commanders' efforts to applaud bravery passed with less notice. Desertions increased. Men grown tired of killing began to walk away from the war that summer, and masses of soldiers surrendered at the height of battle. For the first time, Russia's commanders ordered their artillery to fire its precious shells upon the positions of peasant soldiers who refused to fight.[32]

Perhaps too readily impressed in earlier months by the quality of soldiers he had seen, General Knox now thought Russia's soldiers "listless, of brutally stupid type," whose officers "slouched along without making any attempt to enforce discipline,"[33] and, at the beginning of June, one of his deputies rated Russia's Third Army as "a harmless mob."[34] The Third Army was more than that, and still capable of determined resistance once *Stavka* assigned it a more energetic and talented commander. But it did not pose the threat to its Austro-German foes that it had three months before, and General von Falkenhayn now proposed to use that very obvious fact to the advantage of Germany and her allies. He and his fellow commanders had only to agree about when and where they should strike their blows.

At their Poznań conference with the kaiser on June 19/July 2, Falkenhayn and Mackensen had challenged Ludendorff's certainty that Kovno was "the cornerstone of the Russian defense on the Niemen," and disputed Hindenburg's contention that a victory on that remote sector of Germany's eastern front could change the course of the war in any significant way. It was *OHL*'s task, Falkenhayn now pointed out loftily, "to work for a victory that would probably have its effect on the whole operation" and not squander precious men and materiel to reap a mere "tactical success of local importance" that would "simply lengthen our line." Hindenburg reluctantly admitted that "it was more a matter of sentiment" than of sound tactical judgment that dictated his preference for an attack north of the Niemen rather than against the Narew sector of the front that spread between Novo-Georgievsk and Warsaw. Falkenhayn therefore insisted that the Austro-German armies concentrate their full weight upon the Polish salient that protruded westward between his armies' advanced positions to the north and south of Warsaw. Without hesitation, the kaiser agreed.[35]

With the imperial seal of approval now imprinted firmly upon the important shifts in tactics and strategy that Falkenhayn had audaciously initiated even before the Poznań meeting, German commanders hurried their forces into position. Taking full advantage of the capacity of German industry to produce great quantities of ammunition, and its own awesome ability to move lavish quantities of materiel to its attacking armies, *OHL* gave commanders all along the eastern front crushing quantities of shell for their heavy guns. Shell reserves rose to a thousand per gun and continued to climb. In the rear, hundreds of thousands more stood ready to be rushed forward if needed to blast through the scantily supplied and poorly armed Russian armies that barred the Germans' way.

In Galicia, Mackensen had begun to reposition his forces even before the conference at Poznań because, as *OHL* recognized all too readily, it might be possible "to capture further ground from the enemy, but it was scarcely possible to inflict any real damage upon him in the broad plains of Volhynia and Podolia," where the retreating Russian armies "possessed unlimited possibilities for retirement."[36] Just before the middle of June, Mackensen therefore had wheeled his armies away from their eastward advance until he had them posed to strike directly northward toward the Russian fortress of Brest-Litovsk, some

175 kilometers directly east of Warsaw. Slowly, he began to drive another "Mackensen wedge" into the channel formed by the natural geographical barriers of the Bug River on his right and the Vistula on his left.

Spread across the 140 kilometers that separated the two rivers, Mackensen's new wedge was more blunted than the one he had used in the Tarnów-Gorlice sector, where his breakthrough had left what General Knox called "a running sore that was draining the vital force of the Russian defense."[37] He now drove his Eleventh and Fourth armies resolutely against the Third, Fourth, and Thirteenth Russian armies that had taken positions before Lublin and Cholm to bar the way to Brest-Litovsk. This time, he did not attempt to repeat his classic strategic *percée,* but pressed more slowly, demolishing Russian troop formations with the heaviest concentrations of artillery he could assemble. "The heavy artillery would take up positions in places which were entirely—or almost entirely—beyond the range of the Russian field artillery, and the heavy guns would start to shower their shells on the Russian trenches, doing it methodically, as was characteristic of the Germans," General Golovin remembered. "That hammering would go on until nothing of the trenches remained, and their defenders would be destroyed."[38] Then, the infantry would occupy the demolished trenches and the artillery would move up to begin the process anew. In one such operation, General Knox saw a group of sixteen hundred raw recruits "churned into gruel" by the German heavy guns as they stood in support trenches "to wait unarmed till casualties in the firing line should make rifles available."[39]

The ill trained and poorly armed Russians stiffened their opposition as Mackensen's armies neared Lublin and Cholm. In two weeks of heavy fighting, his left managed to advance nearly fifty-five kilometers, but his right moved less than thirty-five. It took the newly-formed *Bug-Armee,* which Falkenhayn added to Mackensen's command on June 25, a full ten days to break through the Russian lines at Cholm, while Mackensen's elite Eleventh Army, which had smashed the Russians' defenses in the Tarnów-Gorlice line two months before, battled for almost a fortnight before it and the Austro-Hungarian Fourth breached the Russian defenses at Lublin on July 23. Still, the Russian defense was not shattered, and their Third, Fourth, and Thirteenth armies fell back to a new defensive line that stretched from Ivangorod to Opalin to Kowel.[40]

While Mackensen inched his wedge forward, Germany's Twelfth Army moved southeast from Soldau, scene of Samsonov's debacle the year before, to concentrate seven full-strength divisions and 860 guns, each with nearly a thousand shells in reserve, on the thirty-five kilometers of front that stretched east and west between Przasnysz and Ciechanów. Hoping to duplicate Mackensen's earlier breakthrough and win his spurs as Germany's newest army commander in the east, the Twelfth Army's newly-appointed commander General von Gallwitz planned to drive his main assault force between the Russian First and Twelfth armies at Przasnysz, where Russians and Germans had met in a bloodily inconclusive battle some five months before. Like Mackensen, Gallwitz

hoped to batter his way forward by weight of superior numbers and overwhelming firepower. To do so, he had the services of Colonel Brüchmüller, perhaps
the most brilliant barrage specialist to be found on the eastern front.

Facing Gallwitz, the combined forces of Russia's Twelfth and First armies
numbered just over 100,000 men, many without rifles and some without bullets.
Only forty kilometers to their rear stood the great fortress of Novo-Georgievsk,
whose walls bristled with sixteen hundred guns and whose arsenal stored tens
of thousands of shells and several million rifle cartridges. In another incredible
oversight, the Russian High Command gave no orders to move guns and ammunition from Novo-Georgievsk toward Przasnysz, and left the Twelfth and First
armies to face Gallwitz with less than four hundred guns, each with the pitiful
daily ration of five shells.[41]

At five o'clock in the morning of June 30, Gallwitz ordered all of his guns
into action at once—a "hurricane bombardment" one observer said—that cost
the Russian First Army nearly one out of every three of its men during the next
five hours. "The Germans literally plow up the field of battle with a hail of metal
and level our trenches and defenses, often burying their defenders in the process," one division commander wrote in describing this method of attack. "They
use up metal," he concluded. "We use up human life."[42] In vain did members
of the State Council insist that Russia's commanders "must [learn] . . . to
substitute lead, steel, and explosives for the power whose source is human
blood."[43] That first morning at Przasnysz the Turkestan I Corps fought forty-
two German heavy guns with only two of its own and suffered predictably
murderous casualties, while the 11th Siberian Division lost half of its men in
thirty minutes.

Held in its trenches by its commander's uncompromising order that there
could be no retreat, the First Army stubbornly repulsed Gallwitz's infantry, and
it was not until evening that the weight of the Germans' superior numbers and
firepower drove its men back to their second line of defense. During the night,
the German artillery moved forward to begin a day-long bombardment of the
new Russian positions while Gallwitz brought up three more divisions to
strengthen his next infantry assault. By the morning of July 2, he had ten
divisions to hurl against the Russians' four and a half. Three more days of
fighting brought him some eight kilometers further into the Russian defenses,
and a week later his divisions crashed across the Narew River to sweep away the
last Russian defenses in the northern sector of the Polish salient.[44] "The Russians are losing all of Poland with all its immense resources," Paléologue reported
sadly. From Bangor, in North Wales, Lloyd George promised his fellow countrymen with an optimism utterly unwarranted by events that, "with their monster
artillery, [the Germans] are shattering the rusty bars that fettered the strength
of the people of Russia.[45]

While the Russian armies in Poland shuddered beneath the combined assaults of Mackensen and Gallwitz, Hindenburg and Ludendorff ordered their

Tenth and Niemen armies to attack along the 250-kilometer front that stretched from the Baltic to a point some thirty kilometers west of the Kovno fortress in a scaled-down version of the plan that *OHL* and the kaiser had rejected at Poznań. On July 1, just one day after Gallwitz first bombarded Przasnysz, the Germans attacked near Szawli, their artillery outnumbering the Russians' by almost two to one. Commanded by the still sick and still stubborn General Plehve, the men of the Fifth Russian Army had orders from *Stavka* to take "not one step backwards," as the Germans unleashed another crushing artillery assault, but Plehve proved more independent—and more sensible—than many of his colleagues, and ignored *Stavka*'s suicidal order. Preferring to let the German heavy guns chew up empty ground rather than the army he would need if he hoped to fight again, Plehve withdrew to more defensible ground. To do so meant abandoning the center of Russia's tanning industry at Szawli where, on July 8, the soldiers of the Niemen Army seized over a million rubles' worth of tanned hides destined to be sewn into desperately needed boots. Plehve, however, had kept his army intact and, with it, kept the vital centers of Riga and Dvinsk from falling into German hands for the better part of two years.[46]

While the Niemen Army pursued Plehve's retreating forces, the Tenth German Army besieged Kovno, where the fortifications were so outdated that Grand Duke Nikolai Nikolaevich once had said that the fortress ought to be renamed *"Govno,"* the Russian word for "shit." Kovno's flimsy brick bastions were no match for the Tenth Army's heavy 420-mm howitzers, and its defense was weakened further by the cowardice of General Grigorev, its aged commander, who fled so precipitously on the day before the fortress fell that, according to one report, even his own chief of staff "did not for some time even know that he had gone." After the Russian gendarmerie finally discovered Grigorev at the Bristol Hotel in the Lithuanian capital of Vilno, Nikolai Nikolaevich sentenced him to fifteen years at hard labor, even though the Russian army usually did not punish commanders for failure. In the meantime, Kovno's defenders, most of whom were raw *opolchenie* troops, fled headlong before the Germans final assault without even destroying any military stores. When the Tenth Army's men entered the empty fortress on August 4, they found more than thirteen hundred guns and almost a million artillery shells.[47]

Kovno was not one of the first but, in fact, one of the last of a half-dozen Russian western fortresses to fall to the Germans. Built at a cost of hundreds of millions of rubles and defended by thousands of the heavy guns that Russian field commanders had begged for so desperately that summer, these great bastions now could neither be held nor emptied of their rich stores of precious weapons and ammunition before the Germans reached them. At best, they could be used to delay the enemy advance for a few days or weeks, but then they would have to be abandoned if Russia's forces were not to be trapped by the German armies that threatened to cut off the salient containing Novo-Georgievsk, Osowiec, Ivangorod, and Warsaw. Incredibly, the Russian High Command never had prepared plans for the general retreat that was now required, evidently sharing the view of General Polivanov, who told the Council of Ministers at the beginning of August that he had always "put my trust in impassable spaces, impenetrable mud, and the mercies of St. Nicholas the Miracle-Worker, the patron of Holy Russia."[48] At best, Russia's senior commanders nourished vague thoughts about repeating the scorched-earth tactics that Field Marshal Kutuzov had used in 1812 when Napoleon had rashly led his armies onward into Russia's vastness until they stood alone, without food or forage, in the midst of lands that had been laid waste. Winter had driven Napoleon's hungry and freezing soldiers from Russia, and Kutuzov had become a national hero for whom popular legend quickly wove the mantle of greatness that allowed future generations to forget that he had spent most of his life loving women, drink, and food in the greatest variety and quantity obtainable.

Tolstoi had immortalized Kutuzov in *War and Peace*, where legend had assumed an aura of reality, and the aged, one-eyed field marshal, in real life a sly voluptuary and corrupt courtier, had stood forth as a captain of genius, the author of the grand plan that had doomed Napoleon from the moment his armies set foot on Russian soil. Tolstoi's account was so brilliantly cast, its drama so immediate, and its detail so intense, that many Russians forgot all too easily that Kutuzov had done little more than respond to desperate necessity in 1812, and that he was not the creator of a timeless strategy that could repel any enemy unwise enough to invade the Russian land. "The difficulties that Napoleon had at that time," Colonel von Hoffmann later wrote of those long-ago days, "have been overcome by modern means of communications and conveyance. If Napoleon had had railways, telephones, motors, the telegraph, and airmen," he concluded with a brash confidence that Hitler would have been wise to question, "he would be in Moscow today."[49]

Whether in Tolstoi's version or Kutuzov's, the tactics of 1812 had little relevance to the Great War, but Russia's generals could think of no other course than to dedicate themselves to a hopeless attempt to transform Tolstoi's fictional mirage into wartime reality. "The retreat will continue as far—and for as long —as necessary, until the day comes when we have enough rifles and munitions to mount a general offensive," Nicholas told Paléologue. "The Russian people," he promised, "are as unanimous in their will to conquer as they were in 1812." Others saw the Tolstoian vision differently, and their faith in the Russian people took another form. "The Russian people will save Russia," one journalist said, "but before that happens, the regime will perish."[50]

Perhaps none had thought less about a general retreat from the Polish salient than General Palitsyn, the man Nikolai Nikolaevich had hoped to name his chief of staff when he had taken command of Russia's armies in July 1914. Just before Gallwitz's divisions stormed across the Narew, the tsar's nephew, Grand Duke Andrei Vladimirovich, discussed with Palitsyn the difficulties of Warsaw's exposed position near the tip of a salient that thrust several dozen kilometers westward from the new Russian positions on its flanks.

"What will the High Command do now?" Andrei Vladimirovich asked Palitsyn.

"I don't know," the general replied.

"But surely you people have something in mind!"

"Of course, we have something in mind. There are some good, very excellent ideas," Palitsyn assured him.

Not surprisingly, the grand duke asked what those "good, very excellent ideas" were, but Palitsyn spoke only in incomprehensible generalities. Andrei Vladimirovich therefore tried another approach, but with results that, he later confessed to his diary, "left my head spinning."

"Well, then, shall we attack?" he asked.

"No, that's absolutely impossible," Palitsyn replied. "We have to save our strength."

"Then you mean that we must make a stand?"

"Good God, no! When we make a stand, they can hit us wherever they wish. We made a stand at Cicchanow and Prasznysz and look what happened there!"

"Then you mean we must retreat?" the grand duke asked again.

"God save us! How can we retreat?" Palitsyn replied. "According to all theorics, you lose much more during a retreat than during an attack."

"So what are we going to do?" Andrei Vladimirovich asked in amazement as he returned to his original question.

"I don't know," Palitsyn said again. "But there are some good, very excellent ideas."

"I felt as if my thoughts had gotten tangled up in a terrible knot, where neither logic, nor will, nor reason could be of any help," Andrei Vladimirovich remembered. "How the army would carry out these 'good, very excellent ideas' when they could neither attack, nor retreat, nor make a stand, I simply could not comprehend," he confessed. "Was this terrible nightmare a dream, or was it, in fact, reality?"[51]

Andrei Vladimirovich had not long to wait for the full force of reality to strike. Less than a fortnight after his discussion with General Palitsyn, the Germans occupied Warsaw, center of Russia's power in Poland ever since Napoleon's final defeat at Waterloo a century before. "The Russian armies have had to abandon Warsaw, the sanctuary of the Polish soul," the archconservative leader Vladimir Purishkevich lamented at a session of the Duma. Quoting Adam Mickiewicz, Poland's greatest national poet, he asked: "Will there be found among us the magic word that can . . . return to us all that is dead?" and promised "an even stronger hatred for our common enemy, an even deeper faith in ultimate victory." Purishkevich looked forward to "that glorious future day [when] unified Slavdom will triumph," but there were only more defeats in store.[52] Along with Warsaw, *Stavka* abandoned Ivangorod, the crumbling fortress that guarded the Vistula crossings to the city's south, and made ready to preside over a new series of defeats that would drive Russia's armies from all of Poland in just a few weeks.

On July 22, the very day Warsaw fell, Ivangorod's commander, General Aleksei Schwarz, destroyed everything in the fortress that could not be carried away. A tough soldier who called *Stavka*'s order to retreat "a horror and a tragedy," Schwarz had dedicated his entire career to the theory and practice of fortification and was one of the rare commanders who successfully withstood a siege during the Great War. He had fought against Japan at the siege of Port Arthur in 1904, had taught the science and tactics of fortification at the Imperial School of Army Engineers until July 1914, and, that fall, had fought off an earlier German effort to capture Ivangorod. Unlike Grigorev, whose precipitous flight from Kovno had left huge stores in German hands, Schwarz methodically destroyed Ivangorod's supplies, weapons, and fortifications. "The citadel was a picture of utter devastation," he wrote some years afterward as he recalled the moments before he and his men turned their backs on Ivangorod forever. Then Schwarz gave his final command, and watched the fortress "disappear entirely in an enormous black cloud" that left nothing for the enemy when they arrived the next day.[53]

After abandoning Warsaw and Ivangorod to the Germans, Nikolai Nikolaevich and his chief advisers at *Stavka* decided to defend the great fortress of Novo-Georgievsk, even though the fall of Warsaw and Ivangorod had left it well behind German lines and its pitifully weak fortifications had been built to withstand only the artillery of the 1880s. Novo-Georgievsk had a garrison of more than a hundred thousand men, with sixteen hundred guns and almost a million shells. To destroy it, the Germans called upon General von Beseler, one of their most renowned siege experts, whose credentials included the destruction of Antwerp at the beginning of the war. With Beseler came a fearsome train of heavy guns, dominated by nine twelve-inch and six mammoth sixteen-inch howitzers, in addition to several batteries of great ten-inch Austrian siege mortars, all of which could fire high explosive shells that stood as tall as a man and weighed hundreds of pounds, from positions well beyond the range of the fortress guns.

Beseler's awesome technological resources were augmented by amazing good fortune. On the first day of the siege, one of his patrols captured the chief engineer of Novo-Georgievsk and an entire set of plans of the fortress's fortifications that he had been stupid enough to carry with him on a last-minute inspection of its outer works. Now Beseler knew exactly where his gunners should place their shells and they did so with their usual precision. "We are under continual fire, which is growing more intense," a radio telegram reported to General Alekseev's northwestern front headquarters at the beginning of the siege.[54] Quickly, Beseler tightened his ring, bringing all his guns to bear on measured sectors of fortifications, reducing them to rubble, and then storming them with his best infantry. By August 5, he was ready for a final assault against the inner works of the fortress, while the Russians burned the last of their supplies and ammunition. "We are being wiped out," a lieutenant colonel telegraphed to Alekseev. "Everything is being swept away, everyone is in a panic." That was the last word either Alekseev or *Stavka* received of Novo-Georgievsk's fate until a lone airman, who had flown out of the fortress at the last moment, brought word that the fortress had fallen on the morning of August 6. All too typical of the Russians' war effort at that point, the airman who brought the news had to fly through Russian anti-aircraft fire as he tried to land at *Stavka*. By that time, thirty Russian generals, ninety thousand prisoners, and seven hundred guns from Novo-Georgievsk were in German hands, and Hindenburg had already escorted the kaiser into the fortress in grand style. "The barracks and other military buildings were still blazing," he later remembered. "Masses of prisoners were standing around. . . . Before the surrender, the Russians had shot their horses whole-sale."[55]

By early August, Russia had lost nearly all of the Polish salient. From Kovno on the Niemen to Zloczów, some nine hundred kilometers to the south, the Austro-German armies needed only to deal with the fortresses of Osowiec and Brest-Litovsk before their advance could continue. Once those had fallen, the Germans could push on to Grodno, Vilno, and, even, Nikolai Nikolaevich's headquarters at Baranovichi. With nothing larger than six-inch guns, but with well-constructed entrenchments that prevented the Germans from setting up proper artillery observation posts, the Russians held Osowiec for nearly a month and delayed the German Eighth Army from breaking into their rear. During those weeks, the commander of Osowiec estimated that the Germans fired almost a quarter-million shells during the siege, and the Russian press wrote stories about its heroic defense against more than two million shells. Actually, the Germans fired only about twenty-five thousand shells, the damage was nothing like that suffered at Novo-Georgievsk, and the Russians managed to withdraw their troops and even some of their field guns before they destroyed the fortress and its stores on August 13. On the same day, Mackensen broke into Brest-Litovsk. By mid-August, all Russian forces were gone from Poland. Except for a very small sector of front that ran from Tarnopol to Khotin in the far south, the eastern front now lay on native Russian soil, where resistance toughened.[56] The German Eighth Army took Grodno only after "violent street fighting," Gallwitz advanced only by "fighting all the way," and the Niemen Army failed to wrest Riga from Plehve's bony grip.[57] After unsuccessful attacks against Dvinsk and Molodechno, the Germans halted. At the beginning of September, the Lithuanian capital of Vilno became the last major center to fall before Falkenhayn ended the fall campaign in the east.

As Russia's armies retreated from Warsaw and Ivangorod, past Lublin, Cholm, Kovno, Brest-Litovsk and Grodno, beyond the Pripet and Pinsk Marshes of West Russia into the lands of Russia itself, they made clumsy efforts to destroy everything that might be of use to the Germans. "Our enemy always took the most enormous pains to destroy everything, especially supplies," Hindenburg later wrote, but in fact the Russians did both more and less damage than that. "Everywhere there was evidence of misdirected or undirected effort," Knox reported on one occasion. "Without an officer to direct them, the gendarmes ran about setting fire to piles of straw, but leaving the crops untouched." Horses were slaughtered, but fodder stayed unburned. Although sometimes extensive, this random damage often solved more problems for the Germans than it created. "The Russian destructions were in many ways an advantage for us," Hoffmann said some years later. "Although [Brest-Litovsk, for example,] was burnt down, we were able to find quarters there, while the 80,000 inhabitants for whom we should have had to provide, were not there."[58]

What Hoffmann rightly thought an advantage for the Germans became an immense disability for the Russians, for their haphazard policy of random destruction turned hordes of peasants and townfolk from all over Poland and West Russia into refugees overnight. "*Stavka* has lost its head completely," the Council of Ministers complained only days after the retreat had begun. "It has no notion of what it is doing, nor does it realize the sort of abyss it is dragging Russia into." As with so much of Russia's war effort, the scorched-earth policy was clumsily and insensitively put into effect from the very first. "People were torn away from their homes with but a few hours in which to settle their affairs. Their stores of food and at times even their houses were burned before their very eyes," one horrified senior official reported. "Homeless and hungry, beset with hardship and privation," he continued, such people were bitter and "intensely hostile to the authorities," whose callous unconcern was astounding. Senior officers' mistresses, their furniture, even their pets, took precedence over refugees in this cruel and heartless world of forced mass migration. "While tens of thousands of people are trudging along the railroad tracks," wrote Arkadii Iakhontov—the tireless stenographer whose notes on the Council of Ministers' secret meetings during the summer of 1915 provide some of the most revealing glimpses we have into what went on in the government's inner circles—"they are passed by trains loaded down with . . . all sorts of useless junk, including even cages filled with canaries belonging to bird-loving supply officers." It was, the Council of Ministers warned on July 30, the "equivalent to condemning all of Russia to calamities of the worst sort."[59]

At the end of July, the disasters were only beginning, their dimensions yet unknown, but they were heart-rending nonetheless. "We saw the most pathetic sights," General Knox remembered about his days with Russia's retreating armies. Along eastern Poland's torn and primitive roads, he saw "whole families with all their little worldly belongings piled onto carts, two carts tied together and drawn by a single miserable horse, one family driving a cow, a poor old man and his wife each with a huge bundle of rubbish tied up in a sheet and slung on the back." First there were thousands, then hundreds of thousands, and, finally, more than a million, as the stream of refugees swelled with each week that the great retreat continued. "Men who had fought in several wars and many bloody battles told me that no horrors of a field of battle can be compared to the awful spectacle of the ceaseless exodus of a population, knowing neither the object of the movement nor the place where they might find rest, food, and housing," the tough cavalryman General Gurko later wrote. There seemed no end to the masses moving east. "I passed twenty continuous miles of such fugitives," Knox wrote in late August. "If asked where they were going, they replied that they did not know. . . . Some struggled even beyond the Urals [into Western Siberia]." From the Baltic to the Carpathians, refugees clogged the path of the armies Russia was trying to save, sometimes making it impossible for reinforcements to move into battle or for defeated troops to move out of it. "Many a time," General Gurko remembered, "our forces had to stop and fight a rear-guard action just to allow the crowd [of refugees] to make room for the troops."[60]

Hordes of desperate and destitute refugees posed problems for which the Russian civil authorities were utterly unprepared. Especially concerned to know if Russia could feed these starving masses, Minister of Agriculture Krivoshein called it "the most unexpected, the most menacing, and the most irremediable" of all the war's crises. "Sickness, misery, and poverty are spreading all across Russia," he reported to his colleagues on the Council of Ministers. "Starving and ragged masses are sowing panic everywhere. . . . They leave behind them a virtual desert, as if a swarm of locusts or the hordes of Tamerlane had passed through."[61] Having left Moscow's streetcars and hospital trains to work as a medical orderly nearer the front, Konstantin Paustovskii looked on helplessly when guards allowed refugees to come near the army's field kitchens. "Men tore basins of food from each others' hands," he remembered, "[while] women with starving children at their breasts hastily crammed pieces of gray stewed pork into their mouths" and trampled others underfoot.[62]

As Krivoshein's colleagues angrily confessed, "to feed, to water, to warm this multitude is impossible." From the very beginning, this flood of refugees had the makings of a tragedy of monumental proportions. "Unburied corpses are strewn along the roads," the ministers continued in their litany of disaster. "Everywhere there is carrion and an unbearable stench. This human mass is spreading over Russia like a vast wave." Russia's ministers confessed that they were helpless. "What can one do when things are on such an unprecedented scale?" Iakhontov asked. "Everything is so mixed up that it will take years to straighten out all this chaos."[63] "What is happening is simply monstrous!" Krivoshein exclaimed a few days later. "No people on earth ever saved themselves with destruction."[64] Into the fall and early winter the tragic human wave flowed eastward. "Life isn't life anymore," an innkeeper told Paustovskii then. "Everything's turned to dust."[65]

Not the consequences of their generals' ineptitude, but the fruits of their nation's economic backwardness, finally came to the Russians' aid at summer's end. Once the Austro-German commanders invaded the primitive regions of West Russia, where there were few paved roads and fewer railroads, they could not concentrate supplies quickly enough, and in large enough quantities, to sustain their advance. Nor could they maintain proper communications with each other or plan where and when to strike with the precision they had used in Poland. "Progress takes ages," one German corps commander complained. "It is not so much [caused by] the enemy's strength as by the complete impossibility of all observation in terrain of this type."[66] The German advance therefore began to bog down in September. Convinced that a decisive victory had eluded him in the east, Falkenhayn already had begun to concentrate his resources to launch an all-out assault against Serbia within thirty days of signing the Pless Convention with Bulgaria on August 24/September 6. Still unable to abandon their vision of entrapping the crumbling Russian armies in a decisive pincer movement, however, Ludendorff and Hindenburg ignored their chief's warnings and launched another offensive in mid-September. This time, they achieved extremely limited results. By then, they faced a different commander, whose unexpected appearance at *Stavka* altered the course of Russia's history.

As Russia's ministers watched their nation's western defenses fall to the Germans, and struggled to find some means to feed and shelter the hordes that *Stavka*'s scorched-earth policy had driven into the heart of Russia, a catastrophe of even greater dimensions burst upon them. During one lengthy discussion of the refugee and Jewish problems in the Council of Ministers on August 6, Iakhontov paused in his hurried note-taking to notice that War Minister Polivanov had been unusually silent. "The usual twitching of his head and shoulder had become especially noticeable," he remembered. "It was very evident that something was bothering him." Perhaps also noticing Polivanov's disquiet, the Council's aged chairman Ivan Goremykin asked him to report on the situation at the front. Iakhontov remembered that Russia's war minister, "speaking in short, abrupt phrases," barely managed to hold his disquiet in check as he charted the course of military events for his colleagues. "Given the present situation at the front and in the rear of our armies," he told them, "one can expect an irreparable catastrophe at any moment. The army is no longer retreating—it is simply running away and its faith in its own strength is utterly destroyed." Polivanov's report, Iakhontov recorded in his notes, was "a sadder picture of military defeat and disorganization than anything we had heard before."[67]

Yet Polivanov's disquiet did not stem solely from the defeats Russia's armies suffered as they faced the combined assaults of the Austro-German forces, for he tended to present his appraisals of conditions at the front in dramatically overpessimistic terms as a matter of course. Just the week before, he had told Russia's assembled ministers that there was "no ray of light in the situation in the theatre of war," and several weeks before that he had described the retreat as nothing less than "a disorderly flight."[68] More than the expectation of impending military disaster accounted for the war minister's strange behavior in council, for a much greater weight had settled upon his shoulders that very morning. "No matter how badly things are going at the front, there is another still more terrible situation that now looms before Russia," he told his colleagues darkly. "I feel obliged to inform the government that, when I presented my report this morning, His Majesty told me that he had decided to relieve the grand duke [Nikolai Nikolaevich] and take over supreme command of the army himself."[69]

For a moment, Polivanov's colleagues sat frozen, stunned by what they had heard. At first, they spoke with a certain listlessness about the obvious problems that Nicholas's decision raised. How could the emperor be shielded from the wrath of public opinion if the army suffered more defeats, or if the moment came when, as Russia's supreme commander, he had to order the evacuation of Petrograd or Moscow? Could the emperor's safety be guaranteed at the front? Could the safety of Aleksandra and their children be assured in the rear? Yet none of these obvious worries touched what stood foremost in the minds of Russia's ministers at that moment, for each faced a dilemma of much greater dimensions than questions of imperial safety and imperial prestige. Should they —or could they—break with centuries' of custom and speak openly critical words of their sovereign's course? Some could not. "I am a man of the old school," Goremykin told his colleagues, speaking with the wisdom drawn from a half-century of service in the tsar's counsels. "For me, an imperial command is law."[70]

Yet, if Goremykin, born in the reign of Nicholas's great-grandfather, and faithful servant of three tsars, would not criticize the decision of his imperial master, his words seemed to provide the catalyst that freed others to do so. Now certain that duty to his country must come before duty to his tsar, Sazonov took a different and daring course. One of the few prewar ministers who still held his post, and the statesman best able to judge the impact of the emperor's decision upon Russia's allies, Sazonov decisively rejected Goremykin's readiness to "bow before the will of our tsar." For a moment after Goremykin had concluded his remarks, Sazonov tried to focus upon the course he intended to follow. "This is all so terrible that my mind is in utter chaos," he exclaimed.

"What an abyss Russia is being pushed into!" Soon, a number of ministers followed Sazonov's lead. "One needs to have special nerves to live through what is happening now," Krivoshein replied. "Russia has lived through much more difficult times, but there has never been a time when everything possible was done to complicate and confuse an already impossible situation." Fearful for the future, the ministers continued their discussions during several more meetings. "I expect the most terrible consequences from this change in the high command," one of them confessed a few days later. "We are sitting on a powder keg. All we need is a single spark to set it off [and], in my opinion, . . . the Sovereign Emperor's assumption of the army's command is not just merely a spark but a whole candle thrown into a powder magazine."[71]

"The temperature is rising," the indefatigable Iakhontov jotted in his notebook on August 10. "The arguments and objections are taking on a tone unusual for the Council of Ministers." Russia's ministers knew all too well that their emperor was utterly ignorant of tactics and strategy, that he was indecisive and weak, and that he usually asserted himself for the wrong reasons at the wrong time. Most of all, they feared the influence of Aleksandra on her weak husband, and the influence of Rasputin upon Aleksandra. Many Russians shared those apprehensions, and the ministers feared the impact that Nicholas's departure from the capital would have on public opinion. "There can be no doubt but that the emperor's decision will be interpreted as the result of the influence of the infamous Rasputin," Russia's recently-appointed minister of internal affairs Prince Nikolai Shcherbatov warned his colleagues. "Revolutionary and antigovernment agitation will not pass up an opportunity as convenient as this." So fearful were Russia's ministers about the consequences of their emperor's plan that, with one or two exceptions, the entire Council took the daring step of sending Nicholas a collective statement warning "that the decision you have taken threatens . . . Russia, You, and Your dynasty with the gravest consequences."[72]

Nicholas would not be swayed. "May God's will be done!" he supposedly told Aleksandra and Vyrubova just before he announced his intention to become Russia's commander in chief. "Perhaps a sacrificial lamb is needed to save Russia, and I am to be the victim."[73] From this distance, we cannot know for certain if that was his view, but no less a statesman than Krivoshein thought the tsar's decision "fully in keeping with his spiritual make-up and with his mystical understanding of his imperial calling."[74] Now Nicholas hurried to take up the terrible burden for which he was so pitifully unprepared. "Our sole aim now must be to drive the enemy from our borders," he announced grandly as he reassigned Nikolai Nikolaevich to command Russia's Caucasus front against Turkey and appointed General Alekseev—a man whose genius for administration and self-effacement in his tsar's presence soon made him the real commander of Russia's armies—to replace Ianushkevich as *Stavka*'s chief of staff.[75] Two days later, he added, "a new clean page begins, and only God Almighty knows what will be written on it." Aleksandra comforted him with reassurances that "it will be a glorious page in yr. reign & Russian history," and assured him that he had nothing to fear because "Our Friend's [Rasputin's] prayers arise night and day for you to Heaven & God will hear them."[76]

Nicholas's first letters to Aleksandra from *Stavka* showed how utterly he had failed to comprehend the full dimensions of the crises he faced. On the way to Mogilëv (*Stavka* had just been transferred there from Baranovichi, which was soon to be occupied by the Germans), he had seen streams of desperate refugees, and sadly reported that "it is quite impossible to restrain these poor people from abandoning their homes because nobody wishes to be left in the hands of the Germans or Austrians."[77] But Nicholas's inability to understand that refugees did not abandon their homes out of love for tsar and country but because they had been driven away at gunpoint by his generals' unwise policies, was far less ominous than other remarks he wrote at the same time. "Think, my Wify, will you not come to the assistance of your hubby now that he is absent?" he asked. "Do not fear for what remains behind," Aleksandra reassured him. "Don't laugh at silly old wify, but she has 'trousers' on unseen. . . . God will give me the strength to help you . . . & you have Gregory's [Rasputin's] St. Nicholas [icon] to guard & guide you."[78]

Thus began the unfortunate and fateful partnership of two ill-starred sovereigns who perceived neither their limitations nor the ominous complexities of the political maelstrom that swirled about them. At the front, Nicholas, the unschooled strategist and inept administrator, commanded Russia's armies, while Aleksandra, suffering from delusions of persecution and convinced that her insight into state affairs stemmed not from "wisdom, but a certain instinct given by God," played the autocrat in Petrograd with Rasputin as her chief counselor and prime confessor. Theirs was a partnership predestined for failure in the most tragic sense, because their unswerving devotion to each other intensified their isolation from the armies they were supposed to command and the people over whom they were supposed to reign. Nicholas had neither the will to command nor the sternness to rule. In place of those vital qualities he had only the petty stubbornness of a weak man unable to take the proper measure of those who served him in order to determine who offered sound advice and who spoke merely to curry favor.

For Nicholas, the world of military command thus always remained one in which contradictory advice could never be evaluated on its merit, but had to be taken or rejected on the basis of how "loyal" he perceived an adviser to be. Sadly, such traits as likeability, glibness, respect for his empress, and a variety of other inconsequential criteria determined his choice of the men in whom he placed his trust, and these could never provide him with the advice and counsel he needed as Russia passed through the most critical moment in her history. At the same time, Aleksandra perceived the complex world of Petrograd politics to be a vast and multifaceted conspiracy directed against her, her husband, their son, and their dynasty. Very quickly, she defined Russia's war effort as a struggle between a handful of "good" people who supported her against a vast and complex configuration of "poisenous" [sic] elements that included the Duma, leading army generals, and any senior statesman who refused to admit the value of Rasputin's advice. "Its a hunt against wify," she wrote. "You must back me up," she insisted, as she went on to lament a few days later that "I am but a woman fighting for her Master & Child, her two dearest ones on earth."[79]

From the moment he left for *Stavka,* Aleksandra flooded her husband with ill-advised, sometimes hysterical urgings. "Get Nikolasha's [i.e., Nikolai Nikolaevich's] nomination [as commander of the far-away Caucasus front] quicker done," she insisted in her fragmented English prose just hours after Nicholas had left. "No dawdling, its bad for the cause & for Alekseev too—& a settled thing quieten minds, even if against their wish, sooner than that waiting & uncertainty & trying to influence you." Within days, she was hard at work to remove the Holy Synod's well-regarded Director-General Aleksandr Samarin, who had made his disapproval of Rasputin well-known. "Samarin seems to be continuing to speak against me," she wrote. "We shall hunt for a successor." Only a week later, she insisted that Prince Shcherbatov should be ousted as minister of internal affairs, and urged Nicholas to interview a nonenity who came highly recommended by Vyrubova's father as a possible replacement. "Have a strong & frank talk with him," she urged. "Put the position of our Friend clear to him from the outset, he dare not act like Shcherb[atov] and Sam[arin and] make him understand that he acts straight against us in persecuting & allowing him [Rasputin] to be evil written about or spoken of."[80]

Such disjointed ramblings proved not the exception but the rule, and they grew more lengthy, more convoluted, more demanding, and more paranoid as Aleksandra flooded Nicholas's desk with complaints, urgings, even orders. "The ministers are rotten & Krivoshein goes on working underhand," she wrote in mid-September. "Repremand them very severely for their behaviour."[81] "It rests with you to keep peace and harmony among the Ministers," Nicholas later wrote. "Thereby, you do a great service to me and to our country." Aleksandra began to send her husband lists of assignments—including a number from Rasputin—with instructions to "keep my little list before you." Such reminders were necessary, she explained, because "you are so terribly hard worked—that you may forget something—& so [I] act as your living notebook." In one such list, she insisted that Nicholas tell the man about to be appointed Russia's minister of internal affairs that he must "listen to our Friend's [Rasputin's] councils." Confidently, she wrote, "a country where a man of God helps the Sovereign will never be lost." She sent Nicholas charms touched by Rasputin and Dr. Philippe, and reminded him on several occasions to comb his hair with Rasputin's comb "before all difficult talks & decisions," in order to strengthen his will in dealing with "evil" advisers. Certain that Rasputin was the source of profound knowledge, she assured Nicholas that he could be confident of "His wonderful brain—ready to understand anything," and explained that "God opens everything to him." Nicholas, she insisted, must always "remember the miracle—our Friend's vision."[82] "I do not know how I could have endured . . . if God had not decreed to give you to me as a *wife and friend!*" Nicholas replied.[83]

Of course, Russians knew nothing of the letters that passed between their tsar and tsarina, but Nicholas's arrival at *Stavka* stirred new apprehensions among them. Common Russians still remembered the tragedy at Khodynka Field, where hundreds of poor folk had been trampled to death as they assembled outside Moscow to celebrate Nicholas's coronation. Nineteen years had passed since then, but they began to speak of it again now. "One disturbing notion deep in the consciousness of the people was widely reflected in the letters examined by military censors," General Denikin remembered some years later. "Everyone considered the tsar to be 'unfortunate' and an 'unlucky fellow.' "[84] Nicholas had indeed been born on the name day of Job the Sufferer, and superstitious peasant women had been quick to remark, in reference to Aleksandra's appearance at Alexander III's funeral not long after she came to Russia, that she would bring misfortune because "she has entered our land behind a coffin."[85] "Ach, they've screwed it all up, and we're stuck with cleaning up the mess!" General Brusilov heard his soldiers mutter at summer's end.[86] A new wave of cynical humor washed over the common soldiers in the trenches. "Now the Emperor is going to fight," General Knox heard some of them say in those days. "Soon the Empress will come too, and then all the women of Russia will follow!"[87]

In a field hospital near the West Russian town of Minsk, Paustovskii lay recovering from a shrapnel wound in his leg. "The dark shadow of a crow's wing," he wrote, "has fallen across Russia."[88]

CHAPTER VI

The Tsar Takes Command

High on the east bank of the Dnepr River sits the White Russian provincial capital of Mogilëv. From the rear windows of its Governor's Mansion, a building once occupied by Marshal Davout when Napoleon's Grand Army invaded Russia in 1812, one can look down the valley, where woods and meadows intersperse with fields of grain in a patchwork of green and gold. Here and there, ancient white churches dot the landscape, their blue onion domes and glinting spires sprouting upward like the fall mushrooms that grow profusely in that part of Russia. Some had been there since the days of King Stefan Batory, when Mogilëv had been part of Poland's eastern lands before Tsar Aleksei Mikhailovich's Cossack legions had made it part of Russia. Mogilëv's people had greeted Peter the Great with their traditional offerings of bread and salt early in his reign, and Catherine the Great had met there with the Austrian Emperor Josef II. Russia's nineteenth-century sovereigns had prayed under the gentle gaze of Mogilëv's Virgin that the artist Borovikovskii had painted in Catherine's time to adorn the town's cathedral, and Nicholas himself had once come there in earlier, happier times. Before the war, there had been many Jews in the city, but they had all been driven away by Ianushkevich's policy of mass deportations, a fact that the officer assigned to prepare press releases about the tsar's experiences with Russia's fighting forces thought important to mention in his official account of Nicholas's arrival at *Stavka*. [1] All told, Mogilëv was a pleasant place for a headquarters, with "a delightful view over the Dnepr and the distant country," Nicholas wrote to Aleksandra soon after he arrived. [2] Its picturesque setting, so close to nature and so alive with history, belied the root of its name —*mogila*—which, in Russian, means "a grave." There, from the moment he

164

arrived, Nicholas began unwittingly to dig the grave of his dynasty. At Mogilëv, General Brusilov sadly remarked, Russia's last tsar "struck the last blow against himself" by becoming supreme commander of Russia's armies.[3]

On September 5, 1915, Nicholas II, Emperor and Autocrat of All the Russias, arrived at Mogilëv to take up one of the most awesome burdens ever to confront a monarch. Russia had just suffered through thirteen of the most terrible months in her history during which she had lost all of Poland and great chunks of her West Russian and Baltic provinces. About a fifth of all the coal produced in the empire before the war had come from these lost territories, as had a tenth of the iron ore, two-thirds of the chemicals, much of the leather, and most of the flax. The Polish industrial cities of Łódz, Warsaw, Radom, and Lublin, all important centers of cotton and machine-tool manufacture, now lay behind enemy lines. That many of their factories remained in or near working order spoke to the world of yet another Russian failure. Some years before the war, Sukhomlinov had warned that these centers might have to be abandoned in the event of a German invasion, but he never had prepared a plan for evacuating their factories and skilled workers any more than he had thought to draft plans to cover the army's general retreat. With no evacuation plans, extremely limited transport facilities, and no means for establishing priorities in time of emergency, very little of Poland's capital machinery had been removed when Russia's armies began their Great Retreat, and not even vital raw materials could always be gotten away from the Germans in time. The Russians had to abandon almost five hundred tons of desperately-needed zinc when the Germans seized the Baltic port of Libau in May, and Plehve had not had time to destroy over a million rubles' worth of tanned hides when he had retreated from the Germans at Szawli in mid-July.[4]

Territorial losses thus had deprived Russia's sagging war effort of vital raw materials and the factories needed to shape them into the implements of war just when the army needed them more desperately than ever before. Although they had been on a war footing for a year, Russia's weapons factories still could replace only one out of every three of the rifles her soldiers lost in combat, and General Staff weapons experts estimated that they needed about two million new rifles just to make up those losses and arm the men being called up that fall. So many Russian soldiers had no weapons at the end of the Great Retreat that General Alekseev thought it an accomplishment worthy of mention when he could report in January 1916 that seven out of every ten of his front-line troops now had a rifle. Not until later in 1916 would Russia's rifle shortage end, and there would not be sufficient bullets until that time either. At the end of 1915, Russian factories still produced only half of the two hundred million cartridges the army needed every month, and foreign shipments did not make a notable impact until the middle of the following year. Russia had begun the war with about five hundred extra cartridges for each rifle stored in her arsenals; her reserves would not approach that level again until mid-1917.[5]

In August 1915, Russia had only a bare handful of antiaircraft weapons, and still had no artillery heavy enough to compete with that of the Germans and Austrians. Just to replace the guns that had been lost when they abandoned the fortresses of Poland and West Russia, Russia's armies needed more than three thousand new weapons ranging from light 76-mm field guns to heavy 152-mm cannon. From the beginning of August until the end of October, Russia's factories delivered less than eight hundred. They still produced only half the 107-mm ammunition, slightly more than a quarter of the 76-mm shells, only a sixth of the ammunition for the army's 122-mm field guns, and a mere tenth of the heavy 152-mm projectiles that *Stavka*'s staff thought necessary to hold the front against further losses of territory. Desperate for more shells and weapons, the War Ministry's Artillery Department placed large orders in France, England, and the United States, but these shipments did not begin to arrive in any significant quantity until after the middle of the following year.[6]

Nicholas thus assumed command of his armies when every sort of weapon, all types of ammunition, and even telephones, telephone wire, gas masks, uniforms, and boots were in critically short supply, and there was no chance that his allies could send help before the new year. "With firm faith in God's mercy, and with unshakeable confidence in our final victory," he took his place at *Stavka*. "We shall fulfill Our sacred duty to defend Our Homeland to the end," he announced on September 5. "We shall not bring disgrace to the Russian land." To Grand Duke Nikolai Nikolaevich, who for thirteen months had borne the burden of commanding Russia's war effort, Nicholas now explained: "My duty to the nation that has been entrusted to my keeping from on high ordains that I must, at a time when the enemy has penetrated deep into the territory of the Empire, take upon myself the Supreme Command of the fighting forces. . . . The ways of Providence are inscrutable," he continued. "Recognizing that, in view of the complicated situation there, your help and counsel are essential to me on our southern front, I hereby appoint you my Viceroy in the Caucasus and Commander-in-Chief of the valiant Army of the Caucasus."[7]

Certainly relieved to be rid of the supreme command that he had never wanted and had asked not to be given, Nikolai Nikolaevich nonetheless feared for his emperor's future. "It is essential that under your command the army be victorious," he replied to Nicholas's announcement. "Although I am convinced that, in the end, victory will be ours, I am not at all certain that the immediate future will see a change for the better." Ever mindful of Aleksandra's warnings that Nikolai Nikolaevich "fears my influence (guided by Gregory [Rasputin]) upon you," and attentive to her complaint that she would not rest easy until "all is done at Headquarters and Nikolasha [is] gone," Nicholas replied to his uncle's concern by sending him to the Caucasus front immediately, and refused to allow him even to visit Petrograd on the way. "Thank God it is all over," he wrote to Aleksandra. "I feel so calm—a sort of feeling like after the Holy Communion!"[8]

Very few shared Nicholas's feelings of comfort, for Russians retained great confidence in Nikolai Nikolaevich and blamed others for their army's defeats. Shortages of weapons and ammunition, Russia's soldiers, politicians, and public opinion all believed, were the fault of the now-disgraced Sukhomlinov, and these same people blamed the cruel and stupid persecutions of the Jews that had added so much unnecessary misery to the Great Retreat on the viciously anti-Semitic Ianushkevich, not the grand duke. Even those Russians who had good reason to doubt the grand duke's abilities knew that the tsar would be a wretched replacement and that, at the very least, he would add substantially to the army's already monumental difficulties. "Those people who failed to use the most decisive measures—including even force—to dissuade Nicholas II from assuming those duties for which he was so ill-suited by reason of his ignorance, inability, utterly flaccid will, and lack of stern inner character are no better than criminals," General Brusilov once said.[9] At the time, Nicholas's nephew Grand Duke Andrei Vladimirovich warned, "the exile of Nikolai Nikolaevich to the Caucasus is one huge mistake."[10]

Except for Goremykin, the aged chairman of the Council of Ministers who believed that "the Imperial Command is law," virtually every statesman, politician, and diplomat in Russia shared Andrei Vladimirovich's view, as did many aristocrats and other members of the imperial family. "This is dreadful," one princess remarked. "It will bring us to revolution."[11] Nicholas's mother, the Dowager Empress Maria Feodorovna, shook her head in disbelief. "Where are we heading?" she asked again and again. "This is not my Nicky, he's not like this." In confusion, this wise and sensible woman whom Russia's statesmen had from time to time begged to speak to Nicholas when he seemed about to do something especially dangerous or stupid, confessed that she no longer could influence her son. "I don't understand anything anymore," she said hopelessly. "It's all too terrible!" One day at tea, Nicholas's younger sister asked Aleksandra how it could be that Nikolai Nikolaevich was being replaced at a time when he was "still so popular." In reply, Aleksandra snapped: "I'm sick of all this talk about Nikolasha. People talk of nothing else, but Nicky is much more popular."[12] Free of his unwanted duties as supreme commander, and now on his way to the Caucasus, perhaps only "Nikolasha" knew the full weight of the burden that his nephew and sovereign had just taken upon his shoulders at Mogilëv.

Certainly, Nikolai Nikolaevich's new post in the Caucasus promised fewer difficulties and greater successes. Ever since the Turks had entered the war on the side of the Central Powers, Russia had held a decisive advantage in the region, although she had pursued it rarely because so many of her resources were committed against Austria and Germany. For the Turks, the war had begun on the morning of October 16/29, 1914, when ships of their navy, supported by the German cruisers *Breslau* and *Göben,* had bombarded the Crimean naval base of Sevastopol and Russia's Black Sea ports of Odessa and Feodosia.[13] Two days later, the Russians had declared war. "The ill-advised intervention of

Turkey will only hasten that nation's downfall," Nicholas had announced that day, while Sazonov had added: "We shall be obliged to make Turkey pay dearly for her error." From the first, Russia's stance against Turkey was to be defensive, Sazonov said. "Before all else," he assured his allies, "we must defeat Germany." Russia therefore would "keep at a minimum the forces we shall use for defense against the Turkish fleet and army."[14]

Urged on by their German allies, who brought much-needed gold to the Ottoman treasury, the Turks at first waged an aggressive campaign, beginning with Enver Pasha's December offensive against Russia's I Caucasus and II Turkestan Corps in the area around Batumi, Kars, and Sarykamysh.[15] A graduate of Germany's General Staff Academy, and advised by a German chief of staff, Enver Pasha hoped to use his XI Corps to hold the Russians' front, while, in a maneuver patterned after the campaigns waged by Schlieffen, Hindenburg, and Falkenhayn, he turned his IX Corps and X Corps against their flank to cut the Tbilisi-Kars-Sarykamysh railway upon which the Russians depended for their supplies.[16] Designed for the flat lands of Europe, the strategy that Enver Pasha borrowed from his German mentors proved ill-suited to the jagged mountains and blinding snows of the Caucasus, and his assumption that the Russian army would never do more than hold its defensive positions proved a fatal mistake.[17] During the fighting on December 22–23, the Russians moved to the attack, smashed the Turks' IX and X Corps, and captured three Turkish divisional commanders as well as the commanding general of IX Corps. By the beginning of January 1915, the Russians had destroyed much of the Turks' Third Army, leaving Enver Pasha with less than a seventh of the ninety thousand men with which he had begun his campaign.[18]

Russia's decision to commit every available division to her battles against Austria and Germany meant that the Army of the Caucasus could do no more than hold the line of positions that stretched from Archave on the Black Sea eastward to Tavriz. Late in the spring, the Turks reformed their Third Army in the vicinity of Erzerum and, in July, attacked northeastward through Kop and Meliazhert toward Kara-Kilis and Alashkert, only to be driven back with such heavy losses that, when Nikolai Nikolaevich took command of the Army of the Caucasus in the middle of September, the Russians had already pushed their lines some fifty kilometers further south into Turkish territory. Nikolai Nikolaevich would press onward from that point, adding the Turkish fortresses and cities of Erzerum, Trapezun, Mamakhatun, Erzindzhan, and nearly all of Armenia and the Anatolian plateau to his list of conquests during the next few months.[19]

While Nikolai Nikolaevich won victories on the Caucasus front, Nicholas turned to what he described to Aleksandra as his "*new* heavy responsibility" at *Stavka*.[20] He did not know what a commander in chief ought to do, and did not seem inclined to learn. He failed to comprehend the complexities of modern strategy and tactics, even in the vaguest sense. Nor could he command men. All who met Nicholas in those days spoke of his never-failing politeness and his reluctance to take a firm stand or give a definite order. "He never liked to dot his i's," General Brusilov wrote in his recollections about how he discovered that his tsar-commander was "inclined to prefer indecisive and indefinite situations." Nicholas lacked utterly that charisma that draws men to their leaders, although Russia's soldiers were desperately ready for a soldier-tsar to stand at their head as Peter the Great had done two hundred years before. "The word 'tsar' still had a magical effect upon the men," Brusilov remembered, but Nicholas never could bring that magic to life because "he could not find those words which could win over men's souls and raise their spirits." Nicholas loved military parades, and insisted upon holding them even within range of the Germans' heavy guns. But when he appeared before his men, his generals found that "he did not know what to say, where to go, or what to do."[21]

With somewhat more than seven hundred kilometers of clogged railways to separate him from ministerial prophets of doom and quarrelsome Duma politicians who unceasingly urged him to grant Russians a greater role in national affairs, Nicholas found life at *Stavka* pleasant and undemanding. "My brain is resting here—no Ministers, no troublesome questions demanding thought," he once wrote. "I consider that this is good for me," he added in a tone of confidence that was hardly justified by the crises he and his empire faced. "The tsar's personality faded into the background," Miliukov remembered. "Neither the intrigues and the gossip of the Petersburg Court, nor the sound of artillery fire at the front reached him there [at Mogilëv]."[22] Nicholas assured Aleksandra that his will became more "strong" and his brain "sounder" at *Stavka* than it had been in Petrograd. Securely wrapped in the snug cocoon of the comfortingly orderly world of petty daily routines that Alekseev wisely created for him, Nicholas quickly concluded that, at *Stavka*, he could "*judge* correctly the real *mood* among the various classes of people," and that his ministers actually knew "terribly little of what is happening in the country as a whole." Very quickly, the unreal world of splendid formations wheeling precisely in review, and of military maps, with the Russian and enemy positions all neatly marked with brightly colored little flags, became war's reality for Nicholas. He thought some of the regiments he reviewed were "amazingly beautiful," and collected what he called "a whole rainbow of impressions" from his inspections. He sent Grand Duke Kyrill Vladimirovich off to distribute crosses of St. George by the thousands, and cheerfully reported to Aleksandra that his younger cousin had been "delighted to have been given such a job." On one occasion, Nicholas solemnly promoted all the officers at a ceremonial dinner, and thought "the effect was tremendous," but he understood almost nothing of men or politics or war. "I do not read newspapers here," he once wrote from *Stavka*, while to Paléologue he exclaimed: "The life I lead here at the head of my army is so healthy and comforting!"[23]

For Nicholas, Russia at war thus simply became a mightily enlarged version of Krasnoe Selo, the great plain outside Petrograd where the Russian army traditionally held its summer maneuvers. As a young grand duke, he had spent some of his happiest weeks there, and he tried to relive those fondly remembered times at *Stavka*. He brought the Tsarevich Aleksei to *Stavka* to share these experiences with him and reported that the presence of a sickly child in a place that issued commands that sent hundreds of thousands of Russians to their deaths, "gives light and life to all of us." Nicholas thought it a delight for his officers when Aleksei "climbed everywhere and crept into every possible hole." While grizzled generals and graying diplomats talked soberly of politics and strategy at lunch, Nicholas thought it a fine thing when his son's noisy chatter disrupted their debate. "It is pleasant for them," he concluded, "and it makes them smile," while Aleksandra promised him that the sight of tsar and tsarevich ("your Sunbeam, your very own Child") would "touch the hearts" of even the cruelest critics. "Minds will be purified & they will carry the picture of you & yr. Son in their hearts with them," she assured him. Proudly, Russia's supreme commander noted how "tremendously pleased" Aleksei had been with his first military parade and spoke of how well the eleven-year-old "Baby" had reviewed the troops. "I shall never forget this review," Russia's wartime commander wrote to Aleksandra. "The weather was excellent and the general impression astounding."[24]

Nicholas's personality, his failings, and his fantasies, made life at *Stavka* unlike that at any other High Command in modern history. Life's pace remained unhurried, and it was even possible for Aleksandra and their children to visit him on occasion. Usually, Nicholas arose at nine o'clock and then went to Alekseev's office for a morning briefing, during which his chief of staff indicated any changes in the Russian or enemy positions by repositioning the little flags on his large map. From eleven o'clock until about half-past twelve, the tsar received ambassadors, advisers, and others who had come to *Stavka* for that purpose. A simple lunch, during which Nicholas sparingly drank one or two small glasses of vodka with the hors d'oeuvres and another glass or two of his favorite port with the main course, lasted for about an hour, after which he returned to his study. "At three o'clock," an aide remembered, "they would bring up the automobile [a Rolls-Royce], and the Sovereign, in the company of four or five of his Suite, drove off for a tour of the countryside." After some two hours of driving, during which he often stopped for a brisk walk in the woods or along the river, Nicholas returned. Dinner was ready by seven, after which he took another walk if the weather was good. Otherwise, he and his staff listened to music or played a quiet game of dominoes. *Stavka*'s chief duty officer devoted long hours to arranging special cinema showings for the tsar, and worried about whether a film in which the hero and heroine kissed passionately and often would offend Aleksandra and her daughters on their next visit. At one point, he obtained *The Secrets of New York*, a detective serial in twenty parts, from Pathé Films, and, to young Aleksei's delight, noted on the day's menu which new installments would be shown that day. Sometimes in the dark winter afternoons and evenings, Nicholas read such popular English novels as Florence Barclay's *The Wall of Partition* or A. W. Marchmont's *The Millionaire Girl* because he found them "soothing to the brain." These evidently replaced dominoes as evening entertainment for a while, although just three days before the revolution broke out in February 1917, he decided to take them up again as a substitute for the games of patience he so enjoyed with Aleksandra on his visits to Tsarskoe Selo.[25]

From time to time, this routine varied, but only slightly. Nicholas occasionally received visitors after lunch, "while Baby is resting," so that he could go out for his daily excursion half an hour early, "because the days after dinner have become so short." At other times, he and Aleksei, who "marched about with his rifle and sang loudly," took a walk before lunch was served, rather than wait until afterward. There were the usual minor discomforts that Nicholas insisted he must not be spared because they were a part of the life of a commander at war. He found his field bed "hard and stiff," but then sternly rebuked himself with the admonition that "I must not complain—how many sleep on damp grass and mud!" How different it all was from the descriptions we have of Falkenhayn's *OHL*, or the High Command headquarters of Hindenburg and Ludendorff. While these hard-eyed technicians of modern warfare planned their attacks with grim determination, knowing war for the life-or-death matter that it really was, Russia's tsar approached his empire's defense with little more disquiet than he had displayed as a junior cavalry squadron commander at Krasnoe Selo in the happy long-ago days when his stern father still lived.[26] "The heart of the Tsar is in the hands of God," he had once calmly assured one of his ministers. "I am prepared at any moment to render Him an account of all my actions."[27]

Obviously, there was more to war than plotting Russian and enemy positions on *Stavka*'s big map and inspecting weary troops who had been called back from their trenches to march for the tsar's edification. Someone had to know when to attack and when to retreat, and, although the fact had to be concealed from him, these were not decisions that Nicholas could be allowed to make if Russia hoped to survive the war. The real task of commanding Russia's fighting forces thus fell to Nicholas's new chief of staff, Infantry General Mikhail Vasilevich Alekseev, the fifty-seven-year-old son of a career noncommissioned officer in the army of Alexander II. After attending the Moscow School for Military Cadets, Alekseev had entered the army in 1876 as an ensign, but had to wait for nine years to command his own company. Only after he had served in an infantry line regiment for eleven years did the plebian Alekseev's career blossom when he passed Russia's General Staff Academy entrance examination and finished at the top of his class. Appreciated by aristocratic superiors as a talented officer whose capacity for work was exceeded only by his dedication, Alekseev had a genius for military organization, for drawing many diverse components together and forming them into an efficient military operation. General Ivanov had so valued Alekseev's efforts as the Quartermaster-General of his Third Manchurian Army during the Russo-Japanese War that he had insisted upon having him as his chief of staff when he became commander of the Kiev Military District in 1908. When Ivanov became commander of Russia's southwest front at the beginning of the Great War, he again would accept none but Alekseev as his chief of staff.

With Ivanov's full approval, *Stavka* had named Alekseev to replace General Ruzskii as commander of the northwest front in March 1915. That had proved to be a mistake, however, for Alekseev had been no more sympathetic to Ivanov's urgings than had Ruzskii, and his refusal to send reinforcements at the time of Mackensen's breakthrough at Tarnów-Gorlice had been instrumental in enabling the Germans to consolidate their initial success. Yet, Alekseev redeemed himself quickly. Given responsibility for the entire Polish salient when the Germans renewed their offensive, this slim and balding officer, with piercing narrow eyes and an upward-swirled gray mustache, proceeded to organize Russia's Great Retreat on a moment's notice and performed that seemingly impossible task remarkably well. Determined that the Germans would have no more successes such as they had enjoyed at Tannenberg and on the Tarnów-Gorlice line, Alekseev used to the utmost his extraordinary ability to organize quickly and work effectively for long hours. Poland could not be saved, and the fortresses with their precious contents had to be surrendered, but he did not lose Russia's army, as might well have happened under a less able general.

Achievements such as Alekseev's did not necessarily bring recognition from the tsar, however, and it still remains unclear how or why Nicholas suddenly chose this man of humble birth to be his chief of staff. It was easily one of his best decisions, for his decision to appoint Alekseev proved as wise as his choice of himself as Russia's supreme commander was foolish.[28] At *Stavka*, Alekseev lived only for his work and his country. He slept little, worked long, and had little time for the social niceties that had been so much a part of *Stavka*'s dinners during the Ianushkevich era. Simple faith and simple patriotism sustained Alekseev, and in times of great crisis he turned first to God, and then to his own inner reserves of confidence in the rightness of his task. General Vasilii Gurko, whose cavalry had led Rennenkampf's advance into East Prussia and who temporarily became *Stavka*'s chief of staff while Alekseev was recuperating from cancer treatments in the fall of 1916, spoke of his "unusual modesty, accessibility, and simplicity," and thought him a "gifted, high-minded soldier."[29] His one great failing was his inability to delegate responsibility. Alekseev always took onto his own shoulders too much that should have been done by his staff, and this eventually clouded his ability to see strategic problems in their broadest perspective.

Such praise as Gurko's might well be expected from a man whose career had benefited from Alekseev's support, but others proved equally enthusiastic. Even after the atrocities of the Civil War had separated them, Mikhail Lemke, the long-time critic of tsarist censorship who dealt with press affairs at *Stavka* for more than a year before he turned to support Lenin in 1917, remained full of admiration for the man who had helped to organize the White armies that had marched against the Bolsheviks in 1918. "When you speak with people who have seen Alekseev on a day-to-day basis during the fifteen months we have been at war," Lemke confessed to his diary in the fall of 1915, "you come to realize fully what a gigantic military force is contained in this man of moderate height and size [who] . . . wants only to serve his homeland with all his heart and mind. . . . If you ever see before you a stern general looking every inch the part, whose countenance reflects his own great appreciation of the grandeur of his position," Lemke concluded, "then you are *not* in the presence of Alekseev."[30]

Alekseev impressed men with his dedication and drew them to him by his unpretentious manner. Lemke found him "accessible to everyone, devoid of all external pomp, and a comrade to his subordinates." Not well-versed in courtly manners, and ignorant of foreign languages, Alekseev provided much grist for Petrograd's gossip mill, and the city's aristocratic salons were full of tales about how he did not know at what point in a meal coffee should be drunk, and how he even left the table before the tsar in order to get back to his work. Often he simply remained away from the table, for he lived a Spartan life, asked little for himself, and even insisted upon paying his own mess charges rather than have them billed to the General Staff. "No 'hidden hands' or high influence had advanced his career," Lemke concluded. "Alekseev owed his position exclusively to his own achievements."

Even Aleksandra was at first taken with this unpretentious man of homely virtue who had remained always aloof from politics, and urged her husband to make every use of his talent. "It is such a comfort to know that you are contented with Alekseev & find work with him easy," she wrote only a week after his appointment.[31] Yet, so far as the empress's favor was concerned, Alekseev's intelligence and good sense worked against him, just as it did with nearly every other man of talent who held high office. No more prepared to admit Rasputin to his inner councils than Nikolai Nikolaevich or the array of civilian ministers whom Aleksandra drove from office in 1915 and 1916 had been, Alekseev could hope to enjoy imperial favor only so long as he could avoid his empress's urgings to consult Rasputin about military affairs.

Far more adept than most, Alekseev managed for almost a year to sidestep Aleksandra's hints that Rasputin be invited to *Stavka*, but, when she approached him directly about it in the fall of 1916, he had no choice but to reject her suggestion. Instantly, the empress struck against him, just as she did whenever anyone criticized the man she thought had been sent by God to save her son, her dynasty, and her adopted country. "If we let our Friend be persecuted we & our country shall suffer for it," she had warned Nicholas even before he took over the supreme command. Now, she repeated her warning in terms reminiscent of her early ragings against Nikolai Nikolaevich after he had threatened to hang Rasputin if he ever set foot in *Stavka*. "A man who is so terribly against our Friend as poor Alekseev is—cannot have blessed work," she wrote. Less than a week later, she admonished Nicholas to remember that "for your Reign, Baby & us you *need* the strength prayers & advice of our Friend." In Aleksandra's mind, Rasputin was more vital to their future, more necessary for Russia's survival, than any general or statesman ever could hope to be, and she insisted that those who failed to heed his urgings would suffer God's wrath. "God sent this illness," she explained to Nicholas after cancer struck Alekseev late in 1916, "to save you fr[om] a man who was lossing his way & doing harm by listening to bad letters & people."[32]

In the fall of 1915, however, still healthy, passionately dedicated to Russia's defense, and secure in the favor of his tsar and tsarina, Alekseev's most immediate tasks were to halt the Austro-German advance, stabilize the front, re-arm and restore Russia's fighting forces to full combat strength, and then plan an offensive to recover the losses of 1915. To do so, he needed support from the tsar, Russia's senior commanders, and Russia's allies. First of all, he had to establish a routine at *Stavka* that would make it appear that Russia's tsar-commander, who knew neither strategy and tactics nor the art of commanding men, was, in fact, supreme commander of the armies. Nicholas therefore presided at the War Council of Front Commanders that assembled at *Stavka* from time to time, although this did little in the way of strategic planning. On a day-to-day basis, Nicholas was prone to meddle, but Alekseev soon found ways to minimize his interference. "The tsar often gets in the way of [Alekseev's] work when it comes to working out strategic problems," Lemke reported. "Alekseev therefore does a great deal on the spur of the moment, that is, he reports to the tsar about things that have already been accomplished and then, of necessity, the tsar must approve—sometimes with a grimace and sometimes without it."[33]

Nicholas almost never opposed Alekseev's major strategic decisions, although that virtue appears to have been dictated mainly by his preference for avoiding firm commitments or decisions for which he would have to take sole responsibility. When, for example, General Ivanov tried privately to convince Nicholas to countermand the decision to allow Brusilov to begin his highly successful offensive in 1916, Nicholas refused. "I do not think it proper for me to alter the War Council's decisions," he wisely told Ivanov. "You'll have to take it up with Alekseev."[34] Alekseev thus usually retained ultimate authority over Russia's military strategy, but there still were times when Nicholas's personal quirks got directly in the way of military operations. In amazement, one of France's military observers reported that a Russian offensive in Bukovina had to be delayed for a fortnight because Nicholas insisted upon reviewing several Guards corps before the operation began but had to postpone the review because the tsarevich was ill.[35]

In the fall of 1915, Alekseev's first strategic concern was to avoid any risks that might bring embarrassing defeats for which Nicholas, as supreme commander, would have to bear responsibility in the eyes of Russian public opinion, and in no case could he allow Moscow and Petrograd to be threatened. Alekseev therefore followed a cautious course that won him few victories but produced no great defeats. When Nicholas abdicated at the beginning of March 1917, Russia's positions on the now divided northern and western fronts remained virtually what they had been when Alekseev first came to *Stavka*, while in the southwest, Brusilov's offensive in the summer of 1916 would have moved Russia's lines from twenty to fifty kilometers further into Polish and Hungarian territory.

Although Nicholas had a sense of his own ignorance which, combined with Alekseev's clever planning, usually kept him from interfering with major strategic decisions at *Stavka*, his inability to estimate men's worth and measure their talents did not prevent him from interfering frequently in command appointments. From the very first, senior combat commanders feared, as General Knox reported, that there would be "a crop of intrigues [and] that advancement would be given to court favorites," and the eighteen months when Nicholas commanded at *Stavka* did indeed see a motley crew of failed and incompetent generals win enviable appointments through the tsar's intercession. Perhaps none figured more prominently in what a French officer once called "the facts of scandalous favoritism," than General Bezobrazov, a man whose company Nicholas much enjoyed and author of the *bon mot* that "order, counter-order, and disorder" reigned in the Russian army. As the commander of a Guards division during the early days of the war, Bezobrazov had refused to obey the orders of his corps commander and eventually had come to blows with him in a railroad station. Nikolai Nikolaevich had removed the belligerent general from his command, but Nicholas had intervened to name Bezobrazov commander of an entire Guards corps. During the Great Retreat, the Third Army commander had ordered Bezobrazov to pull back his Guards, only to receive the arrogant reply (sent from a position far in the rear of the men under his command) that Russian Guards never retreated. Bezobrazov's absurd pride had cost thousands of lives, not only in his corps, but in another whose flank he carelessly allowed to be exposed. Again the authorities relieved Bezobrazov from his command; again the tsar intervened. Some two months later, Nicholas created a new Guards Army, and bestowed its command to Bezobrazov, who had loftily assured his tsar in the meantime that none ought to command in the Guards but men of exalted birth.[36]

Bezobrazov was only one among scores of incompetents who profited from Nicholas's favoritism as Russia's supreme commander. General Aleksei Kuropatkin, who had managed to overshadow his lackluster performance as war minister by leading Russia's armies to an inglorious defeat at the battle of Mukden in the Russo-Japanese War, took advantage of what one observer called the tsar's "grievous influence" and received command of the northwest front for a time. Likewise, General Dimitriev, who as commander of Russia's Third Army during the winter and spring of 1915 had not bothered to move his artillery from month to month and had forgotten to order his troops to build reserve entrenchments during the weeks before Mackensen's breakthrough in the Tarnów-Gorlice sector, received command of Russia's Twelfth Army from Nicholas. At least two generals received army corps to command after they had lost entire armies, and General Ruzskii managed to win and lose command of the northern front no less than three times during Nicholas's eighteen months at *Stavka*. Another of Nicholas's appointees had the distinction of registering his artillery on his own infantry's trenches just to be prepared in case the Germans overran them, and General Pleshkov used his artillery to so little effect that commentators called one of his major bombardments (in which he neglected to use forward observers) "General Pleshkov's *son et lumière*."[37]

The list of misappointments lengthened further down the chain of command as Aleksandra bombarded her husband with requests and reminders to place various flatterers and members of her friends' families in command positions. During just one three-day period in the fall of 1915, she reminded Nicholas that "Kussov waits for a regiment," urged that he must "get work for Groten," and asked "does Dobriazin get a brigade?" It "would be lovely," she continued, if "Arsenev may receive command of our brigade," and begged that "if General Vesselovsky receives a brigade, please let his successor be Sergiev." Of course, there were good appointments too, and Gurko and Brusilov were sterling examples of obviously brilliant generals who rose even higher under Nicholas's command. But the number of incompetents restored to duty and promoted through the favor of the tsar and tsarina was dangerously large. Soon, Russia had to pay a dear price in lives and weapons lost for Nicholas's refusal to allow Alekseev to appoint men of his own choosing to carry out his strategic decisions. "The tsar's decisions are often in direct contradiction to Alekseev's recommendations," one military attaché reported to his superiors on France's General Staff, while conflicts between various cliques who had patrons at court and on Russia's General Staff made the situation even worse.[38]

While Alekseev struggled to limit Nicholas's meddling in command appointments, he also tried to rebuild his nation's shattered army. Russians needed weapons from their allies, and their desperate need increased the bitterness they felt toward France and England who, many thought, had given them too little support during the campaigns of 1915.[39] Paléologue readily dismissed frequently-heard bitter statements that "France is allowing Russia to carry the entire burden of the fighting" as "a theme of German propaganda,"[40] but such feelings were very real among Russians and far more significant than Paléologue allowed himself to believe. Convinced that their terrible defeats at Tannenberg and the Masurian Lakes had enabled France to halt the Germans at the Marne, the Russians had waited throughout 1915 for their allies to match their sacrifices. Allied help never came. Aside from the Joffre's limited offensive in Champagne which, in any case, came too late, the French and British never opened the great offensive that could have done for Russia in 1915 what the Russians believed their armies had done for their western allies the previous year. "While most of the German forces, and almost all of the Austro-Hungrian army, hurled themselves against us, our allies in the West did nothing," the influential journal *Novoe vremia* declared. History would despise England and France for "having 'sat still like rabbits' month after month in the Western theatre, leaving the whole burden of the war [in 1915] to be borne by Russia," one embittered Russian general complained to General Knox over a bottle of vodka one evening. During the first fifteen months of the war, he went on, "Russia had grudged nothing" to the cause of victory, while "England [only] gave money freely but grudged men."[41]

The everyday brutality of war had cut into the lives of each and every Russian in a way that Knox's countrymen, who sat secure in their island kingdom, could never know. "Do you think it is easy for us to look on those long columns of fugitives fleeing before the German advance?" one Russian officer asked Knox. "We know that all the children crowded into those carts will die before the winter is out."[42] Death in the winter; death of all the children before spring. It was a theme Russians knew too well, and one of the first that Nicholas noted in his letters to Aleksandra after he became Russia's supreme commander. "The children suffer very acutely," he wrote of the refugees he had seen on his way to *Stavka*. "Many of them, unfortunately, die."[43] To many Russians, it began to seem that they must pay for their industrial backwardness with the lives of hundreds of thousands of their young men and children, while their allies' great industries and overflowing treasuries supplied the weapons and the gold. "England and France," some of them said bitterly, "have made up their minds to carry on the war until the Russian soldier has shed the last drop of his blood."[44]

Strong sentiments against Russia's western allies among Alekseev's countrymen, and the Allies' grave misgivings about Russia's ability to bear a fair share of the war effort after the losses she had suffered in the Great Retreat, sowed the seeds of contention among them, and these were nurtured further by persistent rumors of the kaiser's efforts to secure a separate peace with Russia. Rumors of contacts between Russian and German agents in Stockholm, Copenhagen, and Athens had fed allied fears throughout 1915 and had led France's diplomats and politicians to demand repeated reassurances that Russia would pursue the war to final victory. "Germany is making energetic attempts to tear Russia away from her allies," France's foreign minister had warned in the spring of 1915, while the Swiss Social-Democratic newspaper *Berner Tagwacht* reported peace negotiations between the Russians and Germans as "reliable information," just a few months later.[45] Nicholas himself had tried to set Paléologue's mind at rest late in July by insisting that Russian opinion was so strongly against a separate peace that "a revolution would break out if I should make a peace today [with Germany]."[46] French (and British) apprehensions had continued even as the Great Retreat neared its end, despite Winston Churchill's bluntly stated conviction that Russia "is a broken Power but for our aid, and has no resource open but to turn traitor—and this she cannot do."[47] At the end of 1915, the editors of the *Revue de France* insisted, Russian ultraconservatives were "intriguing" with Germany to end the war.[48] Again, Sazonov tried to calm the French by telling Paléologue that Nicholas had refused to reply to German peace feelers under any circumstances, insisting that "any response, however discouraging" would be improper unless the kaiser sent his suggestions to Russia's allies as well.[49]

Such mistrust and failed confidence had to be overcome and the Russians rejoined to their allies in renewed bonds of common purpose if they were to hope for victory in 1916. They found these bonds in expanded war aims, all of which were inflated dramatically that winter. "I want France to emerge from this war as great and as strong as possible," Nicholas told France's emissary General Pau. "I shall agree in advance to anything your government wishes. Take the left bank of the Rhine. Take Mainz. Take Coblentz. Go still further if you think it useful."[50] In return, Russia demanded Constantinople and the Straits, which guaranteed access from the Black Sea into the Mediterranean. "Constantinople must be yours," Britain's King George V told Nicholas, while Paléologue assured Foreign Minister Sazonov that he could "count on the goodwill of the French Government when it came to resolving the questions of Constantinople and the Straits in accordance with Russia's views." While France reaped her rewards in Germany's provinces along her borders, Britain was to receive Egypt and large portions of the oil-rich Middle East.[51] At the highest level, the Allies thus found renewed strength in their common dedication to stripping more territory and greater sums of gold from their enemies. They now had to agree about the best way in which to pool their strategic and material resources in order to achieve those goals.

Alekseev's early efforts in this direction were made more difficult because Grand Duke Nikolai Nikolaevich had curried the favor of Western diplomats, military observers, and correspondents with great care from the very first days of the war, and they did not at first take kindly to his dismissal. Ambassador Buchanan remembered that his government received word of the grand duke's dismissal "with considerable apprehension," and Paléologue thoroughly shared that view.[52] Western military attachés were seduced by Nikolai Nikolaevich's vigorous soldierly bearing and the confidence with which he always set down his views. French observers thought him "an incomparable commander," and reported that "the grand duke had truly become the idol of the army and of the people in Russia." Nikolai Nikolaevich's "extraordinary aggressive spirit," which denied the enemy even "a single moment of respite," impressed the French, who credited him with having saved the Russian army from total destruction during the Great Retreat. Alekseev, whom General Knox described as a man "of simple, unassuming manners," could not hope to win such acclaim through his personal charm. Nor was he blessed with the charismatic aura that Nikolai

Nikolaevich projected among the Russians. Only by demonstrating his ability as a planner and a strategist could Alekseev prove himself to his countrymen and the representatives of Russia's allies.[53]

Ironically, Alekseev's dedication to his soldierly profession made it difficult at first for him to win the approval of those who doubted his ability to replace the popular grand duke. Determined to restore Russia to full partnership in the alliance against Austria and Germany, Alekseev carefully followed his subordinates' progress in replenishing the ranks of their shattered divisions. He insisted upon knowing in detail how quickly recruits were trained and how well they were armed, and his inquiries required far more extensive communications with the army's corps and division headquarters than Nikolai Nikolaevich had ever thought necessary. Each day in March 1915, Nikolai Nikolaevich's *Stavka* had transmitted an average of just over fourteen thousand words spread over 140 dispatches. Alekseev's detailed inquiries, many of which he drafted himself, doubled those figures in October and by March 1916, doubled them again.[54] He therefore found little time for the many courtesies that the grand duke had lavished upon foreign diplomats, correspondents, and senior Allied staff officers. Anxious to get on with rebuilding Russia's armies, Alekseev often chose to leave his foreign guests unattended.

To rebuild Russia's army required reconstruction on an unprecedented scale. In terms of soldiers killed, wounded, and taken prisoner, the battles of 1915 had drained more than two and a half million men from Russia's army, while only slightly more than two and a quarter million appeared at the front to replace them.[55] Clearly, Russia, whose reserves of manpower had always been thought inexhaustible, did not have enough men at her training depots to fill the decimated ranks of her shattered army. "What had been thought to be impossible was now a stark reality," one expert wrote recently. "The giant Russian 'steamroller' was running out of fuel."[56]

Looked at in historical perspective, Russia's sudden manpower shortage in the fall of 1915 was not as surprising as it might seem. In the four decades since the empire had introduced Universal Military Service, her conscription laws had become a hopelessly tangled web of contradictory regulations, exemptions, and exceptions. Young men whose fathers, grandfathers, or uncles had died in military service, who had a brother or a father in the army, or could claim to be only sons or only grandsons, all were exempted from the draft with openhanded generosity. The pool of draftees then shrank further as a result of even broader exemptions, the most amazing of which freed large numbers of ablebodied young Russians from the draft merely for being "breadwinners." Each year, these hundreds of thousands, along with all the others whose family status allowed them to claim exemption, were placed into a second category of the *opolchenie,* the Home Guard or territorial militia, which, according to law, could not be drafted for front-line service even in wartime. Public opinion and military planners all had assumed that these men never would be needed for their nation's defense and that they would, in fact, never be mobilized even for rear-echelon duty.

Russia's annual contingent of draftees thus was assembled every spring from the ranks of able-bodied unmarried nineteen-year-old males who did not qualify for any of the exemptions that allowed so many to escape. Half of those summoned by the authorities appealed their conscription, and nearly three out of every four won a release. "An army," cynical old soldiers in Russia often said, "is an assembly of people who have failed to evade military service."[57] So few draftees remained after exemptions and appeals had taken their toll that the authorities were able to reject only one out of six for physical reasons, as compared with one out of two in Austria-Hungary and one out of four in Germany, where higher standards of living produced generally better health. As a result, many Russian draftees suffered from physical defects or poor health, and this weakened not only the army's present quality but its future fighting ability, because these unhealthy men made up the bulk of the ready reserve after they had completed their term of service as front-line troops. When the former soldiers who made up the reserves at the beginning of the Great War (now in their thirties and early forties) were called back into active service in the middle of 1915, they proved a poor match for their healthier counterparts in the Austro-German armies, not just because they were less healthy, but because there were considerably fewer of them. According to the best estimates, out of every ten men released from active duty at any given time after Universal Military Service had been introduced, only slightly more than six were still alive or not seriously incapacitated a mere decade later.[58]

Russia's immense losses quickly consumed her scant reserves of trained manpower, and even before the end of 1914, her General Staff planners had to look elsewhere for men to fill the army's shattered ranks. All reservists and all men in the first category of the *opolchenie* who were under thirty-two years of age were drafted before Christmas and, in April 1915, the authorities raised the age limit to thirty-eight. All young men due to be called up in 1916 and 1917 were drafted before mid-1915, but the ranks of the army continued to shrink as casualties soared beyond the numbers of men inducted. Convinced that "there should be one man [in the rear] for every three or four at the front," Alekseev was appalled to find that "for every man at the front there are two men in the rear."[59] Numbers of these did nothing to serve the war effort; in fact, some of them got in its way, as became clear from reports about the "extreme development of pillaging in the rear of the army."[60] At every opportunity, Alekseev pleaded with rear-echelon commanders to reduce their forces and send more men to the front. He did so in vain.

Nor was the Russian army's grossly enlarging rump the only waste of its fighting forces. In most modern armies, burgeoning sick lists indicate large numbers of men trying to escape combat assignments, but the poor physical condition of the men conscripted in 1915 meant that many Russian soldiers—

perhaps as many as one out of eight—were truly not fit for combat duty.[61] At the same time, desertions climbed steadily as the Great Retreat continued and as more raw recruits were sent to face the enemy without weapons. That August, Krivoshein complained of "the hordes of soldiers wandering through towns and villages, along the railroads, and, in general, across the face of the entire Russian land." Military authorities insisted that these men were on leave, but some clearly were deserters and their numbers obviously were on the rise.[62] "Many men are running away," a soldier in the headquarters battalion of the 2nd Siberian Rifles wrote to a woman in Moscow. "In one platoon of sixty-five men, there now are only thirty because thirty-five have deserted, while [another] ten have run off from our platoon."[63] "The number of soldiers traveling without documents on the railroads is increasing," General Ivanov reported in September, and that very month Nicholas ordered extra surveillance at railroad stations and along all major roads to apprehend runaways.[64]

Desertion undermined discipline. "The very existence of the army," General Ivanov warned that fall, now depended upon the "firmness and energy" with which commanders maintained discipline over the diminishing numbers of troops under their commands at a time when experienced officers were in desperately short supply and the numbers of younger officers whose "sense of duty, unfortunately, is poorly developed" was on the rise.[65] Men had to be found who could be promoted to officer rank and trained to maintain discipline, and some way had to be found to fill the gaps left in the army's ranks by the millions of men killed, wounded, taken prisoner, fallen sick, or run away.

As the Great Retreat neared its end, Russia's military planners thus had to face the inevitable and draft those hundreds of thousands of breadwinners in the second category of the *opolchenie*, whom the government had sworn never to put into the front line. Born of desperation, it was a decision fraught with peril, for Nicholas, his generals, and Russia's politicians all knew that these men would go neither willingly nor peaceably to the front. Now in their mid-twenties to early forties, most would leave behind families from whom they had never expected to be parted. "Men [already] are hiding out in the woods and unmown grain fields," Russia's minister of internal affairs warned his colleagues as they discussed how this previously forbidden source of manpower might be mobilized. "Given the present mood, I am afraid that we shall not get a single man."[66] Only if the Duma supported the new mobilization of breadwinners, Russia's senior statesmen realized, was there any chance to avoid widespread violence. For perhaps the first time in Russia's history, the chief figures in her government admitted openly that the Duma could speak for Russia's people and could do so effectively.

A number of Duma leaders thought that if a mobilization of breadwinners such as the Council of Ministers proposed was carried out in the fall of 1915, it would be possible to give most of the new levies extensive training before they were sent into battle the next spring. Urged on by Russia's ministers and generals, they rushed the appropriate law through the Duma in August and, by the time Nicholas assumed supreme command of Russia's armies, the authorities had already begun to call up some of the men who had never expected to wear an army uniform.[67] In vivid and frightening contrast to the levies of the Great War's first year, this new mobilization brought violence and bloodshed to villages, towns, and cities all across Russia. In the industrial center of Rostov-on-the-Don, previously exempt breadwinners demonstrated against the police and tried to refuse induction, while in Astrakahan their women and children mobbed assembly points. Joined by their men, these women and children looted dozens of shops, and much blood flowed before the police restored order. "Go ahead and shoot!" angry men and women shouted at the police. "Better that we should die here than be sent to war without bullets!"[68]

All across Russia it was the same. In Petrograd, hundreds of new conscripts turned upon the police with cries of "bloodsuckers, into the army with you!" while in Iaroslavl they threw bricks and cursed the police as "swine." As they were marched to waiting trains, numbers of these men continued their protest simply by walking away. With utter disdain for military authority, others jumped off the trains that carried them to distant training centers and returned to their families.[69] "As the train approached the station at Uvarovka, the new draftees began running away. Our escort opened fire, but the fugitives paid absolutely no attention," one veteran wrote to a friend in Moscow. "When we finally stopped at Uvarovka, it turned out that 372 out of 800 had escaped."[70]

Those who remained made less than biddable fighting men. Especially the older ones—called the *borodachi* (the bearded ones) in contemporary accounts —helped to drag the army's morale to an even lower level as a new wave of war-weariness set in the next year. "This is a dog's life, and I'm totally sick of it," one of them wrote. "Everyone's completely fed up with this war," another added. "How can we keep fighting? But how can we manage to make peace?"[71] Those *borodachi* who survived their first year of military service went on to spread dissent in the Russian army in 1917, when desertions soared beyond a million and soldiers began to make their own peace by walking away from the war. For the moment, however, the conscription of almost a million previously exempt breadwinners gave Alekseev the manpower he needed to rebuild the army after the losses of the Great Retreat.

Alekseev's efforts to rebuild and re-arm the Russian army late in 1915 were aided by the Central Powers' decision to support their new Bulgarian ally in an attack against Serbia, which made it necessary for them to reduce their pressure against the Russian front that fall. True to the promise he had made when he signed the Pless Convention in August, the chief of Germany's General Staff, General von Falkenhayn, sent the undefeated Mackensen to attack the Serbs from the north within thirty days. From the first, the Germans overwhelmed their enemy, for Falkenhayn had sent 330,000 elite German troops, including the crack Alpine Corps, against less than 200,000 Serbs. As the Serbs fell back before Mackensen's heavily supported attacks, six double-strength Bulgarian divisions struck them in the flank. Now pressed from both north and east, the Serbs fell back toward the Kossovo Plateau and the rugged land beyond.[72] Here, there was only "wild mountain country," in which, Falkenhayn remembered, "the Serbian Army moved quickly to its fate." With a victor's unvarnished satisfaction, Falkenhayn noted that "only a few miserable remnants [of the Serbian army] escaped into the Albanian mountains, losing the whole of their artillery and everything else that they could not carry."

By the end of November, Falkenhayn could report that "there was no longer a Serbian Army."[73] From the German point of view, the campaign had cost little and achieved much, for it ended the Serbian threat to Austria's southern flank, won the strong support of the Bulgarians, and seriously weakened the Entente's position in the Balkans. What Falkenhayn did not see at the time was how well Alekseev would recover his balance and strengthen his forces during the weeks when the Germans' attention was focused on Serbia. In a special report to the kaiser at Christmas 1915, Falkenhayn assured him that the "offensive powers [of Russia's armies] have been so shattered that she can never revive in anything like her old strength."[74] On several occasions within the next six months, Alekseev and his field commanders proved how badly Falkenhayn had overestimated the impact of his armies' victories during Russia's Great Retreat.

Despite the monumental obstacles he faced, Alekseev's efforts during the fall and winter brought impressive results. By the beginning of 1916, greater numbers of Russian conscripts were being better trained and better armed than ever before. Sunk to a dismally low ebb at the end of the Great Retreat, the army's morale improved as soon as its soldiers had guns and bullets in their hands once again. "The men's spirits rose and, fully expecting to defeat the enemy, they began to say that they now were ready to go on the attack," General Brusilov remembered. Hope and optimism returned. "We had every reason," Brusilov continued, "to expect to smite the enemy and drive him from our land."[75] Military censors found that almost a third of soldiers' letters reflected "good spirits," while two-thirds indicated their writers' "calm faith in the final victory of the Russian army," and only one in fifty expressed attitudes that could be classified as "depressed" or "pessimistic" about the war's outcome.[76]

Now more confident of his men and more certain of his allies, Alekseev planned a broadly conceived stroke that could enable Russia, France, Britain, and Serbia to coordinate their attacks against the Central Powers. At the end of 1915, he therefore urged the Inter-Allied Conference held at Chantilly to coordinate an Anglo-French thrust north from Salonika with a Russian general offensive in the Carpathians. Supported by the tattered remains of the Serbian army that had escaped from Albania to Corfu in December, these two assaults would converge in eastern Hungary to join in a united thrust against Budapest. Alekseev's plan aimed at cutting the Germans off from their Turkish and Bulgarian allies, defeating Austria-Hungary, and leaving Germany to face the combined forces of the Entente alone. Boldly conceived, and with a chance of success, Alekseev's plan was condemned to oblivion by the parochial strategies of his allies.[77]

Although his passion for detail eventually obscured his strategic vision, Alekseev's initial aims probably were less blurred than were those of his French and British allies, who insisted that the western front always had been—and must always remain—the main focus of Allied military operations. Blindly certain that the decisive blow against Germany could only be struck in the west, the French and British insisted that Russia must again attack unsupported from the east in order to divert enough German forces to allow their armies to break through on the western front that spring and summer. Until then, France wanted to remain on the defensive. General Zhilinskii, sent to represent Russia at Chantilly after his appalling failure as commander of the northwest front in August and September 1914, protested in vain. "In exchange for all [the aid] we are receiving, they'll take the shirts right off our backs," Alekseev bitterly remarked at one point. But Russia was in no position to do more than issue "a calm and dignified protest" against what Alekseev called "Joffre's crude opinion that only France is fighting the war."[78] "The armies of France," Marshal Joffre insisted with a certainty every bit as weighty as his great bulk, "must be preserved until the beginning of the decisive offensive" that he planned for the summer of 1916.[79] Never allowing it to be forgotten that he, not the Russians, had triumphed at the Marne, Joffre remained insistent that the Allies could follow no other course. "The enemy won't let Joffre get the better of him so easily," Alekseev warned. "He'll attack as soon as the weather and road conditions permit."[80]

As Alekseev predicted, Russia's allies did not complete preparations for their offensive on the Somme before the Germans seized the initiative. Certain that "Russia's internal troubles will compel her to give in within a relatively short period," and that Italy's "internal conditions will soon make her further active participation in the war impossible," Falkenhayn urged Germany to focus her main effort in 1916 against France. "France has almost reached the breaking point," he explained to the kaiser that Christmas. Germany now should act to "open the eyes of [France's] people to the fact that, in a military sense, they have nothing more to hope for." If that could be accomplished quickly, he argued, "that breaking-point would be reached and England's best sword knocked out of her hand." A way must be found to make France pay for every German life she took with the lives of at least two of her soldiers, and Falkenhayn proposed to achieve that end by a brutal campaign of coldly calculated attrition.[81] "There are objectives for the retention of which the French General Staff would be compelled to throw in every man they have," he told his kaiser. "If they do so, the forces of France will bleed to death."[82] "To inflict on the enemy the utmost possible injury with the least possible expenditure of lives on our own part," he chose Verdun, the ancient portal through which warriors from the German lands had stormed into Gaul in the long-ago days of the Nibelun-gen.[83] Now a depleted fortress from which Marshal Joffre had already drained most of the guns and men, Verdun was to become "a piece of flypaper on which the French army would be caught" if Falkenhayn had his way.[84]

As one observer remarked many years ago, "the German offensive [at Ver-dun] was to be based on fire-power rather than man-power."[85] Founded, as Winston Churchill later concluded, "upon Falkenhayn's appreciation of French psychology and German artillery," the German strategy was to make Verdun "the anvil upon which the remaining force of the French army would be pulverized in successive relays by the German heavy howitzers."[86] At precisely 7:15 on the morning of February 21, 1916, the German offensive began with an artillery barrage that rained a raging storm of metal and high explosive upon the French positions for more than nine hours. Heavy shells shredded barbed-wire barriers in an instant, and entire trenches disappeared beneath the up-churned earth. "The craters made by the huge shells gave to all the countryside an appearance like the surface of the moon," one observer reported,[87] and, for the next several days, the French defenses simply washed away before the unceasing fire of the massed German guns. Fearful that Verdun's defenders might yield to the crushing power of the German artillery before they could be reinforced, Joffre insisted that there could be no retreat. "Every commander who . . . gives an order for retreat," he telegraphed from his headquarters, "will be tried by court-martial."[88]

Acting on Alekseev's instructions, Zhilinskii had proposed at the Inter Allied Conference at Chantilly that the Allies agree to launch separate offensives in response to any attack that the Germans might make. At the time, he had encountered what he called "the strongest opposition" from the arrogant Joffre, who, with exquisite rudeness, implied that Russia ought not to keep asking the armies of France to come to her aid every time she feared a German attack.[89] Yet, once the German artillery began its ceaseless pounding at Verdun, Joffre was not at all reluctant to ask Alekseev to do for the French precisely what he had so readily refused to do for the Russians. "On the basis of the resolutions made at the Chantilly Conference," Joffre telegraphed to Alekseev less than a fortnight after the Germans began their attack, "I request that the Russian army begin urgent preparations for an offensive."[90]

Although his armies were not yet ready, and the weather was at its worst, Alekseev responded to Joffre's plea by ordering an offensive at the beginning of March. Beginning on March 5, 1916, and continuing for a fortnight, 350,000 Russians of the Second Army, supported by nearly a thousand guns, attacked less than 100,000 Germans in the region of Lake Naroch, somewhat to the east of Vilna and south of Dvinsk. With a reserve of more than a thousand shells per gun, the Second Army boasted considerably greater artillery superiority over its German foes than Mackensen had enjoyed at Gorlice the previous May. Yet the Naroch Operation quickly became one of the most mismanaged of the war and cost nearly 100,000 Russian lives. Ordered to advance through swamps that had thawed since commanders had drawn up their attack plans, Russian soldiers had to attack through near-freezing water that stood knee-deep. Left in the open at night, more than twelve thousand of these sodden, weary men suffered frostbite in less than two weeks.[91] All across the Naroch front, the first spring thaws mired soldiers, animals, and machines in slush and mud. Communications broke down, generals feuded (Alekseev had no confidence in several of the corps commanders, but Nicholas would not allow his favorites to be removed), and the artillery and infantry worked at cross-purposes. One general complained that his headquarters had to receive and transmit more than three thousand telegrams a day during the height of battle, and one expert estimated that more than half of the orders issued were so confused that they had to be countermanded.[92]

Every phase of the Naroch Operation, *Stavka*'s press officer later wrote, was "nightmarishly unsuccessful," as Russia's divisions gained almost no ground despite their terrible losses.[93] Still, Alekseev had managed to win another desperately needed respite for the French. Surprised by his unlooked-for offensive, the Germans halted their attacks on Verdun for more than a week until they were certain that they would not have to commit large numbers of men against an all-out Russian advance on the eastern front. The delay cost them their advantage, and forced Falkenhayn to abandon his hopes that Verdun could become the Churchillian "anvil upon which the remaining force of the French army would be pulverized." A little more than three months later, on June 18/July 1, 1916, the British launched their long-planned offensive on the Somme and opened yet another gore-stained floodgate through which the blood and limbs of nearly a million more German, French, and British soldiers would swirl without purpose into eternity.

No-man's-land, those once-fertile fields of northern France that now lay between the Allied and German trenches, became Europe's new killing ground, where men took lives and surrendered their own for shell-torn gains that could be measured only in feet. Into these nether regions, where terror and suffering beyond all human imagining lurked in a tangle of barbed wire and corpse-filled shell craters, the generals of Europe drove men by the hundreds and watched in morbid fascination as they returned by scant tens and twenties, their minds and bodies forever scarred by what they had done and seen. "We have just come out of a place so terrible that a raving lunatic could never imagine [its] horrors,"[94] wrote an Australian officer who survived one of the many battles that formed part of what generals now exalted as the strategy of "methodical progress," in which commanders repeatedly hurled large numbers of men against strongly-fortified enemy positions in an effort to erode their defenses.

For the men sent again and again into battle for the merest of gains, the time of hope and optimism had long since passed, and those who would survive the desperate struggles of 1916 would never see the world in the same way again. Only for the Russians would there be a fleeting moment of hope as their army, briefly reborn as a victorious fighting force under the brilliant command of General Brusilov, surged to the thunder of their new-made guns into Austria's Galician lands that summer. "Our losses are staggering," Nicholas confided to Aleksandra as he sent her one of a pair of maps that showed every detail of the army's new positions, "but, against this, the success is prodigious!"[95]

Brusilov's short-lived victorious march in the summer of 1916 reflected a military situation very different from the terrible days of the Great Retreat. Equally dramatic was the change in political conditions within Russia, as representatives of the people stepped forward to play a crucial role in the affairs of their government and their nation for the first time in Russia's history. Proclaiming that "without the support of our social institutions our armies cannot be victorious,"[96] Russian politicians in mid-1915 tried to plot a course that would place constitutional politics at the service of a renewed national war effort, and allow the Duma, as the moderately conservative Aleksandr Guchkov said, to "take the revolution into its hands" before it fell victim to the uncertain mercies of "uncontrolled forces."[97] Men of talent and foresight must be found to replace those who had failed Russia so miserably during the war's first year. For, if government and Duma could not work together, then defeat and revolution might follow quickly in the terrible wake of the Great Retreat.

Even before the Great Retreat ended, thoughtful men and women had concluded that "the tyranny of tsarism . . . [had] brought Russia to the brink."[98] Uncertain that their nation could be pulled back in time, men of sound and sober judgment nonetheless made ready for a final attempt during the hot mid-summer days of 1915. The government could not merely exchange "some crows for others," Zinaida Gippius warned then.[99] The only hope, the Kadet lawyer and politician Vasilii Maklakov* insisted, was for "the right men," to be put "in the right places."[100] Government, Pavel Miliukov added sagely from his position as the Kadet's leading spokesman, must be by "men whom the country knows and respects," for "the abnormal relationship between the government and the public"[101] that had marred Russia's war effort during the dark spring and summer of 1915 could not continue. In the words of one of Miliukov's colleagues, the government could not continue to "spurn all who have offered their help."[102] New ways of politics must be explored, Russia's moderates warned, before their government and its statesmen were "doomed to moral impotence"[103] and before men on the extreme left and far right succeeded in exploiting their nation's crises for their selfish ends.

---

*Under no circumstances should Vasilii Maklakov, the brilliant lawyer who spoke so often and with such eloquence in defense of parliamentary government in Russia, be confused with his reactionary younger brother Nikolai who, as Minister of Internal Affairs from 1913 to 1915, had connived to prevent any public participation in state affairs.

CHAPTER VII

New Ways of Politics

Never in Russia's history—not even during the dark days when Napoleon had marched upon Moscow—had war pervaded the lives of her people as it did during the first year of the Great War. From the Baltic coast to the Rumanian frontier, the Great Retreat of 1915 spread a zone of devastation across Poland into the lands of West Russia. By the hundreds of thousands, old men, children, and women of all ages streamed eastward, "going they know not where, wherever the whim of fate sends them," one soldier wrote to a friend at home. "Almost every woman had a babe in her arms and several more little ones grasping at her skirt and getting tangled up in her legs as she tried to walk," a government observer added.[1] Worn by hunger and disease, these refugees had turned their backs upon the labors of a lifetime and abandoned their lives' poor possessions to the invader's mercies. Wherever they went, these wandering folk were ever-present grim reminders that many suffered and few benefited when rulers and statesmen unsheathed the swords of nations.

Far from the devastation of battle, a sense of the war's desperation first entered Russia's long-silent peasant villages in semiliterate, hastily scrawled messages sent from the front by loved ones who had no weapons, bullets, or winter clothing. "Reinforcements come here without rifles and only get them after men have been killed or wounded," one soldier wrote.[2] "Where is that Russian wealth they have in mind when they say that 'Russia is richer than anyone else'?" another asked bitterly. "It is the Germans who have all the artillery."[3] Some angry soldiers urged families and friends not to pay taxes until the war ended.[4] "If the emperor thinks it's necessary to fight," others muttered darkly, "let him do the fighting himself."[5]

War seemed all the more fearsome when some of the young men who had marched away in the late summer of 1914 began to make their way home. With limbs lost, minds shattered, and bodies torn and poorly mended, these victims brought living evidence of war's terrors to peasant hamlets that had known nothing of war's suffering in nineteenth-century wars because wounded serf soldiers had never been allowed to return home. These men now spoke not of great deeds and war's glory, but of the army's exhaustion, their comrades' apathy, and their desire to leave the killing and return to their fields.[6] That summer, the peasants stirred within their villages. Instantly, the authorities grew fearful. Rowdy assemblies mushroomed into "revolts" in official reports, and one district police chief actually reported "mute dissatisfaction" among the peasants, although he could find no overt expression of it.[7] "The worst thing about all this is that [it is developing] by itself, without organization, plan, or system, and therefore is all the more fearsome," one Russian nobleman told a friend. "It can lead . . . to terrible consequences."[8]

Refugees and peasants were not the war's only critics during the days of the Great Retreat. Between April and September 1915, almost eight hundred strikes idled nearly 400,000 workers and cost Russia's industrialists more than a million lost working days. In June, the police shot striking workers in Kostroma and drew the first blood to be shed on the home front since the war had begun.[9] Then, in August, strikes for higher wages in the mills of Ivanovo-Voznesensk turned into open protests against the war. "Let's make a clean sweep of all the butchers, from the tsar and his ministers right down to our chief of police," some four thousand striking workers shouted. "Rather than die fighting against *their* enemies, we'd rather die on the barricades!" Angry workers converged on the city's central square to free several of their leaders from jail. Their jackets opened to bare their chests in a symbolic gesture of defiance, they shouted "Go ahead and shoot!" to two hastily assembled companies of militia. Under a hail of stones from the crowd, two companies of soldiers killed thirty workers and wounded another fifty-three. "A bloody fog blankets Russia!" revolutionary handbills proclaimed. "Now death lurks on every corner! The murderers be damned! Down with the war!"[10] In Petrograd, more than twenty thousand defense workers struck in sympathy, and men and women all across Russia soon took up their gesture. More than twice as many workers went on strike in August as in the month before.[11]

As workers' protests shifted from rising prices to the war itself in mid-1915, politicians, industrial leaders, and prominent private citizens called for a new government of ministers who held the people's trust and would allow the participation of their representatives in making decisions that might rescue Russia's crumbling war effort from collapse. "We had given the government the opportunity to unite all classes and nationalities [at the beginning of the war]," *Kadet* leader Pavel Miliukov later explained, but the government had failed to do so.[12] Led by Prime Minister Ivan Goremykin, whom a leading French official

once called "a disaster" (this same official had branded War Minister Sukhomlinov "a catastrophe" in the same breath),[13] Russia's first wartime cabinet had
tried to order a national war effort into being without calling upon anyone
outside the bureaucracy for advice. "This single fact," the Octobrist leader
Aleksandr Guchkov later confessed, "cost us the entire first year of the war" in
terms of time lost in mobilizing Russia's industrial establishment for her defense.[14] Guchkov did not speak as a man whose political sentiments made him
prone to criticize his government and tsar, for the Octobrist Party that he had
led since its founding in December 1905 had always taken its stand squarely
upon the Manifesto by which Nicholas II had granted Russia a constitutional
government on October 17, 1905. A party of merchants, industrialists, and
provincial landowners, Guchkov's Octobrists stood moderately on the Right to
defend tsar and government against attacks from the Left, and their defense
included opposing some of the more extreme demands that Miliukov's Kadets
had brought forward from time to time. The grandson of a serf whose family
had grown rich in the manufacture of woolen textiles, Guchkov had personally
tried to mediate between government and public opinion as president of the
Third Duma. Yet in the Russia of Nicholas and Aleksandra, neither Guchkov's
loyalty nor his genuine concern for defending the institutions of monarchy in
Russia won him much thanks in high places. "Oh, could one not hang Guchkov?" the Empress Aleksandra exclaimed as this public-spirited industrialist
struggled to mobilize public opinion and private resources to support the war
effort. "A strong railway accident in wh[ich] he alone w[ou]ld suffer w[ou]ld be
a real punishment fr[om] God & well deserved."[15]

Even though it had been allowed to meet only briefly in July 1914 (to
approve war credits) and for two days in January 1915, men like Guchkov hoped
that the Duma might somehow take a more direct part in Russian wartime
affairs. Desperate to divert their nation from its disastrous course, but with no
immediate prospect that the tsar would call the Duma back into session before
the first anniversary of the war's outbreak, politicians representing a broad
spectrum of Russian opinion began calling for a government of "national confidence"[16] early in the spring of 1915 as they sought a forum for their views in
the *Zemstvo* Union and the Union of Municipalities. Certain, as the *Zemstvo*
Union's Prince Lvov warned, that Russia was "truly in danger,"[17] men whose
views ranged from conservative to strongly liberal set out to form an alliance
that, in Lvov's words, would be "strong enough to persuade the government to
adopt a new course."[18] Among them stood the Duma's president Mikhail
Rodzianko, a full 270 pounds in weight and with a stentorian voice to match,
for whom every problem became a crisis and every crisis a national calamity, but
whose courage remained unwavering. Outraged at the "inefficiency of the highest military authorities," and convinced that "the masses will rise up as one
. . . if they see that all the suffering they have borne and all the blood they have
shed has been for naught," Rodzianko pressed his colleagues in the still-recessed

Duma to take action.[19] "Without wasting a single moment, we must mobilize all industry and adapt all factories and mills to serve only the requirements of the war," the Moscow industrialists' influential daily *Utro Rossii* insisted as it took up Rodzianko's urgings.[20] Industrialists, bankers, politicians, and workers must set aside their differences and join together to mobilize Russia's industry in support of the war. "All Russia," Lvov concluded, "must be welded into one military organization."[21]

"The mobilization of industry as a civic movement was born at the Ninth Congress of Representatives of Trade and Industry that assembled in Petrograd on May 26–29 [1915]," War Minister Aleksei Polivanov reported in his diary not long after he took office. At that meeting, the publisher of *Utro Rossii,* Pavel Riabushinskii, whom some called "the mighty voice of the Russians," proclaimed that "Russian industry was called upon to join together to meet all of the war's needs." One of nine sons in a wealthy industrialist family known as much for its patronage of the arts and sciences as for its prominence in the textile industry, Riabushinskii had long urged the government to allow private citizens a larger voice in national affairs. Only two generations removed from the peasant forebears whose talents as traders and entrepreneurs had launched his family upon its path to wealth and fame, he shared their long-standing suspicion of government officials and their belief that individual initiative could accomplish more than bureaucratic caution.[22] Riabushinskii therefore proposed a Central War Industries Committee made up of representatives from Russia's scientific-technological establishment, various industrial organizations, railroad lines and steamship companies, the *Zemstvo* Union, and the Union of Municipalities to weld all of Russia's industries into a nation-wide war production effort. Guchkov, his political ally and friend, became its chairman.

Guchkov liked nothing better than to stand at the center of history's stage, for fate had endowed him with gallantry and daring in generous measure. Certain that this venture must either pave the way for more public participation in Russia's government or weaken the nation's war effort even more, he warned that "if we do not succeed . . . then all our efforts, all our sacrifices, all our enthusiasm, and all our ardor will fly up the chimney like a puff of smoke."[23] "The historic moment has arrived," *Utro Rossii* added. "Much of Russia's fate depends upon how skillfully and intelligently we use it."[24]

Led by some of their most prominent colleagues, including the magnates of the textile, mining, sugar, and oil industries, Russia's industrialists organized local war industries committees in 1915 to assemble raw materials and allocate orders to the factories best able to produce war goods in more than two hundred towns and cities. Their progress was slow and their development cumbersome, for these committees quickly took on all the burdensome routines and outright corruption of the government bureaucracy they so despised. The Central War Industries Committee, whose administrative personnel ballooned from two hundred to two thousand in less than a year, was among the worst, as it blatantly channeled orders to the factories of industrialists already heavily involved in military production rather than bring smaller factories into the war effort.

Almost half of the Central War Industries Committee's orders went directly to the great industrialists of Petrograd and Moscow. Late in the summer of 1915, it divided a single lucrative order for seventy-three million rubles' worth of artillery shells among eleven great factories and sent orders for more than eight out of every ten mine throwers made in Russia to three favored firms. Such did little to aid the war effort. Too many industrialists were too anxious to take handsome government advances (usually based upon deliciously inflated profits) but too slow to meet contract deadlines.[25] The greater significance of their effort lay in the extent to which they brought private citizens, politicians, and politics into Russia's war effort which, in earlier times, the tsar and his advisers had jealously screened from public participation and public opinion.

In any case, the Central War Industries Committee could not address the broader problem of Russians' participation in their nation's political and economic destiny. Even before the men from Moscow had called for a national mobilization of industry, a loose coalition of political parties whose outlooks ranged from moderately conservative to unabashedly liberal urged Nicholas to call the Duma back into session. Yet Prime Minister Goremykin feared that a new session of the Duma would stir more turmoil than the government could control, and warned Nicholas to resist any demand for greater public participation in government. A man who had insisted loudly and often that all in Russia were mere servants of the tsar, Goremykin had honed his political instincts to perfection during a full half-century's apprenticeship in the upper reaches of the Russian bureaucracy, where an incautious remark could bring instant ruin, and a carefully planted insinuation could reap great rewards. With all his sharply-tuned senses keyed to political survival and personal advancement, he feared that the rising wave of bitterness between government and governed would overflow the moment the Duma was called into session, and hoped to avoid it at any cost. "The first sight of [the especially despised minister of internal affairs Nikolai] Maklakov will call down the anger of all factions," he confessed to Rodzianko, "and then I'll have no choice but to dissolve the Duma."[26]

Nicholas would have liked to follow the course Goremykin proposed, for he thought that "the time has come to cut down the powers of the State Duma." Yet he also sensed that he might do irreparable harm to his throne and to Russia's war effort if he did so. He therefore swallowed his dislike for parliamentary government, and, in a rare act of independence, ignored Aleksandra's urgent warnings of dire things to come and instructed Goremykin to announce that the Duma would assemble "no later than August."[27] In the meantime, in a major concession to public opinion, he decided to dismiss the men whom progressive Russians had branded as "odious ministers."[28]

The first to fall was Nikolai Maklakov, minister of internal affairs since 1913 and an unabashed advocate of autocratic government, who had only contempt

Imperial family, 1914.

Rasputin.

*Grand Duke
Nikolai Nikolaevich,
Supreme Commander
of Russia's Armies,
1914–1915.*

*War Minister
General Sukhomlinov.*

*Petrograders standing in the Palace Square hear Nicholas II announce Russia's
declaration of war against Germany.*

Nicholas II holding an icon before his soldiers.

Nicholas II (with binoculars) and Grand Duke Nikolai Nikolaevich.

Russian soldiers before an attack.

Russian infantry on the march.

Cossacks.

Russian howitzer battery.

A Russian field kitchen just behind the front.

Russian soldiers surrendering at Tannenberg.

Russian soldiers in the trenches, 1915.

Russian prisoners in 1915.

Refugees in a forest near Riga.

A Russian war casualty.

General Brusilov.

German soldiers with mounds of Russian corpses dragged away from in front of German trenches.

Workers inside the Putilov Works.

Demonstration on Znamenskaia Square at the beginning of the February Revolution.

Russian women's battalion.

Aleksandr Kerenskii.

General Kornilov.

Lenin.

Red Guards.

Vladimir Antonov-Ovseenko.

Vladimir Maiakovskii.

Feliks Dzerzhinskii.

Lev Trotskii as Commissar of Foreign Affairs.

House where Brest-Litovsk Peace Conference held its meetings.

for the minor concessions Russians had won during the Revolution of 1905. As different from his liberal brother Vasilii as night from day, this Maklakov rejected any thought of public participation in government for any reason. "One leg had been lifted" when the Duma had been created, he told his captors after the Revolution of 1917, and ever since that time Russia's political life had resembled "a drunkard's walk, tottering from wall to wall."[29] Politics, he insisted, must remain the province of the tsar and his officials, whose orders all Russians must obey, and his widely-known public statements had made him the object of hatred among all who did not share his reactionary outlook.

Although personally much taken with Maklakov, Nicholas removed him just before the Great Retreat began, when it became clear that his mere presence would stir bitter opposition the moment the Duma resumed its meetings. A week later, War Minister Sukhomlinov left office amid the scandal of Colonel Miasoedov's widely-publicized trial and execution, and not long afterward Director-General of the Holy Synod Vladimir Sabler and Minister of Justice Ivan Shcheglovitov followed. One of the first in a long stream of wartime ministers who curried the favor of Rasputin, Sabler had built his career as a bureaucrat in a variety of religious institutions before he had won appointments to the Senate and the State Council. His most notorious act, according to well-substantiated rumor, was that he had actually knelt before Rasputin on at least one occasion,[30] and he certainly had little to recommend himself as a member of Russia's wartime cabinet except for his ardent political conservatism. Likewise, Shcheglovitov, an unabashed careerist who had masqueraded as a liberal on more than one occasion when conditions seemed to warrant it, shared Sabler's personal dishonesty and gross lack of principle. A furtive figure who like Sabler remained out of public view, he was nonetheless widely regarded as one of the most pro-German statesmen in Russia. When Nicholas sent both into retirement later in June, few mourned their passing. Ecstatic at the removal of the men its patrons had come to despise, *Utro Rossii* proclaimed: "The voice of the government now speaks in unison with the voice of the Russian land!"[31]

Fearful that her husband might allow the Duma a greater voice in Russia's affairs, Aleksandra poured forth warnings. "We are not ready for a constitutional government," she insisted. "They will try to mix in & speak about things that do not concern them." Ominously, she reminded him that Rasputin was against summoning the Duma. "I always remember what our Friend says & how we do not enough heed his words," she had written a few days before. "When he says not to do a thing & one does not listen, one sees one's fault always afterwards."[32] Certainly, she was right to expect that the Duma would not remain quiet. "Przemysl fell and Maklakov left the government—Lwów fell and out went Sukhomlinov—Warsaw is about to fall and Goremykin is about to go too!" exulted Aleksandr Kerenskii, now spokesman in the Duma for an amorphous conglomeration of peasant socialists who called themselves Trudoviks. The government now "was ready to work loyally and amicably with the State Duma," the sensationalist *Birzhevye vedomosti, The Stock Exchange Gazette,* told its readers.[33] As one commentator later remarked, "public opinion had concluded that it now had the opportunity not only to 'dismiss' ministers but also to 'appoint' them."[34] Should the tsar not concede such unfounded assumptions, the men who had clamored for a new session of the Duma were prepared to challenge him. Proclaiming that "we only want to help the government and, for that purpose, we are expressing the will of the people," the coalition that had urged Nicholas to summon the Duma at the beginning of the summer now searched for further common ground and made ready to oppose the government if it refused them the strong voice in national affairs they now thought was theirs by right of public opinion.[35]

On the first anniversary of the Great War's outbreak, Russia's State Duma reassembled in the White Hall of the Taurida Palace, originally a gift from Catherine the Great to her former lover and long-serving counselor Prince Potëmkin in the 1780s and site of the Duma's sessions ever since its birth in 1906. As they took their places for their first meeting since January, every deputy must have thought about the great changes that the past months had brought in Russia's military fortunes. At the war's beginning, they had assembled in optimism, certain that the Austro-German armies would be defeated before year's end. Now they faced foes that seemed everywhere victorious. In less than a year, enemy soldiers had occupied tens of thousands of square miles of Russian territory. Germany's armies were at Warsaw's very gates, and, within a fortnight after the Duma reassembled, they would seize the great fortresses of Novo-Georgievsk, Ivangorod, Osowiec, and Kovno. As they had for some months past, Duma members and prominent private citizens spoke of "dark forces" that had brought Russia to such a terrible state. "Who controls Russia at the present moment?" Pavel Riabushinskii asked. "To answer that question seriously," he went on, "we would have to say: 'We don't know.' "[36]

In this atmosphere—Rodzianko called it "tense" and remembered that everyone "expected stormy speeches"—the Duma assembled at one o'clock on the afternoon of July 19.[37] From his vantage point in the special gallery reserved for foreign ambassadors, Paléologue saw that all "faces seemed electrified." Rodzianko opened the session, but it was Goremykin, as president of the Council of Ministers and prime minister, who gave the first speech. He spoke first about the war and the need for national unity. "Sooner or later, victory will be ours," he promised his listeners. "For Russia at war, there can be only one party —the Party of War to the End. There can be no party program but one—the Program of Victory." "Bravo!" cheered the men on the Right. "Announce an amnesty for all political prisoners!" their opponents on the Left shouted in reply. Quickly, War Minister Polivanov took up Goremykin's themes to remind the delegates that Russia had a moral duty to fight against "the crude, mindless yoke of Teutonic power," and promised that "the war will be pursued to a victorious end." Foreign Minister Sazonov and Minister of Finance Petr Bark spoke in a similar vein. "Great Mother Russia will never be anyone's slave," Rodzianko replied for his Duma colleagues. "We will fight to the end, to the very last soldier. We must be strong, with firm faith in the strength of Russia's heroes."[38]

Similar ringing words of heroic resolve had embellished the Duma's brief sessions at the Great War's beginning and in January 1915, but they now served only as part of the ritual that opened an urgent debate about Russia's war effort. "The entire administrative organism of Russia must be reformed," men who had begun to look into the failings of Russia's war effort told Paléologue at the time.[39] Beyond that, many of them now were convinced that the tsar could not continue the war successfully unless he agreed to a government of national confidence. "Only a government having the confidence of the country behind it can have the strength to demand the supreme effort from the people," the *Zemstvo* Union's Prince Lvov insisted, while Kadet Party leader Miliukov quickly explained that any such government must include "men whom the nation knows and respects, in whom it has long placed its confidence." Others spoke much more bluntly. "We applaud the change [of ministers] that has [just] taken place," the leader of the Progressists—a group who spoke for the most liberal of Moscow's industrialists—announced. "But the system itself has not changed. . . . The Duma still cannot say to the people: 'Here are your true leaders. Believe in them. Place all your hopes in them. Follow them without fear or doubt.' " Count Vladimir Bobrinskii, founder of the moderately conservative Nationalists, provided the final formula upon which many Duma members seized to frame their program. Only "a close union between the entire country and a government that enjoys its complete confidence," Bobrinskii insisted, could assure victory over Austria and Germany.

Still, the Duma did not speak with one voice, for men who sided with the tsar and his ministers spoke out against those who urged public participation in state affairs. "Is it possible that we are really going to try to settle accounts with the government at the very moment when it is straining every nerve to repulse the enemy's attack?" asked Nikolai Markov, leader of the right-wing Union of the Russian People. "Is this an act of patriotism?"[40] "The debates in the Taurida Palace are becoming more and more lively," Paléologue reported. "All the failings of the bureaucracy are being criticized and all the vices of tsarism brought into the daylight. The same conclusion recurs like a refrain: 'Enough lies! Enough crimes! More reforms! More punishments! We must transform the system from top to bottom'!"[41] "If events continue to evolve with that tragic necessity with which they have developed thus far," one of Guchkov's colleagues remarked at a meeting of the Central War Industries Committee, "the time will come when we can ask for anything and can be certain that we'll get everything we ask for."[42]

Just a few days later, Riabushinskii's *Utro Rossii* reported that a proposal for a "Cabinet of National Defense," presided over by Rodzianko and including several prominent politicians in addition to Krivoshein and Polivanov from the present government, was circulating in the Duma.[43] For the moment, there was no chance that any such cabinet could be formed. Still, the diverse party affiliations of the Duma men mentioned as candidates for such a cabinet reflected an unprecedented union of almost three-quarters of the Duma's 420 deputies into an alliance that called itself the Progressive Bloc. Never in the brief decade of their existence had Russia's political parties succeeded in uniting on such a broad front. It was almost a revolutionary act, carried out by sober men whose devotion to the existing order was so obvious that even the ultrareactionary Markov readily noted the sharp distinctions that separated them from the handful of "Reds" in the Duma. "It might be more appropriate," Markov mused aloud, "to call it a Yellow Bloc." "Not yellow," the irate Nationalist Vasilii Shulgin shot back, "but tricolored."[44]

An immensely complex alliance constructed by some of Russia's most talented politicians, the Progressive Bloc was made possible only by the terrible strains of the war, popular outrage at the War Ministry's scandalous failure to provide enough ammunition and weapons, and the intractable refusal of Nicholas and Goremykin to bring any members of the Duma into the Imperial Cabinet. Miliukov remembered that "the spirit of the bloc was already evident," as the Duma men had listened to the speeches of Goremykin, and his fellow ministers on July 19,[45] but the ephemeral spiritual sympathy of men distressed at Russia's crushing military crisis could not alone establish a concrete fleshly union, and it required almost a month of negotiations before the bloc emerged full-blown. On August 11, seventeen Duma deputies therefore held a series of secret meetings with nine members of the Council of State, the conservative upper chamber of the Duma, to highlight the common ground they shared and to find ways to gloss over their differences. That Baron Aleksandr Meller-Zakomelskii offered his lodgings as a meeting place was itself a dramatic statement of the breadth of opposition to Goremykin's cabinet, for the baron was well known for his ruthless suppression of a mutiny among the sailors at Sevastopol in November 1905, and an even more brutal punitive expedition that he had led along the Siberian Railway to crush the revolution's last dying embers the following January. In the past, when Meller-Zakomelskii had dealt with enemies of the regime, they had become, to use his own phrase, "as quiet as water and as meek as grass."[46] Now, he stood ready to join men who opposed the policies of the government he had so long supported.

For some weeks, men had been speaking of a government of "national confidence," but that term had been left purposely vague and the Progressive Bloc had yet to agree upon what sort of government it had in mind. "Would it be a government of bureaucrats or of civic leaders and politicians?" one of them asked. Obviously, there were going to be sharp differences of opinion about specific points and, as another of them pointed out, "we cannot be completely agreed upon everything from A to Z." The important thing, most of them thought, was that serious and responsible men of diverse views, "united by our desire for victory . . . [could] join together and define the ways and means for achieving our goal." As one of Guchkov's Octobrists pointed out, such a goal was easy enough to define, but immensely difficult to achieve. "Here is the most fundamental question of all," these men concluded. "How can we reshape the state power without brushing the Supreme Power aside in some fashion?"[47]

Convinced that "only a strong, firm, and active state power can lead our homeland to victory," and that "such can only be a state power that enjoys popular confidence," nine of the men who met at Meller-Zakomelskii's in August 1915 drafted a program that emphasized their areas of agreement and called for "the formation of a united government of men enjoying the confidence of the country." They urged broader public participation at all levels of government and called for the abolition of all laws that oppressed the empire's Jews and national minorities, as well as those wartime regulations that had limited the civic freedoms that Russians had won during the Revolution of 1905.[48] Sharply opposed to Goremykin's views, the bloc's program posed serious dilemmas for his government when it was first published on August 25.

From the very first, Goremykin and his cabinet could not agree upon a course of action against the bloc. A number of less rigid senior statesmen thought their prime minister was not the man to lead them through Russia's first great parliamentary crisis and several hoped he would retire or resign. Anxious to avoid the onerous appointment that he had himself twice refused, but convinced that someone more vigorous, more clear-headed, and less antagonistic to the Duma must take Goremykin's place, Krivoshein urged Nicholas to replace him with the energetic and efficient War Minister Polivanov. Krivoshein came away from his weekly meeting with the tsar certain that a new government was about to be formed,[49] but, although immensely skilled at interpreting the subtle signs that foretold the course of events at the highest levels of Russia's government, he had failed to read the tsar's mind correctly. Determined to remain with Russia's armies at the front, Nicholas was equally determined to leave Goremykin ("dear old Goremykin," Aleksandra called him) at the head of the government in Petrograd.[50] As a key figure in that inner circle of mediocre men and women in whom they placed their confidence, Nicholas and Aleksandra were determined to defend Goremykin to the utmost for the time being. "I fear the old man cannot continue working when all are against him," Aleksandra wrote the day after Nicholas reached *Stavka*. "Please tell him," Nicholas quickly replied, "that as soon as the Council of State and the Duma finish their work they must be adjourned."[51] Given an opportunity to choose, Nicholas much preferred to dispense with the Duma, not Goremykin.

As a bureaucrat of the old school, Goremykin had always believed that he must devoutly serve, but never dispute, his imperial master, and he therefore saw the formation of the Progressive Bloc as a direct challenge to his tsar. "As far as the bloc itself is concerned," he told his colleagues on the day its program was published, "I consider any conversations with it impermissible for the government." His diplomatic instincts always alert for any opportunity that could temper conflict and yield agreement, Foreign Minister Sazonov was aghast at Goremykin's stubbornness. "This is a huge political mistake!" he exclaimed. "The government cannot remain suspended in a vacuum and depend only upon the police!" Ignoring the prime minister's accusation that "the bloc has been created [merely] to seize power," Sazonov insisted that "in essence this bloc is a moderate one and needs our support." To Goremykin's arrogant assurance that "it will break up and all its participants will end up squabbling with each other," Sazonov replied: "If it falls apart, we will get one that stands much further to the left." Sazonov clearly thought Goremykin's reaction unjustified and unjustifiable. "Why aggravate relations [with these men] for no good reason when they are bad enough already?" he finally asked. Quickly, Russia's new minister of internal affairs Prince Nikolai Shcherbatov took Sazonov's side in the direct and honest fashion for which he had become so well known in his home province of Poltava. "It is lamentable," the prince began, "that, up to now, the government has not spoken with the bloc." Ominously, he warned: "Our silence, our efforts to ignore this bloc, could affect in a most undesirable way the mood of the great majority of the Duma that is well-disposed to us."[52]

Although Goremykin dismissed the bloc's call for "a united government of men enjoying the confidence of the country" as nothing more than "fancy words on which we shall not waste our time," he reluctantly agreed to send four ministerial representatives to a meeting with the Progressive Bloc the next day.[53] Sazonov proclaimed the meeting "a great step forward," and reminded his colleagues that the Duma, "if not pacified in time, can give rise to serious conflicts that can adversely affect the country and the progress of the war." Goremykin's reply was that of an insensitive bureaucrat who had lost all touch with the political pulse of the nation he claimed to serve. "It's all nonsense," he retorted. "Aside from the newspapers, the Duma doesn't interest anyone and everyone has gotten bored with its idle chatter." Angrily, Sazonov warned that "without good relations with its legislative institutions, no government, no matter how self-confident, can govern a country," and reminded Russia's recalcitrant prime minister that he would do well to remember that the "mood of the deputies influences the psychology of the public."[54] Irritated, Goremykin remained unimpressed.

When Krivoshein urged his colleagues to warn the tsar that "the general internal situation of the country requires a change in the cabinet and in our political course," Goremykin stubbornly insisted that any such recommendation amounted to "the resignation of the Council of Ministers and [a demand for] a new government." In short, he snapped, it was "an ultimatum to the tsar." Now furious, Sazonov demanded an apology. "We are not presenting, and do not intend to present, an ultimatum to His Imperial Highness," he said. "We are not some sort of seditious plotters, but are as loyal subjects of our tsar as Your Excellency. I beg you," he concluded indignantly, "not to mention such words in our discussions." Reluctantly, Goremykin agreed to report their views to the emperor. "The crisis is ready to burst," the council's secretary Arkadii Iakhontov jotted in a personal comment at the end of the notes he had taken at the end of August. "I do not think that it will be easy to get through it."[55] Nor did Aleksandra when she heard about the ministers' dispute. "I long to thrash all the ministers and quickly clear away *Shcherb*[*atov*]," she wrote to Nicholas. "I hope to send you a list of names to-morrow for a choice of decent people."[56]

Goremykin's desire to hold his position was part of his reason for urging Nicholas not to name a new cabinet, but he also had stern and sober personal beliefs that held him to that course. "Let them curse and blame me," he told Iakhontov the next morning. "I'm already old and haven't much longer to live anyway. But, while I'm still alive, I shall fight for the inviolability of the tsar's power. Russia's only strength lies in monarchy. Otherwise, you'll end up with such a mess that everything will be lost." We cannot know precisely what Goremykin told Nicholas when they met at *Stavka* the next day, but he returned to tell his irate colleagues that Nicholas had ordered him to close the Duma until November and "for all ministers to remain at their posts." Instantly suspicious, Polivanov asked: "*How* did you report our views to the emperor?" And Sazonov demanded rudely, "*What*, exactly did you tell him?" "I find it intolerable to be addressed in this fashion," Goremykin replied. "I find it unnecessary to reply to such questions."[57]

Certain that the worst was yet to come, Russia's angry ministers continued their bitter discussions. "The situation in Moscow [where the *Zemstvo* Union and the Union of Municipalities were meeting at that moment] is very serious," Shcherbatov told his colleagues. "Everything there is seething, agitated, irritated, and has a clearly antigovernmental character about it." Unmoved even by Shcherbatov's warning that "an outburst of disorders is possible at any moment," Goremykin replied: "I shall fulfill my duty to Our Sovereign Emperor to the very end, no matter what opposition and lack of sympathy I encounter." With some of the ministers in an absolute rage, Goremykin prepared to adjourn the meeting with the vague assurance that "His Majesty the Emperor said that he will come here himself and see to everything." "Yes, but then it will be too late," Sazonov shot back. "Tomorrow, blood will flow in the streets and Russia will plunge into an abyss!" At that point, Iakhontov noted, Polivanov was "overflowing with bile and, it seemed, was ready to bite," while Sazonov, now "almost at the point of hysteria," stormed from the room. Fearful that Russia's Foreign Minister might faint, Iakhontov followed him into the anteroom just in time to hear him shout as he left the building: "That old man is utterly mad!"[58]

Clearly a majority in Russia's Council of Ministers had turned sharply against Goremykin, but an even larger storm was brewing, for their conflict involved more than policy disagreements. Certain that Goremykin's unyielding refusal to recognize the just concerns of the Duma and public opinion "fatally" threatened Russia's very foundations, Sazonov and Krivoshein had convinced several of their colleagues to join them in telling the tsar that their "irreconcilable differences" with the prime minister had made them "lose faith in the possibility that we can usefully still serve you and our homeland."[59] If Goremykin were to remain, they said, they could not, in good conscience, continue to serve under him. As if it were not enough for Nicholas, at the height of the Great Retreat, to have to confront an irate Duma in Russia's first parliamentary crisis, he now had to deal with the threat of mass resignations from his ministers. No Russian emperor had ever faced that danger before. To Russia's misfortune, he dealt with both crises in such a way that he drove away even more of the handful of competent men that remained in his government.

A fortnight after their last bitter confrontation with Goremykin, Nicholas summoned his angry ministers to *Stavka* to settle the conflict between them and their prime minister. By that time, Goremykin had convinced both of his royal defenders that the ministers' lack of confidence in him as prime minister was a direct challenge to the tsar's authority, and Aleksandra, especially, was in a raging mood. "The ministers are rotten," she wrote. Krivoshein, she reported, was "working underhand," while Sazonov was "a pancake" and "a milksop." Other ministers, she insisted, were "insolent," "stupid," "cowards," and "liars," and Nicholas must "be the master & lord" if he hoped to deal with them. "Show y[ou]r fist," she urged. "You are the *Autocrat* & they dare not forget it." On September 16, the day of the showdown at *Stavka*, Aleksandra reminded her husband to "several times comb y[ou]r hair with His [Rasputin's] comb before the sitting of the ministers."[60] "I will show them," Nicholas replied, while Aleksandra hurried to send further instructions and complaints. Unable, or perhaps simply unready to comprehend the full dimensions of the domestic political crisis his empire faced, Nicholas rejected his ministers' pleas and refused to discuss any of the disagreements that had brought them to such a decisive parting of the ways with Goremykin. Anxious to see their conflict in the most simple terms, Nicholas called their criticisms of Goremykin's policies "a ministers' strike" and reprimanded them all sharply.[61]

To their great credit, some of the men who had opposed Goremykin insisted on being heard despite their emperor's displeasure. No matter how well built, any dam, they warned, would eventually burst unless it had outlets to relieve the pressure. It made far more sense, they argued, to provide legitimate outlets for Russian public opinion rather than dam up all opposition indiscriminately until its pent-up force broke loose. "Unless there is cooperation between the government and people," Krivoshein concluded bluntly, "we shall not be able to conquer our enemies." These men spoke wisely, but with no effect. "I told them my opinion sternly to their faces," Nicholas proudly wrote to Aleksandra, while Goremykin smugly told Iakhontov that "everyone surely caught hell from our Sovereign Emperor."[62] Very quickly, Nicholas replaced Krivoshein and several of his allies with more biddable, less able men, although Sazonov still remained at his post. "The reactionary influences around the emperor are becoming stronger every day," Paléologue warned his superiors in Paris. "Under these conditions," the statesman Vladimir Gurko later remarked, "a compromise with the Progressive Bloc was out of the question. A split between government and public was inevitable."[63]

Nicholas dealt with his ministers sternly on September 16 in part because the Duma's closing had already convinced him that public anger was not as intense as they claimed. Thanks in no small measure to Rodzianko, the Duma's closing on September 3 had proceeded without difficulty. When Duma deputies first learned of Goremykin's intention to announce their adjournment, Rodzianko remembered, "feelings ran very high, and some of the speeches were almost revolutionary." With the help of Miliukov, he delayed the final meeting to give raging tempers time to cool,[64] and, only after the better part of a day had passed did he read the imperial decree stating that the Duma was adjourned but would be summoned back into session "not later than November 1915 depending on extraordinary circumstances."[65] At that point, "according to custom," Rodzianko proclaimed: "For the Sovereign Emperor, Hurrah!" Sazonov and his embittered colleagues would have been surprised to hear most of the delegates join in a second "Hurrah!" and then leave the chamber.[66] The entire task that Goremykin's enemies had so feared had taken less than two minutes. However, neither Nicholas nor Goremykin realized how thin that veneer of loyal calm had become, nor did they know that the split Gurko predicted lay just beneath its surface.

If Nicholas and his advisers would not cooperate with public opinion in the Duma, they had already begun to do so in an arena less open to public scrutiny. As a result of a suggestion that Rodzianko had made to Nikolai Nikolacvich at the height of the shell crisis in April 1915, the grand duke had convinced Nicholas to establish what he called a special conference "for strengthening the artillery supplies of the army in the field," that would bring together members of the Duma and a handful of Russia's leading armaments makers to work with representatives from the artillery department and other military agencies. Far more ready to allow experts from the Duma and weapons factories into government chambers, where their meetings were less open to comment and public scrutiny, Nicholas had given the proposal such surprising support that, before the middle of May, Nikolai Nikolaevich had been able to order Russia's War Ministry to use "the utmost energy" in getting discussions underway, warning that "every day this is put off can have the most irreparable consequences."[67]

Clumsily, and in an atmosphere of confrontation, the War Ministry's generals faced Russia's politicians and industrialists, each prepared to believe the worst of the other. Utterly inexperienced in working together, and not yet able to perceive how to cooperate with former antagonists, these men accomplished little at their first meetings. "Alarmism," Miliukov once said, "was part of [Rodzianko's] nature,"[68] and it was that alarmism which now made it difficult for him and his allies to take full advantage of the opportunity for cooperation with the government that his urgings had brought into being. At the same time, the War Ministry's artillery experts overreacted to Rodzianko's criticism, and they also raised some very legitimate objections to the astronomical prices that the men from Russia's private sector seemed willing to pay for weapons and shells.

Certainly, Rodzianko and Petrograd armaments magnate Aleksei Putilov, who sat with him on the special conference, had no hesitation about cost when it came to negotiating contracts for weapons and ammunition. For every three-inch cannon his factories produced after February, Putilov arranged to receive a bonus of two thousand rubles. At about the same time, these men urged the special conference to award Putilov's vast network of armaments enterprises a contract for three million artillery shells at a price that was almost a fifth higher than that bid by other industrialists. On that order alone, Putilov stood to make an extra profit of eighteen million rubles! Not many months later, the special conference's industrialists and Duma representatives proposed that the Artillery Department place an order for 2,500 three-inch cannon with the Russian Joint-Stock Company of Artillery Factories in which a banker who had been invited to take part in their discussions by Rodzianko was a major partner, even though the company had not yet built the factory that was to produce the guns. On one level, at least, there were some hard truths behind War Ministry efforts to brand these men "domestic enemies" of Russia.[69]

Putilov and his allies managed to use the special conference for unscrupulous personal gain that spring and summer partly because Rodzianko had invited none outside the tiny circle of Petrograd's armaments magnates to attend its meetings. Led by Deputy Chairman of the Central War Industries Committee Aleksandr Konovalov, industrialist-publisher Riabushinskii, and a number of Kadet politicians, Moscow's textile magnates and Petrograd's liberal Duma deputies campaigned to enlarge the conference, broaden its authority, and control Putilov's corrupt practices. By mid-August, the Duma, State Council, Council of Ministers, Admiralty, and War Ministry all had agreed to establish three new special conferences to deal with the problems of food supply, fuel shortages, and freight transport. More significant, they proposed a fourth conference—this one to be the most important—to deal with everything that related to national defense. With about forty members that included Duma deputies, government officials, military experts, and influential private citizens, this new Special Conference on National Defense represented the broadest invitation ever issued to the citizens of Russia to take part in planning their nation's defense. It was, as one expert wrote some years ago, "a small parliament" dedicated to Russia's war effort.[70]

General Aleksei Manikovskii, one of Russia's first technocrats, and the War Ministry's leading artillery expert, claimed that the Special Conference on National Defense was too unwieldy to work efficiently, but the fact remains that it combined great political and bureaucratic authority with equally impressive technical expertise. By November 1915, its members reported that Russia's monthly production of three-inch shells had risen to more than a million and a half, and by early 1916 she was producing almost twelve times as many artillery shells in a month as she had at the beginning of the war. The Special Conference matched its achievement in spurring shell production with similar success in producing weapons, so that on the eve of the Revolution of 1917, the Russians actually produced more field guns each month than did either the British or the French.

Such production "miracles" did not free Russia from her dependence upon foreign weapons, nor could she hope to fight the battles of 1916 and 1917 without ammunition shipments from abroad. What the Special Conference on National Defense did accomplish in late 1915 and 1916—and that most impressively—was to increase armaments production so dramatically that foreign supplies became less significant than anyone had anticipated at the end of the Great Retreat, when Nikolai Nikolaevich's soldiers had half a million too few rifles and their artillery (despite the shell shortage) was wearing out faster than it could be replaced.[71] Nor did the Special Conference limit its achievements to improving weapons and ammunition production. As one scholar noted recently, although it "began as an engine for utilization of free enterprise," the Special Conference's struggles to control supplies of labor, metals, fuel, and food across war-torn Russia meant that "it ended by inaugurating the Soviet [planned] economy."[72]

As Russian industrialists and politicians began to work effectively with government officials and military experts, they began to take a more conciliatory tone toward the government in public, and their sense of having at last won a voice in state affairs became even more certain when Nicholas appointed four prominent industrialists, including both Guchkov and Riabushinskii, to the Council of State in the fall of 1915.[73] Moreover, the wave of strikes that broke out in Petrograd and Moscow right after Goremykin recessed the Duma made the men who directed Russia's booming industries especially fearful and all the more anxious to make common cause with the government.[74] In this atmosphere of reasonable reconciliation, the *Zemstvo* Union and the Union of Municipalities held a national congress in Moscow at the beginning of September at which more than three-quarters of the delegates urged compromise with the government. "We must become unified and organized as quickly as possible," Guchkov insisted at a meeting on September 7, his quiet voice conveying all the authority of his long experience at the pinnacle of national politics. "It is not for the purpose of revolution that we urge the state power to grant the demands of society. Rather, it is to strengthen that power in order to defend our homeland from revolution and anarchy."[75]

Behind Guchkov stood an impressive rank of political leaders ready to heed his call to support the government in its moment of crisis in return for greater benefits at a later time. "Each of our wars has ended with the victory of society and the defeat of reaction," Kadet leader Andrei Shingarev explained. "Just remember [the war of] 1812, the Sevastopol campaign [i. e., the Crimean War] and, finally, the Japanese War. Now the decisive moment has arrived," he concluded. "For a brilliant future just one more friendly bit of pressure is needed and then we shall win the very things that the best Russians have always dreamed of."[76] An upright and clear-headed man who had risen from a modest position in Russia's provincial public health service to national prominence, Shingarev believed in moderation, not confrontation, and his "friendly bit of pressure" involved neither direct opposition nor even a firm demand for a responsible ministry. So anxious were he and his allies to set aside past differences and make common cause with the government that they suggested only that "in place of the present government there ought to be summoned men who enjoy the confidence of the people."[77] The congress therefore chose a delegation of six moderate and sober men led by Prince Lvov, as head of the *Zemstvo* Union, to carry their recommendations to the tsar.[78] "We, the chosen representatives of the *zemstva* and municipalities of Russia, are sent to tell you the real truth," their message began. "The people are ready for any sacrifice ... [and] everyone in Russia, down to the very last man, is prepared to offer himself for feats of valor." The state need only "lead the nation to victory [but] it must be made strong and powerful by the nation's confidence and respect." For that, Nicholas must only heed the people's will. "Renew your power," they urged. "Place its heavy burdens upon the shoulders of men made strong by the nation's confidence ... [and] open for your people this singular path to victory. ... In your hands lies the salvation [of Russia]."[79]

Hopeful that the moment had come to establish a true alliance between tsar and people, Lvov and his delegation asked for an audience at *Stavka*. On September 18, an urgent telegram summoned them from Moscow to the headquarters of the Ministry of Internal Affairs in Petrograd, where a very reluctant Prince Shcherbatov sadly told them that the tsar "did not find it possible to receive a deputation of the congress to discuss matters that did not relate directly to the tasks of the *zemstva* and municipal unions." Nicholas had thought their suggestion that his autocratic authority needed to be "made strong and powerful by the nation's confidence and respect" presumptuous. He therefore had instructed Shcherbatov to tell them that "the Sovereign Emperor considers this intrusion into state politics by trying to evade regular government channels very abnormal," and to remind them that they should address any comments or complaints they might have directly to the proper offices in Petrograd.[80]

"Thus the circle had been closed," one writer remarked wryly. "Those complaining about the government were told to address their complaint directly to the government itself."[81] Nicholas had squandered yet another precious opportunity to repair his dynasty's crumbling foundations. Like Goremykin's his views remained the product of an era when simpler problems had called for less complex solutions, and he therefore continued to reject any genuine working relationship with the men who represented the economic might and industrial power of Russia's emerging modern age. Certainly a man of Shcherbatov's honesty and good sense could not easily follow such a course. Before winter's first snows swirled across the Neva's lead-gray waters, Nicholas replaced him with an intriguer who seemed to share his royal master's desire to hold back the sands of time at precisely the moment when Russia needed to move forward more urgently than ever.

Aged forty-three when Nicholas named him minister of internal affairs on September 27, 1915, Aleksei Nikolaevich Khvostov proved to be as ingratiating as he was unscrupulous, and his physical appearance gave ample testimony to his greed in all things. Grossly overweight (Aleksandra thought "his body collosal . . . but his soul light and clear"), his appetite for food was matched only by his lust for power, and he watched his star soar as his waistline swelled. While still in his thirties, Khvostov had been the governor of Vologda and Nizhnii-Novgorod provinces, and he had used those positions to curry favor in the capital, where his talent for playing upon the base desires of any who stood above him served him well. Always confident of his course, he moved smoothly through those salons and anterooms in which men peddled influence and favor. Only once had his step faltered, and then only briefly, when a dirty monk, with steely-blue eyes that cast a hypnotic stare, came to Nizhnii-Novgorod to "look at [his] soul" and spoke to him of becoming Russia's minister of internal affairs. Khvostov had not recognized Rasputin that day and had sent him off with the curtest of farewells. Too late did he realize his mistake, but he moved quickly to make amends. Determined to regain his lost opportunity, he went to St. Petersburg in 1912 as a member of the Duma and spoke very loudly in favor of extreme right-wing causes. Soon, he took up with a pair of notorious schemers, the hack police official Stefan Beletskii and the sinister Prince Mikhail Andronikov, who once described himself as "the Lord God's adjutant."[82]

For many years, Andronikov had dedicated himself to currying favor by being effusively pleasant to people at court, not the least of whom were Rasputin, Vyrubova, and the empress, and he had soon managed to win the fullest confidence of the latter two. In that way, he eventually brought Khvostov to the attention of Vyrubova and Aleksandra. Both thought him a rare find and sang his praises. "He looks upon me as the one to save the situation whilst you are away," Aleksandra wrote delightedly as she urged Nicholas to name Khvostov minister of internal affairs. "Shcherbatov . . . plays fast and loose," she warned. "Please take Khvostov in his place." Aleksandra thought her new discovery might well become the savior to guide Russia into calm waters through the turbulent currents stirred by the Great Retreat. "He is very energetic, fears no one & is colossally devoted to you," she scribbled in a long, disjointed note. A few days later, she explained that "he thinks he can manage to set all to rights." Happy at last to have found a true champion—"a man [who wore] no petticoats," she said—Aleksandra took added heart from Khvostov's promise that he would "do all in his power to stop the attacks upon our Friend [Rasputin]."[83] "I shall see him immediately on the day of my return [to Petrograd]," Nicholas hastened to promise.[84] Ten days later, on September 28, he named Khvostov to replace Shcherbatov.

"This is a person absolutely inexpert in his work, one who by character is entirely unsuitable," the new minister's uncle warned Goremykin. "I expect no good to come from it," he concluded ominously, "and in some ways I expect even harm."[85] The elder Khvostov knew his nephew's failings all too well, but his remark that he was "absolutely inexpert in his work" was an overstatement calculated to add urgency to his warning. Sir Bernard Pares, a perceptive Englishman who spent much of the war years in Russia, was certain that neither Khvostov nor Beletskii, whom Khvostov elevated to the post of deputy minister of internal affairs in charge of the police, were without ability. Ruthlessly in quest of personal gain, even at the cost of their nation's well-being, both men took on the duties of their new offices so energetically that Beletskii's assistant later remembered them as "two volcanoes."[86] If properly exploited, even the most modest success in coping with Russia's monumental problems could bring vast rewards, and both were determined to reap more than their share if any way could be found to do so.

Leaving police affairs to Beletskii, Khvostov turned first to deal with the press, intent upon rescinding the limited freedom to comment upon domestic policy that Shcherbatov had permitted. Proclaiming that "compromises are both necessary and essential," Khvostov summoned several prominent newspaper editors to a meeting, where he assured them that he was "a stern man and therefore [had] . . . no fear of compromises." Compromises, however, turned out to be unyielding dictates that left the press little room for comment or maneuver. "In wartime," Khvostov explained loftily, "general formulas are subject to replacement by real and categorical instructions." There could be no more talk about the Progressive Bloc's "government of confidence," he warned. So long as Russia's armies continued to fight the more deadly battles of guns and bullets at the front, two concerns must stand before all others at home. All Russians must dedicate themselves to liquidating "German domination" and to winning the "war against inflation" that must be waged at home to bring the skyrocketing food prices under control. After all, he assured his listeners, his expansive vest stretched tightly over his yet more expansive girth, "politics depends on the stomach, and one must think about one's stomach before all else."[87]

The theme of "German domination" in Russian industry and society was one that Khvostov had first raised in a demagogic Duma speech soon after he arrived in the capital, and it had become an ongoing feature of the xenophobia and spy mania that gripped Russia throughout the war. With more justification, many Russians feared the influence of German financiers upon their nation's war effort, and there is some evidence to suggest that Rasputin's more unsavory acquaintances in Petrograd's world of international banking had very real German connections that brought German gold from Stockholm into their secret coffers in exchange for favors done or promised.[88] Khvostov had played upon these fears to attack the honest policeman Vladimir Dzhunkovskii in his first Duma speech, and it seems that he did so to curry favor with the empress who, the sinister Andronikov knew, despised the deputy minister of internal affairs who spied on Rasputin and had recently denounced his lewd behavior in a well-documented report to the tsar. That most fears of "German domination" proved groundless did nothing to lessen the hysteria. Countless loyal Russians, including Generals Rennenkampf and Danilov, not to mention the empress herself, continued to be suspect as spies, and these fears grew stronger as the chain of Russian defeats stretched toward the end of 1915.

If the threat of "German domination" often proved more imagined than real, the dangers of inflation and food shortages were more real than most senior statesmen dared to admit during the war's first year. Such comfortable aristocrats could not even begin to envision the crushing impact that soaring prices had upon Russia's lower classes, and they allowed the laws of supply and demand to operate freely at first. By the time Khvostov took office, however, even the picture painted in official reports was so thoroughly grim that few could ignore it. Petrograd needed more than four hundred freight cars of provisions a day, but received just over a hundred at the Great Retreat's end. "Supplying the capital with provisions—at all times a troublesome question—has gotten much worse in the past month," a top-secret secret police report stated at the beginning of winter. "Two-thirds of the meat shops have no meat at all . . . [and] we now see women being turned away from butchers' shops and then going in crowds of five or six hundred to others . . . [where] they curse the butchers, spit on them, and even come to blows." But meat was only one of many necessities in short supply. "During the present year," the same report went on, "more than a third of the bakeries have closed as a consequence of shortages of oil and flour." By December 1915, women had to stand for hours in subzero weather to buy pitifully small quantities of sugar and flour. There were shortages of candles and soap, not to mention kerosene and firewood. Of 659 cities surveyed at the beginning of October, over 500 reported shortages of one sort or another, and over half did not have enough rye or wheat flour, even though a good harvest had just been brought in.[89]

True to his creed that "politics depends on the stomach," Khvostov tried to use Russia's soaring prices and looming shortages to win the masses' loyalty. If hungry workers received food from those very people whom revolutionary agitators continued to urge them to despise, he reasoned, it would weaken the Left's support among the men and women who had grown most obviously disaffected in recent months. Khvostov therefore arranged for Russia's notorious archreactionary Black Hundreds to organize consumer cooperatives that distributed food among the workers in Petrograd's factory districts. Born as a buttress for autocracy amidst the turmoil of the 1905 Revolution, the Black Hundreds had marched forth again and again as champions of tsar, church, and country to wreak vengeance upon any who supported constitutional monarchy, broader political rights, or social justice. Their ranks swelled by small shopkeepers, tradesmen, priests, and minor officials, they had always been enemies of Russia's proletarians. Yet, because of their blind loyalty to the throne, they could be trusted to do Khvostov's bidding as no other group in Russia could, even if it meant aiding men and women they regarded as enemies.

At the same time, as men widely known to despise Russia's Jews, the Black Hundreds could be relied upon to carry out another key element of Khvostov's plan. This called for refocusing the masses' rising anger against Russia's factory owners and government by turning it upon the Jews.[90] In a flurry of orders to all police and local officials, Khvostov accused the Jews of trying to assure Germany's victory by "spreading rumors among the masses that the Russian government is going bankrupt." They, as a single evil group, he insisted, had caused the food shortages that had driven food prices so precipitously upward.[91] But the Black Hundreds' brutality alone could not protect Russia against the threats she faced. To combat the combined dangers of inflation, food shortages, and the threat of "German domination," all of which could be laid at the doorstep of Russia's Jews, Khvostov explained to a group of specially chosen correspondents, the government must place the nation's best interests before everything else. That meant postponing the Duma's next meeting until the new year, and canceling the joint national congress that the *Zemstvo* Union and the Union of Municipalities had planned to hold in Moscow during December to continue their campaign to convince the tsar to bring "men who enjoy the confidence of the people" into the government. Only in this way, he warned, could Russia's government be free to devote all its strength to winning the two-front war against inflation and the armies of Germany and Austria.[92]

Khvostov's grandiose promises to Russia's emperor and empress were one thing, but results lay in another realm entirely. Despite his best efforts, the average cost of staples continued to rise by more than 1 percent every week.[93] But, if he could not actually curb inflation or end the food shortages in Russia's cities, Khvostov enjoyed more success in undermining the opposition's base of popular support. For a few brief months that fall and winter, events worked in the government's favor, as Germany's armies outran the railroads of Poland and halted their offensive. Once it was clear that the Germans were not going to burst into Russia and seize Petrograd and Moscow, the government faced less pressure to compromise, and Khvostov had been quick to exploit that advantage.

By late fall, the urgent sense of impending doom that the Progressive Bloc had tried to exploit in its effort to wrest a "ministry of confidence" from Nicholas and Goremykin in August had faded. "*Then* we could say that the government was leading us to defeat, and we could call for a ministry of public figures," the brilliant Kadet lawyer Vasilii Maklakov remarked in late October,

when the bloc's leaders renewed their meetings at Meller-Zakomelskii's lodgings. "Moscow won't be taken [by the Germans now]," one of his listeners
added, "and that is infinitely more important [in the masses' view] than who will
be minister or when the Duma will be summoned back into session." Like many
of his colleagues, the prominent Moscow liberal jurist-turned-politician Nikolai
Astrov now sensed that public opinion had turned against the bloc. "[Khvos-
tov's] effort to discredit us [with the masses] has succeeded," he stated bluntly.
"The anger of the population has [now] been brought down upon the elective
institutions of government."

Most of Astrov's listeners at Meller-Zakomelskii's apartment agreed that the
time had come to change tactics, yet Khvostov's success at turning public
opinion to his advantage had left them few options. Flatly, Guchkov stated the
problem: despite inflation, soaring prices, and the terrible losses of the Great
Retreat that had crippled Russia's war effort for the time being, they had failed
to force the mediocre and corrupt men who ruled Russia to take them into their
inner councils.[94] Dared they even think of taking their cause into the streets?
Certainly the thought crossed their minds, but each drew back instinctively from
that awesome step. "We have come to a fateful frontier across which constitutional public opinion cannot cross," Astrov warned his colleagues. "We cannot
become revolutionaries."[95] Wryly, but with a sense of irony that cut to the very
heart of the dilemma they faced, Prince Lvov remarked that perhaps the bloc's
new slogan ought to be: "Not struggle, but self-defense."[96]

Certainly, any political group that stood to the left of the reactionary Black
Hundreds needed to think seriously about self-defense as Khvostov continued
to form societies whose names promised commendable social and charitable
action, provide them with funds from his ministry, staff them with staunch
allies, and use them against his opponents. In the Society for Fighting Against
Inflation and in the Russian Society for the Care of Refugees of the Orthodox
Faith, the sinister policeman Beletskii served as a trusted adjutant to make
certain that prominent men from the Black Hundreds could turn its offices into
centers of political support at a moment's notice. At the end of November, these
right-wing organizations met in Petrograd for a national congress in which
Russia's ultrareactionaries sternly warned that "any monarchist who supports
the call for a ministry of public confidence is not a monarchist," and loudly
accused the foundering Progressive Bloc of spreading a "bacchanalia of lies"
across Russia.

The progressives protested to no avail, pointing to their many efforts on
behalf of the war effort. "Do Not Interfere!" one prominent conservative
warned in a stern article that appeared in the right-wing *Moskovskie vedomosti*,
The Moscow Gazette, just before Christmas. Support for Russia's war effort
from such public organizations as the *Zemstvo* Union, these reactionaries insisted, had proved to be of far too little consequence to merit all the publicity
and accolades heaped upon them by the progressive press.[97] That these organi-

zations were known to have engaged in questionable fiscal practices from time to time embellished the negative image that the rightists set out to project, and their case was strengthened still more by the sleazy behavior of tens of thousands of men who served in the ranks of the *Zemstvo* Union's public service organization. By late 1915, these so-called *zem-gussary* (*zemstvo* hussars)—once described as men "whose gaudy uniforms, ultramartial bearing, and assiduous attendance at night clubs . . . aroused amusement not unmixed with irritation" —had become notorious across the length and breadth of Russia as draft evaders whose privileged status spared them the dangers of combat.[98]

By the end of 1915, Nicholas and his advisers took heart from the fading laments with which the Progressive Bloc greeted the Right's mounting assaults and turned again to the question of the Duma. When should it be reconvened? Or, should it be reconvened at all? Its Austrian counterpart had not assembled since the beginning of the war and there were those who noted that Russia might follow that path, especially since the Fourth Duma's normal five-year term was about to expire in somewhat more than a year.* But Nicholas now felt confident enough to seek reconciliation with the Duma—"to round off the sharp corners" he said—rather than reject it utterly.[99] One of the sharpest remaining corners was the aged Goremykin, who still reigned as prime minister and president of the Council of Ministers and whom the progressives still had every reason to criticize as the "most undesirable" of all the ministers left over from the previous summer.[100] Although closely tied to him by long-standing ties of personal friendship, Nicholas knew Goremykin's physical and mental limitations and so did Aleksandra, who now suggested that he "get the old man out" because "the dear old man" could no longer "grasp, alas, everything."[101]

The biting criticisms that Rodzianko leveled against Goremykin in a private letter that he sent to him on New Year's Day (1916) showed the depth of public resentment against the prime minister who had so stubbornly resisted all efforts to bring men who could claim to hold the people's confidence into Russia's government. Rodzianko cursed the "catastrophic condition of rail transport [and] . . . the prospect of hunger that menaces the people of Petrograd and Moscow." Sternly, he warned of "a general disintegration of all aspects of national life," and concluded with an unprecedented demand for Goremykin's resignation: "If you, Ivan Longinovich, do not feel that you have the strength to bear this heavy burden and cannot use all means at your disposal to help our nation emerge on the road to victory, then have the courage to admit the fact

---

*According to the Fundamental Laws of 1906, the tsar was obliged to convene the Duma for annual meetings, although he had the right to dissolve it at any time before its term ended provided he announced the date on which its successor would be summoned. Russia's First (1906) and Second (1907) Dumas had met only briefly before their unyielding opposition caused Nicholas to order their dissolution. As a result of changes in Russia's election laws, the far more conservative Third Duma (1907–1912) served out its entire term. The full term of the Fourth Duma (1912–1917) was cut short by a few months by the outbreak of the February Revolution.

and yield your place to someone of more youth and greater energy."[102]

Clearly, Goremykin's removal from office at the age of seventy-six could strip the bloc of one of its few remaining potent criticisms, and therefore at the end of January Nicholas announced his prime minister's retirement. Paléologue, who confessed that he would "miss the skeptical and malicious old man," thought it a wise move, for he was certain that, even if Goremykin's "powers of observation, criticism, and prudence remain intact," a point about which many disagreed, he nonetheless was "woefully lacking in authority and energy."[103] Members of the Progressive Bloc saw Goremykin's removal as a clear effort at reconciliation by the tsar and, although cautious, reacted warmly. "If a hand is being extended to us, we shall work together," Meller-Zakomelskii declared a few days later.[104] For a brief moment, it seemed that the tsar and the Duma's progressive majority might begin working together in the common cause of victory.

Nicholas's gross inability to judge men destroyed that precious opportunity for cooperation. Just as the hope for a union of tsar and Duma arose, on January 9, the tsar had written that he wanted the announcement of Goremykin's replacement to "come like a clap of thunder,"[105] and his choice undisputably produced that result and more. Widely regarded as an abject servant of Rasputin and a fawning lackey of the empress, Boris Vladimirovich Stürmer was only a decade younger than Goremykin, and a flatterer whom the Council of Ministers' former president Count Kokovtsev once called "an incapable and vain man."[106] Many others shared that estimate of Stürmer's character. "For the last three days, I have been collecting information about the new president of the Council [of Ministers] from various sources," an appalled Paléologue confided to his diary, "and I have no reason to congratulate myself about what I have learned." Paléologue described Stürmer as "worse than a mediocrity, with limited intelligence, mean spirit, low character, questionable honesty, no experience, and no idea of statecraft."[107]

Compared to his predecessors, Stürmer had shockingly little experience in state affairs. Although a former governor of Novgorod and Iaroslav, his highest position of responsibility in the central government during a career of more than forty years had been the directorship of a department in the Ministry of Internal Affairs. Aleksandra thought Stürmer well fitted for the awesome responsibility of ruling Russia's ministers in council nonetheless. "Sweety, are you seriously now thinking about Stürmer?" she asked at the beginning of January. "I do believe its worth risking," she continued, "as one knows what a right man he is." A few days later she added that Nicholas should "think of Stürmer [because] his head is plenty fresh enough," and, most important of all, "he very much values Gregory wh[ich] is a great thing." The next day, she urged him to have Stürmer "quietly come to Headquarters . . . and have a quiet talk together."[108] By the middle of the month, Nicholas had made his choice; Stürmer would preside over the Imperial Council of Ministers although he never had been a minister himself and had managed to sit utterly unnoticed on the State Council for an entire decade.

A man with little experience, no talent, and even less honesty, Stürmer entered his exalted office surrounded by men of the most dark and corrupt connections, including the sinister Manasevich-Manuilov, who became his personal secretary and dutiful envoy to Rasputin.[109] This long-time *agent provocateur* and Okhrana informer, whom Paléologue described—after knowing him for almost twenty years—as "a stool pigeon, spy, sharper, swindler, chiseler, forger and ruffian," who used a "quick and crafty mind" to satisfy his "passion for high living, pleasure, and *objets d'art,*" quickly turned Stürmer's anteroom into a cesspool of influence-peddling and crooked deals.[110] Stürmer not only tolerated Manasevich-Manuilov's behavior but evidently encouraged some of his schemes, while he dutifully made weekly reports to Aleksandra and listened attentively to the rambling advice she gave in meetings that sometimes went on for an hour or more. "Stürmer is honest and excellent," she wrote to Nicholas, as she happily pointed out that he disliked all the people she despised and "completely" believed in Rasputin's "wonderful God-sent wisdom."[111] Was this at last to be the beginning of that "glorious page in your reign & Russian history" that Aleksandra had promised to Nicholas in her letters?[112] Rasputin had recommended Stürmer warmly, and Aleksandra was certain that "our Friend" spoke God's words to them both. "It is not for nothing [that] God sent Him to us," she assured Nicholas, "only we must pay more attention to what He says."[113] Long at odds with their empress's belief in Rasputin, knowledgeable Russians now disputed her faith in Stürmer. Historians have been almost unanimous in thinking that the appointment of this base conniver marked the beginning of the end for the tsar whose tsarina knew so little, yet insisted so urgently that he appoint to key positions in Russia's government selfish, inept, and dangerous mediocrities, whose only talent was to speak convincingly and often of the sanctity and wisdom of Rasputin.

PART THREE

1916

CHAPTER VIII

The Roots of Upheaval

A pall of depressed indifference hung over Petrograd during February and March 1916. For men and women who once had thought Russia capable of great deeds, dreams soured and life seemed to slip further out of focus as Stürmer's corrupt band of schemers replaced—in Guchkov's words—Goremykin's "regime of favorites, sorcerers, and buffoons."[1] Fatalism turned into cynicism, and a longing for the war's end on any terms extinguished the few scattered embers remaining of that stern patriotic resolve with which Russians had met the defeats of 1915. "Oh, if a Zeppelin would come,/and smash the whole of Petrograd!" a popular ditty began.[2] Men and women no longer hoped to shape history or direct the course of events, and many feigned disinterest in what the future held in store. "We do not take defeat amiss,/And victory gives us no delight," an anonymously written rhyme explained. "The source of all our cares is this:/Can we get vodka for tonight?"[3] Expecting the worst, but still uncertain what that might be, many Petrograders waited restlessly for events to overtake them.

On the surface, such indifference created a sense of false calm at the very time when new seeds of discord were about to sprout. That spring, these set their roots firmly and began to flourish in the soil of Russian discontent. This time, the nourishment that helped them grow strong and deep did not come from the educated, articulate men and women of Petrograd and Moscow. Nor did it stem from Duma politicians appalled at the incompetence and gross self-interest they perceived in high places. Disgust and indifference could stir criticism and, if the criticism went unheeded, despair. But this could never truly nourish what Russians call a *perelom*— a sudden and irreparable fracture of the established order—that would turn the nation and its people onto new and uncharted paths. A *perelom* called for deep, abiding outrage at social and economic injustices borne by too many for too long, and this could not stem from the scandal of Stürmer's appointment or the low maneuverings of Rasputin and his sinister friends. These left no mark upon the men and women who labored in Russia's factories. Nor did they touch the lives of the tens of millions of peasants who tilled the vast grain fields of the Ukraine, New Russia, and the *chernozem*— the region of the Black Earth—from which Russians had drawn their sustenance since time immemorial. For the toiling men and women of Russia, life had other concerns that overrode the gross stupidity of scandal and the deep disgust that came from seeing high offices bestowed upon base and unworthy men. In crumbling huts and wretched tenements, the struggle of poor folk for life's necessities continued unabated, made all the more desperate by the shortages and inflation that the Great War had brought to Russia.

For Russia's peasants, too often thought to be the least of the tsar's subjects, winter's approaching end in 1916 signaled a return to life and labor. "The Russian peasant knows for certain . . . that Nature allows him very little time in which to complete his work in the fields," the great nineteenth-century historian Vasilii Kliuchevskii had told those who came to hear his popular "Course of Lectures on Russian History" some years before. "He therefore must work at top speed in order to get a great deal done in the short time [before winter returns]."[4] Every peasant knew that the *ottepel*— the great spring thaw that cleared the snow from Russia's fields and turned her primitive, rut-filled roads into smooth avenues of knee-deep mud—was the signal to begin the season of workdays that stretched from sun-up beyond sunset. Within a few short weeks, fields must be plowed, harrowed, and sown, livestock put out to pasture in the common meadow, and lesser tasks too numerous to count accomplished without delay. Nothing could stand in work's way, for everything not done before the first snows flew must be set aside for another year. So closely were their lives tied to nature's course that the peasants' ways had become as predictable as the seasons themselves.

Not even a conflict of the Great War's dimensions, nor scandals as appalling as those stirred by Stürmer's low followers, could alter the peasants' yearly routine, for centuries of experience had shown them all too often that if the crop failed, they would starve, and, even if the crop proved moderately good, *golodnyi khleb*— the bitter and unwholesome breadlike concoction made from such hardy weeds as goosefoot that peasants used to eke out dwindling grain supplies —was likely to appear on many of their tables long before spring. In the years before the Great War, meat remained a rarity in peasant diets, and many did not taste eggs, milk, and sugar from one month to the next. For such folk, famine always stood only a small step further away than the other scourges of poverty and disease.

Most peasants still lived in cottages which they called *izby,* made of rough logs and crumbling thatch, often without windows or chimneys, and shared with pigs, lambs, and chickens throughout the winter. Hardly more than a decade before the Great War, shocked public health officials—and the physician-turned-Duma-politician Andrei Shingarev stood prominently among them—had learned of entire villages without a single privy, where peasants left their excrement in the *izba's* outer passage in winter to be consumed by foraging fowl and hungry animals. Bedbugs, cockroaches, and beetles carried disease into every corner, and deadly epidemics followed an annual course that seemed nearly as unvarying as the peasants' daily lives. Influenza came with the new year, to be followed by malaria in the spring and cholera in the summer. Fall and winter brought diptheria, and the year closed with a new wave of influenza that carried over into the new year to begin the cycle anew. Pure chance often dictated survival. In this dark world of ignorance, poverty, and disease, life hung in such precarious balance that only a handful of peasants managed to live past fifty.[5]

As they had for a millennium past, Russia's peasants made ready to resume their labors once again in the spring of 1916, but this beginning, so much like those that had come before, was also different, for families not only went to the fields without their menfolk, but they went with the certain knowledge that many of the men who had marched away to war never would return. Russia's peasants now knew the burden of war's grief all too well and were destined to know it better still. During the first thirty months of war, nearly one out of every two peasant men between the ages of nineteen and forty-five had been called to the colors, and so many more had gone to work in war industries that by the time Nicholas lost his throne, more than six out of every ten men of working age no longer tilled Russia's fields. The great fertile grain fields of South Russia felt the loss even more, for less than one man in ten remained to follow the plow that spring. In 1916, for the first time in Russia's history, the peasant crowds who set out to break open the frost-closed fields for spring planting were mostly women. Wives without husbands, mothers without sons, daughters without fathers, and sisters without brothers began anew the age-old cycle upon which Russia's food supply depended.[6]

Russia's villages had not begun to feel the economic consequences of losing almost ten million men until well into the second year of the Great War. Usually the enemy of primitive agricultural societies, overpopulation had worked briefly to the peasants' advantage at the war's beginning, because some villages in Central Russia had more workers than there was land for them to till. The first troop levies of 1914 and early 1915 therefore merely reduced villages' oversupply of farm workers, in many cases without cutting into much of the essential village work force.[7] The departure of the war's first millions for the front therefore brought grief to their loved ones, but caused no serious shortage of labor in their families' and neighbors' fields before the spring of 1916.

From the very first months of the war, Russia's rich peasants and manor lords had not shared the brief good economic fortune of their less prosperous neighbors, for the loss of surplus peasant field hands struck them much more quickly. Because family holdings laid first claim upon the labor of the men not yet taken by the recruit levies of 1914 and 1915, only a sixth as many households had men to spare for hired work, and the agricultural entrepreneurs who grew much of the grain to feed Russia's cities found it difficult to hire the legions of field hands needed to till their fields.[8] Complaints echoed throughout the corridors of Petrograd's chanceries as rich and powerful men cloaked their self-interest with false concern for their nation's welfare and spoke ominously of reduced acreage and impending grain shortages. Government officials reacted to their warnings with uncharacteristic haste, for they feared that crowds of hungry city workers might erupt all too easily into demonstrations, riots, and even revolution. Even though war loans had driven Russia's foreign debt to astronomical levels, Minister of Agriculture Count Aleksei Bobrinskoi, whose long years of dedication to scientific farming had convinced him that the machines offered the only cure for Russian agriculture's ills, begged to order fifty million rubles' worth of farm machinery from the United States to offset the "shortage of agricultural workers" and increase farm production.[9] "The agricultural sector of Russia's economy, comprising as it does the main source of its economic strength and the source of provisions for all civilians as well as the army," Bobrinskoi warned, "is experiencing very serious difficulties as a result of the present war."[10]

Bobrinskoi spoke not for Russia's small peasant farmers but for her great producers of wheat, rye, and sugar beets, and theirs were the complaints and warnings he passed to his colleagues in the Council of Ministers. In a frantic effort to replenish Russia's shrinking supply of hired farm workers, senior officials in Petrograd's Ministry of Agriculture recruited more than 400,000 prisoners of war and moved about 100,000 families of refugees from clogged urban relief centers into villages where they could live from their work in the fields.[11] At the same time, Bobrinskoi convinced Russia's military authorities early in 1916 to send over a quarter of a million militia and reservists into the fields to help plant the spring crops.[12] For a few brief weeks, commanders even released soldiers from the regular army to help with the local sowing, and one general on the southwest front actually postponed all military construction in the army's rear areas "in view of the great importance of getting the planting done on time."[13]

As refugees and prisoners of war set out to help till the fields of Russia's great estates in the spring of 1916, statesmen and politicians took comfort from the hope that food shortages might still be avoided. Total farm acreage had declined by a modest amount to be sure, but Russia's fields actually had produced more bread grains in 1915 than in the years before, and her reserves had risen even faster than her production because almost none of the surplus could be sold abroad. Toward the end of the 1890s, the Polish financier Ivan Bliokh had predicted that any attempt to cut off Russia's exports in wartime would serve only to increase her food reserves, and a number of well-known economists published more sophisticated versions of his conjectures soon after Turkey joined the Central Powers and closed the Straits in October 1914. Unlike much else connected with Russia's war effort, these predictions proved correct, and that fifth of her crops usually marked for export remained for domestic consumption.[14] By the middle of 1916, Russia's estimated grain reserves therefore had doubled to more than sixteen million tons.[15] Clearly, the Russian Empire had grain to spare as the Great War's second year neared its end.

Although grain was plentiful in Russia's villages, a food crisis was in the making nonetheless. When the government had fixed grain prices at low levels at the beginning of the war, Russia's peasants had responded by hoarding their grain in the hope that greater demand from the cities would drive prices higher. As grain began to disappear from Russia's urban markets, fear of famine spread quickly, and even officials from provinces that produced grain surpluses worried about hunger in the fall of 1915. Russia's cities thus began to suffer food shortages in the midst of bountiful harvests. With the specter of urban hunger riots looming darkly on the horizon, nervous senior officials in the Ministry of Agriculture showered low-interest loans upon town and city governments all over Russia to encourage them to lay in larger grain reserves before winter.[16]

This flood of unexpected wealth set off a wave of frantic grain buying all across Russia. Fearful that there would not be enough grain, agents of the *Zemstvo* Union, the Union of Municipalities, and the army all competed to buy as much as possible, and their well-intentioned efforts drove prices higher still. Higher prices heightened fear of famine, as these agents continued to buy grain at any price throughout the winter and into 1916, and their effort created real market shortages that made famine all the more likely. When spring planting began in 1916, Russia actually faced serious food shortages even though 1915's bumper crop had followed a plentiful harvest the year before and all of that grain had remained in Russia. If peasants hoarded grain again that fall, if the harvest in 1916 were poorer than average, or if the empire's overburdened railways continued to haul fewer tons of nonmilitary freight, hunger was certain to strike Russia's cities before winter.[17] Misguided efforts by the government and well-intentioned private citizens to spare Russia from famine had created a tragic situation in which food shortages had become all but inevitable. As an official in the Ekaterinoslav office of the State Bank warned his superiors, "the grain business [now] is taking on the appearance of a struggle between the producers and the government."[18]

Food shortages were not the only dangers officials faced at the beginning of 1916. As the *ottepel* ended, tension between peasants and authorities flared into open conflict. With cash earned from selling a part of their grain reserves to overanxious army quartermasters and city agents, peasants set out to make long-postponed purchases of consumer goods, only to find that the war's demands had driven prices to astronomical levels or removed goods from the market altogether. Even before spring passed into summer, mobs of hundreds

—sometimes thousands—of peasants in the grain-growing provinces of Kiev, Kherson, Tula, Kharkov, and Nizhnii-Novgorod looted shops for goods that had become scarce. At the end of May, riots flared all across the Kuban region of the south, and continued into the middle of June. Only nine times in 1915 had peasants protested against soaring prices and shortages of goods in this manner; by the end of 1916, the number of their protests had risen to 263.[19] Clearly, the patience of Russia's peasants had worn dangerously thin.

While peasant anger smoldered in Russia's far-flung villages, city factory workers, themselves often recent arrivals from the countryside, also protested rising prices and demanded higher wages and better working conditions. For these men and women, urban life was cruel, and their misery was all the greater because they had been drawn to the city by the lure of dreams that almost never came true. Cities throbbed with excitement and smelled of success, a heady draught for rural folk worn by long generations of poverty, their lives deadened by the boredom of existence in villages where time stood still. To men and women from the country, the city promised the fulfillment of hopes, and gave birth to visions of a better life that could not even have been dreamed in Russia's sleepy hamlets. Beyond that, the city's very existence offered hope of a sort that could never be analyzed or even described. "We provincials somehow turn our steps toward Petersburg instinctively," wrote Mikhail Saltykov-Shchedrin, Russia's master satirist and chronicler of nineteenth-century provincial life, in one of his many timeless portraits. "It is as if Petersburg all by itself, with its name, its streets, its fog, rain, and snow, could resolve something or shed light on something."[20]

With their mystical promise of a future that somehow would be brighter than the present, Russia's cities continued to exert an almost magical attraction upon rural folk. Concluded a provincial official at the turn of the century, Russians young and old seemed compelled "to rise with the spring tide and ride the crest of the huge wave moving toward St. Petersburg, Moscow, and other centers."[21] During the Great War, this movement became so vast that, by 1916, armaments factories had drained nearly as many people from the villages as had the army. Moscow's factory work force grew by about a tenth during each year of the war, and Petrograd's burgeoned by more than a fifth. On the average, almost 1,700 new workers came just to those two cities every week, and most of them—almost 200,000 in all—became metalworkers, *metallisty* who had earned the highest wages and had been the most radical ever since the first days of Nicholas's reign.[22]

Not just dreams but hard economic realities had drawn peasants to the cities ever since the onset of Russia's Industrial Revolution, and they continued to do so during the Great War. Ever since the emancipation of 1861, the more daring and desperate had left Russia's hamlets to seek work in her new factories, and these proletarian pioneers had sent back glowing tales of lives filled with excitement, opportunity, and higher wages.[23] Although the first waves of Russia's urban pioneers often went to the city with the idea of returning home once they had saved a certain sum of money, many stayed on, married, raised children, and spent their entire lives in the factories' shadow. Often, they or their families still held claim to fields in their native villages, but many peasant proletarians had never set eyes on this land and knew it only as the birthplace of parents or grandparents because they themselves had been born in the city. Stubbornly, they held onto the land as a hedge against hard times and as an expression of their hope that their fondly cherished dreams of eventually abandoning Russia's sprawling, raucous factory slums and returning to a life of rural peace might still one day come true.

Ties to distant villages also had real practical value for newly arrived proletarians in Russia's cities, for peasants from certain villages, districts, or provinces tended to cluster together and draw new arrivals into their midst. These *zemliaki,* as men and women from the same village or district called themselves, shared a common rural past that bound them together in their less familiar urban present to make city life less fearsome.[24] Many years later, Petr Moiseenko, a former serf who had bravely set out from his small country village to become a weaver in St. Petersburg, remembered how *zemliaki* sheltered him and his brothers during their first days in the city. "All day—from morning till night—we walked about the city," he wrote of their first day in Russia's capital. Anxiously, the young men searched for work, but found none. Because they had "not even a single kopek left," they had no food or shelter and faced the coming night hungry and alone. Moiseenko never forgot how *zemliaki* came to their rescue, fed them, took them in for the night, and schooled them about how to look for work and where to find it.[25]

Moiseenko and his brothers were proletarian pioneers of the 1870s, and it was partly the result of good fortune that they had stumbled upon *zemliaki* in the midst of the anonymous crowds that thronged St. Petersburg's streets. By the turn of the century, however, the networks that drew new arrivals to the doors of city *zemliaki* had become far more refined, and much more effective. *Zemliaki* usually lived in the same section of the city, and those who had been in the city longest used their network of connections with foremen and skilled workmen to find places at their factories for newcomers. More than half of all the men and women in Moscow's Tsindel Cotton Mill came from the province of Riazan, as did one out of every four workers at the Prokhorovs' Trekhgornaia Mill, the city's largest.[26]

As *zemliaki* found work and housing for each other, they banded together into food cooperatives that helped men and women who still received desperately low wages to survive in the urban world of high-priced food and services by purchasing and preparing food in bulk. As in the villages, black bread, buckwheat groats, and *shchi*— a watery soup made from sour pickled cabbage, with bits of potato and meat added on holidays—were the standard fare. Spoiled food often had to be used to save money, and even in the best of times only the poorest grades of meat and vegetables ever made their way into the workers' soup pots. "You'd come in from work and, long before you got to the dining room, the smell would clog your nose," Tamara Dontseva, a worker at the Prokhorov's mill, told the Soviet sociologists who interviewed her after the revolution. "Then they'd pour out that repulsive *shchi,*" she remembered, "and you'd gulp it down as fast as you could and leave." In bitter jest, some workers called their groups "cooperatives for the disposal of spoiled food." Others remembered trying to make a cup of souring milk feed several children for at least two days, and still others recalled the hordes of maggots that infested the always-scarce meat on their tables. But if the quality of the food was poor, and its nutritional value less than adequate—animal protein ranged from a quarter to one and a half ounces each day—the quantity, at least, was substantial. Food cooperatives gave poor men and women more to eat than if they had bought food on their own, and sometimes the men and women who ate together came to share radical ideas and revolutionary dreams.[27]

Russian factory workers' housing at the turn of the century was no better than their diet. Even on the eve of the Great War, some proletarians still slept on their workbenches, although a number of factory owners had built barracks for their workers in an effort to dilute some of the worst stains that soiled the fabric of urban life.[28] In the late 1890s, Sergei Gvozdev, a government factory inspector whose reports are one of the most revealing sources ever compiled about the conditions under which Russia's proletarians lived and worked, warned that "the housing problem at our factories becomes more critical with each passing year." Gvozdev described barracks without heat or light, in which men and women slept in eight-hour shifts on plank platforms that were divided into sections about two feet wide and six feet long and stretched from one end of the building to the other.[29] "These plank platforms were infested with bedbugs and lice," Fëdor Dovedenkov told Soviet sociologists who later questioned him about his life before the revolution in the barracks at the Prokhorovs' mill. "One could have filled up a whole convoy of wagons with the filth and dirt taken from this barracks," his fellow worker Ivan Kukhlëv added. "That's why there were all sorts of vermin crawling everywhere."[30] Others recalled how the walls of the barracks at St. Petersburg's Thornton Factory "were green with mold," with air "so stale that the lamps kept flickering and almost going out."[31]

Here the least fortunate of proletarian children spent the few childhood years that intervened between their birth and the day when they too took their places in the factory. With kitchens and latrines their only playgrounds, these children's chances of surviving childhood stood at less than two out of five.[32] Disease and hunger were their worst enemies, but accidents also took a heavy toll, and every working man or woman could recount tales of terrible disasters that had befallen unwary children. "Just after noon, a four-year-old lad fell into one of the latrines on the fourth floor [of the Prokhorov mill barracks]," a Moscow weekly reported in October 1909. "The boy was dead by the time they managed to pull him out."[33] Others died from burns suffered in kitchens, from diseases brought by rats and vermin, and from consuming poisonous substances. For factory children at the time of the Great War, life at home proved to be nearly as dangerous as it was for their fathers and uncles at the front.

By the time of the Great War, many workers lived in private lodgings only marginally better than factory barracks. In St. Petersburg, workers crowded into cellars never meant for human habitation, with walls that were perpetually damp from water that had drained through mounds of human feces piled in the courtyard. "The rooms are filthy and the walls and ceilings are thick with soot," one contemporary wrote. "Two rows of cots line each room and two men are obliged to sleep in each one."[34] There were reports of a hundred people living in a single large room, and a government survey at the turn of the century had discovered that there was only one sleeping space for about every two and a half workers in the city.[35] During the Great War, it remained common practice for as many as eighty people to share a single kitchen in buildings where every drop of water used for cooking, drinking, or washing had to be drawn in pans or buckets from a spigot in the courtyard or on the street. Even at the end of the Great War, more than two out of every five homes had no sewage systems or supplies of running water, and epidemic diseases that stemmed from polluted water and poor sanitation accounted for almost half of all the deaths in the city.[36] So wretched were workers' living conditions that many workers in Moscow kept their youngest children in the country with relatives rather than subject them to the too-often fatal risks of city life.[37]

Men who worked in the coal mines of the Urals and in South Russia's Donbas lived in even worse circumstances. One observer called the barracks at the Briansk mines dens of "parasites, filth, fetid air, and drunkenness," while another reported in 1913 that the miners still "drank more than factory workers, trying to erase all awareness of their hard and bitter lives with vodka." One crusading journalist described miners' lives underground as being akin to "a sentence of penal servitude, a place of hard labor, danger, and fines [for every minor infraction], while, above ground, there was only bad food and rotten, stinking barracks for these men. For unmarried men," this writer continued, "there was homelessness, and time spent in drunken stupor, with stinking, infected whores, while for those with families there was terrible overcrowding, deprivation, sick and querulous wives, and puny, emaciated children."[38] When the Great War's demands for weapons and ammunition brought more factory jobs to Russia's urban centers under conditions that were better than in the mines, these men marched away in droves, leaving desperate foremen struggling to find enough miners to do the work. As in agriculture, prisoners of war augmented the miners' ranks, and well over 100,000 had been put to work in the distant mines of the Urals by the time of the revolution. To these were added what Russian officials crudely referred to as "yellow labor," imported, with great misgivings on the part of the tsar's advisers, from China and Korea at the beginning of 1916. By the time of the revolution, these two new sources had produced almost half a million workers for Russia's mines and railroad construction gangs. Beyond that, women, who had not been seen in the mines before, now began to work underground, drawn by the promise of better pay than they could earn above ground.[39]

In the mines and the working class districts of Russia's cities, vodka was everywhere, and many of the lower classes drank to excess. Vladimir Giliarovskii, a Moscow writer who made the city's slums his life's study, quickly found that "drunken ten-year-old whores were not a rarity," as these children whom fate and Russian society had treated so cruelly deadened the pain and indignity of their lives with alcohol.[40] Many workers spent as much as a tenth of their meager wages on vodka so raw that it seared the throat and scorched the stomach, and the government's ban on the sale of spirits in wartime had almost no effect on workers' drunkenness.[41] So much vodka was available in Russia's cities during the war that by 1916 peasants had even stopped selling their infamous *samogon* — the equivalent of moonshine—there. "Vodka is seen in cities and villages, in the rear and at the front," the Petrograd Soviet of Workers' and Soldiers' Deputies lamented the next spring in one of its first decrees. "Drunks have appeared on the streets, in railway carriages, in factories, and in barracks."[42] When asked how they spent their leisure hours, more than three-quarters of all the workers surveyed in southern Russia at the beginning of 1915 said they "read" or "held conversations."[43] However, much of the time claimed to be spent in reading and conversation actually was probably given to drinking if the contemporary studies which showed that workers spent up to six times more money on vodka than on books and other cultural pursuits are accurate.[44]

While proletarian men used vodka to ease the pain of daily life, women turned to religion. For centuries, Russia's women had formed the bulwark of village religious life and they now brought their belief in the magical power of Russian Orthodox rituals with them into the factories and mines. Many used their primitive faith to sustain them in the ordeals of urban proletarian life, and some had no other interests beyond work, family, and prayer. Until well after the revolution, female workers in Russia thought religious holidays more important than secular ones, and observed them far more often. On these days, women prayed still more fervently while their men used them as an excuse to guzzle more vodka.[45]

Proletarians who came to work drunk faced greater risks of accident than did those who were sober, but every worker had to contend with the threat of injury and sickness because of the dangerous and unhealthy conditions under which they worked. On the eve of the Great War, proletarians' working conditions were as grim as those under which they lived, and the war did little to better them. Russian employers always had been careless about workers' health and safety, rarely kept accident records, and reported only mishaps so serious that they could not be concealed, while they tried to disclaim any responsibility for them. Just before the turn of the century, observers reported seeing prominently displayed signs in factories that read: "In the event of an accident, the owner and the director of this factory assume no responsibility." During the Great War, even Russia's limited factory regulations made such blatant disclaimers of responsibility impossible, but frequent injuries remained the rule, for even the best designed safety systems could not protect drunken workers. The coal mines of the Donbas continued to suffer one of the highest accidental death rates in all of Europe, while most other industries continued to report more frequent accidents than did their European counterparts. On the eve of the Great War, Petrograd's factories alone reported over fourteen thousand accidents a year.[46]

Desperate to supply Russia's armies, statesmen and senior officials listened sympathetically to factory owners' pleas for longer working days, reduced safety regulations, and relaxed child and female labor laws, while they did little to compensate workers for added hardships. Even though Russian factory workers had worked twelve to fourteen hours a day before the war, managers now imposed compulsory overtime that lengthened work days by as much as a half. In Moscow, printers, tailors, seamstresses, woodworkers, and workers who processed animal by-products all saw their real wages decline significantly between 1914 and 1916. The *metallisty*, leather workers, chemical workers, and men in the building trades saw their real wages rise somewhat, but such modest increases did little to offset the greater difficulties in obtaining life's necessities for proletarians who had to stand longer hours in lines to purchase fewer and poorer quality goods and food at higher prices.[47]

In response to the army's desperate calls for more guns, ammunition, supplies, and men, Russian factory owners hired more workers to manufacture the goods of war, and the number of men and women who worked with metal rose well beyond a million before the end of 1916. Although mainly a textile center, Moscow alone added more than forty thousand *metallisty* to her labor force before the end of 1916, and in Petrograd, where the great Putilov and Obukhov works manufactured three out of every five field guns produced in Russia during the war, the number was four times greater.[48] Russia's armaments factories in particular suffered a chronic shortage of workers, and factory managers left few stones unturned in their unending search to fill empty places in production lines. Skilled workers evacuated from the industrial centers of Poland and western Russia during the Great Retreat provided one source, but their numbers were far too small to replace the men taken by conscription, even though military authorities had exempted more than two million skilled workers from the draft during the war's first two years.

Employers then turned to city women whose husbands had been mobilized, war widows, and adolescents from urban proletarian families.[49] When these still could not fill their needs, they looked to the countryside where peasant women, their ties to life in the village weakened by the loss of so many of their menfolk, stood ready to begin new lives in the cities. More than 100,000 women between the prime working ages of fifteen and twenty-four moved to Moscow alone during the war years, and another 70,000 (one in ten of whom were war widows) went to Petrograd. In Petrograd, the number of women metalworkers increased by more than 600 percent, so that one out of every five *metallisty* was a woman when revolution broke out in February 1917. In 1914, there had been sixty-three women for every hundred men in Russia's factory labor force. By 1916, there were seventy-eight.[50]

Unskilled men and women could not be trained for high-precision work in a few short days or weeks, and the proportion of skilled workers to unskilled mill hands therefore dropped precipitously. In turn, productivity plummeted from a level that already stood well below that of other industrialized European nations. "Everywhere in Russia, in factories as in offices, little work is gotten through in the day," lamented a foreign engineer who worked in Moscow during the Great War.[51] Unenergetic prisoners of war workers only made matters worse. The number of workers in the Donbas coal mines rose by 75 percent between 1914 and late 1916, but annual production per worker dropped by almost a fifth. Production slipped by almost 10 percent in the iron mines of Krivoi Rog, while the output of each worker in the iron smelting industry fell by more than 40 percent, and that of the oilfields at Baku plummeted by a disastrous 65. Even in munitions factories, productivity dropped precipitously. Production in some sections of the giant Putilov Works fell so sharply that by January 1915, Okhrana agents became convinced that German secret agents and revolutionaries had found some way to slow Russia's production of cannon and artillery shells.[52] By the beginning of 1916, more workers worked longer hours to produce fewer weapons, ammunition, and supplies than ever before. Only mass hirings that bloated the industrial work force far beyond its earlier dimensions made it possible for factories to increase their total output.

More workers crowded into Russia's cities meant more severe shortages of food, housing, fuel, and medical services, and a second winter of cold, hunger, longer hours, and poorer working conditions left Russia's proletarians in a sullen, angry mood.[53] Crowded together in large factories, they could express their anger more readily than could peasants scattered across Russia's vast countryside. In Petrograd, nearly four out of every five factory hands worked in the company of at least five hundred other workers, and, in some factories, the work force stood at many times that number.[54] The gigantic Putilov Works employed over thirty thousand workers, most of them *metallisty*, and at least four other weapons and munitions factories—the Fuse Works, the Cartridge Works, the Obukhov Works, and the Okhta Explosive Works—each had over ten thousand hands. Over eighty thousand men and women, many of them newly-arrived from peasant villages, and most of them single and between the ages of fifteen and thirty, worked in just these five enterprises alone. They had less to lose than did older men and women with families to worry about, and they quickly became prime targets for revolutionary agitators.[55]

Once it became clear to Russia's proletarians that their war effort, like that of their village breathren, would demand much sacrifice and bring little reward, they returned to the labor protest they had abandoned at the beginning of the war. More than 100,000 workers went on strike in September 1915, and as the number of strikers swelled, bloody confrontations with the police became more frequent. That month, soldiers shot nearly a hundred workers who had joined a crowd in front of the town hall in the mill town of Ivanovo-Voznesensk to shout "Down With the War!"[56] After that, hardly a day passed during the next year that did not find well over a thousand workers on strike somewhere in Russia, as the elite *metallisty*, long known for their willingness to join labor protest, led the rising wave of discontent that would cost Russia's war effort more than four and a half million lost working days before fall came again. By then, Russia had seen the first incidents in which soldiers joined strikers rather than shoot them. On October 17, 1916, the eleventh anniversary of the October Manifesto by which Nicholas had conceded a constitutional government to Russia, the soldiers of the 181st Infantry Regiment joined a crowd of almost thirty thousand workers at the Finland Station in singing the *Marseillaise*. In the protests and countermeasures that followed, Petrograd's workers sent no weapons, ammunition, or supplies to the front for two weeks. Given no other choice, the government had to back down and meet their demand to release more than a hundred men and women being held on political charges. With the *metallisty* playing a leading part, Petrograd's workers had won their first victory over the government since 1905.[57]

Not only was the political consciousness of the *metallisty* the most highly developed among Russia's workers, but they were among the most literate. Nearly nine out of every ten could read and write, and most had spent three or four years in school.[58] Generally, younger *metallisty* were more literate than their elders, and this meant that the young and unmarried men and women most likely to take part in labor protest also were the ones who had the education to develop their revolutionary consciousness to the highest degree. These men and women more clearly understood the world in which they lived than did their illiterate peasant breathren, and they had a better sense of their place in it. Anxious to share the new horizons that their reading opened, they organized clubs to discuss their new ideas, and these gave them a sharper sense of identity as an elite group during the years before the Great War.[59]

In an age when the printed word remained the only means of mass communication, Russian working men and women who knew how to read and write could enter a vibrant world of information and ideas from which illiterates were excluded. Because time and space posed no barriers in the world of the printed word, they could extend their horizons as far as their reading and dreams could reach, and develop a sense of unity with other workers in Russia and abroad. During the late nineteenth and early twentieth centuries, such proletarians began to frequent the handful of city school libraries that opened their doors to nonstudents, as well as the scattered public libraries (about one for each hundred thousand residents in Petrograd) in city working-class districts. At first they read only to escape the world around them, pored over such scandal sheets as *Gazeta-kopeika, The Kopek Newspaper,* and lost themselves in adventure tales or cheap romantic novels. Many never pursued their interests any further, but some went on to read the great works of Dostoevskii, Tolstoi, and Turgenev, tried to study political economy and natural sciences, and took courses at night schools.[60] "When I came in from work, I did not lie down to sleep immediately," a weaver from Ivanovo recalled after the revolution. "Instead, I picked up a book, lit a candle that I had bought with my savings, and read until I no longer could keep my eyes open." At first, he read tales of adventure and scandal, but soon turned to the writings of men who described a world of equality and social justice utterly unlike the one in which he labored. Enthralled by this new vision, this young worker dedicated himself to bringing it to Russia. Books helped to prepare him and showed him the way. Most important of all, he later wrote, "books taught me how to think."[61]

Nurtured by the attentions of revolutionary agitators who began to court them in earnest in the 1890s, such workers found it easy to dedicate themselves to the revolutionary cause, and their numbers grew quickly during the first years of the twentieth century. The Revolution of 1905 tested them and gave them their first taste of revolutionary politics, but, as the tsar's police broke their ranks apart in the aftermath of the December uprising, few remained free to carry on the work of revolution after the year's end. Proclaiming that "I believe in Russia's brilliant future with all my heart," the forceful Petr Stolypin became prime minister early in 1906, as the revolutionary movement of 1905 shattered against the strengthening forces of order. Straightforward and unambiguous in his political principles, he dealt severely with Russia's revolutionaries. "What you want are great upheavals," he told them scornfully, "but what *we* want is a GREAT RUSSIA!" Sternly, he ordered the Okhrana into action, and decimated the revolutionaries ranks with an unprecedented series of raids that sent most of their leaders and many of their followers to prison, Siberia, or foreign exile.[62]

A bearlike man of forty-four who had little patience with the slow workings of Petersburg's bureaucracy, Stolypin had moved resolutely to drive his nation into the twentieth century even before he finished dealing with Russia's revolutionaries. The key, he insisted, must be the best of Russia's peasants, and he set out to transform the most energetic and able among them into a new class of prosperous private farmers by allowing them to assemble the scattered strips of land they had used as members of Russia's peasant communes into consolidated farms that they owned outright. Insisting that Russia's hope lay in "the wise and strong, not the drunken and weak" among the peasantry, Stolypin therefore set out to stem the flood of revolutionary propaganda in hundreds of thousands of villages all across Russia by giving Russia's peasants a stake in the existing order.

"This is our great task," he announced, "to create a strong private property owner as the most reliable bulwark of the state and our culture.[63]

Between 1906 and 1911, Stolypin's reform programs transferred the title to tens of millions of scattered strips of land to millions of peasants all across the empire as he declared war on the peasant commune that had for so long held the land, discouraged initiative, and penalized enterprise among Russia's peasant farmers. In doing so, he broke with the combined forces of tradition and the wisdom of some of Russia's leading government officials who continued to insist that the commune must be preserved to buttress their government's power throughout the empire. Stolypin's first efforts thus had promising beginnings, but he was not destined to see their results. His political success had long since made him a target for terrorists, and his daughter had suffered two broken legs when his revolutionary enemies had blown up his summer house outside St. Petersburg early in his term as prime minister. In the fall of 1911, his enemies struck again. On the evening of September 1, Dmitrii Bogrov, a terrorist in the pay of the Okhrana, shot Stolypin while he attended a performance of Rimskii-Korsakov's *Tale of Tsar Saltan* at the Kiev Opera House.

"We had just left the box," Nicholas wrote to his mother a few days later, "when we heard two sounds as if something had been dropped. I thought an opera glass might have fallen on somebody's head, and ran back into the box to look." As he looked into the orchestra, the tsar's polite concern for a theater-goer's discomfort turned to horror. Before him stood Stolypin. Blood had already begun to stain the watered blue silk sash that draped his great barrel chest from left to right, but, for a brief moment, the prime minister's hand obscured it from Nicholas's view. When he saw his tsar in the box above him, Stolypin raised his hand and made the sign of the cross. "Only then did I notice that he was very pale and that his right hand and uniform were bloodstained," Nicholas remembered. At that moment, four days of hopeless agony stood between Stolypin and death. Before that came to pass, his assassin died mysteriously in his jail cell, before a full investigation could be made of the crime. Because Russia's high-ranking reactionaries had come to oppose Stolypin's progressive policies with a passion only slightly less than the hatred of his antirevolutionary policies borne by his enemies on the far left, a number of contemporaries speculated that the Okhrana itself may have helped to bring about his murder.[64]

Stolypin's wager upon the "wise and the strong" began a surge of agricultural production that brought Russia's farm output in 1914 to a level that would not be reached again until the 1960s. At the same time, only a comparative handful of radicalized workers remained at their looms and benches in Russia's factories as the average workers' enthusiasm for revolutionary visions faded, and almost none seemed willing to join new battles against factory owners and the government. A mere handful of workers went on strike in the months after Stolypin's death. "It would be empty and stupid democratic phrasemongering to say that the success of such a policy [as Stolypin's] in Russia is impossible," Lenin warned his Bolsheviks. "It *is* possible!"[65] For the revolutionaries the future looked decidedly bleak.

As so often happens in history, however, a single chance event transformed workers' apathy into rage and gave new life to Russia's withering revolutionary movement. In April 1912, soldiers and police shot more than two hundred workers who had assembled to protest the terrible working conditions in Siberia's Lena goldfields. Unrepentant, Minister of Internal Affairs Aleksandr Makarov then announced that soldiers and police would continue to fire at any crowd that ignored orders to halt. Arrogantly, he concluded: "That is how it has been and that is how it will be in the future."[66] Makarov's apparent callousness galvanized workers into action. During the next two years, a new wave of strikes, in which workers' political demands overshadowed their economic concerns, swept over Russia. Once again, the glory of labor protest seized workers' imaginations so that, on the eve of the Great War, more of the men and women who worked in Russia's mines and factories began to choose the revolutionary path than ever before.

Many of these aspiring proletarian revolutionaries had tried to live by a new and rigid morality ever since the first radicals appeared in Russia's mills in the 1870s. "We . . . were of the opinion that a good socialist ought not to drink vodka," the *metallist* Ivan Babushkin recalled. "We insisted that a socialist must stand as the best example to people in all things, and we tried to set such examples ourselves."[67] This puritanical code of ethics was the first thing to set worker radicals apart from their fellow workers. Whether employed as *metallisty* or engaged in some other trade, these radical men and women as a group were usually more highly skilled, more literate, and better paid, and these qualities, which brought them the respect of their fellows, tended to set them apart even more. Well before the Great War, an elite had emerged from the rank and file of Russia's factory workers, and that elite would play a central role in revolutionary wartime protest.

If elite worker-proletarians shared a stern morality, a heightened revolutionary consciousness, and a hatred for their employers, they agreed about little else. Contention and strife had often cut deep channels through the ranks of workers' social democracy all across Europe during the nineteenth century as its members had debated whether to work for political power and economic and social justice within the framework of existing political systems or to overthrow them and create others in their place. Even more than its European counterparts, social democracy's Russian version, the Russian Social Democratic Workers' Party— abbreviated as RSDRP in Russian—was scarred by schism and conflict almost from its birth. Should uncontrolled spontaneous acts continue to dominate Russia's revolutionary movement? the first Russian Marxists asked. Or should the spontaneity that had been so much a part of Russia's early revolutionary history now be suppressed in favor of fostering disciplined revolutionary consciousness?

Always, Russia's revolutionaries had looked to the peasants as the bearers of the nation's revolutionary standards, and, in the days of the great Cossack peasant wars of the seventeenth and eighteenth centuries, that vision had held true. But such wise autocrats as Catherine the Great and her grandsons Alexander I and Nicholas I had transformed the peasants' Cossack leaders into defenders of the established order by bestowing special rank and privilege upon them, with the result that Russia's peasants had remained passive onlookers when revolutionary populist agitators had tried to stir them to action in the decades after the Emancipation of 1861. Was it now time for revolutionaries to turn away from the peasants and concentrate their efforts upon the urban proletariat that Russia's industrial revolution had brought into being? Russia's early Marxists thought that time had come, but few could agree on anything beyond that single premise. Who ought to lead and who ought to follow? Must a party of working men and women be organized from the bottom up or from the top down? How should its members be chosen? With their passion for seeking absolute answers to impossible-to-answer questions, Russian social democrats debated these issues violently and found themselves in complete disagreement. Wracked by dissension and plagued by arrests that condemned some of their leaders to prison or exile, they tried to form a national organization as the nineteenth century drew to a close, only to have their effort shattered by their leaders' refusal to compromise.

Nine delegates attended the first RSDRP national "congress" on March 1, 1898, in the West Russian city of Minsk, but social democracy's leading Russian theoretician, the peasant populist-turned-Marxist Georgii Plekhanov, could preside only from his distant exile in Switzerland, where he had spent the better part of two decades, while its two most dynamic younger leaders, Iulii Tsederbaum (later to lead the Mensheviks under the revolutionary name of Martov) and Vladimir Ulianov (soon to become the most famous of all Russian revolutionaries as leader of the Bolsheviks, under the name of Lenin) both were in exile in Siberia. Neither Lenin nor Martov therefore could exercise much influence upon the birth of the workers' organization in which they were destined to play such a major part and which their conflicts would split into Bolshevik and Menshevik factions within half a decade.

Although both had joined the Russian revolutionary movement early in the 1890s, Martov and Lenin had come from very different backgrounds and had chosen the revolutionary path for very different reasons. Born into a prosperous Odessa Jewish family in 1873, Martov had come to know the raw terror of Russians' anti-Semitism at the tender age of seven when a particularly violent wave of pogroms had raged through Russia's southern cities in the months after Alexander II's assassination. His family's wealth had spared Martov from injury, but he could not escape the knowledge of the horrors that less fortunate Jews suffered during those terrible months. Long before he entered St. Petersburg University in 1891, Martov knew the cruel fate that awaited Jews in the Russia of Alexander III, and he had concluded that his people could not hope for freedom in a nation where hundreds of thousands suffered for having been born Jews, and tens of millions more suffered nearly as much for having been born peasants and workers. Passionately dedicated to winning freedom and a better life for all who bore the scourges of poverty and persecution, Martov threw himself into the revolutionary movement. Arrested and expelled from the university within months, he was arrested again and sent to one of Siberia's most remote regions at the beginning of 1896.[68]

While Martov had lived his childhood and adolescence in the fearsome shadows cast by Russian anti-Semitism, Lenin had known only security and comfort. Born in 1870, the second son of a provincial schoolteacher who had been ennobled for his service as an inspector of schools in the province of Simbirsk, Vladimir Ilich Ulianov suffered none of Martov's childhood torments and knew no fears for the safety of those he loved. Simbirsk, the provincial capital in which Vladimir and his five brothers and sisters grew up, was in those days the most backward of the provincial backwaters that lined the Volga river. It had no railroad and no university such as Kazan boasted, was younger and poorer than such bustling entrepôts as Nizhnii-Novgorod and Astrakhan, and could not even rival Samara or Saratov, its sister cities to the immediate south, as a river port. Its history thus far had been bathed in stagnant obscurity—its peasants too poor to rebel and its nobles too poor and ignorant to create any notable cultural life—and so it would remain until 1960, when it burst renamed as Ulianovsk into the world's headlines, the place where one of America's high-flying U-2 pilots had been shot down by a Soviet guided missile.

Surprisingly, between 1866 and 1881, Simbirsk gave birth to three men who played a central role in the Revolution of 1917. Born the earliest and of highest birth, Aleksandr Protopopov's ineptitude as Imperial Russia's last Minister of Internal Affairs helped to bring on the first Russian revolution in February 1917. Born last, the son of the director of Simbirsk's high school, Aleksandr Kerenskii grew up to become a crucial link in the chain that connected the revolution of February with that of October. Born between them, young Vladimir Ulianov seemed destined to lead a comfortable, conservative, uneventful life as the son of a successful civil servant and dedicated churchman, who held a noteworthy place among Simbirsk's singularly unnoteworthy citizens.

Then, in just a few months, Ulianov's comfortably predictable world crumbled, and his life turned onto the path that transformed him into Lenin. Just months after their father's death, Lenin's older brother Aleksandr joined a plot to assassinate Alexander III, was arrested, and hanged in May 1887. Unable to face the disdain of her Simbirsk friends, Lenin's recently widowed mother moved her family to a tiny village where her seventeen-year-old second son began the ideological journey that eventually carried him to the writings of Karl Marx, and the political journey that brought him to St. Petersburg in 1893. Within two years, he and Martov had begun to recruit St. Petersburg's workers into the ranks of Russian social democracy but, in December 1895, the police arrested Lenin and forty of his comrades. As it had for Martov, arrest for Lenin meant a sentence of three years in Siberia. Thanks to his mother's tireless pleading with the authorities, however, Lenin spent the time in a more temperate part of Siberia and under far less trying conditions than Martov had to endure.[69]

With Plekhanov in Switzerland, Lenin and Martov in Siberia, and some five hundred of its most active members (including the rest of its Central Committee) in prison, the RSDRP found itself dangerously adrift as the "heresies" of Economism and Revisionism, cursed by Plekhanov as "dirt, triviality, and stupidity," threatened to drive it away from Marxist orthodoxy onto dangerous ideological shoals as the nineteenth century drew to a close.[70] The former insisted that the economic struggle must take precedence over all political concerns, while the latter, based on the revisionist views of the German Social Democratic Party leader Eduard Bernstein, condemned the Social Democrats' most cherished precepts—the labor theory of value, the iron law of wages, and the notion of class struggle—and argued that "the transition from capitalism to socialism would take the form of a gradual socialization of capitalist society."[71]

Still in Siberian exile, Lenin tried to provide Russian Social Democrats with the theoretical weapons needed to combat these heresies. The champions of Revisionism and Economism must not be allowed to "peddle old bourgeois ideas under a new label," he insisted, for "the fundamental task of social democracy" must continue to be "to instill socialist ideas and political self-consciousness in the mass of the proletariat and to organize a revolutionary party inseparably linked to the spontaneous workers' movement." Revolutionaries must be single-minded, dedicated, and stern. Grandly, he announced: "We must prepare men and women who will dedicate not merely free evenings, but their entire lives, to the revolution."[72] Men and women prepared to dedicate their entire lives to the revolutionary cause would become the backbone of Lenin's Bolsheviks and be guided by his unyielding precepts. But must those whose personal commitment was not so deep, or whose dedication was not so single-minded, be excluded from the revolutionary cause? On this vital point, Lenin and Martov parted company for all time and split the RSDRP into two irreconcilable halves.

Lenin and Martov clashed about how the RSDRP should be organized, what its tasks should be, and what role it should play in the future of Russian social democracy. Always, Lenin's answers remained straightforward and uncompromising. "The task of social democracy must center upon diverting the workers' movement from the spontaneous strivings of trade-unionism under the wing of the bourgeoisie, and bring it under the protection of social democracy," he explained in *What Is to Be Done?*, the famous pamphlet that he published in Stuttgart in March 1902. The Party, "a strong organization of professional revolutionaries," must lead the way, direct workers' strivings in the proper direction, and merge *"into a single whole* the spontaneous destructive force of the crowd and the conscious destructive power of the revolutionaries' organization."[73] In the vision of the RSDRP that Lenin developed, as an exile in Switzerland, Germany, and England, there could be no place for any who hesitated to dedicate themselves totally to the revolutionary struggle. Nor could there be any who were not prepared to follow the Party's dictates without question. Martov, on the other hand, argued that the Party's organization must be looser, its decision-making more democratic, and its requirements for membership less uncompromising. Such disagreements could not go unresolved. The Party itself, both men demanded, must decide once and for all which course to follow when the Second Congress of the RSDRP assembled in Brussels during the summer of 1903.

With Plekhanov in his self-appointed role as sage and teacher, the forty-three delegates to the RSDRP's Second Congress assembled in an abandoned flour warehouse in the Belgian capital on July 30, 1903. From the beginning, there was an air of supreme detachment about their debates, in part because some of the leading figures in the RSDRP had only the vaguest sense about what life in Russia was like. Forced to flee from St. Petersburg and Moscow in 1880 in order to escape arrest for his revolutionary activities, Plekhanov had founded the first Russian Marxist party, the five-member Emancipation of Labor Group,* three years later in Switzerland to become the "father of Russian Marxism" at the age of twenty-seven. Almost completely cut off from Russia during the 1880s, he therefore had concentrated on Marxist theory, not its revolutionary applications, and his writings and speeches always remained vague about how such precepts, the truth of which he believed so absolutely, could be applied to Russian conditions. Plekhanov thus was destined to live out his entire life arguing questions of revolutionary theory, his ironic smile often masking an amused contempt for opponents less versed in Marxism's complexity than he. Now approaching fifty, his somber countenance and sad eyes set off by a neatly trimmed Vandyke beard that was just beginning to show its first streaks of gray, Plekhanov had not set foot in his homeland for twenty-three years, while Vera Zasulich, who had grown to a matronly middle age in exile, had not been to Russia for a quarter of a century.

Points of theory, not bolder questions of revolutionary practice, therefore held the deputies' attention during the first days of the Second Congress. Their numbers still were few, and they had only a fragmented organization in Russia, yet their most intense debates centered upon the regime they would establish after their revolutionary victory. Could they permit a democratically elected constituent assembly to have a voice in determining Russia's course after the revolution? If so, how could that be squared with Lenin's and Plekhanov's

---

*In addition to Plekhanov, the Emancipation of Labor Group included Pavel Akselrod (his close friend in exile and companion in his conversion to Marx), Vera Zasulich (one of Russia's first female terrorists who had shot St. Petersburg's governor-general in 1878), Lev Deich (who had been instrumental in instigating a rare peasant protest in the province of Kiev in 1877), and a young man by the name of Ignatov who died soon afterward from tuberculosis.

insistence that "the essential condition for the social revolution is the dictatorship of the proletariat"? Defense of the revolution must always be first and
foremost, Plekhanov told them in reply. *"The health of the revolution,"* he
insisted, *"is the supreme law."*[74]

While Plekhanov reigned as the RSDRP's master theoretician, Lenin and
Martov vied to become its commanders in the field. Both defined their positions
differently and neither would permit any compromise. For more than a week,
both men jockeyed for position, carrying their intrigues into all-night cafés and
private meetings. When the Belgian authorities succumbed to pressures from
the Russian government and expelled the entire congress after ten days of
meetings, the squabbling delegates crossed the English Channel to resume their
meetings in a London church where Martov urged the delegates to allow anyone
sympathetic to the goals of social democracy to have a place in a party whose
looser organization would permit everyone a voice in its decisions, while Lenin
once again explained his plans to make the RSDRP a party of dedicated, highly
disciplined, professional revolutionaries. "Our task is to protect the steadfast-
ness, firmness, and purity of our Party," he insisted. "We must strive to raise
the title and importance of a party member higher and higher."[75]

Martov warned his comrades that Lenin wanted a party that would be
"divided into those who sit and those who are sat upon."[76] At first, a slim
majority heeded his words and took his side. Then, an angry Lenin for a brief
moment seized control of the Party's Central Committee and the editorial board
of the Party newspaper *Iokra, The Spark,* only to have his victory slip away the
next instant. Yet it was from this momentary victory that Lenin began to call
his followers Bolsheviks (those in the majority), while he dubbed those who had
supported Martov the Mensheviks (those in the minority), even though they
always outnumbered his Bolsheviks right down to the moment when he seized
power in October 1917. Few realized at the time that the rift between Lenin
and Martov had split Russian social democracy so irreparably, and many refused
to believe Lenin's declaration that there was "absolutely no hope for peace."[77]
Rigidly and sternly insistent that revolutionaries must be a carefully chosen elite,
Lenin eventually profited most, for the split, and his uncompromising attitude
about any reconciliation with his former comrades, made him the central figure
in Russian social democracy. "[The name of] Lenin rings all the time in your
ears," one of Martov's revolutionary allies exclaimed at the beginning of 1905.
"Whatever one may say to you, you can only comprehend it in terms of *pro* or
contra Lenin."[78]

Neither the opportunity to put decades of revolutionary theory into practice
in 1905 nor the common threat they faced from Stolypin and the Okhrana in
the wake of that revolution's failure could heal the breach between Bolsheviks
and Mensheviks, and they remained bitterly at odds even within the broader
framework of the international socialist movement. In the spring of 1915,
Martov urged European socialists to convoke an international conference at

which deputies from all nations could form an international peace movement. The conference met at Zimmerwald that September only to find Martov and Lenin now in conflict about how socialists ought to work to end the war. As one of his close friends later remembered, Martov's main aim was "the rallying of the proletariat and the laboring masses in general under the banner of a struggle for the soonest possible liquidation of the 'imperialist' war and the conclusion of a general peace 'without annexations or indemnities, on the basis of the self-determination of nations.' "[79]

In April 1916, the Zimmerwald delegates met again at Kienthal where Martov insisted that a revived international social democratic movement must work "to end the war, to prevent annexations by the victors, [and] to democratize their governments and military forces."[80] This was the only way, he said, to prevent a "shameful war" from ending with a "shameful peace," for all belligerents were equally imperialist, since all were "carriers of grabbing, imperialist tendencies."[81] Postwar governments must have new values and new aspirations. "The new era in the history of the labor movement," he wrote, "must open with the struggle for peace [and] the erection of a proletarian front against all belligerent states."[82]

In sharp contrast to Martov's efforts to construct the broadest front of socialist opposition to the war, Lenin led a small minority of delegates at Zimmerwald who insisted that the primary objective of international socialism ought not to be peace but "the transformation of the imperialist war into a civil war" in order to bring about the workers' seizure of power in all belligerent countries.[83] As with his position about Party membership, organization, and discipline, Lenin's views remained more uncompromising than his opponents. In this case, his position also proved to be more advantageous. "This war position of Lenin's," his Menshevik adversary Fëdor Dan noted ruefully, "had the advantage of allowing him to make his central agitational slogan an immediate, though 'separate,' exit of Russia from the war."[84] Committed to the view that the world socialist revolution would take the form of a series of brief civil wars in Europe's imperialist countries, that would hand power over to "an international staff of "self-restrained and trained' revolutionary socialists,"[85] Lenin could look for a rapid transformation from capitalism to socialism without the long intermediate transitional stage that Plekhanov and Martov expected.

Scattered across the globe, and divided on questions of membership, organization, tactics, and how best to achieve peace, Russian Social Democrats found it difficult to direct the mounting wave of labor protest that arose in their homeland early in 1916. Lenin, Martov, and Plekhanov all were in Switzerland, while the thirty-seven-year-old Iosif Dzhugashvili, now known to his comrades as Stalin, had been sent to spend most of the war years on the edge of the Arctic Circle in far-away Turukhansk after escaping from less remote places of exile twice in a single year. Lev Bronshtein, better known by his revolutionary name of Trotskii, and, like Martov, a Jew driven into the revolutionary movement in response to tsarist persecution, was about to join a handful of Bolshevik comrades in the United States after being driven from Switzerland, France, England, and Spain in little more than two years.

Such men could not even hope to sense the true pulse of the Russian proletariat across the thousands of miles that separated them. In those days, Lenin even went so far as to confide to his lieutenants that he did "not expect to live to see the revolution."[86] Nonetheless, despite the absence of Russia's leading revolutionaries and despite even Lenin's deep pessimism, a confrontation between Russia's government and proletariat seemed to be approaching in the spring of 1916. There was a chance that it might be averted—but only a chance—if Russia's military fortunes shifted dramatically.

That spring, allies and enemies alike looked upon Russia as the great imponderable in their plans for victory. Had the crushing Austro-German victories of the previous summer and fall been as decisive as Hindenburg and Ludendorff had hoped? Or would Russia draw renewed strength from the new weapons sent by her allies and produced by her own factories, again take the field, and, as Martov once wrote, make the Germans "defend themselves against a Cossack invasion"?[87] Even Russia's leading generals could not be certain of the answer.

At a Council of War held at *Stavka* in mid-April, Russia's High Command argued that their nation could expect nothing but disaster if her armies went on the offensive that summer. But there was one who took a different view. Newly appointed to replace Ivanov as commander in chief of the southwest front, General of the Cavalry Aleksei Brusilov shocked his fellow commanders by urging that Russia attack without delay. "As far as the southwest front is concerned," he reported, "in my opinion we not only can, but must, go over to the offensive, and I am personally convinced that we have every chance for success."[88] Some of those who sat around the table looked at Brusilov and shrugged their shoulders "in pity," Brusilov remembered, while others simply were appalled. "Whatever made you expose yourself to such dangers?" one senior general asked after their meeting broke up. "Had I been in your shoes, I would have done anything in my power to avoid taking the offensive!"[89] Yet Brusilov's daring was based not on reckless hope but upon a healthy dose of common sense and a carefully calculated decision that Russia's steadily increasing reserves of weapons, ammunition, and supplies made a new offensive possible. Whether or not his plan would succeed even he could not tell for certain. But he was willing to make the attempt, and his willingness to accept the risk that all other senior commanders feared to take gave Nicholas and his regime one more chance to survive at a moment when they did not even realize how tenuous their position had become.

CHAPTER IX

The Brusilov Offensive

By late 1915, Russia's military fortunes had reached their lowest ebb. All of Poland now lay in Austrian and German hands. Still elated by victory, their armies had marched confidently into several of Russia's key western provinces. "We had taken a further and great step towards the final overthrow of Russia," Ludendorff wrote. "The German soldier was justly convinced of his unquestionable superiority over the Russian."[1] Yet, from the victors' point of view, things seemed not to be quite as they ought to have been. "There was something unsatisfactory about the final result of the operations and encounters of this year," Hindenburg confessed later. "The Russian bear had escaped our clutches, bleeding no doubt from more than one wound, but still not stricken to death."[2]

In those days, Hindenburg thought particularly of Napoleon's advance into Russia, and sternly reminded himself to remember "the tragic conclusion of that bold campaign," in which the Russians had suffered huge losses of men and materiel only to rise and conquer "the proud armies of the great Corsican" after his victory had seemed certain. How strong was the Russian bear after the battles of 1915, Hindenburg wondered. "Would he be able to show that he had enough life-force left to make things difficult for us again?"[3] Russia's losses in men could be repaired, and, as a French military attaché reported as early as mid-September, the shortage of weapons and ammunition from which Russia's armies had suffered during the Great Retreat already was "beginning to lessen."[4] In the equations that Allied and enemy generals tried to balance as they made their plans for the coming spring, Russia's will, and, most of all, the will of her commanders, remained the key, and that will had weakened far more than either the Germans or Austrians dared hope. Too many of Russia's senior generals thought their nation's cause lost, and too few dared stake their reputations on coming to her rescue. "From the point of view of morale, [the Russian Army] has but a single fault, and that is the lack of confidence that reigns in the High Command and the General Staff," one of France's military observers reported to his superiors in the War Ministry in mid-1916.[5] Such timid generals, the French assumed, had lost the "*faith* [that was] so indispensable to the success of any military undertaking."[6] In fact, parts of Russia's High Command suffered from a far more ignoble malaise, in which senior commanders struggled to keep the stain of failure from their service records, even at the cost of their country's welfare.

Tragically, such self-interest ruled the strategies of men who commanded armies and, even, entire fronts. None allowed his actions to be guided more closely by such base considerations than General Aleksei Kuropatkin, the dull-witted commander who had led Russia's armies in retreat during the Russo-Japanese War. In retirement ever since his disastrous failure at the battle of Mukden, this disciple of tactical indecision and strategic immobility, who had the rare good fortune to be nearly half a head shorter than a tsar whose commanders often stood above him, had returned to active service to command Russia's Grenadier Corps in the fall of 1915. Proclaimed to be "one of the few good men available" by the tsar himself, Kuropatkin had received command of Russia's northwest front before the fighting resumed in 1916, even though he had not known victory since his youthful campaigns in Turkestan, Samarkand, and Kokand some forty years before.

"The Minister of the Imperial Court [Count Frederiks, an old man of little ability and even less intelligence, whom Nicholas and Aleksandra loved and trusted deeply] does not forget his friends," France's military attaché remarked as he searched for reasons to explain the tsar's surprising new appointment.[7] Anxious to avoid staining the last command of his career with an even greater failure than he had suffered against the Japanese, the sixty-eight-year-old Kuropatkin dedicated the spring and most of the summer of 1916 to avoiding any offensive at all costs. Insisting that he would attack only when he could "guarantee" victory, he announced that it was "utterly impossible to breach the German Front," and that it was "hardly imaginable" that any Russian offensive could succeed.[8] Russia's only chance to escape defeat, Kuropatkin insisted, was to remain always on the defensive.

Yet Napoleon had been driven from Moscow not by faint hearts but by a nation united in arms against a foe who had defiled the Russian land. General Brusilov, whose exploits had won him fame since the early days of the war, looked for the tragedy of the Great Retreat to rekindle that same sense of national purpose once again. "Until this moment, the army alone has made war as best it can with the limited means at its disposal," he remarked when the Great Retreat finally came to a halt. "Now, the entire nation will do it."[9] Educated as a courtier in St. Petersburg's elite Corps of Pages, and married to

Stolypin's cousin, Aleksei Alekseevich Brusilov had been born in the Georgian capital of Tiflis, the son of a fighting general in the army of Nicholas I. With his mustaches twirled, his hair clipped short, and his body slim and straight, Brusilov still remained the dashing cavalry officer he had been during the Russo-Turkish War of 1877, although he was in his early sixties when he replaced Ivanov as commander of Russia's southwest front. Well connected with both groups that had struggled for control of Russia's military establishment in the years before the Great War, he had favored neither cavalry, artillery, nor infantry, but saw them all as units with specialized functions whose effective interaction had become vital to the success of modern warfare.

As commander of Russia's Eighth Army on the southwest front, Brusilov had won victories while other commanders suffered defeats. He had done so by assembling a staff of winning generals and dedicated military technicians who were truly expert in their art and whose minds were unfettered by the conservative wisdom of conventional military tactics. These men developed the complex entrenchment systems from which Brusilov launched his attacks in the spring and summer of 1916 at the same time as they erected proper emplacements for his artillery. This was especially important because Brusilov had become convinced, as few cavalrymen were, that "an artillery commander must direct his fire in much the same manner as a conductor directs an orchestra, and his role [in any offensive] is of the greatest importance." Brusilov therefore had assembled one of the best artillery staffs in the Russian army and used them with brilliant effect.[10] All of these men—engineers, infantry, cavalry, and artillerymen—dedicated their art to Russia's defense outside the constraints of politics. They served their tsar loyally until his weakness and ineptitude destroyed him despite their best efforts to bring him victory. Eventually, they went on to win high rank in the Red Army, and the youngest among them, the fortifications expert Dmitrii Karbyshev, became a Hero of the Soviet Union during the Second World War.[11]

In the dark days of the war's second winter, when most of his disheartened colleagues desperately sought to find ways to keep the stain of defeat from marring their careers, Brusilov took heart from an inner sense which told him that the cumbersome weight of Russia's industry was, at last, about to make a telling impact upon her war effort. "We began to receive rifles—of various types and calibres to be sure—but in quantity and with sufficient cartridges nonetheless," he remembered some years later. "Artillery shells, especially for the light guns, also began to arrive in great quantities," he added. "They increased the number of machine guns, and organized in every unit so-called grenadiers who were armed with hand grenades and bombs." With weapons in their hands, the morale of Russia's soldiers soared. "The men were in excellent spirits," Brusilov wrote. "They began to say that it was indeed possible to fight and have every hope of defeating the enemy under such conditions."[12]

Of course, as General Alekseev pointed out, Russia still could not hope to sustain a full-scale war effort without aid from her allies, and her shell output, "particuliarly for the heavier calibres, composed a quite hopeless picture."[13] Still, by the beginning of 1916, the long months of effort by the empire's industrialists and military planners had at last begun to bear fruit. Rifles, bullets, light shells—even heavy guns—began to pour from the assembly lines of her weapons factories as Russia's war production soared beyond all expectations during the second winter of the Great War. After beginning the war with nearly a million too few rifles and a weapons industry that had produced a mere forty-one rifles during the entire first seven months of 1914, Russia had raised her military output to a moderately impressive level well before the end of the second year of fighting. By the spring of 1916, yearly rifle production stood at one and a third million, and cartridge production had risen to more than a billion and a half. Every rifle in use in the Russian army now had an average reserve of four hundred rounds.

Between 1900 and 1914, a period that included the entire Russo-Japanese War, Russia's factories had produced an average of 1,237 field guns each year; in 1916, they produced more than 5,000, and repaired more than seven times as many as they had the year before. Ordnance officers reported similar accomplishments in almost every other area of military production: from a million and a half shell fuses in 1914 to eighteen million in 1916; from eleven hundred machine guns in 1914 to eleven thousand two years later; from somewhat less than eighty thousand shells in 1914 to over twenty million in 1916, and so on.[14]

Weapons shipments by Russia's allies made these figures even more impressive, for the comparative calm of the western front in 1915 had allowed England and France to accumulate large enough reserves of guns and ammunition that they could spare some for Russia, while the factories of the United States had begun to complete some of the orders they had accepted from desperate Russian procurement officers the year before. By July 1916, the White Sea port of Arkhangelsk had received almost fifty airplanes from France, nearly eighty armored cars from England, more than eighty-five thousand rolls of barbed wire, a hundred and fifty heavy guns, almost five hundred trench mortars, over two million hand grenades, more than a hundred and seventy million rifle cartridges, and a third of a million trench bombs, not to mention some two and a half million pounds of explosives: cannon and rifle powder, fulminate of mercury, pyrolyline, and picric acid.[15] During the winter of 1915–1916, General Knox had seen over a million rifles arrive at Arkhangelsk—Winchesters from America, Gras-Kropatscheks and Lebels from France, and Vetterlis from Italy. By the beginning of April, he estimated that the Russian army had over three and a third million rifles, 6,000 machine guns, and 6,356 pieces of field artillery, all in working order. "Generally speaking," he observed, "the Russian military position had improved by the commencement of the summer of 1916 far beyond the expectations of any foreign observer who had taken part in the retreat of the previous year."[16]

Russia had men trained and ready to use these new weapons, and foreign observers were especially impressed with her military manpower at the beginning of 1916. As an admiring French attaché had reported that fall, Russia's army had remained intact throughout the Great Retreat, despite the images of impending catastrophe that had prevailed everywhere. "The Russian Armies are neither demoralized nor disorganized," this officer announced at the beginning of his report to France's minister of war, as he went on to speak in detail of what was perhaps Alekseev's greatest achievement during his first months as *Stavka*'s chief of staff. "From everything that I have seen personally in Russia during the course of this mission, and from all that has been reported to me," France's observer continued, "I have concluded that . . . the conservation of all major units, the auspicious partial resumption of the offensive, the manner in which the retreat was carried out, and the evacuation of the abandoned territories all indicate that the Russian armies remain, from the material point of view, a force whose momentary weakening is exclusively the result of a lack of means and not of a disorganization stemming from four months of struggle and retreat."[17]

Much more had been accomplished during the war's second winter beyond keeping Russia's bleeding army intact. Painfully aware of the army's desperate shortage of manpower, Russia's energetic war minister, Aleksei Polivanov, had drafted about a million men with still another million called up and held in reserve. "Unfortunately," he later wrote, "some portion of these two million had to be sent directly to the front lines to replace casualties, but the remaining mass was given more extensive training for four, five, or six months . . . in order to have a new and formidable force ready for the moment when we would have a sufficient quantity of guns, shells, rifles, and bullets at hand."[18]

The first of these so-called *Polivanovtsy* therefore reached the front with no training to speak of, but enterprising field officers trained them under the very shadow of the German and Austrian guns while Europe's eastern front stood immobile under a blanket of snow and ice. "A good part of the training was actually given in front of the [Russian] line in our spacious No Man's Land," Britain's Sir Bernard Pares reported later.[19] Even the German High Command was struck by the flurry of activity on the Russian side during those crucial months. "Activity was uncommonly lively in the enemy's rear areas" that winter, Hindenburg remembered. "Deserters complained of the iron discipline to which the divisions drawn from the lines were subjected, for the troops were being drilled with drastic severity."[20] By winter's end, one and three-quarter million Russian troops, better trained and armed than at any time since the first days of the war, faced just more than a million Germans and Austrians along Europe's eastern front. More than three-quarters of a million more were ready in reserve. All were armed, with sizeable reserves of rifle and artillery ammunition within easy reach.[21] "We had every reason to expect to smash the enemy," Brusilov wrote, "and drive him beyond our frontiers."[22]

Yet, if Polivanov had restored Russia's war effort by improving the army's reserves of weapons and manpower beyond all expectation, he had failed to curry favor with the empress, in whose eyes he had committed one unpardonable act after another. An enthusiastic supporter of the Special Conference on National Defense, Polivanov had been equally willing to work with the representatives of the Duma to win their support for the army's requests. For those reasons alone, he had incurred Aleksandra's stern displeasure. Far more serious, however, he had outraged her by his open hostility toward Rasputin's meddling in army affairs and by calling Stürmer to account for some of his corrupt schemes. "Polivanov [is] simply treacherous," Aleksandra exclaimed in one of her letters to Nicholas as she pressed her attack against the man who dared question Rasputin's judgment. After a few more letters and a few more accusations, Nicholas needed no more convincing. Despite Polivanov's amazing achievement in restoring Russia's fighting forces, he rudely dismissed him before the middle of the month with the ungrateful remark that "after P[olivanov]'s removal I shall sleep in peace."[23]

Dramatic as Russia's military regeneration seemed, serious problems still remained. Arkhangelsk lay nearly two thousand kilometers to the northeast of the nearest part of the war zone, while Vladivostok stood some ten thousand to the east. Neither port had the facilities to unload and trans-ship large quantities of military supplies to the front quickly. At one point, General Knox found "an enormous accumulation of stores" in Arkhangelsk, including "no less than 700 automobiles in wooden packing-cases." So desperate did the Russians become to move these vital supplies from the White Sea coast that the Council of Ministers approved the building of a new railway from Murmansk, on the ice-free Kola Bay, to Petrozavodsk, whence supplies could be shipped to Petrograd via the Ladoga Canal.

It was a Herculean undertaking. Of more than a thousand kilometers of rails, at least one out of every four had to be laid through shifting, unstable marshes, and bridges had to be built for one kilometer out of every fifty. For part of the year, men labored unceasingly under the perpetual light of the midnight sun until the earth shifted on its axis at the solstice and the days began to shorten. Soon darkness reigned at noon, and the mercury fell far below zero. Then men worked by the light of "enormous torches, which had to be used," Knox explained, "because lanterns proved useless in the strong north-east wind and continual snowstorms." At times, the number of men in the work-gangs rose beyond fifty thousand, including prisoners of war and imported Chinese laborers.[24] The task proved to be another of those astounding feats of which Russians have shown themselves to be capable time and again in their history. Just three months before the February Revolution in 1917, the work gangs completed the vital rail link to bring weapons and supplies more quickly to Europe's beleaguered eastern front.

Construction of the Murmansk-Petrozavodsk railway had only begun when Alekseev began to lay out his plan of operations for the spring and summer of

1916. Early in April, he urged the tsar to permit either a two-pronged attack on Vilna, or to consider some other action that would "forestall any new offensive on the part of the enemy, compel him to respond to our initiatives, and destroy his plan of operation."[25] When they assembled at Mogilëv for the year's most important Council of War on April 15, *Stavka*'s senior generals and Russia's front commanders discussed these options. Nicholas remained a passive observer throughout, doing nothing to direct the debate or probe his generals' motives. "The tsar sat in silence the whole time," Brusilov later wrote. "Not once did he even venture an opinion."[26] The day before, while Russia's senior generals and their chiefs of staff made their final preparations for the Council of War that might determine their nation's military fortunes that year, Nicholas had "read from morning to night" to get through A. W. Marchmont's *The Man Who Was Dead* and Florence Barclay's *Through the Postern Gate*, a touching tale about Little Boy Blue. That night, Nicholas shed tender tears over Little Boy Blue's adventures. "I had to resort to my handkerchief several times," he confessed to Aleksandra. "I find [his adventures] so pretty and true!"[27]

Whether Nicholas's thoughts wandered to Little Boy Blue's adventures as his senior generals tried to gloss over their timidity, incompetence, and self-interest with insincere statements about Russia's welfare, we of course cannot tell. Suddenly, however, the tone of the Council of War shifted. After the commanders of the northwest and western fronts finished explaining in the most sage and sober terms that any Russian attack must be doomed to failure, Brusilov insisted that Russia must go on the offensive immediately. He already had urged that course upon the tsar the previous week, when Nicholas had inspected some of the troops under his command at Kamenets-Podolsk, and it was then that the uneasy commander of Russia's southwest front had discovered that his nation's irresolute Supreme Commander "never liked to dot his i's." While Russia's other front commanders protested that they "could not guarantee success" in the sort of offensive Brusilov proposed, Russia's southwest front commander tossed their reservations aside. "No matter what the time or the place, no general could do that, not even if he were a thousand times greater than Napoleon," Brusilov retorted. There was every chance of success, he insisted, and that alone made an offensive worth the risk. Alekseev approved Brusilov's plan but warned that he could spare no extra men, artillery, or munitions, for those must be held for the western front where Russia's main offensive was to be launched later that summer. That night, Brusilov left *Stavka* to return to his southwest front headquarters at Berdichev.[28] Scantily supported by his superiors and scorned by his fellow front commanders, Brusilov was about to launch what one expert later called "the most brilliant victory of the war."[29]

What distinguished Brusilov from Russia's other senior generals was his unwillingness to accept conventional military truths and his ability to find alternatives to long-accepted strategic precepts. Ever since the positions of the warring nations had become fixed in the Great War, Europe's generals had attempted to assemble the great weight of men and munitions needed to achieve the decisive breakthrough that would bring the enemy to his knees. Mackensen had set the example in Galicia in the spring of 1915 when he had amassed overwhelming quantities of men, ammunition, and heavy artillery, to smash through the already-weakened Russian lines in the Tarnów-Gorlice sector. Mackensen and his fellow commanders refined this tactic, sometimes called the "Mackensen steamroller," into a system of slow, inexorable movement, which demanded relentless support by the army's heaviest artillery. Using crushing quantities of shell, Mackensen's forces simply battered their way forward:

> The following is a rough picture of Mackensen's main attack [wrote Russia's General Nikolai Golovin some years later]. Creeping like some huge beast, the German army would move its advanced units close to the Russian trenches, just near enough to hold the attention of its enemy and to be ready to occupy the trenches immediately after their evacuation. Next the gigantic beast would draw its tail, the heavy artillery, toward the trenches. That heavy artillery would take up positions which were almost or entirely beyond the range of the Russian field artillery, and the heavy guns would start to shower their shells on the Russian trenches, doing it methodically, as was characteristic of the Germans. That hammering would go on until nothing of the trenches remained, and their defenders would be destroyed. Then the beast would cautiously stretch out its paws, the infantry units, which would seize the demolished trenches.[30]

Time and again during the remainder of the war, opposing commanders assembled all the men and metal at their disposal in a vain effort to duplicate Mackensen's success. Yet the result always proved the same: thunderous barrages hurled tons upon tons of heavy shell onto the enemy's barbed wire defenses and machine gun emplacements; divisions of infantry then stormed across the new-made wastes to occupy trenches from which all but the dead and dying had fled, only to face murderous barrages from the enemy guns that remained on their flanks. Under the crushing weight of counterattacks by forces that the enemy always held in reserve, such attacks invariably crumbled. The only hope of success lay in surprise, but it always proved impossible to assemble in secret enough men, weapons, ammunition, and supplies for the initial breakthrough. Inevitably, the enemy learned of the impending attack, and massed the men and weapons needed to meet it, which meant that no tactical success could ever be exploited to its fullest and converted into a major advance. Thus, strategists of the Great War found themselves caught in a hopeless contradiction: the only way to achieve a breakthrough seemed to be to hurl enough men and heavy shells upon the enemy positions to crush them; yet, the very act of assembling such quantities of men and weapons inevitably warned the enemy and showed him where to assemble the reserves needed to defeat the attack.

Brusilov had studied these dilemmas and had tested some tentative solutions during the second winter of the Great War. "In every operation that we undertook during the winter, we searched for the very best means for resolving those new problems that [the conditions of] the present war had created," he explained later to a correspondent from the *Times* of London.[31] Together, Brusilov and his staff of engineering and artillery experts learned that surprise —so ardently sought and so rarely enjoyed by attacking commanders—could be achieved if they spread their preparations along an entire front and delivered a series of simultaneous attacks rather than concentrating upon assembling a single crushing blow. "I considered it absolutely vital to develop an attack at many different points," so that the enemy could not know where to concentrate his reserves, Brusilov later explained.[32] Inevitably, such an offensive would bring heavy losses, but the southwest front's new chief considered losses justified if they produced results. "An offensive without casualties may be staged only during manoeuvres," he once wrote. "To defeat the enemy, or to beat him off, we must suffer losses, and they may be considerable."[33]

The sort of offensive that Brusilov envisioned in the spring of 1916 required intense training and the construction of complex offensive entrenchments all along the front to conceal from the enemy the points at which his lines would be assaulted. Vast *places d'armes* had to be built very near the front as staging areas for reserve troops, enemy entrenchments had to be charted with absolute accuracy, and there had to be the closest possible cooperation between the previously antagonistic aristocratic artillery and plebian infantry. If all of these complex and contradictory requirements could be brought completely into harmony, Brusilov saw a chance for success where his counterparts on the western and northwest fronts envisioned only failure.

As Ivanov's successor in command of the southwest front, Brusilov at first found it difficult to impose his new regimen of intense training and preparation upon subordinate army commanders.[34] With only two months in which to make his preparations, he therefore frequently visited the most advanced positions at the front, personally overseeing training and surveying entrenchments. As the great Suvorov had done on the eve of his assault against the Turkish fortress of Ismail in 1791, Brusilov built full-scale models of Austrian entrenchments and trained his troops to assault them so that they would feel "at home" in the enemy's trenches once the offensive began. Already high by winter's end, his men's morale rose higher still. Bernard Pares met a battalion chief of staff who reconnoitered the enemy's lines "with the enthusiasm of a schoolboy," and remarked upon the disappointment of the troops who had no part in the real offensive. These were a vastly different breed of soldier from the ones that one of Knox's deputies had called a "harmless mob" less than a year before. Observers now thought them "keen and spirited," the "first-fruits of [Polivanov's] excellent system of training."[35]

Although Alekseev had at first hesitated to approve Brusilov's offensive, unforeseen events in the west soon added a sense of urgency that made him especially anxious to begin the attack. As a result of the Allies' promises that she would receive the Trentino, the Tyrol, Trieste, and some of the Dalmatian islands, after the defeat of Germany and Austria if she entered the war on their side, Italy had declared war upon her former partners in the Triple Alliance in 1915 and had opened a front against Austria in the Trentino. For the rest of the year, the Austrians had remained on the defensive, their resources of men and weapons totally committed to the war in the east.[36] Only after spending some six months in buttressing dugouts, trenches, and machine gun emplacements all along Europe's eastern front had the Austrians turned their attention back to the south and west in the spring of 1916 to what one expert termed a "fatal diversion." Now confident that they could withstand any force the Russians might hurl against them, the Austrians had opened an offensive against the Italians in the Trentino and quickly had overwhelmed their advance units.[37]

As the Austrian armies broke through Italy's Trentino front, the Italians frantically begged Alekseev to relieve the pressure against their armies by an offensive against Austria's forces in the east "at the earliest possible moment." Although the French and the British urged him to postpone any offensive until they had completed preparations for their attack on the Somme, Alekseev took a middle course. "When will preparations for an offensive against the Austrians be complete?" he telegraphed Brusilov a week after the Italian front had begun to crumble. When Brusilov replied that his armies could attack in seven days, Alekseev urged him to do so as the "opening act of a general operation," in which an even heavier assault would be launched on Russia's western front a week later.[38]

Before the end of May, Brusilov was ready. He centered his offensive against Lutsk, but planned to support it with other large-scale attacks all along his front. He had sixty-five divisions—well over 600,000 men—poised to fall upon the Austrian line from Rafalovka, just south of the Pripiat marshes, to Czernowicze, some 350 kilometers to the south. He had nearly five hundred more pieces of light artillery than did his enemies, but, in heavy artillery, the Austrians still outnumbered him by more than three to one.[39] Nonetheless, Brusilov looked for success, hoping that the strategies he had developed that winter would enable him to conceal the time and place of his main attack from the enemy. Much to her irritation, even Aleksandra remained in the dark about his final plans. "Are you ready to go on the offensive?" she asked at a private meeting she had arranged on the very eve of Brusilov's attack. Brusilov cautiously replied that he was "not yet fully prepared," but that he hoped "to be able to crush the enemy before year's end." "*When* do you plan on starting your attack?" the empress persisted. Brusilov said he "did not yet know," that the final date "depended on conditions that could quickly change," and concluded with the pointed remark that "such information was so secret" that he "could not at the moment call it to mind." After a few perfunctory comments, the empress took her leave.

"After a cold welcome," Brusilov remembered, "she gave me an even colder farewell.[40]

Brusilov now had far more pressing problems to contend with than the empress's sullen displeasure, for he was almost certain that Alekseev's "general operation" would never materialize, and that General Aleksei Evert, commander of the western front from which Russia's major offensive of 1916 was to be launched in support of his own, would somehow avoid going on the attack. An officer who in 1916 shared Kuropatkin's passion for defense and who already had warned Nicholas that any offensive had "no chance of success,"[41] the fifty-nine-year-old Evert had behind him a record of accomplishment that belied the timidity he showed that summer. Wounded in the Russo-Turkish War of 1877–1878, in which so many of the men who commanded Russia's armies in the Great War had received their baptism under fire, Evert had, for the next quarter-century, avoided staff work and preferred field command. Although by no means a brilliant strategist, he nonetheless proved a competent tactician, and his record in the Russo-Japanese War reflected that fact. Left in command of the Irkutsk Siberian Military District when the Great War began, Evert found himself on Russia's southwest front within a month with orders to take command of the shattered Fourth Army that the incompetent Baron Zalts had marched headlong into a superior Austrian force.

Yet, if Evert had proved a competent chief of the Fourth Army, he far outstepped his abilities when he took command of Russia's western front at the end of the Great Retreat. Rigidly conservative, his heavy beard topped by a long, drooping mustache in the style of the Zabaikal Siberian Cossacks he had once commanded, he had no trace of the daring that had made such men famous. Now, like Kuropatkin a disciple of a strategy that offered no chance for victory but minimized the danger of defeat, he desperately wanted to avoid joining Brusilov's offensive in May 1916. In reply to Alekseev's repeated orders, he bombarded *Stavka* with warnings and excuses. Even as Brusilov reported his first victories, Evert postponed his attack for the third time in a week because "enormous enemy forces and great quantities of heavy artillery had been concentrated" at precisely the point where he had planned to launch his major assault. Boldly announcing that "defeat was a certainty" if he did not change his plans, Evert quickly secured permission from Russia's timid supreme commander to cancel his attack so that he could make preparations to strike in an area where, "in his opinion, he could expect success." Furious because he was certain that Alekseev "understood perfectly the criminality of Evert's and Kuropatkin's actions" but was trying "to conceal their lack of initiative" because they once had been his superior officers, Brusilov made ready to fight alone.[42]

Russia's most stunning victory of the Great War—officially designated the "Offensive of 1916 on the Southwest Front," but more often remembered as "the Brusilov Offensive"—began, in Brusilov's own words, "at dawn on May 22, with a thunderous artillery barrage all along the southwest front in all designated

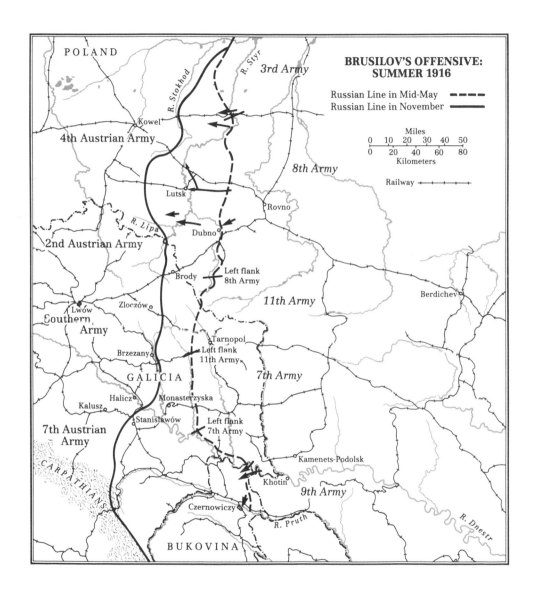

POLAND

R. Stokhod

R. Styr

3rd Army

Kowel

4th Austrian Army

**BRUSILOV'S OFFENSIVE:
SUMMER 1916**

Russian Line in Mid-May ▬ ▬ ▬
Russian Line in November ▬▬▬

Miles
0 10 20 30 40 50
0   20   40   60   80
Kilometers

Railway ┿━┿━┿━┿

8th Army

Lutsk

Rovno

R. Lipa

Dubno

2nd Austrian Army

Brody

Left flank
8th Army

Zloczów

Berdichev

Lwów

11th Army

Southern
Army

Brzezany

Tarnopol
Left flank
11th Army

G A L I C I A

7th Army

Halicz

Monasterzyska

Kalusz

Stanisławów

Left flank
7th Army

7th Austrian
Army

Kamenets-Podolsk

C A R P A T H I A N S

Khotin

9th Army

Czernowiczy

R. Pruth

R. Dnestr

B U K O V I N A

sectors." For the entire day and on through the night, his well-trained gunners worked, methodically directing their fire and carefully observing its impact. For two hours, from 12:30 until 2:30 a.m. on May 23, they lifted their barrage so that scouting parties could assess the damage, and then resumed their bombardment in some sectors for another entire day. With "complete success," Brusilov said, they managed to "totally obliterate" the first belt of Austrian trenches and barbed wire so that his infantrymen found "nothing but ruins and shattered corpses" when they advanced at 2:00 a.m. on the morning of May 24.[43] Within two hours, they had occupied the enemy's first and second lines of trenches. "The entire zone of battle was covered by a huge, thick cloud of dust and smoke, often mixed with heavy explosive-gases," an Austrian officer wrote. "[This] allowed the Russians to come over the ruined wire-obstacles in thick waves into our trenches."[44] Only in the best-fortified dugouts did Austrians remain alive and Brusilov's grenadiers dealt with them quickly. "A grenadier with a single grenade would stand at the entrance," Brusilov explained. If the enemy refused to surrender, the grenadier simply tossed the grenade into the dugout and killed everyone inside. Unable to escape, and unwilling to die, as Brusilov remarked, "to no good purpose," Austrian soldiers surrendered in droves.[45]

Brusilov's plan thus enjoyed brilliant tactical successes from the very outset. Within forty-eight hours, his grenadiers and infantrymen took more than forty thousand Austrian prisoners; in three more days they had shattered Archduke Josef-Ferdinand's Fourth Army and broken through the Austrian defenses all along a seventy-kilometer front. "The word 'victory' has been used for the first time in the official communications," a joyful Nicholas wrote to Aleksandra, while in Petrograd even the ever-skeptical Paléologue confided to his diary that "General Brusilov's offensive . . . is actually beginning to take on the momentum of victory." By the end of the first week, the Russians had seized several hundred machine guns, grenade launchers, and field guns and had captured nearly seventy thousand prisoners. In the second week, the successes mounted. "Our troops are attacking and pressing the enemy," Nicholas wrote on the offensive's eighth day. "The number of prisoners exceeds 100,000 and one general."[46]

The Eighth Army's assault against Lutsk formed the spearhead of Brusilov's offensive and was the first prize to fall to Russian arms since Przemysl more than a year before. Although he had strengthened even his reserve positions with concrete reinforcements, the Archduke Josef-Ferdinand had forgotten to fortify the Krupy heights, from which the Russian guns could fire directly into Lutsk's defenses. By May 25, the Russians were in the city with the Austrians in full retreat. "Masses of testimony from prisoners drew a hopeless picture of the Austrian flight [as] hordes of unarmed soldiers from various units fled in panic through Lutsk, throwing everything away as they went," the Eighth Army's quartermaster general reported. "Many prisoners told us that their officers were the first to flee to the rear, leaving only noncommissioned officers in command."[47] "It's a complete debacle," the headquarters of one Hungarian corps confessed. "We can do nothing."[48]

Although individual units sometimes encountered heavy resistance, Brusilov's Eleventh, Seventh, and Ninth armies moved with speed and fortune only slightly less dramatic than the Eighth's. Although so short of heavy artillery and shell that its commander had at one point begged to postpone the attack (he had only a sixth of the heavy shell reserve that French experts set as the absolute minimum), Russia's Ninth Army overwhelmed the Austrian Seventh Army on the fourth day of the offensive and drove them across the Dnester River. Within another week, the Austrian Seventh was completely shattered, with its commander reporting losses that exceeded 100,000 men. Combined with the Eighth Army's victory at Lutsk, the Ninth's successes helped to carry forward the rest of Brusilov's forces so that, by the ninth day of the offensive, Brusilov's four armies had captured almost 3,000 officers, 200,000 men, over 200 field guns, and 645 machine guns. They had seized so much rifle ammunition that the War Ministry ordered all Russian firms to stop manufacturing ammunition for the tens of thousands of Austrian rifles that Russian troops had taken from Przemysl the year before.[49]

Brusilov's first week of victories had cost Austria-Hungary more than half of the men on her eastern front, and left her with no hope for victory against Russia. "Peace must be made in not too long a space," Conrad confessed, "or we shall be fatally weakened, if not destroyed."[50] Yet Conrad and the men who stood with him in urging peace with the Russians knew that any but a victorious peace could weaken their Habsburg masters irreparably and destroy the empire. Austria therefore had no choice but to stand in league with Germany, commit herself to the total war the Germans wanted to fight, and transfer control over the remnants of her shattered ground forces to the German High Command. There must be a "wholesale change" in which Hindenburg "would take over the command of the whole Eastern Front," including all Austrian forces, Ludendorff said, and it was "imperative to act quickly." In mid-July, the Austrian General Staff therefore relinquished to the Germans control over all its forces on the eastern front except for the Archdule Karl's Army Group in southeast Galicia and Bukovina.

"God help you and our Fatherland," Falkenhayn remarked as he gave Hindenburg his new command. Hindenburg must weld the two forces together so that they could fight against the Russians as one. Austria's fighting men needed more than leadership. Like so many of Russia's senior generals, they required massive transfusions of confidence to give them new will to fight. The task that lay before Hindenburg therefore was immense. "If ever an army needed one controlling and resolute will, one single impulse, it was this army," he recalled some years later. "Without them, the best blood would run feebly in such an organism and be poured out in vain." But Hindenburg's new appointment remained a month in the future as Brusilov calculated the extent of his first week's victories. "Bitter experience was needed," Ludendorff later wrote, "before this change was effected." That would come only after Brusilov pressed his attack to Austria's eastern gateway at the foot of the Carpathian passes. Beyond these, only the plains of Hungary stood between the Russians and Budapest. If Brusilov reached Hungary's ancient capital, only the easily-crossed Danube would bar his path to Vienna.[51]

As Austria's armies crumbled before Brusilov's attacks, the German General Staff faced its most anxious moments of the war. Against all expectations, the French still held at Verdun, and the Germans well knew that the British were about to launch their offensive on the Somme. No German divisions could be diverted from the west to shore up Austria's collapsing eastern front, especially because repeated efforts to break through at Verdun now cost the Germans greater losses than they inflicted upon the French defenders. All German forces sent to stiffen the crumbling Austrians therefore had to be drawn from other parts of the eastern front, and Hindenburg and his staff realized that they might well be forced to stretch their resources to the breaking point if Evert or Kuropatkin had the foresight to open a second offensive. "For a moment," Hindenburg confessed, "we were faced with the menace of a complete collapse."

Frantically, Ludendorff formed reserve battalions from raw recruits that had just arrived at Germany's eastern conscript depots, although he realized at the time that these could do little against the Russians. "If the Russian's had a really great success at any point," he later admitted, "these units would [have been] but a drop of water on a hot stone." As the Germans knew all too well from their victories of the previous summer in Poland, raw recruits could slow the advance of trained soldiers only briefly at best. Hindenburg and Ludendorff thus could only hope that Evert and Kuropatkin would not launch the all-out offensive they feared. Especially on Kuropatkin's northwest front, the Germans anticipated an assault against their positions near Riga which, Hoffmann later confessed, "was the most sensitive spot on the whole of the Northwest Front. If the Russians could succeed in breaking through there," he once wrote, "the whole of the front would have had to retire."[52]

If the Germans and Austrians found themselves in dire straits, the Russians had problems of their own, for Brusilov's commanders had had to pay a heavy price for their first victories. The Eighth Army alone had lost thirty-five thousand men, and much of the southwest front's shell reserve had been spent in the first days' assaults. Because he had turned most of the horsemen under his command into infantry in order to be free from supplying fodder for tens of thousands of horses, Brusilov now had not enough cavalry to press the fleeing Austrians. As he consolidated the gains of his first fortnight's victories, he hoped to send the Eighth Army against Kowel and the headquarters of the Austrian Fourth Army at Vladimir-Volynsk. With its infantry already far ahead of its supply lines, the Eighth Army had not the men, the munitions, the supplies, nor the transport to move so far so quickly, yet, just to its north, Evert commanded a million men, two-thirds of the Russian army's artillery, and great stores of heavy shells that held the key to victory. To follow up his early successes, Brusilov needed Evert either to send him reinforcements or take up the major offensive that the War Council at *Stavka* had assigned him in April. Without that, Brusilov would expose his northern flank to German counterattack the moment he moved further westward.[53]

As he continued to shift the focus of his main attack and find excuses to postpone it, Evert was no more inclined to take the offensive in June than he had been at the Council of War in April. Too many German reserves had been brought up to make success possible in the Lake Naroch area, he announced, and victory would be far from certain if the offensive were launched toward the Grand Duke Nikolai Nikolaevich's former headquarters at Baranovichi. Then the attack on Baranovichi had to be postponed for a week, and then for an entire month, first, because Evert thought "it would be unseemly to attack on Trinity Sunday," and then because he discovered that the Pinsk Marshes had not yet fully dried. By mid-June, he claimed that he could not launch a full-scale offensive under any conditions, and that his effort to support Brusilov had to be confined to "frontal blows, promising only very slow progress with the greatest of difficulty." Therefore, while Brusilov struggled to press on with his offensive, the great masses of men and artillery under Evert's command made only a handful of feeble thrusts. On the northwest front, Kuropatkin's troops, which outnumbered the Germans by more than two to one, remained utterly passive.[54]

Brusilov was furious: furious at Kuropatkin, even more furious at Evert, whom his men now thought a traitor, and equally furious at Alekseev, who seemed reluctant to force his old Russo-Japanese War chiefs to launch the second offensive that Ludendorff and Hindenburg so feared. Evert's timidity would "turn a won battle into a lost one," Brusilov stormed at Alekseev, as the Eighth Army had to halt its advance for want of replacements and supplies.[55] In response, Alekseev transferred four corps to the southwest front, despite strong objections from Kuropatkin, who insisted that there could be no question of attack and that he needed at least another corps just to strengthen his defenses. "What is the basis for your demands?" Alekseev asked Russia's former minister of war. "Right now, you have 420,000 men against [the Germans'] 192,000. Surely those figures ought to tell you something!" His patience nearly exhausted, Alekseev curtly informed Kuropatkin that he must place the "success [of Russia's total war effort] ahead of all personal self-interest."[56]

Finally, after endless prodding from Alekseev, Evert gave the order to attack on Russia's western front on June 19. Utterly oblivious to the lessons of Brusilov's success, he concentrated his assault on a front of less than seven kilometers near Gorodishche, some fifteen miles north of Baranovichi. A thousand guns with a thousand shells apiece prepared the way but did little damage because, although he had only the merest handful of airplanes equipped with wireless, Evert had chosen to attack in terrain where only airborne observers could properly adjust artillery fire. To make matters worse, all of Evert's backing and filling during the past weeks had left his commanders unready for battle, and despite his preachings about the need for careful preparation, they had less than two weeks in which to bring their forces, supplies, and reserves to bear on their objective.

Evert's assault therefore bore all the marks of poor preparation and botched command. Communications became hopelessly tangled the moment the infantry moved forward. Telephones broke down and five cavalry divisions (a full 20 percent of Evert's entire force) clogged supply lines in the rear. It took four hours for the general whom Evert had entrusted to command his main assault to receive word that one of his key divisions had broken through the enemy's lines. By then, its men had been driven back with extremely heavy losses by the inevitable German artillery counterattack which they had had to face without reinforcements or artillery support. Such tragedies repeated themselves for several days as Evert followed barrage after barrage with waves of massed infantry assaults against heavily-fortified Austro-German positions. Very quickly, his many dire predictions of inevitable defeat became self-fulfilling prophesies. In less than a week, the Russians paid for sixteen thousand Austrian and German casualties with more than eighty thousand of their own, some of whom simply slipped from sight in a bottomless bog that did not appear on their outdated, crudely-drawn maps.[57]

Such losses slowed Brusilov's advance but did nothing to relieve the pressure against his front. "If we had had a real supreme commander, and if all the front commanders had acted according to his orders," he once wrote bitterly, "then my armies . . . would have moved so far forward that the enemy's strategic position would have become so desperate that he would have had to withdraw to his frontiers even without a battle. The course of the war would have taken a very different turn and its end would have come much more quickly."[58] As it was, Kuropatkin's stubborn refusal to advance and Evert's costly and stupid failure left Brusilov with a deepening sense of futility about his own effort. "No matter how great our successes, my front alone could not win the war that year," he remembered in angry frustration. "It simply was not possible to advance far enough so as to give any serious strategic benefit to the other fronts."[59]

By late June, Alekseev had come to share Brusilov's disgust at the stubborn and arrogant ineptitude of Kuropatkin and Evert, and even Nicholas, so long their defender, grew disillusioned. "Many of our commanding generals are silly idiots who, even after two years of warfare, have not been able to learn the first and simplest A. B. C. of the military art," he complained to Aleksandra as Evert's effort at Gorodishche collapsed.[60] Perhaps sensing that the tsar would not oppose the idea, and in any case certain that the only hope for a major victory in 1916 lay in the southwest, Alekseev tardily began to shift reserves of men and ammunition to support Brusilov's offensive on the largest possible scale. Yet, the very promise of unlimited resources dulled Brusilov's brilliance, and he reverted to more orthodox tactics. As his Ninth and Seventh armies took Brzezany, Monasterzyska, and Stanislawów on the left of his front, he turned his attention to Kowel on his extreme right where he hoped to use the Third Army that Alekseev had just transferred from Evert's command to support the Eighth in an assault that would open the way for Evert's armies to join his advance.[61] Yet the men Brusilov now sent into battle were not the equals of those he had used to begin his offensive, for his losses of more than a half-million killed, wounded, and missing since the middle of May had decimated the well-trained *Polivanovtsy*. Fearful that raw recruits might have to be brought into the line if such heavy losses continued, Alekseev added to Brusilov's command Russia's Guards Army, the one elite force he had assiduously withheld from all front commanders until that moment.

Tall, strong, and well-trained, the Imperial Guards had always been the force in which Russia's emperors had taken their greatest pride. The creation of Peter the Great, the Guards had been the arbiters of Russia's succession during the eighteenth century, and had placed Alexander I and the Empresses Elizabeth and Catherine the Great on the throne. Every autocrat since Peter the Great had been the honorary colonel of the Preobrazhenskiis, the Guards' most elite regiment, in which every man, dressed in the bottle-green dress uniform first chosen by Peter himself, stood over six feet tall, a living monument to the best that Russia had to offer. Russia's rulers committed these prize units to combat only with the greatest reluctance, but Nicholas had sent them into the line not long after the Great War began. After the Great Retreat, during which they had suffered extremely heavy losses, Nicholas had withdrawn them all, refilled their ranks, and consolidated them into a Guards Army of sixty-five thousand men that he designated as his personal reserve.[62]

An assault force of striking quality—General Knox called them "physically the finest human animals in Europe and all of the best military age"[63]—the Guards Army was led by men of high birth whose ability was utterly unworthy of the men they commanded. As its commander, Nicholas chose none other than General Aleksandr Bezobrazov, notorious for his bad judgment, who had repeatedly been relieved of lesser commands for incompetence and outright insubordination. Brusilov thought him a man of "limited intelligence and unbelievably stubborn," and Knox, who certainly was no friend of Brusilov, agreed that he was "a difficult old man."[64] Bezobrazov's two chief deputies were no better than he. Brusilov called I Corps' Commander Grand Duke Pavel Aleksandrovich "one of the most noble of men," whose only failing was that "he knew absolutely nothing about military affairs," while Duma President Rodzianko was appalled to find that "the grand duke does not carry out even the orders of his immediate superiors." II Corps was commanded by an even less accomplished general who, Brusilov pointed out, "suffered from one great failing in wartime [which was] that, whenever he found himself in danger, he lost all presence of mind and was unable to conduct operations because his nerves could not stand the sound of riflefire."[65]

Combining unsound tactics with criminally poor strategy, Bezobrazov and his two chief subordinates slaughtered the flower of Russia's army in a few days. Behind them stood one of the heaviest concentrations of heavy artillery yet seen on the eastern front, but their gunners had to plot targets on maps that had not been corrected since 1897 because Russia's General Staff had never thought that their army might have to fight so deep inside Russian territory. When Brusilov ordered the Guards into action in the middle of July, Pavel Aleksandrovich unhesitatingly disobeyed his orders and sent his men to attack through a swamp over which German planes hovered to rake them with machine-gun fire while they struggled through chest-deep water and mud. While General Knox watched in horror, German airmen returned again and again to strafe the thousands of trapped Guards. "The wounded sank slowly in the marsh, and it was impossible to send them help," Knox confided to his diary a few hours after the grand duke's murderous performance. "The Russian Command for some unknown reason," he concluded, "seems always to choose a bog to drown in."[66] Absurdly unaware of the slaughter into which his personally chosen commanders had led his prize troops, Nicholas wrote to Aleksandra that Pavel Aleksandrovich's Guards were "attacking and performing miracles."[67]

Nor was I Corps commander alone in his fatally inept performance. Not to be outdone by his imperial colleague in the matter of disobeying orders, his sensitive nerves shattered by the sound of gunfire, the commander of II Corps exposed the flanks of the divisions he was supposed to support to brutal enemy fire when he retreated after being ordered to attack. "We are willing to give our lives for Russia, for our Motherland," a young lieutenant said bitterly, "but not for the whims of generals."[68]

In less than a fortnight at Kowel, the senior officers of the Guards Army, the very men whom Nicholas had chosen personally, lost more than 80 percent of Russia's finest fighting force.[69] As complaints against Bezobrazov poured into *Stavka,* Nicholas agreed to dismiss him, but only with the greatest reluctance. "What an honest and well-bred man he is!" he exclaimed to Aleksandra. "I have given him leave for two months. . . . [Then] I have promised him that if . . . some vacancies occur in one of the Guard Corps to appoint him there! He was very good as the head of a Guard Corps."[70] It was another tragic example of Nicholas's chronic inability to take the proper measure of men. As a Guards Corps commander in the summer of 1915, Bezobrazov had been relieved for insubordination after he had ignored his superior's urgent order to retreat. On that occasion, Bezobrazov's belligerent stupidity had caused thousands of his men to be slaughtered to no purpose except to prove the truth of his arrogant statement that Imperial Guards never retreated.[71]

Brusilov's right flank bogged down with slaughter of the Guards Army, but

Alekseev had become so committed to taking Kowel that he now insisted on still heavier attacks. These required still more men and weapons, but none could be taken from the southwest front where Brusilov needed all his reserves to support the Seventh and Ninth armies further south. Evert's western front was the best source, but the only way Alekseev could be certain that Evert would send the men and munitions needed was to place the Kowel operation under his command. On July 29, just days after the Germans had mauled the Guards Army, Alekseev returned command of the Third Army to Evert and added to it the battered Guards Army, now called the Special Army.[72] As the Third and Special armies continued to attack the Austro-German positions around Kowel, they did so according to the method whose certain failure Evert had proved time and again. In his worst performance of the summer, Evert sent eighty-six battalions of the Third Army against an enemy force one-fifth their size with no result except for huge Russian casualties.[73] Kowel—what the Eighth Army's quartermaster general once called "the Kowel pit"—became a deadly drain on Russia's manpower until winter put an end to Evert's hopeless attempts to bludgeon his enemy into submission.[74]

While Evert's battering-ram tactics bogged down against Kowel, Brusilov orchestrated the advance of his Seventh and Ninth armies with greater success. By launching less concentrated attacks along a longer front, these forces took Halisz and Brody, cleared all enemy forces from Bukovina, and, at two points, actually advanced onto Hungarian soil. As the autumn rains rendered impassable the marshes where so many of the Guards Army had perished, and as seventeen divisions brought in from the western and Italian fronts stiffened the Austro-German defenses, Brusilov's offensive finally ground to a halt.[75] Clearly, he had won the Russians' greatest victory of the war, for the enemy's losses had exceeded a million and a half men, more than 400,000 of whom had been taken prisoner. Almost 600 field guns, 1,795 machine guns, and untold quantities of military supplies had fallen into Brusilov's hands and his divisions had seized some twenty-five thousand square kilometers of enemy-held territory in less than three months. Yet he had paid a high price for his victories, for he had lost half a million of the troops that Polivanov had trained with such care the previous winter.[76] How soon replacements could be properly trained was a question that none could answer that fall. Upon that single factor, the success of Russia's future war effort would in large measure depend.

At the end of August, Brusilov's final victories in Bukovina and Galicia convinced the government of Rumania, which, for more than two years, had hesitated to declare itself, to enter the war on the side of the Entente. The Allies now seemed to have every opportunity to open another front against the Central Powers in the east at precisely the moment when the Russians were pressing the Austro-Hungarian armies the hardest. "It was the moment, if ever, for Rumania to march," Britain's Ambassador Buchanan wrote,[77] and Hindenburg, just named to replace Falkenhayn, agreed. "It is certain," the new chief of Germany's High Command later wrote, "that so relatively small a state as Rumania had never before been given a role so important, and, indeed, so decisive for the history of the world at so favorable a moment."[78] All of Europe shared that view. "By making the necessary arrangement with Rumania to crush the Austrian army, we should compel Germany to make an additional effort which may well be beyond her immediate resources," France's President Poincaré wrote to Nicholas. "I think it my duty to inform Your Majesty of the very great interest which the General Staff of France attaches to the conclusion of an agreement with Rumania at the earliest possible moment."[79] For one moment more, Rumania hesitated—to be certain that her choice would allow her "to fly to the aid of the conqueror," a sharp-tongued French diplomat remarked—and then, on August 27, declared war on Austria-Hungary.[80]

The Rumanian plum, which all of Europe's belligerent powers had been so eager to pluck, proved sour fruit indeed. Rumania's army numbered 620,000 ill-trained men, led by an officer corps which one sympathetic observer described as "utterly devoid of any sense of military morality," and which at the time of its mobilization had to be reminded that none below the rank of colonel had permission to wear make-up.[81] Largely illiterate, and with no experience in using the machines of modern war, the Rumanian army did little to realize its allies' hopes. Leaving one lone division to guard against a Bulgarian attack in the Dobrudja—the very lands they had seized from a still-resentful Bulgaria in 1913 —the Rumanians hastily launched the rest of their armed forces into Transylvania, their so-called "natural objective," where they were quickly overrun by an Austro-German force under the redoubtable Mackensen.[82] Quickly the Allies felt the dead weight of this ineptly-led, primitive fighting force, on which Russia had to lavish precious weapons, munitions, and manpower. Rumania was better thought of as "Ruritania," Bernard Pares caustically remarked at one point, and the moment he saw the Rumanians in combat, the commander of Russia's relief force begged to be released from what he called "a punishment for some crime that I did not even know I had committed. . . . I have to fight more with the Rumanian troops than with the enemy," he continued. "I feel almost certain that, when the first Bulgarian shell lands, the Rumanians will scatter in all directions."[83] His fears proved all too well-founded. By Christmas, Mackensen had occupied Bucharest, driven the remnants of the Rumanian army into the borderlands of Moldavia, and begun to harvest the spoils of victory: a million tons of oil, two million tons of grain, 200,000 tons of timber, and over a quarter-million cattle, goats, and swine, all of which the Central Powers needed almost as desperately as the Russians had needed weapons and munitions the year before.[84]

As the war's third winter approached, the Rumanian debacle and the torrents of blood that Evert's clumsy tactics continued to shed in the marshes before Kowel began to tarnish the brilliance of Brusilov's victories. Men wearied of being sent again and again through terrain that could not be crossed to attack enemy strongholds that could not be taken, and Russia's soldiers began to feel a sense of betrayal by senior officers who sent them so readily and so often to their deaths but gave them no sense of the purpose in their dying. This was particularly so because the battles of the Great War had slaughtered the junior officers who traditionally had led Russia's armies into battle. "The Russian soldier requires leading more than any soldier in the world," General Knox once wrote. "The average Russian soldier was indeed very much like an Indian buffalo," he remembered some years later. "He would go anywhere he was led or driven, but would not wander into uncomfortable places on his own." The numbers of men left to lead Russia's soldiers now had shrunk to a comparative handful. To save officers, regimental commanders sent only two per company into action at a time and held the rest in reserve to replace casualties.[85]

Without officers to lead them into battle without question or hesitation, men grown weary of the war's killing began to ask why the war must continue. For men condemned to stay in the line until they were killed or the fighting ended, that question took on deadly urgency that fall, as the toll of killed and wounded rose beyond the War Ministry's ability to calculate. The deaths seemed unending. In the marshes before Kowel, so many rotting Russian corpses littered no-man's-land in September that their comrades asked for a truce to clear them away.[86] How many had died? "No one knows the figures," Hindenburg wrote after the war ended. "All we know is that sometimes in our battles with the Russians we had to remove the mounds of enemy corpses from before our trenches in order to get a clear field of fire against fresh assaulting waves."[87]

That fall, too many men too long deprived of hope began to look for avenues other than desertion or self-mutilation to escape war's endless horrors. A sense of disquiet that filtered to the front from Russia's villages and factories stirred soldiers' restive bitterness, and a handful of mass protests began to replace individual acts of insubordination. Wildly apocryphal rumors about starvation in Petrograd, about special cemeteries for the legions of civilians perishing from hunger, and about bread that cost a ruble a loaf raced through Russia's trenches and fanned smoldering fires of discontent.[88] Too long separated from loved ones from whom they had too little news, soldiers added fear for family and friends to their resentment at seeing comrades killed to no apparent useful purpose. For some, the disquiet grew too intense to bear, and more than a dozen regiments mutinied during the last weeks of 1916.[89] "Give us boots and warm clothing first!" some shouted when officers ordered them forward, while others insisted that "We'll hold the front, but we won't attack."[90] As the snow began to swirl across Eastern Europe's blood-soaked no-man's-land for the third time, soldiers turned away from the weak and incapable tsar who allowed himself to be ruled by selfish, incompetent men and a neurotic and paranoid empress. "The tsar surrounds himself with Germans who are despoiling Russia!" some of the mutineers cried.[91] Others so detested the war's killing that they no longer even feared death. "Take us and have us shot," one company telegraphed with fatalistic resignation to Nicholas, "but we just aren't going to fight any more."[92]

The army that had been the tsar's one great buttress against revolution in 1905 now seemed about to crumble. Even the Knights of St. George, who wore Russia's highest medal for valor, began to think that no good could come from the war now. "Somehow, I have a feeling that, after all this is over, we are not going to be thanked for all the hardships and privations which we are going through now," one general told his men on their patron saint's feast day that December. "Rather [I feel] that all this is going to be held against us."[93] Far more quickly than anyone could have imagined in June, Russia's last promise of victory had soured. As they dined at Donon's elegant restaurant on the Moika Embankment one evening late in September, former Prime Minister Kokovtsev, who had lost his emperor's trust and friendship for speaking out against Rasputin, warned Paléologue: "We're heading for revolution." Aleksei Putilov, their immensely wealthy dinner companion, whose Petrograd mills had produced at such great profit so many of the weapons with which Brusilov had won his victories, thought differently. "We're heading for anarchy," he told his two dining companions. "There's a vast difference. The revolutionary has the intention to reconstruct; the anarchist thinks of nothing but destruction."[94]

Not long after Paléologue, Kokovtsev, and Putilov had dined together, the Union of Municipalities sent Konstantin Paustovskii, who had left his duties as a medical orderly in order to work on a Moscow newspaper, to the countryside beyond Moscow to test the feelings of the masses. Uncertain what he would find, Paustovskii went with pen and sketchbook in hand to "moss-covered old Russia," to a backwoods backwater of "impassable roads, decayed old settlements, ancient peeling churches, little horses with manure stuck to their hides, [and] . . . cemeteries with overturned crosses above the graves." Instantly, he felt a chill more cutting than any stirred by winter's winds. "This land was freezing over," he remembered, "darkening the spruce palisades with scarlet stripes of late fall sunsets, crackling with the first ice crystals, covering the fields with smoke from freezing villages." Fascinated, Paustovskii listened to the sound of the people and the land. "I heard drunken and sober talk, timid and desperate talk, talk full of submissiveness and full of hatred," he wrote. Everywhere, people spoke against the war, with all their anger "centered in the west, with the army." Surrounded by anger too long suppressed and bitterness too long unrelieved, Paustovskii felt a great tension building. The dark shadow of impending apocalypse, stark and unrelieved, hung over the land that winter. "It was as if no one, anywhere, expected any happiness," he decided. "Everyone, languishing, was waiting for the drama's denouement."[95]

CHAPTER X

"Dancing a Last Tango"

The tense waiting that Paustovskii found in Russia's villages at the beginning of the war's third winter reflected not only the masses' deadening war-weariness, but also a deepening sense that some titanic event lay just beyond the horizon. Few were certain of its form, but all awaited its approach, perhaps hoping that in some way it would create order out of the contradictions that now seemed to litter every corner of Russian life. For no good reason, able statesmen fell quickly from office, while sycophants of no talent rose to high positions. At the front, men with weapons had no officers to lead them into battle. Artillery had shells but no way to chart the impact of its fire on enemy positions because the Russian army had almost no airplanes or barrage balloons. Generals feared to attack when they outnumbered the enemy by more than two to one, but were willing to launch hopeless assaults against his strongest defenses. War contracts promised profits beyond industrialists' wildest dreams but brought declining real wages and worsening working conditions for factory workers. Bountiful harvests produced shortages of food in towns and cities.

Not long before he went to watch and listen in Russia's villages, Paustovskii had seen a procession of Uzbeks being driven through Moscow on their way "to build an Arctic railroad [from Murmansk to Petrozavodsk] and to die." At their destination, they would find that Russian military engineers had transformed the unending Arctic winter nights into eerie torch-lit days to finish the tracks that would carry Allied munitions from Murmansk to the front for the campaigns of 1917. This "procession of the doomed,"[1] of men sent to the frigid Arctic from the torrid lands of Central Asia, somehow epitomized the stark conflicts that seemed about to overwhelm the Russians as the Great War entered its third year. Where were events leading and what were they coming to? A sense of impending Apocalypse gripped Russians now. As the shape of the Apocalypse grew more concrete, their visions darkened. "Now the time had come for a reckoning," the brilliant young conservative politician Vasilii Shulgin realized. "And there we were," he remembered as he recalled the strange intensity of those days some years later, "dancing a 'last tango' on the rim of trenches filled with broken corpses."[2]

Broken corpses, gut-shot men screaming as they lay impaled upon the barbed wire of no-man's-land, trenches plowed into level fields by the shells of heavy guns, their occupants vanished without a trace—these were the reality that overwhelmed the grandiose imprecise visions of the Apocalypse that had fascinated Russia's writers and poets ever since 1900, when Vladimir Solovev had first proclaimed that its advent would be marked by Asiatic hordes more ruthless than any led by Genghis Khan storming out of the East to impose a new Mongol yoke upon Russia and the West. Spiritual father of the brilliant culture that blazed across Russia in the two decades before the Great War, Solovev made no predictions about when the West would face the onslaught of Asia's Yellow Peril. His vague warnings therefore were heard at first only by a few, but they took on the false garb of prophesy in 1905 after Japan's armies defeated the forces of Russia in the Far East. Solovev's "prophesies" held Russians' imaginations with a force akin to the power of Tolstoi's preachings, and they responded with a flood of apocalyptical writings more numerous and more intense than any that appeared anywhere else in the world.

A lonely gentle thinker, whose long hair and haunting, deep-set eyes heightened his ascetic appearance, Solovev was once described as a man who "spoke like a prophet, lived the life of a monk, and yet, like a child, could not resist the temptation of a bonbon or a jelly tart."[3] He may have been Dostoevskii's model for Alësha in *The Brothers Karamazov*, and, like his fictional likeness, he chose to live in the real world even though he remained apart from it. A monk without a monastery, he dreamed of a "free theocracy" in which the Russian tsar reigned as the ideal Christian prince and where men and women would strive to achieve a state of "total-oneness" through sexual love and artistic expression. Always, he emphasized symbols, not reality, and, for an intelligentsia grown weary of the dictum that art must first serve some useful social or practical purpose, his preachings promised long-awaited liberation.[4]

Ready to accept art for art's sake after more than a half-century's dedication to realism, Russia's writers and artists seized upon Solovev's preachings as new gospels that opened the way to higher harmonies and greater beauties. Under the influence of his vision, they wrote of Armageddon, the world's approaching end, and the Antichrist's triumph. Proclaiming that sexual pleasure was "the anticipation of the Resurrection of the flesh," Dmitrii Merezhkovskii greeted the twentieth century wrapped in a greenish haze of cigar smoke, his physical ugliness contrasting vividly with the striking auburn and ivory beauty of his wife, the poet Zinaida Gippius, as they presided at the *Dom Muruzi,* their *avant-garde* Petrograd salon. Convinced that "the principal thought of Christianity is the thought of *the end of the world,* " and fearful (or, perhaps, hopeful) that "the period of crisis and of the Last Judgment is beginning," Merezhkovskii clearly believed that modern civilization was approaching an apocalyptic climax. He pursued that theme at the turn of the century in *Christ and Antichrist,* a vast and complex trilogy of historical novels in which he recounted the western historical experience in terms of a titanic struggle of the gods. Beginning with the death of the gods under Julian the Apostate, Merezhkovskii's work told of their rebirth under Leonardo da Vinci and concluded with the final confrontation between Christ and Antichrist, which he depicted as the struggle of Peter the Great against tradition and his son Aleksei. Merezhkovskii envisioned the coming of the Kingdom of the Holy Ghost, the time foretold in Revelation 11:15, when "the kingdoms of the world are become the kingdoms of our Lord, and of his Christ." Yet his vision remained always out of focus, for he never spoke clearly of who (or what) would be defeated at Armageddon, nor did he locate the Apocalypse in time and space.[5]

The poet Valerii Briusov spoke with more originality and his writings bore a more dramatic message than Merezhkovskii's. Grandson of a serf and son of a wealthy Moscow merchant, his high-cut cheek bones and Mongolian eyes an ever-present reminder of the conquerors who had stormed out of Asia in centuries past, Briusov took up Solovev's theme that the Yellow Peril would herald the approaching Apocalypse, exclaiming: "Where are you, you marching Huns? I hear your iron tread!" Briusov's Huns were the new conquerors from Asia, the men who would drive "thinkers and poets" into hiding "in catacombs, in deserts, and in caves," but they remained a distant rumble "on the still undiscovered Pamir Mountains" and posed no immediate danger to Russia or the West. He therefore shouted a welcome ("You who will destroy me, I welcome with a hymn of greeting") to those who would come "to revive our prematurely decrepit bodies with a wave of burning blood."[6] For the moment, Briusov felt no clear and present danger, merely the delicious anticipation of danger in the distant future, and that made the vagaries of his vision all the more appealing.

Other apocalyptic symbols pressed upon Russians with greater urgency, as the Great War neared. Its throbbing factories and roaring machines a living monument to the achievements of modern science and technology, the modern city readily became a symbol of the coming Apocalypse, as Russia's poets and painters transformed factory chimneys into the Beast's "red fingers" which threatened to "rip out of the soil the onion domes of the faithful."[7] Briusov's famous "Pale Horse" moved apocalyptic images even more deeply into the vitals of Russia's urban life, where "the street was like a storm. Crowds passed by./ As if pursued by fate's inevitability." In the midst of this turmoil, Briusov saw Death. Astride a pale horse as foretold in Revelation 6:8, the specter flitted for an instant across his lines and vanished, as quickly and as uneventfully as it had come, while "Omnibuses, cabs, and automobiles rushed on past,/ And the surging torrents of humanity flowed on endlessly."[8] Briusov had moved his vision of the Apocalypse into the center of the modern city. But he drew back from adding the details that could reveal its true nature or locate his vision more precisely in time.

Briusov's reticence about his vision had less to do with his perception of the Apocalypse than with the longing of Russia's *avant-garde* writers to break free from the bonds of time and space in the years before the Great War. A number sought refuge in the famous "Tower" of Viacheslav Ivanov, a poet who published his first book of verse at the age of thirty-seven, and who, some said, "loved not so much poetry as power over poets."[9] In appearance, Ivanov resembled more a prophet of the Old Testament than an oracle of modernism, and the learned treatises on philosophical and philological questions that he had written during his twenties and thirties gave no hint of his future importance among Russia's artists and poets. With his wife Lidia Zinoveva-Annibal, a minor writer best known for her lesbian novel *The Thirty-Three Abominations,* "Viacheslav the Magnificent" came to preside over a salon in a penthouse overlooking the famous Taurida Palace where the Duma met. There, he and St. Petersburg's *avant-garde* poets set out to erase all sense of time and space from the Tower's rooms: "Square rooms, rhomboids, and sectors," the poet and novelist Andrei Belyi once said, "where thick rugs swallowed up the sounds of footsteps . . . [and] day became night and night became day." Belyi found it easy to "forget what country you were in and what time it was" in the Ivanovs' Tower. "You'd blink," he remembered, "and a month would have passed."[10]

To the Tower, people came—some for a few hours, others for days, even weeks—and then, according to their inclinations, continued on their way. Many came to listen to the few who came to speak from the small circular space that Ivanov insisted upon reserving for that purpose no matter how many crowded into the Tower's rooms. Tall, broad-shouldered, his gray visionary's eyes radiating poetic warmth, Aleksandr Blok declaimed his verse from the Tower's roof to an audience whose unconcealed delight led young Kornei Chukovskii to remark many years later that "Blok's poetry affected us as the moon affects lunatics."[11] Andrei Belyi, at times Blok's bitter foe, at others his closest friend, appeared often, and once stayed for five weeks, while Mikhail Kuzmin (a minor poet known for his dedication to erotica and devotion to beautiful adolescent youths) stayed for more than a year.[12] There, Georgii Chulkov once proclaimed his belief that "Life—is love."[13] On other occasions, the middle-aged Vasilii Rozanov, married to Dostoevskii's former mistress, and consumed by what a critic once called "his simple- and single-minded concern with the mystique of sex," held forth.[14] A former provincial schoolteacher who launched anti-Semitic campaigns while he wrote of his love for Judaism, Rozanov insisted that "each soul is a Phoenix," and that "each soul must burn, and the great bonfire of these burned souls creates the flame of history." Piling sincerity upon hypocrisy, this churchgoing sinner who aspired to saintliness and thought that God might safely co-exist with the Devil, continued his contradictory path. Proclaiming that "my works are blended with neither water nor blood, but with semen," Rozanov eventually concluded that "a bull mounting a cow" expressed more theology than all the seminaries in Russia, and then, as if to confound his critics beyond the point of confounding, took refuge in the famous Troitse-Sergeevskaia Monastery where he wrote *The Apocalypse of Our Time,* his last and most renowned work, at the end of the Great War.[15]

By the time Ivanov closed the doors of his "Tower" and went to live abroad in 1912, more urgent forces had begun to press upon Russia's *avant-garde* writers. In the three years before the Great War, their sense of the approaching Apocalypse, gained substance from a series of events that seemed destined to build toward an apocalyptic climax, although these events still did nothing to clarify their visions of when and how the Apocalypse might come to pass. The first was the long-awaited reappearance of Halley's Comet in May 1910 which, Blok thought, might bring universal catastrophe and reconciliation in its wake. For a brief moment, he focused his image of the Apocalypse clearly in space and time as he greeted the impending catastrophe. "You threaten us with the last hour," he wrote. "But ruin is not terrible for the hero while his dream still raves."[16] Other writers including Merezhkovskii and Briusov also connected visions of Apocalypse with the comet, but no one, not even Blok, seriously thought that Halley's Comet would strike the earth, or that its passing threatened an immediate catastrophe.[17] The comet's appearance nonetheless added to the sense of impending doom that weighed more heavily upon Russia's *avant-garde* with each month that passed.

That same year, three giants of Russian culture—Lev Tolstoi, the longest-living and perhaps greatest of Russia's nineteenth-century novelists, Mikhail Vrubel, Russia's first individualist painter, whose tortured mind produced paintings of rare imaginative intensity, and the actress Vera Kommissarzhevskaia, thought by some to be the "poetic image" of the Symbolist theater—all died within the space of a few months. Tolstoi's death was not unexpected, for he was over eighty, but Kommissarzhevskaia at forty-six was at the peak of her beauty and artistic power when smallpox struck her down in faraway Tashkent, in Central Asia. Perhaps there was something particularly terrifying in knowing that the beauty of the woman who was sometimes thought to be "the emblem of the times" had been ravaged so suddenly, and so utterly, by disease, for her death—Blok called it an "agonizing, but young, but pre-spring death"—left an especially deep scar on Russia's literary and theatrical world.[18] Coming less than two months later, Vrubel's death and the circumstances surrounding it added to this deepening sense of doom. The first of Russia's painters to reject realism in favor of art for art's sake, Vrubel died stark raving mad in an asylum, "having reduced himself to utter exhaustion," Blok wrote his mother, "because he had concluded that God would give him emerald eyes if only he could remain standing for seventeen days."[19]

Such experiences with death and celestial turmoil tormented Russia's writers throughout the fall and winter of 1910–1911. "I've decided to get away from here," the aspiring poet Elizaveta Kuzmina-Karavaeva told Blok, with whom she was having a brief affair. "The only thing left to do here is to lie down and die."[20] Blok agreed, but with no intention of following her path, for he "already sensed the smell of burning, blood, and iron in the air."[21] That spring the historian and Kadet leader Pavel Miliukov spoke soberly about "An Armed Peace and the Reduction of Armaments," while others began to write about "The Nearness of the Great War." Public opinion split bitterly in a new wave of anti-Semitism that surged over Russia in the wake of frenzied efforts by the minister of justice to convict the hapless Mendel Beilis* of ritual murder even after his innocence was discovered.[22] That summer, Blok remembered, the "heat was so intense that the grass burned to its very roots,"[23] an earlier symbol of impending catastrophe destined to reappear in the vast peat and forest fires whose smoke, Gippius and Paustovskii recalled, lay heavy upon St. Petersburg and Moscow in the summer of 1914.[24] People seemed to suffer from "a sort of sickness in the soul," and the crisis seemed to be approaching very rapidly now. This was a time of "drunkenness without wine, of food that does not satisfy," Kuzmina-Karavaeva confessed as she prepared to break away from the overwhelming sense of "ineradicable decay" that she felt in St. Petersburg.[25]

Foreign crises combined with these domestic tragedies to convince Russia's *avant-garde* writers that the world stood at the brink of a cataclysmic conflict that would transform their native land forever. Certain that "[Kaiser] Wilhelm is looking for a fight, and by all the signs he *will get his war,*"[26] Blok pleaded with Belyi to break off his foreign travels in Europe and North Africa and "come back to Russia. It may turn out that there is not much time left in which to know her as she is now!" he urged.[27] For a moment, the place and time of the Apocalypse seemed more clear and immediate, but it was Belyi, not Blok, who captured the deep sense of foreboding that gripped the men and women who once called themselves "the children of Russia's dreadful years,"[28]—in a brilliant novel of startling apocalyptic imagery.

Belyi began the novel, which he called "The Lacquered Carriage," in 1911, the same year that Blok decided that "the governments of all nations finally have gone too far,"[29] and that a shattering cataclysm lay just beyond the horizon. After three months of frenzied work (he had written his first novel, *The Silver Dove,* in only five weeks), Belyi had it finished, only to find that his publisher, the general editor of Russia's leading semipopular journal *Russkaia mysl, Russian Thought,* thought it "immature," filled with "nonsense," and "preten-

*See Chapter 5, pages 139–40.

tiously and carelessly" written.[30] For more than a year, Belyi labored to recast his sprawling text until one of Russia's leading publishing houses agreed to publish it in installments as *Petersburg*, just a few months before the Great War began.[31] In a radical departure from the work of Russia's great nineteenth-century writers, Belyi set his novel in the midst of a world of sound, sometimes resonant, often dissonant. The ticking of a time bomb, concealed in a sardine can and set to explode in twenty-four hours, dominated its pages as a sometime-revolutionary son struggled with the party's orders to kill his reactionary father. "Pepp Peppovich Pepp is a party bomb," Belyi wrote in one of his most memorable uses of sound in prose. "Pepp Peppovich Pepp will expand and expand. And Pepp Peppovich Pepp: will burst!"[32] Inescapable, never-ending, the cadence of the bomb's ticking penetrated to the novel's very depths, just as the vision of impending catastrophe—at various times in the form of war, Armageddon, revolution, or Apocalypse—stood riveted in the deepest reaches of the consciousness of Belyi and his friends.

Belyi, an ardent Muscovite, saw St. Petersburg as the "un-Russian-but-nonetheless-capital city" that still guided the destiny of an empire in decay. A terrible and terrifying modern industrial city like the apocalyptic image Briusov and Blok had described earlier, Belyi's Petersburg remained an alien realm of "bodies, bodies, and more bodies: bent, half-arched, bent hardly at all, and not bent."[33] It was a city of specters, of "gray human streams" with "utterly smoke-sodden faces" and "greenish faces," dehumanized by the "many-chim-neyed factories" in which they labored. A city of "saffron murk," "greenish murk," "yellow-green fog," and "white-hot fog," Petersburg remained "ashy and indistinct," "a dot [which] is the place where the plane of being is tangential to the surface of the sphere and the immense astral cosmos." Petersburg had no sun, Belyi said. Only after the titanic battle between East and West, would the sun "rise in radiance." "As for Petersburg," he concluded in a statement of bitter disenchantment, "it will sink."[34]

Frequently disturbing, sometimes fearsome, Belyi's visions of the approaching cataclysm nonetheless remained as spectral as his images of Russia's capital. "Suffering and weeping [and] . . . divided in two until the final hour,"[35] Russia would at last achieve the universal reconciliation of which Blok had dreamed as he pondered the approach of Halley's Comet. Like Kuzmina-Karavaeva, who thought that "the last act of the tragedy of estrangement between the people and the intelligentsia"[36] had been played out, Belyi saw Russia poised on the brink. "There will be a leap across history," he wrote. "Great shall be the turmoil [and] . . . the very mountains shall be thrown down by the cataclysmic earth-quake." Returning to Solovev's warnings about the Yellow Peril and to Briusov's theme of marching Huns, he spoke of "yellow hordes of Asians" whose struggle with the West would "encrimson the fields of Europe in oceans of blood" in a final titanic struggle, "the like of which has never been seen in this world."[37] Belyi's vision of Armageddon thus was of a great set-piece confrontation between Russia and her Asian foes. No images of modern technology entered the picture. Belyi cast his vision of the "final days" in terms of the battles of olden times, not of modern warfare, to herald the events that would usher in Christ's Second Coming.

Overwhelmed by a sense of decadence and sterility, on the eve of the Great War many of Russia's intelligentsia shared Belyi's growing impatience for the cataclysm they hoped might cleanse and regenerate the society in which they lived. "The literary circles in Petersburg have reached the last stages of putrefaction," Blok wrote to Belyi not long before the war. "They are beginning to stink."[38] The "children of Russia's dreadful years" now were very far from childhood, and life on the eve of the Great War was far from being very dreadful for them in any material or cultural sense.[39] For many, the war's approach coincided with the onset of middle age. Gippius, Merezhkovskii, and Ivanov all were approaching fifty when the war began, and all passed that mid-century milestone well before its end. Briusov was over forty, while Belyi and Blok—the *enfants terribles* of the Symbolists—were in their mid-thirties. As the Great War neared, the Symbolists' raging declarations of disenchantment somehow rang off key, for the world in which they awaited Armageddon's cleansing fire was the best that Russia's intelligentsia had ever known. Although there had been rare exceptions before, theirs was the first generation of writers and artists to live comfortably from their brushes and pens. In a manner unknown to earlier generations of Russia's intelligentsia, literary accomplishment on the eve of the Great War offered upward mobility and material prosperity to men and women of humble origins.

Perhaps Maksim Gorkii, orphaned at the age of eight and forced to live as a rag-picker, river boat scullion, and cobbler's apprentice before the age of fifteen, was early-twentieth-century Russia's most striking literary success story. A man who combined a passion for understanding mankind with a passionate dedication to serving humanity, Gorkii spent the 1880s wandering through Russia, living by his wits and manual labor, and meeting the thousands of men and women who later served as the raw material for the novels and plays that catapulted him to literary greatness on the eve of the twentieth century. Gorkii's first literary success came after an acquaintance locked him into a room and forced him to write down some of his adventures. After his first tale appeared in the Tbilisi newspaper *Kavkaz, The Caucasus,* in the fall of 1892, he went on to write for a number of provincial dailies until he published his first story in the influential national monthly *Russkoe Bogatstvo, Russian Wealth.* From month to month, Gorkii's popularity soared. Now almost always dressed in boots and peasant shirts as a monument to his plebian origins, he became almost instantly known across the length and breadth of Russia, his morose face always contrasting sharply with his warm blue eyes wherever he appeared. When the first edition of Gorkii's collected stories was published in 1898, the writings of this man who had had no more than five months of formal schooling in his entire life sold on a par with those of Tolstoi, Russia's best-selling author until that time.

There were other writers whose success paralleled Gorkii's. All profited from the soaring number of readers in Russia in the 1890s[40] and from the moderately improved political climate of the early twentieth century that made possible unique cultural achievements. In the wake of the Revolution of 1905, the virtual abolition of censorship left them free to explore new themes and experiment with genres in ways undreamed of by such giants of the nineteenth-century literary scene as Pushkin, Turgenev, Dostoevskii, and Tolstoi. Secure in their material comforts and free to write as they chose, they looked into the future and, to their dismay, found a murky vision of the Apocalypse illuminated by the dreamlike hope that Armageddon's fire could cleanse them of the stains of modern life.

Such dreamlike visions could not survive in a world of real war, where Armageddon seemed destined to be fought again and again (and with no conclusive result) on the fields of East Prussia, Poland, Belgium, and Northern France, and such titanic struggles as Tannenberg, Tarnów-Gorlice, Verdun, Kowel, and the Somme had made the awesome destruction of modern warfare too well-known for men to dispute its raw brutality or glorify its suffering. "In war, there is no room for dreams," wrote Ilia Ehrenburg, a Russian Jew who lived most of the war years in Paris. *Real war was very* deadly and very cold, its scenes painted only in flat monochromes, again belying the Symbolists' visions of Armageddon's cleansing fires. "In war," Ehrenburg explained as he recalled the days when he had written reports about the war on Europe's western front for Russia's newspapers, "guns and soldiers' faces lose color. Straight lines, planes, drawings like blueprints, an absence of anything arbitrary, of any thing that is lovably irregular" made war purely and simply "a well-equipped factory for the annihilation of mankind."[41]

Into this "well-equipped factory for the annihilation of mankind" that reality had deprived of all romance and glory, tens of millions had marched during the first two years of fighting, and millions had not returned. Glory had no place in their deaths, nor had it much meaning for those who remained among the living. For Russia's younger generation, born after the great famine of 1891 and obliged to watch friends die while still in the flower of early adulthood, the cataclysm of the Great War became as awesome and brutal as it was terrifying. No longer could the Apocalypse be awaited in fearsome anticipation from afar. Unlike the "children of Russia's dreadful years," the young men and women of the Great War era knew that the Apocalypse—ugly, brutal, and bitter—was upon them and that it promised not rebirth but only a very cold and certain death. Escape could not come through the cleansing fire of Armageddon, but only, perhaps, through a revolution that would wipe away the old world and usher in a regenerated new society.

Among these young men and women, none spoke more brilliantly or more forcefully than Vladimir Maiakovskii, son of a forester and just twenty-one when the war began. Unlike his Symbolist predecessors, Maiakovskii was deeply and seriously interested in politics, joined Lenin's Bolsheviks at the age of fifteen, and was arrested three times before his sixteenth birthday. After spending his sixteenth birthday and several months more in solitary confinement in Moscow's notorious Butyrki Prison, Maiakovskii concluded that poetry could not express his deepest feelings, decided that he should paint instead, and joined a group of artistic rebels in a "fight for futurism," which they launched by painting their faces with wildly-colored designs and parading through Moscow in outrageous clothing. Together they painted, wrote, and lectured to win a hearing for their works. "The publishers wouldn't print us," Maiakovskii remembered. "The capitalist nose smelled dynamite in us."

From Moscow, these young artists carried their cause into the provinces—Maiakovskii called their journeys his "Golgotha of auditoriums"—to confront (or, at least, affront) the public and win a hearing.[42] "Throw Pushkin, Dostoev-skii, Tolstoi, etc., overboard from the steamer of modernity," they challenged their listeners. "From the height of skyscrapers we look down on their insignifi-cance."[43] Always, Maiakovskii appeared in a tunic with alternating black and vivid orange-yellow stripes. Young Boris Pasternak first saw him then and found an "irrevocable" quality about his intensity that made him look "hugely unbent and unconstrained."[44] By the time he had finished his provincial "fight for futurism," Maiakovskii had become a serious poet. Like a number of his poet-painter contemporaries (Guillaume Apollinaire led a similar group in Paris), he continued to paint and even exhibited his cubist paintings along with the work of such renowned modern Russian artists as Larionov, Goncharova, Chagall, Tatlin, and Malevich in the fall of 1915. But by then, his paintings had become pale shadows of his poetry.[45]

The outbreak of war found Maiakovskii briefly in Moscow under secret Okhrana surveillance as a man of dangerous views. In a surge of patriotism that he quickly regretted, he tried to enlist, was rejected as politically unreliable, produced a few posters to support the war effort, and then moved to St. Peters-burg—now Petrograd. "I greeted the war with agitation," he later confessed. "In the beginning, I only noticed its decorative and noisy qualities."[46] Perhaps more quickly than most Russians, he came to understand the terrible side of war and, by early 1915, he raised his voice against it. By then, Petrograd's poets no longer proclaimed their poetry to the stars and the city's cultural elite from the heights of Ivanov's Tower. Now, they aimed in the opposite direction, declaiming their verses to the general public in a cellar cabaret on Mikhailovskaia Square called the "Stray Dog."

In vivid contrast to "Viacheslav the Magnificent," who had presided over the Tower's nightly discussions of art, politics, and poetry, a half-decade before, the Stray Dog's proprietor Boris Pronin possessed finely-tuned entrepreneurial instincts, and used them to attract to his crumbling establishment crowds of politicians, war profiteers, and similar philistines who wanted to feel in touch with the city's cultural pulse. With remarkable virtuosity, Pronin exploited writers' and artists' thirst for public acclaim and the public's passion to feel a part of their achievement. "Like hot dishes, prepared in another part of the city and carried here in a thermos," one of Maiakovskii's friends remarked snidely, "newly won triumphs were brought here when one wanted to prolong them, and taste their flavor for a second or third time." On occasion, Pronin's sleazy patrons paid as much as twenty-five rubles (nearly a week's pay even for one of the elite *metallisty*) to partake of the Stray Dog's newest offerings, while they drank coffee or water flavored with pineapple juice in patriotic homage to wartime prohibition and "squeezed each other's lustful knees under the table."[47] Anna Akhmatova, perhaps the greatest of Russia's woman poets, appeared at the Stray Dog dressed in black silk, a large oval cameo at her waist, and a bright yellow stole draped around her shoulders, as living proof that Nathan Altman had not exaggerated her classic beauty in the portrait he had done earlier that year. In a small side room Akhmatova intoned, her sad, dark eyes cast upon an audience who hung upon her every word: "We are all sinners, we are all whores."[48] Blok often stood nearby, "with his stony, impenetrable expression," while Maiakovskii observed the cabaret's comings and goings "half-lying in the position of a wounded gladiator, on a Turkish drum."[49]

The Stray Dog stank perpetually of urine, sweating bodies, and stale smoke from the searing makhorka tobacco favored by its habitués. Pronin's patrons greeted each performance with shouts of *"Homage! Homage!"* and examined "thoughtfully, though inconclusively" Akhmatova's latest work. Throughout it all, Maiakovskii brooded stormily, "with very dark eyes which had an insolent and sullen provocative look," banging on his drum whenever a fellow Futurist appeared.[50] Always dressed in his orange-yellow and black tunic, he listened but spoke little. Then, one night in February 1915, he rose to confront the men and women who sat before him. *"Vam!"*—"To You!"—he roared at his expectant and unsuspecting audience. *"Vam!* who only live from orgy to orgy. . . . *Vam,* who love only women and food,/ Why should I give my life for your convenience?!/ I'd be better off serving pineapple water/ To the whores at the bar!"[51] The next moment, the Stray Dog exploded. Men in *Zem-gusar* uniforms, who had never heard a shot fired in anger and had no intention of ever doing so, cursed in patriotic outrage, while women burst into tears. "We would consider coming here [again] below our dignity," some of the older artists and writers told Pronin disdainfully. According to the newspaper account that appeared the next evening, Pronin turned away in disgust and tossed the words "Good riddance!" over his shoulder as he did so.[52] With a mere seventy-seven words—sixteen brilliant lines—of poetry, Maiakovskii had stripped away the sleazy aura of false heroism with which men who peddled patriotism for profit surrounded themselves to become a resolute critic of Russia's war effort.

By a stroke of bitter irony not uncommon in Russia at the time, the authorities had rejected Maiakovskii the would-be patriot when he had tried to enlist in the fall of 1914, only to draft Maiakovskii the bitter war critic a year later. For a brief moment, his poetic question: "Why should I give my life?" took on note of special urgency. But Maiakovskii was more fortunate than many of his contemporaries. Gorkii, then at the peak of his fame as chronicler of life in Russia's lower depths, proclaimed him "a poet of great caliber" and came to his rescue. "There is really no futurism, there's just Maiakovskii," he remarked after their first meeting.[53] Anxious to spare men of talent from the war's carnage, Gorkii arranged to have Maiakovskii assigned to Petrograd's auto school along with a number of other young writers including Viktor Shklovskii. After a helpful engineer spent several nights teaching him how to use drafting instruments to make blueprints of automobiles, Maiakovskii became one of the school's "experienced" draftsmen.[54]

Partly influenced by Gorkii's conviction that the war must be ended in any way at any price, Maiakovskii wrote *Voina i Mir (War and the World)*, interspersing poetry with bars of popular tango music in what became in a very real sense another "last tango." Although the title held the same words as Tolstoi's *War and Peace*, since the Russians used the same word for "peace" and "world," Maiakovskii's long poem had little in common with Tolstoi's epic work. "Listen!" he exclaimed in its dedication. "Each person,/ Even someone who is of no use,/ Has the right to live./ You can't,/ Simply cannot/ Bury him alive/ In trenches and dugouts—/Murderers!" For a moment, his anger turned even upon the cities he had held up as symbols of progress: "Where the woodland was—now is a square with a hundred-housed Sodom./ Whorehouse after whorehouse/ With six-story high fauns darting in dances," he raged. As his tango's tempo changed to staccato drumbeats, he exclaimed, as if to someone directing a spectacle: "Stage Manager!/ The hearse is ready!/ Put more widows in the crowd!/ There still aren't enough there." And then, the drum beats quickly shifted to become part of an Orthodox mass for the dead. "Lay to rest, O Lord, my soul," the liturgy began, while, against its backdrop, Maiakovskii proclaimed: "No one ever asked,/ That victory be/ Inscribed for our homeland./ To an armless stump left from the bloody banquet/ What the hell good is it?"[55]

In striking contrast to the Symbolists' vision, Maiakovskii's poem revealed Armageddon in all its raw horror, stripped bare of any glory or poetic romance. Yet, if he saw the brutality of war more vividly than did his predecessors, so too did he speak more clearly of the world that might emerge from Armageddon's carnage: "All around!/ Laughter./ Flags./ In a hundred colors./ Pass by./ Rising high./ By the thousands." With a joyous shout, unlike anything ever heard from the Symbolists who had lived in the easier times of Russia's "dreadful years," Maiakovskii, the product of Russia's bitter war years, proclaimed: "People! Beloved,/ Not loved,/ Familiar,/ Unfamiliar,/ Pour through these doors in broad array./ And he,/ The free man,/ About whom I shout/ Will come/ Believe me!/ Believe me!"[56] Maiakovskii already had written that he could see, "approaching in the thorny crown of revolutions, a certain year,"[57] but he did not name it until 1918, for his vision of revolution remained less real than his sense of Armageddon's horrors. "Many of us felt that revolution was imminent," his friend Shklovskii remembered. "But many of us thought that we were outside space, that we had established our own kingdom of time."[58] Revolution still remained a vision, clearer than before, but a vision nonetheless, while the war, had become a reality so stark that none in Russia could turn aside from it. By 1916, all Russians had been forced to see the war as it was: not a noble quest for greatness and glory, but a searing, savage struggle that in time could consume them all.

If the Symbolists sought the cleansing fire of Armageddon, while Maiakovskii and his friends feared its brutality, there were others who responded to all warnings of its approach with pure Russian fatalism combined with supreme nonchalance. Belyi called this *tryn-travizm,* a term he derived from the slang phrase meaning "What the hell, who cares?" with which certain segments of Russia's youth rejected all responsibility for society, their actions, and themselves. On the eve of the Great War, such men and women lived only for the here and now, determined to experience all sensations before the opportunity slipped away or life's fragile thread was severed. These were Russia's first sensualists, whose preoccupation with sex promised immortality in the sense that one generation of men and women utterly unconcerned with the future mindlessly begat another as an incidental byproduct of their *"tryn-travistic"* quest. *Sanin,* a brief novel in which Mikhail Artsybashev insisted that "conceptions of debauchery and purity are merely as withered leaves that cover fresh grass," stated their credo.[59] Put into practice in a variety of sex clubs and love shops on the eve of the Great War, Saninism glorified vulgar self-gratification, carnal excess, and sexual experimentation as Russia's youth heeded the urgings of *Sanin's* leading characters to "plunge into the stream of sexual enjoyment," and to "enjoy absolute sexual liberty, allowing women, of course, to do the same."[60] Following *Sanin's* example, other novels featured similar themes, insisting that women need "find nothing more worth remembering about where, under what circumstances, and with whom they satisfied a natural [sexual] function than they would about a chance dinner companion and the menu in the restaurant."[61]

Despite such daring pronouncements, the men and women who heeded Artsybashev's preachings remained far from equal, for, although Russia's early sensualists practiced sex freely, they did not share its more predictable and immediate consequences equally. In *Sanin's* world, men moved happily from boudoir to boudoir, while women could not escape the narrow path that carried them from boudoir to kitchen to nursery. *Sanin's* women endured tragic pregnancies, suffered the pain of abandonment, and all too often found themselves entrapped by the obligation to care for the byproducts of their attempted sexual liberation. To enjoy emancipation equal to their men, women must be freed, in

Lenin's words, from "the need to spend three-fourths of their lives in smelly kitchens,"[62] and liberated from the burden of child rearing (as opposed to childbearing) that entrapped them in a traditional environment in which their chief functions were to serve and to be possessed. The cause of women's sexual and personal liberation needed a champion, but none could be found among Russia's liberals, or even among the leaders of the revolutionary movement. Many revolutionaries, in fact, assigned such a secondary place to the women's question that, just a few years before the Great War, the Bolsheviks' Petersburg Committee actually posted a sign that read: "Tomorrow, Meeting for Men Only."[63]

Russian women found their champion in Aleksandra Kollontai, the daughter of a wealthy Finno-Russian woman and a tsarist general who traced his genealogy to a prince who had ruled the principality of Pskov in the thirteenth century.[64] Passion for political and social issues ended her first marriage to Vladimir Kollontai, and drove her to spend most of the late 1890s at the University of Zurich where she applied Marxist theoretical principles to a study of Finnish workers. Quickly, she found a place among Lenin's Bolsheviks and took up the cause of winning true liberation for the millions of women who toiled in Russia's fields and factories, insisting that Russia's Social Democratic movement must dedicate special attention to their needs. She found little sympathy for her cause. With both the men and the older women in the Social Democratic ranks arrayed against her, Kollontai persisted until she found some measure of support among Petersburg's largely female Union of Textile Workers, and established the nucleus of a women's movement among them.[65]

Always elegantly dressed and immaculately groomed amid the working women to whom she dedicated her entire life, Kollontai remained a rare aristocrat among revolutionaries. From the moment she first committed herself to radical causes at the turn of the century, her stunning appearance and well-known aversion to the bonds of traditional marriage added to her notoriety and led various Western contemporaries to portray her as a woman dedicated to sharing her bed and body with legions of men. In fact, Kollontai's private life and public views were very much the same, and both were far from the notorious *tryn-travizm* that scandalmongers attributed to her. "Sanin, with his over-simplified psycho-physiological make-up," she wrote very soon after Artsybashev's book appeared, "would be a poor partner in 'game-love,'" the less-than-permanent liaisons between men and women which she defined as "the difficult 'school of love' where the emotions are refined" so that "the human psyche [can] develop its potential for loving." Kollontai thought that a "monogamous union based on 'great love' still remains the ideal" for which men and women should strive, but argued that society must "learn to accept all forms of personal relationships" so long as they did not involve incest or prostitution. In all relationships, Kollontai insisted, "the self-preservation of the individual" remained of paramount importance. There must never be "the loss of one's personality in the waves of passion," and people must be led from the "blind-alley" in which love was "either a tragedy that tears the soul apart or a vulgar vaudeville."[66]

Kollontai's opposition to conventional marriage thus stemmed not from any perverted desire to have an inordinate number of men use her body, but from her unshakeable belief that the institution, as traditionally practiced, prevented women from enjoying lives that realized their fullest potential. "Only by break-ing the domestic yoke will we give women a chance to live a richer, happier, and more complete life," she once proclaimed in the hearing of the American journalist Louise Bryant, who later called her "the only articulate voice of the new order for women which has been so greatly misunderstood outside of Russia."[67]

If Russia's women were to be free, Kollontai insisted, they must be liberated from the kitchen and nursery, the two greatest symbols of the servitude that traditional marriage imposed upon women. "The separation of the kitchen from marriage," she announced, would be a "great reform, no less important than the separation of Church from State."[68] Likewise, the burden of child rearing must be lifted from women's shoulders. "Maternity," she explained, "is 'sacred' " and therefore it was the duty of society to "arrange all forms and kinds of 'aid-stations' for women, that . . . [would] give them moral and material support" when they were pregnant and after they became mothers.[69] "Every mother must be convinced," she proclaimed, "that, once she fulfills her natural function . . . , the collective will love and attend to her and her child."[70] Because children were one of society's most precious resources, Kollontai thought that their care and upbringing required professional attention. "To mend shoes, one has to pass through an apprenticeship," she pointed out, while "only a mother's instinct" traditionally had been sufficient "to guide such a delicate creation as the spirit of a child." Clearly, children in need of rearing deserved as much consideration as shoes in need of repair, and it must become the duty of society to provide government-run nurseries in which children could be raised "in a hygienic, morally pure atmosphere."[71]

If Kollontai hoped to transform the *tryn-travizm* of *Sanin*'s disciples of free love into socially useful forms by calling for a revolution that could liberate women from marriage-imposed servitude, others sought to cope with increas-ingly concrete visions of the Apocalypse and Armageddon by turning the Sani-nist cult into pure sexual escapism that touched broader (and higher) strata of society. Prince Feliks Iusupov, scion of one of Russia's wealthiest noble families who, as the prima ballerina Anna Pavlovna once exclaimed, had "God in one eye and the Devil in the other," became such a notoriously successful transves-tite that he actually managed to perform as a female singer on six occasions at Petrograd's well-known Aquarium cabaret before anyone discovered his true gender.[72] And the Duma's vice-president Aleksandr Protopopov, a politician whose appointment as Russia's last minister of internal affairs Aleksandra insisted "God will bless," was widely believed to be an ardent necrophiliac.[73]

The paths of both these men crossed that of the *starets* Rasputin, for his hand had guided Protopopov's rise to ministerial office, and Iusupov later became his assassin. Rasputin also espoused sexual excess, but unlike any of his aristocratic friends or enemies, he justified it as an instrument of salvation, and did so with such passionate conviction that his following nearly assumed the dimensions of a cult. Rasputin's early preachings about the ease with which sins of the flesh could help one enter upon the path to salvation had been popular from the first, but they found an especially ready hearing in Petrograd during the war years, as an endless stream of admirers and petitioners—from princesses and countesses to shop clerks, cleaning women, and whores—made their way to his lodgings to seek a favor, a blessing, or guidance. Often he used his influence to aid petitioners in need; from time to time, he would ravish one of them crudely and quickly before sending her on her way with vaguely muttered assurances that "now, Mother, everything is in order."[74]

In those days, Rasputin lived at No. 64 Gorokhovaia Street, in a third-floor walk-up flat that overlooked the inner courtyard of one of the many nondescript apartment buildings that had sprung up to mar the city's pristine architectural regularity during the last decades of the empire. The *starets* shared his stairway with several people of modest means, including the masseuse Utilia, and the seamstress Katia, both of whom spent nights with him on occasion. The sour smell of *shchi,* rancid butter, and hot sheep's cheese, all perpetually trapped in the cloud of steam that arose from the battered tin samovar in the porter's lodge, clogged the nostrils of anyone who stepped across the building's threshold.[75] There in flat No. 20, where Prince Iusupov thought that "the whole contents, from the cumbersome sideboard to the crowded and abundantly stocked kitchen bore the stamp of bourgeois well-being and prosperity," while the "lithographs and badly painted pictures on the walls were fully in keeping with the owner's taste,"[76] Rasputin entertained the *rasputinki,* the women infatuated with his ability to guide them toward "salvation" through sexual excess or some other means. The meals Rasputin shared with his *rasputinki* always were simple: a fish soup such as *ukha* or *solianka,* black bread, boiled eggs, and desserts, all eaten without knives or forks and washed down with Rasputin's favorite Madeira. The *rasputinki* took it as a compliment when their host offered them bits of half-eaten food from his plate, or when he showed one of them particular favor by plunging his fingers into the jam pot and allowing her to suck the residue from them. "Humble yourself," he would urge. "Lick it clean, lick it clean." On occasion, he would launch into disjointed preaching, sometimes ordering one of his visitors to write down his words, as he spoke of sacred love and profane passion in such a way that his listeners readily confused the two.[77] Often, as he described in crude detail how the stallions mounted the mares on his father's farm, he would draw one or another of his listeners to him with the words: "Come, my lovely mare."[78]

Some of the *rasputinki* thought he possessed special holy powers, and there is some evidence that Aleksandra was beginning to think of him as a reincarnation of Christ.[79] Certainly, the empress thought him sent directly from God to aid her, her husband, her son, and Russia, and would allow no word to be said against him. When the Princess Zinaida Iusupova insisted upon telling her "everything she had come to say" about Rasputin in the fall of 1916, Aleksandra coldly dismissed her with the remark: "I hope that I never see you again." Even when her saintly sister, the Grand Duchess Elizaveta Feodorovna, tried to warn her, the empress sent her away. "She dismissed me like a dog!" the grand duchess exclaimed afterward. "Poor Nicky, poor Russia!"[80] Utterly unmoved by such warnings, Aleksandra remained calmly confident. "A country, where a man of God helps the sovereign," she assured Nicholas a few days after she had sent her sister away in tears, "will never be lost."[81]

Some thoughtful Russians found it difficult to bear the excesses they saw around them. "We have a chorus of rebellious spirits, dissatisfied with life and love," Vasili Shulgin wisely confided to his diary then. "In their search for the 'keys to happiness' some among them throw themselves into mysticism while others turn to debauchery. . . . Grishka [Rasputin] ties these two currents together. Holding in one hand a hysterical female mystic, and in the other a hysterical nymphomaniac, he adorns the ballet of Petrograd with his two-faced countenance—that of a sorcerer and a satyr."[82]

Rasputin flourished in Petrograd society, and his influence soared during the Great War in part because Russians found it so difficult to gain access to their tsar and tsarina. In Russia, the saying that "every law has an exception" applied very literally because, until the great judicial reform of 1864, judges had been permitted only to apply the law, never to interpret it. For those cases in which the rigid application of the law seemed patently unjust or obviously foolish, the tsar traditionally had the right and the duty to make an exception, and, for centuries, he had tempered the laws of his land with his mercy. "You need a superior mercy so as to soften the law, and this only can come to us in the form of absolute monarchy," the widely misunderstood Nikolai Gogol explained in the 1840s. "A state without an absolute ruler is like an orchestra without a conductor."[83]

When Gogol wrote these words, the application of the autocrat's mercy had become institutionalized in the tsar's personal chancellery, to which men and women in need of imperial intervention directed a steady flow of petitions that always began with the apology that, "had it been permissible under the law, I should never have taken it upon myself to trouble His Imperial Majesty."[84] By the time of the Great War, Nicholas's chancellery received more than 70,000 such pleas every year, or more than 250 every working day.[85] Of course, no tsar could hope to deal with such a flood of importunities, and most of them therefore were attended to by his personal staff. Still, there always had been those particularly touching or difficult cases where the petitioner felt it necessary to speak directly with the tsar or to have someone who saw the tsar frequently —a minister of state, a favored courtier, a member of the imperial family, or, perhaps, even the empress—present their case.

Great moments of national crisis, and wars most of all, increased the number of special pleas the Russians directed to their tsar, yet, by the time of the Great War, this special avenue to imperial intervention had been closed to most Russians. More outraged at their subjects' disloyalty than fearful for their personal safety, Nicholas and Aleksandra had isolated themselves at Tsarskoe Selo's Alexander Palace ever since the Revolution of 1905. Ministers of state, and even members of the imperial family, saw them rarely, and they almost never appeared in public in such a way that they could be approached with special pleas. As one of the rare people who saw them often, and who also appeared often in public, Rasputin came to occupy a very special position in Petrograd society. It was natural and expected that he would be besieged by hordes of petitioners seeking his intercession with the tsar and tsarina, as well as by women seeking his personal "assistance" in gaining salvation. Women wanting to have a loved one transferred from the front, returned from Siberia, or given a promotion all flocked to his lodgings, and there was even one who appeared in a low-cut gown and begged him to have her made a *prima donna* at the Imperial Opera. From some, Rasputin took gifts or money, from others he asked nothing at all, and from still others he demanded sexual favors in return for his intercession.[86] No complaints against such behavior were ever given even the slightest credence at Tsarskoe Selo, as Nicholas and Aleksandra dismissed accusations from ministers of state and even detailed police reports with the simple phrase: "He is hated because we love him."[87]

In those days, shady characters desperate to seize everything within reach for fear that it—or they—would not be there on the morrow gravitated into Rasputin's orbit. Numbers of them lurked in several Petrograd salons where the host or hostess was reputed to enjoy some close connection with the imperial court, most usually through Rasputin or one of his unsavory associates. Perhaps the salon whose connection with the Alexander Palace remained the most mysterious was that hosted by Baroness Evgeniia Rozen, a woman who, Minister of Internal Affairs Khvostov later reported, "had no visible means of support whatsoever," but always dressed with exquisite taste. The baroness, Khvostov said, presided over a salon noted both for its elegance and for the "never-ending orgies and drunkenness" that took place in the wee hours of the morning. Between midday and early evening, a visitor might well find any number of grand dukes and grand duchesses, high-ranking army officers, and ministers of state in the baroness's drawing room or at her table, while the late night hours brought an array of fashionable courtesans, journalists, and actresses. "Imagine it," Khvostov exclaimed. "[General] Dobrorolskii and Grand Duke Boris Vladimirovich would dine there with their ladies, and at breakfast you'd meet none other than [Boris] Rzhevskii," a notorious yellow journalist and Okhrana spy. Rzhevskii was known to maintain very questionable connections with Rasputin's archenemy, the monk Illiodor, who devoted part of the war years to writing a scurrilous pamphlet entitled *The Holy Devil*, which portrayed the *starets* in particularly unflattering terms. At one point, Rzhevskii also was rumored to be involved in an unsavory scheme to publish letters that had been passed between Aleksandra and Rasputin in the hope of being well paid for not doing so.[88]

Rumor had it that the baroness Rozen's elegant dinner parties, late-night suppers, and fashionable late-morning breakfasts all were financed by a mysterious engineer whose identity not even Khvostov's secret police could discover. Khvostov himself often visited the salon in the hope of learning more about it, because he suspected that if his deputy Beliaev could "learn a lot that is interesting from the grand dukes [there], so could any number of German agents." Clearly the baroness seemed to know someone who enjoyed the trust of the imperial family, for the confidential information she shared with her visitors almost always proved accurate, just as the requests and petitions she forwarded usually received prompt attention. This seemed all the more certain in 1916, when Rzhevskii's visits suddenly grew few and far between, and Princess Obolenskaia, who, Khvostov reported, used her position to "conduct various business . . . that people would have thought rather fishy" had she not enjoyed the protection of her husband's princely name, became a frequent guest. In search of wealth and position, Obolenskaia had contrived to marry a scion of one of Russia's oldest noble families; in an instant, her beauty drew Rasputin to the salon where men and women dealt in government contracts and sold other sorts of favors.[89]

Countess Rozen was but one among a number of influence peddlers who flourished in the *tryn-travistic* atmosphere of wartime Petrograd. Nikolai Burdukov, the tsar's equerry, whose kitchen and cellar were kept up by the wealthy financier Ignatii Manus, presided over another salon that was widely known for the quality of its food and drink. An especially close friend of Vice-Admiral Nikolai Sablin, commander of the imperial yacht, and the "old sea-bear" Admiral Konstantin Nilov, both of whom stood among the favorite aides-de-camp of Russia's emperor and empress, Burdukov seems to have been more than willing to sell his influence to anyone willing to pay the price. Thus, like that of Baroness Rozen, Burdukov's salon was connected with Rasputin and was thought to be another center of German espionage in the capital, especially by Khvostov during his months as minister of internal affairs.

Many suspected Burdukov's friend Manus of being a particularly influential German agent, and that suspicion grew stronger toward the end of 1916. A Jew whom Aleksandra knew well enough to mention in her letters to Nicholas, Manus had long proved more than willing to serve anyone who paid enough and stood ready to betray anyone who did not.[90] A strong supporter of Prince Meshcherskii's virulently anti-Semitic newspaper *Grazhdanin, The Citizen,*

Manus even wrote for it from time to time under the pseudonym *"Zelenyi"* ("Green").[91] For some months, this shady swindler dined once a week with Rasputin and the equally shadowy Stefan Beletskii, whom Khvostov had named deputy minister of internal affairs in charge of Russia's police, and used the information he gathered on those occasions to considerable financial advantage.[92] "He is the distributor of German subsidies," Paléologue said of Manus. "He secures the relationship with Berlin, and it is through him that Germany hatches and maintains her intrigues in Russian society."[93]

Far more unsavory than the gatherings at Baroness Rozen's and Burdukov's was the salon of Prince Mikhail Andronikov, the patron of Khvostov, friend of Beletskii, and inveterate intriguer who was thought by some who knew him well to be a homosexual.[94] Although in the pay of the Ministry of Internal Affairs, Andronikov held no fixed position, and made his way by peddling favors. "I am a man, a citizen," he once said, "who tries to make himself as useful as possible."[95] Everywhere he went, he carried a bright yellow attaché case, always giving the impression that it held important information, but which, according to Okhrana reports, he had stuffed with old newspapers.[96] He was apt to turn up anywhere, "always mixing in matters that didn't concern him," Nicholas's minister of court Count Frederiks complained.[97] Many in high society feared his ready pen and malicious wit, for Andronikov was notorious for publishing flattering portraits of friends and vicious satires of enemies for the amusement of those whose favor he courted in high places. Aleksandra placed such great trust in him that on at least one occasion she called him "my Andronikov."[98]

By 1915, Andronikov had become friendly with the present and future ministers of internal affairs Khvostov and Protopopov, and dealt frequently with such disreputable characters as Stürmer's chief agent Manasevich-Manuilov and Rasputin's unscrupulous secretary Aron Simanovich (who evidently curried favor with the empress on the side by allowing her to buy jewels from him at much-reduced prices). Rasputin himself visited Andronikov from time to time, as did Vyrubova. Most of all, Andronikov spent his evening hours with Beletskii, whose chief occupation was to dirty his hands in unsavory police affairs. All of these sinister hangers-on prospered in the squalid atmosphere of intrigue, betrayal, and counterbetrayal that flourished under the petty and meddlesome interference of Aleksandra and Rasputin in state affairs in wartime Petrograd. Even taking into account that each of them vied to outdo the others in condemning former associates after the revolution, the testimony that these men gave before the Supreme Investigating Commission in 1917 reeks from the stench of the base and ill-concealed schemes in which they wallowed, always immune from retribution because their empress held them in high esteem.[99]

Like the peasants whose talk drew Paustovskii to their villages, many Petrograders felt a sense of foreboding which deepened as Russia made ready to face her third winter of fighting. Among them was Aleksei Tolstoi, a young writer who later made his peace with the Bolsheviks and won the Stalin Prize for the huge novel about the Great War and Revolution that he called *Purgatory*. One of his friends remembered how he seemed to fill the entry of the Stray Dog that winter as he stormed in, wrapped "in a raccoon fur, beaver cap or top hat, with a haircut *à la moujik.*"[100] Although not yet dedicated to the Bolsheviks, Tolstoi anxiously awaited the revolution that winter. Petrograd was "seething and satiated" in those days, he wrote some years later. Its people "tormented by sleepless nights, stupefied and deadened by wine, wealth, and lovemaking without love," the city seemed about to explode. "The spirit of destruction pervaded everything," Tolstoi remembered. "Destruction was thought to be in good taste, and neurasthenia to be a sign of refinement."

A certain desperation touched almost everyone then, as the poor struggled against food shortages, and the wealthy struggled against the ennui that came from discovering that life held no more mysteries or challenges for men and women who had tried all things. Everywhere—from the fine restaurants and opulent hotels, where wealthy industrialists and war profiteers entertained their ladies, to the ballrooms of the rich and famous, to the cellar cabarets, to the "People's Houses" that provided culture for the masses—Tolstoi sensed "the heartrending, impotent sensuality of the strains of the tango" wailing in the background. In its native habitat, it had been a song of love and passion; transported half a world away from its Latin roots, the tango seemed so strangely transformed that Tolstoi called it "a hymn of death." Against its compelling, strangely sinister, backdrop, he wrote, "the city lived, as if awaiting that one terrible and fateful day [of reckoning]."[101]

Even Minister of Internal Affairs Protopopov, now kept in office only by the support of Rasputin and the empress, felt the weight of the impending catastrophe. "The Supreme Power had ceased to be a source of life and enlightenment," he told the Supreme Investigating Commission the next spring. "It had become the captive of stupid forces and stupid influences [and] could give no direction or leadership." Protopopov understood that Russia had stood on the verge of defeat at the beginning of the war's second winter and had survived only to find herself on the verge of domestic collapse a year later. "No one was happy," this fawning favorite concluded, as if echoing the conclusions Paustovskii had drawn in his distant and isolated peasant village that winter. "The former tsar felt this instinctively, he explained, "but none of his ministers understood the depths of the dissatisfaction."[102]

Relieved to be away from Petrograd at his headquarters at Mogilëv, Nicholas continued to rely upon the counsels of Aleksandra, who, in her turn, now depended almost entirely upon Rasputin and his agents for advice. Inevitably, Russians thought the worst about the empress's deepening trust in this fraudulent holy man whose shocking conduct had become known all across the empire. "It is a terrible dilemma," one loyal Russian confessed to his diary just before the year's end. "The tsar offends the nation by what he allows to go on in the [Alexander] Palace . . . while the country offends the tsar by its terrible suspicions. The result is the destruction of those centuries-old ties which have sustained Russia. And what is the cause of all this? The weakness of one man and one woman. . . . Oh, how terrible an autocracy is without an autocrat!"[103]

An autocracy without an autocrat, an army without a real commander in chief, a nation, in fact, without a leader: this was Russia as the Great War entered its third winter. Rarely in history had "the weakness of one man and one woman" threatened such dire consequences for any nation. No real statesmen now remained at the top of Russia's government, for Aleksandra and Rasputin had removed them all in such rapid succession that the long-loyal archconservative Vladimir Purishkevich cynically dubbed the process "ministerial leapfrog."[104] Fearful and aghast at what they saw happening all around them, Russia's leading politicians demanded a greater role in their nation's affairs. But, despite their good intentions, talent, and, in some cases, even brilliance, they lacked broad experience in government and a clear sense of the crisis their nation faced.

In any case, so long as Nicholas refused to heed their urgings, these men could not make their views felt in government, no matter how loudly they raised their voices in protest. Such concessions as a "responsible cabinet" were not necessary, Aleksandra assured Nicholas again that winter; the advice of "our Friend who leads through God," and raw, naked force, were enough. "Russia loves to feel the whip," she wrote less than a fortnight before Christmas. "Crush them all under you."[105] During the war's third winter, the tsar's refusal to make a place in his government for these men, and their failure to convince him of the necessity for doing so, removed the last barrier that stood between Nicholas and Aleksandra and revolution.

CHAPTER XI

"Is This Stupidity, or Is This Treason?"

Nicholas II's visit to the State Duma of the Russian Empire on the afternoon of February 9, 1916, was an unprecedented event. Traditionally, Russia's autocrats had governed as they thought best, and, when they had seen fit to hear their subjects' views, they had summoned them into their presence. In medieval times, tsars sometimes had convened the *Zemskii Sobor*, the Assembly of the Land, to tell them its opinions about matters of great import, but after Peter the Great had done away with that body, autocrats had consulted with their subjects much less often. Even the Duma's emergence after the Revolution of 1905 had not changed the traditional pattern of the tsar summoning his people's representatives. On the rare occasions when Nicholas had addressed the Duma, he had done so with great pomp and circumstance in the Winter Palace. At the beginning of February 1916, in the wake of Russia's first victory in many months (against the Turks at Erzerum) and the "retirement" of Goremykin, the prime minister whose stubborn opposition to the Duma's very existence had been a festering thorn for too long, Nicholas made ready to meet his people's elected representatives, on their own ground for the first time. Russia's tsar, autocrat, emperor, and supreme commander would not summon the Duma to the Winter Palace as he always had done before; he would go to their chamber in the Taurida Palace. Some called it "a step of great resolution . . . [which] will effectively purge the poisoned currents of domestic policy."[1] Certainly, it was a serious attempt to ease the rancor that had poisoned relations between tsar and Duma for more than a year.

Ever since Catherine the Great had commissioned Ivan Starov to build the Taurida Palace in 1783, it had been a landmark in Russia's capital. Its Domed

Hall and Gobelin Drawing Room were among the most splendid in the city. Its magnificent gardens began at the end of the Furshtatskaia, the short street on which the American Embassy stood (and where its present-day Consulate General occupies the same building), and spread over more than eighty acres. In the days before the war, Blok and Belyi often had looked down into the gardens from the heights of Ivanov's Tower, while Petrograders frequently strolled through the less elegant southern sections that were open to the public. In 1906, Nicholas had turned the entire Palace over to the Duma for its first session, and it had met there ever since, its four hundred and forty seats arranged in a newly-designed chamber that once had housed the palace's winter garden. Now the tsar was to visit the assembled Duma for the first time in history.

Precisely at 1:55 p.m., Nicholas arrived at the Duma, where its president Rodzianko and his deputies awaited him on the Palace's portico. Openly skeptical and aloof, the victim of what one contemporary called "his too perfect thinking apparatus" that produced a "superabundance of formal logic" but flawed political judgments, Pavel Miliukov stood so far aside that he could not hear clearly the tsar's remarks, but Rodzianko greeted the imperial gesture of reconciliation with all the warmth and enthusiasm that Nicholas could have hoped for. With the deputies and a small suite of courtiers in tow, Nicholas attended a Te Deum in honor of Nikolai Nikolaevich's triumph at Erzerum and the Duma's opening. There, beneath the icon of St. Nicholas the Miracle Worker, and surrounded by his ministers and a handful of adjutants, the tsar bent to kiss the cross while the delegates and the members of the State Council looked on. Rodzianko remembered that "the Sovereign was deathly pale, and his hands shook from nervousness" when he first entered the hall, but that he grew calm and "his agitation gave way to a look of tranquility and satisfaction" as the service went on. Solemnly, the deacon intoned: "May all who have laid down their lives in battle be remembered eternally." Although it was customary to remain standing for the remainder of the service, Nicholas suddenly knelt to express his gratitude for the millions who had sacrificed their lives in Russia's defense, and the dignitaries, delegates, and visitors all hurried to follow his example.[2]

According to Rodzianko's recollections, his Sovereign arose as the Te Deum ended, turned to him, and said: "Mikhail Vladimirovich, I would like to say a few words to the members of the Duma. Do you think it best that I speak here or would you suggest some other place?" When Rodzianko replied: "I think it best to do it here, Your Imperial Majesty,"[3] Nicholas turned to the assembled deputies and addressed them briefly. Some years later, Rodzianko remembered that the tsar spoke "calmly, clearly and firmly," but Ambassador Paléologue, who recorded his impressions in his diary that very day, reported that "he could scarcely force his voice out of his throat," and that, "after each word, he stopped or stumbled."[4] Still, even if the words were not well spoken, no one could doubt their meaning. "I rejoice to be with you," Nicholas began. Then he spoke about the victory at Erzcrum and of his hope that the deputies would be guided in their work by their "burning love for our Motherland." He added only a few words more. "With all my heart," he concluded, "I wish the State Duma fruitful labors and every success."[5]

Perhaps it was, as Miliukov reported, "a colorless but well-meaning" speech, but the deputies were more than prepared to give their tsar every benefit of the doubt.[6] Although their remembrances of Nicholas's oratory differed sharply, Paléologue's diary and the official Russian account of the meeting agreed that a thunderous "Hurrah!" greeted Nicholas's words, and the crowd of deputies and dignitaries spontaneously sang "God Save the Tsar!"[7] As the last strains of Russia's national anthem faded, Rodzianko's stentorian bass voice swelled forth. "Your Imperial Majesty!" he began. "What a joy it is for us to see our Russian tsar here in our midst. . . . May the All Highest Lord God bestow his blessings upon you! Long live the great sovereign of all Russia!" More cheers, more handshakes, more congratulations, and then the tsar turned to leave. "Make the most of this shining moment, Your Imperial Highness, and proclaim here and now that you will grant a responsible ministry," Rodzianko begged in a low voice as Nicholas began to move toward a side door. "You would write a glorious page in the history of your reign if you did so." "I shall give it some thought," Nicholas replied. "This has been most pleasant," he added as he favored Rodzianko and his deputies with his famous warm and charming smile. "This is a day I shall always remember."[8] Then he was gone, leaving Russia's newly appointed prime minister, the corrupt and already much-despised Boris Stürmer to speak to the Duma in his name.

Stürmer's pallid address instantly drained the tsar's visit of its conciliatory effect. He spoke "in lengthy and muddled phrases," Rodzianko remembered, and the fact that "he said nothing about the government's intentions," left "a depressing impression" upon all but the extreme rightists in the Duma. When Stürmer finished, the Duma members sat in stony silence, which contrasted sharply with the warm reception they gave War Minister Polivanov and Foreign Minister Sazonov. "A waterfall of eloquence," quipped the ultraconservative Purishkevich, as he hastened to compare Russia's new prime minister to Chichikov, the vulgar, banal hero of Gogol's *Dead Souls*, who tried to curry favor by saccharine manners and meaningless flattery before he set his crooked schemes in motion.[9] Miliukov cynically dubbed Stürmer's cabinet "the ministry of confidence of the Union of the Russian People," perhaps the most reactionary of all Russian right-wing political organizations, and the one to which Nicholas long ago had promised his personal protection.[10]

Yet, no matter how distasteful they found Stürmer's regime, the Duma could do little in protest. Men like Purishkevich were not yet ready to go against their tsar, no matter how much they resented the growing influence of the empress and Rasputin in their nation's politics, while the Progressive Bloc had lost much of its unity and sense of common purpose in the course of the winter. Even

Miliukov, the lion of the previous August and September, seemed only a shadow of his former self as he confessed his impotence before his Duma colleagues. "I know where the exit is," he told them in a speech the day after the tsar's visit. "But I don't know how to get to it."[11] So fragmented had the Progressive Bloc become that one radical deputy exclaimed: "The Progressive Bloc is dead. Long live the regressive Bloc!"[12] Ministers even moderately sympathetic to the cause of reform disappeared without a murmur that spring, as both Polivanov and Sazonov fell victim to campaigns waged against them by Rasputin and Aleksandra. In their stead, Stürmer flourished, his cynical mediocrity winning him supporters in high places.

Stürmer's corrupt scheming and gross incompetence made men angry, but they grew even more outraged as they realized how helpless they were to prevent him and his unsavory allies from dragging Russia beyond the revolutionary brink from which there could be no turning back. Such feelings gripped many during the first days of the Duma's new session, and they came out even more sharply when the Kadets held their sixth national congress in Moscow some ten days after Nicholas's visit to the Taurida Palace. All spoke of their fears that Russia stood on the brink of catastrophe, and all lamented their powerlessness to deflect their nation from that path. "The mood of the country is such that it is terrible to think of the immediate future," Andrei Shingarev exclaimed, his many years of experience as a provincial physician warning him that "the anguish and indignation of the masses has reached its limit."[13] Indeed, the nation's mood was no secret. People wanted revenge, an Okhrana report stated, "revenge not only from the present government in the sense that it included a particular cabinet of ministers, but from the supreme power itself."[14] Clearly, revolution was in the air. "Only the rapid resignation of Stürmer and the minister of internal affairs [Khvostov]," some of Shingarev's Kadet colleagues at the congress insisted, "can, if not avert, at least weaken somewhat, the force of the explosion that will come at the war's end."[15]

None of the men who assembled in Moscow to discuss the Kadet program, or who defended it so weakly in the Duma, imagined that the fall of one of their enemies was very near. Nor did they sense that his long-hoped-for disgrace would make their position even worse. Khvostov and Stürmer had risen to high office as two of Rasputin's favored creatures, but, unlike the man whom Aleksandra so often referred to as "the tail" in her correspondence with Nicholas (*khvost* means "tail" in Russian), Stürmer was satisfied not to strike out on his own without the company of his protector. Rasputin was quick to perceive that crucial difference between the two men, and Stürmer's appointment as president of the Council of Ministers—in effect, Russia's prime minister—meant that Aleksandra had firmly sided with him against the man who, she once had insisted, was "very energetic, fears no one, and is colossally devoted." Somehow, "the devil got hold of him," she now explained to Nicholas. "As long as Khvostov is in power and has money and police in [his] hands—I honestly am not quiet for Gregory and Ania [Vyrubova]."[16]

Aleksandra was quite right to be concerned for Rasputin's safety at the beginning of 1916. More than a month before the Duma met, Khvostov had tried to seize Goremykin's place for himself, only to fail because Rasputin (and therefore Aleksandra) had supported Stürmer. To even the score, Khvostov decided to have his former ally killed and set several plots in motion at the very moment when Russia desperately needed all of the resources of his powerful Ministry of Internal Affairs to shore up her crumbling war effort.[17] To murder Rasputin, Khvostov recruited some of the most sinister characters ever to set foot in Russia's central government offices—Manasevich-Manuilov, General M. S. Komissarov, and that unsavory intriguer Boris Rzhevskii—all of whom were so dishonorable that none could keep faith with any of the others.

Most eminent among this ignoble band was Manasevich-Manuilov, who had at different times been a dedicated practitioner of Judaism, Lutheranism, and Russian Orthodoxy, and had served and doublecrossed any number of prominent officials in the Okhrana to support a style of life that cost far more than he could obtain through legitimate means. At one point not long before the Russo-Japanese War, he had betrayed agents of the reactionary Minister of Internal Affairs Plehve to those who served the more progressive Minister of Finance Witte, and then stole secret documents from Witte's safe and turned them over to Plehve. When the honest Stolypin refused to allow him to work anywhere in the government, Manasevich-Manuilov made a living by betraying Okhrana agents to Russian revolutionaries living abroad, took up blackmail, and peddled favors on a scale that far exceeded his ability to influence the course of human events. So accustomed was he to treachery that even when Stürmer appointed him to head his personal chancery, he evidently seemed sufficiently open to other enticements that Khvostov tried to buy his services against Rasputin along with those of Komissarov—who pilfered documents from Allied embassies and at one point poisoned Rasputin's cat—and Rzhevskii, who functioned at a lower, more disgusting echelon in the "dirty tricks" trade.

Soon, a shocked Aleksandra learned of Khvostov's plans from her ever-faithful Vyrubova. As frightened as she was angry, she immediately ordered General Mikhail Beliaev, the harried chief of Russia's General Staff who was working night and day to repair the terrible damage of the Great Retreat, to take personal responsibility for Rasputin's safety. While Beliaev desperately searched for rifles, bullets, field guns, and shells, and tried to assemble more than a million new soldiers to fill the gaps that the massive firepower of the Austro-German armies had torn in Russia's defensive lines during the Great Retreat, he also had to worry about the safety of a peasant-turned-holy man to whom Stürmer himself assigned especially fast military cars so that he could evade police surveillance on his nightly prowls.[18]

Rasputin thus escaped Khvostov's embittered plotting, but only at considerable cost to his imperial protectors. At the very least, Aleksandra had committed a shocking abuse of her prerogatives when she used resources so vital to the Russian Empire's war effort to protect such an out-and-out scoundrel, and she soon faced the scandal that her intemperate actions inevitably caused. In a predictably ill-fated attempt to forestall controversy, Nicholas prohibited any reference to Rasputin, Khvostov, or his deputy Beletskii in the press, but before his order reached the offices of the gossipy evening edition of *Birzhevye vedomosti, The Stock Exchange Gazette,* its editor published a confidential interview with Beletskii who, in an effort to better his standing in the Rasputin-Stürmer camp, betrayed his chief with a flood of lurid—and in some cases fanciful—details. By that time, Nicholas had relieved Khvostov from office, but the entire capital was awash with rumors that made the whole offensive mess seem even worse than it was.[19]

As always, controversy did nothing to sharpen Nicholas's political instincts, and he continued to forge ahead, making more enemies every step of the way. Stubbornly insensitive to the anger that Stürmer's appointment had stirred at the end of January, he stupidly named him to replace Khvostov as minister of internal affairs at the beginning of March, and insisted that he also continue as chairman of the Council of Ministers. Stürmer, the former provincial governor whose experience in directing affairs in the central government had been limited to supervising a department with less than a hundred employees, now held more power than any man had wielded since Stolypin. Stolypin's appointment a decade before had signaled a brief era of great progress; Stürmer's elevation promised only disaster. "I thought him a man who was utterly false, two-faced, not particularly intelligent, and never believed a word he spoke in the Council of Ministers," Russia's minister of justice once said. Perhaps Krivoshein's successor in the Ministry of Agriculture best summed up the dilemma that knowledgeable Russians faced when their emperor placed his trust in a man of Stürmer's low character. "People have a well-known name for it when an ordinary person does not tell the truth," he remarked in disgust that summer. "But, when a man who holds such a position and is entrusted with such authority [as Stürmer] tells lies, what does one call *him?*"[20]

As if to confirm everyone's worst suspicions, Stürmer hurriedly slipped a resolution establishing a secret fund of five million rubles for his private use into a pile of papers that his colleagues customarily signed at the end of each Council meeting. Minister of Education Count Pavel Ignatiev, the tsar's friend ever since they had trained together as junior guards officers, remembered that he and his colleagues usually signed these resolutions "mechanically," because they were supposed to contain only decisions that they had approved at the previous meeting. Just by chance—Ignatiev's glance fell upon Stürmer's unapproved resolution, and it caught his attention. "What's this?" he asked. Since the sum was charged against the budget of the War Ministry, Ignatiev called Polivanov over to explain the matter: "Here, look at this," he insisted. "Do you need such an appropriation for the War Ministry?" Amazed that a resolution would be drawn to allocate funds over which only he had control, Polivanov exclaimed: "No, that's got nothing to do with me!" and turned to Stürmer for an explanation. Stürmer could only mumble that it was needed for "espionage" costs, and that he had submitted it only to satisfy one of the tsar's special requests. Unmoved, Polivanov insisted that such a course was "utterly impossible." "After that," he remarked later, "I had the distinct feeling that my days in office were numbered."[21]

The great confidence Nicholas and Aleksandra placed in Stürmer at the very moment when all who knew him well denounced his appointment as a shocking disaster showed that both of Russia's sovereigns had grown far out of touch with reality. "I am going away this time with greater peace of mind," Nicholas wrote to his wife as the train carried him back to *Stavka* a few days after Stürmer's appointment. "How glad I am that you have got Stürmer now to rest upon," Aleksandra wrote the next morning. For her, nothing else mattered so long as Stürmer had the full support of Rasputin, whom she now identified more closely with Christ than ever. "During the evening Bible [reading] I thought so much of our Friend and how the *bookworms* and pharisees persecute Christ," she wrote a few weeks later. "He [Rasputin] lives for His Sovereign & Russia and bears all slanders for our sakes. How glad I am that we went all with Him to Holy Communion in the first week in Lent," she concluded, capitalizing the pronouns that referred to Rasputin as one customarily did only when referring to God, Christ, or the tsar.[22]

Clearly, Nicholas and Aleksandra had chosen a path so far outside the bounds of common sense that there remained virtually no hope of bringing them, their people, and their people's representatives in the Duma into agreement. Nicholas had succumbed almost completely to Aleksandra's far stronger will. To defend Rasputin Aleksandra would pay any price and dispense with any statesman, no matter how vital he might be to Russia's war effort. Dzhunkovskii, Kokovtsev, Krivoshein, and Grand Duke Nikolai Nikolaevich all had been driven from their posts for that reason. Now Sazonov and Polivanov would follow them before the summer of 1916 passed its zenith. To Polivanov, Nicholas denied even the pro-forma letter of thanks that he usually bestowed upon ministers defeated by the empress and Rasputin. Instead, he heaped unjust complaints upon the man whose effort to rebuild Russia's forces after the terrible losses of the Great Retreat had been little short of brilliant. "The activities of the war industries committees do not inspire me with confidence," Nicholas wrote to Polivanov less than a week after he had given him a warm reception at *Stavka,* "and I find your supervision of them to be insufficiently assertive."[23]

Although the tsar's strongly-expressed dissatisfaction took Polivanov by surprise, it certainly was true that Nicholas had always despised the war industries committees as hotbeds of liberalism, and it irked him beyond measure to see

Polivanov try to work with them. For that reason alone, court circles viewed Polivanov with distrust. "He is very intelligent," the elderly minister of court Count Frederiks once said, "but very dangerous because he is too much inclined to sympathize with parliaments."[24] The war minister's "liberalism" thus gave clear cause for alarm and conflict, but his main sin lay elsewhere. Here again, the opinion that counted most was Aleksandra's, and that was uncompromisingly negative. "Oh, how I wish you could get rid of Polivanov," she wrote at the beginning of January.[25] "He is simply a revolutionist," she went on two months later in one of the many rambling, disjointed letters with which she flooded Nicholas at *Stavka.* "Lovy mine, dont dawdle, make up yr. mind, its far too serious, & changing him at once, you cut the wings of that revolutionary party; only be quicker about it—you know, you yourself long ago wanted to change him—hurry up Sweetheart, you need Wify to be behind pushing you. . . . Promise me that you will at once change the M. of War, for your sake, your Sons & Russias."[26]

That very day, Aleksandra received the word she had awaited so impatiently. "I have at last found a successor for Polivanov," Nicholas wrote. "All the ministers will feel relieved."[27] Again, Nicholas and Aleksandra had diverged utterly from the views of Russia's best statesmen and her Allies' leading representatives. General Knox thought Polivanov "undoubtedly the ablest military organizer in Russia," and called his dismissal "a disaster," while the only word that Rodzianko could find to describe the mood of Russia's politicians and educated public when they received word of his dismissal was "despondent."[28] These men chose their words with good reason, for Polivanov's loss was irreparable, and not merely because his successor, General Dmitrii Shuvaev, the untalented son of a serf soldier, was weak. Perhaps Paléologue best summed up the significance of the war minister's unwarranted disgrace when he explained that "he appeared to be one of the regime's last defenders, a man capable of defending it both against the follies of absolutism and the excesses of revolution."[29]

Nowhere did Aleksandra's and Rasputin's misappointments cast a darker shadow than among those men of the Duma who continued to hope against hope that Russia could somehow be shielded from the revolution that now loomed above her. Dedicated to Russia's victory over Germany, such men drew back instinctively from any domestic political confrontation that might weaken her war effort. For the moment the government therefore had nothing to fear from them, despite the urgent campaign they had waged for a "ministry of public confidence" a few months before. "One can say with certainty that, at the present moment, and until the end of the war, the government has nothing to fear from any special complications in our domestic life" a smug Okhrana report pointed out at the end of February 1916.[30] Miliukov put it more bluntly. "If someone told me that to organize Russia for victory meant organizing her for revolution," he exclaimed on the eve of Khvostov's ouster, "then I would have to say that it would be better to leave her as she was, unorganized, for the remainder of the war."[31]

Miliukov's statement reflected a general weariness among the leaders of the Progressive Bloc that stemmed in part from the successful efforts of Goremykin and Khvostov to thwart the campaign they had waged for a "ministry of public confidence" the previous fall. It also reflected a momentary physical weakening in their ranks that deadened their political activism fully as much as their desire not to undermine their nation's war effort. That winter, pneumonia muted Riabushinskii's powerful calls for Russians to be given a greater voice in their nation's destiny and kept him from taking his recently acquired seat in the Council of State for several weeks, while Aleksandr Kerenskii, the Duma's firebrand young radical lawyer from Simbirsk, had to undergo a major operation to remove a tubercular kidney and endure a long recuperation that stretched from winter into summer.

As a man who lived to speak, and one who shaped his words to accommodate his listeners' shifting moods with amazing virtuosity, Kerenskii had held the trust of the masses ever since he had first entered the Duma at the age of thirty-one in 1912. By 1915, his outspoken defense of their interests had convinced the Okhrana that he stood at "the center of revolutionary work in Petrograd," and that he, more than any of the men who figured prominently in the revolutionary camp, was most likely to weld the revolutionaries' fragmented ranks together and lead an assault against autocracy.[32] Even more than the temporary loss of Riabushinskii, Kerenskii's long recuperation denied men critical of the government one of their strongest voices and added to the gloom that lay upon those who felt condemned to look on helplessly as Stürmer and his corrupt hirelings continued their course.

Most of all, sober and responsible Russians felt the loss of Aleksandr Guchkov, the Octobrist leader and powerful chairman of the Central War Industries Committee, who suffered a serious heart attack at the end of 1915 that kept him convalescing in the Crimea for almost six months. Guchkov had ridden along China's Great Wall during the Boxer Rebellion, fought with the Boers against the British in South Africa, and stayed on after Kuropatkin's defeat at Mukden to transfer Russia's helpless wounded to the care of the victorious Japanese in 1905. His adventurous spirit and willingness to face danger had endowed him with an aura of indestructibility that made his sudden disappearance from the political turmoil of Moscow and Petrograd seem strange.[33] This gave rise to a flood of disquieting rumors when people recalled Aleksandra's pathological hatred for the man who had struggled so valiantly to make a place for public opinion in Russia's war effort. Certainly, Aleksandra thought Guchkov despicable[34] and she had taken few pains to conceal her feelings from those who came and went from her now-famous mauve boudoir, with the inevitable (and predictable) result that his heart attack now provoked grim whisperings that he had been poisoned by "Rasputin and company."[35]

Such apprehensions deepened the gloom that lay upon Petrograd at the beginning of 1916 and made Stürmer's appointment especially loathsome to the many dedicated men and women who had placed Russia's welfare ahead of their own interests for so many weary months. Although politicians from Petrograd still tried to defend the tsar and tsarina, who seemed bent on ruining themselves, their dynasty, and their empire, others urged more radical solutions. Men from Russia's provinces, who saw all too readily the damage that inept government policies had wrought in their native regions, disputed the course charted by their parties' leaders. Early in 1916, they challenged the Duma's political establishment, and nowhere did their challenge prove more formidable than among Miliukov's Kadets.

Provincial rebels from Kaluga had raised the standard of rebellion among the Kadets even before the Great Retreat ended, and others soon swelled their ranks. By the fall of 1915, men from Samara, Kiev, Kostroma, and Moscow had joined the movement to turn their party away from collaborating with Russia's reactionary government. Angrily challenging Miliukov's stubborn insistence that the victory over Germany and Austria must take precedence over any attempt to force concessions from the government, the rebels used *Russkie vedomosti, The Russian Gazette* (in Moscow), and *Kievskaia Mysl, Kievan Thought* (in Kiev), to insist that their party must follow a new course. "We cannot stand aside from the popular movement," one of the Moscow insurgents explained. Lest the "forthcoming popular day of judgement against the criminal government become disorganized, chaotic, and senseless," the Kadets must use all resources at their command to "try to play the leading role in it."[36] Escalating their demands far beyond the previous fall's plea for a "ministry of public confidence," these men now called for "a ministry responsible to the Duma," and urged that there be no more cooperation with the government until one was granted. So many flocked to the rebels' standards that fall that Miliukov's moderate principles seemed doomed. In one last desperate attempt to restore unity, he summoned a meeting of Kadets from all over Russia to discuss the party's program and tactics.[37]

The rebels launched their attack against Miliukov's views the moment the Kadet Party congress convened in Petrograd on February 18, 1916. "It is said of Miliukov that his tactics are leading the party into a dead end, into a swamp," one of their Moscow leaders exclaimed. "He sees only the Duma Kadet fraction, persons completely [immersed] in the Petrograd atmosphere, and he judges all Russia by them."[38] Others pressed further the theme that the Kadets in the Duma had lost contact with their constituents. "The subtleties of Duma politics," one of Kiev's representatives explained, "in no sense satisfy our colleagues in the provinces."[39] Hoping to exploit the breach they had opened, the provincial rebels urged the Kadets to demand a "transfer of power to a ministry supported by a majority in the Duma." No longer should they place victory over Germany ahead of politics. "The party," they now insisted, "must not hesitate at open conflict with the government."[40] The rebels' proposals left Miliukov aghast. "For God's sake, do not get caught in the government's provocation," he warned. Peace would bring "Russian liberalism's victory complete and unquestioned," but there must be no conflict with the government until then. "All we can do now," he said, "is to tolerate everything and swallow the most horrible pills."[41]

Although their rebukes stung, the rebels' attacks contained more sound and fury than substance. They, not Miliukov, swallowed the first bitter pills of defeat, for this able politician, who had planned his political strategy with meticulous care, left the congress with even greater control over his party.[42] Convinced that Russia's European allies would force the tsar to grant a responsible ministry once the war ended, and fearful that a revolution would destroy any chance for victory, Miliukov carried his program of "moderation until victory" to the national congresses of the Union of Municipalities and the *Zemstvo* Union in Moscow. "I cannot be sure that the government will lead us to defeat," he confessed at one point, "but I know that a revolution in Russia will, without fail, lead us to defeat." Avoid striking "a spark which will ignite a great conflagration," he begged the rebels at the Moscow congresses. At the same time, he promised the Poles and Finns among the delegates that after the war they would find his assurances of freedom from the tsarist yoke "not empty words, but a sacred debt of honor." In the meantime, revolution must be avoided at all costs. Allied victory, Miliukov insisted with all of the professorial wisdom at his command, would bring Russia's moderates the political triumphs they had awaited for so long, while a revolution would bestow power directly upon the masses. "Never, perhaps, did a political leader descend to such blindness for the sake of a temporary parliamentary combination," exclaimed one of the spokesmen for the Duma's left wing. Miliukov was "the first of the opposition ranks," this prominent radical later said, "to make a speech of a clearly counter-revolutionary hue when the country had not yet pronounced the word revolution."[43]

Miliukov's increasingly conciliatory stance toward the government reflected widely-shared apprehensions about impending revolution that drove sober politicians toward the Right during the spring and summer of 1916. All feared that a mass revolution would wrest power from them even before they had got it firmly into their hands. "The day after peace is declared, we shall face the bloodshed of a civil war," Aleksandr Konovalov, Guchkov's deputy on the Central War Industries Committee, warned at one point. "When that happens," he went on, "we shall find ourselves face to face with the workers—and, at that point, their strength and our utter lack of it will become very clear."[44] The men of the Progressive Bloc thus had to choose between supporting reactionary government by such incompetents as Stürmer in the hope that their western allies could somehow force the tsar to yield them power at the war's end, and courting the masses, who saw little difference between them and the reactionaries who ruled for Aleksandra and Rasputin. For the moment, these thoughtful and moderate men very reluctantly chose the former and hoped that events would not push them too quickly into a confrontation with the latter. Their parliamentary allies in France and England looked on in amazement. "Russia must be very rich and very confident of her strength to be able to permit herself the luxury of a government such as yours, where the prime minister is a disaster, and the minister of war [Shuvaev, who had just replaced Polivanov] is a catastrophe," France's deputy minister for artillery and munitions remarked to Rodzianko at the beginning of May.[45]

Throughout the spring and summer of 1916, Stürmer easily held the Duma at bay. "The political achievements of the [Progressive] Bloc," *Birzhevye vedomosti, The Stock Exchange Gazette* reported at the beginning of April, "have been minimal or, more precisely, nonexistent," while Guchkov's allies lamented to their still-recuperating leader that "this Duma poses absolutely no danger to Stürmer's government."[46] To make matters worse, Miliukov and several other prominent Duma figures visited England, France, and Italy that spring as part of a special parliamentary tour arranged by the ambassadors of Russia's allies, and party discipline crumbled while they were out of the country. At least three major revolts shook the Kadet ranks during their brief absence, and the Progressive Bloc quickly polarized into extreme radical and conservative wings. Perhaps even more than the men of Stürmer's government, the Duma deputies seemed relieved in mid-June when Nicholas prorogued the Duma until November.[47] "All hopes had been centered in the Duma," Rodzianko sadly remarked, "but it, unfortunately, proved to be powerless."[48]

A weak Duma made weaker by its government's disdain and its leaders' desperate efforts to avoid any confrontation that might provoke the masses intensified Russia's domestic difficulties during the spring of 1916. Soaring prices, food shortages, industrial turbulence, and sullen unrest in the countryside all were as well-known to the Okhrana's guardians of public order as was the Duma's apparent inability to threaten Stürmer's government. Certainly, these problems demanded serious attention. While Brusilov's victorious armies pressed on toward the Carpathians, Nicholas experimented with other ways to resolve the crises his empire faced at home. A detailed report drafted by *Stavka*'s chief of staff General Alekseev at the beginning of June turned the tsar's thoughts back to October 1905, when Count Witte had insisted that he either must grant Russians some form of constitutional government or appoint a dictator to crush the forces of revolution. Now fearful that Russia's mounting domestic crises would undermine her war effort, Alekseev urged Nicholas to think again in terms of a "dictator" to deal with his empire's most pressing domestic problems. "Just as all power is concentrated in the hands of the supreme commander in the theater of military activity," he explained, "so it should be that, in the internal provinces of the empire . . . all power should be placed in the hands of one single plenipotentiary, who might well be named Supreme Minister for State Defense [and would] . . . unite, lead, and direct by his will alone the activities of all ministries and all government and civic organizations that function outside the theater of military action."[49]

Rarely did Nicholas pursue any suggestion as single-mindedly as he did Alekseev's. "It is imperative to act energetically and to take firm measures, in order to settle these questions once and for all," he wrote. "As soon as the Duma is adjourned," he told Aleksandra in the middle of June, "I shall call all the Ministers to this place for the discussion of these problems, and shall decide upon everything here."[50] To implement some form of Alekseev's plan while the Duma was in recess, he hoped to move rapidly, decisively, and in secret, but secrecy proved the scarcest of all commodities in shortage-ridden Petrograd that summer. Despite the bureaucracy's penchant for working behind closed doors, the shroud of military secrecy that zealous officers drew around much that transpired between the front and the rear, and the Okhrana's intense surveillance, very little remained secret for very long in Petrograd during the last months of the old regime. Within days, Duma President Rodzianko not only learned of Alekseev's scheme, but even obtained a copy of his top secret report.

Outraged that the emperor planned to usurp the Duma's prerogatives, Rodzianko hurried to *Stavka*. According to Rodzianko's best remembrance, the tsar was "attentive" at times during their meeting, but otherwise somewhat "indifferent." "What measures would you propose for setting affairs in the rear in order?" Nicholas asked when Rodzianko spoke out sharply against any sort of "dictator." "Your Highness, I can propose only one way out of the situation," Rodzianko replied. "Grant a responsible ministry." Nicholas assured the Duma's president that he would "think it over."[51] He had absolutely no intention of doing so. "It goes without saying that Rodzianko talked a lot of nonsense," he wrote Aleksandra after the Duma president had left *Stavka*. "Of all the foolish things which he said, the most foolish was the suggestion of replacing Stürmer."[52] Rather than think seriously about a "responsible ministry," Nicholas already had decided that Stürmer should become the "dictator" Alekseev had suggested.

Moving with what Stürmer later called "such speed that it made your head spin," Nicholas summoned the entire Council of Ministers to *Stavka* at the very end of June to tell them his plans. Utterly insensitive to his ministers' muted opposition, Nicholas forged ahead even though Stürmer himself pleaded that he "absolutely could not hold three posts at one time."[53] In reply, Nicholas ordered the unambitious but honest Aleksandr Khvostov (not to be confused with Aleksei, his nephew, whose brief tenure as minister of internal affairs had been so turbulent during the previous fall and winter) to take over the Ministry of Internal Affairs in order to lighten Stürmer's burden. Then, in a particularly short-sighted move, he named Stürmer to replace Sazonov as minister of foreign affairs a few days later. Here again Aleksandra had gotten her way. Insisting that Sazonov's continued service "w[oul]d be Russia's ruin," because he was "such a coward towards Europe and a parlamentarist," she had long urged Nicholas to remove him. Almost anyone would be better, she insisted, especially since she thought that Sazonov's replacement as director of Russia's diplomatic corps really "need not be a diplomat."[54]

In the five months since he first caught Aleksandra's fancy, Stürmer thus had parlayed his limited experience and minimal talent into appointments as Russia's president of the Council of Ministers, minister of internal affairs, minister of foreign affairs, and supreme minister for state defense. He had no notable experience in any of the areas of government over which he now reigned, nor had he any conception of the magnitude of the tasks he must accomplish.[55] Moreover, Stürmer's months in office had done nothing to mend his reputation for corruption and dishonesty. Ambassador Paléologue once described him as "ceremonious and cloying," while Ambassador Buchanan bluntly warned his superiors at the Foreign Office that they ought never to have "confidential relations with a man on whose word no reliance can be placed."[56]

Precisely because Stürmer's incompetence was so widely known, perhaps no appointment that Nicholas made ever stirred more anger. From the very beginning, he proved utterly unable to direct Russia's government, as serious and informed officials resisted his underhanded and deceitful methods of administration. Within a month, he had provoked Rodzianko's Special Council on State Defense to fury, and took to making grandiose gestures rather than working toward any substantial achievement.[57] Clearly, he could not bend Russia's central administration to his will except by personal appeals to the tsar in every instance where he came into conflict with the government agencies he was supposed to direct. Safely away from Petrograd's turmoil at *Stavka,* Nicholas had no desire to be as directly involved in Russia's day-to-day administration as Stürmer's failings made necessary. Worst of all, Stürmer had no success in solving the ominous food crisis that had been a major reason for Alekseev's suggestion that Nicholas consider a "dictatorship" in the first place. What Miliukov so aptly termed "a paralysis of authority" therefore undermined Stürmer's "dictatorship" from the moment it began.[58] "There was no one to put things into order," one of his associates remarked. "Everywhere there were people in authority giving orders, and there were a lot of them, but there was no directing will, no plan, no system."[59]

Officials issuing orders, officials providing solemn assurances that everything that could be done was being done, officials trying to give the impression that something important was being accomplished—this was the essence of a "dictatorship" whose colossal bluster fully matched its colossal failings. Nowhere did these stand out in sharper relief than in Stürmer's hopelessly botched attempts to deal with the food shortages that threatened Russia's cities that summer. By mid-1916, Russia's urban food supply crisis had replaced the army crisis of the previous year as the major worry for Nicholas's government and the Duma. What made the new crisis even more dangerous than the problems of 1915 was that no one understood at the time its real causes and therefore it could not be dealt with in the direct and effective manner that Polivanov had used to rebuild Russia's shattered armies the year before. For Russia's food shortages were not brought on, as some thought, by reduced grain production caused by the war's troop levies, nor were they the work of evil Jewish speculators (as anti-Semites claimed) or the result of failed government efforts to institute crude price controls, as some of the leading economists in the *Zemstvo* Union so urgently insisted. Not even the failure of Russia's railroads to move goods and commodities on time—so often the explanation given for shortages during the war years —could legitimately account for the food crisis.

To state the hard and simple truth, there was more than enough grain to feed everyone in Russia in 1916, but Russia's peasants simply refused to sell their grain at the low prices fixed by the government while the cost of rapidly disappearing consumer goods soared. They ate more themselves, fed more to their livestock, distilled grain into illegal vodka, or stored it away in hopes that food prices would rise and consumer goods prices would fall enough for them to sell their reserves at a greater advantage.[60] In the meantime, Russia's urban masses faced hunger, perhaps even starvation.

To end the urban food crisis, Russian planners needed to drive down the prices of consumer goods and produce enough to meet peasant demands at the very moment that Brusilov's armies were calling for increased production of guns and shells. Ending the crisis might well have proved beyond the skill of even a talented "dictator," but there is no doubt that it called for far greater ability than Stürmer possessed. Utterly at sea, he quickly made the crisis worse by establishing several impotent government agencies that worked directly at cross-purposes with ones that already were in place.[61] Even worse, when the never-failingly corrupt Manasevich-Manuilov tried to bilk a Moscow bank out of thousands of rubles, Aleksandr Khvostov, the honest new minister of internal affairs who proved to be so unlike his scheming nephew, ordered his arrest. All too quickly, the scandal rubbed off on Stürmer, especially when people began to whisper that he did his private secretary's bidding, not the other way around, and every remaining shred of his support disappeared. No one would cooperate with Stürmer because no one trusted him.[62]

In the midst of Petrograd's corruption and scandals, Aleksandr Khvostov's honesty was all the more striking at that particular moment because rumor had it that he had just refused Aleksandra's request to release Sukhomlinov from prison. "If I order you to do so, why can't you do it?" Russia's irate empress was said to have asked. According to the tale that passed through Petrograd's salons, Khvostov had replied: "Your Highness, my conscience will not permit me to free a traitor even at your command."[63] If Khvostov had the courage to oppose Russia's empress, he had more than enough to oppose her "dictator." For Stürmer and his corrupt inner circle, Manasevich-Manuilov's arrest thus signaled disaster. That he could not loosen Khvostov's grip on the man whom Paléologue once called a "man utterly lacking in moral scruples" caused Stürmer to lose credit with his underhanded accomplices. As such men are wont to do, they began to make new alliances and plot new avenues of escape.[64]

By summer's end, Stürmer therefore faced problems even more vexing than the food crisis, Manasevich-Manuilov's arrest for extortion, and Khvostov's honesty. Most dangerous from his scandalously self-interested perspective, his stock with his sovereigns had slipped badly. Nicholas, after all, had moved to *Stavka* to escape the irritation of attending to government affairs in Petrograd, and his "dictator's" continual appeals for his personal intercession demanded precisely the involvement that he so much wished to avoid. Further, Aleksandra now aspired to play an even greater role in Russian domestic politics, and, to carry out her will, she had found a man who thought Rasputin even more holy than did any of the others she had championed for high office. Stürmer, she now explained to Nicholas urgently, was nothing but "a big act." For several months, he had failed to consult Rasputin and, for that reason, had "lost his footing." The man in whom they now must place their faith without hesitation must be Aleksandr Protopopov, the elegant, impeccably dressed vice-president of the Duma thought by some to have all "the over-done politeness of a tradesman." This man, she insisted, "venerates our Friend & will be blessed."[65]

Urged on by Rasputin, Aleksandra concluded that Protopopov was far better able than Stürmer to deal with the pressing shortage of foodstuffs in Russia's cities, and she badgered Nicholas to appoint him in place of honest Aleksandr Khvostov as minister of internal affairs. "Gregory begs you earnestly to name Protopopov," she wrote at the beginning of September. Could such a man resolve the crisis that now loomed over Russia's cities? Aleksandra thought they need not worry. "Our Friend begs you not to too much worry over this question of food supply," she wrote to her husband. "I fully believe Protopopov can cope with this question."[66] At first Nicholas hesitated. "Our Friend's opinions of people are sometimes very strange," he mused. "All these changes make my head go round. In my opinion, they are too frequent." The next day his resistance collapsed. "It shall be done," he wrote to Aleksandra, although he did not announce Protopopov's appointment until later in the month.[67]

"Ministerial leapfrog" was now in full swing. Putting his thoughts into doggerel, Vladimir Purishkevich wrote: "The pace of it is simply mad;/ Whatever else we lack, there's none/ Could count the Ministers we've had . . . You'll see your Minister uncrowned/ Within a month—or rarely two./ By minutes now we count their term;/ They go and leave a sulphurous smell;/ Only Rasputin still holds firm."[68] Conservatives not blessed with Purishkevich's intelligence, wit, and gift for satire shared his anger nonetheless. "They ought to hang up a sign that reads: 'Piccadilly—the show changes every Saturday,' " one of them muttered in disgust. "You'd hardly get settled in your seat before the program changed again." New appointees thought their positions so tenuous, Purishkevich confessed to his diary, that they no longer even bothered to move into the government residences to which their high offices entitled them.[69]

Although several of her ministerial choices had stirred widespread outrage during the year since Nicholas had taken over supreme command of Russia's armies, Aleksandra's choice of Protopopov won wide acclaim. Much of the Duma and the press seconded her choice, and the Stock Exchange greeted the news by posting one of its largest gains since the war's beginning. *Birzhevye vedomosti, The Stock Exchange Gazette* compared Protopopov's appointment in significance with the ouster of the much-hated "odious ministers" the previous year, and proclaimed that "a representative of public opinion now wields political power," while one of its rivals assured its readers that Protopopov's appointment "guarantees that plans for political and economic progress will be given first priority."[70] Perhaps Aleksandr Konovalov, Guchkov's deputy on the Central War Industries Board and vice-president of Moscow's Stock Exchange, was most enthusiastic of all. "Capitulating to public opinion, the supreme power has made a colossal and unexpected leap forward," he exclaimed. "The very most that we could have expected was the appointment of some bureaucrat who would play the liberal to some extent. And suddenly—we have the Octobrist Protopopov, a man utterly alien to the bureaucratic world." Now looking forward to the time when "in a few months we shall perhaps see a cabinet including Miliukov and Shingarev," Konovalov urged his colleagues in the Duma to set aside their pessimism and ennui. "Everything depends on us, everything is in our hands," he announced to a group of liberal politicians, industrialists, university professors, and journalists whom he had invited to his home in Moscow. "The upcoming session of the State Duma can witness our decisive assault against the bastions of power; it can be our final storming of the bastions of the bureaucracy."[71]

Buoyed by Protopopov's appointment, moderate and progressive Duma delegates looked forward to November 1, 1916, when Nicholas had promised to summon them back into session. But they had grossly misunderstood both Aleksandra's motives in proposing Protopopov and Nicholas's hopes in appointing him. Hating "those rotten people" in the Duma more than ever, Aleksandra had urged Nicholas to approve Protopopov's appointment in hopes that it would "make a great effect amongst them & shut their mouths."[72] She did so confidently because only she and a handful of close confidants knew that Protopopov had long been part of Rasputin's circle. "Doctor" Zhamsaran Badmaev, the notorious Tibetan herbalist who, on occasion, had "prescribed" concoctions for the tsarevich's hemophilia, had first brought Protopopov and Rasputin together at some point during the thirteen-year period in which he treated Protopopov for an illness whose hallucinations, recurrent paresis, and "tubercular" leg ulcers bore an unmistakable resemblance to the symptoms of advancing syphilis. Badmaev was famous in court circles for his herbal infusions which, Rasputin once explained, could make one's greatest cares "seem like petty trifles" after just one tiny draught. "You'll become happy," Rasputin promised, "so-o ha-appy, and sil-ly that you won't worry about anything."[73] Greedy for power and mentally unstable (Paléologue, Rodzianko, and Buchanan all agreed later that he was insane),[74] Protopopov quickly proved unjustified the optimism with which so many had greeted his appointment.

During the six weeks that separated his appointment from the Duma's reopening, Protopopov let no one doubt his allegiance to the empress and to the corps of unprincipled hangers-on who surrounded her, and even went so far as to appear before the Duma Budget Commission garbed in the uniform of the Imperial Gendarmerie, long-standing symbol of repression in tsarist Russia.[75] When his former Duma colleagues took offense at his tactless behavior, questioned his worrisome intimacy with Aleksandra, Stürmer, and Rasputin, protested his readiness to set free Sukhomlinov and Stürmer's corrupt henchman Manasevich-Manuilov, and criticized his plan to seize control of all food supplies not held by the military, Protopopov struck back with a vengeance. There could be no more public meetings, he insisted, including hitherto private joint sessions of the Union of Municipalities and *Zemstvo* Union, unless police officials attended. Even before the Duma reassembled, September's fleeting moment of optimism had passed. "A man who works with Stürmer, frees Sukhomlinov, whom the entire country considers a traitor, liberates Manasevich-Manuilov, and persecutes the press and civic organizations cannot be our friend!" Miliukov exclaimed in righteous indignation.[76]

Miliukov carried his outrage directly into the Duma when it reassembled. Sensing that something unpleasant was in the wind, Stürmer and his cabinet left the Taurida Palace as soon as Rodzianko finished his brief opening remarks. Then Miliukov rose to speak. "With this Government we cannot legislate, any more than we can, with this Government, lead Russia to victory," he said angrily. Stürmer's government had weakened Russia at home and undermined her relations with her allies so as to make victory impossible, he continued. Sukhomlinov's inexcusable failure to prepare Russia for war had been treason. Was it the same with Stürmer? "Does it matter, practically speaking, whether we are dealing with stupidity or treason," Miliukov asked his colleagues. "While the Duma everlastingly insists that the rear must be organized for a successful struggle against the enemy," he went on, "the government persists in claiming that organizing the country means organizing a revolution . . . and deliberately prefers chaos and disorganization." Speaking in measured tones, to emphasize every word, Miliukov asked: "Is this stupidity, or is this treason? Choose either one, the consequences are the same."[77] Such a state of affairs could not continue, he warned. Stürmer and his band must be driven from office "for the sake of the millions of victims and the torrents of blood poured out [and] for the sake of our national interests."[78]

Miliukov offered no hard evidence of Stürmer's treason, nor was there any. Like Sukhomlinov, Stürmer's crimes were greed, incompetence, and stupidity, not betrayal of his country. Miliukov later confessed that when he posed the alternatives of stupidity and treason, he had been inclined to think the former, but that the country as a whole preferred to assume the latter. Certainly, his speech was, as he later wrote, "a storm signal for the revolution."[79] Quickly, others leaped into the fray. The conservatives' brilliant rising young star Vasilii Shulgin announced that Stürmer was "a man without convictions, ready for anything, who understands nothing about state affairs," while Miliukov's colleague, the silver-tongued lawyer Vasilii Maklakov, used his brilliant gifts to curse the government as "a cabinet without a program [and] ministers without opinions . . . which paralyzes and enfeebles the strength of all Russia."[80]

Boiling with outrage, Stürmer set out to prosecute Miliukov as a political criminal, only to find that most of his cabinet would not support him for fear of provoking the Duma to still greater anger. He then demanded that the Duma be recessed. This time, Nicholas had had enough. While Alekseev sternly warned him against closing the Duma as Stürmer insisted, his mother and several of his cousins and uncles told him very bluntly that the influence Rasputin exercised through Aleksandra threatened his dynasty's future. The tsar would not admit openly the justness of their complaints, but their urgings had enough cumulative impact to ruin Stürmer, who, perhaps more than anyone at that moment, personified all the various failings that had brought Russia to ruin.[81] "Nobody believes in him," Nicholas wrote to Aleksandra. "It is much worse than it was with Goremykin last year." Stürmer's fate was sealed. Despite Aleksandra's assurances that "Our Friend says Stürmer can remain still some time as President of Council of Minister," Nicholas decided to "give him leave for the present."[82]

During the first days of the Duma's November session, Miliukov played the role of virtuous knight to Stürmer's archknave, while a nationwide audience cheered their approval. His speech had lanced what he later called "a pus-filled boil" that revealed the gaping, suppurating sore which sapped the vital forces of the Russian state. Miliukov had spoken firmly and in measured tones to an audience of less than five hundred, and the police had moved quickly to censor all accounts of his speech in the press. But he had spoken the words that had lain too long in the hearts of too many Russians for them to be suppressed. Word of his speech, and excerpts put together from the notes and recollections of those who had heard it, swirled across the empire. While the Duma stood recessed for a fortnight, the people of Russia became what Miliukov later called "a megaphone" for his words.[83] "If, by some miracle, it would have been possible to gather the entire population of Russia into the White Hall of the Taurida Palace and Miliukov could have repeated his speech before those tens of millions," Shulgin later wrote, "the applause that would have greeted his words would have drowned out even the drumfire of the hundred and fifty great ammunition dumps that General Manikovskii had assembled [for the next year's campaigns]."[84]

Now fearful of the wide-ranging consequences of Stürmer's failure, Nicholas appointed Aleksandr Trepov, an honest conservative bureaucrat and long-time student of European legislative institutions, to preside over the Council of Ministers and even spoke of dismissing the incompetent Protopopov. Words tumbled from Aleksandra's pen by the thousands during the three days after she heard the rumors of Protopopov's possible dismissal, but this time, she used her letters to Nicholas only to gain time until she could get to *Stavka* to plead her case in person. "I entreat you dont go and change Protopopov," she wrote when she first heard the rumors. "Protopopov is honestly for us. Oh, Lovy, you can trust me. I may not be clever enough—but I have a strong feeling & that helps more than the brain often." Insisting that only men who held Rasputin's trust could save Russia from ruin, she begged Nicholas to "remember that for your reign, Baby & us you need the strenght prayers & advice of our Friend." Therein lay Trepov's failing and Protopopov's great virtue, she insisted. Trepov "does not trust me or our friend," she complained, while "Protopopov venerates our Friend & will be blessed."

No longer content to press advice upon Nicholas as she had in the past, Aleksandra now gave instructions that came close to being direct orders. "For me dont make any changes till I have come, tell Trepov you wish to think it over a day or two and tell him you do not intend sending Protopopov [away]," she wrote. That spring, she had come close to identifying Rasputin with Christ. Now she spoke of Rasputin and God in the same breath. "I pray so hard to God to make you feel and realise that He [Rasputin] is our caring, were He not here, I dont know what might not have happened. He saves us by His prayers and wise councils and is our rock of faith and help." Clearly, God spoke through Rasputin, and the men who refused to heed his words must not be allowed to continue their dangerous course. Nothing less than the fate of autocracy hung in the balance. *"The Czar rules and not the Duma!"* Aleksandra exclaimed at one point. "I am fighting for your reign & Baby's future." As she made ready for her journey to *Stavka,* she concluded: "Be firm [and] don't listen to men, who are not from God but cowards."[85]

After a fortnight at *Stavka,* Aleksandra got her way, although Nicholas must have been more stubborn than usual. "Forgive me if I was moody or unrestrained—sometimes one's temper must come out!" he wrote just hours after she returned to Tsarskoe Selo. "You were so strong and steadfast—I admire you more than I can say." In reply, Aleksandra counseled him to have "just a little more patience & deepest faith in the prayers & help of our Friend—then all will go well." Protopopov had been saved, but her victory had not been complete, for Trepov still remained. Fearful that Russia's new prime minister might undo all she had done, she urged Nicholas to "let no talks or letters pull you down—let them pass by as something unclean & quickly to be forgotten." Now she wanted her tsar to be firm, to assert the will that, in fact, was her will. "Show to all, that you are the Master & your will shall be obeyed," she insisted. "Now comes your reign of will and power, & they shall be made to bow down before you & listen your orders & to work how & with whom you wish."[86]

While Aleksandra exhorted Nicholas to end the "time of great indulgence & gentleness" and to be the autocrat his father had been, Trepov set out to strike an alliance with the Conservatives in the Duma that could make it possible for them to work with his government. His first efforts left the Duma's moderate right and center in a quandary. Did his conciliatory tactics mean that the tsar and his ministers had decided to tread softly, to perhaps seriously seek the common ground on which they could stand together against the volatile masses? Or was this merely "a reseating of the musicians," with no significant implications for the future? On one thing they agreed: Stürmer's fall could not alone save Russia from defeat or revolution. Protopopov must follow Stürmer into disgrace before anything could be done.[87] Anxious to clear the way for a strong conservative government that could lead Russia to victory, and hopeful that his tsar was ready to follow his urgings, Trepov enlisted his Duma allies to help oust Protopopov from Russia's largest and most important ministry. Neither he nor the Duma would succeed. In the battle over Protopopov, Aleksandra won the day, for she had built her defenses so strongly that he would remain so long as she and Nicholas occupied Russia's throne.

When the Duma reassembled on November 19, thoughts about Protopopov, the empress, and the ominous "dark forces" around the throne filled the minds of progressives and conservatives alike, but none expected to hear them expressed in such forceful and dramatic terms as those chosen by Vladimir Purishkevich, stalwart dean of the Duma's right wing at the age of forty-six, and a staunch defender of the throne. Just two weeks before, Purishkevich had dined with Nicholas at *Stavka* where he had been appalled to find that the "cowards" on the tsar's staff refused to tell him the true state of affairs in Russia for fear of damaging their careers. Apprehensive and self-centered, these men had urged Purishkevich to speak rather than shoulder the burden themselves. "Tell him about Stürmer," they urged. "Let him know about Rasputin's fatal role in things." Shocked at the vulgar careerism and "self-love" of such men, Purishkevich had spoken to Nicholas that evening, but with no appreciable effect.[88]

As a leader of Russia's reactionary forces and a dedicated anti-Semite who was soon to become known as one of Russia's first Fascists, Purishkevich always had held public opinion in low esteem and believed that only an autocrat could govern Russia. What he called his "limitless allegiance" to his sovereign had survived even his visit to *Stavka*, but the terrible experience of seeing his tsar surrounded by lickspittle "caliphs for an hour" had shaken the very foundations of his faith in the men who stood closest to the throne. Now convinced that only public pressure might divert the tsar from his ruinous course, he arose in the Duma on November 19 to make the speech that every loyal servant of a monarch hopes never to make. Proclaiming himself to be, "perhaps, the most extreme of the Rights," Purishkevich warned that, despite the rightness of Russia's cause, "the hour of victory [would] be postponed for a long time" because of the "consistent and systematic internal disorganization of the state." He spoke of the stupidity and capriciousness of Russia's censors, of the "marauder-profiteers" in the army's rear, and of "a hidden hand" working against Russia's efforts at victory. With Protopopov looking on from the ministerial box, Purishkevich turned and spoke bluntly of his failings and hinted broadly at his crude self-interest. Noting that Russia's minister of internal affairs once had claimed to have lost several pounds from overwork, Purishkevich exclaimed to the deputies' cheers and laughter: "Now I think that Aleksandr Dmitrievich Protopopov has lost not merely three or four pounds of weight but, in fact, has lost all his weight, all his authority, in the eyes of Russian public opinion."

But Purishkevich had larger game in his sights than such "small-fry" as Protopopov. Now he felt compelled to insist that loyal Russians, and loyal ministers most of all, must "throw themselves at the sovereign's feet, beg the tsar for permission to open his eyes to horror of the current state of affairs, and plead with him to deliver Russia from Rasputin and all his corrupt band." All too aware of the dangerous ground upon which Purishkevich had begun to tread, Rodzianko admonished him to "remember of whom and what you are speaking." But Purishkevich, who, as a young man of thirty-five had sworn to defend the throne at all costs during the Revolution of 1905, and had founded the Union of the Russian People to terrorize Jews and shatter the foundations of constitutional government in Russia, was not to be stopped from using that very forum he had once sworn to destroy to indict those whose greed and stupidity had placed Russia in mortal danger.

Ignoring the Duma president's warning, he plunged on to deliver the most crushing indictment of all. Together with all who sat in Russia's legislative chamber, he announced, the tsar's ministers must join forces and "beg the Sovereign to remove Grisha Rasputin as the director of Russia's domestic and public affairs."[89] Now it had been said. Others on the Left had hinted broadly at Aleksandra's absurd reliance upon Rasputin's counsel, but the voice that now spoke the indictment against the "filthy, depraved, corrupt peasant" whose advice Russia's empress listened to above all others', stood on the Duma's far Right.[90] Russia's autocracy seemed about to lose the very last of its loyal defenders. Russians' "journey to Golgotha," one of Purishkevich's allies on the Right said, had begun.[91]

Unknown to his colleagues, Purishkevich was prepared to go to far greater lengths to rid Russia of Rasputin than merely denounce him from the Duma's rostrum. On the morning of November 21, Purishkevich received a visit from Prince Feliks Iusupov, recently married to the dazzling Grand Duchess Irina Aleksandrovna, whose mother was Nicholas's favorite sister. Not yet thirty, and as handsome as his wife was beautiful, Iusupov moved with equal ease in Petrograd's most elegant salons and most dissolute bordellos, and was thought self-centered and spoiled by many. Yet Purishkevich, who possessed the rare ability to take the proper measure of what lay within a man, immediately saw in Iusupov "a man of great will and character." He had come to talk of Rasputin, whose influence, he had told Purishkevich on the telephone the day before, was "an embarrassment."[92] Iusupov was convinced that Rasputin "was the cause of all Russia's misfortunes," and that if he were removed, "that satanic power which envelopes our sovereigns," would vanish.[93] Purishkevich had dared to confront Rasputin's evil influence by speaking out; Iusupov was ready to use other means.

Although neither man knew the other well, they talked for almost two hours that morning. According to Purishkevich's diary account—the only firsthand record we have—Iusupov came directly to the point.

"You know, your speech is not going to accomplish what you are hoping for," he began. "The Sovereign has an intense dislike for any attempt to put pressure on him, and one has to expect that Rasputin's importance now is going to increase, not diminish, because of his totally unlimited influence on Aleksandra Fedorovna."

"Then what's to be done?" Purishkevich asked, posing yet once more the fateful question that had confronted so many Russians for so long. For a moment, the two men looked at each other in silence, each quite probably trying to take the final measure of the other before deciding whether he dared continue.

An "enigmatic smile" played across Iusupov's lips, but it never touched his eyes, which grew hard as they held Purishkevich's own with an unblinking stare. Now certain of his man, Iusupov spoke two words more: "Eliminate Rasputin."

"That's easy enough to say," Purishkevich replied, perhaps forgetting in his confusion that no one had dared to say it so bluntly until that very moment. With a nervous laugh, perhaps because he now wanted to gauge Iusupov one last time, he asked: "Who's going to do it, when there are no resolute men left in Russia, and when the government, which might ordinarily do such a thing itself, and even do it skillfully, is controlled by Rasputin himself?"

"You're right to say that the government cannot be counted upon," Iusupov continued, "but one can find such men in Russia nonetheless."

"Do you really think so?"

Iusupov replied without a moment's hesitation. "I am certain of it," he said. "One of them stands before you at this very moment."

Purishkevich's almost instinctive first reaction was to chide Iusupov not to "make jokes about this," but it took only a moment for him to realize that the young prince spoke in deadly earnest. Feliks Iusupov, perhaps the empire's wealthiest prince, whose family had served Russia's tsars since the time of Ivan the Terrible, stood ready to become an assassin in his country's cause.

If Iusupov was ready, so was Purishkevich. "If you are willing to take part in an attempt to finally rid Russia of Rasputin, then here's my hand. Let's sit down and discuss all the possibilities for such an operation."

"Two others have already agreed to join us," Iusupov replied. "Come to visit me if you happen to be free this evening. Then you can meet them both."[94]

For someone who had just claimed that no truly resolute men remained in Russia, it must have been disconcerting for Purishkevich to learn that there were three who had been prepared to go to greater lengths than he to rid their land of Rasputin. And he must have been even more surprised to find that one of them was none other than Grand Duke Dmitrii Pavlovich, the favorite nephew to whom Aleksandra was thinking of marrying her eldest daughter, the Grand Duchess Olga. Tall, slender, just-turned twenty, and a frequent visitor in Nicholas and Aleksandra's family circle, Dmitrii Pavlovich, who on occasion sat next to the tsar when they dined at Mogilëv, awaited Purishkevich at Iusupov's palace that evening with a Captain Sukhotin, about whom we know nothing except that he was on medical leave and recuperating from a wound when he became the fourth of the conspirators. Together they laid plans to lure Rasputin to the palace by using Iusupov's beautiful wife as bait, to kill him with poison, and then to slip his weighted corpse under the ice so that it would sink to the bottom of the Neva. To make certain that the poison would not fail, they agreed that Purishkevich should recruit his close friend and physician Dr. Lazovert. They allowed themselves a month to complete their preparations. If all went well, Rasputin would meet his death in the early hours of December 17.[95] None relished the course he had chosen, but all were certain that Russia's salvation or ruin hung in the balance. "Aleksandra Fedorovna administers Russia as she does her boudoir," Purishkevich wrote in his diary then. "No one can be certain what tomorrow will bring."[96]

The night of December 16, Purishkevich remembered, was one of those warmish northern winter nights when the mercury stood just below freezing and the heavy, wet snowflakes held together barely long enough to reach the pavement. Just after midnight, the conspirators met at the great palace that Iusupov's family had built on the banks of the Moika more than a century before. On its walls hung a dazzling array of paintings by old Dutch masters, including several by Rembrandt, along with valuable works by Teniers and Claude Lorraine. Beyond a grand ballroom that rivaled that of the Winter Palace in splendor stood a complete theater, decorated in gold leaf and red velvet. There, in the old days, the Iusupovs' serf actors and serf orchestras had performed for their masters' pleasure, as Iusupov's illustrious ancestors had entertained the ambassadors of Europe's kings along with Russia's great lords and ladies. A place rich in history, the palace overflowed with memories that reflected the greatness of Russia and the Iusupovs' loyalty to her tsars.

In the palace's semi-basement, Iusupov's workmen had labored for more than a week to hang tapestries, arrange furniture, and transform gray walls and granite floors into an intimate dining room and cozy salon that radiated intimacy and warmth so that the prince could entertain Rasputin without taking him into the main part of the palace. Three huge Chinese vases of red porcelain had been placed in niches cut into the walls to complement the rich and rare furnishings that Iusupov had chosen for the occasion. In one corner stood a miniature Oriental armoire of ebony, with bronze fittings and many secret drawers and panels. Upon it stood a large silver-ornamented cross that an artist had carved from a single block of crystal during the long-ago days of the Italian Renaissance. A large rug of white bearskin covered the floor. Intricately carved chairs cut from rare woods, and sofas done up in costly fabrics, added to the room's warmth. In the center, Iusupov's servants had placed the table "where Rasputin was to take his last glass of tea." Not until eleven o'clock on the evening of Rasputin's visit had Iusupov's servants put the last touches on their work. Just before the other conspirators arrived, Iusupov surveyed the scene with satisfaction. A samovar steamed in the middle of the table, surrounded by plates of Rasputin's favorite cakes and pastries. On a chest nearby stood bottles of Rasputin's favorite Madeira and Marsala. After careful thought, Iusupov had decided that these, and petits-fours filled with Rasputin's favorite rose-flavored cream, would hold the cyanide that was to kill him.[97] "I was absolutely amazed at the skill with which this cellar had been transformed in so short a time into something resembling an elegant bonbonnière," Purishkevich confessed. "The room [that so recently had been a semi-basement] was completely unrecognizable."[98]

Just after midnight, Purishkevich, the Grand Duke Dmitrii, Captain Sukhotin, and Dr. Lazovert sat briefly with Iusupov in his newly-arranged dining room and disordered its table settings so as to make it appear that several people had just finished dinner. Then Iusupov donned a fur coat, pulled a fur hat low over his eyes, and drove to Rasputin's apartment on Gorokhovaia with Dr. Lazovert, now masquerading as his chauffeur. They found Rasputin ready, dressed in his finest, for his long-awaited meeting with the Princess Irina Iusupova. All of Petrograd's high society knew that she was in the Crimea; during the past month, it had been Iusupov's task to convince Rasputin that his wife had planned a brief secret visit to Petrograd in order to speak with him. The *starets*'s appearance that evening showed that Iusupov had been masterful in his deceit. "Rasputin was dressed in a silken shirt embroidered with cornflowers, gathered in at the waist by a heavy, twisted, raspberry-colored cord," Iusupov remembered. "His wide trousers of black velvet looked new, and so did his boots. He had slicked down his hair and trimmed his beard with particular care. When he came near, I detected a strong odor of cheap soap." They chatted briefly as the *starets* made ready to leave. "Enough chatter," Rasputin exclaimed. "Let's be off!"[99]

Once at Iusupov's palace, Rasputin suddenly grew suspicious. The disordered table showed that other people were in the house, and he could hear the strains of "Yankee Doodle" being played on a victrola. It was all part of an elaborate scheme to explain the Princess Irina's absence. His wife had been entertaining a few close friends who had stayed later than expected but were about to leave, Iusupov said in an off-handed way. In the meantime, they would wait in the basement salon, in order to keep Rasputin's visit a secret as he had wished. To support Iusupov's tale, "Yankee Doodle" played again in the distance. "The miniature armoire, with its many tiny drawers held his attention particularly," Iusupov remembered. "He amused himself with it like a small child, opening and closing it, and looking at it from every angle." As they settled comfortably into the salon, Iusupov offered his guest some of the rose-cream cakes. "I don't want any. Too sweet," Rasputin insisted. For a while he sat taking neither food nor drink. Then Iusupov pressed a glass of poisoned Madeira upon him. Rasputin drank it at a gulp and helped himself to another. Seeing that neither glass had any effect, Iusupov offered him a third as they waited for the upstairs to grow quiet as a sign that the "guests" had left. Rasputin helped himself to one of the rose-cream cakes. To Iusupov's amazement, it too had no effect. For some terrible minutes, it seemed that Rasputin could devour poison with impunity.

Iusupov hurried upstairs to consult with Purishkevich and the others. "Wait a bit longer, Feliks," Grand Duke Dmitrii urged. "Go back and try another dose." His nerves clearly set on a very fine edge, Iusupov returned to his guest. During the next half-hour, Rasputin drank more wine and ate more cakes, while his host played gypsy ballads on his guitar. Still, the cyanide made no impact. Now almost beside himself, Iusupov returned to his friends a second time. His patience exhausted, he turned to Purishkevich.

"Vladimir Mitrofanovich, you wouldn't have any objections if I just shot him, would you?" he asked. "It would be quicker and simpler."

"Do as you wish," Purishkevich replied. "The main thing is to finish him off and to do it tonight."

"I had scarcely uttered these words," Purishkevich wrote, "before Iusupov strode quickly and resolutely to his writing desk, took out a small caliber Browning pistol, and rapidly went downstairs."

"Why are you looking so intently at the crucifix?" Rasputin asked when the prince returned.

"You would do well to look at it yourself, and say a prayer," Iusupov replied.

For a brief moment, Iusupov was paralyzed. "Should I aim at his head or his heart?" he asked himself. He chose the heart and fired. Rasputin let out a roar and fell, "like a broken puppet," upon the bearskin rug.

Their deed done, the conspirators rejoiced. It was almost three o'clock in the morning, and it had been a full two hours since Rasputin had taken the first glass of poisoned wine. But their moment of relief was brief. Suddenly their victim struggled to his feet and staggered to a side door that led into a courtyard that opened directly onto the Moika Embankment. He tore open the door and stumbled into the night, with Purishkevich in pursuit. With Rasputin several meters away, Purishkevich raised his heavy Sauvage pistol and fired twice. An expert marksman who often practiced at the Semenovskii Guards' pistol range, Purishkevich expected to see his man fall. To his amazement, Rasputin continued his flight, and was almost at the embankment gate, when Purishkevich raised his revolver again. Two more shots brought Rasputin down in a heap. "This time, I was certain," Purishkevich wrote in his diary a few hours later, "that his swan song had been sung and that he would never arise again."[100]

Purishkevich's wish could not make it so; Rasputin was unconscious but not yet dead. Only when the conspirators gathered up his body and dumped it beneath the Neva's ice did he breathe his last. According to the autopsy done after the police found Rasputin's corpse the next day, the man whom Aleksandra thought had been sent to her by God to save Russia had met his death by drowning. Even before the police had determined the cause of Rasputin's death, the assassins' identity had become an open secret in Petrograd, but the question of their arrest raised some awkward issues. Purishkevich's status as a Duma member left him immune from arrest for the moment, while Russian law regarded all members of the imperial family as sacrosanct and specified that they could be taken into custody only on direct orders from the tsar. Almost without hesitation, Aleksandra usurped the emperor's powers and ordered a colonel in the Imperial Gendarmes to place Iusupov and the Grand Duke Dmitrii under house arrest.[101] Purishkevich, however, remained free to make a scheduled inspection tour of the front.

During the weeks after Rasputin's death, Nicholas and Aleksandra recklessly destroyed the last pillars of their political support. Without a moment's hesitation, the tsar abandoned an important military council, for which commanders from all fronts had been summoned to *Stavka,* in order to rush to Tsarskoe Selo. Because Alekseev was on sick leave in the Crimea, the tsar's hasty departure left Russia's military affairs in the hands of General Vasilii Gurko, *Stavka*'s recently-appointed deputy chief of staff, who could do nothing without direct approval from Alekseev or Russia's supreme commander. A small, dapper general who had fought in Central Asia's Pamirs and, as Russia's military attaché, had been captured by the British during the Boer War, Gurko at the age of fifty had led a division of Rennenkampf's cavalry into East Prussia. Since then, he had risen to command a corps and then an army before he had been appointed as Alekseev's temporary replacement in the fall of 1916. An officer of rare ability, Gurko nonetheless sat at *Stavka* with his hands completely tied, for when he abandoned his army to be with his empress, Nicholas left Gurko without any authority to act on his own. Any German or Austrian offensive could have brought disaster. To Russia's good fortune, her soldiers had none to face at that moment.

The same good fortune did not hold for Russia's domestic affairs. As the first weeks of the new year brought a desperate food supply crisis to Petrograd and Moscow, Aleksandra placed still greater faith in Protopopov who, according to one account, claimed to enjoy frequent nocturnal visits from Rasputin's ghost and dutifully reported the instructions and warnings that fell from that specter's lips.[102] At the same time, Nicholas replaced every minister who had a shred of talent or independence with men whose cause Aleksandra had pleaded during the weeks just before Rasputin's murder. Anxious for Trepov to bear the stigma of closing the Duma, Nicholas had written to Aleksandra some days before that, "after Trepov has done the dirty work" of closing the Duma, he intended to "kick him out."[103] A week after Rasputin's murder, he did so, and forced the premiership upon the sick and aged Prince Nikolai Golitsyn, who begged to be spared from an assignment he thought himself "utterly incapable" of performing.[104] Although a senator and a member of the State Council, Golitsyn, who had not actually served in any central government office in forty years, had unwittingly won Aleksandra's favor for his work as chairman of a charitable committee that worked to better the lot of Russian prisoners of war in Germany and Austria. At the same time, Nicholas removed War Minister Shuvaev, a barely competent administrator with whom Aleksandra had grown disenchanted, and replaced him with the slow-moving General Mikhail Beliaev, whose talents Aleksandra had touted even before Polivanov's disgrace. Again at Aleksandra's urging, Nicholas removed minister of justice Makarov, who stubbornly still refused to withdraw charges against Manasevich-Manuilov, and replaced him with Nikolai Dobrovolskii, another of Rasputin's favorites. Now thoroughly disgusted, Minister of Education Count Pavel Ignatiev, who had given the tsar his loyal friendship without question for thirty years, begged to resign his post to be spared "the unbearable burden of serving against the commands of my conscience."[105] At first, Nicholas made no reply. A week later, he relieved Ignatiev without so much as a word of thanks for his years of diligent service. With these last remaining able men removed from office, no one who could hope to deal with the social and political tensions that threatened to overwhelm Russia stood near the head of her government.

If Nicholas's and Aleksandra's actions during the days after Rasputin's death drove away the handful of statesmen and politicians who still might have supported them, Aleksandra's illegal order to arrest Iusupov and the Grand Duke Dmitrii infuriated most of the imperial family. "This means open revolt," the Grand Duke Andrei Vladimirovich exclaimed. "Here we have the war, with the enemy threatening us from all sides, and we have to deal with this sort of nonsense. How can they not be ashamed to stir up all this fuss over the murder of such a filthy good-for-nothing!"[106] Without exception, Russia's grand dukes and grand duchesses urged clemency for Iusupov and Grand Duke Dmitrii. "All my deepest and most tender prayers surround you because of your dear son's patriotic act. May God protect you all,"[107] Aleksandra's sister, the Grand Duchess Elisaveta Fedorovna telegraphed to Iusupov's mother just a few days before she and fifteen other members of the imperial family signed a joint letter begging Nicholas not to punish Rasputin's assassins. Nicholas insisted otherwise, for "no one," he replied sternly, "has the right to commit murder."[108] Purish-kevich's political fame had risen so high that Nicholas feared to touch him, but he immediately exiled Prince Feliks to the Iusupov estate of Rakhitnoe. Then, he sent away not only Dmitrii Pavlovich, but all those who had spoken on his behalf.

Their relatives driven away, their government in the hands of incompetent and corrupt men, and their army left under the command of a deputy chief of staff who had no authority to act without its absent commander's approval, Nicholas and Aleksandra went into near seclusion to console each other over Rasputin's loss. Together they buried "the unforgettable Gregorii" in a remote corner of the park at Tsarskoe Selo.[109] Aside from his infrequent meetings with his new appointees to high office, Nicholas spent all his nights and days with Aleksandra now, reading, listening to music, and playing patience, for they were as alone among their hundred and eighty million subjects as they had been when they had stood together in the cold gray morning mist before Rasputin's open grave.

"Break down the barrier that separates you from your people and regain their confidence," Ambassador Buchanan urged Nicholas a day or two later. "You have, Sire, but to lift up your little finger, and they will once more kneel at your feet as I saw them kneel after the outbreak of war, at Moscow." Haughtily, Nicholas stared for a long moment, and Buchanan knew he had asked the impossible. "Do you mean that *I* am to regain the confidence of my people," Russia's tsar asked coldly, "or that they are to regain *my* confidence?"[110]

The confidence of Russia's tsar in his people, and of Russia's people in their tsar now remained only as memories of better days long since past. "This cannot go on for long," Nicholas's cousin and childhood friend the Grand Duke Aleksandr Mikhailovich warned. "Discontent is mounting rapidly and, the further it goes, the more the abyss deepens between you and your people. . . . It is the government which is preparing the way for a revolution," he concluded. "We are watching a play with an unheard-of plot: we are looking on while revolution comes from above, not from below." By the end of 1916, no one questioned if revolution would come. It had become only a question of when. "We are living through the most dangerous moment in the history of Russia," one of the grand dukes wrote on Christmas Day. "You are exaggerating the danger," Aleksandra replied haughtily a few days later. "When you are less excited, you will admit that I knew better."[111]

Certainly Minister of Internal Affairs Protopopov shared Aleksandra's view. As Russia began her final steps toward the revolutionary precipice, he busily went on arranging choice appointments for friends and friends of friends, and ardently championed the cases of unworthy acquaintances who sought undeserved promotions. Judging by the correspondence files and logbooks kept in his personal chancery, his main worry at the beginning of 1917 was the flood of Christmas and New Year's greetings he had received from petty minions and petitioners in all corners of the empire, for he wanted to answer each one personally and bestow a small reward upon each correspondent. Carefully, Protopopov arranged all of their warm wishes in neatly-ordered files, compiled careful lists of the favors each requested, and noted what they already had received. As the crisis in Russia deepened, Protopopov resolutely set to work to answer his petitioners' greetings.[112]

"Anything is preferable to the state of anarchy that characterizes the present situation," Paléologue wrote to France's minister of foreign affairs. In a flash of candor such as an ambassador can allow himself only in a moment of extreme crisis, he concluded: "I am obliged to report that, at the present moment, the Russian Empire is run by lunatics."[113] "On the surface, everything still seems smooth enough, but underneath everything is seething," a Moscow newspaper editor warned Konstantin Paustovskii. "The tighter the government screws down the top on the cauldron, the bigger the explosion is going to be."[114]

PART FOUR

1917

CHAPTER XII

"Down with the Tsar!"

Although Aleksandra had dismissed the stern warnings of Russia's grand dukes as absurd exaggerations, overwhelming evidence supported their opinion against hers. According to the Okhrana's best estimates, the cost of a pair of boots had quintupled during the first two years of war, coal prices had quadrupled, and the cost of a cheap meal in a workers' café had increased by as much as seven times.[1] Even those shocking increases paled in comparison to the inflation that ravaged Russia at the beginning of 1917. After climbing at a frightening rate throughout the first thirty months of the war, food prices now soared to catastrophic heights. Every week saw Petrograd's bread prices rise by more than 2 percent, potatoes and cabbage by 3, and milk by 5, while ham and sausage rose by 7 percent, and sugar and chocolate by more than 10. Most workers had long since stopped eating eggs, meat, sugar, milk, and fruit.[2] More and more, city dwellers had to settle for watery *shchi* to go with whatever bits of bread they could find. At the end of 1916, firewood had become so expensive that workers often had to choose between being warm but on the verge of starvation, or being only moderately hungry but freezing. "Children are starving in the most literal sense of the word," an Okhrana agent reported. "A revolution, if it takes place," he predicted grimly, "will be spontaneous, quite likely a hunger riot."[3]

More and more workers saw their marginal standard of living plummet into outright poverty. Even the *metallisty,* whose skills were so vital in armaments production, now began to feel the bite of inflation. Until the war's third year, these elite workers had enjoyed the highest standard of living among the proletarians of Petrograd, but as 1917 opened, their long-standing upward real wage spiral suddenly turned downward, and added a note of real urgency to their demands.[4]

With these grim economic conditions came deepening bitterness and rising frustration as the government spoke stern words but did nothing. "Protopopov fancies that he is Bismarck and threatens everyone with a mailed fist," Rodzianko's wife wrote to a friend, but everyone knew that he had failed utterly to rescue Russia's cities from the onrushing crisis of food shortages.[5] It was worst in Petrograd, where the surrounding areas produced little grain, and nearly all food products had to be shipped over rail networks that were clogged with military supplies and short of rolling stock. During the last ten days of January, Petrograd received only a sixth of the minimum quantities of flour and grain that its inhabitants consumed each day, only to have those shipments cut in half before the middle of February.[6] Then the temperature fell to almost fifty degrees below freezing in parts of European Russia, and the crisis deepened. "Sixty thousand freight cars with fuel, provisions, and fodder lay buried beneath the snow drifts," Protopopov later confessed.[7] Many people lived in buildings where the temperature never rose above twelve degrees centigrade, even at midday.[8] "Since the war began, Russia has not faced such a critical moment as the one we are living through right now," Guchkov exclaimed, while Rodzianko's wife nervously reminded a friend that "even the tiniest spark could start a conflagration."[9]

No matter how grim their predictions, none of these people overstated the danger. "Every day, the masses are becoming more and more embittered," an Okhrana agent reported. "An abyss is opening between the masses and the government."[10] Again, Rodzianko prepared a lengthy report about the difficulties Russia faced and took it to the tsar at Tsarskoe Selo. Factories had no fuel, weapons plants had no iron and steel, and cities had no bread. Everything seemed out of step with everything else. "We have grain at flour mills that have no fuel, flour where there aren't any freight cars to move it, and freight cars where there is no freight for them to carry," Moscow's Mayor Chelnokov wrote in despair.[11] "The final hour is beginning to strike," Rodzianko warned as he presented his conclusions to the tsar. "The time is very near when any appeal to the masses' reason will be too late and useless."[12] "Hurry up and finish [your report]," Nicholas interrupted crossly. "I can't waste time on this." When Rodzianko begged to continue, the tsar cut him short. "I already know everything that I need to know," he snapped, "and your information simply contradicts mine."[13] Aleksandra called such efforts to ward off the approaching storm "high treason," as she ranted about wishing to "hang Trepov for his bad counsels" and send Miliukov, Guchkov, Polivanov, and Prince Lvov to Siberia. "My soul and brain tell me it w[ou]ld be the saving of Russia," she exclaimed, as she urged Nicholas to "crush them all" and "go forth like a lion . . . in the battle against the small *handful* of brutes and republicans."[14]

In those days, one of the Duma's leading progressives watched his colleagues "walking around like underfed flies," powerless to resolve the crisis their nation faced, and unable even to confront its major issues. "No one believes in anything," he lamented. "Everyone has lost heart. They all realize how powerless they really are."[15] Still dedicated to his conservative ideals, Vasilii Shulgin was struck by how much "alarm and melancholy overflowed" in Russian society.[16] Rich and poor alike sought escape through suicide. In just one morning, the Okhrana reported in its typically flat prose, the "daughter of a colonel, E. V. Kritskaia, aged thirty-three [killed herself] by drinking liquid ammonia. At 8 A.M. a noble[man], Stanislav Iakovich, aged fifty-five [took his life] by shooting himself in the head with a revolver. At 9 A.M., a staff-captain, P. Gehring, aged twenty-eight, [suffered death] by shooting himself in the right temple with a revolver. At 12 noon, a student of the Women's College, Gebeksmann, aged twenty-seven, [was] found dead by hanging herself."[17] "The enormous increase in suicides shows that the forces of disintegration are silently at work in the very heart of Russian society," Paléologue concluded.[18] Already, those forces had bored deep into Imperial Russia's vitals. "The revolution was ready," Shulgin remembered, even though "the revolutionaries were not yet prepared for action."[19]

Still fearful that any revolutionary action might harm Russia's war effort, Miliukov continued to insist that "our word is already our action."[20] Few agreed. "No one believes in words any longer," Shulgin wrote to Grand Duke Nikolai Mikhailovich, perhaps the most politically astute of the Romanovs, whom Nicholas had exiled to faraway Kherson province for urging clemency for Iusupov and the Grand Duke Dmitrii.[21] In the Duma, Aleksandr Kerenskii, now more than ever the voice of the people, railed against those "superficial thinkers" whose calls for restraint were "just a pretext to stay safely in warm armchairs."[22] By early 1917, even most of Miliukov's Kadets had abandoned hope for any domestic peace, as Russians turned increasingly to radical measures. "We stand before the unknown and unexpected," one of them remarked in January. "Around us the gloom is thickening and we know not what lies two steps ahead." Certain that "the country is listening to the Left, not to us," another concluded that it was "too late" to save the regime by parliamentary efforts, while their best orator Vasilii Maklakov spoke even stronger words of warning. "The revolutionary path of struggle is inevitable," Maklakov concluded. "The only question is when to start the fight."[23]

Petrograd's cold and hungry proletarians thought the time to start the fight was then and now. On January 9, nearly 150,000 workers from more than a hundred factories poured into the streets to observe the twelfth anniversary of the Bloody Sunday massacre that had claimed the lives of several hundred of their comrades. Led by close to forty thousand *metallisty* from the Putilov and Obukhov works, they marched in protest against the war, soaring prices, and their employers' refusal to raise their wages.[24] All across Russia, workers did the same. That day, over thirty thousand working men and women from fifty-one

Moscow factories, fourteen thousand from the oil fields of Baku, ten thousand from Kharkov, and thousands more from Tula, Tver, and Nizhnii-Novgorod, took to the streets. Before the end of the month, more than a quarter-million striking workers had cost Russia's industrialists nearly three-quarters of a million lost working days. But the danger was far greater than the loss of working days for Russia's still struggling war effort. Sounding a new and foreboding note for the old regime, nearly one out of every three striking workers pressed political demands upon the authorities. A number of their banners urged: "Down with the War!" Some bore the words: "Down with Autocracy!"[25]

The January demonstrations marked only the beginning of a powerful strike wave that surged across Russia during the next month. Elated by their success on January 9, revolutionary groups planned more demonstrations for February 14, the end of the Shrove-tide holidays and the day when the Duma was to reopen after its two-month recess. In Petrograd, nearly ninety thousand workers from more than fifty factories, including virtually all of the major armaments plants, demanded political change, labor reform, and economic relief. Already they talked of revolution. "Down with the Government of Traitors!" their slogans proclaimed. "Long Live the Second Revolution! Long Live the Democratic Republic!"[26] Fearful that the "second" revolution might become a fact (the Revolution of 1905 was the "first"), the authorities sent Cossacks to break up a large rally at the Izhora factory. The Cossacks did their work, but an Okhrana observer noted apprehensively that "there was an impression that the Cossacks were on the side of the workers."[27]

Half-hearted Cossack assaults against workers could not stop the rising tide of labor protest that struck Russia's factories in the days after the Duma reopened. During the next week, another 200,000 workers took part in over 150 strikes. Banners reading "Down with the War!" and "Down with the Government!" became commonplace. Among them another appeared, the simplest and perhaps the most ominous of all, bearing only one word: "Bread!"[28] Bread became the key. Rapidly radicalizing proletarians might protest against the war and the government, and strike for higher wages, better hours, and heated latrines (as they had at Petrograd's Izhora Mill). But for bread, Petrograd's workers would launch a revolution.

Thursday, February 23, dawned cold, sunny, and clear. Although they did not arise much before midday, Petrograd's intelligentsia spoke of nothing but the "gorgeous revival" of Lermontov's *Masquerade* that had opened at the Alexander Theater the night before. The play itself had been the product of its author's own revolutionary yearnings in the 1830s when he had struggled against constraints imposed by the censors of Nicholas's great-grandfather, Nicholas I, who eventually suppressed the writings of Tacitus, Plato, Shakespeare, and even Catherine the Great. "Stupidity and perfidy, Lermontov had written then, "are what make the world turn," and the imperial censors had banned the production of his plays for many years "because of the violence of [its] passions and charac-

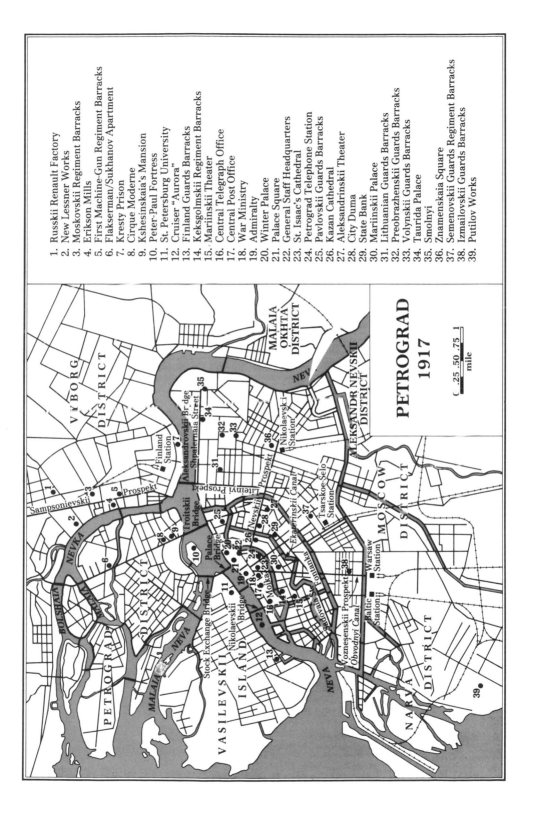

1. Russkii Renault Factory
2. New Lessner Works
3. Moskovskii Regiment Barracks
4. Erikson Mills
5. First Machine-Gun Regiment Barracks
6. Flakserman/Sukhanov Apartment
7. Kresty Prison
8. Cirque Moderne
9. Kshesinskaia's Mansion
10. Peter-Paul Fortress
11. St. Petersburg University
12. Cruiser "Aurora"
13. Finland Guards Barracks
14. Keksgolmskii Regiment Barracks
15. Mariinskii Theater
16. Central Telegraph Office
17. Central Post Office
18. War Ministry
19. Admiralty
20. Winter Palace
21. Palace Square
22. General Staff Headquarters
23. St. Isaac's Cathedral
24. Petrograd Telephone Station
25. Pavlovskii Guards Barracks
26. Kazan Cathedral
27. Aleksandrinskii Theater
28. City Duma
29. State Bank
30. Mariinskii Palace
31. Lithuanian Guards Barracks
32. Preobrazhenskii Guards Barracks
33. Volynskii Guards Barracks
34. Taurida Palace
35. Smolnyi
36. Znamenskaia Square
37. Semenovskii Guards Regiment Barracks
38. Izmailovskii Guards Barracks
39. Putilov Works

PETROGRAD
1917

0 .25 .50 .75 1
mile

ters." Now, in late February 1917, Lermontov's work emerged as a revolutionary artistic statement under the hand of the *avant-garde* director Vsevolod Meierhold, whose dazzling new techniques fused the diverse artistic streams of Symbolism, Futurism, and Expressionism into revolutionary new theatrical perspectives.

Although the production sometimes struck Britain's aristocratically aloof Ambassador Buchanan as having a "richness and extravagance [that] went oddly in company with a distressed and impoverished Russia," its view that the world of the Russian court, high society, and government was nothing but a false masquerade, in which vile intrigues, debauchery, and all lack of decency guided the life of Russia must have seemed especially appropriate to the times in which Petrograd's theatergoers found themselves.[29]

Certainly, the opulence of Meierhold's production stood in dramatic contrast with the stark masses of gray and worn working women who had arisen before three o'clock to stand in bread lines. That day, the lines were longer than usual, for Petrograd seethed with rumors that rationing soon would be imposed, and the city's workers were anxious to buy up any extra bits of bread to make *sukhariki,* those dried husks that Russians had prepared since the Middle Ages whenever they faced hard times.[30] Bread would spoil in a few days; *sukhariki* could last for weeks.

Grim-faced and stolid, Petrograd's women then waited by the tens of thousands to buy bread, knowing that the most fortunate among them would be able to acquire only a fraction of what their families needed to stave off hunger, while the rest would fail even to reach the bakeries' counters before the sign that announced "No More Bread" would be posted. Even foreigners, so often the objects of privileged treatment in Russia, received no special consideration that day. "My cook waited for four hours in the freezing dawn that day to buy two rolls," one French correspondent reported in amazement. One of Ambassador Buchanan's deputies reported that the women were "patiently waiting," and opined that their "exemplary patience" would "probably continue." Paléologue feared otherwise. Where his British colleague saw "exemplary patience," he saw a "sinister expression" that made him anticipate the worst in a country where, he said, the authorities suffered from "a congenital inability to organize."[31]

This "congenital inability" affected Russia's radicals as well as her government. For radical working men and women, February 23 was International Women's Day, a minor holiday that some Russian Socialists had used in the past as a pretext for distributing antigovernment and antiwar propaganda. This year, however, the rising wave of labor protest in January and February had taken up their attention, and the holiday came upon them before they had made any plans. Only a handful of radicals even bothered to distribute leaflets, and none wanted demonstrations that day. Before renewing their strike activities, Russia's revolutionaries intended to build up the strength of their organizations in Petrograd's factories. Poorly organized strikes, they insisted, could only sap vital energy that should be directed toward recruiting and training more activists. On International Women's Day, they hoped the workers would wait, not demonstrate.[32]

The anger that seethed in Petrograd's streets on International Women's Day in 1917 thus came not from inflammatory propaganda or even revolutionary oratory. Frustration at working longer hours for wages that bought less food, the deadening burden of trying to support families while their menfolk were away at the front, and, most of all, the pain and weakness of real hunger—these were the forces that stirred the suppressed rage Paléologue had sensed behind women's "sinister expressions" that day. "Every day one hears complaints that people have not had bread to eat for two or three days or more," a police officer had reported to his superiors the day before.[33] Another added that he had seen women cross themselves and weep with joy simply for being "fortunate enough to be able to buy two loaves of bread."[34] At that moment, "daily life had disintegrated to the point that brought people to the bitter realization that they just couldn't live that way any longer," one of Petrograd's angry workers remembered. "There occurred one of those moments," he added, "when the real masses, without any inducement from the outside, suddenly threw aside their passivity and forced their leaders to take action."[35]

Working women who toiled especially long hours for notoriously low wages in Petrograd's cotton mills led the way, not the men who worked for much higher pay in the city's weapons and munitions plants. Well over seven thousand of them left their looms in Petrograd's Vyborg District that morning and poured into the streets. To the New Lessner Works, the Nobel Plant, the Russian Renault Factory, and, finally, the Erikson Mills, they marched. "Bread!" they called out. "Bread!" By ten o'clock, twenty thousand had joined them. Before noon, their ranks had swelled to more than fifty thousand. Some of Petrograd's elite metal and munitions workers joined their ranks in the afternoon, and the strike spilled into the neighboring Petrograd District. As it did so, it took on political overtones. "Down with the War!" banners now proclaimed. "Down with the Tsar!" Cries for "Bread!" still rang out, but revolutionary marching songs began to swell above them: "Bravely onward, comrades. Our spirit shall be tempered in the fray!"[36]

Late that afternoon, Petrograd's police received the first reports that women and teenagers had begun to break into food shops in a surge of violence to which the authorities were unaccustomed. According to various Okhrana and city police reports, somewhere between eighty and ninety thousand workers had gone into the streets by nightfall. On the other side of the city, in the industrial suburbs that lay along the Peterhof Highway to the southwest, legions of workers had been locked out of the gigantic Putilov Works the day before because of a labor dispute. All day, the Vyborg workers had hoped to hear the shout that would have proclaimed that the thirty thousand elite Putilov metalworkers had joined their protest: *"Putilovtsy idut!"*—"The Putilov men are on the move!"

That day, the *Putilovtsy* did not come, but they were less than ten miles away: ready, waiting, disgruntled, and hungry.[37]

A large number of confrontations between demonstrators and the authorities produced little bloodshed and very few arrests on International Women's Day. Clearly, the police had kept the upper hand, and, by seven o'clock that evening, quiet reigned in most of Petrograd. No one thought that anything much out of the ordinary had occurred, and few expected anything untoward on the morrow. True, the workers had pushed on with demonstrations despite their leaders' urgings to the contrary, and Petrograd's radicals now had to decide whether to support the rank and file or try to convince them to return to work and save their strength for another day. Certainly, none of Petrograd's socialists expected to see much result from another day of demonstrations beyond some modest propaganda benefits. Many agreed with Aleksandr Shliapnikov, the cautious thirty-three-year-old head of the Russian Bureau of the Bolsheviks' Central Committee in Petrograd, that Russia's would-be revolutionaries ought not to squander valuable personnel or resources on further demonstrations.[38] "Similar disorders had taken place scores of times," one of them explained later. Aside from what some thought might be a momentary lack of resolve on the part of the authorities, who seemed to be trying to avoid violence whenever possible, none of Petrograd's revolutionaries had any reason to expect that the events of February 23 had been much different from any that had come before.[39]

The authorities reached a similar assessment. That evening, General Sergei Khabalov, an officer with almost no combat experience who had been given command of the Petrograd Military District, ordered that nagaikas—the vicious weighted whips capable of tearing the flesh from a man's back that Cossacks used to disperse crowds—be issued immediately, and summoned the city's key police and military officials to a special meeting. After some discussion, they agreed to deploy troops to guard key points throughout the city, and ordered that more flour be sent to some of the largest city bakeries.[40] Khabalov and his deputies thought that such measures would be sufficient, for none of them expected the next day to bring anything beyond a few more demonstrations.

The day before, Nicholas had finally ended his two-month absence from *Stavka* and returned to his armies, grateful to escape even the minor interruptions that civilian ministers had brought to his cloistered life at Tsarskoe Selo. "My brain is resting here," he told Aleksandra the day after he had resumed his duties as Russia's supreme commander. To his delight, he found "no Ministers, no troublesome questions demanding thought" at Mogilëv.[41] Reluctant to disturb their emperor's peace, none of the authorities in Petrograd thought the demonstrations even worth mentioning in their reports. "In general, nothing very terrible has happened," Aleksandra's faithful Protopopov wrote in his diary. For the moment, he and Khabalov thought it would be enough to post handbills spreading false assurances that there were large reserves of flour still available in the city.[42]

Hardly anyone else in Petrograd regarded the events of the day as anything much out of the ordinary either. Zinaida Gippius guessed that the day's incidents probably were just an "ordinary sort of hunger riot like those that have been happening in Germany," although she thought that "the putrefaction of the government" was a "huge fact" that might have some unforeseen impact upon the course of events.[43] Count Louis de Robien, an elegant young French diplomat recently assigned to Paléologue's staff, took his dog for a walk at dusk and returned to report that "everything is perfectly calm."[44] Without a second thought, Paléologue went ahead with plans to have twenty-six guests for dinner, including the recently dismissed Trepov. Although Paléologue found Trepov's comments about the government's ineffectual efforts to bring more food into Petrograd "anything but reassuring," the rest of his guests preferred to discuss the large party that the Princess Radziwill was planning for Sunday evening and to debate the merits of the Imperial Ballet's three prima ballerinas. Concern about the day's events seemed far from everyone's thoughts. "Women's Day had passed successfully, with enthusiasm and without victims," Trotskii noted. "But what it concealed in itself, no one had guessed even by nightfall."[45]

As if to confirm the authorities' best estimates, the first cold moments of foggy daylight on Friday morning found Petrograd's center calm, and Protopopov congratulated himself that the disorders had come to an end. In the workers' quarters, however, trouble beyond Protopopov's wildest dreams already was brewing. All through the night, strike organizers had been at work, like an army of industrious termites gnawing at the foundations of Russia's old order, even though some of their leaders still opposed further demonstrations. "We must go ahead and solve our problem by force," one of them shouted to the *metallisty* who assembled at the Stetinin Factory early that morning. "Only in this way will we be able to get bread for ourselves. . . . Arm yourselves with everything possible—bolts, screws, rocks," he urged. "Start smashing the first shops you find!"[46] With grim determination, angry workers began to move toward the Aleksandrovskii Bridge, the artery that would take them across the Neva to Liteinyi Prospekt and into the city's center. As they marched, some forty thousand workers from the Vyborg District joined them. Before nine o'clock, they faced five hundred soldiers, Cossacks, and mounted police who barred the bridge's approaches.[47]

To pass such a barrier raised frightening specters of thundering Cossack horses' hooves and slashing nagaikas tearing away exposed flesh. Understandably, the demonstrators hesitated. Then, one Bolshevik worker remembered, working women once again led the way. "You also have mothers, wives, sisters, and children," they pleaded to the horsemen who stood before them. "We are only asking for bread and an end to the war!"[48] Tense moments followed. Then, a young Bolshevik *metallist* by the name of Kaiurov remembered, "several of the Cossacks grinned, and one of them winked slyly at the workers." To some, at least, it now seemed clear that the Cossacks would not attack them, and

Kaiurov heard "shouts of 'Hurrah!' for the Cossacks burst from thousands of breasts."[49] Many workers now bypassed the Cossacks by crossing the Neva on the ice. Some of the most daring simply breached the Cossack barrier by darting beneath their horses' bellies while they stood rigidly at attention. Once across the river, the workers faced still-loyal mounted police detachments, who attacked them without mercy, but a beginning had been made and it could be exploited. "A remarkable incident!" Trotskii remembered a few years later. "The revolution . . . made its first steps toward victory under the belly of a Cossack's horse."[50]

Despite the Cossacks' sympathy for the working women they faced, the viciously determined mounted police detachments standing in reserve would normally have been more than sufficient to quell the forty thousand workers who marched on the Aleksandrovskii Bridge that morning. But the strike organizers who had scurried through the predawn hours had done their work well. Like so many forest fires, strikes and labor protests burst forth all across the city, and took the authorities by surprise. In the Vyborg District, 74,842 workers left their jobs, while another 22,596 followed their example in the adjacent Petrograd District. On Vasilevskii Island, 23,248 workers lay down their tools, while, in the Moscow District, the number of strikers reached 19,506. In Petrograd's three central districts, home of aristocrats, prosperous middle classes, and most government offices, 16,421 workers went on strike, and, even in the Narva District, where troops already occupied most of the large factories, another nine thousand men and women walked resolutely away from their machines. All told, well over 160,000 Petrograd workers joined the strike by nightfall. Where General Khabalov and Protopopov had expected to see fewer workers in the streets that day, their numbers, in fact, doubled.[51]

What made the situation even worse that day was that some forty thousand strikers carried their protest from Petrograd's working class districts directly onto the Nevskii Prospekt in the very heart of the city. Not since the revolutionary events of October 1905 had striking workers converged on Petrograd's central avenue in such numbers. Some marched to the square surrounding the Kazan Cathedral less than a kilometer from the Admiralty and Winter Palace. Others proceeded along the Liteinyi to a mass political rally around the ponderous statue of Alexander III that stood, heavy and unmoving, on Znamenskaia Square at the far end of Nevskii Prospekt.[52] Confronted by new Cossack formations, the workers cheered and the Cossacks bowed in reply. Revolutionary speeches rang out. "One brave soul, his head bare and his coat hanging open, scrambled upon others' shoulders to gain a foothold on the statue's pedestal," one of the demonstrators remembered. "Then, this daredevil stood to his full height and raised his right fist. 'Comrades!,' he shouted. 'From beneath the mindless iron heel [of her rulers], Russia has poured out her blood. This senseless slaughter undertaken by the [tsarist] camarilla has brought the masses to utter ruin!' "[53]

Others followed, their speeches showing much anger but little clarity. "Logic and coherency were not needed by the people," one worker explained, "[to understand] the first free speech under the open sky in front of the massive crowds in full view of the Cossacks and the police."[54]

No one interfered. It was as if the authorities dared not challenge the good feeling that reigned between Cossacks and crowd for fear that both might unite against them. At some point such a test must come if the demonstrations continued, but General Khabalov desperately wanted to avoid it. That evening he called Petrograd's top officials to another meeting, where they agreed to prevent the city's bakers from selling flour on the black market, arrest any revolutionaries whom the Okhrana could locate in Petrograd, and strengthen the garrison by calling in units of the Guards' Reserve Cavalry that had been billeted in Novgorod's Krechevitskii Barracks, some six hours distant by train. According to Russia's foreign minister, Protopopov could not be present at Khabalov's meeting because he spent the entire evening trying to contact Rasputin's ghost in order to ask its advice.[55]

Whatever their sources of information, neither Protopopov nor Khabalov realized that the demonstrations of the previous two days were about to become a general strike. While these men laid plans to secure the city's center, strike organizers made ready to confront them on the morrow. Overnight, the men and women who had emerged to lead the city's rank and file during the past two days transformed factories from work places into meeting places that rang with political demands that would have been unthinkable two days before. No longer did the slogans "Down with the Government!" "Down with the War!" "Down with the Tsar!" "Long Live the Revolution!" burst from anonymous throats in the midst of faceless crowds. Now, men and women well known to those who stood before them pronounced them loudly and hopefully.[56] "The atmosphere was tense," a worker at the New Parviainen Works remembered. "As one, the toiling masses were gripped by an impulse, by an irresistible desire and resolution to live or die in the struggle."[57] Now ready to confront the authorities head-on, these men and women now numbered well over 200,000.[58]

Convinced that Cossacks and soldiers might be won over to their cause, Petrograd's workers made every effort to avoid hostile confrontations and, again, proletarian women led their effort. "More boldly than men," Trotskii remembered, working women would go up to detachments of soldiers, "take hold of the rifles, [and] beseech, almost command: 'Put down your bayonets—join us!'" More and more often during the next several days, such efforts coaxed soldiers away from their officers. The police, however, were another matter. These "Pharaohs," Trotskii remembered, were "fierce, implacable, hated and hating foes," and the best way to deal with them, he insisted, was "to beat them up and kill them."[59] To face the nagaikas of these enemies, Petrograd's workers reinforced the backs, shoulders, and collars of their jackets with heavy padding made from rags, towels, and old blankets. Beneath their jackets, they carried knives, sharpened pieces of metal, spikes, and broken bottles to slash at the police's horses as they charged past.[60]

As on the day before, a squadron of mounted police barred the workers' way at the Aleksandrovskii Bridge. Implacable in his hatred for the men and women who stood before him, their chief charged as he had on days past, but this time the workers were ready. Now protected against his slashing nagaika, they surrounded him and drew him into the midst of the crowd before his men could reach his side. Desperately, he tried to fight his way out, and the brief moments of his last struggle before the workers seized his revolver and shot him, transformed the workers' protest from a demonstration to a rebellion.[61]

As if to emphasize that point, on the other side of the city, the Putilov workers at last went into action. Locked out of their plant since the day before the demonstrations had begun, they now stormed the factory's gates. Inside, a strange peace reigned, the courtyard's new-fallen snow and the cloudless sky untainted by the grime that usually poured from the factory's belching chimneys. Hastily the workers formed Petrograd's first "provisional revolutionary committee" to "lead the struggle against the police, to organize fighting detachments, and to establish the revolution in the streets." It had taken the Putilov men three days to decide. Now they would fight violence with violence, gunfire with gunfire.[62]

While the Putilov workers occupied factories along the Petrograd Highway, other workers made their way along Nevskii Prospekt to continue their previous day's protest on Znamenskaia Square. As they marched to the stirring revolutionary refrains of the *Marseillaise* and the *Varshavianka,* hungry workers unfurled red flags that proclaimed "Down with the War!" and "Down with the Tsar!" Again they massed around the statue of Alexander III to hear workers and students demand an end to the war, autocracy, and oppression, while several squadrons of mounted police and Cossacks closed a tight ring around the entire square. Led by a young lieutenant, the police charged the crowd, their nagaikas cutting left and right, concentrating their attacks against the marchers who carried red flags and revolutionary banners. For the moment, the Cossacks stood aside. Then several workers approached them, heads bare, caps in hand. "Brothers—Cossacks," they began. "You see how the Pharaohs treat us, hungry workers. Help us!"[63] Suddenly the Cossacks charged, their sabers seeking out the police, not the workers. As the police fled, the Cossacks returned to their positions and stood calmly, as if nothing had happened. On the pavement, the police commander lay dead. His blood drained onto the snow from several gunshot wounds and saber cuts. As reinforcements arrived, a new shout rang out: "Comrades! pile up posts and scaffolding! Build a barricade against these beasts!"[64] A symbol of revolution ever since they had clogged the narrow streets of Paris in 1789, barricades now appeared in Petrograd.

The incidents at the Aleksandrovskii Bridge and on Znamenskaia Square showed that a number of Khabalov's Cossacks sympathized with Petrograd's workers, but none had yet actually joined their ranks. The first defection of Petrograd's garrison came just a few hours later near the intersection of Nevskii Prospekt and Sadovaia Street, where demonstrators faced bayonets, not sabers. With the soldiers' bayonet tips almost touching their breasts, a rank of working women pleaded: "Put down your bayonets. Join us!"[65] The soldiers hesitated for a moment and then slung their rifles over their shoulders and disappeared into the crowd. Would others join them? And would those who remained apart be willing to fire upon workers if their officers ordered them to do so? Late that afternoon, events seemed to answer both questions in the affirmative. Not far from the spot where the first soldiers had joined the workers on the Sadovaia, others fired into the crowd, killing and wounding more than a score, including at least two women. Strangely, neither the authorities nor Petrograd's top revolutionary leaders yet realized that these were revolutionary events. "What revolution?" the Bolsheviks' Aleksandr Shliapnikov exclaimed angrily that day when one of his subordinates from the Vyborg District remarked that a revolution had begun. "Give the workers a loaf of bread and the movement will be gone!"[66] Events soon proved him wrong.

While Khabalov's mounted police faced sullen and hostile crowds, Nicholas had spent a pleasant Saturday at *Stavka*. He arose late to a leisurely breakfast, and then spent an hour and a half listening to General Alekseev's report on conditions at the front. After lunch, he jotted a brief note to the empress and then went to a nearby monastery to offer prayers for them all before an especially revered icon of the Holy Virgin. He made certain that the brooch Vyrubova had pinned to the icon some weeks before remained in place and took care to report that he had "touched it with my nose when kissing the image." Nicholas then went for the afternoon walk that he never missed unless he were ill and returned to *Stavka* to find a letter from Aleksandra that said nothing of the disorders beyond the fact that a crowd had attacked Filippov's "bakery" (actually a fancy patisserie), and that Cossacks had been called out. That evening, Nicholas attended vespers in the Mogilëv Cathedral, had a light dinner, and settled down to a quiet evening of staff work. By that time, Alekseev had received detailed reports from Khabalov and Protopopov, although it is not clear just how much of the information about the situation in Petrograd he passed on to Nicholas, for the tsar did not see fit to mention the "disorders" in his diary until two days later.[67] Perhaps unaware how serious the danger really was, Nicholas sent brief instructions to Khabalov. In Russian, the text of his telegram amounted to a mere sixteen words, the most fateful message any tsar ever sent:

I ORDER YOU TO BRING ALL OF THESE DISORDERS IN THE CAPITAL TO
A HALT AS OF TOMORROW. THESE CANNOT BE PERMITTED
IN THIS DIFFICULT TIME OF WAR WITH GERMANY AND AUSTRIA.

NICHOLAS[68]

As the conflict between workers and government now approached its final confrontation, Nicholas did not even realize that the fate of his dynasty hung in the balance.

"What was I to do? How was I to bring the disorders to a halt?" Khabalov asked rhetorically when he was questioned a month later by the Provisional Government's Supreme Investigating Commission.[69] The only way to end the disorders "as of tomorrow," as Nicholas had ordered, was to declare war upon Petrograd's workers and attack them with rifles, machine guns, and even heavy artillery, just as the elite Semenovskii Guards had done when they had crushed the armed uprising of Moscow's workers in December 1905. Now, the perils of such an attack far exceeded those of 1905, and Khabalov understood its terrible risks all too well. Should the soldiers disobey their officers' orders to fire, the workers' protest would become a revolution. More time, negotiation, and concessions might make it possible to avoid such a confrontation, but Nicholas had left Khabalov no room in which to maneuver.

Feeling "absolutely finished," Khabalov met that evening with Petrograd's chief military and police commanders to give them their orders for the next day. "If the crowd is small, if it is not at all aggressive, and if it is not carrying [red] banners," he told them, "then use your cavalry detachments to disperse it. But if the crowd is in any way threatening, and if it carries banners, then you are to act according to regulations. Give three warnings, and then—open fire."[70] Khabalov placed extra military units on alert, moved machine guns and armored cars into position, and proceeded to arrest every revolutionary that the Okhrana had been able to locate during the past forty-eight hours.[71] At the same time, he ordered two proclamations to be posted all across the city before dawn. One announced that any workers who did not return to work within two days would be sent directly to the front. The other forbade all Petrograders from gathering in the streets, and warned that the authorities would break up any assemblies with gunfire.[72]

For a city that had been on the brink of armed warfare between its citizens and their government as dusk fell, Petrograd seemed uncharacteristically peaceful and calm that night. Paléologue took the Vicomtess du Halgouët to a concert at the Mariinskii Theater and found it nearly empty. This struck him as particularly strange, for he once had insisted that "one might as well ask the Spanish to do without their bull fights" as to ask Russians to "give up their theater, music, and ballet," no matter how difficult the times. As they drove across the deserted theater square after the performance, the vicomtess felt the cold desolation so keenly that she asked: "Are we perhaps witnessing the last night of the regime?"[73]

Just off the Nevskii Prospekt, less than a mile away, the Aleksandrinskii Theater was performing Lermontov's *Masquerade* once again. As the performance ended on a note of high pathos, one theatergoer found himself remembering that Nicholas's reign had begun with the tragedy at Khodynka Field, where more than a thousand Russians had been trampled to death as they celebrated the new tsar's coronation. "A bloody baptism" to begin the reign, he thought. Was this "splendid masquerade" going to serve as "a magnificent theatrical funeral" to mark its end?[74] As Petrograders made their way home from the theater, the city grew more quiet still. "It's like some sort of gigantic corpse is suffocating," Gippius confided to her diary before she went to bed. "Apparently someone, somewhere, something, is giving orders," she wrote. "A strange sensation."[75]

Gippius awoke the next morning to find the weather "cold and sunny," but thought it "an extremely harsh day" nonetheless. No newspapers appeared, but diligent policemen had posted Khabalov's proclamations everywhere.[76] "It [will be] . . . interesting to see how he is going to send all of us to the front," someone heard a worker remark as he glanced at a notice that had been posted on a street corner. "Who is going to produce supplies for the army? He himself?"[77] Overnight, Petrograd had become an armed camp, an army readying itself for battle. Machine-gunners had taken up positions where their guns could sweep strategic intersections. Infantry platoons guarded key buildings, and Cossack squadrons patrolled the streets. Stores, cafés, and restaurants had shut down, and there were no horse cabs or street cars. Anyone who moved in the city's center needed a special written pass.[78] For the first time in its two hundred year history, Petrograd was an occupied city, occupied by its own army against its own people. Order reigned. By mid-morning, Khabalov breathed more easily. "Today, February 26," he telegraphed to *Stavka*, "the city is entirely peaceful."[79] He spoke too quickly. At that moment, workers were massing in Petrograd's industrial suburbs to march again on the Nevskii Prospekt.

Late that morning, workers converged upon Petrograd's center from all directions, their routes chosen to intersect the Nevskii Prospekt at several key points. They did not know Khabalov had assigned special training detachments —units in which the Guards trained their noncommissioned officers from men chosen for their bravery and loyalty—to guard those very intersections. This time, when the marchers ignored orders to halt, the training detachments opened fire. Terrified workers dived for cover in doorways and courtyards or fell to the ground in the middle of the street. By midafternoon, casualties numbered in the scores, as officers stood among their men with pistols drawn to make certain that they did not fire into the air. On Znamenskaia Square, the Volynskii Guards Regiment again faced the crowd, and again the soldiers drew back from shedding the people's blood. "The first volleys claimed no victims," one of them remembered. "By tacit agreement, the soldiers fired into the air." Then an enraged officer strode into their midst. "Each man must fire in turn so that I can watch him shoot!," he shouted. "Aim at the heart!" Then, the moment they all had awaited with such trepidation arrived. "Suddenly a machine gun that had been aimed at the crowd by the officers began to chatter," the soldier-memoirist wrote. "And the blood of workers stained the snow."[80]

Certain that unarmed demonstrators could not stand against machine gun fire, many of Petrograd's worker-revolutionaries thought the end had come. But there were others who insisted that they must arm themselves and continue the struggle. "Let's not be naive any longer!" one of them exclaimed. "In the arsenals and weapons factories we shall find revolvers, rifles, and cartridges. . . . Comrades! It's now or never!"[81] As Petrograd's workers retreated from the blood-stained pavement of the Nevskii Prospekt that afternoon, they took with them a stern resolve to return with rifles and pistols instead of rocks, knives, and clubs, to continue their struggle against whatever forces the government sent against them.

Would the resolve of the army and police be as firm? Could they be counted upon to continue killing workers in Petrograd's streets? Upon those questions hinged the survival of Nicholas, Aleksandra, their children, and Russia's autocracy. That very afternoon, disturbing signs hinted that the answers might not be what the Romanovs and their counselors expected. Three companies of the Pavlovskii Guards had been sent out to fire upon the crowds while a fourth remained nervously in reserve, shut away in their barracks without access to newspapers, telephones, or visitors from the city. Word traveled quickly in Petrograd's streets nonetheless, and, at some time between four and six o'clock that afternoon, someone burst in with word that "the *Pavlovtsy* are shooting the people!" Anxious to stop the shooting, about 150 men of the fourth company stormed into the street, ran to their arms room, seized about thirty rifles, and hurried toward the Nevskii Prospekt. A company of mounted police barred their way and opened fire. Instantly, the few *Pavlovtsy* with weapons replied.[82] "A terrible breach in the stronghold of tsarism," the Menshevik Nikolai Sukhanov called it in his unparalleled eyewitness account of the revolution, "the first revolutionary act of the military against the armed forces of Tsarism."[83]

Although Nicholas's regime still survived, and Khabalov remained in command, men such as Sukhanov saw particular importance in the "Pavlovskii rebellion," for it had breached the impenetrable wall of bayonets that the army had raised between Russia's tsar and his enemies since the time of Peter the Great. That night, the military authorities sent nineteen *Pavlovtsy* to the Peter and Paul Fortress and another sixteen to the guardhouse, while they did everything they could to keep the news concealed from other military units.[84] "At the moment, everything depends on conduct of the military units," an agent of the Okhrana concluded. "If they do not go over to the side of the proletariat, then the disorders will recede rapidly. If the troops turn against the government, then nothing can be done to save the country from revolutionary upheavals."[85]

Everyone assumed that the next day's events would decide the troops' loyalty once and for all. That Russian superstition held Monday to be an unlucky day must have made them all—police, army, workers, and revolutionaries—even more nervous. As the last hours of February 26 passed, each therefore awaited the morrow in his own way: some fearful, others expectant, still others overcome by fatalistic resignation. As Paléologue's car turned into the Fontanka Embankment somewhat after eleven o'clock, France's ambassador noticed that the Princess Radziwill's much-discussed party was "in full swing," and that at least one Romanov grand duke was there, perhaps resigned to letting Fate decide the course of Monday's events. In this moment of crisis, Paléologue thought the sight of the grand duke's car at the curb of the Radziwill Palace an ominous sign.[86] Even more ominously, as the clock passed midnight, dogs howled on the Nevskii Prospekt, now dark and heavily patrolled. That night Gippius wrote of "tomorrow's revolution" in her diary. "A leaderless revolution," she exclaimed. "Poor Russia!"[87]

While Paléologue noted ominous signs and Gippius worried about a "leaderless revolution," some of the soldiers who had been forced to fire upon workers on Znamenskaia Square set out to transform Sunday's sporadic violence into a full-scale insurrection. What would they do, some of the men from the Volynskii Regiment asked each other, if their officers ordered them to kill more unarmed workers the next day? "Fathers, mothers, sisters, brothers, even brides, are begging for bread," Sergeant Kirpichnikov, one of many men whom the events of 1917 raised from obscurity to influence the course of Russia's history, told his men. "Should we strike them down? Have you seen the blood running in the streets? I propose that we not march against them again tomorrow. I personally don't want to."[88] Although barely out of his teens, Kirpichnikov acted resolutely. Quickly forging the company commander's name to an order for cartridges from the regimental armory, he seized the company's rifles and machine guns, and waited for the morrow. "Enough blood has been shed," he announced to his men about an hour before reveille. "Now it is time to die for freedom." When their company commander appeared, another sergeant flatly stated that he and his men would "no longer shoot" at civilians. Now certain that they no longer could control their men, the Volynskii officers hastily left the barracks. As they did so, the rebels ran to the windows, opened fire, and killed their commander.[89]

With their first volley, Kirpichnikov's men crossed the revolutionary Rubicon. Now they and the revolution must triumph or face the hangman as rebels. Aware of the dangers, the other companies of *Volyntsy* at first refused to join them, but Kirpichnikov and his supporters won over men from regiments quartered nearby: the Preobrazhenskii and Lithuanian Guards Regiments, and the Sixth Military Engineers Battalion. With still more machine guns, rifles, and cartridges in their hands, and with the Sixth Engineers' band playing at their head, these rebels from four elite army units now joined Petrograd's worker-insurgents. By early afternoon, they had seized the Liteinyi Arsenal with forty thousand rifles, thirty thousand pistols, and at least four hundred machine guns.[90] Sunday's general strike against food shortages, the war, and the government had now become an armed rebellion against the autocracy.

But it was not only soldiers unwilling to shoot unarmed workers who led the way to armed rebellion that day. In the dark hours between Sunday night and Monday morning, a Bolshevik tinsmith commanded a handful of workers from the outlying Lesnoi District in an attack against another nearby armory. A few hours later, other workers attacked the Petrograd Cartridge Works, and, by noon, armed workers fought their first battle with the army when the Moscow Regiment's training detachment tried to bar their way at the Aleksandrovskii Bridge. In a battle that lasted only a few minutes, the workers shot the detachment's commander and broke through to join Kirpichnikov's rebels on the other side of the Neva. United, they marched back to the Vyborg District, their bayonets festooned with red ribbons. Over a quarter-million Petrograders now stood against the government. One in ten among them was an armed soldier; the fall of the Liteinyi arsenal and the Lesnoi armory meant that another three in ten were armed workers.[91]

Arms and men alone could not make a revolution in Petrograd, however. No matter how bitter their opposition to the government, insurgent workers and soldiers needed to focus their attack, not wander from one part of the city to another celebrating the mere fact of their rebellion. Not far from the Vyborg District's Finland Station stood Kresty Prison, dark and lowering on the Neva's north bank, symbol of the government's power to repress dissent and long an object of the workers' hatred. Not long after noon, some of the insurgents attacked it, burned its police records, and freed its 2,400 common criminals and political prisoners. In close succession, workers stormed the Military Prison, the Women's Transit Prison, and several police headquarters, destroying files and liberating prisoners as they went.[92] These victories gave Petrograd's insurgent workers and soldiers a sense of common purpose and liberated the revolutionary leaders they so sorely needed to guide them.

By late afternoon on Monday, Petrograd's insurgents had weapons, ammunition, and leaders; by nightfall they even had armored cars, including an armored cannon car that the British had sent to strengthen the tsarist army's coming offensive. Some of the armored car crews joined the insurgents of their own accord, but others needed persuasion. That afternoon, an armored car had stopped near a crowd in which a veteran Bolshevik weapons-maker stood. "Comrades!" the driver shouted. "Who knows how to fire this machine gun?" The weapons-maker leaped into the gun turret as the driver sped off, careening through the streets toward the Armored Car Division garage to recruit its men for the revolution. At first, no one inside the garage responded to the insurgents' summons. "I fired a long burst along the [building's] iron roof [to help them reach a decision]," the machine-gunner remembered. "From inside the garage a loud 'hurrah!' burst out when they heard the shots."[93] The men of the Armored Car Division had joined the revolution.

Similar scenes occurred all across the city, as armed soldiers nervously asked each other, "Are you for the people?" On hearing an affirmative reply, strangers hugged and kissed each other, sensing, perhaps, their growing strength and unity of purpose. "We all kissed a lot in those days," the twenty-three-year-old armored car driver Viktor Shklovskii remembered. Of Russian-German-Jewish origin, Shklovskii met the revolution with all the enthusiasm to be expected of one whose early literary work was as revolutionary as his politics. Later to become a leading figure in Russian Formalism and a major literary critic, Shklovskii joined without question the movement that Kirpichnikov and his small band of rebels had begun that morning. The next day, the Petrograd garrison joined the insurrection.[94]

By that time, Khabalov and his deputies had completely lost their heads and panic reigned in their headquarters. One observer remembered that Petrograd's chief of police was seized by "an uncontrollable, senseless laughter of the kind that affects people in fires and earthquakes," while Khabalov screamed out orders for airplanes to bomb the city.[95] With the Petrograd garrison melting away like snow on a summer's day, Khabalov's resolve to launch an all-out attack against the city's insurgent workers and soldiers collapsed. "I beg you to inform His Imperial Highness that I am not able to carry out his instructions about the restoration of order in the capital," he telegraphed to *Stavka* late on Monday evening. "The majority of army units, one after another, have betrayed their oaths, refusing to fire upon the rebels. Other units have joined with the insurgents and have turned their weapons against the troops still remaining loyal to His Highness."[96]

Khabalov's telegram arrived at *Stavka* just after midnight to find Russia's High Command in turmoil. Until late that afternoon, it had never occurred to anyone to doubt Khabalov's ability to deal with the disorders in Petrograd, and even when he had called for reinforcements earlier that day, no one had any notion that his defenses were about to collapse. But *Stavka* had not been unwarned. As tsar and supreme commander, Nicholas simply had chosen to ignore the warnings that men concerned for Russia's welfare had pressed upon him. Less than a fortnight had passed since he had rejected out of hand Rodzianko's dark predictions because they contradicted the information he received from men anxious to place Russia's approaching crisis in the most optimistic light. Now, as unarmed protests in Petrograd turned into armed revolt, he turned away from the hard truth once again. "The capital is in a state of anarchy," Rodzianko had telegraphed on Sunday. "It is necessary that some person who enjoys the confidence of the country be entrusted at once with the formation of a new government. There must be no delay."[97]

Admittedly, Rodzianko overstated the extent of the crisis at that moment, but everything he predicted came to pass in less than twenty-four hours. Certainly, his estimates of the situation proved far more accurate than those of Khabalov or Protopopov, but Nicholas simply rejected them because of his intense personal dislike for their author. "Again, this fat Rodzianko has written me lots of nonsense, to which I shall not even deign to reply," he exclaimed to old Count Frederiks when the Duma president's telegram had arrived late on

Sunday evening.[98] The next day, as Guards regiments mutinied and workers broke into arsenals to arm themselves, Rodzianko tried again. "It is impossible to have any confidence in the troops of the [Petrograd] garrison," he telegraphed just before noon. "Do not delay, Your Majesty," he begged. "The hour that will decide your fate and that of our homeland has come. Tomorrow it may be too late."[99] This time, General Alekseev himself took the telegram to Nicholas. "I did everything possible to convince him to take the road to salvation at last," he later told the chief of *Stavka*'s diplomatic office. "Again, I ran against a stone wall."[100]

Rodzianko's pleas for a responsible ministry were repeated that same evening by Grand Duke Mikhail Aleksandrovich, Nicholas's younger brother. Tall, slender, and looking remarkably young despite his receding hairline, Mikhail Aleksandrovich had once insisted upon marrying a twice-divorced commoner, and his stubborn refusal to heed Nicholas's and Aleksandra's arguments against the marriage had for some years stood between them. Until the war, he had been obliged to live abroad because of their displeasure, but their mutual concern for Russia's welfare had repaired the breach sufficiently for Mikhail Aleksandrovich to return to Russia and take up a command at the front. Because of his morganatic marriage, and the imperial family's refusal to receive his wife, this grand duke lived more among the Russians, and had a better sense of their feelings, than did any of the other Romanovs. Certain that the tsar must be taken away from his wife's influence, and equally certain that the army's supreme commander should be at his headquarters, he had urged Nicholas to leave Tsarskoe Selo and return to *Stavka* just a few weeks before the disorders had broken out.[101] Now certain that Russia stood at the brink, Mikhail Aleksandrovich tried for one last time to warn Nicholas of the danger. At about 10:30 on Monday night, he begged to communicate directly with the tsar by telegraph and, when Nicholas insisted upon speaking only through Alekseev, made his plea in any case, but to no avail.[102]

Even before his brother's telegram, and several hours before Khabalov confessed that he could not restore order in Petrograd, Nicholas had decided to grant no concessions. Rather than agree even to the "responsible ministry" that Rodzianko hoped might yet soothe the tensions that gripped the capital, Nicholas summoned General Ivanov who, since Brusilov had replaced him as commander of the southwest front, had become something of a fixture at *Stavka*, and ordered him to march against Petrograd. Insisting that he must have an "absolutely loyal, although not necessarily large force" if he were to crush the revolt, Ivanov received command of the St. George Battalion, all battle-hardened heroes who wore the coveted Cross of St. George. Beyond that, Alekseev promised to send four infantry regiments, four cavalry regiments, two machine-gun units, and four batteries of artillery from front-line units to rendezvous with Ivanov at Tsarskoe Selo. From there, the general would launch his assault against the turbulent capital. As further reports about the collapsing situation in Petrograd came in, Alekseev added other reinforcements to Ivanov's command: "the most reliable battalion" of the Vyborg Fortress Artillery, the "two most reliable battalions" of the Kronstadt Fortress Artillery, and three Guards battalions, which he ordered Brusilov to hold ready on the southwest front to dispatch at a moment's notice.[103] *Stavka* and Nicholas were at one in their resolve to commit whatever military forces were needed to restore order in Petrograd.

As it turned out, most of the forces Alekseev promised never reached Ivanov's command. As revolutionary events closed around Nicholas and the men at *Stavka*, their orders grew garbled, even contradictory, and Ivanov found himself, first at Tsarskoe Selo and then at nearby Vyritsa, awaiting orders that never came. As revolution spread from Petrograd's garrison to the soldiers of Kronstadt, Luga, and Moscow, as the forces demanding Nicholas's abdication grew more insistent, and as events pressed Alekseev to defend the Duma's seizure of power in order to keep the revolution from falling into the hands of true revolutionaries, the crushing counterrevolutionary thrust that had seemed so certain on the night of February 27 no longer appeared desirable or even possible. As events convinced Alekseev to avoid any bloody battle that might drive more army units to unite with the revolutionary forces still at large in Petrograd, he recalled the reinforcements he had promised and ordered Ivanov to return to *Stavka*. [104]

No one could have predicted such an outcome when Ivanov left *Stavka* on the morning of February 28, because no one realized the strength of the forces unleashed by the previous day's insurrection nor did they perceive the speed with which they would move. Within hours after the soldiers' insurrection began, the leading figures of Nicholas's government became fugitives, fearful for their lives, fearful even to admit they knew each other as they fled into the night from the Mariinskii Palace where they had gathered to demand Protopopov's resignation. As these men scurried away, each in search of a haven for the night, the cabinet in which Nicholas and Aleksandra had placed all their faith slipped into oblivion.[105]

That very day, Petrograd's workers and soldiers organized a soviet of workers' deputies, as the revolutionary leaders just freed from Kresty Prison and socialists from the Duma led by the ever volatile Kerenskii and Nikolai Chkheidze, the Georgian schoolmaster-turned-Menshevik deputy to the Duma whose heavy accent added an exotic aura to his speeches, tried to forge the instruments of revolutionary power. "To assume that the bourgeoisie, in the form of the Progressive Bloc and the Duma Committee, would take up and support the revolution and unite with it, even provisionally and formally, was impossible," the revolution's Menshevik chronicler Nikolai Sukhanov warned. "If the revolution was to be continued, consummated, and fortified, the democracy had to be ready to take on *itself alone* the burden of this feat."[106] By evening, representatives of "democracy"—soldiers, workers, and the radical intelligentsia—gathered in the Taurida Palace for the Soviet's first meeting. Yet, even if they had the means to defend the revolution because they held the soldiers' loyalty, these men had neither the knowledge nor the experience to displace the crumbling government of Nicholas II's regime. For that, they and their allies turned to the Duma, although they remained contemptuous of the "cowardice, flabbiness, and reactionary nature" of even its most liberal figures.[107]

To govern Russia meant, in fact, to recognize the insurrection of February 27 as an accomplished revolution, and that thorny issue posed terrible dilemmas for men certain that evolution, not revolution, should chart Russia's future. Not yet ready to move against their tsar, the Duma men now understood that they could not leave the revolution—if, in fact, it became that—in the hands of Petrograd's soldiers and workers. Clearly, the Duma must play a part, but to do so meant, in effect, joining the revolution they all had struggled to avoid. "What could we do?" Vasilii Shulgin asked later. "Stand to one side? Wash our hands of the entire business? Leave Russia without a government?" Convinced that leaving the revolution in the hands of the masses could bring only disaster, Shulgin exclaimed: "If we do not take power, then others, who have already chosen some villains in the factories [as delegates to the Soviet] will do so!"[108]

In order to satisfy "the needs of the moment," as Miliukov explained, yet not "predetermine anything for the future," the Duma men searched frantically for a way to traverse the impossibly treacherous terrain that separated loyalty from disloyalty. By five o'clock on the afternoon of February 27, they decided that a Provisional Committee "for the restoration of order and for relations with individuals and institutions," could serve their purpose best. Among its eleven members, Rodzianko, Miliukov, Prince Lvov, Shulgin, Chkheidze, and Kerenskii proved to be the most influential.[109] As members of the Executive Committee of the newly-formed Soviet, Chkheidze and Kerenskii immediately became a conduit between the two groups and reveled in what Kerenskii ecstatically called "the spirit of unity, fraternity, mutual confidence, and self-sacrifice welding all of us into a single fighting body."[110]

Standing on the Right, the loyal conservative Vasilii Shulgin saw things very differently. Contemptuous of "His Majesty, the Russian masses," he insisted that "only the language of machine guns could be readily understood by the crowds in the streets. Only hot lead," he concluded, "could drive this terrible beast, that somehow had burst free, back into its den."[111] Nor did the prowlings of Shulgin's "terrible beast" remain confined to Petrograd. Before the end of the month, the soldiers, sailors, and workers of Kronstadt, Tsarskoe Selo, Oranienbaum, Vyborg, Helsingfors, Reval, Pskov, Dvinsk, and Riga—an area which the forces of order soon dubbed "the rotten triangle"—all joined the revolutionary ranks.[112] So too did those of Moscow, where the old regime fell late on March 1. Such overwhelming victories for the revolutionary forces made the tsar's abdication inevitable.

With his customary blindness to the political realities of his empire, Nicholas was the last to perceive that the course which he, Aleksandra, Rasputin, and

Protopopov had charted had left him at the mercy of events over which he could exert no control. At the first reports of danger in the capital, he decided to join Aleksandra at Tsarskoe Selo, placing the natural worries of a father and husband far ahead of his duties as Russia's emperor and supreme commander. Nothing and no one—not even Alekseev's warnings that the distance between Tsarskoe Selo and Mogilëv could create fatal communications difficulties—could alter Nicholas's stubborn resolve to abandon his headquarters, the protection of his front-line army, and the advice of his best generals, at the very moment when the greatest crisis any monarch could face burst upon him. Now hopelessly out of touch with the events that swirled around it, the imperial train plunged into the emptiness of Russia's winter landscape, isolated and without reliable sources of information, where any stationmaster who sympathized with the revolution could divert its path. Its way to Tsarskoe Selo soon barred by revolutionary workers at the Dno railway junction, Nicholas's train shunted aimlessly across the Russian countryside for almost two days until it came to Pskov, headquarters of the northern front, a mere six hundred kilometers to the north and west of Mogilëv.[113]

During Nicholas's last hours at *Stavka,* Alekseev and Ivanov had begged him to grant the ministry of public confidence that Miliukov and Rodzianko had urged upon him for more than a year, but he had stubbornly refused them both. A day later, it was too late. Only a ministry responsible to the Duma, the Duma's Provisional Committee now insisted, could prevent the revolution from falling into the hands of the proletariat. Certain that Nicholas could not waste a moment if he hoped to keep his throne, and unable to reach the imperial train as it slowly made its way to Pskov, Alekseev telephoned General Ruzskii, now holding his third appointment as commander of the northern front, and convinced him to urge their emperor to grant the Duma's demands. Therefore, when Nicholas arrived at Pskov on the evening of March 1, he found Ruzskii, "bent, gray, and old, wearing rubber galoshes and the uniform of the General Staff," awaiting him with news of more revolutionary victories. Moscow had fallen, the sailors at Kronstadt had gone over to the revolution, and the Baltic Fleet had recognized the authority of the Duma's Provisional Committee.

One of the generals who traveled with Nicholas remembered that Ruzskii's face looked "pale and unhealthy, while his eyes, peering out from behind his glasses, looked unfriendly."[114] Certainly he did not relish the task that awaited him, and the fact that he knew Nicholas disliked him made it no easier. Nor could he have been pleased that he had to wait for more than an hour to see the emperor, when every minute lost made the fall of monarchy in Russia more certain.[115] "I always believed that the emperor could never govern a country as vast as Russia," Ruzskii once said. "His character was too unstable."[116] Now fate had allotted him the task of trying to save the crown of the sovereign he thought unfit to reign.

"For the first and only time in my life," Ruzskii remembered, "I had the opportunity to tell the emperor what I really thought about the men who had held responsible posts during the past few years, and to outline what seemed to me to be the gravest failings in the general workings and activities of *Stavka.*" Without trying to soften his words, Ruzskii then went on to insist that the revolution's new victories left the tsar no choice but to concede a ministry responsible to the Duma. "Calmly, coolly, and with a feeling of deep conviction," he reported, the tsar refused. "I am responsible to God and to Russia for everything that has happened and for what is happening now," Nicholas explained, and he did not try to shift the blame. "He wanted nothing for himself," Ruzskii remembered, but he insisted that, in all good conscience, he could not grant a ministry responsible to the Duma.[117]

As Ruzskii cast about for other arguments to support his case, an aide delivered a telegram from General Alekseev. "The information we are receiving here," Alekseev reported, "gives us some ground to hope that the Duma men, led by Rodzianko, can still halt the general breakdown." But the support of these men could be secured only with concessions, and these must be granted without delay. "Each hour we lose," Alekseev warned, "diminishes these last opportunities for restoring order and makes the seizure of power by the extreme radicals more likely." *Stavka*'s chief of staff therefore begged Nicholas to grant a responsible ministry and even included the text for the necessary manifesto with his telegram.[118] If Alekseev and Ruzskii, who, Nicholas once said, "never agreed about anything,"[119] held the same opinion, the tsar realized that the entire army command must share their view. Alekseev's final pleadings therefore convinced him that he must give way, and he authorized Ruzskii to dispatch the proper messages to Rodzianko and to *Stavka* to announce that Russia henceforth would be governed by ministers responsible to the Duma, not the tsar.

Long past midnight, Ruzskii fell into bed, exhausted but certain that Russia had been calmed and the monarchy saved. Some two hours later, he learned that his effort had been in vain when an aide awakened him to say that Rodzianko asked to communicate with him through the Hughes apparatus, a primitive teletype machine that connected each front headquarters with the General Staff in Petrograd. Proudly, Ruzskii announced that the tsar, "in an effort to meet the wishes of the legislative institutions and the people halfway," had agreed to grant "a ministry responsible to the legislative chambers." Rodzianko's response, which took a full fourteen minutes to transmit with the Hughes apparatus printing at its maximum speed of twenty words a minute, took him completely aback. "Evidently His Majesty and you have not received word of what is going on here now," Rodzianko began. "One of the most terrible revolutions has broken out here, and it will not be easy to overcome." During the past forty-eight hours, he went on, "popular passions have become so inflamed that one can scarcely hope to contain them." It was too late to concede a responsible ministry. "The dynastic question," Rodzianko concluded, "arises point-blank."[120]

To Ruzskii's query about "what sort of resolution of the dynastic question is taking shape?" Rodzianko explained that there was very strong feeling in Petrograd that Nicholas must abdicate in favor of his son. Still, Rodzianko avoided firmly taking that position himself. Instead, he launched into a bitter litany about the failings of the past three years, when "the masses, in the form of our valorous army, poured out their blood and bore incalculable losses." Nicholas, his empress, and their advisers had "made a mockery" of these sacrifices, and had made no effort to ease the people's suffering. "Remember the liberation of Sukhomlinov," he said. "Remember Rasputin and his entire clique, remember Maklakov, Stürmer, and Protopopov." When Ruzskii insisted that some compromise still must be possible, Rodzianko self-righteously pointed out that, "unfortunately, the manifesto [granting a responsible ministry] is too late. . . . Time has been lost and there is no way of turning back the clock." By that point, their conversation already had gone on for the better part of two hours as the Hughes apparatus clicked out the mournful tale of the revolution's spreading violence with painful slowness. With Rodzianko reminding him that "revolutions can be voluntary and completely painless for all concerned if everything is over and done with in just a few days," Ruzskii sent his final words: "May God assist you!"[121]

Even before Ruzskii and Rodzianko finished their conversation, Ruzskii's chief of staff began to transmit its contents to General Alekseev, pointing out that "the dynastic question has been posed point-blank," and that Rodzianko had reported that "the crowds and the soldiers [had] . . . brought forward a demand for [the tsar's] abdication in favor of his son under a regency of [Grand Duke] Mikhail Aleksandrovich."[122] For three hours, Alekseev considered what course he, as chief of staff of Russia's Supreme Headquarters, should follow. His first thought was for his nation's war effort. By mid-morning he had concluded that only Nicholas's abdication could save the army and Russia from complete collapse.

After more than a year as *Stavka*'s chief of staff, Alekseev knew his Sovereign's mind better than almost any man in Russia. To convince Nicholas to abdicate, Alekseev knew that the army must present a united front, with all its senior commanders insisting upon the same course. To all of Russia's front commanders, he therefore telegraphed that Rodzianko had informed Ruzskii that "one of the most dreadful revolutions" had broken out in Petrograd, that "the dynastic question has been posed point-blank, and that the war can continue to a victorious end only with the fulfillment of the demands that have been set forth for an abdication from the throne in favor of his son under a regency of Mikhail Aleksandrovich." Alekseev reminded the front commanders that it was "essential to preserve the field army from disintegration, continue the struggle against the foreign enemy to the end, and save Russia's independence," and then asked each of them to transmit his views with the greatest speed. "I repeat," he concluded, "every minute that is lost can be fatal for the existence of Russia. . . . Among the chief commanders of the field army, there must be established a unity of thought and purpose in order to save the army."123

With speed befitting the urgency of Alekseev's request, Russia's front commanders sent their replies within the hour, showing precisely the "unity of thought and purpose" needed to convince Nicholas that abdication remained the only course.

> —At the present moment, the only way out that can save the situation and make it possible to fight on against the foreign enemy . . . is to renounce the throne in favor of the Sovereign Heir the Tsarevich under a regency of Grand Duke Mikhail Aleksandrovich.—General-Adjutant Brusilov [Commander, southwest front].
>
> —I implore your Imperial Highness, in the name of the salvation of Russia and the dynasty, to decide, in agreement with the statement of the President of the Duma, as expressed by him to General-Adjutant Ruzskii, as evidently the only way to end the revolution and save Russia from the horrors of anarchy.—General-Adjutant Evert [Commander, western front].
>
> —I, as a loyal subject, in accordance with the obligation and spirit of my oath of allegiance, consider it essential to beg Your Imperial Highness, on my knees, to save Russia and Your Heir, knowing Your sacred feeling love for Russia and for Him. Making the sign of the cross, pass on Your inheritance to Him. There is no other way.—[Grand Duke] Nikolai [Nikolaevich, Commander, Caucasus front].124

While Alekseev and his two chief aides spoke with Russia's chief commanders, General Ruzskii took the full transcript of his conversation with Rodzianko to the tsar. As Nicholas read the document from beginning to end, he learned for the first time that the crowds in Petrograd demanded his abdication. Ruzskii always remembered the "terrible moment of silence" after Nicholas laid the transcript on the table. The tsar first remarked that he was "firmly convinced that he had been born for unhappiness and that he had brought unhappiness to Russia," Ruzskii later reported. Then, Nicholas calmly spoke of abdication: "If it is necessary for me to step aside for the welfare of Russia, then I am ready to do so." An aide brought word that Alekseev was asking the front commanders for their opinions. "What do you think?" Nicholas asked Ruzskii. "The matter is so important and so terrible that I beg Your Highness's permission to think about this dispatch before I answer," Ruzskii replied. "Let's see what the front commanders say. Then the situation will be more clear."125

At two-thirty in the afternoon, Nicholas received a telegram from Alekseev in which *Stavka*'s chief of staff included the front commanders' replies to his query about abdication. "Most humbly having reported these telegrams to Your Imperial Highness, I beg You to decide without delay in the manner that the Lord God inspires You," Alekseev added. Thus far, the army had been spared "the illness that had seized Petrograd, Moscow, Kronstadt, and other cities," but

Alekseev gave no guarantees for the future. Instead, he begged the tsar to think of Russia's welfare. "For the sake of [our nation's] unity, its independence, and for the cause of victory, permit yourself to take the decision that can bring this gravest of situations to a peaceful and safe conclusion."[126] Alekseev had not actually pronounced the word abdication, but no one failed to grasp his meaning.

A few moments later, Nicholas summoned Ruzskii to the salon car of the imperial train. "In view of the extreme importance of the moment," Ruzskii remembered some years later, he took his two chief aides with him.[127] For the next half hour, Nicholas read all the telegrams that Ruzskii had brought to buttress his recommendation that he abdicate. While their aides tried to remain inconspicuously in the background, the tsar and Ruzskii smoked cigarette after cigarette, filling the unventilated car with a heavy haze of tobacco smoke. Did Ruzskii's aides share the front commanders' views, Nicholas asked. With the air as heavy with their own sense of awkwardness as it was with burning tobacco, both assured him that they did. "There came a general silence lasting for one or two minutes," one of them remembered. Then Nicholas spoke: "I have decided. I shall renounce the throne." With that, he made the sign of the cross, and all three generals, overwhelmed by the awesomeness of the moment, followed his example.[128]

His decision made, Nicholas drafted two brief telegrams. To General Alekseev, he announced: "In the name of the welfare, peace, and salvation of our deeply beloved Russia, I am prepared to abdicate in favor of my son. I beg you all to serve him faithfully and without hypocrisy."[129] On another blank, he wrote to Rodzianko: "There is no sacrifice that I would not bear in the name of the true well-being and for the salvation of our Mother Russia. Therefore I am ready to abdicate in favor of my son, under a regency of my brother Grand Duke Mikhail Aleksandrovich, providing that he can remain with me until he comes of age."[130] He handed these two fateful messages to Ruzskii, who left, only to return a few moments later with a telegram announcing that two representatives of the Duma's Provisional Committee were en route to Pskov. For a moment, Nicholas had a change of heart, asked Ruzskii not to send the telegram to Alekseev, and even took back the one he had drafted to Rodzianko. Of the two men en route from Petrograd, he despised Guchkov thoroughly for his work on the equally despised Central War Industries Committee, but might not Shulgin, the brilliant advocate of conservatism and life-long defender of Russia's monarchy yet find a way to avoid abdication? About fifteen minutes later, at 3:45 p.m. according to Ruzskii's records, Nicholas set such false hopes aside, decided that there could be no other course, and instructed the commander of Russia's northern front to send both messages.[131] Nicholas's abdication was not yet accomplished, but his final moments as tsar were not far away. Within minutes after the tsar's telegram arrived at *Stavka*, Alekseev asked his expert on diplomatic affairs to draft an abdication manifesto. By early evening, the document had been telegraphed to Ruzskii's headquarters.[132]

Neither Shulgin nor Guchkov knew what to expect at Pskov, and both feared that the tsar and the High Command still intended to suppress the revolution by force. To their surprise, an escort immediately ushered them into the emperor's salon car, and not long afterward, Nicholas entered, with "no trace of agitation," dressed in his favorite gray Cossack uniform.[133] Along one wall stood a small table on which hors d'oeuvres—the famous Russian *zakuski*—were served before meals. After polite and perfunctory greetings the men arranged themselves along its three sides, Nicholas at one end, his long-trusted Count Frederiks at the other, and Guchkov and Shulgin on the long side between them.[134] Guchkov launched into a pompous speech that Shulgin thought was both "long" and "superfluous."[135] As he reached the point of condescendingly explaining to Nicholas that "if too much time is wasted, we shall not be in a position to give you any advice,"[136] Ruzskii entered and whispered loudly to Shulgin that the tsar had already decided to abdicate. "So as to save as much of the Sovereign's prestige as possible," Ruzskii later explained, "I wanted it to be known that he had made the decision to abdicate of his own free will before their arrival and not under pressure from them."[137] Taking advantage of Ruzskii's interruption, Nicholas informed Shulgin and Guchkov that he had decided earlier that day to abdicate in favor of his son. Since then he had reconsidered. "I have come to the conclusion," he told them, "that because of his illness, I must abdicate at the same time for my son as well as for myself."[138]

No one—not Shulgin, not Guchkov, not Ruzskii—knew for certain if the law allowed a tsar to abdicate for his son as well as himself. Shulgin and Guchkov were doubtful and, since Guchkov seemed inclined to dispute the matter, Shulgin asked for a fifteen-minute recess. Certain that the volume of *The Complete Collection of Laws of the Russian Empire* that included the Emperor Paul I's carefully-drafted "Fundamental Laws" covered even this situation, Ruzskii asked how the law read and was outraged to learn that neither Shulgin nor Guchkov had thought to bring a copy of that precious volume with them. Guchkov continued to quibble, but Shulgin, having learned that Ruzskii had just received reports of armored cars moving on Pskov from Petrograd, wisely noted that "every moment is precious" and insisted that they conclude their negotiations. "If there is some juridical irregularity—if the Sovereign in fact cannot abdicate in favor of his brother—then let there be an irregularity!" he exclaimed. Perhaps more to himself, he thought, "perhaps this will gain some time. Mikhail will reign for a brief time, and then, when everything has calmed down, it will become clear that he cannot reign and the throne will pass to [Nicholas's son] Aleksei Nikolaevich."[139]

As the clock approached midnight, Guchkov returned with Ruzskii and Shulgin to tell Nicholas that the Provisional Committee would accept his abdication for his son as well as for himself. In a few moments, the last scene in this final act reached its climax as Nicholas signed the elegantly-worded manifesto sent from *Stavka* to announce to the world that neither he nor his son would ever reign again in Russia:

> In these days of great struggle against a foreign enemy who has been trying for almost three years to enslave Our Nation, the Lord God has seen fit to send upon Russia a new and terrible ordeal. This new domestic turmoil threatens to have a fatal effect upon the outcome of this hard-fought war. The fate of Russia, the honor of Our heroic army, the well-being of the people, and the entire future of Our Nation require that this war be continued to a victorious end. Our cruel enemy is now drawing upon his last resources and the hour is already near when Our valiant army, together with Our glorious allies, finally will be able to smite the enemy. In these decisive days in the life of Russia, Our Conscience imposes upon Us the duty to draw Our people into a close union and to rally all the forces of Our people for the most rapid attainment of victory. Therefore, and in agreement with the State Duma, We recognize that it is necessary to abdicate from the throne of the Russian State, and to resign the Supreme Power. Not wishing to be separated from Our much-loved Son, We bequeath Our inheritance to Our Brother, Grand Duke Mikhail Aleksandrovich, and We give Our blessing to His accession to the throne of the Russian State. We ask Him to govern in full and indestructible unity with the representatives of the people assembled in legislative bodies and to swear His inviolable oath to them. In the name of Our deeply beloved Nation We summon all true sons of the Motherland to fulfil their sacred obligations to Him who has taken up the duties of Tsar at this very difficult moment of national crisis and to help Him, together with the representatives of the people, to lead the Russian State along the path of victory, prosperity, and glory. May the Lord God help Russia!
>
> —NICHOLAS[140]

So that it would not appear that the abdication had been wrung from the tsar by the Duma deputies, all agreed that the manifesto should be dated at 3:00 p.m. that day, March 2, 1917.

"These parting words were so noble," Shulgin thought as he read through the text, "that I felt that He, perhaps even more than we, loved Russia most."[141] Yet the simplicity that so moved Shulgin disappointed Guchkov. "Such an important act in the history of Russia," he later lamented, "and it all took place in such a simple, everyday manner." He seemed offended that Nicholas appeared to have "no feeling for the tragedy of his position," and wondered if he was dealing with "a normal person."[142] Certainly, Nicholas could not be expected to give a man he so despised the satisfaction of seeing his inner torment, but his self-control was all the more noteworthy because even those who knew him well did not see through it. "He renounced the Russian throne as simply as if he were turning over command of a cavalry squadron," an aide-de-camp who had been with him ever since he had assumed command at *Stavka*, re-marked.[143] Only to his diary, the life-long companion in which he had recorded the orderly passing of his days without embellishment or emotion for almost fifty years, did Nicholas confide the true depths of his feelings that day. "At one o'clock this morning, I left Pskov with a heart that is heavy over what has just happened," he wrote that night. "All around me, there is nothing but treason, cowardice, and deceit!"[144]

Nicholas's only thought now was to reach his wife and family at Tsarskoe Selo, but the forces of the revolution kept them apart until March 9, when they all were united as prisoners in their private quarters at the Alexander Palace, its gates now locked from without, not from within. "Nicholas Romanov!" a slovenly guard at the palace gate announced to identify the man whose auto awaited the revolutionary duty-officer's permission to enter the palace grounds that morning.[145] A scant fortnight before, Nicholas had passed through that very same gate as Supreme Commander of Russia's Armies, Nicholas the Second, Emperor and Autocrat of All the Russias, Tsar of Poland, and Grand Duke of Finland. Now, he had returned, his wealth, titles, and throne all stripped away, to take his place with his wife, children, and a handful of faithful friends as Nicholas Romanov, private citizen and prisoner of the Russian Provisional Government and the Petrograd Soviet of Workers' and Soldiers' Deputies. To many Russians, this inept and shattered sovereign already had become the symbol of evil incarnate, although his greatest failings had been his weakness of character and his inability to fend off his wife's absurd advice. Now Russians thought of revenge although they would not exact its fullest measure until the early morning hours of July 18, 1918, when agents of the Ural regional soviet shot Nicholas, Aleksandra, and all their children in the far-away Siberian city of Ekaterinburg.[146]

So intense was the popular hatred against Nicholas and Aleksandra that nothing of Russia's ancient autocracy remained twenty-four hours after Nicholas's abdication. "An immense red flag floats over the Winter Palace," the Comte de Robien wrote on March 3. "Gangs of soldiers, tired of hunting policemen, are now busy removing eagles and other imperial emblems."[147] On Liteinyi Prospekt, a cheering crowd delightedly threw piles of Russia's famed two-headed imperial eagles into the fire that had gutted the District Court building.[148] Again and again, Petrograd's workers struck at the symbols that had oppressed them. "Death to the two-headed eagle!"/Sever its long-necked head/ With a single stroke!/ So that it can never come to life again," the poet Maiakovskii exclaimed with shameless delight. "We have triumphed!/ Glory to us all!/ Glo-o-or-r-y to us all!"[149]

None of the members of the Duma's Provisional Committee dared guarantee the Grand Duke Mikhail Aleksandrovich's safety, and most feared terrible bloodshed if he dared attempt to right his brother's toppled throne. Unwilling to take the risk or bear the burden, he renounced the throne just hours after word of his brother's abdication reached Petrograd. "Your Royal Highness, you have acted nobly and like a patriot. I assume the obligation . . . to defend you!" Kerenskii exclaimed at that moment. "You have magnanimously entrusted to us the sacred cup of your power," he continued a few moments later. "I swear to you we shall pass it on to the Constituent Assembly without having spilled a single drop from it!"[150] His statement was shallow, theatrical, and false. Just as soon as the representatives of the Provisional Government had drafted the final text of the grand duke's renunciation, Shulgin remembered, Kerenskii "rushed headlong from the room to the printing office" to announce to all of Petrograd and Russia that "Nicholas abdicated in favor of Mikhail; Mikhail has abdicated in favor of the people!"[151] Within hours, every lamp post in the city bore a copy of Kerenskii's placard and then, that very afternoon, Petrograd's workers and soldiers resolved to place the grand duke under the "surveillance of the revolutionary army."[152] Four months later, Kerenskii himself sent the prince he had sworn to defend to prison on a series of crudely concocted charges of plotting a counterrevolution.

CHAPTER XIII

New Men and Old Policies

To the very end, Miliukov and Guchkov had fought for Russia's monarchy, but the tide of revolution had surged so powerfully that none could stem its flood. Shulgin recalled the "iron Miliukov," already five days without sleep, "his face ashen from fatigue, completely hoarse from speaking in the barracks and at meetings," hauling his weary body to its feet one more time to make one last ardent plea for the monarchy on the morning of March 3, when the delegates of the Duma's Provisional Committee met with Mikhail Aleksandrovich to learn if he would accept Russia's crown. "If you refuse the crown it will mean anarchy! —Chaos! —Bloody turmoil! . . . The monarch is our only focus. . . . Without that, it's impossible. . . . There won't be a state, not even a Russia! Just nothing!" Throughout it all, Miliukov "croaked like a crow," but Shulgin the life-long monarchist, thought "he croaked wise, prophetic words, the greatest words of his life."[1]

Miliukov's pleadings alone could not save Russia's monarchy, and he knew full well that Kerenskii spoke the truth when he warned of "a bloody disintegration" and civil war "at the very moment when Russia needs complete unity to face the foreign foe," if the grand duke took his brother's crown. "The crown went skidding away," Shulgin wrote. "How mournfully it clanked on the granite pavement!"[2] By one o'clock that afternoon, the old regime had come to an end. In its place stood men who had never dreamed of Russia without a tsar but who now had to determine who would wield power in their nation, and how. For the moment, they had nothing to guide their steps but the urgent need to rule the lands and peoples of what had been the earth's most vast empire. "Russia *was* an empire," Shulgin mused as he walked home to sleep in his own bed for the first time in five days. "But what is she now?" The answer that came to him as he crossed the Troitskii Bridge, the new iron span that had been built to commemorate the silver wedding anniversary of the fallen tsar's parents, was far from comforting. "Neither a republic nor a monarchy," Shulgin concluded. With no tsar and no formally declared alternative, Russia had become "a state formation without a name."[3]

What would be the source of power in this nonmonarchy, nonrepublic? Who would govern? To whom would the men who did so be responsible, and how would they be held accountable? These were the most vital questions of all, more important even than the issue of war and peace, for the answers would define the nature of Russia's revolution. Would it be, as the Duma leaders hoped, a revolution in which Russia's businessmen, industrialists, and professional men and women would lead their nation along the political paths that European countries had taken after the revolutions of 1830 and 1848, or would the revolution of February 1917 become merely a stopping place on the road to a "socialist" revolution that would elevate what Shulgin once had called "His Majesty, the Russian masses" to become Russia's new rulers? Until these questions were settled, Russia and the Russians must live from day to day, uncertain of the morrow, and with no guarantees for the future.

From the moment the workers' demonstrations had become an armed rebellion, the Russian revolution had two centers of authority—the Duma's Provisional Committee (soon to become the Provisional Government) and the Soviet of Workers' (and, beginning on March 1, Soldiers') Deputies—each with its own vision of what the revolution ought to accomplish, where it ought to end, and what course Russia was destined to follow. When revolution had first broken out in Petrograd, Russia's most dynamic revolutionary leaders all had been in prison or exile, and the less daring men who remained to act in their stead in the Petrograd Soviet of Workers' Deputies had feared power, the masses, and the forces of counterrevolution, all at the same time. Uncertain of their course, they at first refused to compete for power in any "bourgeois" government and insisted upon remaining aloof from it. "Soviet democracy should entrust power to its class enemy, the privileged elements," one of the Soviet's early leaders had insisted. "At the moment, it cannot succeed in governing under these terrible conditions of disintegration."[4]

Surging revolutionary crowds, angry, embittered, and not easily led—the proletariat "in the full reality of its emaciated flesh,"[5] as one eyewitness remarked later—quickly proved far more fearsome than anyone had anticipated during the days when the Okhrana had kept order and radical propagandists had struggled against its network of spies to make their voices heard. It was one thing, these men and women now discovered, to instruct small groups of factory workers in the fundamentals of revolutionary theory, and quite another to confront tens of thousands of them giving vent to their pent-up rage in the streets. At the same time, the very thought of a counterrevolutionary assault filled the revolutionaries in the Soviet with terror. "No one knew . . . how to defend the revolution and its citadel, the Taurida Palace," Nikolai Sukhanov wrote as he recalled the moment when he had joined the Soviet's Executive Committee for its first meeting. "If Cossacks, or any sort of organized assault force had actually attacked us, even in the most negligible numbers, they could have conquered the revolution with their bare hands, for we had nowhere to turn for protection."[6] A bespeckled, thirty-five-year-old journalist with strong sympathies for Martov's Mensheviks, Sukhanov was a far cry from the hard-eyed revolutionary strategists who would dominate Russia's political landscape later in 1917. Ever since his radical student days at Moscow University, he had preferred to do battle with his pen. Now events thrust him forward, along with a former school-teacher, a handful of radical lawyers, and a motley assortment of publicists, to lead the masses as a member of the Executive Committee of the Petrograd Soviet. Fearful that they had not the skill to navigate the difficult course that lay before them, and, in any case, still prisoners of an ideology that decreed a lengthy period of bourgeois government between the fall of Russia's monarchy and the advent of a socialist democracy, such men were willing, even anxious, to concede the terrible burden of power to the "privileged elements" of the Duma during the desperate days of February and March 1917. Beyond those considerations, as Sukhanov later explained, "our revolution did not at that point have the practical strength to bring about the rapid socialistic transformation of Russia, nor were the conditions yet ripe."[7] As men whose loyalties lay mainly with the Menshevik and Socialist Revolutionary parties, Sukhanov and his comrades therefore held to a moderate course.

Long used to disputing the intricacies of revolutionary theory, but far less accustomed to revolutionary practice, the men who stepped forward to direct the Executive Committee of the Petrograd Soviet asked for little more than theoretical and political guarantees for "complete political freedom . . . and absolute freedom of organization and agitation" from the Duma in exchange for delivering Russia's instruments of power into their hands.[8] Yet, to hungry workers and war-weary soldiers, the concessions so ardently yearned for by the persecuted revolutionaries of the old regime seemed poor protection against a privileged class whose first three "revolutionary" acts had been to defend the Romanov dynasty, urge the disarming of the city's workers and soldiers, and form a government designed to halt the revolution in mid-course. By no means so willing as their revolutionary mentors to exchange their first victories for theoretical concessions, Petrograd's workers and soldiers demanded stronger guarantees. Sensing that the revolution needed real defenses, not theoretical bastions, they took firm steps to erect them even before Nicholas abdicated.

From the first days of the revolution, Petrograd's civilian and military masses had placed their trust in the Soviet of Workers' Deputies, not the Duma. They had forged it themselves as their own revolutionary instrument in the revolutionary crucible of 1905, and it was to the Soviet, reborn on February 27, 1917, that they always gave their first allegiance. Certain that Rodzianko's order (signed just after midnight on February 28 in the name of the Duma's Provisional Committee) for "all officers to return to their units and take whatever measures are necessary to restore order"[9] was an attempt by the "privileged classes" to disarm the revolution, Petrograd's soldiers turned instantly to the Soviet. Irate, they burst into its chambers just after noon on March 1, angrily demanded that the *soldatskii vopros*—the issue of the soldiers—be discussed at that very moment, and allowed none but soldiers to speak.

Anxious to avoid this explosive issue, the Soviet's Executive Committee tried to postpone its discussion until it had time to strike an agreement about transferring power to the Duma's representatives, but Petrograd's soldiers refused. "The soldiers must not give up their weapons, and, in political questions, must obey the Soviet," the first soldier-orator to come before the Soviet insisted, while another exclaimed: "We're the masters along with the workers! We won't let [the privileged elements] get the upper hand!"[10] Quickly, they resolved that "the soldiers should organize themselves into a soviet of soldiers' and workers' deputies," and that their "weapons should not be turned over to their officers, but only to [elected] battalion committees."[11] The Petrograd Soviet now changed its name to the Petrograd Soviet of Workers' and Soldiers' Deputies, and launched into a discussion about how to abolish the abuses that soldiers had suffered at the hands of their officers under the old regime. Clearly, officers still must command and soldiers still must obey, but the soldiers in the Soviet insisted that their officers must command "politely, without using mother curses,"* as so many of them did as a matter of course. One soldier-orator sagely pointed out that, if the Soviet removed every officer who used "mother curses," there would be no officers left in the army, but the others insisted that their dignity had been abused long enough.[12] Beyond that, the soldiers demanded that they and their officers be treated as equal citizens when they were not on duty. No longer could the uniform of an army private deny civil rights to the man who wore it.

Petrograd's soldiers insisted that the resolutions they had rushed through the Soviet, now swollen to more than a thousand members, be published that very day. Quickly, they forced the Executive Committee to end its temporary adjournment and draft the appropriate order. With it triumphantly in hand, the soldiers returned to the full session of the Soviet to vote their final approval of PRIKAZ No. 1—"Order No. 1"—the most fateful document of the February Revolution. "The garrison of the Petrograd Military District, all soldiers of the guard, army, artillery, and fleet," Order No. 1 announced, must "immediately form committees of representatives elected from the lower ranks." These com-

*For centuries, so-called "mother-curses"—elaborate forms of profanity that employed a variety of picturesque variations of the phrase "mother-fucker"—had been widely used by Russians from all walks of life. As such, they hardly constituted anything like the grave affront to their sensitivities that the soldiers claimed in 1917.

mittees would moderate all disputes between officers and men, punish all cases of officers' "rude behavior toward soldiers of all [lower] ranks," and control all weapons, including "machine guns and armored cars." Most important of all in determining the revolution's future course, Order No. 1 instructed all soldiers and sailors that orders of the Duma "shall be executed only in those cases where they do not conflict with the orders and resolutions of the Soviet of Workers' and Soldiers' Deputies."[13] Thus, before the unpracticed revolutionary intellectuals in the Soviet's Executive Committee could obey the dictates of theory and hand control of the "bourgeois revolution" over to the "privileged elements" in the Duma, the revolutionary soldiers and workers of Petrograd made certain that the real source of power in the city—the armed garrison of Petrograd—would remain under their control to defend *their* revolution. "One had to see the faces of the soldiers to understand the revolutionary meaning the order carried," one onlooker recalled some years later. "The soldiers were beside themselves with joy."[14]

While Petrograd's soldiers boldly forced Order No. 1 through the Soviet and its Executive Committee, their "class enemies," the Duma representatives of the "privileged elements" led by Miliukov, had begun to form a provisional government. More than a decade later, Shulgin recalled the dramatic moment on March 1 when Miliukov, "between endless discussions with thousands of people who kept plucking at his sleeve, meetings with deputations, speeches at unending meetings in the Catherine Hall, frantic hasty trips to various regiments, discussions of telegrams received by direct line from *Stavka,* and squabbles with the increasingly insolent Executive Committee of the Soviet . . . sat down briefly at a table off in one corner, and drew up a list of ministers."[15] Such a scene hardly rivaled the Soviet's raw revolutionary fervor, but, for the Duma men, whom events had finally forced to turn against a tsar who had not yet abdicated and a dynasty that had not yet relinquished the throne, Miliukov's act was every bit as revolutionary as the soldiers' struggle for Order No. 1. "We were first of all the loyal opposition," Shulgin remembered. "Within us, respect for the throne was intertwined with a protest against the course that the tsar had set, a course which, we realized, was leading straight to the brink."[16]

Still, the moment was neither as spontaneous nor as ill-planned as Shulgin's remembrances might indicate. Ever since the Progressive Bloc had first called for a ministry of confidence in August 1915, the Duma men had maintained some consensus about who among them could best direct their nation's affairs. Therefore, when Miliukov jotted down the names for a cabinet, few men in the Duma were much surprised. Even the absence of Rodzianko's name on the list occasioned little comment, for although he was president of the Duma, he had not been mentioned as a potential cabinet member since August 1915. Only Kerenskii, whom Miliukov urged to become minister of justice, had not been included on earlier lists of names for positions in either a ministry of confidence or a responsible ministry.[17]

At about three o'clock the next afternoon, Miliukov announced to a large crowd assembled in the Taurida Palace's great columned Catherine Hall the names of the men who would guide Russia until a Constituent Assembly could be elected. "I went out to the crowd . . . in a very elated mood," he later wrote. "My words somehow strung themselves together all by themselves." A Provisional Government, in which he would serve as foreign minister, Miliukov explained to the assembled masses in his best professorial manner, had taken power at this critical moment to guide Russia through the turmoil of war and revolution. To any who dared question the new government's right to speak for the nation, he proclaimed: "The Russian Revolution itself has chosen us!" But the revolution's choice could not be the final one. Miliukov insisted that it must be tested by national elections, "on the basis of universal, direct, equal, and secret vote, as soon as the danger [of war] is past, and a firm peace is established." These elections would determine what form of government, and what sort of men, would rule Russia. "We shall not," he promised his listeners, "hold this power for a single moment beyond the time when the freely elected representatives of the people tell us that they want to see others, more deserving of their confidence, in our places."[18]

From the first words of his speech to the last, Miliukov had trod upon uncertain and dangerous ground, and the responses of his listeners showed the limits of their trust in the Duma's "privileged elements" all too clearly. Obviously, common men and women did not trust them to defend the revolutionary victories they had just won in Petrograd's streets. For that reason, they cheered loudest when Miliukov named Kerenskii, champion of the "nonprivileged classes," as Russia's new minister of justice. Of all the men in Russia's first Provisional Government, Shulgin remarked with reluctant admiration, "only Kerenskii knew how to dance upon the revolutionary quagmire."[19] Nowhere was his immense talent for bending revolutionary crowds to his will more evident than in the way he manipulated the Soviet, its Executive Committee, and the Petrograd masses so that he could accept the position Miliukov offered.

To Kerenskii's dismay, just as the Duma men decided to include him in the government, the Executive Committee voted that none of its members could hold any post in any "bourgeois" Provisional Government. Unlike his colleagues on the left, Kerenskii desperately wanted power, and refused to wait for the supposedly inexorable workings of the Marxian historical dialectic to deliver it into his hands. He justified his personal desires in terms of the revolution's best interests. "I was compelled to fight continually against the academic dogmatic socialism [of the Executive Committee] of the Soviet, which from the very beginning tried to thwart the normal development . . . of the Revolution." To that, he added an urgent moral imperative. Not only must he save the Revolution; he must save the hapless statesmen of the old regime who now languished in prison. "Could anyone else, could any bourgeois minister of justice, save them from lynching and keep the Revolution undefiled by shameful bloodshed?" he asked rhetorically. Certain that "under the circumstances, no one but myself could do this," Kerenskii decided that he must at any cost become "the hostage of revolutionary democracy" in the camp of the "bourgeoisie,"[20] and took his case directly to the full Soviet, where a spokesman for the Executive Committee had begun to present a lengthy report about the conditions under which he and his colleagues had agreed to transfer power to the Provisional Government.

As the speech neared its end, the ever-present Sukhanov remembered, "Kerenskii leaped to his feet as if he had been stung, and hurled himself toward the rear of the hall." None escaped his sense of desperate urgency, as he turned, "as pale as snow, agitated to the point of total shock, to choke out short, abrupt phrases, interspersing them with long pauses."[21] Pouring all of his great oratorical skills into this one effort, Kerenskii leaned toward his audience:

> "Comrades!" he began in a confidential stage whisper. "Do you trust me?"
> "We trust you!" the audience roared back. "We trust you!"
> "Comrades! I speak from the very depths of my soul. From the bottom of my heart. And if I must prove this—if you don't trust me—then I am ready to die—Here, before your very eyes! . . . Comrades! At this very moment, a Provisional Government, in which I am to occupy the post of minister [of justice], has been formed. Comrades! I had less than five minutes in which to reply and therefore I could not seek your approval. . . . Comrades! The representatives of the old regime were in my hands. And I could not let them out of my hands now!"
> "Right!" his listeners cheered, punctuating their cries with stormy applause. "Bravo!"

Nor would Kerenskii let the Soviet's delegates out of his hands before he had everything he desired. Not only would he take up a ministerial portfolio in the Provisional Government; he intended to keep his powerful position in the Soviet, and to do so at the masses' urging. "Because I accepted the post as minister of justice before I could obtain your formal consent," he announced grandly, "I hereby divest myself of my position as vice-president of the Soviet of Workers' Deputies. But . . ." For a long moment, Kerenskii paused dramatically until he was certain that he held Petrograd's proletarians firmly in his grasp. Then he continued. "But . . . I am ready to take this title from you again if *you* think it necessary."

Again, the representatives of Petrograd's workers and soldiers replied with cheers of approval. "Yes! Take it! Take it!" they shouted. "Long Live the Minister of Justice!"

Kerenskii had won his victory. Now minister of justice and vice-president of the Petrograd Soviet of Workers' and Soldiers' Deputies by popular acclaim, he would not allow his audience to escape his words' relentless grip until he had wrung one more pledge of confidence from them. "Can I trust you even as I would trust myself?" he asked in a voice that demanded confidence, yet promised the confidentiality of confession.

"Trust us, Comrade! Trust us!" his audience begged.

"I cannot live without the people!" Kerenskii confessed, as he took the twelve hundred men and women who now crowded into the hall into his confidence. "If the moment ever comes when you doubt me—then kill me!" As his listeners screamed their approval, he concluded with one last flourish: "I enter the new government with your permission, then! I go only as your representative!"

For a moment, the shouts of "Long Live the Minister of Justice!" faded. Then, Petrograd's workers roared: "Long Live Kerenskii!"[22]

The thunder of the assembled delegates' applause drowned the sound of the new minister's steps as he stormed from the hall to become "the representative of the people . . . the representative of democracy" in revolutionary Russia's first Provisional Government.[23]

Although Kerenskii now spoke for them in the "bourgeois" camp, the Soviet's Executive Committee began to have second thoughts about surrendering all power to the "bourgeois" Provisional Government. Most of all, they feared Miliukov and Guchkov, who had returned from presiding over Nicholas's abdication to take up the powerful post of war minister. Both were men of strong will, who could sway the masses with the force of their words and their personalities, and both still insisted that a Constituent Assembly, not the Petrograd crowd, must determine the future of the Romanov dynasty. Ever mindful of Miliukov's insistence that only delegates chosen in national elections could decide what form of government ought to rule Russia,[24] the Executive Committee hurried to qualify their approval of the Provisional Government and agreed to support its policies only "in so far as they correspond to the interests of the proletariat and of the broad democratic masses of the people."[25] "We agreed to leave it to the government . . . to deal with the Romanov monarchy," Sukhanov wrote of himself and his colleagues on the Executive Committee. "But we categorically declared that the Soviet, for its part, would immediately wage a broad struggle for a democratic republic." Ominously, he emphasized, *"our 'agreement' was an agreement about the conditions for a duel."*[26]

Combined with Order No. 1, which pledged the Petrograd garrison's first loyalty to the Soviet, the Executive Committee's hesitant partial support obliged the Provisional Government—"a revolutionary cabinet that contains not a single revolutionary other than Kerenskii," Gippius archly remarked[27]—to make policy, but allowed it to enforce only those policies that the Soviet sanctioned. "The Provisional Government possesses no real power," Guchkov confessed to General Alekseev a few days later. "Its orders are executed only in so far as this is permitted by the Soviet of Workers' and Soldiers' Deputies, which holds in its hands the most important elements of actual power, such as troops, railroads, postal and telegraph services."[28] This situation resulted in a stalemate—*dvoev-lastie*—dual power the Russians called it—in which both Provisional Government and Soviet, now "locked in a struggle to the death," to use Sukhanov's phrase,[29] diverted vitally-needed human and political resources from the war effort in a desperate attempt to prevent each other from pursuing its revolutionary vision to a conclusion. *Dvoevlastie* became the revolution's curse, and prevented the Provisional Government from pursuing any policies that stirred the disapproval of Petrograd's masses. "The old government is in prison," one observer remarked. "And the new one is under house arrest."[30]

As the new government struggled into existence, the revolutionary crisis spread into Russia's provinces and began to move toward the front. Everyone everywhere worried, wondered, and argued, as they peered into the murky future and tried to define their course. "Day and night, a never-ending disorderly meeting went on across the entire country," Paustovskii remembered. "Mobs of people shouted on city squares, around monuments, in railroad stations smelling of disinfectant, in factories, in villages, at markets, in each courtyard, and on every stairway of any building that was inhabited." Once begun, there seemed no end to the debates, the challenges, the accusations. Russia, the peoples' Russia—the dark and silent Russia of shabby-booted, bark-shoed men in greasy sheepskins, of women in gray woolen shawls, shapeless smocks, and heavy padded jackets, their fingers grown thick from hard labor begun in earliest childhood—had found its voice. "In the course of a few months Russia spoke out about everything it had been silent about for centuries," Paustovskii continued. "All of a sudden, a whole host of babblers sprang up. . . . Cheap demagoguery flourished in well-manured market places."[31] In Petrograd, Viktor Shklovskii thought "it was like Easter—a joyous, naive, disorderly carnival paradise."[32]

Neither carnival nor demagoguery restrained those who spoke or those who acted upon their words. "There existed not only no governmental power," Kerenskii said, "but literally not one policeman."[33] Law and order had perished during the first days of the revolution, and there seemed no way to get events under control. To Paustovskii, it seemed that "the state was falling apart like a hunk of wet mud." As he watched the revolution's course from Moscow, he was certain that "provincial Russia was no longer ruled by Petrograd, and no one knew what they lived on or what seethed inside them." During the first weeks of the revolution, he remembered Kerenskii, "with his lemon-colored puffy face, his reddened eyelids, and his close-cropped, thinning, grayish hair," laboring, it seemed, beyond the limits of human endurance, to "knock the broken pieces of Russia back together with his ecstatic eloquence."[34] But Kerenskii's stirring speeches could not settle the staggering problems that faced Russia. "The Provisional Government," he explained some years later, "inherited nothing from the autocracy but a terrible war, an acute food shortage, a paralyzed transportation system, an empty treasury, and a population in a state of furious discontent and anarchic disintegration."[35] All too easily forgotten during the excitement of the revolution, the impending catastrophe brought on by the shortsighted and stupid policies of Protopopov's last weeks in office remained still poised to strike. "There is much more that is absurd than is grandiose," Maksim Gorkii mused as he surveyed the scene from the editorial offices he shared with Sukhanov. "We cannot go back," he wrote to his wife, "but we are not going forward very far."[36]

Most of all, the war—ever-present, ever-bitter, draining the substance of enemy and ally alike—barred the way. The war stood in the way of desperately needed domestic reforms, it lay at the root of the deadly spiral of inflation that sucked more and more workers into poverty, and its costs had brought Nicholas's government to the brink of bankruptcy. More than four million Russians lay dead because of it; once the 1917 campaign began, hundreds of thousands more were certain to follow. From his winter quarters near the foothills of the Carpathians, the commander of Russia's Seventh Army jotted wearily in his diary at the beginning of March: "The common soldier today wants only one thing—food and peace, because he is tired of the war."[37]

For months before the revolution, Russia's allies had feared she would leave the war because her people had turned so vehemently against it, and Paléologue already had begun to pay out handsome sums to induce senior officials to place materials favorable to the Allies' war effort in influential Russian journals and newspapers.[38] "Time is not working for us on the Russian front," he warned France's delegates when they arrived in Petrograd for the Inter-Allied Conference in mid-January. "Russians no longer care about the war."[39] Paléologue was doubtful, even fearful, of Russia's ability to continue the struggle. "What is the use of sending her guns, machine guns, shells, and airplanes, which would be so valuable to us [on the western front], if she has . . . [not] the will to use them?" he had asked just a month before revolution broke out in Petrograd.[40]

Such fears loomed even larger in the minds of Russia's allies as the Provisional Government took its first uncertain steps:

"Today it is necessary to save Russia by carrying on the war to victory," Miliukov assured Paléologue on the day the Provisional Government took office. "But the passions of the masses are so exasperated and the difficulties of the situation are so terrible that we must agree immediately to make major concessions to the national conscience."

"I have no doubts about your personal feelings," Paléologue replied. "But the direction of Russian policy is henceforth at the mercy of new forces. . . . I should like to be certain that the new Russia considers herself bound by the oath of her former tsar!"

"You can be sure that I shall do everything in my power," Miliukov assured him at their next meeting. "But you cannot imagine how difficult our socialists are to handle! And we must avoid a break with them at all costs. Otherwise, it means civil war."

"Whatever reasons you may have for sparing the hotheads of the Soviet," Paléologue retorted, "you *must* understand that I cannot allow any equivocation about your determination to maintain the alliance and continue the war!"[41]

Certainly, Miliukov wanted to give such assurances, because he was committed to the aims of tsarist foreign policy and because he wanted the Allies' recognition of Russia's Provisional Government to come as soon as possible. "My fundamental idea," he explained some months later, "was that there are no such things as the 'tsar's diplomacy' and the 'Provisional Government's diplomacy,' that what exists is a mutual diplomacy of the Allies."[42] To better ensure continuity during Russia's revolutionary transition, Miliukov kept all of his ministry's personnel at their posts. "I knew that among those serving under me there were people who did not share my views about current questions of foreign policy," he later wrote, "but I relied upon their conscientiousness as professionals."[43] In an effort to tie his government securely to the Allies' mutual diplomacy, and to assure them of Russia's resolve to fight on to victory, Miliukov issued his first official foreign policy statement on March 5, the day of his second meeting with Paléologue. As a "victim of premeditated aggression . . . which has aimed at . . . subjecting Europe of the twentieth century to the shame of domination by Prussian militarism," he promised that Russia would fight on by the side of her allies "against the common enemy until the end, without cessation and without faltering."[44]

Miliukov's statement so outraged the Soviet that the Provisional Government had to take a more moderate tone in its first official public declaration "to the citizens of Russia" the next day. Now seeking safer ground, Miliukov offered much less than he had led the French and British to expect. "The Government will sacredly observe the alliances which bind us to other powers, and will unswervingly carry out the agreements entered into with the Allies," he promised. But, he continued, "while taking measures to defend the country against the foreign enemy, the Government will, at the same time, consider it to be its primary duty to open a way to the expression of the popular will."[45] The Allies wanted more. "Determination to continue the war until complete and final victory isn't even mentioned!" Paléologue exclaimed in an outburst of righteous indignation. "Even Germany isn't mentioned!"[46]

Was Russia prepared to press on against the Germans and Austrians or not? A month before the revolution, General Gurko had told the assembled members of the Inter-Allied Conference that Russia's armies were still in the throes of a serious reorganization that would prevent them from taking the offensive before late spring. Until then, he had warned, the Russians would find it "impossible to undertake any serious operations," and the Allies must expect them to do nothing more than remain on the defensive.[47] At best, this was far less than the Allies had hoped for, especially in return for almost eight thousand cannon, five million artillery shells, nearly twenty thousand machine guns, one and three-quarter million rifles, and a half-billion rifle cartridges that the Russians demanded before their forces could go on the offensive.[48] Now, before they drained their arsenals to support Russia's war effort, the Allies desperately needed clear and public assurances that the revolution had not weakened Russia's will to fight. Desperately, Miliukov tried to satisfy them yet not enrage the Soviet, upon which Order No. 1 had conferred the power of final approval in all military questions. "I must have an assurance that the new Government is prepared to fight the war out to a finish and to restore discipline in the army," Ambassador Buchanan insisted.[49] Paléologue spoke even more bluntly in private. "Yes or no?" he demanded of Miliukov. "Will Russia continue fighting at the side of her allies until final and complete victory, without weakening and without ulterior motives?"[50]

While Russia's European allies pressed Miliukov for stronger public assurances than he dared give before granting official recognition to the Provisional Government, her American ally, the United States, took a different course. "I request respectfully that you promptly give me authority to recognize the Provisional Government, as the first recognition is desirable from every viewpoint," Ambassador David Francis urgently telegraphed to the secretary of state just after the Provisional Government was formed. "This revolution is the practical realization of that principle of government which we have championed and advocated—I mean government by consent of the governed. Our recognition will have a stupendous moral effect, especially if given first."[51] At Francis's urging, the United States moved with extraordinary speed so that he could recognize Russia's new government on March 9. France, England, and Italy now had little choice but to do likewise, and did so two days later, even though Miliukov could do no more than tell them privately of his nation's unaltered resolution to pursue the war to victory.[52] As Buchanan, Paléologue, and Italy's Ambassador Carlotti announced their governments' recognition, Miliukov declared that "Russia would fight till her last drop of blood" was shed in the Allied cause. "I have no doubt that Miliukov would," General Knox remarked after he left the ceremony. "But can he answer for Russia?"[53] Paléologue, too, had doubts. "In the present phase of the revolution," he warned his superiors at the Quai d'Orsay, "Russia cannot make either peace or war."[54]

Through Miliukov, the Provisional Government had spoken. But what of the Soviet? Just three days after the Allies gave their blessings to Russia's new government, the Soviet declared "To the Peoples of the World," that the two centers of *dvoevlastie* stood very far apart on questions of foreign policy. Partly in response to Miliukov's dedication to tsarist foreign policy, and partly in reply to greetings they had received from the American Federation of Labor and the socialists of France, the Petrograd Soviet addressed the "proletarians and toilers of all countries" for the first time. Socialists and labor leaders in the West all had urged caution upon their Russian brothers. In clumsily-chosen words whose patronizing tone could not help but sting, the Americans insensitively lectured the Soviet that "it is impossible to achieve the ideal state immediately" and loftily explained that freedom "cannot be established by a revolution only [but] . . . is the product of evolution." At the same time, France's socialists called upon Russia's workers and soldiers to "work together to destroy the last and most formidable citadel of absolutism, Prussian militarism."[55]

In reply, the Soviet appealed "to all people who are being destroyed and ravaged in this monstrous war," and announced that "the time has come for the people to take into their own hands the decision of the issue of war and peace." To the "toilers of all countries," they proclaimed: "Refuse to serve as an instrument of violence and conquest in the hands of kings, landlords, and bankers." Then, they promised, "by our united efforts, we will put an end to this frightful carnage."[56] This declaration, Sukhanov explained, "launched the struggle for peace against the piratical efforts not only of [Kaiser] Wilhelm, but also of Miliukov and his allies against that policy, which placed their obligations to Anglo-French imperialism higher than their duty to democracy." It would not threaten Russia's security, but could hasten peace. By joining with socialists in other nations, Russia's toilers and soldiers hoped to force Europe's warring ruling classes to end the conflict.[57]

Although "fundamental differences" separated the Soviet from the Provisional Government and its allies in their attitudes about the war, the Soviet itself could not at first decide about how to conduct what Sukhanov once called "Miliukov's frenzied imperialism."[58] "We are striving not to take territory from other peoples, but to help them attain liberty," the Petrograd Soviet's official newspaper, *Izvestiia, The News,* announced a few days after it published the appeal "To the Peoples of the World." "We are striving for final victory not over Germany, but over her rulers," its editors explained. "As soon as the people of the Austro-German Coalition compel their rulers to lay down their arms and to give up the idea of conquests, we will also lay down arms."[59] Such statements paid lip service to the misty dreams of Europe's international socialists, but they offered no concrete course of action to counter Miliukov's vow that "Russia would fight till her last drop of blood" in exchange for the Allies' continuing promise that she would receive Constantinople and the Dardanelles—the Straits —as her reward at the war's end. Almost from the moment that Catherine the Great had first gained for Russia a permanent foothold on the Black Sea in the 1770s, the specter of Turkish-controlled Straits had haunted tsarist statesmen. At any moment, Russia's traditional southern enemy could allow enemy ships easy access to Russian coasts, prevent Russian warships from sailing forth to defend their nation's interests in the Mediterranean, or, at the very least, close off Russia's vital grain trade with the West. It therefore was understandable that Miliukov, the historian who once had written a book about Russia and the Straits, should remain faithful to the traditional demands of his predecessors and insist that control of the Straits must be Russia's reward for her loyalty to her allies when the Central Powers were defeated. "The possession of the Straits is the protection of 'the doors to our home,'" he insisted at a press conference on March 23. "It is understandable that this protection should belong to us."[60]

If the Soviet hoped to counteract such policies, it needed to agree upon a well-defined program to replace its vague appeals to international socialist unity and fervently-phrased promises to extend "the hand of brotherhood across the mountains of our brothers' corpses, across rivers of innocent blood and tears, over the smoking ruins of cities and villages, [and] across the shattered treasuries of culture."[61] But, before that could be accomplished, the dispute between those socialists who had agreed to support their nations' governments in a "defensive" war and those who insisted that "the slogan 'defense of the fatherland' and its variations . . . are only a device to mask the rapacious pretensions of the ruling classes . . . and turn the worker into a blind instrument for their imperialist ambitions"[62] had to be set to rest. This debate had kept the socialists of Russia and Europe in disarray throughout the war, and the Soviet's first broad statements about the war had been necessarily vague precisely because its Executive Committee included men whose opinions ranged across the entire spectrum of those views. Until someone found a middle ground, the Soviet could not hope to take a firm stand on the conditions under which Russia should seek war or peace and counter Miliukov's very clearly-stated war aims.

Irakli Tseretelli, a practical, realistic, clear-thinking Georgian aristocrat "with a pale and elongated face like a painting by El Greco," one contemporary remembered,[63] set out to transform the Executive Committee's vague statements of principle into concrete articulations of policy that could rival Miliukov's precisely formulated statements. Friend and foe alike thought him "a man of high moral standards," and Miliukov later spoke warmly of him as "a natural-born peacemaker who became a remarkable expert on inter-party relations." Even Trotskii, whose ascerbic characterizations of men and events during 1917 left few unscathed, called him "a distinguished orator" known for his "desire to pursue a consistent policy."[64]

After having spent ten years at hard labor in Siberia for his revolutionary activities, Tseretelli returned to Petrograd on the morning of March 20 to enthrall the Soviet and its Executive Committee with his brilliant but sensible words. "Full of burning enthusiasm, but always even tempered, elegantly restrained, and calm," one observer remembered, Tseretelli held his listeners' attention with well-structured speeches that had "something sharp and decisive about them" and stood dramatically apart from the "confused pathos" of speakers who usually addressed the Soviet.[65] "On the first day [after he returned from Siberia], he modestly refused to venture an opinion because he had not yet fully examined the situation," a revolutionary memoirist who saw him often during those days wrote. "Then, on the next day, he delivered a wide-ranging speech," which seized the imaginations of nearly all factions on the Executive Committee. "On the third day," Tseretelli's associate concluded in wonderment, he "emerged as the confident leader of the [Executive] Committee and the Soviet."[66]

Tseretelli "knew what he wanted, had a definite plan, [and] believed firmly in it," one of his close friends later wrote.[67] The moment he arrived in Russia's revolutionary capital, he urged its workers and soldiers to "unite the struggle for peace with the defense of the revolution against a foreign military threat," and pressed the Soviet to agree upon "a constructive policy of peace and defense."[68] Insisting that "either we shall become a responsible force or be an irresponsible opposition," he argued that the Soviet must do nothing to weaken the army's ability to defend Russia's frontiers but must dedicate itself to seeking "universal peace without annexations or reparations."[69] Nation and revolution must be defended equally, and a lasting peace could come only if the Soviet followed what he called "a constructive policy of peace *and* defense."[70]

Only after bitter opposition from Miliukov, who branded Tseretelli's proposal "a German formula that they endeavor to pass off as an international socialist one,"[71] did the government and the Soviet agree to a formal statement about Russia's war aims. "The purpose of free Russia is not domination over other nations, or seizure of their national possessions, or forcible occupation of foreign territories," the Soviet's declaration announced on March 27. "The Russian people does not intend to increase its world power at the expense of other nations, [and] it has no desire to enslave or degrade any one."[72] It seemed for the moment that Tseretelli had succeeded. "A torch [has been] thrown into Europe, where it will burn with a brilliant flame," he exclaimed. Yet Miliukov had retreated, not capitulated. "I refused to throw this 'torch into Europe,' that is, to issue this declaration as a direct communication to our Allies," he later wrote. "I insisted [instead] that it be directed only to Russia's citizens, that is, that it be given only domestic circulation."[73]

To Russia's allies, Miliukov insisted that "we by no means renounce the securing for Russia of the vital interests stipulated in the respective agreements" that would give her ownership of Constantinople and the Straits.[74] To emphasize the point, he spoke publicly as if his government's declaration of war aims to its citizens had no force abroad. "Russia must insist on the right to close the Straits to foreign warships and this is not possible unless she possesses the Straits and fortifies them," he told the *Manchester Guardian's* correspondent.[75] That his words were widely quoted in England as evidence that the revolution had brought no significant change in Russia's foreign policy particularly enraged Viktor Chernov, who had spent most of the Great War in exile in Switzerland and England and now saw at first hand, as he made ready to return to Russia, how widely the Allied press reported Miliukov's views and how completely it ignored the Soviet's demands for peace "without annexation or indemnities."[76]

A very ordinary-looking man and, by most accounts, a poor public speaker who rarely looked his listeners in the eye while he hurled tendentious and pretentious phrases at them, Chernov was the founder of Russia's Socialist Revolutionary Party. A large and loosely organized revolutionary group that, on occasion, showed itself less than strong on first principles, the Socialist Revolutionaries demanded that private landownership be abolished and insisted that each peasant family be given the use of as much land as it could farm without hired labor. As champions of the peasants' interests, the Socialist Revolutionaries therefore found their main support in the countryside, not among city factory workers, and Chernov, the provincial from Tambov, remained their chief spokesman and theoretician throughout 1917. Even more than Sukhanov, he relied upon his pen in revolutionary battles, and he used that weapon to deadly effect against Miliukov from the moment he reached Petrograd on April 8.[77]

Convinced that Russia's foreign minister would "persevere in his doctrinaire stand until history shows him the door," Chernov vowed to drive "Miliukov-Dardanelskii" from office.[78] Together with Tsereteli, he set out to force Miliukov to state the war aims of peace "without annexations or indemnities" to her allies in a formal diplomatic note. Stubbornly, Miliukov resisted until France's socialist minister of munitions, who had just arrived in Petrograd for meetings with the Soviet, suggested that he could satisfy both Soviet and Allies by adding a covering note to underline Russia's resolution to continue the war.[79] On April 18, Miliukov therefore assured Russia's allies of his nation's firm resolve "to bring the World War to a decisive victory." Then, unwisely, he added a sentence of fateful impact at the end. "The leading democracies," he concluded, "will find a way to establish those guarantees and sanctions which are required to prevent new bloody encounters in the future."[80] With those words, Trotskii later wrote, Miliukov "touched the match to the fuse."[81]

Was Miliukov speaking of "indemnities and annexations"? Certainly the Soviet thought so. Tsereteli thought the note "was nothing but a repudiation of the basic principles of the Soviets' foreign policy," while Nikolai Chkheidze, Tsereteli's more volatile Georgian comrade, who had spent the past decade as one of the Mensheviks' rare representatives in the Duma, muttered ominously that "Miliukov is the evil genius of the revolution." Tsereteli and several others on the Soviet's Executive Committee counseled restraint nonetheless. Clearly, they could not allow a "genius of tactlessness" like Miliukov to "substitute imperialistic slogans for those of the Russian Revolution," but they preferred to avoid open confrontation and to settle the issue by negotiation. Mass demonstrations and street violence, they insisted, would create more problems than they could solve.[82] "What is the point of demonstrating," one of them asked the assembled Soviet rhetorically? "Against whom will you use force? After all, the only force is you and the masses who follow you!"[83]

Less tractable revolutionaries refused to mute their opposition so easily. Alone among Russia's revolutionary leaders, Lenin had insisted from the very beginning of the Great War that the proletarians of Russia and Europe must not allow politicians to use "false phrases about patriotism" to divert them from the ever-urgent tasks of class struggle and "distract their attention from the only genuine war of liberation, namely, a civil war against the bourgeoisie."[84] Only Lenin had urged Russia's workers to resist fighting in a war that was not their affair and not in their interest. The real enemy, he insisted, could never be the workers and peasants of Germany and Austria-Hungary who had been deceived by their governments and the socialist leaders they trusted into taking up arms in an "imperialist" war. "The conversion of the present imperialist war into a civil war is the only correct proletarian slogan," he told them. "It is only along this path that the proletariat will be able to shake off its dependence on the chauvinist bourgeoisie."[85]

From his refuge in Switzerland, where he had settled after the Okhrana had driven him from Russia and Poland, Lenin had looked on in disgust as Europe's leading socialists rushed to vote war credits, end labor protest, and urge workers to increase their productivity in the name of patriotism and national defense in August 1914. Unwavering in his certainty that the workers of Russia and Europe must be taught "to use weapons not against their brothers, the wage slaves in other countries, but against the reactionaries and bourgeois governments and parties of all countries,"[86] he insisted that Europe's proletarians now must be summoned anew to wage "a merciless struggle against chauvinism and . . . the bourgeoisie of all countries without exception."[87] They now must be taught that "there is no escape from barbarism, no possibility of progress in Europe, without a civil war for socialism."[88]

Throughout the Great War, Lenin remained unmoved by the efforts of Europe's socialists to justify their "defensist" position and refused to soften his uncompromising stand. These men, he insisted, had destroyed the international socialist movement by their "betrayal," but he found reason for optimism in the political, social, and economic crisis of unimagined proportions that the war they supported had produced. "The conflagration is spreading," he exulted. "The smoldering indignation of the masses, the vague yearning of society's downtrodden and ignorant strata for a kindly ('democratic') peace, the beginning of discontent among the 'lower classes'—all these are facts." Precisely these facts, he felt, made it clear that Europe and Russia had entered "a revolutionary situation." If "the proletariat's revolutionary consciousness and revolutionary determination could be sufficiently aroused" by the efforts of dedicated socialists, it might be possible to turn them toward "revolutionary action." Would this lead to revolution? Lenin insisted that "nobody can know" for certain, but he thought that the chances were better than ever before.[89] "Europe is pregnant with revolution," he insisted. "We of the older generation may not live to see the decisive battles of this coming revolution," he told a group of young Swiss workers at the beginning of 1917, but he promised that it would happen within their lifetimes.[90] Not quite two months later, with Nicholas's abdication less than twelve hours away, a young friend burst into Lenin's Zurich flat just as he and his wife and loyal assistant, Nadezhda Krupskaia, were finishing lunch.

"Haven't you heard anything about it?" the young man exclaimed. "There's a revolution in Russia!"[91]

The news took Lenin completely by surprise. For two days, he struggled to get his bearings, to construct from afar an accurate picture of the events in Petrograd. So much remained unknown, and so much of what little he did know seemed unclear. Of one thing he remained confident: "The new government," he wrote on March 4, "cannot give the peoples of Russia . . . either peace, bread, or full freedom." Its program and its promises must be viewed with the "deepest distrust," not only because it represented the bourgeoisie, but it was "composed of avowed advocates and supporters of the imperialist war." Thus, "the revolutionary proletariat" must continue its "fight for a democratic republic and socialism," he telegraphed a group of Bolsheviks about to leave Norway for Petrograd. "Only a workers' government that relies, first, on the overwhelming majority of the peasant population, the farm laborers and the poor peasants, and, second, on an alliance with the revolutionary workers of all countries in the war," Lenin concluded, "can give the people peace, bread, and full freedom."[92]

Such comments said very little about the actual state of affairs in Russia and even less about the tactics that the Bolsheviks must pursue. Desperate to know more than "the scanty information available in Switzerland" about Russia's new course, Lenin chafed to return to Petrograd, but could not.[93] With Russia's allies sworn to arrest him if he dared set foot on their soil, and threatened by arrest in Austro-German territories as the national of an enemy country, Lenin could only envy each and every one of his comrades who had been fortunate enough to be in neutral Sweden and Norway, or even in Siberia, at the time of Nicholas's abdication. For several weeks, he toyed with the most fantastic plots and schemes including, Krupskaia remembered, trying to make the journey from Zurich to Petrograd by airplane. Wisely, the sober and sensible Krupskaia dissuaded him from schemes that "one could think about only in a state of nocturnal semi-delirium."[94]

For twenty-five days, Lenin remained cut off from Russia's revolution, marooned in Zurich while others directed the course of events in Petrograd. Then, in mid-afternoon on March 27, he, Krupskaia, and twenty-seven fellow Bolsheviks boarded a train—what historians have since called the "sealed train"— on which the secretary of the Swiss Social-Democratic Party had arranged for them to travel through Germany to Scandinavia. Ready to support anyone or anything that might weaken the Allied war effort, the Germans had been only too glad to assist Lenin in returning to Petrograd, provided that he and his comrades refrained from spreading their revolutionary ideas among German workers as they passed through the German territory. They therefore insisted that no one could enter the Bolsheviks' railroad car, and none of them could leave. During the entire journey across German territory to the Baltic port of Sassnitz, Lenin and his comrades therefore would be kept apart, quarantined from any contact with German citizens or even German officials.[95] Unwilling to remain in the West a moment longer than necessary, Lenin was the first to sign the statement assuming "full political responsibility" for his "participation in the journey" that the Germans required from each traveler. As for those socialist "defensists" who had betrayed the Second International to support the bourgeoisie in the imperialist war that still raged in Europe, Lenin saw their moment of retribution approaching. "We are firmly convinced," he announced as he and Krupskaia said their last farewells, "that this filthy froth on the surface of the world labor movement will be soon swept away by the waves of revolution."[96]

After an uneventful journey across Germany and Sweden, Lenin and his fellow travelers entered Russian Finland on Easter Sunday, traveling in what Krupskaia called "those dear, wretched little third-class [Russian] railroad cars"[97] until they reached the Finno-Russian border at Beloostrov and began their triumphal railway journey to Petrograd. As they traveled, the city's Bolsheviks hurried frantically to prepare their best welcome, even though almost everything had closed down for the holiday. On a few hours' notice, thousands of workers, including some two thousand *Putilovtsy*, assembled on the platform of the Finland Station. With an honor guard of sailors from Kronstadt lining the platform, the crowd raised red banners and cheered while bands blared out the *Marseillaise*. On the evening of April 3, exactly one week before his forty-seventh birthday, Vladimir Ilich Lenin, the long-time exile and bitter foe of tsarism, capitalism, and imperialism, stepped down from the fifth car to stand once again among the Russians.[98]

Even before he reached the reception that loyal followers had arranged for him in the private waiting rooms of Russia's deposed emperors, Lenin made his views clear. "I know for certain that, when . . . [the spokesmen for the Provisional Government] offer you sweet speeches and great promises, they are deceiving you just as they are deceiving the entire Russian people," he said as he turned to address the honor guard of revolutionary sailors who had come from the Kronstadt naval base that lay just beyond Petrograd in the Gulf of Finland. "The people need peace! The people need bread! The people need land!" The revolution had begun, but a beginning was not enough. "Comrade sailors!" Lenin concluded. "We must struggle to the very end, to the final victory of the proletariat. Long Live the World Wide Socialist Revolution!"[99] It had taken Lenin only a moment to repudiate the moderate "defensist" position favored by those Bolsheviks who had arrived earlier on the front lines of Petrograd's revolution. No longer could Bolsheviks call for Russia's workers and peasants to "remain steadfastly at their posts, answering bullet with bullet and shell with shell" so long as their nation's enemies continued their advance.[100] Loyal Bolsheviks, Lenin insisted, must follow a different course.

The next day, Lenin detonated an ideological and political bombshell whose repercussions would reverberate across Russia for the next seven months. "Because of the capitalist nature of our present government," he insisted in the first article of the famed "April Theses" that he read to a group of Bolsheviks who were about to attend a meeting to discuss reunification with the Mensheviks, the war remained an "unconditionally predatory imperialist war." The masses therefore must be shown, "thoroughly, persistently, and patiently . . . [that], without the overthrow of capitalism, it . . . [would be] *impossible* to end the war with a truly democratic, unoppressive peace," and that support for Russia's war effort in any form remained as "intolerable" as ever. Neither the Provisional Government nor the war deserved the support of Russia's workers and soldiers. Only a "revolutionary war," a war against capitalism, Lenin insisted, could be supported by the "class-conscious proletariat." Loyal Bolsheviks must oppose the Provisional Government at every opportunity and dedicate their energies to transferring all power to the Soviet in whose hands all power and authority in Russia must rest.[101]

Lenin's April Theses stunned his fellow Bolsheviks. Such uncompromising tactics seemed clearly out of keeping with the obvious minority status that enabled their party to command the vote of merely one delegate in twenty at the Petrograd Soviet,[102] all the more so since their potential allies on the Left thought Lenin's speech the "ravings of a madman" and dismissed his urgings as pure "clap-trap." Outraged Mensheviks and Socialist Revolutionaries assured each other that Lenin was a "has-been," whose compulsion to speak the "super-annuated truths of primitive anarchism" showed the sterility of his views.[103] Nor were only Socialist Revolutionaries and Mensheviks outraged. Thirteen out of the sixteen members of the committee that directed the Bolsheviks' activities in Petrograd rejected Lenin's views and made their vote public, while the Bolshevik newspaper *Pravda, The Truth*, hastily explained to its readers that the April Theses represented Lenin's "personal opinion" only, and assured them that "Lenin's general scheme appears to us unacceptable."[104]

Yet, as Sukhanov later remarked, "Lenin displayed such amazing force, such superhuman power of attack,"[105] that he often bent the most resolute opponents to his will. During the next fortnight, Lenin pressed his views upon comrades anxious to seek common cause with the government, arguing always that there could be no compromise and no retreat. The stormy debate about the meaning of the diplomatic note that Miliukov had sent to Russia's allies on April 18 added fuel to the fire and won a more sympathetic hearing for Lenin's "primitive anarchism," as he insisted that only "the revolutionary proletariat, together with a revolutionary army in the person of the Soviet of Workers' and Soldiers' Deputies . . . [could] bring the war to a rapid end by a truly democratic peace." When he called for "peaceful" mass protests and demonstrations to demand Miliukov's resignation on April 20, Petrograd's masses moved readily into the streets.[106]

By mid-afternoon on April 21, a mass of workers from the Vyborg and Petrograd districts with banners proclaiming "Down With Miliukov!" and "Down with a Policy of Conquest!" joined some twenty-five thousand armed soldiers on the square in front of the Provisional Government's headquarters in the Mariinskii Palace. "I spent two years in the trenches and I know what war is," a passing student heard a worker tell a bystander who urged support for the war. "If the bourgeoisie wants it, then let them go and fight it. We've had enough. We've already spilled enough blood for them. Take your fat belly, go there, and see how you like it. Then you can go around shouting: 'Long Live the War!' "[107] Others who shared such views moved to the Nevskii Prospekt, stern and grim in their opposition to the Provisional Government's foreign policy. "Their faces amazed me," a journalist wrote. "All those thousands had but one face, the stunned, ecstatic face of the early Christian monks."[108] At the same time, other crowds carried banners bearing the slogans: "Long Live Miliukov!" "Confidence in Miliukov!"[109] Danger lurked in Petrograd's streets and courtyards as these groups confronted each other. For a brief moment, the threat of civil conflict hung over the city.

"I did not fear for Miliukov so much as I feared for Russia," Miliukov wrote when he recalled those days and the crowds that demanded his resignation.[110] Certainly, nearly everyone feared something. Quite rightly, the Provisional Government feared for its safety, and Prime Minister Prince Georgii Lvov hurried to announce to the Soviet that "since we no longer have your trust, we are ready to resign."[111] The Soviet feared equally that the government might attack the revolution from the right and plunge Russia into civil war. Perhaps most immediately, the mere presence of armed men in large numbers threatened everyone's safety, and a number of Petrograders met untimely deaths during the so-called "April Crisis" when undisciplined armed civilians fired their weapons indiscriminately. Clearly, armed workers must be kept under control, yet they must remain under arms to defend the revolution against attack from the Provisional Government. "Only the armed working class," a proclamation "To All Workers of Petrograd" insisted the day after the April disorders ended, "can be the actual defenders of the freedom we have fought for. . . . The business of the Revolution is not yet finished," the proclamation continued. "The Red Guard, as the true sentry of the Revolution, will be its defense in the hour of need."[112]

The "Red Guard" of armed and organized revolutionary factory workers would not become a major factor in determining the direction of Russia's revolution for some weeks yet, for Provisional Government and Soviet, each hoping not to provoke the other's all-out attack, still sought ground upon which to compromise. "Experience has shown that the Soviet, without merging with the government, retains the greatest possible influence upon the most inflammable section of the population," Tseretelli insisted. "So long as we maintain this position," he warned, "we shall be able not only to check the growth of extremist tendencies in the masses, but also to exercise a real influence upon the government in the direction of a democratization of its policies."[113] Therefore, the Soviet's Executive Committee much preferred getting the Provisional Government to amend Miliukov's note rather than take the government's responsibilities upon themselves. "Who will take the place of the government [if it resigns]?" one speaker asked the Soviet. "We? But our hands tremble."[114]

The Soviet's reluctance to press for the government's resignation made it possible for its Executive Committee and the Provisional Government to avoid open conflict during the April Crisis. But this no more than masked the symptoms of the disease that threatened the government's very life. Behind the tattered curtain drawn to cover the conflict between Soviet and government, the most deadly dangers still remained. "Before Russia rises the terrible apparition of civil war and anarchy," the Provisional Government reported as it made a final plea "For the Support and Cooperation of All the Vital Forces in the Nation" later that week.[115] Still, Miliukov would not give way, and stubbornly insisted in a *New York Times* interview that "our policy remains unchanged. We have conceded nothing," even though his government had just issued another note to explain that his earlier reference to "guarantees and sanctions" meant "the limitations of armaments, international tribunals, and so forth," not "annexations and indemnities."[116]

Clearly, the situation demanded major alterations. "The salvation of Russia is at stake," the newspaper *Novoe vremia, New Times,* warned its readers. "A choice faces us: either, following the example of other revolutions, to come— via anarchy—to the restoration of the old despotism, or immediately to turn the events onto the road which reason, conscience, and duty demand."[117] That course, Kerenskii and Prince Lvov now agreed, must be for "representatives of the toiling democracy . . . [to] no longer stand aside from responsible participation in ruling the state."[118] The Soviet must relieve Kerenskii from his lonely outpost as the sole hostage of the revolution in the Provisional Government. Other socialists must join him in a true coalition government in which "the personnel of the Russian diplomatic corps abroad must undergo a radical change."[119]

Neither the moderate Miliukov nor the more conservative Guchkov could go so far as their colleagues in accepting the Soviet's new role, for they felt that to involve the Soviet more directly in Russia's government could only spawn further crises. "Our fatherland is on the verge of ruin," Guchkov exclaimed as he warned an anniversary convocation of Russia's Duma deputies of a "mortal ailment which undermines the very life of the country." The wounding process of disintegration, he went on, had far outstripped the healing process of consolidating Russia's "new social bonds," leaving the nation in a "sickly state of fermentation." The cure demanded the sternest measures. "Only a strong government, homogeneous within itself and in union with the nation," he concluded, "can create the mighty and vital creative center, in which lies the salvation of the country."[120]

Guchkov spoke with fervor, but mainly in generalities. Miliukov saw the situation in the much clearer terms of two very stark and simple choices. "On the one hand," he remembered, "one had the possibility of using the government's firm authority, but, in that case, it would be necessary to do without Kerenskii . . . and be ready to oppose the Soviet's active efforts to seize power. On the other hand, one could agree to a coalition government, . . . the further weakening of the government's authority, and the disintegration of the state."[121] To his mind, only the first course offered any chance of solving Russia's problems, but most of his colleagues had already chosen the path of coalition. For Miliukov, the burden of that course was too heavy, and his conflict with the Soviet too bitter. "For the first time, Miliukov seemed to be shaken in his courageous optimism and his confidence and will to go on," Paléologue reported. "The dull tones of his voice and his haggard look showed all too clearly the distress he held within himself."[122] Democratic Russia's first foreign minister had not yet resigned, but the forces pressing upon him had become all but impossible to resist.

The men who had spoken for Russia's allies throughout the war shared Miliukov's view. "The first weeks of the revolution have seen a crescendo of disorder," France's General Staff concluded from Paléologue's reports. "It has been a period of naive intoxications and of brutal awakenings."[123] Paléologue now reported that "anarchy is spreading through all of Russia and will paralyse her for a long time,"[124] an opinion that Ambassador Buchanan endorsed wholeheartedly. "Russia is not ripe for a purely democratic form of government," Buchanan warned the Foreign Office, and predicted that "a series of revolutions and counter-revolutions" probably would be the lot of England's ally for some time to come. "The Russian idea of liberty," he added in a private letter, "is to take things easy, to claim double wages, to demonstrate in the streets, and to waste time in talking and in passing resolutions at public meetings."[125] To such sane and sober men, Russia's immediate future looked hopelessly grim, despite Prince Lvov's ecstatic reassurances that "the great Russian revolution is truly miraculous in its majestic, quiet progress" and that "every day that passes renews the belief in the inexhaustible creative power of the Russian people."[126]

Guchkov so disagreed with Lvov's saccharine portrayal of the conditions under which Russia must be ruled that he abruptly left the government on May 1, despite Miliukov's pleas to wait until they could resign together. Bluntly criticizing the conditions "which threaten the defense, freedom, and even existence of Russia with fatal consequences," Guchkov insisted that he no longer would "share the responsibilities for the grievous sin which is being carried on against the fatherland."[127] Now, if Soviet and Provisional Government failed to reach an agreement quickly, Russia would have no government. Convinced that "only with the active participation of the democratic forces will it be possible to form a new government . . . capable of eliminating the economic disorganization, organizing the defenses of the front, and hastening the conclusion of peace on an international scale," the Soviet voted on May 2 to reverse its prohibition against its members' participation in the cabinct after a debate that lasted well into the early hours of the morning.[128]

Only hours after the Soviet's vote, its delegates met with Russia's ministers to form a coalition government. The negotiations "grew more complicated with every hour that passed," one of the negotiators later wrote. "Each step of the negotiations, every proposal, every correction made a break in the negotiations necessary so that the members of the Government and the representatives of the Committee could reach an agreement among themselves. . . . They could not find ministers for some portfolios, nor portfolios for some ministers." For three days and four nights, the meetings continued, with enemies, rivals, and even friends shouting and cursing in a desperate effort to reach agreement. Viktor Chernov struggled bitterly to take the Ministry of Agriculture away from Andrei Shingarev, a man of respectable and respected views and the Kadets' leading agricultural expert. No ministry seemed quite right for Tseretelli, although both Government and Soviet insisted he must havc a place. "No longer were questions even discussed," one onlooker remembered. "Everyone simply spoke—or, more precisely, shouted—from his corner." Then, when Tseretelli and Chkheidze, who had been friends for years, began to shout at each other, Kerenskii, in another of the dramatic gestures for which he was becoming famous, burst in from another room to announce that a solution had been found. He proposed no one whose candidacy had not already been disputed at length, but his decisive statement changed the mood. Suddenly, men wanted desperately to agree, and by late evening on May 5, Russia's Provisional Government had its first coalition cabinet, with Prince Lvov again named to stand at its head.[129] "I did not withdraw, but was ousted," Miliukov remarked privately. "I stood at my post until a large majority of my comrades told me that I ought to go."[130]

Miliukov's replacement as foreign minister by the young Ukrainian sugar magnate Mikhail Tereshchenko, a man whose energy and talent could not offset his utter inexperience in foreign affairs, signaled the final failure of the old tsarist foreign policies that Russia's first Provisional Government had struggled so desperately to perpetuate. "Revolutionary democracy expected Tereshchenko to . . . revise the diplomatic heritage of autocracy," Chernov explained.[131] Russians now demanded that their nation's foreign policy reflect the new sense of national purpose that had been forged in the turmoil of the February Revolution. "Deep down in their hearts, almost no one thought that the new order of February marked the end of the revolution," Paustovskii remembered. "The confused, almost unreal, condition of the country could not continue for very long," he went on. "The people's life demanded clearness of purpose and the precise application of effort."[132] In foreign policy, such feelings demanded that Russia's leaders pay serious attention to the Soviet's demand that their government end the war "without annexations or indemnities, and based on the rights of nations to decide their own affairs." The Western Allies, Petrograd's Soviet now insisted, must revise their war aims in accordance with Russia's example "in the firm conviction that the fall of the regime of Tsarism in Russia and the consolidation of democratic principles in our internal and external policy will create in the Allied democracies new aspirations toward a stable peace and the brotherhood of nations."[133]

Other men guided by other policies now set out to lead Russia forward, and socialists, for so long Tsarist Russia's most notorious political criminals, now headed some of the new government's most powerful ministries. As Russia's new minister of war and navy, Aleksandr Kerenskii quickly emerged as the coalition government's leading figure, the one man apparently able to control the awesome forces that Petrograd's revolutionary crowd had unleashed. By the spring of 1917, he had come to represent the revolution in the minds of many Russians, and he had fostered that image above all others, proclaiming "I am sent by the Revolution!" as he marched time and again through angry crowds, his left arm upraised "like the flaming torch of revolutionary justice," one commentator remembered, to rescue men like Protopopov and Sukhomlinov from the masses' wrath. Perhaps more than anyone in Russia, this unimpressive looking man, who hurried from meeting to meeting looking for all the world like a wounded hero because he carried his right arm in a sling to relieve the pain of chronic bursitis, thrived in the revolution's atmosphere of crisis.[134]

Kerenskii's immense personal charisma and dazzling oratorical gifts thus became a force in shaping the revolution's course, for none could hear him speak and remain untouched by the force of his words. "A simple reading of his speeches offers no clear idea of his eloquence, for his physical personality is perhaps the most effective element of the power he has to fascinate a crowd," Paléologue once wrote in describing Kerenskii's effect. "From time to time a mysterious or prophetically apocalyptic inspiration transfigures the speaker and radiates around him in magnetic waves. The burning intensity of his features, the sometimes halting, sometimes passionate flow of his words, the sudden fits and starts of his thoughts, the somnambulistic deliberation of his gestures, the fixed intensity of his pupils, the twitching of his lips, and the bristling of his hair all make him look like a monomaniac or one possessed."[135] Others shared Paléologue's estimate. Britain's Vice-Consul Bruce Lockhart once saw a general "who had served the Tsar all his life and who hated the revolution as the pest, weep like a child" when Kerenskii spoke at Moscow's Bolshoi Theater. In later years, Lockhart remembered Kerenskii's "epic performance" as being "more impressive in its emotional reactions than any speech of Hitler or of any orator I have ever heard."[136]

Kerenskii brought these gifts, not a well-defined program, to his office when he replaced Guchkov as Russia's minister of war and navy at the beginning of May. With them, he urged, cajoled, demanded, even terrorized friend and foe to do his bidding as he took command of Russia's destiny in a cabinet in which the unassertive and nonpolitical Prince Lvov had become increasingly unable to assert his authority as minister-president and minister of internal affairs. With charisma as his only offering and words as his only instruments, Kerenskii set out to restore the army's fighting capacity, a task of the greatest and gravest dimensions. Only a leader with the utmost personal magnetism could even contemplate such a task. "For the army to fight again meant to conquer anew the animal in man, to find anew some sort of unquestionable slogan of war that would make it possible again for everybody to look death in the face calmly and unflinchingly," Kerenskii later explained. "For the sake of the nation's life it was necessary to restore the army's will to die."[137]

Guchkov had lost precious months precisely because he could do none of these things, and Kerenskii launched his campaign to restore the will and the discipline of Russia's crumbling armies even before Guchkov left office. "Is it really possible that Free Russia is only a country of mutinous slaves?" he exclaimed at a meeting of officers summoned to hear Guchkov's resignation.[138] "You are the freest soldiers in the world," he exhorted units at the front. "Our army under the monarchy accomplished heroic deeds. Will it be a flock of sheep under the republic?"[139] For a brief moment, Russians at home and at the front responded to his words. As Kerenskii took the helm, thoughtful Russians felt a dim ray of hope that his fiery words could shape order out of chaos.

CHAPTER XIV

"To Keep the Swing from Going Over the Top"

From the moment they drove Nicholas and Aleksandra from the throne, Russians sensed that their nation's fate was somehow different, that destiny had singled them out to follow an untried path. "With the 'broad Russian nature' experiments are possible that could not be tried in Western countries," one private soldier exclaimed when the Provisional Government first took office. "Russia will find a Dostoevskii, not a Napoleon!"[1] Full of hopes, needs, and dreams to be fulfilled, Russians abandoned themselves to an orgy of meetings, endless debates, and meaningless speeches that spring. "Across the entire country, an unending, disorderly meeting went on day and night," an eyewitness wrote, as his countrymen assembled in their factories, villages, and trenches to speak the words they had been forbidden to speak for so long.[2] "I found 'meetings' going on everywhere," Paléologue reported from Petrograd in early March. "One of the group gets up on a stone, or a bench, or a mound of snow, and talks on and on, gesturing wildly. The crowd gazes fixedly at the orator and listens in a kind of trance," he continued in an attempt to describe the captivating magic of the moment. "As soon as he finishes, another takes his place and immediately gets the same fervent, silent, and concentrated attention."[3] The incredibly rich, ever-changing Russian language even created the new verb *mitingovat'*—"to meetingify"—to describe the phenomenon as it spread across the nation.[4] In Moscow, "the entire city was on its feet," Paustovskii later said. "Their voices rasping, people spoke at meetings the whole night through, sleepily loitered in the streets, then sat down and argued in public squares or on the sidewalks."[5] Moscow's thieves held a huge meeting at Nikitin's Circus to discuss "liberty, equality, and brotherhood," and, a few days later, the city's whores did the same.[6] Days and nights blended into a continuum of thoughtless, timeless confusion. "Yellow street lights burned day and night," Paustovskii remembered. "People simply forgot to switch them off."[7]

Russians everywhere shared a sense that everything was possible and that nothing remained out of reach. "Everyone is overcome by a recognition that a miracle has occurred, and, consequently, that more miracles will follow," Aleksandr Blok wrote just before Easter. "Freedom is extraordinarily majestic!" he exclaimed, as he told his mother of his "extraordinary consciousness that everything [now] is possible."[8] "Events still have not even begun to unfold in their entirety," he promised a week later. "Almost everyone feels this."[9] Yet few had any sense of the stern discipline and outright self-denial that Russia's new course demanded. "Such things could not be born on their own to the clash of cymbals and the rapturous cries of the citizenry," Paustovskii explained. "The establishment of justice and freedom was going to require a great deal of hard work and even some brutality."[10] "What was wanted," General Knox remarked acidly, "was a little narrow common sense."[11]

Neither in town nor country did the common sense that Knox thought so necessary dampen the revolution's first euphoric weeks. "We have suffered 300 years of slavery," one Petrograd soldier exclaimed the day after Nicholas abdicated. "You cannot grudge us a single week of holiday!"[12] Everywhere, Russia's peasants and workers wanted to taste the revolution's fruits quickly, even though more sober men knew that limits must be imposed and Russia's path more clearly defined. Wracked by domestic crises, Russia remained at war. A powerful enemy stood at her very gates, threatening at any moment to burst into lands that had not felt an invader's heel since the time of Napoleon. These dangers had to be faced, and the terrible reality of the war recognized, without denying the nation's toiling masses the immediate gratification of their most cherished revolutionary expectations. Russia's Provisional Government, the poet-briefly-turned-revolutionary Aleksandr Blok insisted, must at all costs find a way "to keep the swing from going over the top, yet not allow the trajectory to decrease." It must guide Russia "along the edge of a precipice, not letting it fall or lose itself."[13]

Euphoric celebrations of freedom elevated the swing's trajectory to a dangerous height, but only briefly masked the desperate crises that loomed on Russia's domestic horizon. As Miliukov and Guchkov left office at the beginning of May, dire shortages showed Russia's working men and women how illusory their newly-won freedom could become. In the coal mines and oilfields of the south, the cotton mills of Ivanovo and Orekhovo-Zuevo, and the weapons factories of Moscow and Petrograd, hundreds of thousands of mill hands and miners struggled not to starve. "We have to pay four, five, six, or even more rubles for what we could buy for a ruble [in 1914]," the workers of Ivanovo wrote in a collective complaint to their new government that spring.[14] Suddenly, it began to matter less that cloth cost ten times what it had in 1914, or that kitchen utensils cost even more.[15] Workers' desperate struggle against hunger eclipsed all other concerns as they found that even if they spent every kopek they earned on food, they could buy only about half of the calories that conservative estimates set as an adult worker's minimum daily needs. Never in Russia's history had inflation cut so deeply into the diet of the lower classes. In August 1914, a worker's ruble had bought 14,000 calories; in May 1917, it bought precisely 168![16]

Not only did the price of food soar beyond workers' reach in the spring of 1917, but even those who could scrape together the necessary cash found food in desperately short supply. Not since the terrible plague years of the seventeenth century had Russia's cities known such dire food shortages. "It is frightful to remember that, at the beginning of March, there were moments when Petrograd and Moscow had bread enough to last only for a few days, and there were parts of the front, with hundreds of thousands of soldiers, where there were only enough reserves of bread to last for half a day," one group of experts reported.[17] Before the revolution was a month old, Russia's desperate minister of agriculture gave in to the "bitter necessity" of rationing bread in Petrograd to one and a half pounds a day for men and women engaged in manual labor, while the rest of the city's adults received only two-thirds of that amount. Bread lines spread again across the city. "There is no street, not even an obscure alley, into which these endless lines for black bread do not reach," the Bolsheviks' *Pravda* reported.[18] As bread lines grew, the government restricted workers' food consumption still more. With sugar supplies almost exhausted, Petrograders received two and a half ounces a week, only to see that meager quantity reduced several times before summer's end.[19] "Hunger is knocking at the door," one city council member reported. "Scurvy poses a serious threat. The children's death rate is rising every day. Catastrophe looms ahead."[20] As a last resort, workers hurried to plant communal kitchen gardens in any odd bit of vacant land they could find, hoping that these would yield up vital green vegetables and root crops by summer's end.[21]

In the few weeks after the Provisional Government took command of Russia's affairs, crowds of hungry workers thus swelled into legions of starving men, women, and children. At the same time, soaring housing costs and shortages of the materials needed for repairs condemned much of Russia's urban work force to even more wretched shelter than they had known at the war's beginning. As more than three million refugees from the towns and cities of Poland and western Russia flooded into already crowded city workers' dwellings, the situation grew even worse.[22] "What one sees in the lodgings of the urban poor completely defies all description," one observer wrote. "The population is simply sinking into dirt and filth."[23] Filth brought disease. Just as inflation drove the prices of medicine to more than fifty times their prewar levels, epidemics of small pox, scarlet fever, and typhus struck.[24] For Russia's factory workers, their country's first months of freedom had brought misery even greater than that which had driven them into the streets in February.

While factory workers struggled to escape the tightening grip of inflation, the Provisional Government, burdened with war expenses that approached sixty-seven million rubles a day and a national debt that cost almost two billion rubles each year in interest,[25] tried to respond to the slogans that had won the workers' loyalty during the February Revolution. At first, it tended to resist far-reaching changes in wartime, and tried to postpone meeting the workers' demands. Quickly, proletarian patience wore thin. "Comrades! Where are those slogans that were inscribed on our banners?" one Moscow worker exclaimed. "Where are the demands and what are their results?"[26] All across Russia, workers insisted that their new government share their sense of urgency. "We ask you, Mr. Kerenskii, to keep in mind that we . . . have addressed our complaints to you," one group of miners wrote in the middle of March, "[because] life has become impossible."[27] In those days, workers called for better working conditions, a voice in the management of their factories, social security, and higher pay, but most of all they sought relief from the grinding routine of long hours of heavy labor. For every mill hand who called for higher wages when Russia's work force returned to their factories in mid-March, three demanded an eight-hour day.[28] "Eight hours for work, eight hours for sleep, and eight hours to spend in relaxation or as one sees fit," Moscow's Bolsheviks insisted. "These [last] eight hours are needed by every worker so that he or she can become a responsible citizen."[29]

During the first days of the revolution, a noted historian once wrote, "the demand for an eight-hour day had an almost sacramental character."[30] Workers all across Russia gave it their ardent support, and saw its realization as the first expression of their revolutionary victory. Within a week of Nicholas's abdication, Petrograd's workers simply began to lay down their tools and walk away from their benches after eight hours' work. Anxious to resume military production, factory owners hastily agreed to their demand for "an eight-hour working day, applicable to all shifts," as "the only measure capable of bringing calm to the turbulent sea" of worker protest.[31] By appealing to workers' sense of patriotic duty, factory managers and government officials in many cases convinced their labor force to work longer hours and merely to accept the eight-hour day as a victorious achievement of the revolution in principle. The men and women in Petrograd's Cartridge Works therefore recognized "the eight-hour day as a fundamental principle," but agreed to work longer hours "in view of the danger that threatens . . . our brother soldiers at the front."[32] The city's women, condemned to stand long hours in lines to buy meager rations for their families, proved far more militant defenders of their victory and refused to work overtime under any conditions.[33] All overtime work, the largely female work force at the Nevskii Shoe Factory insisted, must be abolished "forever," while the women of Petrograd's Moscow District insisted that it must be "absolutely forbidden."[34]

Although Petrograd's factory owners regarded the eight-hour day as an "historically necessary measure . . . capable of ensuring the future spiritual development of the working class,"[35] other Russian industrialists proved more obstinate, even to the point of insisting that "this demand of the workers be replaced by . . . a longer working day."[36] Less than a fortnight later, Moscow's workers joined their brother proletarians in Petrograd and instituted the eight-hour day despite angry protests from the city's industrialists. "Citizens!" the Moscow Soviet of Workers Deputies announced, "It is vital for all people [of Russia] that the worker not toil for more than eight hours. . . . We are forging in the fires [of revolution] a new and free Russia," they concluded. "Support the working class in its struggle for an eight hour working day!"[37] From Moscow, the movement spread outward across Russia: to Ivanovo, Nizhnii-Novgorod, Kiev, Kazan, Kharkov, Rostov-on-the-Don, Ekaterinburg, and Perm before the end of the month, and into the mines and foundries of the Donbas, Krivoi Rog, and the Urals by the middle of April. Along with a democratic republic, the Petrograd Soviet's *Izvestiia* explained, "an eight-hour working day . . . [is an] essential condition for making the worker feel that he is a human being, and a citizen."[38] By the time Miliukov and Guchkov left office, even the oil fields of Baku and such distant Siberian and Central Asian centers as Irkutsk, Omsk, and Tashkent had taken their stand. "The spontaneous establishment of the eight-hour working day in almost all factories and mills," the Bolsheviks' *Pravda* exulted on May 6th, "is a vital victory for the Russian Revolution."[39]

No sooner had Russia's factory workers won this "vital victory" over their employers than they launched an equally aggressive campaign for higher wages.[40] Insisting that each of them must be able "to live in a fashion suitable for a worker and a free citizen," Russia's proletarians demanded that even the least able among them receive a minimum wage that would guarantee sufficient food, clothing, and shelter.[41] Again, the workers of Petrograd led the way, in some cases increasing their wages by as much as 50 percent within a fortnight after Nicholas's government collapsed. A month later, the city's workers won a minimum daily wage of five rubles for men and four for women, although skilled workers received more than twice that amount. Quickly, the movement spread. All across Russia, wage increases of as much as 100 percent became common. In some cases, demands went far beyond even that amount, with workers at the huge Nikolskaia textile mills in Orekhovo-Zuevo leading the way in insisting that their wages be raised sixfold in the month of May alone.[42]

Workers' ability to force such concessions very soon proved to be a dangerous two-edged sword. Astronomical wage increases generated astronomical production costs, and prices climbed as never before. Well before the middle of 1917, industrialists calculated that higher wages and higher costs for raw materials and fuel (themselves driven skyward by inflated wage agreements) would absorb many times their gross profits and could only be offset by rapid price increases that quickly robbed the workers of all their gains.[43] "The bony hand of hunger," Moscow's millionaire publicist and political activist Pavel Riabushinskii warned in a remark often cited at the time for its supposed callousness, "would grasp by the throat the members of the different committees and Soviets" who continued to demand higher wages with no thought for their broader impact upon the nation's economy.[44] All too quickly, the elementary laws of economics proved Riabushinskii correct. By midsummer, workers began to negotiate payments of food and clothing in place of higher wages in a last desperate effort to escape the starvation that lay in inflation's deadly coils.[45]

Riabushinskii and his colleagues foresaw only too clearly the consequences of the workers' arrogant demands, but they dared not oppose proletarian men and women still basking in the first flush of revolutionary victory, especially because Russia's workers had organized factory committees during March and April which gave them a collective authority that far outmatched the strength of their industrialist adversaries. These appealed to Russia's workers far more than trade unions, which, some radicals insisted, "only hold us back from the struggle [against capitalism],"[46] and it was in these newly-formed committees that Lenin's Bolsheviks found their easiest access to the workers' ranks. Even though the Provisional Government tried to limit their numbers, more than 300,000 workers organized almost four hundred committees in Petrograd alone.[47] Quickly, these became involved in all parts of workers' lives, arbitrating labor disputes, organizing food supplies, regulating working conditions, and struggling valiantly against drunkenness, that curse of Russia's working class which drove men to consume not only alcohol, but "methylated spirits, varnish and all kinds of other substitutes," lowered their productivity, and "suffocated" their revolutionary class consciousness. Beyond these critical problems, the committees' concerns ranged downward into the realm of trivia, even to the point of discussing workers' demands that scented soap be provided in factory washrooms.[48]

While workers had explained their demands for shorter hours and higher pay in terms of self-interest, they justified their factory committees in terms of self-defense and national concern. "The factory committees at the present time willy-nilly are forced to interfere in the economic life of their mills, because otherwise they would have stopped long ago," one worker explained. "It is up to the workers to demonstrate initiative in the areas [of procuring raw materials and fuel] where the industrialist-enterprisers do not."[49] After noting "the utterly wretched mismanagement of the factory administration," workers from Moscow's famous "Dinamo" factory announced that they had been driven to interfere in the factory's management because "the closing of the factory at the present moment would pose an irreparable threat to the national defense."[50] "The disorganization and dislocation of industry and transport," the workers at the Sestroetsk Weapons Plant added, "is the result of the unsystematic and anarchistic capitalistic structure."[51] "The path of escape from disaster," the First Petrograd Conference of Factory and Mill Committees concluded, echoing the views of the Bolshevik majority on its central committee, "lies only in the establishment of effective worker control over the production and distribution of [manufactured] goods."[52]

The spring of 1917 was not the first time that Russia's proletarians had spoken of establishing "worker control," but it was the first time they had found the instruments to make it effective. Like the factory committees, the first detachments of revolutionary workers' militia were born during the turbulent days of late February and early March when Petrograd's insurgents seized weapons and ammunition from the city's arsenals. Before the middle of March, the ranks of the workers' militia held some twenty thousand men, all armed and sworn to defend the revolution as each saw fit. As they took upon themselves the task of restoring order in Russia's cities, they insisted that theirs must be the new authority in revolutionary Russia to protect their fellow proletarians against their class adversaries.[53] Until "new cadres of the revolutionary militia" were organized and armed, the Bolsheviks' *Pravda* insisted, "the revolution [would] not be completed," and its very existence threatened. "It is impossible to wait," one of the Bolsheviks' military experts insisted in mid-March, as he warned *Pravda*'s readers that "reaction never sleeps," and that the revolution could never be safe until it was protected by "a genuine people's, that is, workers' army."[54] Vladimir Bonch-Bruevich, close friend of Lenin and brother of the tsarist general who became a Red Army commander, named this "workers' army" the "Red Guard of the Proletariat" a few days later.[55]

For the next month, workers seemed more occupied with questions of wages and hours until the April Crisis showed revolutionary leaders how seriously the men and women in Russia's factories had taken the idea of establishing Red Guards units. "Only the armed working class can be the actual defenders of the freedom we have fought for," a maverick Menshevik printer insisted. The Red Guard must become the "strong battle organization that must be the bulwark of the revolutionary working class."[56] Without the Red Guard to "struggle against the counter-revolutionary intrigues of the ruling classes [and] defend with arms in its hands all conquests of the working class," another Bolshevik warned, the revolution stood in mortal danger.[57] Throughout the spring and summer, working men (and a few women) from the textile factories of Ivanovo and the vast mills at Orekhovo-Zuevo joined their proletarian brothers in Petrograd and Moscow in arming themselves despite government opposition and the urgent protests of moderates, who denounced the Red Guards as "unnecessary and harmful."[58]

Debates about the Red Guards' value as a revolutionary instrument did not prevent these armed "insignificant fellows,"[59] as a labor organizer once called them, from supporting the protests of Russia's workers against the Provisional Government's failure to satisfy their revolutionary expectations. One worker later called them "our eyes, which helped us become the masters of the factory,"[60] and Red Guards units unquestionably gave revolutionary workers the means to impose their will upon arrogant managers and factory owners who had dominated their lives for so long. Especially during the early months of the revolution, militant workers drove long-hated managers, engineers, experienced technicians, and foremen from Russia's mills and mines and gave their places to men more ready to support their cause. As one association of coal mine owners reported, such workers began by using "blackmail and threats of violence . . . to extort large sums of money" which they called "back pay," and there was at least one report that workers ordered a factory's directors to fill up several large sacks with cash—calling it "war profits of three years"—or else be put into the sacks themselves and thrown into the river.[61] Soon, workers demanded control of day-to-day operations, even to the point of insisting that mines and factories be run by "an elected body of workers."[62] Only such bodies, the First Petrograd Conference of Factory Committees insisted that spring, could successfully "interfere in the economic life of revolutionary Russia and, by doing so, extricate her from the blind alley into which she is becoming more deeply enmeshed every day. . . . Those who stand at their benches," these delegates concluded, "must save revolutionary Russia."[63]

The First Petrograd Conference's vow to save "revolutionary Russia" did not extend to defending the Provisional Government against labor unrest. February's huge wave of labor protest had crested on the eve of the revolution, with ninety thousand Petrograd defense workers going on strike on February 14 alone.[64] Almost no workers went on strike during the first part of March, but as the first euphoric visions of the February Revolution faded, their protests mounted once again. "The proletariat was the chief motive force of the revolution," Trotskii wrote. "At the same time the revolution was giving shape to the proletariat."[65] As had been the case before the revolution, the most skilled workers in the highest-paid industries led the way, ready and willing to risk a temporary loss of earnings in order to win still higher wages and better working conditions.[66] Before the end of March, strikes hit at least eighteen of Russia's provinces, and quickly spread during April. By the end of May, the number of affected provinces more than doubled as the number of men and women on strike again approached 100,000. In June, the number doubled again, and labor protest even moved into such outlying Central Asian regions as Kokand.

Workers' victories faded as the strikes spread further. Russia's working men and women could claim victory in nine out of every ten strikes in March and April, but by July, management had begun to win three encounters out of every ten, and September saw that figure double.[67] Frustrated by factory owners' growing intransigence, workers' protests turned violent. "Strikes were especially stormy among the more backward and exploited groups of workers," Trotskii remembered, as laundresses, dyers, woodworkers, and shoemakers followed their more affluent brothers from the highly-skilled metal-working trades into the arena of labor conflict.[68] That spring and summer, Sukhanov reported "lynchings, the looting of homes and shops, [and] violence against officers, provincial authorities, and private persons," as he observed "acts of revenge numbering in the tens and hundreds" from his vantage point on the Petrograd Soviet's Executive Committee.[69] People now spoke readily of anarchy, fearful that the revolution had slipped out of control. "Russia has turned into a sort of madhouse," the Kadets' leading newspaper announced sadly, while a representative of thirty thousand *Putilovtsy* exclaimed: "We will have our way! *Let there be no bourgeoisie!*"[70] "Having forever left the shores of the bourgeois world," Viktor Chernov explained as he recalled the turbulent days of summer 1917, "a dangerous exaltation was growing in the workers." Fearfully, Chernov, the leader of the Socialist Revolutionary Party, saw "a peculiar heroism of despair" grow among his nation's toilers. "Although despair is a bad counselor," he warned, "it is irresistible because it creates fanatics."[71] Before summer's end, the most ominous warning of all came from the owners of the Donbas mines, upon whose dwindling production of coal Russia's factories and mills depended for fuel: "General madness," they reported, "is about to grip the masses."[72]

Perhaps the first signs of that "madness"—an elemental rage against all property owners and forms of established authority—stirred among the faceless tens of millions of men and women who remained in the dark reaches of Russia's countryside. Ever since the late Middle Ages, that anger—raw in nature and elemental in its fury—had lain concealed beneath the filth and poverty that blanketed Russia's villages. For decades, even centuries, it lay dormant, only to burst forth in times of political tension or social stress. During those terrible upheavals known as the Time of Troubles that had seized Russia at the end of the sixteenth century, when the nation had no tsar and lay crushed beneath the military occupation of the Poles and the Swedes, peasant anger had flared against all who tried to sit upon Moscow's throne. The late 1660s and early 1670s had seen that anger burst forth again, as Stenka Razin drew more than a hundred thousand of Russia's oppressed peasants and dissident Old Believers into the ranks of his Cossack army to wage the first great peasant war against the Romanovs' authority.[73]

Only after several years of struggle had the Romanovs' generals crushed Razin's protest and driven the peasants' anger into hiding once again beneath the deceptive calm of rural life. It surfaced again under the Cossack rebel Kondratii Bulavin during the reign of Peter the Great and exploded in its greatest fury during the 1770s, at the very moment when Catherine the Great had assured her correspondents among Europe's *philosophes* that "there was not a single peasant in Russia who could not eat chicken whenever he pleased, although he recently has preferred turkeys to chickens."[74] Catherine told Europe that the Cossack renegade Emelian Pugachëv was "a mere highwayman,"[75] but his peasant war encompassed an area larger than France, and pitted half again as many rebellious subjects against Russia's armies as had Razin's a century before. Kazan, Russia's great Volga entrepôt, fell before Pugachëv's raging rebels, and two of Catherine's senior generals lost their armies before he finally was brought down in defeat.[76] Russians even coined a word—*pugachëv-*

shchina—to describe the violence of Pugachëv's rebellion, and every tsar afterward lived more in fear of it than of any opposition that might come from his nation's aristocrats or urban proletarians.

After Pugachëv's barbarous execution in Red Square on January 10, 1775, Catherine had wasted no time in transforming the rebellious Cossacks into defenders of the establishment by granting them special status and numerous privileges, so that the *pugachëvshchina* never returned to Russia. Within a century, Pugachëv's descendants had become such ardent defenders of the throne that the government could rely upon them to break up strikes and workers' protests, just as they did on the first day of the February Revolution. Without Cossacks to give focus and leadership to their outrage, Russia's peasants had become helpless victims, their anger seething as they bore the abuse of masters and government alike. Their rage flared briefly in Saratov province in 1902, and burst across the empire at the end of 1905. Then it calmed, as the peasants stolidly shouldered the monstrous burden of the Great War that was destined to claim their lives by the millions.

First to be called to war, the peasants were the last to hear of the revolution. Rightly fearful that word of the February Revolution would stir once again the raw anger that had seethed in their breasts since the days of Razin and Pugachëv, the men who still held the tsar's commissions in the provinces tried desperately to conceal the revolution's existence from them.[77] Then, as word of Nicholas's abdication seeped into Russia's remote villages, the authorities' worst fears came true. Peasants rejoiced to be freed from the greatest of all traditional restraints. "Now that we have no tsar," they exulted, "we can do anything and not answer for it!"[78] During just five days in mid-March, reports of arson and looting streamed into the Ministry of Internal Affairs' Central Militia Department from desperate landowners in no less than ten provinces, and continued for the rest of the month.[79] Sometimes the peasants burned barns. At other times, they killed livestock, stole firewood and timber, or carried off grain. As their rage grew more difficult to contain, they did all of those things and more. "Peasants from neighboring villages destroyed my estate," one terrified landowner telegraphed to Prime Minister Prince Lvov. "Grain, reserves of food, seed, thoroughbred horses, cattle, swine, fowl, machines, implements, and all other property was seized and all the buildings burned. I just barely escaped with my life."[80] "We are completely at the mercy of the local population," wrote an outraged country gentlewoman who refused to stand idly by while the peasants looted her estate in Riazan and complained directly to the minister of agriculture. "My two sons, who could give us some defense, are now at the front, and here, at the estate, are my two grandchildren, three and four years old, whose fate frightens me more than my own."[81]

The primitive and simplistic terms in which Russia's peasants viewed the world made it especially difficult for them to comprehend the events that burst upon them in the spring of 1917. "No one ever comes to our little village, no explanations or instructions ever reach us, and there aren't any reliable people to set us on the right path and explain the new ways to us!" some of them told government investigators. "We feel as if we've come out of a dark dungeon to stand in God's own light, not knowing where to tread or what to do," others added. Anxiously, some villages sent emissaries to Petrograd and Moscow, to see things for themselves and to bring back reports about what was happening. When they returned, the villagers besieged them with questions. "What did you hear while you were there?" they asked anxiously. "I've forgotten!" some of these confused travelers exclaimed, as much amazed at themselves as at what they had seen. "I've forgotten everything I heard. I heard so much that, in the end, I can't remember any of it at all!"[82]

The political debates of the revolution's first months thus swept over an uncomprehending peasantry. Unaccustomed to abstract thought, or even to thinking in impersonal concepts, Russia's many illiterate peasants thought only in terms of people they knew and actions they performed.[83] Peasants therefore had to be dealt with more directly and personally than did factory workers, and the mass meetings that city proletarians found so exciting had little meaning for them. "A peasant understands only when he asks questions directly," the authors of a report on rural conditions wrote, while another observer explained that "the peasants say candidly that they . . . are waiting for people who would explain to them how to act and what to do."[84] "Send us orators who can formulate the idea [of nationalizing the land] for those among us in whose minds it now resides only in a subconscious form," one group of peasants begged Moscow's Soviet of Workers' and Soldiers' Deputies at the beginning of April.[85] "For whom should we vote—for the tsar or the students?" This cautiously whispered query overheard by one observer in peasant villages in May showed the abysmally primitive level of the peasants' political understanding.[86]

Land and peace—translated in their minds into the fertile fields of nearby estates and the return of long-missed fathers, husbands, uncles, and brothers— were what Russia's peasants wanted most that spring.[87] Insisting that "there ought not to be exploitation of the poor and forgotten by the powerful and the rich,"[88] Russia's countryfolk anxiously awaited the "universal permission" to take the land, for there, in the fields and meadows of the great lords who once had ruled their lives as masters, they saw the long-sought key to well-being and prosperity. The "universal permission" need only be given, one of them told Paustovskii, and they would "drive away the great lords and all the little lords too with our spades under their backsides."[89] Some could not wait, and insisted that the land now must be theirs. "They say that it's anarchy when the peasants seize the lord's lands," a peasant soldier wrote bitterly to the Petrograd Soviet, "but it wasn't anarchy when they took our fathers and sons from among us, the poorest class of all, and sent them to the front!"[90] Yet, like so many of the peasants' age-long dreams, the panacea of the lord's land too often proved an empty hope. "At this moment, the peasant in many places thinks not in terms of reality but of slogans," one expert warned. "If one were to divide all the lords' lands [in Orël province, for example] among them, then each would get about a third of an acre," hardly enough to better the standard of living in even the poorest peasant household. Elsewhere, the chances were better, but statisticians still insisted that the supply of arable land outside the peasants' control in 1917 could not, by itself, solve Russia's land-hunger.[91] Only by bringing more land under cultivation and by producing higher crop yields could that be accomplished.

Even more deeply rooted than the peasants' land-hunger, were the seasonal patterns that had ruled village life since ancient times, and these conditioned the peasants' response to the revolutionary events that swirled around them. After the first great outburst of violence in Russia's villages, the age-old imperatives of rural life took firm control over the peasants' revolutionary acts. Food supplies in Russia's villages always reached a low ebb in March and April as winter neared its end, and at that point the peasants hurried to seize neighboring lords' stocks of grain. As the last gusts of winter's winds reminded them of repairs needed for the coming spring, shortages of firewood and timber drove them into nearby woodlands with their saws and axes at the same time as they seized the estate owners' grain. The great spring *ottepel* slowed their assault, as melting snow flooded roads and fields. Then spring's arrival plunged them into their annual cycle of frenzied plowing and planting to wring every possible advantage from the short Russian summer before winter came again. The cramped and scattered areas they had to plow and plant that spring whetted the peasants' appetites for more land, and they seized the fields of nearby estates and crown properties as the planting began. The opinion of experts and statisticians to the contrary, no power on earth could have convinced Russia's land-hungry peasants then that the lands just beyond their communes' boundaries could not be tilled to better purpose under their plows than under those of laborers hired by nobles, rich merchants, or bureaucrats from government offices. More land to till required more implements and beasts of burden and peasants took those from nearby lords as well.[92] "You [always] felt that there was another man's hand in your pocket," one nobleman wrote after he had reached the safety of exile in the West. "This will all be ours," the peasants explained whenever anyone questioned them. "It's all the same, it'll be ours."[93]

The peasants did all this with the greatest urgency, each fearful that the land would go to others before he took his share. At the front, soldiers worried and grew restless. First sporadically, and then in ever-larger numbers, they drifted away from the war, anxious to claim the lands that they and their ancestors had so long thought theirs by tradition and God's will. They must claim "their" land before the end of September, some thought, or it would be too late. Others believed that Kerenskii had given orders to seize the land, and that they were only carrying out the instructions of the man who, in their minds, personified Russia's government from early March until late October. "The *muzhiki* [peas-

ants] are destroying the squires' nests so that the little bird [a euphemism for
large estates] will never be able to return," one old man muttered, while others
spoke with contempt of the *"burzhui"*—any one of the bourgeoisie—to whom
fortune or talent had given more land, livestock, or buildings than they possessed
themselves.[94]

Until Russia's country folk found more solid footing on the shifting sands
of revolutionary politics, the insistence with which their brethren among the
urban labor force pressed their demands obscured the urgency of the peasants'
needs. Any number of contemporaries agree that after March's outburst of
violence had subsided, Russia's villages became less turbulent than her cities.[95]
Calls for an eight-hour day, higher pay, and better working conditions therefore
took precedence over less focused rural pleas for peace and land during the
revolution's first weeks. Because the city crowd stood so near at hand, ready to
fill the streets of Petrograd at a moment's notice, Russia's new statesmen at first
forgot that the most ardent calls for land and peace came not from the mouths
of tens of thousands but from the hearts of tens of millions. The new government's first declaration "To the People of Russia" had made no mention of the
land question or agrarian policy.[96] "It would be superfluous to speak of the
Provisional Government's program, which is clear and obvious to everyone,"
Prince Lvov had added the next day. Revolutionary Russia, he promised, would
dedicate herself to "liberating the people from all bonds which enchained it.
. . . The main thing is not to think or act in the old manner," he continued.
"The old regime has been overthrown forever."[97]

As they faced a seething peasantry and an exploding proletariat during
March and April, the Provisional Government struggled to find a policy and the
means to implement it. Clearly, the very disparate aspirations of the Soviet of
Workers' and Soldiers' Deputies conflicted with those of the moderate Kadets
who held most high government offices, and rivalries among government agencies, whose place in Russia's administrative structure had been dislocated by
February's revolutionary upheavals, heightened the tension. In any case, Russia's
masses could not be stilled by the instruments available to the new government.[98] "Not having in its local administration those organs of authority which
could quell disorders by applying physical force," Minister of Agriculture Andrei
Shingarev explained at the end of April, "and not recognizing it to be possible
under the conditions of the times through which we are now living to use such
means [even if they were available], this ministry can do no more than bring
moral pressure to bear on the population."[99]

During more than a quarter-century spent as a *zemstvo* physician in "dying
villages" all across Russia's Middle Volga provinces, Shingarev had witnessed
raging malnutrition, poverty, and disease that killed two out of every three
infants before their first birthday and left the others to face a life that rendered
them old at forty-five. "There are some families who do not see milk [even once]
in the course of an entire year!" Shingarev had exclaimed in bitter amazement

as he described the deprivation of life in villages where peasants rarely consumed even so much as an ounce of animal protein a day. Acutely aware of the desperate conditions under which Russia's peasants lived, Shingarev explained the spring disorders of 1917 as "a natural result of the existing state of upheaval,"[100] and searched for ways to better the peasants' standard of living, entice them to produce more grain, and, above all, convince them to sell what they produced to the government in order to stave off hunger at the front and in Russia's volatile industrial centers.

Once again—and not for the last time—the specter of urban food shortages dictated Russia's policy toward her peasants. While his nation's rural masses begged for land, Shingarev urged them to set aside their revolutionary dreams and be satisfied with promises of future benefits. Land reform, the "most cherished dream of many generations of the nation's peasants," he insisted, must be considered the "most fundamental of all fundamental questions," but this could "not be resolved by any sort of seizure." Fearful that land seizures could threaten the coming year's harvest, he sternly condemned "violence and looting" as "stupid and dangerous." Announcing that "only the enemies of the people can incite the masses to such a disastrous course," he insisted that the "land question must be decided on the basis of law," and explained that it required "serious preparatory work."[101]

Promising a Central Land Committee with branches spread all across Russia to assemble the data needed for new land reform laws, Shingarev warned that "the socialization of the land cannot be confused with arbitrary seizure of it for personal advantage."[102] "And so, the mountain had given birth to a molehill," one commentator remarked later. "The government had forbidden any spur-of-the-moment resolution of the land question and firmly announced that all this would be decided by a Constituent Assembly."[103] Even Chernov, the peasants' champion between 1905 and 1917, now urged his Socialist Revolutionary party to replace its time-hallowed cry of "Land and Freedom!" with the slogan: "Resolution of the agrarian question by means of the Constituent Assembly."[104] "They're leading the peasants around by the nose!" Lenin exclaimed indignantly. "On the land question, they say: 'Wait for the Constituent Assembly.' The Constituent Assembly is going to say: 'Wait for the end of the war.' When the end of the war comes, they're going to say: 'Wait for total victory.' This is what's going to happen. The capitalists and landlords are laughing up their sleeves at the peasants, because they have a majority in the government."[105]

Such sober bureaucratic bodies as the district and provincial land committees that Shingarev established in April dared not speak the words that more than a hundred million Russian peasants expected to hear. Not these assemblies of earnest and dedicated experts, but the All-Russian Congress of Peasant Deputies that assembled more than thirteen hundred strong in Petrograd at the beginning of May, spoke first and loudest for the peasants. Despite the best efforts of its Socialist Revolutionary majority to bar him from the podium, Lenin addressed the Congress on May 22, and seized the initiative immediately. Condemning "landowning by squires" as "the greatest injustice," Lenin demanded the abolition of all private landownership, and urged the representatives of Russia's land-hungry peasants to seize the lands that lay beyond their village boundaries.[106] "We feared that the ground would slip away beneath our feet," one of Lenin's opponents confessed after his speech.[107] Partly in response to his urging, the Congress resolved that "the right of private property in land is abolished forever," and urged that all privately-held land be "taken over without compensation as the property of the whole people." All Russians "who desire to cultivate it with their own labor, with the help of their family, or in a cooperative group," these delegates insisted, must have "the right of using the land."[108]

Despite these clearly stated resolutions, the statesmen of Russia's new coalition government remained as insensitive to the urgency of the peasants' pleadings as the Provisional Government's first cabinet had been, and continued to counsel their impatient countrymen to set their hopes aside until the war's end. Nor did they understand the dreams of Petrograd's factory workers much better. Convinced that the working men and women of their nation would willingly accept further sacrifices in order to win total victory against Germany and Austria, they took up several themes of their former "bourgeois" opponents. "To save the country," the long-acclaimed workers' champion Tseretelli announced, "workers must be ready to work more than eight hours a day," while the Menshevik minister of labor warned that, so long as workers and peasants continued to demand more land and higher pay, "the entire country, its entire economic life, is heading toward derangement."[109]

Were these the same men who had sat on the Soviet's Executive Committee since the first days of the revolution? Were these the same radical leaders who had defended the workers' hopes for a "democratic peace" in their bitter struggles against Miliukov and Guchkov at the end of April? "Do we go backward or forward?" Lenin demanded angrily.[110] Looking back upon his days at the helm of the War Ministry, Kerenskii claimed he had seen "a steady diminution of revolutionary chaos and the development of political strength and wisdom" among his countrymen.[111] "The nation had been shaken out of its normal routine by the Revolution," he later explained, but her "recovery after the fall of the Monarchy was extraordinarily rapid" and she had "gained strength at a tremendous rate."[112] With the help of America and her European allies, even victory now seemed within reach. When their armies went on the offensive for the first time since the revolution in mid-June, the optimism of Russia's politicians soared higher still. "Again the sun rises!" the Kadets exclaimed. "Again one wants to live and believe. . . . Everything is saved: freedom, our dignity, our national honor." Russia, Miliukov concluded, had reached a "great turning point."[113]

But hungry and impoverished proletarians were not so willing to postpone their revolutionary dreams as their insensitive leaders believed, nor were they even prepared to accept Marx's admonition that "you must endure 15, 20, or 50 years of civil and international wars . . . [so] that you yourselves may . . . become capable of taking over political power."[114] Russia's working men and women demanded immediate benefits from their revolution, and as 1917 passed its summer solstice, they began to understand that, as in tsarist days, they could best realize their dreams when they ignored their government's urgings and defied its orders. When the Provisional Government had disappointed their long-cherished hopes for an eight-hour day in law, they had implemented their dream in their factories and mills on their own authority. While Petrograd's statesmen and politicians had urged them to postpone their dreams for land until the Constituent Assembly met, Russia's peasants had seized the land upon which they had gazed with longing for so long.

However, such disjointed acts could not institutionalize the revolutionary order for which Russia's workers and peasants hoped. For that, the locus of political power had to shift from the bourgeoisie to the representatives of the proletariat. To that end, Russia's proletarians began to support their arbitrary acts with ballots. At the beginning of June, Petrograd's Municipal Council elections saw the socialists outpoll the once-favored Kadets by almost four votes to one; less than a fortnight later, they widened their margin in Moscow. Where the Kadets had commanded more than six votes out of every ten a decade before, they now won barely one out of five.[115] Strangely oblivious to the meaning of these dramatic political setbacks, the Kadet leaders explained that the "Russian people now thirst for authority and a statesmanlike organization of life."[116]

It seemed that the growing caution of Russia's new leaders had lowered the trajectory of Aleksandr Blok's metaphorical revolutionary swing far below the horizontal at the very moment when the rising anger of the nation's workers and peasants forced it sharply and dangerously upward. "There is increased dissatisfaction with the Provisional Government," the commander of the Petrograd Military District warned in mid-June. "And there is increased support for the [Bolshevik] slogan: 'All Power to the Soviet of Workers' and Soldiers' Deputies!'"[117] Certainly, Petrograd's workers made no effort to mask their feelings. "For the last time we announce: patience has its limit," a group of them warned ominously. "We simply cannot live in such conditions." Would Russia's leaders respond to such plainly-stated warnings? Convinced that they would not, Lenin sneered: "They are the government. They are all together in a bloc. And they will see that nothing comes of it."[118]

Lenin's view seemed confirmed when the First All-Russian Congress of Soviets of Workers' and Soldiers' Deputies—in which delegates from all over Russia met in Petrograd from June 3 to June 24—voted to stage a peaceful mass demonstration as a show of socialist unity. "Revolutionary democracy wants to express its desire for universal peace without annexations and indemnities, self-determination of all peoples, and the preservation of unity in the revolutionary movement of workers, peasants, and soldiers," the Petrograd Soviet's *Izves-*

tiia announced.[119] Yet, although conceived as a statement of unity, the demonstration that the All-Russian Congress staged on Sunday, June 18, showed little support for the Mensheviks and Socialist Revolutionaries who continued to control Petrograd's Soviet, and a great deal of sympathy for the Bolsheviks who did not. "In general, the Menshevik-Socialist Revolutionary bloc at that time was a prime example of a decomposing government, frozen in its self-confidence, self-satisfaction, and blindness," Sukhanov remembered. "On the other hand, the Bolsheviks were feverishly active in the very entrails of the proletarian capital, breaking new ground and organizing their followers into militant columns."[120]

The Bolsheviks' dedication to organization and detail brought handsome rewards on the afternoon of June 18—"a magnificent day and already hot"— when nearly a half-million soldiers and working men and women marched across Petrograd's Field of Mars, for more than a century the parade ground of tsarist armies, in close-knit, orderly ranks. Above them, red banners bore Bolshevik slogans, broken only rarely by "official" Socialist Revolutionary or Menshevik ones. In vivid contrast to such verbose and clumsy sentiments as Chernov's recently conceived "Resolution of the Agrarian Question by Means of the Constituent Assembly!" the simple, direct slogans of Lenin's Bolsheviks bored relentlessly into the people's minds and hearts that day: "All Power to the Soviets!" "Down with the Ten Capitalist Ministers!" "Peace for the Peasant Huts, War for the Palaces!" Here on the Bolsheviks' banners blazed words that promised life to hopes too long suppressed, and spoke directly to common men and women too long ignored. Almost against his will, Sukhanov watched, seduced by the rhythm of their march. "Like Fate itself, like the fatal Birnam Wood, they marched toward us," he remembered. "The sight of these measured, advancing militant columns of the revolutionary army made it seem that the coalition already had sung its swan song, that it already was formally liquidated."[121]

That a political faction which held no offices in government and commanded only a small minority of votes in the Petrograd Soviet and at the First All-Russian Congress of Soviets could seize control of Petrograd's masses so completely posed once again the dilemmas of power that the Russian revolution had not yet squarely faced. If power still lay in the hands of the Soviet *and* the Provisional Government, how could it be consolidated? And, into whose hands should it finally be placed? Moderate Russians still believed that, once convened, the anxiously-awaited Constituent Assembly would resolve the dilemmas of dual power, while the Mensheviks and Socialist Revolutionaries still envisioned some means by which the growing political strength of the proletariat would eventually transfer power peacefully into socialist hands. Again at odds with his fellow socialists, Lenin thought otherwise. "Power is not transferred," he insisted. "It is taken with guns."

Lenin therefore explained that Bolsheviks must prepare the proletariat for "an armed uprising, if not in days, if not in the coming weeks, then in any event in the near future." While most of Russia's revolutionaries struggled to decide *if* they should take power and how, Lenin and his Bolsheviks, the only political group that stood ready and willing to take power at any time, insisted that the fundamental question was not *how* power would be taken, but *when.* The seizure of power in Russia would demand the sort of careful preparation and meticulous organization of which only the Bolsheviks were capable, and they had no intention of sharing it with other socialists if they succeeded. Much more needed to be done, however, before they could count on success. "We must concern ourselves with organization in the most intensive way," Lenin counseled. To move too quickly could be fatal, but Russia's workers and peasants must be convinced "of the impossibility of gaining power by peaceful means."[122]

Carried away by a false sense of their rapidly mounting strength, other Bolsheviks thought Lenin too cautious. "We are now faced with death either in the trenches in the name of interests that are foreign to us or on the barricades for our own cause," exclaimed his volatile lieutenant Gregorii Zinoviev on the day before the great June 18th demonstration. In Trotskii's view "far more bold and unbridled in agitation than any other Bolshevik," and "able to catch out of the air whatever formulas are necessary," Zinoviev could turn the masses' heads with his words.[123] Even within the upper reaches of the Bolsheviks' Petrograd organization, others anxious for action took up his theme during the next several days. "It is time we remembered that we represent not only socialism, but . . . *revolutionary socialism,*" one of his comrades insisted, and soldiers from the First Machine Gun Regiment, ordered on June 21 to send two-thirds of its men and weapons to the front, agreed. Accusing the government of deliberately launching a new offensive "to destroy the proletariat," angry machine-gunners demanded the "transfer [of] all power into the hands of the Soviet of Workers' and Soldiers' Deputies," and urged new elections "so that only Bolsheviks will be in it."[124] For a moment, a stern reprimand from the Executive Committee of the Soviet dampened their spirits, but, as the always militant sailors of Kronstadt added their voices to the soldiers' call for action, it seemed that an armed rebellion might break out in Petrograd at any moment.

Fearful that his comrades had focused too readily on his statement that an armed uprising was "inevitable," and certain that they had forgotten his warnings that the Bolsheviks could seize power only after much more careful organization and preparation, Lenin spoke out sharply against those among them who had set aside the restraints of caution and common sense. "One wrong move on our part can wreck everything," he cautioned. In words that one of his followers likened to a "cold shower," he warned the Bolsheviks to remember that "we are [still] an insignificant minority," that the stark reality of that fact "cannot be brushed aside," and that, even "if we were now able to seize power, it is naive to think that, having taken it, we would be able to hold it." For the moment, it seemed that Lenin had convinced the Bolsheviks to set aside any thought of immediate armed action in favor of propaganda and careful organizational preparation. "The proletariat will prepare for the new stage in the revolution," the cautious Lev Kamenev explained, "not by anarchistic demonstrations and disorganized partial endeavors, but through renewed organizational work and unity."[125] Satisfied that his counsels had won the day, Lenin left the seething cauldron of Petrograd revolutionary politics to spend a few days in Neivola, Finland, at the summer cottage of Vladimir Bonch-Bruevich,[126] the loyal friend and journalist who had been one of the first to join the Bolsheviks when the Russian Social Democrats had split almost fifteen years before.

While Lenin and his sister Mariia lay in the sun and swam in Neivola's nearby lake, a number of hot-headed Bolsheviks chafed at the restraints his caution had imposed upon their allies among Petrograd's soldiers and workers. Throwing caution to the winds, and violating every canon of the rigid party discipline that had become the trademark of Lenin's followers, the editors of *Soldatskaia Pravda, Soldier's Truth,* the daily newspaper that Petrograd's Bolsheviks distributed to Russia's front-line troops, called for immediate action. Urged on by soldiers who had abandoned their front-line positions in favor of the safety of Petrograd, *Soldatskaia Pravda* opened an intense propaganda campaign against the offensive that Kerenskii had just launched. "The time has come not to sleep but to act," its front-page exclaimed on July 2nd. Power belonged only in the hands of Russia's soldiers, workers, and peasants, and they must be ready on a moment's notice to "protect the freedom that has been won" and "remove from power the bourgeoisie and all its sympathizers."[127] "The capital seethed," Sukhanov remembered. "Everywhere, on every corner, in the Soviet, in the Mariinskii Palace, in people's apartments, on the public squares and the city's boulevards, in the barracks and in the mills, everyone talked about some sort of 'demonstration,' if not today, then tomorrow."[128] Writing with all the advantage of hindsight, it seemed to Trotskii that already "the July Days were casting their shadow."[129]

At the very moment when, as Sukhanov said, "the city felt itself on the verge of some sort of explosion,"[130] policy disputes tore the first coalition government apart. For centuries, the Russian Empire had held within its boundaries a complex conglomeration of subject nationality groups, each sheltering unique political and cultural visions and dreaming of the day when its people could declare themselves independent of their Russian masters. None were more ardent in their dreams or more dedicated in their quest than the Ukrainians. While the government in Petrograd counseled patience, the members of the Central Rada—the National Assembly of the Ukraine—openly challenged the Provisional Government and proclaimed a Ukrainian Republic in Kiev at the end of June.

Desperate to prevent a political crisis in their armies' rear at the height of their new offensive, Kerenskii, Tseretelli, and Tereshchenko raced to Kiev to arrange a compromise. After three days of desperate negotiations, they succeeded. In triumph, they returned to Petrograd to find that, at Miliukov's urging, four of the Kadets who held ministerial portfolios in their government would permit no concessions. Condemning the compromise as "a criminal document" because all members of the government "were obligated by oath to preserve the sanctity and indivisibility of the Russian state," these four ministers followed the instructions of their party's Central Committee and left the government.[131] On the evening of July 2, just as some of Petrograd's most disgruntled soldiers and workers prepared to move into the streets to wrest power from the Provisional Government, the coalition that had governed Russia since the beginning of May collapsed.

While the ministers of the Provisional Government's first coalition cabinet struggled with the Ukrainian secession crisis, the soldiers of the First Machine Gun Regiment made ready to disobey the new orders that would send them to the front. Urged on by a handful of Anarchist-Communists and some Bolsheviks who had concluded that it would be better to lead an armed uprising than let it occur without them, these men assembled late on the morning of July 3 for a meeting that Trotskii thought "was a storm from the first moment." Trotskii remembered especially a young American-Jewish anarchist by the name of Bleichman, "his shirt open at the breast and curly hair flying out on all sides" as he urged his listeners, "his Jewish-American accent sharp as vinegar," to confiscate food and money from the "bourgeoisie," seize Russia's mills and mines "immediately," and "overthrow the Provisional Government just as it overthrew the Tsar." Neither well-laid plans nor careful organization had any place in Bleichman's revolutionary struggle. "It is necessary to come out with arms in our hands!" he exclaimed. "The street will organize us!"[132]

His past no less obscure than his future, Bleichman stood at history's center stage for one shining moment. At his urging, the machine-gunners sent delegates hurrying to other regiments and to some of Petrograd's largest factories where they found other radical orators already at work. Led by their Red Guards detachments, all of the *metallisty* in the Vyborg District marched into the Sampsionevskii Prospekt where they joined ranks with the machine-gunners late that afternoon. By early evening, the light of the city's legendary White Nights still strong upon their columns, other workers began to swell their ranks.[133] On the other side of the city, after a marathon seven-hour meeting that began in mid-afternoon, twenty-five thousand *putilovtsy* voted unanimously to join the "demonstration" despite the efforts of loyal Bolsheviks to defend Lenin's prohibition against any attempt to seize power at that moment. With the cry: "Into the streets! Standards in the lead!" the *putilovtsy* marched upon the center of Petrograd.[134]

Now past eleven o'clock in the evening, the sky had become overcast, the light of Petrograd's legendary midsummer White Nights at last having faded into dusk. Not one street lamp burned. "One could hear only the measured tread of many thousands of feet," one writer remembered. "At first they sang, but the closer they came to the city center, the more concentrated and silent the marchers became." Everyone wondered if gunfire would greet them when they reached the Nevskii Prospekt. There they found the same silence, the shops barred, their lights out. Past the Nikolaevskii Station they marched, across Znamenskaia Square, and on to Ligovskii Prospekt. Now their route lay along Ligovskii as they made their way to the Taurida Palace. At three o'clock in the morning, the sky now brightened by the early midsummer dawn, they reached their destination. Together with their wives and children (some of whom stood at their side at that very moment) the Putilov men demanded the immediate expulsion of all "capitalist" ministers from the government. Already assembled when the *putilovtsy* arrived, some thirty thousand Vyborg workers and soldiers added their voices to their demand.[135]

Not far away, at their headquarters in the mansion they had seized from the fallen tsar's first love, the famous ballerina Kshesinskaia, a number of leading Bolsheviks once again had to consider their course, now that the workers and soldiers had taken the lead. Among them was Iosif Dzhugashvili, now known by the revolutionary name of Stalin, who, in a decade, would overshadow all rivals in the struggle to become Lenin's heir. Like Lenin a scant five feet four inches tall, but nearly ten years younger than he, Stalin had been born in the humblest circumstances in the ancient Georgian village of Gori. Swarthy and pockmarked from the ravages of smallpox, his left arm slightly withered from a near-fatal bout with blood poisoning he had suffered as a child, Stalin had entered adolescence as a seminarian and left it a full-blown revolutionary pursued by the police.

After several escapes from Siberian exile, and several brief stays abroad, Stalin had been arrested in 1913 and sent to the remote Turukhansk region, where the brutal Arctic climate made escape impossible. He had spent the entire first thirty months of the Great War there, but had managed to return to Petrograd a full three weeks ahead of Lenin to take control of the Bolsheviks' *Pravda* and to urge a more pliant attitude toward the Provisional Government and Russia's participation in the war than Lenin favored. Now, as the *putilovtsy* and their worker-soldier allies from the Vyborg District filled the square in front of the Taurida Palace and demanded immediate action, Stalin and his Bolshevik comrades had to decide whether to continue Lenin's policy of restrained waiting or to try to direct a political demonstration that seemed about to burst out of control. "Did the Party really have a right to wash its hands of a 'demonstration' by the proletariat and soldiers and stand off to one side?" he later asked rhetorically. "We had no right to do so," he concluded. "As the party of the proletariat, we were obliged to become involved in its 'demonstration' and to impart to it a peaceful and organized character."[136]

But a "peaceful and organized" demonstration could be no more than a fond hope then, for armed workers and soldiers had assembled all across Petrograd.

"Young people appeared on the streets distributing bullets, procured no one knows where, to persons who had arms," one eyewitness remembered.[137] Machine-gunners took up their assigned positions. At the Mikhailovskoe Artillery Academy, armed soldiers and workers marched away with several field guns despite the protests of the Academy's senior officers. In Kshesinskaia's master bedroom, the Bolsheviks made a fateful decision. While the Soviet Executive Committee's Menshevik and Socialist Revolutionary majority tried in vain to convince more than sixty thousand armed workers and soldiers massed before the Taurida Palace to return home, and with Lenin still at Bonch-Bruevich's isolated lakeside cottage at Neivola, Petrograd's Bolsheviks decided to seize the lead before it was too late. "Was this, perhaps, the beginning of the proletarian revolution?" asked Mikhail Kalinin, the peasant metal worker destined one day to become titular head of the Soviet Union. Not even Lenin knew the answer. "We shall see," he replied when he arrived in Petrograd the next morning. "Right now, it is impossible to say!"[138]

Lenin arrived back in Petrograd just before midday on July 4, disheveled as a result of his hasty early-morning departure from Neivola, and still uncertain about what had happened during his absence. He found not only armed soldiers and workers in the city's streets, but a contingent of more than ten thousand armed sailors who had arrived from Kronstadt earlier that morning to announce that they had come "to restore order because the bourgeoisie has gotten too far out of line here."[139] As they had in late February and late April, Petrograd's workers, soldiers, and sailors headed for the Nevskii Prospekt that morning to carry their message of protest from the workers' slums in the Vyborg and Petrograd districts into the heart of the opulent bourgeois and aristocratic quarters of the city. As something over sixty thousand men, women, and children marched along the Nevskii, again en route to the Taurida Palace, a sad combination of trigger-happy tension, frustration, and outright anger caused a series of armed clashes that left scores of dead and wounded along their path. In a perverse turn of events, armed clashes in which both sides thought themselves the revolution's defenders had begun to produce most of the revolution's casualties.

Petrograd's militant proletarian marchers were in such a rage by the time they reached the headquarters of the Soviet in mid-afternoon that they arrested Chernov the moment he appeared to address them. Fearful for Chernov's safety, several of his comrades on the Soviet's Executive Committee tried to speak, but the crowd drowned their words with its angry shouts. "As far as the eye could see, the mob seethed," Sukhanov wrote as he recalled his fears that "all of us, including perhaps even Trotskii, might have been torn apart" that afternoon. One of the last revolutionary leaders to return to Petrograd, after having been interned for a month in a British prison camp in Nova Scotia as he journeyed from New York to Petrograd, Trotskii had captivated Petrograd's crowds with his dazzling revolutionary oratory from the moment he had arrived in early May. Now hoping to take advantage of the enthusiasm with which he had been received by Russia's revolutionary soldiers and sailors in the past, he leaped onto a nearby auto and tried to speak to the Kronstadt sailors who stood nearest the Palace steps. "Red Kronstadt has again shown itself to be a leading champion of the proletariat's cause!" he shouted. "Long live Red Kronstadt, the pride and glory of the revolution!"

After several anxious minutes, Trotskii managed to free Chernov, but the crowd remained, growing larger and more hostile as time passed.[140] Clearly, it had the will to wrest power from the Provisional Government and pass it to the Soviet. It lacked only leaders and some clear signal from the Soviet itself that it stood ready to take the power that these armed proletarians hoped to offer. "Take power, you son-of-a-bitch, when they give it to you," one of them had snarled at Chernov.[141] Not Chernov and his Socialist Revolutionaries; not Tseretelli and his Mensheviks; not even Lenin and his barely-restrained Bolsheviks were ready to take that step. Ever so slightly and subtly at first, the balance shifted as July 4th passed into the early morning of July 5th. On the third day of the riots, the forces of order took control of Petrograd's streets.

Shortly after midnight on July 5, the Izmailovskii Guards marched to the Taurida Palace and openly took the side of the Soviet. "Suddenly a noise was heard in the distance," Sukhanov remembered. "Nearer and nearer it came, until the measured tread of thousands of feet echoed in the surrounding halls." At first, the Executive Committee froze in apprehension. A new threat to the revolution? After the soldiers of the First Machine Gun Regiment, the *putilovtsy*, and the Kronstadt sailors, who else could take the other side? Then Sukhanov and his associates learned the truth. "Regiments loyal to the revolution have come!" someone exclaimed. Behind them came regiments of the Preobrazhenskii and Semenovskii Guards. Together, soldiers and Soviet sang the *Marseillaise*. For the first time in almost two days, the men of the Executive Committee breathed more easily.[142]

A number of important factors had turned the tide in the Soviet's favor. For one thing, Petrograders had wearied of violence, which had claimed close to four hundred killed and wounded in forty-eight hours. Likewise, the city's garrison, staunch champions of the revolution though they claimed to be, had no taste for defending it against the battle-hardened combat troops that now were en route to Petrograd at the Soviet's request. Finally, the minister of justice's emphatic charge that Lenin was a German secret agent, presented (with a few scraps of authentic-looking documentation) to several regiments who had declared their neutrality in the events of July 3 and 4, brought key army units over to the government's side, and disoriented the men and women who had taken to the streets to demand that the Soviet strip the Provisional Government of its power.

Supported by a hastily-published broadside entitled *No Needless Words*, the charges that Lenin was in the pay of the Germans caused a sensation in Petrograd's barracks, factories, and streets on July 5. "All down the Nevskii, [the broadside] sold like hot cakes," one witness reported.[143] But the facts were few and the suppositions as numerous as they were unfounded. At best, the flimsy evidence showed that Lenin and several other Bolsheviks had accepted funds channeled from the German government through agents in Sweden to support their avowed purpose of overthrowing Russia's government. Even the most creative efforts of the Provisional Government's jurists offered no substantial proof (and none has ever been found) that Lenin became a German secret agent to carry out the commands of superiors in Berlin, or that his decisions ever were motivated by any considerations other than a ruthless certainty that Russia's Provisional Government must be replaced by a revolutionary proletarian one. Nonetheless, this charge, cleverly presented and supported by the testimony of corrupt witnesses, momentarily broke the Bolsheviks' tightening hold on the Petrograd crowd and forced Lenin into hiding. Now more confident, the government issued warrants for the arrest of other leading Bolsheviks on charges that, "for the purpose of aiding the enemy countries at war with Russia," they had, among other things, agreed "to organize in Petrograd, from July 3 to 5, 1917, an armed insurrection."[144]

By midday July 5, troops loyal to the Soviet and Provisional Government had cleared Petrograd's streets of demonstrators, wrecked *Pravda's* offices, and driven the Bolsheviks from their headquarters in Kshesinskaia's mansion. Insisting that the only "power that is active is the military dictatorship," Lenin refused to appear in court to answer the government's charges. "It is not a question of 'courts,' but of *an episode in the civil war!*" he exclaimed as he went into hiding once again. Several comrades managed to follow his example, but police and military agents jailed Trotskii, along with a number of other leading Bolsheviks who had remained in the city.[145]

Reaction set in. Outraged that the Left had engaged in "sacrificing national and moral values to the masses in the name of demagogy,"[146] Prince Lvov resigned as head of the government, leaving Kerenskii amid the shambles of the previous days' street violence to form a new government. Proclaiming that "a fateful hour has struck," and demanding from all Russians complete "readiness to give everything—all their strength, possessions, their very lives—for the great cause of saving the country," Kerenskii took charge.[147] His would be the "Government to Save the Revolution," the Soviet's Executive Committee announced.[148] But immense difficulties lay ahead for which, even the Soviet confessed, only "heroic remedies" would suffice.[149] At the front, Russia's new offensive had crumbled, and the Germans had broken through. In the rear, the nation's economy now lay in shambles. Inflation soared, shortages grew worse, and hunger more intense.

Beneath all of these terrible obstacles lay another, "a more evil and stronger enemy," of which only Maksim Gorkii dared speak aloud. This, Gorkii warned from his offices at the newspaper, *Novaia zhizn,* he edited with Sukhanov, was "the oppressive Russian stupidity" that the Petrograd mob had demonstrated so shamelessly during the terrible forty-eight hours of the "July Days," the "disgusting scenes" of which, he lamented, "will remain in my memory for the rest of my life."[150] From the Winter Palace, where he labored to compile the testimony of the old regime's public servants, Aleksandr Blok felt the same fear. "If the proletariat is going to take power," he wrote to his mother, "we are going to have to wait a long time for 'order.' Maybe we'll never live to see it come again."[151]

CHAPTER XV

Kerenskii Takes Charge

Not merely the violence of the July Days, but the apparently mindless brutality of the masses that surfaced again and again that spring appalled Maksim Gorkii, who insisted that the revolution must have a higher moral purpose. "How monstrously cold our attitude toward man is," he exclaimed at one point. "We do not know how to love, we do not respect one another, our consideration for man is not developed." Warning that "our beautiful dreams will never come true in the soil of destitution and ignorance," and that "a Garden of Eden cannot be grown in a rotten swamp," Gorkii asked: "Will we not choke in the mud which we so diligently produce?" Violence, the "chaos of dark, anarchistic feelings," the compulsion to "prove something with bullets, bayonets, or a fist in the face," all threatened what Gorkii saw as the revolution's chief mission, which was to "develop and strengthen a social conscience and a social morality in man."[1] But, as one tsarist general who joined the Bolsheviks understood all too well, there could be no turning back, and the revolution's excesses must in some way be integrated into Russia's new government and society. "The old system was gone, never to return," he concluded. "The wheel of history would not turn backward."[2]

As history's wheel moved forward that spring, it crushed centuries of Russian military traditions beneath its rim. "The army is living through a sickness," a high-level secret conference at *Stavka* had warned only a fortnight after Nicholas's abdication, and by the time Kerenskii took over Guchkov's War Ministry at the beginning of May, the situation had become much worse.[3] Elated by their first taste of "democracy," soldiers now refused to obey orders and drove away their officers. "In the majority of cases," a group of Duma investigators reported in mid-April, "the officers who are the most suspect are the best combat commanders. This is explained," they concluded, "by the [soldiers'] subconscious fear that good officers may force them to attack."[4]

The Socialist Revolutionaries' decisively indecisive leader Victor Chernov, a failure from the moment he entered the government as minister of agriculture in May, thought that the army's crumbling will could be restored by discipline based "not on fear, but on conscience, a discipline reinforced by a fresh consciousness of civil duty." But the hard truth remained that few men in the ranks found revolutionary ideas sufficient incentive to face German and Austrian bullets once the winter lull in the fighting had passed. While socialists in the Petrograd Soviet struggled to force the Provisional Government to implement a policy of "peace without annexations or indemnities," Russia's front-line peasant soldiers thought that "Anneksiia" and "Kontributsiia" (the Russian words for annexation and indemnity) were cities somewhere deep in enemy territory.[5] Neither held much interest for men whose horizons remained confined to distant villages. To such simple peasants, who had never understood Russia's war aims anyway, the reasons for returning to a life of peace at home now seemed more compelling than ever, and the war's purpose slipped even further out of focus. "What the devil do we need another hilltop for," they asked in bewilderment, "when we can make peace at the bottom?"[6]

That spring, as the army began to discuss rumors of a coming offensive, one soldier was heard to ask: "What's the use of invading Galicia anyway, when back home they're going to divide up the land?"[7] Clearly, ringing words of patriotism could have little meaning for men uninterested in Russia's historic mission and ignorant of the meaning of such terms as "Prussian militarism," the "Turkish yoke," and "the national aspirations of the peoples now oppressed by Austro-Hungary and Turkey," with which revolutionary politicians filled their speeches and manifestoes.[8] These men saw the war in far simpler terms. "It's not life here, but just pure Hell,/ And I'd like to head for home," a familiar front-line ditty began. "But you can't just take off for places dear," it went on. "So you become a deserter and disappear."[9] First by the thousands, then by the tens and hundreds of thousands, these men began to walk away from the war, anxious to reap the revolution's promised rewards as the spring planting began at home. "It seems that we finally have lost the war," a junior officer wrote from the front in the middle of May.[10] Stern, devoted to the old army, and destined to remain at heart a Cossack to the very last, General Anton Denikin shared similar fears as he set out to take up his new appointment as *Stavka*'s chief of staff that spring. "The ringing of the Easter bells has lasted too long," he wrote after seeing the thousands of soldiers who sought to escape front-line service by "deepening" the revolution in Petrograd. "They would have done better to ring the alarm bell."[11]

Certain that "a revolutionary army requires no less severe discipline than an old regime army," and equally certain that "an army can exist only as a monolith," Chernov thought that a "militant, revolutionary spirit" could replace the age-old precepts of tsar, faith, and motherland for which Russia's soldiers had fought and died under the Romanovs. But any revolutionary revival of the army's will to fight in the spring of 1917 depended upon finding officers who could press those new values upon the men under their command. "Commanders had to identify themselves with the democracy of the army," Chernov explained, "or make way for a new commanding body produced by that democracy."[12] Yet their dismay at the revolution's excesses, and their frequent insensitivity to the feelings of the men under their command, prevented many tsarist officers from developing any such identification. As *Stavka*'s new chief of staff, Denikin found that even commanders willing to serve the new Russia "felt themselves to be stepsons of the Revolution and were unable to hit upon a proper tone in dealing with their men" under the revolution's prohibitions against corporal punishment and the death sentence.[13] Unable to adjust to the new patterns of relations between officers and men that the revolution seemed to dictate, one colonel in the elite Izmailovskii Guards organized a "Regimental University" at which he delivered enthusiastic lectures to uncomprehending soldiers about such topics as the "Psychology of the Masses" until he began to write his orders of the day in verse and had to be relieved.[14]

Perhaps nothing proved more galling to former tsarist officers, or more awkward to fit into the army's regular command structure, than the soldiers' committees that the revolution spawned at the front. Like the soviets of workers' and soldiers' deputies that emerged in Russia's cities and industrial centers during the revolution's early days, soldiers' committees appeared spontaneously. Russia's peasant soldiers suspected that their officers had concealed vital truths about the revolution from them and hoped that elected representatives might in some way obtain accurate information about the new government's plans. Beyond that, even though the Petrograd Soviet had applied Order No. 1 only to the Petrograd Military District and not to front-line units,[15] soldiers at the front had wasted no time in forming groups to discuss everything from war and peace to "insulting remarks" made about them and the revolution by their officers. "Now that 'freedom' had been won, it was not surprising that the general ambition was to elect, or rather to be elected," England's acid-tongued General Knox reported. "No one who had any acquaintance with the Russian character was surprised to find that 'freedom' was interpreted as freedom to talk without end and to do nothing."[16]

Almost without exception, Russia's senior officers at first opposed elected soldiers' committees, and as acting supreme commander during the days after Nicholas's abdication, Alekseev fired off a barrage of furious telegrams to his superiors in Petrograd's War Ministry to insist that they halt the Soviet's meddling in military affairs. But Alekseev, as a leading Soviet commentator once wrote, was "the captive of an illusion" that the Provisional Government actually had the authority to control the Soviet and the masses who swore their first allegiance to it.[17] On March 10, a candid and depressing letter from Guchkov, who usually marched bravely across terrain upon which others feared to tread, convinced him that he had been mistaken. "The Provisional Government has no real power whatsoever, and its authority exists only to the extent that the Soviet of Workers' and Soldiers' Deputies, which holds all the most important elements of real power—the army, the railroads, post, and telegraph—in its hands, permits it," Guchkov confessed. "I can tell you very frankly," he concluded, "that the Provisional Government can continue to exist only so long as the Soviet of Workers' and Soldiers' Deputies allows it to do so."[18]

Instantly, Alekseev shifted course. "The Soviet of Workers' and Soldiers' Deputies," he telegraphed to Guchkov the next evening, "is made up mainly of moderate elements, who understand the situation and recognize the necessity for continuing the war to victory and the importance of preserving discipline, order, and proper subordination in the army." He now proposed that special committees chosen by the government and the Soviet be established "in each army or on each front" and urged that the government make every effort to get officers elected to them "in order to get control of events and direct them."[19] Alekseev's was a crude attempt to limit the spread of the soldiers' committees or, if that proved impossible, to control them after they came into existence. Yet, within a fortnight he changed his tactics again as he realized that these awkward bodies, filled with men who fumbled to address issues they did not yet comprehend, might provide the only possible adhesive to cement his crumbling army together. "The widespread establishment of committees of officers and soldiers not only is desirable," he telegraphed to the commander of the Rumanian front on March 24, "it is imperative."[20]

Even before Alekseev reconciled himself to dealing with soldiers' committees, Russia's common soldiers already had begun to send delegations to Petrograd to take their grievances and their hopes directly to the Provisional Government and the Soviet. Earnestly trying to comprehend the revolution, and perhaps testing their new freedom to see if, as one of them wrote, it were true that "each humble little peasant can speak freely,"[21] these soldiers sought out Russia's chief ministers. Why did the new government not send enough weapons, they asked. Why were there shortages of ammunition, food, clothing, and equipment at the front? Why were throngs of soldiers allowed to take refuge in the rear to "defend" the revolution, while they still had to face German shells, machine guns, and poison gas? None of Russia's harried ministers found these confrontations pleasant, and all tried to dismiss their questioners by thanking them "warmly," assuring them that they "fully shared such views,"[22] and urging them to visit meetings of the Soviet, where they received a more sympathetic hearing. "If now we are not giving the army the full extent of everything it needs, then this is *not because we do not want to, but because we are not able to do so,*" Kerenskii explained to the Seventh Army's delegation a few weeks later. As he tried to find in the power of words the key that would let him escape and attend to more pressing business, he concluded: "It is impossible for the new power to create everything out of nothing immediately."[23]

Despite the war-weariness that lay so heavily upon Russia's soldiers when the revolution broke out, these first delegations from the front demanded not peace at any price but asked only that they not be left to face the war alone. A few delegations in which staff officers had won control even called for all-out support for the war effort, urged the proletariat to "bend all its strength toward making shells and weapons," and warned the Soviet "not to upset the army with extremist resolutions." Even some delegations controlled by the rank and file supported the Provisional Government's refusal to sue for an immediate peace, although none apparently shared Miliukov's dedication to the annexationist war aims of his tsarist predecessors. At first willing to support a "defensive" war that would defend Russia against German and Austrian attack but not invade foreign territory, these men from the front turned their anger against armaments factory workers who would not make guns, and the despised *belobiletniki*, men who had arranged to be exempted from front-line service and led comfortable lives in the rear while they starved and froze in vermin-ridden trenches. The government, the men from the front said, must force Russia's factory workers to get on with producing ammunition and weapons and send the men who hid in the rear to the front. "We will die for freedom and country," the First Army's elected assembly told the Petrograd Soviet, "but . . . work to straighten out the rear . . . [and] do not stir discord in our midst with hasty decisions."[24]

Contact with the Soviet, and with the more politically conscious Petrograd factory workers and garrison, began to draw these awkward visitors from the nether reaches of no-man's-land away from their support for a limited defensive war as spring approached. They arrived, in Sukhanov's words, "an unprepared, ignorant, hesitant mass," their ingrained peasant suspicions piqued by everything they saw and heard.[25] "A crowd of soldiers would invade a munitions factory at night to check on the work of the night shift," wrote one man who represented the Executive Committee on a number of such occasions. "Workers would be assembled in the court[yard], nervous and angry, facing grim, suspicious, and often arrogant soldiers."[26] Tension stood at a fever pitch. "Just wait," Sukhanov remembered the soldiers saying on similar occasions. "We'll put one of our comrades with a rifle next to every one of the loafers in your workshops."[27]

Clearly, the soldiers from the front blamed the workers in the rear for the shortages that had cost so many of their comrades their lives. Yet they were not certain of their ground, nor had they closed their minds. They approached Petrograd's factories with the skepticism of peasants made cautious by centuries of deception, but they came prepared to listen, to learn, and to understand. Gradually, Petrograd's more politically conscious workers won the soldiers to their cause. "Step by step," Sukhanov remembered, "the Petrograd workers— in private conversations, meetings, in the Soviet, in the barracks, and in the factories—explained to the soldiers the true state of affairs."[28] "We won't forget the front," one of the *putilovtsy* assured a group of soldiers. "We are working hard to turn out shells. But there's no coal, and only enough oil and raw materials for four days. The old government left us in a mess."[29]

What they saw and heard in Petrograd made Russia's front-line soldiers more sympathetic to the views of the Soviet and more critical of the government's determination to press on to "final victory." Even before Guchkov and Miliukov resigned at the beginning of May, soldiers at the front began to turn against even the "defensist" war they had supported during the first weeks of the revolution. General Knox was appalled to hear men exclaim: "The devil with the sixteen Governments [i.e., provinces]" that had fallen to the enemy during the Great Retreat, when he urged a group of soldiers to continue fighting in order to win back the territory that Russia had lost.[30] "The general mood of the army is becoming more strained with every day that passes," Fifth Army's commander reported at the end of March. "Politics, now involving all strata of the army, involuntarily has turned everyone's attention from the front to what is happening in Petrograd," he went on. "This makes the entire rank and file desire only one thing—the end of the war and the opportunity to return home."[31] Just over a fortnight later, General Alekseev warned Guchkov in a report on "the moral condition of the army" that the situation in the army was "getting worse every day," and that "a pacifist frame of mind" was developing in the armies under his command."[32]

By mid-April, Russia's front-line soldiers began to fraternize with the enemy all along the front despite stern orders from their commanders that "all enemy attempts to make personal contact with our troops, must always be met with only one answer—the bayonet and bullets."[33] When their officers tried to end their meetings with enemy soldiers by calling in artillery fire, the men took matters into their own hands. "Infantrymen often cut the telephone lines between artillery batteries and their forward observers [and] warn the artillery that if it begins to fire upon the enemy, they will hoist its gunners on their bayonets," two Duma investigators reported just after the middle of the month.[34] Even the Petrograd Soviet's pleas to all "comrade soldiers" to "defend Revolutionary Russia," and its warnings about the indignities that "German imperialism" would inflict upon them all "if the Russian army sticks its bayonets into the ground and announces that it doesn't want to fight any more,"[35] could not chill the moments of crude warmth that followed the cries of "We won't if you don't!" that rang out from the Russian trenches as men sick to the death of killing replied to the clumsy German and Austrian pleas of "Rus! Don't shoot!"[36] In the office of *Novaia zhizn* Gorkii took heart for a moment. "It is apparent that the accursed war, begun by the greed of the classes in command, will be ended by the power of the common sense of the soldiers," he exulted. "If this happens, it will . . . give man the right to be proud of himself—his will shall have conquered the most abominable and bloody monster, the monster of war."[37]

Russia's senior generals did not share Gorkii's joy at the sight of Russian soldiers drinking and singing with their Austrian and German enemies. Now commander in chief of all Russian land forces, General Alekseev summoned his front commanders to Petrograd at the beginning of May for a meeting with representatives of the Provisional Government's new coalition cabinet to warn that the collapse of discipline threatened Russia's survival. "We have no well-disciplined troops," General Brusilov announced decisively as the meeting opened. "An enemy success may easily become a catastrophe." The Fifth Army's commander, General Avraam Dragomirov, who had been entrusted with defending the important northwestern city of Dvinsk, reported that the desire for peace had become so strong that "reinforcements refuse to accept equipment and arms and say: 'They are no good to us as we do not intend to fight.' " All around the table, others added similar reports.

One after another, Russia's senior generals begged the government to give them the power to force instant obedience from their men. "You have taken the ground from under our feet," the fiery Dragomirov exclaimed. "Will you kindly restore it?" General Gurko, the hot-headed cavalryman who had acted as *Stavka*'s chief of staff during Alekseev's illness late in 1916, spoke in the same vein. "You must help," he pleaded. "It is easy to destroy, and if you know how to destroy—you should also know how to rebuild." As he summed up his commanders' remarks, Alekseev's words took on a menacing tone. "The Army is on the brink of the abyss," he began. "Another step and it will fall into the abyss and will drag along Russia and all her liberties, and there will be no return." Gurko, who resigned his command in outrage just a few days later, had the last word. "If you do not cease to revolutionize the army," he warned the men who had just taken over the reins of Russia's civil government, "[then] you must assume power yourselves." The generals left, "fully conscious," *Stavka*'s Denikin wrote, "that the last card had been beaten."[38]

Not only had the generals' last card been beaten, but, from their point of view, far worse things lay in store that were directly connected with Kerenskii's move to the War Ministry. Less than a week after Nicholas's abdication, the Soviet had taken up the thorny question of soldiers' rights, and had insisted that "soldiers are to enjoy all the rights of a citizen," that "soldiers have the right to organize themselves," and that "a soldier cannot be subjected to punitive measures or prosecution without a trial."[39] To work out the details of a formal declaration of soldiers' rights, and to revise Russia's military regulations to reflect the new revolutionary order, the Soviet had established a special commission of experts which, at the end of April, had prepared draft regulations that Guchkov had thought so revolutionary that he chose to leave office rather than sign them. Citing "conditions which I am powerless to alter," and warning that these "threaten the defense, freedom, and even the existence of Russia with fatal consequences," he resigned on May 1 so as not to "share the responsibilities for the grievous sin which is being carried on against the fatherland."[40]

Ten days later, Kerenskii issued the famous Order No 8, "The Declaration of Soldiers' Rights," as one of his first official acts as Russia's new minister of war. Russia thereby bestowed full personal and political rights upon her soldiers, which allowed them to take part in all political activities (including antiwar propaganda) when off duty, and permitted them unlimited access to antiwar and antigovernment literature. Worst of all, from the point of view of the army's senior commanders, only "elected army organizations, committees, and courts" could punish soldiers except in cases of direct insubordination in combat where officers still could use armed force to compel obedience. Ominously, however, the Declaration did not guarantee the army's support to an officer who forced men into battle, for its provision that an officer could do so only "on his own responsibility" left him open to arrest or demotion if any one of the army's many elected authorities questioned his actions at a later date.[41]

"Let the freest Army and Navy of the World prove that there is strength and not weakness in Liberty," Kerenskii reportedly exclaimed as he published Order No. 8. "Let them forge a new iron discipline of duty and raise the Armed Power of the country."[42] Determined that "an unshakeable, almost automatic conviction of the inevitability and necessity of sacrifice" must be reborn in the hearts of Russia's fighting men, Kerenskii now stormed to the front to "make it possible again for everybody to look death in the face calmly and unflinchingly." Again and again he spoke. "Forward to the battle for freedom!" he exhorted Russia's weary soldiers. "I summon you not to a feast but to death!"[43] Everywhere, he scored stunning oratorical triumphs as he launched his campaign to restore Russia's will to fight. At the front, where thousands of troops assembled to greet him, Kerenskii's words lifted his listeners to unheard of peaks of hysterical patriotism. "The *word* created hypnosis and self-hypnosis," one general explained as he recalled the magic moments when Kerenskii urged Russia's battle-weary soldiers to take up their arms once again, this time in the name of the revolution.[44] For a brief moment, it seemed that words could overcome the problems that too many words, spoken too often and too persistently, had created. But sound and fury could not alone bring lost dreams of victory back into focus. What Gorkii called "the jingling copper of empty words" that fell so readily from the lips of "wordmongers [who] beget wordmongers" might stir men's souls, but they could not induce them to face death without good reason.[45]

Among Russia's revolutionary armies, Kerenskii's words briefly stirred intense enthusiasms, but they failed almost universally to generate what he later called "a new will to action" needed to restore the "unshakeable, almost automatic conviction of the inevitability and necessity of sacrifice" that he knew Russia's newly-liberated soldiers must have to launch the offensive that he and the nation's generals had planned for mid-June.[46] Nor could his words heal the festering sores of domestic crisis that the war had aggravated so intensely that spring. "Our motherland is absolutely turning into some kind of madhouse, where the madmen take all the initiative and are in command, while people who have not yet gone insane crouch against the walls in fear," the Kadets' newspaper *Rech* exclaimed in mid-May.[47] "The war was becoming more and more unbearable," Sukhanov wrote. "The elemental forces against the war, against its support, and against its entire organization—were accumulating drop by drop, day after day."[48]

Kerenskii's impassioned pleadings could not breach this rising wall of antiwar sentiment. "The rank and file welcomed him enthusiastically, promised him everything he could have wanted, and then broke every one of their promises," General Brusilov later wrote in disgust, while a sergeant on another front looked on in amazement while "the identical crowd which gave an enthusiastic welcome to Kerenskii would accord a similar reception to a Bolshevik or Anarchist agitator an hour later."[49] "They trust no one," a young officer explained in a letter to Gorkii. "God knows who this Kerenskii is [they say], maybe he wants to fight another three years."[50]

In Russia's peasant villages, in her cities, and, most of all, at the front, men and women turned against the war as Kerenskii's exhortations lost their effect. In a movement so vast that it "recalled to mind a vast migration of peoples," Sukhanov thought as he looked on in amazement, men left the front. Day after day, week after week, Sukhanov watched as "a huge flood of soldiers took off for home without any sort of permission whatsoever. They clogged all the railroads," he remembered, "terrorized the authorities, threw passengers off the trains, threatened the entire transport system, and, in general, created a civic disaster."[51] Many turned their backs on the war, but a few swore to dedicate themselves anew to the struggle for victory to which Kerenskii continued to summon them. On the southwestern front, in the days just before he replaced Alekseev as Russia's commander in chief, General Brusilov began to organize such men into special "voluntary revolutionary battalions for the formation of shock attack groups."[52]

Despite Brusilov's optimistic support for such shock battalions, Alekseev remained skeptical about their value, especially when he learned that the southwestern front's commander hoped to recruit some of them from the revolutionary garrisons of Petrograd and Moscow.[53] With the help of the newly-formed "All-Russian Central Committee for the Organization of a Voluntary Revolutionary Army," Brusilov persisted nonetheless. At a number of public assemblies—England's young Bernard Pares thought the ones he attended were reminiscent revival meetings[54]—carefully-chosen speakers recruited volunteers to whom they promised glory in exchange for an oath "to believe that my death for our homeland and for the freedom of Russia is happiness."[55] "Forward!" the Executive Committee for the Formation of Revolutionary Battalions from Volunteers in the Rear proclaimed in support of Brusilov's effort. "Everyone who holds dear the fate of our motherland, everyone—workers, soldiers, women, cadets, students, officers, officials—who treasures the great ideals of the brother-

hood among peoples, join us under the red banner of the volunteer battalions!"[56] Thousands responded, but Pares, like Alekseev, remained unconvinced
that their dedication would survive even the first days of army training. "Conver-
sions to discipline," he remarked, "were, in public, fashionable,"[57] but he rightly
suspected that many who answered the call to march to a heroic death changed
their minds when it came to signing the stern "Oath of a Revolutionary-
Volunteer."

For one woman in particular, the promise to liberate Russia from the enemy
or die became a sacred mission. Maria Bochkareva, daughter of a former serf,
had grown up in Siberia and was almost twenty-eight when the revolution broke
out in Petrograd.[58] A natural-born leader, she had left home before her fifteenth
birthday, took the nickname of Iashka, and spent her early adult life as a foreman
of male construction gangs that spread asphalt in the boom towns that were
springing up in the wake of the newly-opened Trans-Siberian Railroad. Accustomed to living a man's life among men, and as tough in mind as she was in
body, Iashka Bochkareva hastened to volunteer for the infantry the moment
news of the Great War's outbreak reached her in the western Siberian city of
Tomsk. "A *baba*—a peasant woman—who wants to enlist!" Her effort provoked
roars of laughter wherever she went. "Women are not made for war," she
remembered them telling her. "They are too weak."[59] Nonetheless, Iashka
persisted until a sympathetic battalion commander helped her petition the tsar
for special permission to enlist. To everyone's amazement, Nicholas granted her
request. By early 1915, *baba* Iashka was in the trenches, fighting in the Fifth
Corps as part of the reconstructed Second Army under General Gurko's command.

A sergeant when the February Revolution broke out, Iashka had spent more
than two years in the trenches, had been wounded several times, and had won
a number of medals for valor. Like Russia's other peasant soldiers, she struggled
with the vexing questions of war and peace, of discipline and democracy. Bitter
at the collapse of the army that spring, she appealed to Duma President Rodzianko to restore discipline to the army. At the beginning of May, Rodzianko
brought her to Petrograd and urged her to express her views to his friends and
to the Soviet. Could the army be restored as a fighting force? And, if so, how?
Those were the questions the Petrograd Soviet faced, and they were the ones
about which they asked Iashka's opinion. Hopeful that "a few women at one
place could serve as an example to the entire front," Bochkareva proposed "to
shame the men" by organizing a Women's Battalion of Death that would
become "an example to the army and lead the men into battle."[60] Her proposal
received a mixed reception, especially from men who cynically raised the inevitable question: "Who will guarantee that the presence of women soldiers at the
front will not yield there little soldiers?"[61] Nonetheless, with support from
Rodzianko, Brusilov, and Kerenskii (who had received a similar proposal to form
the Black Hussars of Death from Valentina Petrova, an equally sturdy female

volunteer in the 21st Siberian Rifles), Bochkareva recruited volunteers for the First Russian Women's Battalion of Death.[62] On July 8, with Bochkareva proudly wearing a lieutenant's stars, the women entered the trenches, ready to join Tenth Army in an attack.

During the next two months, Bochkareva's Women's Battalion of Death served as an example "to shame the men" only once and it cost them nearly 80 percent casualties. Men's battalions of death suffered similar losses, and the Reval Battalion of Death actually lost all but fifteen of its three hundred members when other Russian infantrymen fired upon it after it had stormed the enemy's trenches.[63] In the meantime, hordes of deserters continued to pour into the rear, utterly sick of war and desperate for peace. "If you were now to go out on the village square and to proclaim that the war will end at once, but only on one condition—that Nikolai Romanov returns to power—every single man would agree and there would be no more talk of a democratic republic," the commander of the First Guards Infantry Division told General Knox at the end of June.[64] As the numbers of deserters climbed to nearly twelve thousand a week in mid-July,[65] battalions of death had to be diverted from the front to round them up. During a single night, one such battalion seized twelve thousand deserters in the vicinity of the West Russian city of Volochisk, which lay along the main highway leading from the front.[66]

Determined not to lose their lives before they reaped the revolution's rewards, these deserters had abandoned the front to escape the offensive Kerenskii had ordered in the vain hope that a call to arms could weld Russia's crumbling army together.[67] "If Russia's present activity at the front and the collapse of the army's strength—its discipline—continued," Kerenskii remembered warning a group of his chief military planners, "the Germans as well as the Allies would lose all respect for us. . . . It was our duty to Russia," he concluded, "not only to stop the decay in the army, but also to weld it together again into an efficient fighting force."[68] Whether that could be done remained doubtful indeed as summer approached. Few could have taken much heart from the southwest front commander's report to *Stavka* that, "although there still is not complete readiness [among the rank and file] to launch an offensive, only certain individual units continue to oppose the idea categorically."[69]

Even had Russia's soldiers been more enthusiastic, it would have been difficult for anyone to have begun a more ill-fated undertaking than the offensive Kerenskii planned for the summer of 1917. During his first month as war minister, he had replaced several senior army commanders, beginning with General Alekseev, whom he replaced with General Brusilov as supreme commander of Russia's armies less than three weeks before the offensive began. At about the same time, Kerenskii replaced the commanders of the southwest and western fronts, from which the heaviest weight of men and weapons would be launched during the coming offensive. On the southwest front, the Eleventh Army changed commanders just five days before the offensive opened, while the

Eighth Army struggled through the same process just two weeks before.[70] These new commanders had to send their armies into battle before they could take proper measure of their subordinates or accurately estimate the state of their armaments, communications, and supply networks. In the meantime, although he had fought the entire war on the southwest front, Brusilov had little time to counsel them. At *Stavka* he had more than enough problems of his own as he struggled to work with a new staff unaccustomed to his methods of command. At the very moment when certainty was a key prerequisite for success, everything had become uncertain. "To tell the truth," Brusilov remarked later, "the government itself did not know for certain what it really wanted."[71]

In vain Kerenskii proclaimed that "Russia, having thrown off the chains of slavery, has firmly resolved to defend, at all costs, its rights, honor, and freedom."[72] The army was falling apart, and even those fragments that might earlier have responded to Kerenskii's characteristically dramatic call to arms now stood unmoved. Eloquent summonses "to die for the eternal ideals of freedom" could not stir Russia's peasant lads now.[73] "What's the use of the peasants getting land if I'm killed and get no land?" one frightened recruit asked Kerenskii in a quavering voice.[74] "In essence, the war was over for us," Brusilov concluded, "for we have no means for getting the soldiers to fight."[75] General Knox put it even more bluntly. "There will be no success," he wrote as the first great barrages of the offensive began. "The one object of the [Supreme] Command seemed to be to avoid bloodshed," he observed in retrospect, "and it was only by bloodshed that discipline could have been restored."[76]

While his commanders struggled to move unreliable men into attack positions in unfamiliar territory, Kerenskii himself passed the early morning hours before the Russians launched their first infantry attacks on June 18 at Seventh Army's main observation post. "We were gripped by a terrible fear that the soldiers might refuse to fight," he later confessed. Then, he and the Seventh Army's commanders saw the men before them begin to advance, "their rifles at the ready, charging toward the front lines of German trenches."[77] Overcome by relief, Kerenskii rushed to proclaim "the great triumph of the revolution," and telegraphed to Petrograd for permission "on behalf of the free people to hand to the regiments who have participated in the battle of June 18 the red banners of the revolution."[78]

Even though most experienced commanders failed to share Kerenskii's optimism during the first days of the offensive, all the statistics at that point promised a decisive victory. The Russian forces outnumbered their enemies by almost three to one. Thanks to weapons shipments from France, England, and the United States, and to the increased production of her own factories during the fall and winter of 1916, Russia's infantry had great concentrations of artillery to support their first attacks, and the southwest front, where Brusilov had planned his main assault, had a well-enough developed rail network so that its commanders could mass infantry in heavy concentrations. Never had Russian soldiers known the luxury of artillery preparations in which their guns stood less than thirty meters apart along a hundred kilometers of front, nor had they ever enjoyed the advantage of outnumbering the enemy's heavy guns by more than five to one. At first, this huge weight of men and metal seemed decisive, as the Russian Seventh and Eleventh armies smashed through the first two or three lines of enemy trenches and seized Brzeżany and Zborów on their drive to retake Lwów. Five days later, the Eighth Army occupied Stanisławów, crossed the Bystritsa River, seized Halicz, and moved on to take Kalusz. Quickly, the Russians pushed forward, opening a breach some thirty kilometers deep in the Austro-German front.

As Brusilov ordered supporting offensives on the western and northern fronts, the armies of revolutionary Russia seemed on the verge of a major victory. Now in command of Russia's western front, General Denikin rained such massive artillery barrages upon the German positions that his infantry took the first three lines of trenches almost unopposed. On June 10, the Fifth Army attacked toward Dvinsk in the north and took the first line of enemy trenches with similar ease. Clearly, the Germans were scarcely better able than the Austrians to stand against the crushing weight of metal that Russia's heavily reinforced artillery now rained upon them. Denikin remembered it as "a fire of such intensity as had never been heard before," and added in one report that "in all the three years of war I had not seen such wonderful work of the artillery."[79] Certainly, Russia's commanders had never before seen artillery fire in such concentrations, or with such massive quantities of shell. War correspondents and politicians began to draw ill-advised and inaccurate parallels with the campaigns of revolutionary France in the press, and some even appeared in official military communiqués. For a brief moment, it seemed that the free draughts of revolution had breathed new life into Russia's demoralized soldiers.[80]

Yet, to those who looked more closely and did not allow wishful thinking to cloud their observations, disturbing signs were everywhere, and danger signals appeared even on the first day of the offensive. Even as Kerenskii hurried from the Seventh Army's observation post on the morning of June 18 to send his ecstatic announcement that Russia's "revolutionary army" had proved "to the whole world its supreme fidelity to the revolution and its love for liberty,"[81] General Knox jotted notes in his field diary that predicted not victory but disaster. "Knox . . . is causing trouble," Kerenskii complained to Russia's foreign minister in a confidential telegram. "[Have him] recalled from the front."[82] But recalling Knox could not do away with the disturbing incidents he had observed in the ranks of Russia's revolutionary divisions. Soon, events proved just how wrong Kerenskii's unrestrained and uncritical optimism had been. Within days, the evidence to support Knox's early warnings became overwhelming.

As the Germans launched their first counterattacks, the Russians harvested not the first plums of victory but far more bitter fruits. Even as their divisions advanced against minor opposition, suffering casualties that were minimal in comparison to those Brusilov's men had sustained the year before, Russia's commanders looked on in dumbfounded amazement as their men turned their backs upon victory. Eyewitness accounts of those terrible days read for all the world like a grotesque catalog of actions by men perversely determined to suffer defeat in the face of victory:

>   —It is probable that a large number of the men hid in the woods and only returned when they got hungry and were sure that the fighting was over. [General Knox's report about the 23rd Division's performance on June 18.][83]
>   —Some elements voluntarily evacuated their positions without even waiting for the approach of the enemy. [Eleventh Army Report.][84]
>   —Some units launched their attacks, crossed two or three lines of enemy trenches as if on parade, paid a visit to the enemy's batteries, removed the gunsights from his artillery, and . . . returned to their own trenches. [General Denikin's description of Tenth Army's advance.][85]
>   —I attended the moving ceremony when the Potiiskii Regiment received [one of] the red banners [of the revolution] and when the soldiers swore by it to fight to the very death. This same Potiiskii regiment retreated fifteen *versty* just one hour before the attack. [General Denikin's description of the preparations for his offensive on the western front.][86]
>   —Many units held meetings and formed committees to discuss the orders they had received to attack, let a great deal of precious time slip by, and then absolutely refused to carry out the orders. [Report on conditions in the Eleventh Army in early July.][87]
>   —The disintegration of our armies continues to develop. I have received word that in some units the officers are being slaughtered by their own men. Today I received a report that, in one division, the chief of staff was murdered in this fashion. [Alekseev's diary, June 10.][88]

All along the front, from Rumania to the Baltic, the breakdown spread. "The German [counter]offensive, which began on July 6 on the front of the Eleventh Army, is assuming the character of a disaster which threatens a catastrophe to revolutionary Russia,"[89] a group of desperate commissars, assigned to pour new fire into the veins of its soldiers, reported to Kerenskii on July 9. "The army is on the run," the Eleventh Army's commander warned *Stavka* three days later. "It is hard to conjecture where the enemy might be stopped."[90] Even appeals from the Petrograd Soviet not to "retreat a single step before the enemy" had no effect. When several delegates from the Petrograd Soviet came to the front to meet with disgruntled units in the Tenth Army, the soldiers attacked one of them and "beat him on the head until he was covered with blood."[91]
Suddenly even the Soviet, for so long a defender of soldiers' rights, under-

stood that a revolutionary army without discipline threatened the very revolution it had sworn to defend, and hurried to condemn the beating as a "manifestation of the dark instincts of the irresponsible masses of soldiers." That a member of its Executive Committee had been the victim of such "senseless, savage behavior," the Soviet's newspaper *Izvestiia* insisted, indicated that energetic measures must be taken to instill "into the very heart of our army a true understanding of the meaning of the new principles on which the Russian revolutionary army is being built." The Soviet all too readily blamed Lenin's Bolsheviks, "who throw slogans at the masses which undermine the authority of the recognized and authorized organs of revolutionary democracy."[92] Others who were wiser placed the blame more precisely where it belonged. "It is time for us to come to our senses," Brusilov wrote to Kerenskii on July 11. "It is necessary to restore iron discipline, in the fullest sense of the term. . . . If we delay even a moment, the army will perish, Russia will perish, and we will sink into infamy."[93] Brusilov did not have long to wait. "The government was faced with a tragic choice either to sacrifice the army to cowards and traitors or to restore the only penalty that could frighten them," Kerenskii wrote. Unanimously, Russia's ministers restored the death penalty, but "only," Kerenskii explained to his allies on the left, "to preserve the priceless lives of heroes who would selflessly die fulfilling their duty to the motherland."[94]

The situation had reached the point of desperation when, at Kerenskii's request, Russia's senior generals met with him and Foreign Minister Tereshchenko at *Stavka* on the afternoon of July 16 to discuss how the army might be restored as an instrument for Russia's defense. Warning that "history shows that there are limits to the degree of freedom that can be permitted in any army," Brusilov asked his commanders to state their views. In addition to Brusilov and Alekseev, Generals Klembovskii (commander in chief of the northern front) and Denikin (commander in chief of the western front) spoke at length, while Shcherbachev (commander of the Rumanian front) and Kornilov (just appointed as commander in chief of the southwest front) added their views in lengthy telegrams. All agreed that the Provisional Government must restore officers' authority to demand obedience from their men and insisted that the death penalty be restored in the rear as well as at the front. Yet these men understood that even these stern measures offered no panacea to cure Russia's ills. "Can we really execute entire divisions?" Klembovskii wondered aloud as his colleagues talked of using firing squads. And could one talk realistically of using field courts-martial either? If all the offenders were brought to trial, Klembovskii warned his colleagues, "half the army would end up in Siberia."

Nothing could be done overnight. "It was easy to destroy the army," Denikin told Kerenskii bluntly, "but time is needed for its restoration." In the meantime, Russia's leaders faced the now-victorious divisions of Austria and Germany with an army which, Alekseev lamented, "remained nothing but human dust."[95] All agreed that the human tide that now poured from the trenches to the rear must be stemmed, but when the generals' meeting broke up just an hour before midnight, none had yet determined how that might be accomplished. Russia faced "difficulties beyond the wildest imaginings," Alekseev said a few days later.[96] Led astray by the fleeting charisma of empty words uttered too often and too irresponsibly, Russia longed for a hero, a man of fire and iron—what Kerenskii once called "a general on a white horse"[97]—whose inflexible will could weld its shattered human and institutional fragments together. Lavr Georgievich Kornilov, the daring son of a Siberian Cossack, seemed to many that summer to be just such a man.

Born, like Lenin, in 1870, Kornilov had grown up in eastern Siberia, studied in the Siberian Cadet Corps and the Mikhailovskoe Artillery Academy, and received an officer's commission in the Turkestan Artillery Brigade at the age of twenty-two.[98] For most of his career, he had served in Asia—in Turkestan, Manchuria and, as a military attaché, in China. A newly-promoted general when the Great War began, and known throughout the army as a man of resolve and daring, Kornilov had fought on the southwest front, and his 48th Infantry Division had spearheaded the Russians' victorious advance through the Carpathians in the spring of 1915. There, his effort collapsed when he could not replace lost weapons or replenish his soldiers' ammunition. Wounded and captured during an Austrian counterattack, Kornilov found himself confined in a small town south of Vienna. After almost a year in Austrian hands, he donned the uniform of an Austrian soldier, slipped past his guards, and made his way to Budapest, where he took a train in the direction of Rumania. Obliged to make the last part of his journey on foot because of watchful sentries, Kornilov spent a fortnight walking from village to village, always moving closer to the frontier. For three days, he lived only on wild berries, as he studied the movements of the Austrian border patrols. Then, choosing a dark night for his final effort, he slipped into Rumania. Safely in Allied territory, he hurried to rejoin the Russian army.[99]

To a nation starved for heroes, Kornilov became a legend overnight. His piercing, slanted black eyes set off by high cheekbones and a thick black mustache, his lean, wiry frame accentuated by quick, catlike movements, Kornilov projected a hero's aura among soldiers and civilians with equal ease. Surrounded by a bodyguard of fierce-looking, machine-gun-carrying Central Asian tribesmen renowned for their horsemanship and their unswerving loyalty to the man they had sworn to protect,[100] Kornilov stirred sharp emotions everywhere. General Denikin later described him as "a fighting general who carried fighting men with him by his courage, coolness, and contempt for death." Not willing to see heroes made so easily, General Alekseev thought him "a man with a lion's heart and the brains of a sheep,"[101] although that description hardly seemed to take into account that Kornilov spoke several Central Asian languages fluently, wrote poetry in Tadzhik, and had published a book and several articles about eastern Persia, Turkestan, Baluchistan, and India.

Kornilov was commanding the XXV Corps, a part of the southwest front's Special Army, when Nicholas' abdication left Alekseev in command at *Stavka*. In search of a "renowned fighting general" whom the masses would respect and obey, the Provisional Government urged Alekseev, "for the salvation of the Motherland, for victory over the enemy, and so that those innumerable victims of this long war will not have been sacrificed in vain on the very eve of victory," to appoint Kornilov to command the Petrograd garrison.[102] Proclaiming that the revolution "insures victory over the enemy," Kornilov set to work to bring Petrograd's disorderly soldiers under control, and on at least one occasion dispersed a crowd of disgruntled troops by driving his staff car into their midst and lecturing them about duty, country, and sacrifice.[103] But Kornilov remained too much the soldier and too little the politician to navigate the murky currents of Petrograd revolutionary politics comfortably. Concerned for the safety of the government during the April Crisis, he had ordered his artillery into the city's center, evidently with every intention of using it to break up crowds if they became threatening. Outraged, the Soviet demanded that the garrison refuse all orders to fire unless they came directly from its chambers. Equally irate at what he considered unwarranted civilian interference in military affairs, Kornilov insisted upon being transferred back to the southwestern front.[104]

The opening of the "Kerenskii Offensive" therefore found Kornilov leading Brusilov's old Eighth Army in victorious assaults against Halicz and Kalusz before the Austro-German counteroffensive broke their ranks and drove them back in disorder. Convinced that neither the Provisional Government nor the Petrograd Soviet had the courage to confront the problem of the army's crumbling discipline squarely, Kornilov moved resolutely. Proclaiming that "an army of maddened, benighted people whom the authorities have done nothing to protect against systematic disintegration and debauchery, and who have lost all traces of human dignity, is in full retreat," Kornilov bluntly announced that "the death penalty will save many innocent lives at the cost of those of a few cowards and traitors."[105] Mercilessly, he ordered the few Battalions of Death that had survived the campaign's first assaults to arrest looters and deserters and to hang them at every crossroads.

Kornilov's draconian orders came at the very moment when he found an unlikely ally in the person of Boris Savinkov, a notorious terrorist and associate of Kerenskii, whom the revolution had elevated to the position of commissar of the southwest front.[106] Once described by Gippius as "a soul choked in blood,"[107] Savinkov had entered Russia's revolutionary movement as a fighter in the Socialist Revolutionaries' Battle Organization at the beginning of the twentieth century. Between July 1904 and February 1905, he had had a hand in the assassinations of Minister of Internal Affairs Viacheslav Plehve and Grand Duke Sergei Aleksandrovich, two of the Russian terrorists' most notorious successes. Arrested and sentenced to execution in 1906, he had fled to the West, where he spent the decade before the February Revolution writing several novels that glorified a deranged cult of assassination, and serving as a volunteer in the French army. Returned to Russia after the February Revolution, Savinkov had risen quickly to responsible positions after the fall of the Provisional Government's first cabinet. As commissar of the southwest front, he envisioned himself as a mediator, a responsible third party able to moderate disputes between commanders appointed by the old regime and the rank and file of Russia's new revolutionary army, and he expressed himself as "almost completely" in agreement with Kornilov's views about restoring discipline.[108]

Seeing Kornilov as the hero who might yet save Russia from defeat, and hoping to use him to advance his own fortunes, Savinkov urged Kerenskii to name him as a replacement for General Aleksei Gutor, the just-named commander of the southwest front. Brusilov thought the proposal ill-advised because "any change in the make-up of the high command, and especially of someone in such a high position as the commander of a front, simply because the soldiers' deputies had asked for it, was bound to have serious repercussions."[109] Still, he did not press his views too strongly and gave in to Kerenskii's and Savinkov's urging soon after the summer offensive began. Unknowingly, he had named not only Gutor's replacement, but his own. Convinced after his meeting with Russia's senior generals at *Stavka* on July 16 (and with some added private urging from Savinkov) that Brusilov could not save the army, Kerenskii appointed Kornilov supreme commander of Russia's armies two days later. That same day, July 18, he named Savinkov deputy minister of war.

If Kerenskii expected to find in Kornilov a pliable instrument, the general's first acts proved him gravely mistaken. Kornilov insisted that he would answer to none but his "conscience and to the people [of Russia]," that he would allow neither government nor Soviet to interfere with his military operations, that he would permit no one else to appoint or relieve senior commanders, and that the death penalty must be applied at the front and in the rear by special courts under his personal control.[110] At that point, only political pressures (Kerenskii had been trying to form another coalition government for almost a fortnight) and the best efforts of Savinkov prevented an open break between these two strong-willed men, each of whom felt certain that he alone held the key to Russia's salvation.[111]

While Kornilov chafed to set his command in order, Kerenskii searched for ways to make his stern demands more palatable to a Soviet made increasingly restive by rumors of something untoward in the wind. During the next three weeks, Kornilov came twice to Petrograd to plead with Kerenskii and his fellow ministers for unfettered authority to impose his will upon Russia's crumbling land forces. A man who preferred action to political considerations, he soon grew contemptuous of his government's hesitation, and more willing to heed the urgings of others anxious to use his hero's image for their own purposes. As Kornilov struggled to win the authority he needed to end the flight of Russia's soldiers from the front, some of Russia's great industrialists publicly placed the "hope and faith" of "all thinking Russia" in him. "May God help you," they proclaimed on August 9, "in your heroic undertaking of reconstructing a powerful army and saving Russia."[112]

The next day, Kornilov came to Petrograd to make his second appeal to Kerenskii and his ministers. Surrounded by his scarlet-coated Central Asian bodyguards—Russians called them the *Tekintsy*—with their swords unsheathed and their machine guns at the ready, Kornilov did nothing to quiet the government's growing fears that he might soon lead a coup d'état.[113] None was more apprehensive at that moment than Kerenskii. "I tried to make Kornilov realize that any attempt at hasty and violent action would produce [an] adverse effect on the army," Russia's nervous prime minister remembered. "[I warned him]," Kerenskii continued, "that if any one should try to establish a personal dictatorship in Russia he would find himself the next day helplessly dangling in space, without railroads, without telegrams, and without an army."[114]

The next morning, Kornilov returned to *Stavka* convinced that mortal danger threatened Russia because the government had lost the will to act. Beyond that, he suspected the government's loyalty. During his stay in Petrograd, Kerenskii and Savinkov had cautioned him about giving details when he reported on the condition of the army, because they feared that someone in the cabinet (Kerenskii blamed his rival Chernov) was leaking secret military information to members of the Soviet's Executive Committee which, they now feared, included men willing to betray Russia to the Germans. Although Kerenskii and Savinkov offered no evidence for their fears, Kornilov evidently jumped all too readily to the obvious conclusion. "That's your Provisional Government for you!" he exclaimed to his chief of staff General Aleksandr Lukomskii. "It's got known traitors sitting right there in its inner councils." His anger rising, he continued: "It's time to hang that whole bunch of German spies and sympathizers, the chief of whom is Lenin," he went on. "The entire Soviet of Workers' and Soldiers' Deputies needs to be broken up—broken up, so they'll never be able to meet again!"[115] Yet Kornilov was not ready to take matters into his own hands entirely, for he had other cards yet to play. The Moscow State Conference, made up of what Kerenskii once called "the very best elements of political, social, cultured, and military Russia," was to open the next day in Moscow's Bolshoi Theater, and Kornilov had returned to *Stavka* to find an invitation for him to address its delegates.[116] "I'll go there," he told his chief of staff, "and I'll finally get my requests acted upon."[117]

Once described as having been "conceived by Kerenskii in late July to familiarize authoritative political figures from all over Russia with the country's grave problems and to mobilize their support for the programs of [his] newly created second coalition [government],"[118] the Moscow State Conference attracted men of sober views from all over Russia. In sharp contrast to the political climate in Petrograd, where the radical Soviet held sway, the speeches at the Moscow Conference emphasized the themes of order and security that appealed more to Russia's bourgeoisie, aristocracy, and army officers than to workers, peasants, and common soldiers. Men of moderation now spoke to an appreciative mass audience for the first time in months and happily heard delegates greet their calls for the government to "put an end to the criminal agitation that is going on among our soldiers"[119] with applause that rose to roars of approval whenever they called for stronger government authority.[120] Clearly, the Russia of well-cut morning coats, starched collars, and carefully-trimmed fingernails was determined to have its say, and, as speaker after speaker mounted the platform, the wave of support for law, order, property, and "war to victory" rose higher still. These were men and women waiting for a hero. On the afternoon of August 13, many thought he had arrived.

Kornilov's arrival that afternoon proved to be the closest to a royal welcome that anyone had seen in Moscow since the revolution.[121] From the moment the ever-vigilant *Tekintsy* leaped from his still-moving train to open a path through the crowd, Kornilov seized Moscow's attention. Elegantly dressed women strewed flowers in his path. Senior officers and Duma politicians, whom the revolution's new proletarian heroes had long since relegated to the sidelines, cheered. And Madame Morozova, renowned philanthropist and redoubtable mistress of the famous "Portuguese" Castle that her family had built within sight of the Kremlin to impress Muscovites with their wealth, fell to her knees as he approached.[122] "You are the symbol of our unity," the Kadet politician Fëdor Rodichev exclaimed. "Save Russia and a grateful people will crown you!"[123] Like a conquering hero, Kornilov rode through Moscow's streets, readily acknowledging the crowd's cheers as he made his way to the tiny Chapel of the Iberian Virgin that stood just beyond the northern edge of Red Square. As so many of Russia's tsars had done before him, he prayed before the icon of the miracle-working Iberian Madonna, her head still framed in a net of pearls and surmounted by a diamond crown as it had been since the day almost three centuries ago when the brothers of Mt. Athos's Iberian Monastery had presented it to Tsar Aleksei Mikhailovich.

From the Iberian Chapel, Kornilov returned to his train to receive a procession of distinguished visitors that continued late into the evening. Aleksei Putilov, whose Petrograd factories produced so many of Russia's weapons, came to pay his respects, as did Miliukov, General Alekseev, and General Aleksei Kaledin, whose short stature and mild, shy manner seemed utterly at odds with his distinguished war record and powerful office as hetman of the Don Cossacks. Purishkevich, the archconservative assassin of Rasputin, along with a host of financiers, industrialists, politicians, and generals, all came and went that evening to see the man whom they hoped might yet spare them the loss of their factories, wealth, and power. Just before midnight, Kerenskii pleaded with Kornilov by telephone to speak only briefly, and only about the situation at the front, when he addressed the State Conference on the morrow. Kornilov refused. "I shall speak as I think fit," he told Kerenskii as he prepared to address the delegates, for he intended to present his case for absolute control over Russia's fighting men to the largest public forum then available.[124] "I thought it essential," he explained later, "to make known to the country the real state of affairs in its armed forces, and to point out how necessary it was to raise their battleworthiness."[125]

The next morning, Kornilov rode again through the streets of Moscow and ascended the rostrum in the Bolshoi Theater to a storm of applause. Yet, any who hoped to hear encouraging words from the man whom Kerenskii introduced to the assembly as "the first soldier of the Provisional Government" were destined to disappointment. "I have no confidence that the Russian army will fulfill its duty to our country," Kornilov began. "Since the beginning of August, the men have become like beasts, no longer resembling anything military." Warning that "the army must be restored at all costs, for without a reconstructed army there can be no free Russia and no salvation of our homeland," Kornilov demanded the authority to inflict ruthless punishments upon any who disobeyed his orders. "Only an army welded together by iron discipline, only an army led by the unified inflexible will of its leaders," he stated bluntly, "can achieve victory." The enemy stood at Russia's very gates. The city of Tarnopol had fallen in the South, and territories of Galicia and Bukovina had been lost. Now, the great Baltic port of Riga might fall at any moment. "If the instability of our army makes it impossible for us to hold our defenses on the Riga Gulf," Kornilov told his listeners, "then the road to Petrograd will lie open." There could be no more delay, no more debate. "We cannot afford to waste time," he concluded. "Not even a single minute can be wasted."[126]

Kerenskii's theatrical lament that "all the flowers of my dreams for mankind" had been "trampled upon and spoken of with disdain" could not drown the chorus of approval that greeted Kornilov's words. "Our nation can be saved from total disaster only by a truly strong government," General Kaledin exclaimed, while Vasilii Maklakov pleaded that Russia now must be governed "without utopias, sins, and unnecessary mistakes."[127] Like the politicians and generals, the newspapers took up Kornilov's warnings and spread them from one end of Russia to the other. "The shield and sword of the State and the people have turned into the knife of a murderer and the axe of an executioner," Petrograd's *Novoe vremia* warned, as it pleaded for decisive action against the revolution's excesses. Only two courses remained, its editors insisted. "Either the restoration of the Russian army along the lines indicated by General Kornilov . . . or its final disintegration, with the unavoidable ruin of Russia and all of her freedoms."[128] In vain did the Petrograd Soviet's *Izvestiia* assure its readers that the revolution "possesses its own curative powers,"[129] and proclaim its "faith that revolutionary Russia will be just as invincible as was revolutionary France."[130] Nor did Kerenskii's plea that he wanted "to follow a middle course"[131] find much sympathy among the men and women who had heard in Kornilov's speech the defense of law, order, and property they had longed for since March. Events of

that spring and summer had made too many thoughtful, articulate, well-to-do
Russians fear the worst. As Kornilov's train carried him back to *Stavka* that
evening, he took with him promises that these people stood ready to provide
millions of rubles to help him bring order and stability back to Russia.[132]

If the Moscow Conference had emphasized articulate and educated Russians' desperate longing for order and stability, the fall of Riga filled them with
fearful certainty that the enemy had broken through one of the last remaining
gates that had barred the way into Russia's hinterland. To be sure, the Germans
had thrown some of their most awesome weapons into their Riga offensive, and
these had included heavy concentrations of poison gas in addition to crushing
artillery bombardments.[133] Nonetheless, Fate had given Russia's Twelfth Army
a unique advantage at Riga, for, as one of her commanders later explained, the
German attack was "one of those rarities in the annals of the history of military
operations in which we could, on the basis of the information we obtained,
define with almost mathematical precision not only the point of the attack but
also the exact moment when it would occur."[134] That fact alone should have
enabled the Russians to mount a more resolute defense. Instead, Russia's peasant infantrymen fled so precipitously that one German general later described
the crossing of the Dvina River by the Eighth German Army as having been
done "almost in play."[135] Certainly, Russia's demoralized Twelfth Army offered
no more than token resistance to the Eighth Army's advance, and the absurdly
exaggerated reports of heroic Russian attempts to halt the German advance that
commissars on the northern front filed with their superiors could not disguise
the very obvious fact that the men in whom they had been ordered to instill a
revolutionary dedication to defend Russia at all costs had, in Kornilov's simple
words, "fled in disarray, leaving their guns behind them."[136]

Combined with the explosion of Russia's great ammunition and weapons
depot at Kazan that destroyed nearly a million desperately-needed artillery shells
and several thousand machine guns, and the murder of several senior officers by
rampaging soldiers, the flight of Russia's Twelfth Army at Riga convinced
Kornilov that decisive measures could not wait.[137] Clearly, the Soviet's urgent
appeals for "all revolutionary forces [to] unite for the salvation [of Russia and
the revolution]"[138] had no more value than the cheap paper they were printed
upon, for Petrograd's workers and soldiers continued to deny the undeniable fact
of the army's collapse and look elsewhere for the causes of their nation's defeats.
Reports about "disgraceful desertion from the field of battle," the Socialist
Revolutionary newspaper *Narodnoe delo, The Peoples' Causes,* insisted, simply
were "false," and its editors demanded that "an authoritative commission [be
appointed] to expose the [real] causes of the Tarnopol and Riga break-
throughs."[139] Such absurd statements only obscured the real issue which was,
as Kornilov had told the Moscow State Conference bluntly, that "a whole series
of legislative measures" beginning with Order No. 1 and the Soldier's Bill of
Rights had "transformed [the army] into a mindless mob, valuing nothing but

its own life."[140] Efforts to find scapegoats could not alter that fact, and no medicine other than the bitter draughts Kornilov prescribed could produce a cure. Riga had fallen. The way to Petrograd lay open for Germany's armies if they chose to follow up their gains, and frantic efforts to dismiss Kornilov's accounts of the German breakthrough at Riga as unprincipled "slander of the army" could not wipe away that undeniable fact.[141]

During the fortnight after the fall of Riga, a succession of incredibly confused events unfolded in which Kerenskii frantically tried to enlist Kornilov's support for declaring martial law in Petrograd, crushing the Bolsheviks, and fully restoring officers' authority over their troops, while in Moscow a group of right-wing financiers, industrialists, and generals plotted to overthrow Kerenskii and place Kornilov at the head of Russia's government. Precisely how much Kornilov allowed himself to be a part of either plot remains to this day unclear. What is certain is that he had begun to move elite units into place just beyond the limits of the Petrograd Military District early in August, that he began to concentrate these troops much more heavily after Riga's fall, and that, when these troops eventually marched against Petrograd, Kerenskii concluded that their chief purpose was to overthrow his government, not to crush the Bolsheviks as Kornilov always claimed.

Supported by two divisions of Cossacks from General Aleksandr Krymov's elite III Cavalry Corps, the renowned Savage Division of mountain tribesmen from the North Caucasus, whose fury in battle had long been legendary, formed the core of Kornilov's Petrograd forces.[142] If the Germans pressed on from Riga, these forces could become the nucleus of a special army to defend Petrograd. If the Bolsheviks resisted the government's efforts to restore order and discipline, these brutal cavalrymen could, as Kornilov later claimed, have "disarmed those units of the Petrograd garrison which had thrown in their lot with the Bolsheviks, disarmed the population of the city, and dissolved the Soviet."[143] Finally, if Kornilov had chosen to do so, he could have used these forces to overthrow Kerenskii's government and establish himself as the head of a military dictatorship supported by the right-wing conspirators from Moscow. The question that no one could answer then, and which cannot be answered for certain even now, is whether Kornilov planned to support a conspiracy to overthrow the Provisional Government or intended to stand with Kerenskii to restore order when he began to move his elite strike force into position toward the end of August.

Certainly Kerenskii and Savinkov, who on August 23 carried to *Stavka* Kerenskii's promise that the government would approve sterner measures to restore discipline in both front and rear, had every reason to think that they had won the supreme commander's support for their cause.[144] Kornilov had begun his meeting with Savinkov by announcing that "I do not trust any longer Kerenskii and the Provisional Government," but, according to Savinkov's report, which coincides very closely with the accounts set down later by Kornilov and his chief of staff General Lukomskii, further discussions soon cleared the air sufficiently for Kornilov to urge Russia's deputy war minister to "tell Aleksandr Feodorovich [Kerenskii] that I shall support him in every way, for the welfare of the Fatherland requires it."[145] None of these men trusted each other, and Kornilov and Kerenskii distrusted each other most of all. Still, it seemed to their deputies that an alliance had been struck, and that both were committed for the moment to restoring discipline in the army and to crushing the Bolsheviks in Petrograd.

Unknown to Kerenskii, Savinkov, or Kornilov, a chain of seemingly innocuous events which none of them could control had already been set in motion, and these not only changed their immediate plans but altered the entire course of the Russian Revolution. Strangely, an interview in which Kerenskii himself had been involved formed one of its central links, although Kerenskii apparently had been utterly unaware of it at the time. The day before Savinkov went to *Stavka*, Vladimir Lvov had insisted upon a meeting with Kerenskii at the Winter Palace. A politician whose liberal views in the Third and Fourth Dumas had won him the post of Director General of the Holy Synod after the February Revolution, Lvov had become known as a blunderer whose inflated aspirations to influence events contrasted as sharply with his ineptitude as did his heavy beard with his hairless head. As we can best determine, Lvov told Kerenskii that his "song was sung," spoke vaguely of representing "influential circles" that had grown displeased, and urged Russia's somewhat startled prime minister to abandon his alliance with the Soviet. "I asked him point-blank to tell me on whose behalf he had come," Kerenskii later reported. "He replied that he was not authorized to tell me."[146] After a few perfunctory remarks, Kerenskii sent Lvov on his way, evidently with some vague assurance that he could relate their conversation to "the people he represented," but almost certainly without any authorization "to enter into negotiations" on Kerenskii's behalf with "all the elements" he thought necessary, as Lvov later claimed. "I thought that the matter would end at that," Kerenskii explained later. "Generally speaking, I did not attach any importance to it."[147]

Kerenskii concluded that his visitor probably spoke for a "group of 'men who have been,' " or, as he described them in another context, "moderate conservative circles in Moscow" with whom he had no desire to make common cause, and whom, we now know, conspired to place Kornilov at the head of Russia's government. Undoubtedly, Kerenskii assumed that his awkward visitor would relay some version of their conversation to these men, but he did not anticipate Lvov's appearance at *Stavka* just a few hours after Savinkov had left. Nor did he expect him to present himself, as some observers reported, as "a man fully empowered to speak for Minister-President Kerenskii."[148] At the very least, Lvov's sudden appearance on such a vital mission, and in such a pivotal role, surprised Kornilov. "I had never previously spoken to him [at length]," he later explained. "I had not seen him since April and knew him only slightly."[149]

Everything about this new visitor seemed suspicious, especially when Kornilov and his staff learned that Lvov carried no written authorization from Kerenskii. "Why didn't Savinkov say anything about this, or even seem to know about it," Lukomskii mused aloud. "And why does Kerenskii entrust such a task to Lvov at the very moment when Savinkov arrives at *Stavka* [as his emissary]?" Strangely, although Lvov's reputation as a meddler and a muddler was widely known, neither Kornilov nor Lukomskii seemed inclined to doubt his honesty, and Lukomskii jumped instead to the conclusion that Kerenskii had concocted some sort of underhanded scheme. "I don't like the look of any of this," Kornilov's loyal chief of staff insisted. "And I am very much afraid of Kerenskii."[150] Heeding Lukomskii's warning, Kornilov therefore spoke mainly in generalities. Most of all, he urged decisive action against Lenin and his Bolsheviks, and warned again that Russia could not survive without a strong government. As they parted, Kornilov offered the safety of *Stavka* to Kerenskii and Savinkov should the Bolsheviks attempt the coup that he evidently expected to occur at any moment. Before he reported Kornilov's offer to Kerenskii, Lvov transformed it into a threat against the minister-president's life.

If Lvov's sudden appearance left Kornilov and Lukomskii profoundly suspicious of Kerenskii's motives, the meddler's return to the Winter Palace late in the afternoon of August 26 propelled Kerenskii's apprehensions about Kornilov into the lower reaches of hysteria. Evidently confident of Kornilov's support, Kerenskii had spent the previous four days making arrangements to crush the Bolsheviks and place Petrograd under martial law as a first step to establishing sterner government control over the revolution's course. At that critical moment, Lvov reappeared at the Winter Palace "strangely agitated," Kerenskii remembered, and gripped by "extreme excitement." Insisting that a Bolshevik uprising was a certainty, Lvov warned that Kerenskii was "doomed" because *Stavka* would do nothing to defend him. Since "no one would guarantee" Kerenskii's life under such circumstances, Lvov reportedly urged him to "get out of Petrograd," and to go "somewhere far away." At first, these statements struck Kerenskii as ludicrously absurd, but something about Lvov's urgency made him reluctant to dismiss the warnings out of hand. "It seemed to me," he later explained, that either Lvov "was insane or something very serious had happened." Lvov's name had never appeared in any of the many reports Kerenskii continued to receive about plots against him and his government. And yet, Kerenskii asked himself, what if Lvov's words were "even remotely in accord with reality?" and Kornilov was, in fact, about to launch a coup d'etat?

In an effort to learn more, Kerenskii pressed Lvov for details. Kornilov insisted that all ministers must resign, Lvov replied, and he demanded that "all military and civil authority shall be placed in the hands of the Generalissimo." Would Lvov put all he had just said into writing, Kerenskii asked, evidently hoping that the prospect of signing such a document would shake Lvov's certainty. Without a moment's hesitation, Lvov agreed, and "the readiness, the assurance, [and] the quickness" with which he performed the task removed Kerenskii's "last doubt."[151] Now certain that Kornilov had betrayed him, but equally certain that the Council of Ministers would not consider Lvov's signed statement sufficient proof of Kornilov's treason, Kerenskii hurried to the Hughes apparatus that connected the War Ministry directly to *Stavka*, intending to use its teleprinted transcript to confirm or deny Kornilov's duplicity. Making it seem that Lvov had accompanied him into the transmission room and was, in fact, asking the question, he asked Kornilov "is it necessary to implement that certain decision of which you asked me to inform Aleksandr Fedorovich [Kerenskii] confidentially? Without your personal confirmation," he went on, "Aleksandr Fedorovich is reluctant to believe me entirely." Without asking Kerenskii/Lvov to make clear what "certain decision" was referred to, and almost certainly thinking that the phrase referred to his offer of protection while his elite cavalrymen crushed the anticipated Bolshevik rising, Kornilov sent his fateful reply. "Yes, I confirm that I asked you [Lvov] to transmit to Aleksandr Fedorovich my urgent request that he come to Mogilëv."[152] On the basis of that single misunderstanding, Kerenskii came away from the Hughes apparatus convinced that Kornilov wanted to lure him to *Stavka* to place him under arrest.

The seven confusion-filled hours before midnight on August 27 were among the most momentous of any that passed between the fall of Nicholas II at the beginning of March and Lenin's triumph at the end of October. Their thoughts a welter of suspicions and their conclusions a tragic accumulation of misunderstandings, Kerenskii and Kornilov now altered forever the course of Russia's destiny. "Did Kerenskii understand at this moment that, by proclaiming himself Kornilov's opponent, he gave himself and Russia into the hands of Lenin?" Miliukov later asked rhetorically. "Did he understand that this was the government's last opportunity to challenge the Bolsheviks and still win?"[153] Unable to see into the future, Kerenskii clearly viewed the right, not the left, as his greatest enemies at that moment and acted accordingly. Largely as a result of his misunderstanding of Kornilov's replies to the questions he had sent on the Hughes apparatus, his suspicions supported by the claims of a meddler who wanted in some way to carve out a place for himself in Russia's revolutionary annals, Kerenskii concluded that Kornilov had set in motion a coup d'etat that would bring the III Cavalry Corps into Petrograd with the fearsome Savage Division rampaging in its vanguard. He therefore summoned his Cabinet and asked for extraordinary powers to deal with the "conspiracy" that he now called a "rebellion."

With the approval of his ministers, Kerenskii arrested Lvov, denounced to all of Russia Kornilov's "demand for the surrender by the Provisional Government of all civil and military power," publicly ordered him "to surrender the post of Supreme Commander," announced that he was "taking all necessary measures to protect the liberty and order of the country," and placed Petrograd under martial law.[154] Enraged at what he in turn perceived as Kerenskii's betrayal, Kornilov dismissed his denunciation as "a lie throughout," accused the Provisional Government of acting "in full agreement with the plans of the German General Staff," and called upon the people of Russia to support him, "the son of a Cossack peasant," to aid "our dying motherland."[155] Determined not to relinquish his grip on the supreme command of Russia's armies, he ordered the Cossacks of the Third Cavalry to ride against Petrograd.

With the lines of conflict thus clearly drawn, few thought Kerenskii could survive. "The entire commanding personnel, the overwhelming majority of the officers, and the best combat units of the army will follow Kornilov," the Russian Foreign Ministry's representative at *Stavka* reported. "It is up to the men now in power either to face the inevitable change . . . or, by their resistance, to accept responsibility for countless new calamities."[156] Yet those who celebrated Kerenskii's demise had not reckoned with the Petrograd Soviet and the prime minister's genius for making common cause with the workers and soldiers of Russia's capital. "Comrades! the hour has arrived when your loyalty to freedom and the revolution is on trial," he proclaimed as Kornilov's elite divisions approached under the command of the hard-eyed right-wing General Krymov, who once had promised "to cleanse Petrograd in two days with one division" provided he were allowed to shed blood freely. "Let them see before them truly revolutionary regiments."[157]

Instantly, the Soviet established the Committee for the People's Struggle Against Counterrevolution, which, by voting to arm the workers, made it possible for the Bolsheviks to revive the Red Guard units that Kerenskii had suppressed after the July Days, and which would become vital to the success of their effort to seize power in October.[158] Yet, even before Kerenskii issued his appeals to the Soviet, revolutionary railroad workers had halted Krymov's troop trains and averted the threat of attack. Without hesitation, most of Krymov's men took the side of the revolution. "We declare that we are in complete solidarity with the political platform of the Soviet," the men of the Savage Division announced. "Long live the Soviet of Workers' and Soldiers' Deputies!"[159]

It remained only for Kerenskii and his deputies to gather up the scattered shards from the shattered Kornilov "rebellion," as the fallen general's enemies now chose to call it. With his Cossacks on the verge of placing him under arrest, the now defenseless Krymov agreed to a meeting with Kerenskii at the Winter Palace where, in what one witness described as "a tumultuous scene," Russia's victorious prime minister berated him for his duplicity. Ordered to give a full account to the Extraordinary Commission of Inquiry that Kerenskii already had appointed, the man who had abhorred the revolution's readiness to allow the masses a voice in determining Russia's destiny, saw no way to extricate himself honorably. "The last card for saving the homeland has been beaten," Krymov wrote on a scrap of paper. "Living is not worthwhile any longer." Then, without further hesitation, he shot himself through the heart. "I congratulate you from the bottom of my heart," Lvov wrote to Kerenskii from his prison cell on the

day of Krymov's suicide. "I am glad that I saved you from Kornilov's hands."[160]
Soon Lvov fled to the West, only to return to the Soviet Union some years later
to become a minor Bolshevik bureaucrat.

With Petrograd out of danger, Kerenskii still faced Kornilov and his loyal
supporters who were rumored to be strongly entrenched at *Stavka*. [161] Yet that
specter proved to have even less foundation than had Krymov's threat to Petrograd. After another day of telegraphic exchanges, General Alekseev, whom
Kerenskii brought out of retirement to serve again as *Stavka's* chief of staff, went
to Mogilëv and placed Kornilov under arrest. When the members of Kerenskii's
Extraordinary Commission of Inquiry arrived on September 2, they easily arrested the remainder of Kornilov's allies. Quickly, the Extraordinary Commission ordered them all to a prison that had been fashioned out of an ancient
monastery in the nearby town of Bykhov. Russians' last effort to stem the
revolution's inexorable leftward march lay in ruins.[162]

Kerenskii emerged from his confrontation with Kornilov with near-dic-
tatorial powers, and bearing the title of supreme commander to go with that of
prime minister. Yet he lost more from Kornilov's defeat than he gained. On the
Right, influential men now shunned him as Kornilov's betrayer; on the Left,
revolutionary activists disdained him as Kornilov's one-time ally.[163] As Russia's
workers and peasants grew more militant, all the crises that Kornilov had hoped
to control by restoring discipline and order intensified. Soldiers now refused to
fight. Workers would not work. And peasants thought only of burning manor
houses and seizing more land. Russia's military, economic, and civic life crumbled. Again, all Russia waited, wondering where and how the next blow would
strike. Aleksandr Blok hoped that "blood, violence, and bestiality" would soon
give way to "pink clover," and urged people to remember that "the heavy
hammer breaketh glass but forgeth steel."[164] Gorkii rejected such optimism.
"All the dark instincts of the crowd, irritated by the disintegration of life and
by the lies and filth of politics, will flare up and fume, poisoning us with anger,
hate and revenge," he predicted as summer turned into fall. "People will kill one
another, unable to suppress their own animal stupidity."[165]

In the wake of the Kornilov "rebellion," Gorkii's ominous prediction seemed
all too likely to come to pass, for Petrograd's workers now were better armed
and more ready than ever to defend their revolution. Soon, one out of every five
of the *putilovtsy* had a rifle or a revolver, and the Red Guards at other factories
armed themselves at a similar pace.[166] Always sensitive to the danger of the
mob, Gorkii wrote fearfully of "rifles and revolvers in hands trembling with fear
. . . [which] will fire at windows of stores, at people, at anything!" Therein lay
the makings of tragedy. "They will fire," Gorkii explained, "only because
. . . [They] want to kill their fear."[167]

Increasingly, the Bolsheviks commanded the loyalty of Russia's working men
and women, as they demanded peace, land, and bread—now the dreams of all
workers, soldiers, and peasants—at every opportunity. At the beginning of

September, Bolsheviks won majorities in the Soviets of Petrograd and Moscow for the first time, adding a new dimension to Lenin's call for "All Power to the Soviets!"[168] Slowly but surely, the pathway to power opened. "Before the Kornilov revolt, the army and the provinces could and would have marched against Petrograd [if the Bolsheviks had seized power]," Lenin wrote from his refuge in Finland to which he had fled when Kerenskii's government ordered his arrest after the July Days. "Now . . . all the objective conditions exist for a successful insurrection."[169] The "objective conditions" threatened Kerenskii most of all. A week after he had crushed Kornilov, he found his government in greater danger than ever. This time, the threat came from the Left, not the Right, and from men far better able than Kornilov to impose their will. Less than two months after Kornilov's arrest, Lenin and his Bolsheviks would drive Kerenskii from office and seize power in Russia.

CHAPTER XVI

---∙⟨∞⟩∙---

Lenin Seizes Power

Believing that "every right-thinking and really honest man must be a revolutionary," Vladimir Ilich Lenin had dedicated his entire adult life to Russia's revolutionary struggle.[1] Ever since he had chosen that path as a youth of seventeen, he had worked for that single dizzying moment when the old order would crumble, and former political criminals would step forward to direct his nation's affairs. Painfully and painstakingly, Lenin had shaped the Bolshevik party into what he once called an instrument for *"bringing closer and merging into a single whole* the spontaneous destructive force of the crowd and the conscious destructive power of the revolutionaries' organization."[2] As a lonely exile in Zurich, far from the people and events whose course he one day hoped to influence, he had struggled to preserve that revolutionary instrument throughout the first thirty months of the Great War. Yet, even though the February Revolution did not offer him the opportunity to use his revolutionary forces freely, Lenin retained his devotion to the cause of proletarian revolution and continued to prepare the Bolsheviks for the moment when the workers would rule in Russia. Dedication and, above all, discipline would in time make his party masters of Russia. Until then, the Bolsheviks must forego short-term gains and await the larger prize: the opportunity to seize control of Russia's destiny. Time and again after his return in April 1917, Lenin therefore had demanded that the Bolsheviks take the path of greatest resistance, arrogantly rejecting compromise with potential allies on the Left.

Of all Russia's political leaders, Lenin shifted his course most dramatically during the first half of September in order to take full advantage of the opportunities that arose in the wake of Kornilov's failure. "The Russian revolution is experiencing so abrupt and original a turn," he wrote on the day the Kornilov revolt collapsed, "that we, as a party, may offer a voluntary compromise . . . [that will support] a government of Socialist Revolutionaries and Mensheviks responsible to the Soviets."[3] Then, he received word that his Bolsheviks had won majorities in the Petrograd Soviet (August 31), in the Moscow Soviet (September 5), and in the Soviet in the Ukrainian capital of Kiev (September 8). Then he learned that the Petrograd Soviet had voted to place a majority of Bolsheviks on its Presidium and Executive Committee (September 9).[4] Abruptly, Lenin changed his position. "Perhaps it's already too late to offer a compromise," he mused in a postscript to his essay "On Compromises." "Perhaps the few days in which a peaceful development [of the revolution] was *still* possible have passed too."[5]

Convinced that the events of early September had completely altered the revolution's prospects, Lenin prepared a new tactical statement about the course the Bolsheviks ought to follow. "The present task must be an *armed uprising* in Petrograd and Moscow, the seizure of power and the overthrow of the government," he announced in a letter to the Central Committee just eleven days after he had counseled his party about the virtues of voluntary compromise with the Mensheviks and Socialist Revolutionaries. The time had come for "decisions and not talk, for action and not resolution writing," he insisted. Bolsheviks now must carry the call for insurrection to the workers themselves. "We must despatch our entire group to the factories and the barracks," Lenin wrote. "Their place is there, the pulse of life is there. There is the source of salvation for our revolution." Insisting that all who refused to share his opinion were guilty of a "betrayal of the revolution," Lenin urged his followers to reject any thought of compromise with those socialists who urged a more moderate course and to remember that "the power of decision . . . lies in the working-class quarters of Petrograd and Moscow."[6]

None in Petrograd's Central Committee shared Lenin's view, and even Nikolai Bukharin, the wiry, red-headed twenty-eight-year-old maverick who agreed more often with Lenin's extreme positions than any other leading Bolshevik in 1917,[7] thought the call for an armed insurrection premature and Lenin's uncompromising and brutal condemnations of his comrades unwarranted. "[Lenin's] letter was written with extraordinary force and threatened us with all sorts of punishments," Bukharin later wrote as he recalled the shocked silence with which he and the rest of the Bolsheviks' Central Committee heard Lenin's accusations.[8] As he had when he returned to Petrograd at the beginning of April, Lenin thus stood alone on the left of his party once again, his views to all appearances shockingly out of focus and utterly at odds with the opinions of those Bolsheviks who insisted that they, not he, saw most clearly the path Russia's revolutionaries must follow.

Still unable to be seen publicly in Petrograd, Lenin at first allowed his anger to cascade across the Finnish frontier. Certain by late September that "the crucial point of the revolution in Russia has undoubtedly arrived," that "the crisis has matured, [and that] the whole future of the Russian revolution is at stake," Lenin insisted that the Bolsheviks must act resolutely or "cover themselves with eternal *shame* and *destroy themselves* as a party."[9] The situation demanded firmness, daring, and dedication. "We must not allow ourselves to be frightened by the screams of the frightened bourgeoisie!" he exclaimed. "No power on earth can prevent the Bolsheviks, *if they do not allow themselves to be scared,* and if they succeed in taking power, from retaining it until the triumph of the world socialist revolution."[10]

While Lenin raged against their timidity, the Bolsheviks' Central Committee continued to resist any thought of an armed uprising and insisted that a transfer of all power into the hands of the Soviet offered the most promising means for strengthening the party's support among the masses. "Only by welding together all the forces of the masses, organized in the Soviets, can the victory of the workers, soldiers, and peasants be accomplished," they announced. "Only with their victory will it be possible to achieve a democratic peace and advance the cause of international revolution."[11] With Lenin and the Bolsheviks' Central Committee so sharply at odds, late September and most of October thus became a time of agitation and preparation. Most Bolsheviks intended to have their efforts culminate in a transfer of power from the Provisional Government to the All-Russian Congress of Soviets that was scheduled to meet in Petrograd toward the end of October. Lenin expected more. If he had his way, October would see the Bolsheviks seize power by an armed uprising in Petrograd.

Around the beginning of October, to shorten the distance that separated him from his comrades and to defend his views more effectively, Lenin moved from his Finnish refuge to the apartment of Margarita Fofanova, a dedicated Bolshevik who lived at the beginning of Serdobolskaia Street on the outskirts of the Vyborg District, a stronghold of Bolshevik sentiment in Petrograd. Comfortably installed in Fofanova's corner bedroom, amid its very "middle class" chintz curtains and flowered wallpaper, with a balcony and drainpipe offering a convenient avenue for escape, Lenin hurled forth a torrent of letters and instructions urging, cajoling, even demanding, that his Bolsheviks press on with the urgent tasks at hand. From a small desk that stood under the room's corner window, he filled in the outlines of the Bolshevik programs on the desperate questions of peace, land, and bread, and, from the same diminutive platform, with its single kerosene lamp, he launched verbal missiles to shatter the defenses that more cautious Bolsheviks erected against his demands.[12]

But, before any assault against the Provisional Government could be attempted, the Bolsheviks' Central Committee had to be convinced to support an immediate armed uprising. Day after day—sometimes several times a day—Lenin sent letters, essays, and homilies, all urging action as he chafed at the arrest warrant that kept him confined to Fofanova's apartment. "Go then to the barracks," he insisted to his comrades at the beginning of October. "Go to the

Cossack units, go to the working people and explain the *truth* to them."[13] The need was urgent, and the magic moment when it would take so little to accomplish so much was in danger of slipping away. "The success of both the Russian and the world revolution," Lenin estimated on October 8, "depends on two or three days' fighting."[14] Any delay at a time when "victory is certain, and the chances are ten to one that it will be a bloodless victory," he insisted, would be "childish," "disgraceful," "dangerous," "criminal," "a betrayal of the revolution," and "fatal."[15] Clear-headed and ready to act without hesitation, Bolsheviks must seize the opportunity that lay before them. "Marx summed up the lessons of all revolutions in respect to armed uprising," Lenin reminded his readers, by quoting the "words of 'Danton, the greatest master of revolutionary policy yet known: *de l'audace, de l'audace, encore de l'audace.*' " Evidently forgetting his promise that the Bolsheviks had at least nine chances out of ten for a "bloodless victory," Lenin concluded his "Advice of an Onlooker" with a passionate summons to victory or death. *"Better die to a man,"* he exclaimed, *"than let the enemy pass!"*[16]

Desperate to be heard, and far from certain that even the most passionate urgings could convert the reluctant Bolsheviks of Petrograd and Moscow to his views, Lenin decided to risk a foray into Petrograd's streets in order to lay his case directly before the Central Committee.[17] The Central Committee chose the night of Tuesday, October 10, for what became one of the most fateful meetings that it ever held. Thick clouds darkened the sky that night, and an icy drizzle soaked the outer clothing of those who ventured outside. Only too ready to seek shelter, Kerenskii's police agents had disappeared from Petrograd's streets long before the Bolsheviks began to make their way across the city to the ground-floor apartment of Galina Flakserman, the fragile, tubercular Bolshevik wife of Nikolai Sukhanov, at No. 32 Karpovka Embankment, on the outer edge of the city's Petrograd district. Because her husband remained one of Petrograd's leading Mensheviks, still a prominent figure in the Petrograd Soviet, and co-editor with Gorkii of the socialist, but often anti-Bolshevik, weekly *Novaia zhizn*, *New Life*, the Central Committee's hostess had made certain that Sukhanov would remain at his editorial offices that night. "Oh, the novel jokes of the merry muse of History!" Sukhanov later exclaimed in wistful lament. "This supreme and decisive session took place in my own home . . . but without my knowledge."[18]

By ten o'clock, Lenin had assembled eleven of Russia's leading Bolsheviks around the dining room table where Sukhanov and his fellow Mensheviks had eaten so many meals in the past. For fear they might be recognized, those who still remained fugitives despite the partial amnesty that had freed Trotskii had shaved their heads or grown beards. Lenin "looked every bit like a Lutheran minister," one member of the Central Committee later wrote, because he had shaved his well-known beard and wore a curly wig.[19] To make certain that the *dvornik*—the Petrograd counterpart of the Parisian *concièrge*—who usually worked for the security police, would not observe the unusual comings and goings, their hostess had covered the room's only window with a blanket so that nothing could be seen from the courtyard. Already tense from the fear of discovery that gripped each visitor, the atmosphere grew positively explosive from the sense of impending conflict that pervaded the room. None could escape the feeling that monumental decisions were about to be made.[20]

While Galina Flakserman and her brother Iurii tended the samovar and prepared sausages, Lenin hammered away at his lieutenants' reluctance to launch an armed uprising. "In the political sense, the situation has become ripe for the transfer of power," he insisted, and under such conditions, the "indifference about the whole question of an armed uprising" that he perceived among Petrograd's Bolsheviks was simply "inadmissible." Now "tired of words and resolutions," the masses demanded action. "The decisive moment was near at hand," and the Bolsheviks therefore must immediately turn their "attention to the technical side" of an uprising and discuss precisely the organizational and logistical problems that needed to be resolved for an uprising to succeed.

Lenin's assurances notwithstanding, a number of the Central Committee doubted if the Bolsheviks could take advantage of such a decisive moment. Soon to become the head of the Cheka, the Bolsheviks' fearsome security police, in Petrograd, Mikhail Uritskii warned that "we are weak, not only in the technical sense, but in all other areas of our work as well." So far the Bolsheviks' had little more than "a mass of resolutions" to their credit. "What forces are we counting on?" Uritskii asked. "The Petrograd workers have forty thousand rifles, but this alone cannot decide the matter. This is really nothing." Others added their comments, warnings, and forebodings. After another member reported on the situation in Moscow, they discussed the situation in Russia more generally. "We have to come to a decision about a definite course of action," Uritskii insisted, and pointed out that, among other things, the Bolsheviks needed to begin a concerted campaign to guarantee the support of the Petrograd garrison. Feliks Dzerzhinskii, who would become Uritskii's superior as the Cheka's founder and first chief, thought that expert agitation by the very best of the Bolsheviks offered the best prospects for advancing their cause at that moment. "For political guidance," he concluded, "a Political Bureau, made up of members of the Central Committee, needs to be organized."[21]

Hours passed. "Soon the charcoal ran out," Iurii Flakserman remembered, "and I had to cut up splinters and heat the samovar with those." As the fresh fuel smoldered, the debate continued to flame. From the kitchen, Flakserman thought it sounded "intense [and] passionate." It was that, and more. Lenin was "saturated with a desire to instill into the objecting, the wavering, the doubtful, his thought, his will, his confidence, his courage," Trotskii later remembered, and his arguments slowly took hold.[22] "Lenin's supporters always felt a surge of strength and resolve when they heard him speak or read his writings, while those who experienced a certain hesitation set aside their doubts," Flakserman later explained.[23] At last certain of his followers' support, Lenin proposed a resolution to formalize their agreement. "Recognizing, therefore, that an armed uprising is inevitable and imminent, the Central Committee proposes that all organizations of the party be guided by this fact and that they discuss and decide all practical questions from this point of view."[24]

Nine of the eleven members of the Central Committee present in addition to Lenin voted for the fateful resolution. Convinced that the risk was too great and that failure could destroy the Russian revolution, Lev Kamenev, Trotskii's brother-in-law and a Jewish convert who had been born and baptized in the Russian Orthodox faith, voted against it. So did Gregorii Zinoviev, a product of more traditional Jewish upbringing, who had shared Lenin's exile in Zurich and had gone with him into hiding after the July Days. "We are deeply convinced," these two dissenters concluded at one point, "that to proclaim an armed uprising now means to gamble not only with the fate of our Party, but also with that of the Russian and the international revolution."[25]

As Trotskii wrote some years later, "no practical plan of insurrection, even tentative, was sketched out" that night,[26] and even Lenin's resolution specified no time for a revolt, or discussed how it ought to begin. Still, the wheels had been set in motion, and there was a sense among the Bolsheviks that a seizure of power must take place sooner, not later. That sense stirred doubts among many, for Zinoviev and Kamenev were far from alone in their apprehensions. Trotskii remembered that two of the Bolsheviks' leading military experts, Nikolai Podvoiskii, the former seminarian who soon would lead the assault against the Winter Palace, and Vladimir Nevskii, whose soft features and pudgy figure concealed his talents as a tactician of the streets, had grave doubts, as did a number of other party leaders who had not been able to attend the meeting in Flakserman's dining room. Could a comparative handful of Bolsheviks seize that one-sixth of the globe's surface stretching from Poland to the Pacific that once had been the Russian Empire? Many feared to contemplate the answer. "Here," one Central Committee member confided to a friend, "we really seem to be pygmies thinking of moving a mountain."[27]

No one proved more crucial to the success of Lenin's effort to prepare Petrograd's workers for an armed uprising than Lev Davidovich Bronshtein, a thirty-seven-year-old Jew from the Ukraine. The son of a tough nonconformist who had carved a prosperous farm out of the rough frontier lands of Russia's southern steppe, Bronshtein already had won a notable place in the revolutionary annals of 1917 under the name of Trotskii well before midsummer. Powerfully built, with broad shoulders and a massive chest, and standing more than half a head taller than Lenin or Stalin, Trotskii had first emerged as a leader among Russia's revolutionaries not long after he turned twenty. Just before his twenty-sixth birthday, he had taken command of St. Petersburg's masses during the Revolution of 1905 when he had captured workers' imaginations with his ringing speeches and daring pronouncements. Fearlessly, Trotskii had stood before the city's tired proletarians to reject Nicholas's promised constitution with the scornful sneer that the proletariat "does not want a *nagaika*, a Cossack whip, wrapped up in a constitution."[28] During the summer of 1917, his fiery words had won him dedicated followers in every factory of Petrograd, and he had become better known to the average factory worker than any other revolutionary leader. Just released from jail in an abortive attempt by Kerenskii's government to win allies among some of those whom they had imprisoned for their part in the events of the July Days, Trotskii became the obvious candidate to replace the faltering Menshevik chairman of the Petrograd Soviet when the Bolsheviks won their first majority at the beginning of September.

From his newly-won vantage point as chairman of the Petrograd Soviet, Trotskii took pride in observing that the seven months since Nicholas's abdication "had educated hundreds and thousands of rough diamonds who were accustomed to look on politics from below and not above." These men and women of the people now became a vital link in the Bolsheviks' effort. "A generalizing formula tossed out in the great amphitheater of the Cirque Moderne by one of the revolutionary leaders," he remembered, "would take flesh and blood in hundreds of thinking heads and so make the rounds of the whole country."[29] These people would become the revolutionary shock troops that Trotskii would soon send to storm the centers of power in Petrograd and the other cities of Russia. Such men and women, now close to a quarter-million strong, Lenin insisted, would provide the manpower to govern Russia "in the interests of the poor and against the rich."[30]

Nonetheless, it was not the Bolsheviks, but the Petrograd Soviet under Trotskii's direction, that forged the weapon that made an armed uprising possible, and it did so in response to fears raised by Kerenskii himself. Ever since Riga and the important islands of Dagö, Ösel, and Moon had fallen, the threat of German attack had hung ominously over Petrograd and Kerenskii evidently intended to use that as an excuse to remove the more radical elements of the Petrograd garrison. His unexpected order to transfer garrison troops to the front, coupled with Rodzianko's sensational public exclamation that "Petrograd appears threatened. . . . [and] I say to hell with Petrograd,"[31] convinced many of the city's workers and soldiers that Kerenskii planned to hand the capital over to the Germans in an attempt to destroy the revolution. Reacting to cries that "The government is fleeing to Moscow! The government is surrendering the center of the revolution!"[32] the Soviet met hastily, ready to act but, for the moment, not certain how.

After a long and stormy debate on October 9 in which the Bolsheviks beat back a Menshevik proposal to cooperate with the government in the war effort, a majority of the Petrograd Soviet adopted a hastily scribbled resolution that Trotskii put before them. "In this moment of deadly danger for the people and the revolution, the Petrograd Soviet of Workers' and Soldiers' Deputies resolves that the government of Kerenskii is ruining the country," the resolution began.

Now certain that, "in league with the bourgeoisie," Kerenskii was about "to surrender Petrograd—the citadel of the revolution—to the Germans," the Soviet authorized its now-Bolshevik-dominated Executive Committee to defend the people and the revolution against this impending onslaught of counterrevolutionary forces by organizing a "revolutionary committee of defense," or, in the Executive Committee's words, "a revolutionary headquarters for the defense of Petrograd."[33]

If Lenin had the strength of will to impose his certainty that an armed uprising was "inevitable and imminent" upon the Bolsheviks' doubting Central Committee, Trotskii alone possessed the sheer genius needed to make the uprising occur. Sukhanov remembered that during the fortnight after the Central Committee met in his apartment Trotskii seemed to be everywhere. "He flew from the Obukhov Works to the Fuse Factory, from the Putilov to the Baltic Mills, from the Riding School to the barracks, and it seemed that he spoke in all places at the same time," Sukhanov wrote a few years later. "Every Petrograd worker and soldier knew who he was and had heard him speak. His influence—among the masses and at headquarters—was overpowering. During these days," this admiring Menshevik, who was then so hopeful for Russia's future and so apprehensive at any thought of violence, concluded, "he was the central figure and the real hero of this remarkable page in history."[34]

No task proved too difficult, and no detail seemed too unimportant for Trotskii in those days. When the Petrograd garrison declared that they would remain in the city despite Kerenskii's orders, Trotskii's influence lay behind every word. When the garrison of Petrograd's great Peter-Paul Fortress refused to admit the delegate sent to them by the Military Revolutionary Committee, Trotskii went alone to speak to them and convinced them to yield up the weapons in the fortress arsenal to the city's proletarians. All across the city, his words rang out the revolutionary tocsin summoning Petrograd's workers and soldiers to stand against Kerenskii's government and declare their loyalty to the Bolsheviks' cause. "The time for words has passed," he thundered at one point. "The hour has come for a duel to the death between the revolution and the counterrevolution."[35] Not even the huge amphitheater of the Cirque Moderne, in the center of the Petrogradskaia District, could contain the throngs that flocked to hear Trotskii as he stirred the proletariat's revolutionary fervor to a fever pitch.[36] "The masses have taken the view that the uprising is inevitable," one observer concluded less than a week later. "If there is a call from above, there can be no doubt that the masses will support it." In vain did the cautious and stubborn Kamenev urge his comrades to wait, pleading that "it is not a question of now or never."[37]

If it was not that, it nearly was, for the preparations for an armed uprising already had moved forward decisively, even though the Military Revolutionary Committee promised in Trotskii's resolution of October 9 was not fully organized for another week. Were the Bolsheviks planning an armed uprising, the

Mensheviks asked at a meeting of the Soviet's Executive Committee. "It is not the Bolsheviks who are preparing an uprising but . . . those who are creating a mood of hopelessness among the masses," a Bolshevik spokesman replied. "If the workers and peasants revolt as a result of the government's policy, we will be in the front ranks of the insurgents." As Kerenskii's government did nothing to stand in their way, the uprising seemed all but inevitable. Concluded one Menshevik mournfully, "there is no turning back."[38]

While Kerenskii's government did nothing, Petrograd seethed with rumors about the Bolsheviks' impending coup, and these became even more inflamed when Kamenev and Zinoviev, still disputing Lenin, published their objections in Gorkii's *Novaia zhizn.* "Not only Comrade Zinoviev and I, but also a number of practical comrades, feel that to assume the initiative of an armed uprising at the present moment . . . would be an intolerable, ruinous step for the proletariat and the revolution," Kamenev wrote. "[We] express ourselves against any attempt to assume the initiative of an armed uprising," he concluded. "Our party . . . has too great a future ahead of it, to take such steps of desperation."[39]

In an absolute rage, Lenin read both men out of the party. "We cannot refute the slanderous lie of Zinoviev and Kamenev without doing even greater damage to the cause," he wrote in a pair of furious letters to the Central Committee. "Kamenev and Zinoviev have *betrayed* . . . the decision of the Central Committee of their Party on insurrection and the decision to conceal from the enemy preparations for an insurrection and the date appointed for it. . . . I demand the expulsion of both the blacklegs," he concluded. "Let Mr. Zinoviev and Mr. Kamenev found their own party."[40] Wisely, the Central Committee resisted Lenin's demand to expel two of its key members. "It is easier to theorize about a revolution afterward than absorb it into your flesh and blood before it takes place," Trotskii sagely remarked. "The approach of an insurrection has inevitably produced, and always will produce, crises in the insurrectionary parties."[41]

Russia had reached another revolutionary crossroads. "A ground-swell of revolt heaved and cracked the crust which had been slowly hardening on the surface of revolutionary fires dormant all those months," wrote the American socialist-come-to-see-the-revolution-with-his-own-eyes John Reed.[42] A native of Portland, Oregon, and a Harvard man who never ceased being impressed with that fact, Reed owed his infatuation with revolution to his never-ending love affair with the life and the radical *isms* of Greenwich Village, where he had first set foot in 1911. Dedicated to all rebels, "bound by no creed," he once wrote, but ready to "express them all, providing they be radical,"[43] Reed had stormed back and forth through the Village for the next six years, with time out to report on the Mexican Revolution of 1913 and to observe the Great War from Italy, France, England, and Germany (in that order) before he first went to Russia for a brief visit at the end of 1915. Everywhere, he made devoted friends and implacable enemies, for his passionate dedication to contradictory ideas and his unquenchable thirst for life stirred strong likes and hatreds. Few could agree, even about his looks. Some thought him "lovable" and "generally breathless," with a "real poet's jawbone." Max Eastman, who liked him, thought he had a face "rather like a potato."[44]

Always and everywhere, Jack Reed searched for that magic revolutionary mixture that combined all the *isms* of prewar Greenwich Village and transformed them into real life. When he returned to Russia at the moment of Kornilov's defeat, he thought his search had ended. Perpetually curious, constantly on the move, with an ever-ready *"Ia amerikanskii sotsialist"* ("I am an American socialist") for whomever he met, Reed absorbed the intoxication of Russia's revolutionary drama through every pore, and he set out to record it in one of the most vibrant revolutionary chronicles ever written.[45] *Ten Days That Shook the World* became Jack Reed's masterwork, and his own revolutionary tribute to the revolutionary cataclysm into which Lenin and Trotskii were about to lead the Russians.

As at the beginning of February, a sense of chilling, yet thrilling, anticipation gripped Russia as people uncertain of the future lived only for the moment. "Hold-ups increased to such an extent that it was dangerous to walk down side streets," Reed wrote of his first days in Petrograd. "On the Sadovaia one afternoon I saw a crowd of several hundred people beat and trample to death a soldier caught stealing. . . . Gambling clubs functioned hectically from dusk to dawn, with champagne flowing and stakes of twenty thousand rubles. In the center of the city at night prostitutes in jewels and expensive furs walked up and down, and crowded the cafés."[46] Newly returned to Russia after serving as a war correspondent in Paris, Ilia Ehrenburg was struck by the "very large breasts" of members of the women's shock battalions he saw sauntering along the Nevskii Prospekt, and by the sense of tense anticipation that seemed even more intense in Moscow than in the capital. "Moscow lived as if on a railway platform, waiting for the guard's whistle," he remembered. "There was cursing everywhere, particularly in the trams which crawled along, hung with clusters of human beings. At the [Hotel] Metropol, despairing intellectuals drank French champagne [and] . . . went on living purely by the force of inertia."[47]

For most folk, life grew still more difficult as the revolution deepened. Food became even scarcer in Russia's cities as fall turned toward winter, and almost everything else seemed to be in equally short supply. On October 15, a government commission reported that well over a hundred horses were dying every day in Petrograd for want of fodder, and that, at any moment, Petrograders' daily bread ration might have to be reduced from three-quarters of a pound to a half. To purchase tobacco, chocolate, sugar, milk, all required long hours spent in the seemingly endless lines that stretched along the city's streets. That fall, as apples ripened in the orchards outside Russia's capital, an average worker had to pay almost half a day's pay for four. In Moscow, where food and commodities were cheaper, a pair of shoes cost a worker eighteen days' pay, and a suit came to a full two months' wages.[48] "The great families are obliged to part with their most precious mementoes in order to live," the young French diplomat Comte Louis de Robien wrote, as he confided to his diary his fears that the Louvre would fail to buy a rare *biscuit de Sèvres* bust of Marie Antoinette belonging to Madame Narishkina and that it might end up gracing "the parlor of a Transatlantic pork merchant."[49] The air crackled with tension. Reed saw "mysterious individuals circulating around the shivering women who waited long cold hours in queues for bread and milk, whispering that the Jews had cornered the food supply— and that while the people starved, the Soviet deputies lived luxuriously." Everyone feared the cold and hunger of the approaching winter. Russia seemed to be on the verge of an apocalyptic rendezvous. "Monarchist plots, German spies, smugglers hatching schemes," Reed wrote. "And, in the rain, the bitter chill, the great throbbing city under grey skies rushing faster and faster toward— what?"[50]

Only Lenin, the Bolsheviks, and, particularly Trotskii, knew the scope of events that were taking shape at Petrograd's Smolnyi Institute, formerly a boarding school for young women of high birth and now the headquarters of the Petrograd Soviet and its Executive Committee. For months the pulse of the revolution had throbbed at Smolnyi as nowhere else in Russia, bringing together revolutionary leaders and the humblest of their followers at one of the rare sources of plentiful food in the capital. Reed became a regular visitor to the cellar dining room "where twenty men and women ladled from immense cauldrons cabbage soup, hunks of meat and piles of kasha, slabs of black bread" to a thousand men and women who crowded around wooden tables, "wolfing their food, plotting, and shouting rough jokes across the room."[51] Then, overnight, the atmosphere changed, as Trotskii and his now fully organized Military Revolutionary Committee drew a tight curtain of military security around their work. One day in late October, Reed looked on in amazement as Trotskii himself, unable to find his pass, found his way barred by a suspicious guard.

"You know me. My name is Trotskii," Reed heard the president of the Soviet tell the sentry.

"You haven't got a pass. You can't go in. Names don't mean anything to me," the obviously unimpressed soldier replied.

"But I am the President of the Petrograd Soviet," Trotskii protested.

"Well, if you're as important a fellow as that you must at least have one little paper."

"Let me see the Commandant," Trotskii went on patiently. Instead the soldier summoned the commander of the guard.

"My name is Trotskii," the exasperated Soviet president began again. Reed looked on in amazement while the commander scratched his head.

"Trotskii? I've heard the name somewhere," the commander mused. "I guess it's all right. You can go in, comrade."[52] And Trotskii, commander-in-chief of the armed uprising about to be set into motion by bitter discontent with "[government] policies that have done so much for the bourgeoisie in the course of seven months of revolution and nothing for the masses,"[53] entered Smolnyi to continue his work.

"People do not make revolution eagerly any more than they do war," Trotskii later wrote. "A revolution takes place where there is no other way out. And the insurrection, which rises above a revolution like a peak in the mountain chain of its events, can be no more evoked at will than the revolution as a whole." Mass insurrection, the only means to produce "the victory of one social regime over another," Trotskii insisted, could occur only in the wake of "some swift growth which has broken down the old equilibrium of the nation." Thus, a revolutionary cadre, no matter how well trained and organized, could not launch a mass insurrection spontaneously even though such cadres were necessary to lead a mass insurrection when the proper moment arrived. At some point, mass frustration and anger required conscious direction and leadership. "Just as a blacksmith cannot seize the red hot iron in his naked hand," Trotskii explained later, "so the proletariat cannot directly seize power."

To seize power required organization, and the Bolsheviks saw the Petrograd Soviet and its counterparts in other industrial centers as "the organization by means of which the proletariat can both overthrow the old power and replace it." Still, although the soviets *could* serve that purpose, it was not inevitable that they would do so. Only when a "revolutionary party" controlled it, Trotskii explained, would "the soviet consciously and in good season strive toward a conquest of power." For that to happen, "organized conspiracy [must be brought into] mass insurrection." Thus, Trotskii stressed again a point Lenin had made at the end of September (quoting Marx erroneously, it turned out, for the phrase belonged to Engels) that "insurrection is an art, and like all arts it has its laws." Yet Trotskii insisted that the precise moment for applying such laws, like the laws governing natural birth, could not be set down too precisely. Just as "physical births . . . present a considerable period of uncertainty," he explained, so too did insurrections. Thus, "revolutionary leaders" must determine the precise moment when they should take their position at the head of the insurrectionary masses. "They must feel out the growing insurrection in good season and supplement it with conspiracy," Trotskii said, just as a midwife must choose the proper moment to interfere with labor pains in complex births so as to bring on the birth of a healthy infant. As a result, Trotskii concluded, "intuition and experience are necessary for revolutionary leadership just as for all other kinds of creative activity."[54]

The Petrograd Military Revolutionary Committee became the instrument for directing the armed insurrection once Lenin's intuition determined that the proper level of mass discontent had been reached, and it was a true measure of Trotskii's multisided genius that he was able to assume command with absolutely no previous experience in military affairs. Certain of support from the Petrograd garrison, Trotskii provoked a direct confrontation with the Headquarters of the Petrograd Military District late in the evening of October 21 by informing its commander that "orders not signed by us are invalid."[55] When the commander threatened the Military Revolutionary Committee representatives with arrest, Trotskii announced in a special order to all units of the garrison that by not recognizing the Military Revolutionary Committee, the Headquarters of the Petrograd Military District had "become a direct weapon of counterrevolutionary forces." "Soldiers of Petrograd!" Trotskii continued. "The protection of the revolutionary order from counterrevolutionary attacks rests upon you under the direction of the Military Revolutionary Committee. Any garrison directives not signed by the Military Revolutionary Committee are invalid. . . . The revolution is in danger. Long Live the Revolutionary Garrison!"[56] Although he was not yet ready to put his authority to the final test, Trotskii and the Military Revolutionary Committee had in fact taken control of the very military units upon which the government would have to rely to crush an armed insurrection. They needed only to continue unopposed along that insurrectionary path to be certain of full control over the city.

To intensify popular hatred against Kerenskii's government, the Bolsheviks dedicated Petrograd Soviet Day—Sunday, October 22—to haranguing workers all across the city. Before a vast meeting at the People's House, Trotskii gave one of his most brilliant performances. "Soviet power," he told his listeners, "will give everything in the country to the poor and the men in the trenches. You have two sheepskin coats, bourgeois—give one to a soldier who is freezing in the trenches," he went on with a dramatic flourish. "Do you have warm boots too? Stay at home. A worker needs your boots." Posing his next statement in the form of a resolution, Trotskii announced: "We shall stand for the workers and peasants to the very last drop of our blood!" Enthralled, Sukhanov listened as Trotskii continued to work his oratorical magic on the crowd. "Who is in favor?" he asked. Sukhanov saw the entire crowd raise its hands. "I saw the raised hands and the burning eyes of men, women, adolescents, workers, soldiers [and] peasants," he remembered. "Let this vote be your oath," Trotskii proclaimed as he reached the end of his speech, "to support with all your strength and at any cost the Soviet which has taken upon itself the great burden of leading the revolution to victory, and giving land, bread, and peace [to us all]!"[57]

"To have at the decisive moment, at the decisive point, an overwhelming superiority of force," Lenin once said, "is the law of political success, especially in that seething and bitter war of classes which is called revolution."[58] Trotskii could inflame crowds of workers and soldiers with his words, but to make certain of an "overwhelming superiority of force," he moved quickly to send reliable Bolshevik commissars to all military units in Petrograd "in the interests of defending the revolution and its conquests against attacks from the counterrevolution."[59] At the same time, Trotskii leveled a propaganda barrage against the

garrison of the Peter and Paul Fortress to assure that they, the bastion's resources, and those of the nearby Kronverk Arsenal would be ready to arm the
insurrection when it came. The day after his appearance at the People's House,
Trotskii therefore delivered what one observer called "not so much a speech as
an inspirational song" to bring the Fortress soldiers, cheering and pledging their
loyalty to the Soviet and its Military Revolutionary Committee, over to the side
of an armed uprising.[60] "The most important task of the insurrection, and the
one most difficult to calculate in advance," Trotskii later wrote, thus "was fully
accomplished in Petrograd before the beginning of the armed struggle." By the
night of October 23, Bolshevik commissars controlled most of the Petrograd
garrison.[61]

Yet, as Trotskii knew all too well, the garrison's support contained more
symbolism than substance. Now composed of political activists with a strong
preference for politics over combat, overaged men, underaged boys, and front-
line soldiers convalescing from wounds, the Petrograd garrison no longer was the
formidable fighting force it had been at the beginning of the year. Trotskii
thought this "soldier crowd," which wanted only to go home and seize land for
itself and its families, "incapable of prolonged military effort," but able, hopefully, "to rally once more . . . and rattle its weapons suggestively before completely going to pieces." But its effort could be no more than a gesture. Certainly, it could not be expected to stand against "the best trained elements in
the army"—that fragment of the garrison made up of women's shock battalions,
military cadets, senior officers, and some scattered Cossack units—which had
remained unmoved by the Bolsheviks' best efforts. "It was impossible to win
these elements politically," Trotskii concluded. "They would have to be vanquished."[62] To do so, the Bolsheviks needed a completely reliable and dedicated
armed force. For that, they looked to the Red Guards, born in April, mobilized
to defend Petrograd against Kornilov, but not yet tested in battle against the
established order.

Militant champions of Soviet power, Petrograd's Red Guards had been
moving steadily toward a confrontation with Kerenskii's regime since the July
Days, when the government's efforts to disarm them had driven them closer to
the Bolsheviks. Called briefly to arms in late August in response to the workers'
urgent demands for an "armed offensive against counter-revolution,"[63] their
ranks totaled about twenty-five thousand that fall. In some factories, they numbered only a few dozen or at most a few hundred, but the vast Obukhov Mills
boasted close to two thousand Red Guards, and the even larger Putilov Works
claimed three times that number. Armed with rifles, revolvers, hand grenades,
and even machine guns, Petrograd's Red Guards trained openly and seriously
during September and October, sometimes for six or more hours a week. When
combined with naval units from Kronstadt, these Red Guards would leave the
Bolsheviks very close to holding that "overwhelming superiority of force" that
Lenin had recommended.[64]

Factories thus became armed bastions of Bolshevik power that fall. In those days, the Putilov Works' official history states, *"Putilovtsy* came to work with rifles, stood them next to their benches, ready at any moment—at the first alarm —to march in defense of the revolution," while at the Franko-Russkii Factory, "turners were at their work-benches with cartridge pouches over their shoulders, rifles standing at the benches. In the locksmith shop, the rifles stood in the corners, and the locksmith also had cartridges over his shoulder," another worker remembered as he looked back on those days.[65] As October passed its midpoint, that sense of preparedness heightened. Some Red Guards went on full alert as early as October 15, and almost all did so within the next five days. "We took to sleeping in our factories, and even ate as a group," one of them wrote, while another remembered how tensely he and his comrades waited for the command to attack.[66] Even the weather added to the sense of an impending rendezvous with destiny. Cold, soaking rain, sheeting sideways on the biting Baltic winds, coated the city's long-unswept streets with slimy muck that tens of thousands of proletarian boots carried into government buildings, private dwellings, and public meetings. "COMRADES! FOR THE SAKE OF YOUR OWN HEALTH, KEEP THINGS CLEAN!" boldly lettered signs pleaded. Nobody seemed to care, or even notice. "Nobody is looking down now underfoot," Trotskii explained. "All are looking forward."[67]

But forward to what? None knew for certain, because no one had ever followed Lenin's path into that unknown realm of open conflict where defeat could transform the hopeful victors of today into the traitors of tomorrow. This path led beyond the point of no return where, as one Bolshevik told Jack Reed, "the other side knows it must finish us or be finished."[68] Here, men feared the unforeseen and unforeseeable, the minor detail overlooked—or not even thought of—that could tip the balance from victory to defeat. Such thoughts and fears filled Lenin's mind as he continued to live in hiding at Fofanova's. For almost a month, he had paced the confines of her four small rooms as he tried to measure the forces whose command circumstances had forced him to entrust to lieutenants. Now, as the Red Guards went into action, he could barely restrain his terrible impatience. As Fofanova hurried back and forth between her apartment and Smolnyi with his deputies' cautious replies to his demands for action, he growled: "What are they afraid of?" "Just ask them if they have a hundred trustworthy soldiers or Red Guards with rifles," he exclaimed at one point. "That's all I need!"[69] The prize, he insisted, lay within their reach if only they dared claim it. "The government is tottering," he announced in a note Fofanova delivered to Smolnyi on the afternoon of Tuesday, October 24. "It must be *given the death blow* at all costs," for the *"salvation of the revolution,* the offer of peace, the salvation of Petrograd, salvation from famine, the transfer of land to the peasants" all depended upon the Bolsheviks' will to action. "We must not wait!" he urged. "We may lose everything!"[70] "Everything now hangs by a thread," he warned later that evening. "The matter must be decided without fail this very evening, or this very night."[71]

Yet, as so often happens in history's great events, an aura of unreality, tempered by a sense of tragi-comedy, enveloped the Bolsheviks' armed uprising in Petrograd from its beginning. Just as Tolstoi's Pierre Bezukhov could not find the set-piece beginnings of a great battle as he wandered near the village of Borodino[72] during the hours when Napoleon and Kutuzov assembled their forces on the morning of August 25, 1812, so too did Jack Reed find it difficult on October 24, 1917, to locate the set-piece armed rebellion in which desperate, rag-clad proletarians stormed public buildings and clashed with the forces of their class enemies on the barricades, trading bayonet thrust for bayonet thrust and answering gunfire with gunfire. At several points along the Nevskii Prospekt, Reed saw squads of soldiers sternly requisitioning private automobiles, but then discovered that "nobody knew whether the soldiers belonged to the Government or the Military Revolutionary Committee."[73] A day of searching drove him to ask the actors in history's drama for directions. "On the corner of the Morskaia, I ran into Captain Gomberg," he recalled. "When I asked him if the insurrection had really happened he shrugged his shoulders in a tired manner and replied, '*Chort znaet!* The devil knows!'" Further on, near the Winter Palace itself, Reed found a barricade heaped up out of "boxes, barrels, an old bed-spring, and a wagon." Hopeful that perhaps here he had come upon the scene of the coming conflict, he questioned some soldiers nearby.

"Is there going to be any fighting?" he asked.

"Soon, soon," one of them replied. Pointing toward the Admiralty, he volunteered: "They will come from that direction."

"Who will?" Reed wanted to know.

"That I couldn't tell you, brother," the soldier replied, spitting indifferently on the pavement, and leaving Reed to unravel things as best he could.[74]

If Reed's elemental socialist fervor could find no heroic confrontations between proletarians and bourgeoisie, neither could the calm moderation of Comte Louis de Robien at the French Embassy. "Everything has been quiet," he wrote on Tuesday, October 24, within hours of the time when Lenin proclaimed that Kerenskii's overthrow "must be decided without fail this very evening." The next morning, the elegant French diplomat awoke to find the Bolsheviks "masters of the capital," and discovered, somewhat to his amazement, that, "to all appearances the town is completely quiet." Like Reed, de Robien saw the Bolsheviks' coup as a series of small events: the sight of "the women's regiment going along the English quay, with bands and machine-guns, beautifully in line"; a friend arrested but quickly released; an English general's Russian driver receiving a generous ration of scarce petrol once he assured

everyone that he had no use for Kerenskii, and the annoyance of having the
banks closed down for a few days.[75] Still, during those days, de Robien lunched
at Donon's and dined undisturbed at Contant's, two of the city's most fashionable restaurants, the latter no more than a few minutes' walk from the Winter
Palace, and he found nothing very disturbing in anything he saw. In fact, the
Bolsheviks' coup touched a sensitive chord in the romantic young Frenchman's
breast. "Lenin at least, like Christ of old, brings something new and talks a
different language," he confided to his diary. "I prefer their dreams," he wrote
of the Bolsheviks, "to the gross realism of the 'get-out-and-let-me-in' people of
the first revolution."[76]

While de Robien pondered the meaning of the day's scattered events on
Wednesday evening, Chaliapin gave a memorable concert at the People's House
and the Imperial Ballet performed as usual in the Mariinskii Theater. At the
Aleksandrinskii Theater, where Meierhold's opulent revival of Lermontov's
Masquerade had marked the beginning of the February Revolution, now, his
revival of Aleksei K. Tolstoi's half-century-old *Death of Ivan the Terrible*, a
monument to the last moments of the sixteenth century tsar whose revolutionary
dreams turned tragically sour, heralded the end of Kerenskii's dreams for a new
and revolutionary Russia. The French journalist Claude Anet dined at the
elegant restaurant in the Hotel Evropeiskaia. Not far away, Reed sat down to
share a more modest meal at the Hotel Frantsiia with Louise Bryant, the former
wife of a dentist whom Reed had rescued from the "unfertilized soil" of Portland, married, and taken with him to Russia—the woman he once acclaimed
as "wild, brave, and straight"[77]—and Albert Rhys Williams, an American
socialist journalist who could be counted upon to provide glib theoretical justifications for any event. After the trio hurried through their meal, they considered
going to the ballet, but decided to walk along the Nevskii Prospekt instead. As
they started up the Nevskii, a "saucy young woman in a shapeless jacket" hurried
up to another who was wearing a fur wrap. "Haven't you worn that long
enough?" Williams heard her ask. "I'll try it for a while!" she exclaimed as she
snatched it from the older woman's shoulders and hurried away into the crowd.
"The whole town is out tonight—all but the prostitutes," Reed remarked. As
they walked along, Williams was struck by the "curious unreality about the
scene." It hardly seemed that anything of great consequence could be happening
that night.[78]

Later, Reed and his companions walked on to Smolnyi, where they joined
a group of Russians who took them careening back down the Nevskii in a
hastily-commandeered truck, joyfully hurling in all directions proclamations
announcing the fall of Kerenskii's government. In all, an "orderly, and even
rather gentle" revolution, Williams thought, and Trotskii, as he looked back on
the scene, felt it all had been "too brief, too dry, too business-like—somehow
out of correspondence with the historic scope of the events. . . . There is no
picture of the insurrection," he went on. "Events do not form themselves into

a picture."[79] Yet, across the time and space of the seven decades that have passed since they occurred, the events that contemporaries thought so uncharacteristically inconsequential as they awaited Petrograd's great revolutionary confrontation take on more precise shape. On closer examination, this plethora of seemingly disconnected incidents welds itself into a more clearly defined portrait of the moment in history which became the triumph that the Bolsheviks named the Great October Revolution.

Of all the unrealities of these October days, perhaps none was more striking than Kerenskii's unprecedented paralysis of will. Ever since the secret Central Committee meeting in Galina Flakserman's apartment, Bolshevik agitators and newspapers had spoken and printed what, at any time and from any government's point of view, ranged between subversion and outright treason. Diligent government agents had assembled files of subversive clippings from the Bolshevik newspapers *Rabochii put, The Worker's Path,* and *Soldat, The Soldier,* and had added to them careful compilations of treasonous public statements in which Bolshevik agitators urged their listeners toward an armed uprising. Trotskii's speeches as president of the Petrograd Soviet more than substantiated charges of treason against the Bolsheviks, and Lenin's writings, especially his raging "Letter to Comrades" of October 17, in which he had openly accused Kamenev and Zinoviev of "astounding confusion [and] timidity" because they had spoken out against an armed uprising, provided even more incriminating material.[80]

Yet, not until the predawn hours of October 24, when Red Guards all across the city had gone on full alert and Bolshevik commissars already had control of most of the Petrograd garrison, did Kerenskii move to defend his government.[81] After closing *Rabochii put* and *Soldat,* he took his case to the Council of the Republic, a recently created body of some 550 members designed to act in an advisory capacity to the government until it should finally summon the long-awaited Constituent Assembly into session. There, Kerenskii assured his listeners that the evidence at hand now prevented anyone from "reproaching the Provisional Government with making false accusations or malicious fabrications." Kerenskii then launched into one of the now-tiresome tirades that led de Robien a few days later to describe him as "a seedy-looking barnstormer."[82] The Bolsheviks (who had boycotted the meeting) were working to "betray the liberty and independence of Russia" and plotting to "open Russia's front to the armored fist of [Kaiser] Wilhelm and his friends," Kerenskii insisted. The very workers and soldiers to whom he had so often turned for support now had become, in his words, a "rabble" that must be brought under control. Concluding with what one commentator later called the "hysterical wail of a bankrupt politician,"[83] Kerenskii provoked a burst of laughter from some of the Mensheviks and Socialist Revolutionaries in his audience when he threatened his enemies with "immediate, final, and definite liquidation."[84]

In another of the unrealities of Petrograd's revolutionary days, two leading

Mensheviks, the Bolsheviks' bitterest enemies on the Left for more than a decade, rose to defend them against Kerenskii's charges. Insisting that "the Bolsheviks have only taken advantage of the real dissatisfaction among the broad masses, whose needs have not been met," the first to take the floor urged the government to "satisfy the people's cry for peace . . . [and] formulate the land question in such a way that no one would have any doubts that the Government is taking firm steps in the direction of satisfying the needs of the people." Calling Kerenskii's speech nothing short of a "challenge to civil war," the Mensheviks' leader Martov rose to speak, his words croaking forth from a throat scarred by the ravages of tuberculosis contracted during his years of exile in the Siberian Arctic. Once he had been Lenin's comrade in the long-ago days when they had struggled to organize St. Petersburg's workers into the Marxist Union of Struggle for the Liberation of Labor. Now he stood among his bitterest enemies, the leader of the Mensheviks who had refused to accept Lenin's demands for rigid party organization and absolute obedience. Still, as a Marxist, he could not allow Kerenskii's threat to go unanswered. "Only a government that is guided by the interests of the [people's] democracy can deliver the country from the horrors of a civil war," Martov insisted. If the masses revolted because of the government's insensitive policies, he warned, he and his followers would not support any attempt "to suppress the uprising."[85] For once, Kerenskii's theatrical performance had fallen far short of its intended result. Rather than grant him the unlimited authority he demanded to crush the Bolsheviks, the Council of the Republic sided with the Mensheviks in a vote of "no confidence."[86]

By his own admission "enraged" at the Council's action, Kerenskii at first threatened to resign, but changed his mind and decided instead to act without the Council's support. In the capital, only his cabinet and an undetermined number of troops stood with him against the Bolsheviks, the Petrograd Soviet, the Red Guards, the Kronstadt sailors, and most of the city garrison. "The Bolsheviks were acting with great energy and no less great skill," Kerenskii admitted some years later. " 'Red' troops were in action all over the city . . . [and] the uprising . . . was developing with tremendous speed."[87] "This is defense, comrades, defense!" Trotskii exclaimed, as he promised a delegation from the Petrograd City Duma that the forces of the Soviet would answer "blow for blow, with steel against iron" if the assault against it continued.[88]

Not even the Bolshevik command post at Smolnyi at first realized how fast the uprising was moving. Late on Tuesday afternoon, two of Trotskii's agents, both of whom had forgotten to bring their weapons, walked into Petrograd's Central Telegraph Office, where the Bolsheviks had not one sympathizer among its three thousand employees, and took command in the name of the Military Revolutionary Committee.[89] For the next several hours, Military Revolutionary Committee commissars carried out other assignments with only slightly more difficulty, causing the ever-observant Sukhanov to remark that, "in general, military operations seemed more like a changing of the guard."[90] "The indifference of the crowds in the streets and on the streetcars is astounding,"[91] a General Staff officer reported as the crowd of theatergoers turned toward home late that evening. No one anywhere seemed to care. "In a word, the darkest, most idiotic, dirtiest 'social revolution' that history has ever seen is upon us," Zinaida Gippius wrote in her diary as midnight approached. "It's boring and disgusting," she added. "The element of struggle is completely lacking everywhere. . . . Oh, well, to hell with them all!"[92]

Gippius's supreme indifference on the night of October 24 stood a pole apart from Lenin's burning impatience. All day, he had paced Fofanova's apartment while she, after begging him not to step outside and risk arrest at such a crucial moment, had crossed and recrossed the city as she carried messages back and forth across the five kilometers that separated Lenin from Smolnyi, which now throbbed with the pulse of revolution in a city otherwise indifferent to the momentous events afoot. Jack Reed remembered how "great Smolnyi, bright with lights, hummed like a gigantic hive" that night.[93] Its gate bristled with machine guns and cannon, and in the courtyard armored cars stood ready to go on the attack, their engines roaring in anticipation.[94] "Like a steady flow of electric current," one Bolshevik remembered, orders flowed outward into the city.[95] As yet, none knew the impact of those orders. It was too early to gauge their success, too early to tell whether the men who sent them would become the villains of Petrograd's revolutionary drama or emerge as its heroes.

Late that evening, Lenin could stand his isolation no longer. At 10:50, Fofanova returned from one of her journeys to find a hastily scribbled note propped up on a plate:

I have gone where you didn't want me to go.

Good-bye.
Ilich[96]

Desperate to be a part of the moment he had awaited for so many years, Lenin had donned his curly wig, wrapped his face in a bandage to distort his appearance still more, and rushed into the night accompanied only by Eino Rakhia, a Bolshevik Finn whose middle-aged appearance belied the fact that he sometimes served as a bodyguard. Then, in one of the supreme ironies of the Great October Revolution, the Red Guards who stood at Smolnyi's doors barred Lenin's path because he had no proper identification. Rakhia's quick wit saved the day. "I began to shout, 'What a mess. I'm a delegate [to the Second All-Russian Congress of Soviets, which was scheduled to begin the next day, October 25, at Smolnyi] and they won't let me through!" he explained in his brief recollections of that night. "The crowd supported me and pushed the two Red Guards aside, and we moved in," he remembered. "V. I. [Lenin] came last, laughing, satisfied with the favorable outcome."[97] Finally, Lenin was at Smolnyi. At last, the commander commanded his forces. Some recalled in later years that the pace of action seemed to quicken and become more resolute at that moment.[98]

Whether due to Lenin, Trotskii, or some combination of forces, the Bolshevik-led armed uprising moved forward in deadly earnest as October 24 passed into the wee hours of Wednesday, October 25. Although gunfire rang out infrequently and claimed victims only with the greatest rarity, Military Revolutionary Committee commissars had added the Electric Power Station, the Main Post Office, and the Nikolaevskii Railroad Station to their seizures by two o'clock that morning. With the exception of the Nikolaevskii Bridge that stood nearest to the Winter Palace, they controlled all the bridges connecting the workers' districts to the central part of the city as well. Less than two hours later, the crew of the cruiser *Aurora* added the Nikolaevskii Bridge to their list. When these heroes of the revolution shined their spotlights on the military cadets assigned to defend the bridge, the cadets fled without firing a shot. The State Bank fell at six o'clock in the morning, and the Central Telephone Exchange at seven. At eight, the Warsaw Station, from which trains ran to *Stavka* and the northern front headquarters at Pskov, rounded out the list of victories won that night by commissars who forgot to bring their revolvers and warships that fired searchlight beams instead of high explosive projectiles.[99] A sense of unreality still seemed omnipresent. "The moonlight created a fantastic scene," one of the soldiers of the Sixth Engineer Battalion that seized the Nikolaevskii Station remembered. "The hulks of the houses looked like medieval castles—giant shadows followed the engineers. At this sight, [the bronze statue of] the next-to-last emperor [Alexander III, that stood in the square outside the station] appeared to rein in his horse in horror."[100]

As the early dawn lightened the sky enough to dispel the specters that had lurked around the Nikolaevskii Station, Kerenskii prepared a last desperate effort to save his government. He had spent the bleak predawn hours on Wednesday morning at General Staff Headquarters with Deputy Premier Aleksandr Konovalov as report after report announced the fall of one key point after another to the Military Revolutionary Committee's commissars. Finally, when no regular troops could be found anywhere in the city to march to the government's defense, even Petrograd's arrogantly optimistic garrison commandant had to admit that the the Provisional Government's situation had become critical.[101] "It was as if the Provisional Government had found itself in the capital of a hostile country that had just completed mobilization but had not yet started active operations," one general reported to *Stavka*.[102] By nine o'clock that morning, Kerenskii evidently concluded that his only hope lay in finding troops at the front who would march with him to wrest Petrograd from the Bolsheviks' victorious hands.

With Petrograd's rail stations under Bolshevik control, Kerenskii's only chance was to escape by car. Yet, as his driver later testified, not one car in working order could be found anywhere in the General Staff, and it took more than an hour of frantic searching to turn up the two cars needed for his flight. Well before noon, a Renault commandeered from the American Embassy and still flying the American flag left the Winter Palace followed by a huge Pierce Arrow touring car with Kerenskii huddled in its back seat. Ironically, Kerenskii's last motorcade took the same route that Germany's Ambassador Pourtalès had taken when he made his fateful evening journey from the German Embassy to the Russian Foreign Office to declare war more than three years before. Now Kerenskii retraced the ambassador's steps, destined, like Pourtalès, never to see Petrograd again.

As the now-abandoned German Embassy rose before it, just beyond the point where the Morskaia ran into Mariinskaia Square, Kerenskii's hurrying procession swerved to the left, passing Baron Klodt's statue of Nicholas I that still towered above the shattered land his great-grandson had ruled so wretchedly. Picking up speed, the cars crossed the famous Blue Bridge, where the serf ancestors of the peasant soldiers now in control of Russia once had been sold at auction, and then drove past the Mariinskii Palace, where the Council of the Republic's Presidium had assembled for its regular morning session. A moment later, Kerenskii and his escort sped off along Voznesenskii Prospekt toward Gatchina and beyond.[103] In the wake of his fleeing caravan, detachments of Red Guards moved toward the Winter Palace Square. As Russia's soldiers had done since the time of Peter the Great, they sang as they marched, but they now sang the cadence of the brave new world they hoped to enter. "Bravely forward, comrades," they roared. "We shall raise the the workers' red banner above the land!"[104] A few hours later, the Bolsheviks closed their ring around the Winter Palace to set in motion the closing act in the drama of their Great October Revolution.

Except for Minister of Food Sergei Prokopovich, who had been detained briefly by a wandering Bolshevik patrol, Kerenskii's fellow ministers had arrived at the Winter Palace early that afternoon to find their prime minister and supreme commander gone. Certainly, the impression of unseemly haste that surrounded Kerenskii's departure did not sit well with his fellow ministers, but the problem of the Bolshevik onslaught still had to be faced. "After a brief exchange of opinions," Minister of Justice Pavel Maliantovich remembered, "[we] agreed . . . that the situation was so serious that the Provisional Government could not fulfill its duty if it did not remain at the Winter Palace in full force, and announce that it would meet in continuous session until the final resolution of the crisis."[105] The Provisional Government thus girded for its last stand in the Winter Palace without Kerenskii and without regular troops to defend it. "All of the exits and passageways leading to the Neva [Embankment] were surrounded by military cadets," a member of the Provisional Government's Supreme Investigating Commission wrote in his recollections of that day. Four hundred of these boisterous adolescents, "making a racket, laughing, and chasing each other up and down along the sidewalk,"[106] plus a Women's Shock

Battalion and a handful of Cossacks, were to be the government's defenders in its last extremity.

None but the Cossacks had ever faced enemy fire, and the cadets' image of war still remained romantically untarnished by reality. "If we should be overpowered, well, every man keeps one bullet for himself," one of them told Reed grandly.[107] Not long afterward, Louise Bryant saw a "little man come out of the palace, walk across the square, plant his tripod." As the Bolsheviks' forces began to assemble across the square, she watched in amazement as he unhurriedly began to photograph the women soldiers building a barricade. "It all had an operatic air, and a comic one at that," she remarked later that day to Bessie Beatty, her fellow American, a dauntless war correspondent of the *San Francisco Bulletin.* [108]

While the Provisional Government found time during its last hours to send out a photographer to record the actions of its women's shock battalion, it never took the time to make certain that all the Palace doors had been locked before the Bolsheviks began their assault. The comedy therefore had in it all the makings of a tragedy, as adolescents who did not yet know the meaning of war made ready to carry out the orders of a government that had decided it should defend itself although it had not the will to do so. "Doomed men, abandoned by everyone, roaming around inside a gigantic mousetrap," Minister of Justice Maliantovich thought as he and his fellow ministers "gathered from time to time either together or in smaller groups for brief discussions." No one believed they could succeed and few really cared. "The regime was about to collapse beneath the weight of universal disgust," a Russian journalist wrote later. "It was clear that no one would raise a finger in its defense."[109]

Outside the palace, the Bolsheviks strengthened their forces. "The chain drew nearer and nearer to the Winter Palace Square, and it grew thicker with every hour that passed," their co-commander Nikolai Podvoiskii wrote.[110] Inside, the defenses grew thinner. At six o'clock in the evening, most of the cadets from the Mikhailovskoe Artillery School decided that they had no further taste for defending the Provisional Government. Not having eaten for most of the day, they left the Winter Palace and went in search of dinner, taking most of their guns with them. Then the Cossacks decided not to fight alongside "Jews and wenches," and marched away. "Half the Government's made up of Jews, but the Russian people stand with Lenin," they told one junior officer who cursed them as traitors and cowards.[111] More cadets began to slip away and then even the Women's Shock Battalion departed. All the while, the ministers turned a deaf ear to all pleas for instructions. "We could not give the order to fight to the last man and to the last drop of blood because we, perhaps, were already defending only ourselves. . . . Nor could we give the opposite order, to surrender," Maliantovich explained. "We gave our defenders the opportunity to choose to link their fate with ours," he concluded. As the Provisional Government had done so many times before, so it did once again in its final moment of extremity. Called upon to make a vital decision at a critical time, it struggled frantically not to act.[112]

Meanwhile, the Bolsheviks had problems of their own, though nothing of the magnitude of those that confronted the government's leaders. That morning, the Military Revolutionary Committee had ordered the commandant of the Peter and Paul Fortress to prepare his fortress artillery for possible action against the Winter Palace only to learn that, of the six guns that could be brought to bear on the Palace, five had not been cleaned in more than seven months. Not wanting to take part in the events to follow, the fortress's artillery officers insisted that these could not be fired in their present state. Unaware that a simple cleaning would solve the problem, the Bolshevik commissars on the scene (they all happened to be infantry ensigns) ordered their men to drag several smaller three inch training guns into position, but then learned that the fortress magazine had no three-inch shells and that, in any case, the gunsights had not been mounted. At that point, the fortress gunners pointed out that the dirty six-inch guns could be cleaned without much time or effort. Finally, eight hours after the order to prepare the guns arrived, they stood ready to go into action supported by the guns on the cruisers *Aurora* and *Amur.*[113] "We propose that the Provisional Government and the troops loyal to it capitulate," one of the Bolshevik commanders informed the Provisional Government's ministers at six-thirty that evening. "This ultimatum expires at 7:10, after which we will immediately open fire."[114] "What threatens the Palace if the *Aurora* opens fire?" one cabinet member asked as they assembled in the great Malachite Hall to discuss the enemy's ultimatum. "It will be turned into a pile of rubble," the Provisional Government's last minister of naval affairs replied.[115] Again, the Provisional Government decided not to act. As the ultimatum's deadline passed, they retired to an inner dining room to await the outcome of events.

Yet the Bolshevik forces did not respond to the passing of the deadline as they had threatened. By now, they had deployed forces in overwhelming strength, including several armored cars, field guns, most of the Keksgolmskii and Pavlovskii regiments, a number of elite Red Guards units, and a detachment of tough sailors from Kronstadt, all of them ready to fight if the order came. But even among these embittered men, well-armed and anxious to use their weapons, the atmosphere of unreality persisted. As the armored cars moved into position, a small boy casually clambered onto one and began to peer down the barrel of its machine gun.[116] Finally, "the bass voice of the *Aurora*" spoke, but only to fire a single blank shot.[117] For about an hour, the Bolsheviks' machine guns took up the refrain, leaving the Palace walls "riddled with thousands of bullets," Ambassador Buchanan reported, while the few remaining defenders replied from the Palace's windows and from behind firewood barricades. Although the city waited anxiously, the final assault still did not begin.[118]

As the *Aurora* sounded its single thundering note, Petrograd's white-bearded mayor Grigorii Shreider, aged but resolved to do his duty to the end, opened an emotion-charged emergency session of the City Council with a warning that a full-scale bombardment of the Palace would begin in "a few seconds." Carried on the crest of a wave of mass hysteria, most of the Council and a number of spectators pledged to march to the Winter Palace to relieve the besieged ministers or perish with them. One after another, each announced himself or herself ready to die. Four abreast, armed against the rain with umbrellas but unarmed against the guns they faced ahead, three hundred well-dressed men and women, many of them former pillars of the tsarist regime, began their march down the Nevskii Prospekt, singing the "Marseillaise" as they went. For two hundred meters they marched. At the Kazan Cathedral Square, a cordon of revolutionary soldiers and sailors barred their way. Together, Shreider and Minister of Food Prokopovich, whose impressive credentials as an economist had done nothing to help him end Petrograd's food shortages, demanded that the guards step aside. The soldiers refused. Jack Reed and Albert Rhys Williams heard the marchers announce that they would continue on in any case, ready to die on the spot if the soldiers dared shoot unarmed men and women.

"No, we won't shoot unarmed Russian people," the commander of the cordon agreed.

"We will go forward! What can you do?" Prokopovich and Shreider insisted.

"We can't let you pass," a sailor said slowly. "We will do something."

"What will you do?" the marchers demanded.

"We will spank you!" another sailor roared, totally destroying the heroic aura with which the marchers had tried to surround themselves. "Go home now!"

There, in the chilling rain, the march ended. "Let us return to the Council and discuss the best means of saving the country and the Revolution," Prokopovich loftily counseled his followers. "We cannot have our innocent blood upon the hands of these ignorant men!"[119]

A few blocks away, in the Winter Palace, the Provisional Government's ministers now awaited the Bolsheviks' final assault alone, each in his own way. Deep in thought, Admiral Verderevskii smoked his pipe, spitting from time to time. Nervous as a cat—"like a caged tiger," one eyewitness said—Foreign Minister Tereshchenko paced up and down, grinding the carpet beneath his feet, while Minister of Trade and Industry and Deputy Premier Konovalov sat on the sofa, nervously tugging at his trousers creases until he finally had the cuffs lifted above his knees.[120] Among their defenders, some had been drunk for hours, already having launched an assault upon the Palace wine cellars that Bolshevik soldiers, sailors, and Red Guards would take up in greater earnest later. "Feasting in the time of plague," a young lieutenant thought as he surveyed the scene. "Feasting in the time of plague. What shame!"[121] A mere handful of defenders, some exhausted, others hopelessly drunk, and still others on the verge of hysteria, made ready to repel the invaders. "We're ready to die for the Government!" one exclaimed. "Just give us the order to fire!"[122]

Across the river, while streetcars continued to crawl slowly across the bridges at both ends of the Palace Embankment, the gunners at the Peter and Paul Fortress fired some three dozen six-inch shells at the Winter Palace. Only two found their mark, but one exploded just above the room in which the ministers had assembled, the sound tearing at nerve ends already raw from tension. In the Square outside, the armored cars' machine guns began their attack again, as soldiers, sailors, and Red Guards with rifles moved closer to the Palace walls. A few among them worked their way into the Palace through obscure doorways left unlocked and windows left unguarded, only to be captured by the cadets still willing to defend the government. Others followed. Soon, the process reversed itself, with the invaders outnumbering the defenders and disarming them. In small groups, growing larger as they went, the attackers worked their way through the palace, exchanging shots with frightened defenders, always moving forward in search of the room in which the Provisional Government had taken refuge. "The noise rose, swelled, and immediately swept toward us in a broad wave," Maliantovich remembered. "It filled us with unbearable anxiety, like a cloud of poisoned air."[123] As the minutes passed, it rose to a crescendo, ever-louder, ever-nearer. "Who is really holding the Palace now: we or the Bolsheviks?" the ministers asked each other.[124] None doubted the answer. All knew that the end stood only moments away.

Now Russia's ministers reached a decision. "The picture is clear!" they concluded. "No more bloodshed! We must surrender!"[125] Quickly, they seated themselves around the table at which Nicholas and his family had eaten their private meals in the years before the Revolution of 1905 had driven them into the Alexander Palace at Tsarskoe Selo. Suddenly, the door burst open. "A little man flew into the room, like a chip washed up by a wave, under the pressure of the crowd that poured in behind him and spread, like water, flowing into all corners of the room," Maliantovich wrote later.[126] This was Vladimir Antonov-Ovseenko, the seedy-looking tactician of the streets who sometimes went by the revolutionary codename of "the bayonet," and who, with Nikolai Podvoiskii, had commanded the assault. Although far from heroic, Antonov's image burned itself indelibly into Maliantovich's consciousness so that he could later describe him in meticulous detail. "He had long rust-colored hair and glasses, a short trimmed reddish mustache, and a small beard," he wrote in his remembrances of the Bolsheviks' attack. "His short upper lip raised up to his nose when he spoke. He had colorless eyes and a weary face." For some time, Maliantovich could not tear his eyes from Antonov's shirt. "On his soft shirtfront a long necktie crept up from his vest toward his collar," he remembered. "His collar and shirt and cuffs and hands were those of a very dirty man."[127]

As Kerenskii's deputy, his composure now restored, Konovalov addressed the intruders:

"The Provisional Government is here," he said. "What is your pleasure?"

"In the name of the Military Revolutionary Committee," Antonov-Ovseenko announced, his voice insistent, "I declare that you—all of you—members of the Provisional Government are under arrest!"

"The members of the Provisional Government submit to force and surrender, in order to avoid bloodshed," Konovalov replied.[128]

Acting Minister of Transportation Aleksandr Liverovskii looked at his watch and noted the time: 1:50 a.m. on the morning of Thursday, October 26. Less than two hours later, he and his colleagues entered the Peter and Paul Fortress. There, he made a final entry in his diary: "5:05 a.m. I am in cell No. 54."[129]

Like besieging invaders let loose into an enemy city, the soldiers, sailors, Red Guards, and commonfolk of Petrograd had their way with the Winter Palace.

The matter of the Winter Palace's wine cellars became especially critical [Antonov-Ovseenko wrote as he recalled the crowd's final assault against them a month later]. . . . The Preobrazhenskii [Guards] Regiment, assigned to guard these cellars, got totally drunk. The Pavlovskii [Guards], our revolutionary buttress, also couldn't resist. We sent guards from various other picked units—all got utterly drunk. We posted guards specially chosen from Regimental Committees—they succumbed as well. We dispatched armored cars to drive away the crowd. After patrolling up and down a few times, they also began to weave suspiciously. When evening came, a violent bacchanalia overflowed. "Let's drink the Romanovs' remains!" This happy slogan seized the crowd. We tried sealing up the entrances with brick—the crowd came back through the windows, smashing in the gratings and seizing what remained. We tried flooding the cellars with water—the firemen sent to do the job got drunk instead. . . . This drunken ecstasy infected the entire city [the appalled Antonov-Ovseenko went on, as he described a scene fit for Hogarth's brush or pen]. . . . The Council of People's Commissars, finally, appointed a special deputy, endowed him with exclusive authority, assigned a strong detachment [of troops] to help him carry out his duties. But this person also turned out not to be very reliable.[130]

As intoxication reached epidemic proportions, the Bolsheviks took brutal measures to bring it under control. "In view of the fact that experience has shown that less decisive measures have not had the desired result," one order decreed, "stores of wine will be BLOWN UP WITH DYNAMITE." To strike greater fear into the hearts of Petrograd's drunken legions, the order concluded: "We announce that no advanced warning will be given before the explosions."[131] At the French Embassy, Comte de Robien was beside himself when he heard the news. "It is sickening to see such good stuff thrown away," he lamented to his diary. "There were bottles of Tokay there from the time of Catherine the Great, and it has all been gulped down by these vodka swiggers."[132]

What would life be like in the brave new world where "vodka swiggers" drank the tsar's best wines? None, not even Lenin, could be certain. "You know," Lenin mused to Trotskii on the evening that the Winter Palace fell, "to pass so quickly from persecutions and living in hiding to power—*es schwindelt* —it makes one's head spin!"[133]

PART FIVE

1918

CHAPTER XVII

Birth Pangs of a New Order

During the crucial days before the Great October Revolution, Lenin had insisted that the Bolsheviks' armed uprising must overthrow the Provisional Government before the Second All-Russian Congress of Soviets assembled for its first meeting on the evening of October 25. Yet, events did not move quite swiftly enough, and the Congress opened at Smolnyi just hours before the Winter Palace fell. While the Red Guards and Kronstadt sailors prepared their final assault against the last bastion of Kerenskii's government, a violent dispute between the supporters of the armed uprising and Martov's Mensheviks tore the Congress apart only minutes after it assembled. With Lenin still behind the scenes at Smolnyi, unable to appear in public until the Bolsheviks made absolutely certain of the overthrow of Kerenskii's government, Trotskii led the Bolsheviks' attack against those who demanded that the uprising stop short of seizing the Winter Palace and that the insurgents negotiate with the shattered remnants of Kerenskii's government. "You are pathetic bankrupts," Trotskii replied, his voice full of contempt for those who had spoken out against the Bolsheviks' "insane and criminal action" in leading the armed uprising. "Go where you belong from now on: into the trashbin of history!"[1]

Stormy applause greeted Trotskii's cold, arrogant words. Desperate to counteract them, Martov rose again. "Then we'll leave!" he said, and angrily led his supporters toward the exit. "Martov walked in silence and did not look back," one of his allies remembered. At the exit, he stopped, turned for a moment, and spoke his last words to the assembled representatives of Russia's soviets. "One day, you will understand the crime in which you are taking part," he said wearily. Then he left the Bolsheviks and their allies among the left wing of the Socialist

457

Revolutionaries to continue the work in which he refused to take part.[2] "This does not weaken the Soviets but strengthens them by purging them of counter-revolutionary elements," Trotskii assured those who remained, and began to read a resolution condemning the "compromisers" who had joined Martov's "pitiful and criminal attempt to wreck the All-Russian Congress."[3]

As Trotskii spoke, word arrived that the arrested Provisional Government ministers had entered the Peter and Paul Fortress,[4] and Lenin hurried to draft an historic message that announced Russia's new course. "Backed by the will of the vast majority of the workers, soldiers, and peasants, backed by the victorious uprising of the workers and the garrison which has taken place in Petrograd, the Congress takes power into its own hands," his chosen representative Anatolii Lunacharskii announced to the men and women of the Congress who had not followed Martov into "history's trashbin." "A slight, student-like figure with the sensitive face of an artist," was the way Reed remembered Lunacharskii in those days,[5] but the man who had been with Lenin in the Russian Marxist movement for more than twenty years spoke well—in a manner fitting someone about to become Commissar of Public Enlightenment in the new government. Russia's new government, Lenin promised through Lunacharskii, would "propose an immediate democratic peace to all nations and an immediate armistice on all fronts," transfer all land held by nobles, church, bourgeoisie, and government to the peasants, introduce "complete democracy in the army," place the workers in control of their factories, "ensure the convocation of the Constituent Assembly," guarantee "the genuine right of self-determination" to all nationalities in Russia, and would "see to it that bread is supplied to the cities and prime necessities to the villages." Thunderous applause greeted Lenin's promises, and the man who spoke for him could barely make his final words heard. "The fate of the revolution and the fate of the democratic peace is in your hands!" he exclaimed. "Long live the revolution!"[6]

While the delegates cheered, exhaustion overcame the men who had directed the uprising for the better part of two days. Several hours before October's murky late dawn lightened Petrograd's leaden sky, Lenin left Smolnyi, now free to walk the city's streets in daylight, in search of a few hours' sleep at the nearby apartment of his close friend Vladimir Bonch-Bruevich. Yet he rested only briefly. While Bonch-Bruevich slept on into late morning, Lenin drafted his Decree on Land, easily one of the most revolutionary documents to come out of the turmoil of 1917. Borrowing extensively from the agrarian program of the Socialist Revolutionaries which, as he had told Fofanova some days before, provided the basis for "a ready-made agreement" between them and the Bolsheviks, Lenin set down the brief text that transferred the lands long held by Russia's noble lords into the hands of her millions of peasants.[7] "Now we have only to see to it that it is widely published and publicized," he told Bonch-Bruevich. "Then let . . . [the bourgeoisie] try to take it back!"[8]

This decree, Trotskii later said, was of vital importance, not only as "the foundation of the new regime but also [as] a weapon of the revolution, which had still to conquer the country."⁹ Therefore, even before he returned to Smolnyi to make his first appearance before the Congress of Soviets that evening, Lenin began to search for ways to send its contents to peasants who could not read in a land where electricity (and hence radio messages) had not yet reached hundreds of thousands of isolated hamlets. Could not demobilized soldiers be sent to announce the decree, explain it, and distribute copies, Bonch-Bruevich asked? Lenin thought for a moment, then shook his head. "There's a problem," he said. "They'll want to smoke, and, when they get out where there aren't any newspapers, they'll roll cigarettes out of the decree." The solution, Lenin suggested, was to give leftover calendars to every soldier they sent into the countryside. "Three hundred and sixty-five sheets in all," he reminded Bonch-Bruevich with a smile. "Enough for a soldier and his friends too, and the decrees will get carried safely all the way to the villages!"¹⁰

At 8:40 on the evening of October 26, the Second All-Russian Congress of Soviets opened its second session. With the armed uprising victorious despite his gloomy predictions, Lev Kamenev rose to report on the actions that the Military Revolutionary Committee had taken since Kerenskii's government had fallen. Capital punishment had been abolished at the front, and all officers, soldiers, workers, and peasants arrested by the Provisional Government for political crimes had been freed from prison.¹¹ Several of those who had stormed out of the Congress's first session the night before returned to issue dire warnings, only to be shouted down by the delegates: "We thought you already walked out last night! How many times are you going to walk out?"¹² Then Lenin rose to speak. His "shabby clothes, his trousers much too long for him," caught Reed's eye, and he thought that this "short, stocky figure, with a big head set down in his shoulders, bald and bulging" was indeed "unimpressive to be the idol of the mob" that he had become that night. Reed was seeing Lenin for the first time, but he already understood that this man, although "without the picturesque idiosyncrasies" that commonly characterized political leaders, possessed one of the rarest of revolutionary gifts—a combination of "the greatest intellectual audacity" and "the power . . . of explaining profound ideas in simple terms"—a gift that raised him head and shoulders above all revolutionaries who had risen to the surface of Petrograd's revolutionary maelstrom since February.

Reed watched Lenin's "little winking eyes travel over the crowd" as he stood before the representatives of Russia's masses for the first time, "apparently oblivious to the long-rolling ovation, which lasted several minutes."¹³ Sukhanov, still the ever-present observer even though the Mensheviks' ranks now lay broken in defeat, watched from a seat in the back of the hall as the Congress celebrated the Bolsheviks' victory. "The entire Presidium, with Lenin at its head, stood and sang [the *Internationale*] with excited, animated faces and blazing eyes," Sukhanov wrote, as he recalled "with heavy heart" the moment when the Bolsheviks and the masses seemed to merge into one.¹⁴ Gripping the edge of the podium, Lenin spoke to the masses before him for the first time. "Comrades," he said simply, "we shall now take up the formation of the socialist state."[15]

"The question of peace is a burning question," Lenin began, as he prepared to read the draft of a Decree on Peace that would enrage Russia's allies. After asserting that "the overwhelming majority of the working class and other working people of all the belligerent countries, exhausted, tormented and racked by the war, are craving . . . an immediate peace without annexations . . . and without indemnities," Lenin went on to declare that Russia would abolish secret diplomacy and "proceed immediately with the full publication of the secret treaties" that her government had signed since February. He then proposed "an immediate armistice to the governments and peoples of all the belligerent countries," urged that it extend for at least three months "to permit the completion of negotiations for peace with the participation of the representatives of all peoples or nations without exception," and called upon "the class-conscious workers of all countries" to join Russia in pressing their governments to make peace. The victorious proletariat must let nothing deter it in its efforts to end the war. "The workers' movement will triumph," he assured his listeners, "and [it] will pave the way to peace and socialism."[16]

With almost no discussion, the delegates at the Congress of Soviets voted unanimously to accept Lenin's proposal. Reed noticed that "a grizzled old soldier was sobbing like a child," and that Aleksandra Kollontai, the Bolsheviks' aristocratic champion of women's rights, "rapidly winked the tears back." For one wonderful, magic moment, Russia's people were at one with their leaders as they rose and sang: "When tyranny falls the people will rise, great and free!" Reed saw a young worker's face radiant with hope. "The war is ended!," he heard him call out. "The war is ended!"[17] Of course, it was not. But Lenin had dared to take the first momentous step. That night, Russia's new leaders had rent the chain that had bound the nations of Europe into a single warring, bleeding mass for more than three years. At 10:35 on the evening of October 26, Lenin and his revolutionary comrades joined Russia's war-weary masses in severing their nation's commitment to fight in the Great War. "Nice friendly Allies we have chosen, haven't we?" one British officer wrote bitterly that night, while another exclaimed in utter disgust: "Damn all these chicken-hearted peoples who have no guts to stick out a war!!!"[18] It was, a future American ambassador to the Soviet Union later wrote, "a bitterly unfriendly move."[19]

Perhaps too willing to play a secondary role in the events that swirled around her, uncommonly wise in the ways of politics, and acutely attuned to world's failings, Lenin's wife Nadezhda Krupskaia once explained that the Bolsheviks "had not organized the revolution and seized power in order to hitch a swan, a pike, and a crab to the Soviet cart and form a government that would prove incapable of pulling together and moving forward."[20] Lenin intended to accomplish a great deal more on this "first day of the Socialist Revolution"[21] than just a dramatic gesture toward peace. Therefore, even as the delegates celebrated the Decree on Peace, he had already begun to read the even more revolutionary Decree on Land that he had completed at Bonch-Bruevich's apartment that morning. "The first duty of the government of the workers' and peasants' revolution must be to settle the land question," he announced. Quickly, Lenin enumerated the principles that must "serve everywhere to guide" that settlement. "Private ownership of land shall be abolished forever," he began, and quickly added that all land held as private property "shall be confiscated without compensation and become the property of the whole people." Hired labor could not be used to cultivate land, disabled or sick peasants would receive government pensions, and, henceforth, "the right to use the land shall be accorded to all citizens of the Russian state (without distinction of sex) desiring to cultivate it by their own labor."

A number of other generally stated principles, most of them taken from the program of the Socialist Revolutionaries, and all in the same vein, filled out Lenin's brief speech. "We want no details in . . . [this decree]," he explained. "We trust that the peasants themselves will be able to solve the problem correctly, properly, and better than we could do it." Peasants, not government bureaucrats, must settle peasant affairs, he concluded. Most of all, "the peasants should be firmly assured that there are no more landowners in the countryside."[22] Typically concerned for the poorest and least fortunate, Krupskaia looked not at the men who stood on Smolnyi's rostrum as the delegates cheered Lenin's pronouncement, but at an elderly peasant who sat not far away. "Gripped by deep emotion, his face shown with a peculiar waxen transparency," she remembered. "His eyes glistened with a certain special light."[23] Lenin had spoken the magic word, land—not in the future, not after endless deliberations by government committees, but now, and according to the age-old peasant belief that the land's bounty belonged to those who labored upon it. The decree promised to make the peasants' oldest and most cherished dream come true. "No power anywhere will be able to take back this decree from the peasants and return the lands to the nobles," Lenin had told Bonch-Bruevich that morning. "This is the most important achievement of our October Revolution!"[24]

As its final task that night, the Congress of Soviets took up the question of forming a new government. Earlier that day, Lenin and a number of leading Bolsheviks had met to discuss how a new government might be organized. Convinced that the title Minister was "a foul, worn-out term," Lenin urged his comrades to find another name for the men who would head Russia's first Bolshevik government. "Commissars, perhaps. But right now we have too many commissars already," Trotskii mused. "How about People's Commissars?" he asked suddenly, adding that the government as a whole might be called "the Soviet of People's Commissars." Lenin seized upon the idea. "That's brilliant!" he exclaimed. "It smells strongly of revolution!"[25] Lenin was to be named chairman of the Soviet of People's Commissars, and Trotskii the people's commissar for foreign affairs. But what of the other posts? Who among men with so little experience in the tasks of government could hold such positions? One of those suggested for a post tried to refuse, explaining that he had no experience that could fit him for the task. Lenin burst into laughter. "Do you think any of us has any experience in this either?" he asked. The main thing, he insisted, was that a people's commissar be someone "closely tied to the masses—a type born in the fires of revolution."[26]

With Lenin as their chairman, fourteen people's commissars became the new government that the Second Congress of Soviets chose to direct Russia's affairs before they broke up in the early morning hours. Taken together, Russia's new leaders had spent far more than a century in prison and exile. Five of them had been in prison for their political activities within the past three months, and Lenin had been freed from life as a fugitive less than two days before. Certainly, the will and dedication of these men had been tempered in the flames of revolutionary struggle. Now, they must turn their fire-hardened dedication to the defense of their revolution. Victorious in Petrograd, the revolution still had, as Trotskii said, "to conquer the country." Even as the Second Congress of Soviets triumphantly concluded its sessions on the morning of October 27, men bent on destroying its just-realized dreams had begun their march. Some had never left the city.

The promise that the infuriated Prokopovich had made to return to the City Council and "discuss the best means of saving the country and the Revolution" when Red Guards had barred his marchers' path to the Winter Palace in the late hours of October 25 had not been an idle gesture. Shouts of "Murderers!" "Rapists!" "To prison!" "To the gallows!" greeted the Bolshevik deputies who appeared in the City Council late the next morning.[27] Once the Bolsheviks had been sent away, Prokopovich joined Mayor Shreider and a number of influential moderate socialists and Kadets to form the All-Russian Committee for Salvation of the Motherland and the Revolution. With their headquarters in the City Council building, these men set out to undo the events of the previous forty-eight hours and regain their lost influence.[28]

Instantly, the Committee for Salvation unleashed a barrage of circulars and proclamations condemning Lenin's new government, which they accused of the "criminal" overthrow of Kerenskii's government. "Citizens of the Russian Republic!" their first proclamation began. "The civil war, begun by the Bolsheviks, threatens to plunge the country into the indescribable horrors of anarchy and counterrevolution." The Committee for Salvation promised that it could remove the terrible threat of a war that pitted brother against brother, but it needed the support of Russia's soldiers, workers, peasants, white-collar workers, and intelligentsia to do so. "Don't recognize the authority of these [Bolshevik] rapists!" the Committee urged. "Don't carry out their orders! Stand up for the defense of the Motherland and the Revolution!"[29] Support poured in. Few people thought the Bolsheviks could hold power for more than a few days.

Others thought they might last a few weeks at most.[30] Many who thought the Bolsheviks could not survive had helped to rule Russia between February and October, but few understood why they had failed to win the loyalty of Russia's masses and establish a stable government. That same blindness now guaranteed their failure in their struggles against Russia's new masters.

Ready to strike an alliance with any anti-Bolshevik force, and willing to match violence with greater violence, the Committee for Salvation prepared to turn the Bolsheviks' tactics against them. On the far right, they opened negotiations with the archreactionary Vladimir Purishkevich, now famous as Rasputin's assassin, who advocated "public hangings and shootings as an example to others" in order to wipe the stain of Bolshevism from the fabric of Russian life. "It's necessary to begin at the Smolnyi Institute," he explained, "and then work one's way through all the barracks and factories, shooting soldiers and workers by the masses along the way."[31] But if they were willing to deal with Purishkevich, sometimes thought of as the "first Russian fascist," the Committee for Salvation also tried to use its aura of middle class respectability to appeal for support against their nation's new government from Russia's allies and deny the mantle of legitimacy to the Bolsheviks in the arena of international politics.[32]

Within Russia, the Committee for Salvation also drew strong support from a source that revolutionary forces had overlooked until then, but which had considerable potential to influence the course of events. In what the émigré historian Sergei Melgunov called "one of the most beautiful pages in the history of the revolutionary era,"[33] Petrograd's civil servants and other white-collar workers went on strike, causing Russia's new People's Commissars to find near-empty offices when they arrived to take command. When one of the Bolshevik leaders arrived at the offices of the former Ministry of Internal Affairs, he found only "a heap of scrap papers, locked desks, and cabinets without keys." Employees at the State Bank refused to recognize drafts signed by the new government, and, when Trotskii appeared at Russia's Foreign Office, he found that the officials there had abandoned their offices, taking all the code books and keys to the safes that held important diplomatic papers with them. All around the world, Russia's ambassadors refused to recognize the Soviet of People's Commissars as their nation's government and declined to represent it in the capitals of Europe and America.[34]

While the Committee for Salvation prepared to attack the Bolsheviks' government from within, Kerenskii marshaled his forces for an assault from without, yet the Committee studiously avoided public association with Kerenskii. "Strangely enough," *Stavka*'s chief commissar wrote some years later, "even as they struggled against the Bolsheviks, everyone tried to avoid being linked with the [Kerenskii] government."[35] Northern front headquarters took that position emphatically. "I should make very clear to you the attitude of the [soldiers'] committees here," one general told *Stavka*. "It is clear that the committees will not support the former government and that they do not want to send troops to support Kerenskii as its head."[36] Even foreign observers easily perceived Russians' widespread disgust with their former minister-president. "Kerenskii is utterly discredited with all parties," Ambassador Buchanan wrote in his diary. "The troops, if they do come to Petrograd, will not fight to restore his Government, but to support the Socialist groups who have turned against the revolution."[37]

By the morrow of the Bolsheviks' victory, none knew better than Kerenskii how little support for his government remained in Russia. At every bend in the road that took him away from Petrograd, he had expected to find the troops he had ordered to march to his defense just before he had fled the capital. He had hoped to halt his flight at Gatchina, some forty kilometers away, but had found there only soldiers loyal to the Bolsheviks and had barely escaped arrest. Cold, and on the verge of nervous exhaustion, Kerenskii pressed on to the southwest another two hundred kilometers to Pskov, where General Cheremisov commanded the northern front, only to learn that Cheremisov himself had canceled his order to send troops to the capital.[38] Cheremisov wanted nothing to do with Russia's former prime minister and had no inclination to support any defense of his fallen government. He therefore refused to allow Kerenskii to carry his search for loyal troops directly to *Stavka*'s General Nikolai Dukhonin and, instead, urged him to leave Pskov with all possible haste.

Broad-minded enough to accommodate himself easily to the February Revolution, the forty-one-year-old Dukhonin had risen from regimental commander to acting supreme commander of Russia's armies in just three years. One key to his success had been a straightforward approach to crises that led him to avoid second-hand reports and seek information at its source whenever possible during the turmoil of 1917. Dukhonin therefore wanted to speak to Kerenskii himself, not have his questions relayed through Cheremisov. "Can you have him [Kerenskii] come to the [Hughes] apparatus?" Dukhonin asked in one of several exchanges he had with northern front headquarters that night. "In his own best interests, it cannot be done," Cheremisov had replied.[39] Unable to contact other senior commanders, Kerenskii, according to a report Dukhonin received the following day, "experienced the tortures of Hell" for several hours at Cheremisov's headquarters as he searched for loyal forces. Then, at about three o'clock on the morning of October 26, General Petr Krasnov, a Cossack commander who had ridden with Kornilov a scant two months before, arrived to report that some of the Cossacks of his III Cavalry Corps, the same one that Krymov had once led in support of Kornilov, were ready to march against the Bolsheviks. Within two hours, Kerenskii and Krasnov left for III Corps headquarters to begin the first armed assault against Lenin's Bolsheviks.[40]

In September, General Aleksandr Krymov had led the entire III Cavalry Corps, including its fearsome Savage Division, in support of Kornilov's cause. Now, as Krasnov and Kerenskii set out, they had something less than a thousand men, a handful of light guns, one machine-gun company, an armored car, and an armored train that had to be run by the commander of a squadron of Enisei Cossacks because no regular engineers would do so.[41] Even with this unimpressive force, they had no difficulty in taking Gatchina before dawn on Friday, October 27. At seven o'clock that morning, just as the elated deputies to the Second Congress of Soviets had begun to drift homeward after hearing Lenin's decrees on land and peace, Kerenskii set up his headquarters in the vast Gatchina Palace, built by Catherine the Great's son Paul I. In one of his first general orders, he named Krasnov Chief of All Military Forces of the Petrograd Region of the Russian Republic,[42] and ordered him to advance directly against Petrograd. "To the soldiers of the Petrograd Military District, to each and every one everywhere!" he proclaimed by telegraph. "I, the minister-president of the Provisional Government and supreme commander of all the armed forces of the Russian Republic hereby announce that I arrived [at Gatchina] this morning at the head of troops from the front who have remained loyal to the Motherland. I hereby order all units of the Petrograd Military District, who, through misunderstanding or mistake, have joined the gang that has betrayed the Motherland and the Revolution, to heed the call of duty without delaying a single moment further." Kerenskii went on to announce that several elite units had joined his advance, and urged all others to do so immediately.[43]

While Kerenskii settled into his new headquarters, Krasnov awaited reinforcements to strengthen his assault columns before he attempted to pass Tsarskoe Selo with its revolutionary garrison of some sixteen thousand men. By late evening, less than three squadrons of Cossacks had appeared from the direction of Novgorod, but Krasnov's own units had begun to grow restive beneath a shower of taunts from the Gatchina garrison. "Kornilovites!" they shouted at Krasnov's perplexed Cossacks. "Stranglers of the Revolution!"[44] With the only force yet marshaled against the Bolsheviks about to melt away, Krasnov moved accordingly. Convinced, as he later wrote, that "civil war is not like regular war, [that] its rules are different, and [that, most of all], it demands decisiveness and pressure," Krasnov therefore resumed his march at two o'clock on Saturday morning.

As he approached Tsarskoe Selo, however, Krasnov's decisiveness turned to timidity and he wasted the better part of the next day in parlays with the town's revolutionary troops. Weary of the delay, a battery of Don Cossack artillery settled the matter at dusk with a few well-placed shells that drove the revolutionary garrison into headlong flight. Krasnov then claimed another triumph for the defenders of the "true" revolution. Yet, if the rules for troop movement in civil war were different, so, too, were the rules for victory. As Krasnov himself later wrote, "victory was ours, but it devoured us utterly." Adrift in a sea of "neutral" troops from Tsarskoe Selo and Gatchina, Krasnov's loyal units threatened to slip from his control, all the more so because Kerenskii, for whom the troops now had no use whatever, insisted upon haranguing them every few hours. Russians' fascination with meetings had run out at last. Now utterly sick of speeches, one old soldier spat in disgust as he turned away. "Everything's so tangled up that I don't understand any of it! To hell with all these orators!"[45]

Pulled in a dozen different directions in half as many months, the loyalties of Russia's soldiers had become extremely tattered by late October, and the tangle of men and messages that followed Krasnov's "victories" at Gatchina and Tsarskoe Selo made it impossible to predict the outcome of Krasnov's approaching confrontation with the Bolsheviks' main forces at Petrograd. None knew that better than Lenin, whose "heightened nervous state" and openly-expressed "fear for the fate of the proletarian revolution" surprised the Bolsheviks who had grown used to seeing him as their unruffled commander.[46] Certainly, others at Smolnyi shared his fears. To comrades exhausted from more than a week of sleepless nights dedicated to the revolutionary struggle, the new people's commissar for internal affairs remarked with cynical fatalism: "Tomorrow maybe we'll get a sleep—a long one."[47]

Fear gripped Lenin, but it did not paralyze him as it did some men. Throughout October 27, he drew upon every available source to assemble that overwhelming weight of force that he thought so vital in political and military confrontations. "It is imperative that we have the strongest reinforcements as soon as possible," he told the chairman of the Helsingfors Soviet on the Hughes apparatus that morning, just moments before he urged Tsentrobalt—the Bolshevik-controlled Central Committee of the Baltic Fleet—to send "everything you can spare" to halt Krasnov's advance. That night, he insisted upon knowing precisely how battleships' heavy guns might be used against Krasnov's forces inland, and then sent a personal emissary to Kronstadt to assemble loyal sailors to man the trenches that Red Guards and work details already had begun to dig beyond Petrograd's suburbs. Within hours after Kerenskii had set foot in Gatchina Palace, five thousand men—"loyal and ready to fight" for the Bolsheviks—had left Helsingfors, due to arrive in Petrograd within twenty-four hours "at the outside," and Tsentrobalt had dispatched two destroyers and a battleship that would be in positions to support Lenin's Petrograd Bolsheviks in "eighteen [hours] at the most."[48] Not content to await the arrival of promised reinforcements, Podvoiskii, Antonov-Ovseenko, and Pavel Dybenko, the *troika* to whom Lenin had entrusted Petrograd's defense, marshaled legions of Red Guards to defend the city. On a few hours' notice, two thousand *putilovtsy* marched, their wives joining the factory's women workers as nurses, trench diggers, and cooks.[49] From all across the city, other factories sent men and women to join the *putilovtsy*, all ready to defend their revolution as valiantly as Krasnov's men were prepared to defend theirs. "An amazing, inspired mass in thin, tattered coats and pinched white faces . . . [poured] out of the factories in a mighty, spontaneous people's army," Louise Bryant wrote as she recalled her impressions of the days she had spent with Reed in Petrograd. "They will run like sheep," a member of the City Duma told her knowingly. "Do you think such ragamuffins can fight?"[50]

Petrograd's Bolsheviks did not face attacks only from outside the city. Anxious to be rid of the Bolsheviks once and for all, the Committee for Salvation planned to act in concert with the Kerenskii-Krasnov advance to overthrow them from within. Still able to command considerable sums of cash from his allies on the far right, Purishkevich offered ten thousand rubles for Lenin's murder at the same time as the Committee, upon learning from Kerenskii that he expected Krasnov's troops to reach Petrograd on Sunday, made ready to seize key points in the city.[51] Their plan, prepared by none other than the same general who had so grossly underestimated the strength of the Bolsheviks just a few days before, called for the cadets of the Mikhailovskoe Artillery School to seize Petrograd's Central Telegraph Office and the Mikhailovskii Riding School, where the armored cars were garaged, while cadets from the Vladimirskoe and Pavlovskoe schools seized the Peter and Paul Fortress and freed the imprisoned Provisional Government ministers.[52]

Within a remarkably brief time on the morning of October 29, groups of well-disciplined cadets seized the armored cars at the riding school, occupied the Astoria (the city's most elegant hotel, where several people's commissars had taken up residence), and took control of the telegraph office. Yet Krasnov's troops did not reach Petrograd as Kerenskii had promised, and overwhelming numbers of counterattacking Red Guards and military units soon drove the cadets back into their schools. Quickly, Red Guard commanders placed these under siege. At the Vladimirskoe School, the Bolsheviks used artillery with crushing effect, and white flags fluttered from the windows in reply. Then, as the Red Guards moved forward in the open, a fusillade of rifle and machine-gun fire cut them down. Grimly, the fallen men's comrades set to work and, by mid-afternoon, their heavier artillery and machine-gun fire had blasted the cadets into submission.[53]

The ranks of its cadet defenders shattered, the Committee for Salvation now pinned its hopes on Krasnov, whose elite Cossack units now faced untried Red Guards on the Pulkovo Heights just beyond the city's suburbs. For men unused to war, the prospect of battle with Krasnov's combat-hardened regulars raised specters of almost certain defeat. "What chance have we?" one Bolshevik exclaimed as the battle began to take shape. "A mob against trained soldiers!"[54] However, not long after the last of the cadets' defenses collapsed on the afternoon of October 28, the picture shifted, when more than a thousand sailors from Helsingfors and Kronstadt joined the nearly ten thousand Red Guard volunteers that the Bolsheviks had assembled.[55] Before nightfall, the two sides seemed almost evenly balanced, and few dared to predict the outcome.

Some Petrograders still stood outside the warring camps of the Bolsheviks and the Committee for Salvation, preferring negotiation to bloodshed. On October 29, the All-Russian Executive Committee of the Union of Railway Workers—Russians called the group by its acronym, Vikzhel—had entered the fray, demanding that the two camps either agree to a compromise government that represented all socialist groups or face a nationwide rail strike that would isolate Petrograd from the rest of Russia and cut off its dwindling food supply. For the moment, Lenin and Trotskii had no choice but to negotiate, for Vikzhel possessed the weapon against which they had no defense. Yet talks brought neither peace nor agreement. Meetings between the Bolsheviks and their Vikzhel opponents continued for several days until Vikzhel's efforts to launch a railway workers' strike collapsed for want of support among the rank and file.[56]

In the meantime, Krasnov and the Red Guards fought their battle on the Pulkovo Heights. On the morning of Monday, October 30, Petrograd's armed workers halted the attack against their revolution, just as they would halt Hitler's victorious armies at the same place in 1941. Short of ammunition and fearful of being outflanked, Krasnov's Cossacks fell back upon Gatchina where, two days later, they agreed to surrender Kerenskii to Bolshevik envoys in return for a guarantee of safe conduct to their homes in southern Russia. Warned at the last minute, Kerenskii fled—disguised as "a rather grotesque-looking sailor," he later said—less than half an hour before a Cossack detachment came to arrest him.[57] Now, there could be no doubt that Kerenskii had fallen.* Perhaps no epitaph better reflected his failings than the caustic assessment of Sir George Buchanan. "He was," Britain's ambassador later wrote, "a man of words and not of action. He had his chances and he never seized them. He was always going to strike and he never struck. He thought more of saving the revolution than of saving his country, and he ended by losing both." Like Miliukov, Kornilov, and, most of all, Nicholas II, Kerenskii had always moved too slowly and acted too late. "Tardiness," Viktor Chernov later wrote, ". . . was the chief disease of the 1917 Revolution. . . . It was, in a way, an echo of a certain retardedness in Russia's historical development."[58]

Nowhere did Russians dispute the Bolsheviks' seizure of power more violently during the last weeks of 1917 than in Moscow, where blood flowed for the better part of a week. The largest city in the country, Moscow was the hub of Russia's textile industry and a center of political ferment that had yielded the Bolsheviks some of their most important early victories.[59] Lenin had once thought of beginning the Bolsheviks' armed uprising there because he expected that its chances would be better in the city where Nikolai Bukharin's faction of young Bolsheviks provided strong support for his uncompromising political views than in Menshevik-dominated Petrograd.[60] Yet, while Petrograd's Red Guards had been training since September and its Military Revolutionary Committee functioning since mid-October, Moscow's Bolsheviks had made no attempt to shape these vital revolutionary instruments until just before noon on

---

*Kerenskii eventually escaped from Russia, using a false passport most probably supplied by the British diplomat Bruce Lockhart. He lived in Paris for the better part of two decades until he emigrated to the United States in 1940.

October 25, when word arrived that Petrograd's armed uprising seemed certain of success.[61] Only later that day did they organize a "fighting center" to launch an armed uprising in support of their comrades in the north,[62] and they did so to defend the soviets against the threat of counterrevolution, not to seize power for the soviets as their comrades had done in Petrograd.[63]

While Moscow's Bolsheviks organized a Military Revolutionary Committee like that in Petrograd, its Socialist Revolutionary mayor urged the City Council to form a Committee for Public Safety "to preserve civic order and defend the safety of [Moscow's] citizens."[64] For almost two days, these two organizations waged a war of appeals, proclamations, and denunciations while they assembled their forces. Combining cadets from Moscow's military schools, a few elite assault troops, and a number of volunteers, all accustomed to military discipline and trained in the use of weapons, the forces of the Committee of Public Safety numbered about ten thousand, including virtually all of the army officers in the city. On the other side, estimates of the forces commanded by the Military Revolutionary Committee ranged as high as fifty thousand, but few of these men were trained. Although at first they did not have enough rifles, the Bolsheviks and their supporters enjoyed the great advantage of controlling almost all of the heavy artillery in Moscow.[65] With greater numbers and more dedication to the causes they had been called upon to defend, both sides stood ready to shed much more blood than anyone had seen in Petrograd although the Bolsheviks at first managed to seize a number of key points, including the Kremlin, whose fortress-like walls and huge arsenal counted as a critical element in the success of any insurgent effort, without bloodshed.

On the evening of October 27, the Commander of the Moscow Military District announced that he would turn his artillery upon the Moscow Soviet unless the Bolsheviks evacuated the Kremlin, surrendered the arms taken from its arsenal, and dispersed the Military Revolutionary Committee within fifteen minutes.[66] When the Military Revolutionary Committee ignored the order, the forces commanded by the Committee for Public Safety recaptured the central telegraph and telephone office, the post office, and several key railway stations in rapid succession, leaving only the Kremlin itself in Bolshevik hands. In reply, the Bolshevik-led insurgents appealed "To All Working Men and Women," urging them to "give everything, to the very last drop of blood, for the defense, preservation, and strengthening of true democracy," and warning that their defeat would mean "the barracks and penal servitude for soldiers, the landlord's lash for peasants, the capitalist's unfettered power for factory workers, and hunger for all."[67]

On the morning of October 28, anti-Bolshevik forces surrounded the Kremlin and ordered its insurgent commander to surrender. Cut off from Military Revolutionary Committee headquarters, he agreed to open the Kremlin's historic Troitskaia Gate on the condition that the Committee for Public Safety spare the lives of his men. "Attention!" an officer of the besieging forces barked, after some five hundred defeated men had marched unarmed onto the Kremlin's Senate Square. As the Kremlin's captured defenders moved to obey, their enemies opened fire with two machine guns. Within moments, dozens of corpses littered the square as the unarmed men fled in panic, shot down as they tried to flee through the arsenal's gate by machine-gunners who supported the Committee for Public Safety. Although their commander made no report of the number killed, the commander of the Kremlin's arsenal later wrote of "a mountain of dead bodies" piled in front of its gate, and others reported that a work detail of forty prisoners worked for several hours that night to clear the square and bury the dead.[68] During its first hours, Moscow's armed uprising had claimed more lives than had the Bolsheviks' seizure of power in Petrograd. Tragically, the deaths of the Bolsheviks in the Kremlin marked only the beginning of Moscow's struggle. "To Arms! To Arms!" the newspaper *Sotsial-Demokrat* proclaimed. "The final, decisive battle has begun!"[69]

While pro-Bolshevik military units had played a key role in Petrograd's events, workers' Red Guards dominated the insurgents' side during the October Revolution in Moscow. As the battle for Moscow began in bloody earnest, the forces of the Committee for Public Safety held the city's center, while the Bolsheviks' controlled its outlying workers' districts. For the better part of a week, Red Guards pressed inward, compressing their enemies' forces further into the city's center. As in Petrograd during Krasnov's advance, Vikzhel tried to impose a cease fire; again, as in Petrograd, it failed to end the fighting. On the same day, the Moscow Soviet's moderate Bolshevik chairman tried to negotiate a truce, failed, and came close to losing his life.[70] Fighting raged from street to street and building to building. Only recently returned to Moscow from the countryside, Konstantin Paustovskii was trapped for six days in a cross fire at the Nikitskaia Gate, and only narrowly missed being shot by a squad of Red Guards as an enemy sniper.[71] Lenin ordered Red Guards from other cities to march to the aid of their Moscow comrades. "Don't forget that Moscow is the heart of Russia," he told a group of sailors. "This heart must be Soviet if we are to save the revolution."[72] From more than twenty-five different towns and cities, Red Guards and soldiers loyal to the Bolsheviks converged upon Russia's ancient capital at the beginning of November. Men from the Baltic Fleet served as gunners for the Moscow Red Guards' artillery; rubber workers from Tushino and weavers from the great Morozov Mills at Orekhovo-Zuevo fought in the Nikitskaia Gate battle that trapped Paustovskii; while men from the factories of Podolsk entered the fray along the walls of ancient Kitai-gorod, just beyond the Kremlin itself.[73]

Not men but the weapons to arm them proved the greatest difficulty for Moscow's Bolsheviks, until a metalworker discovered freight cars holding over forty thousand rifles on a siding at Sokolniki.[74] With these, the balance shifted quickly in their favor. Always moving toward the Kremlin, the Red Guards and their allies fought from alley to alley and from street to street, inexorably tightening their lines and their artillery inward along the Tverskaia toward Red Square, and from Liubianka Square toward the Metropol Hotel and Kitai-gorod beyond. By the evening of November 2, only the Kremlin remained to be taken, and the Red Guards opened their final assault. Blasting open the Borovitskii Gate with mines and artillery, they burst into the Kremlin, scattering its exhausted cadet defenders before them.[75] Before midnight, the battle for Moscow had ended, and what Lenin had called the heart of Russia had become Soviet. Men and women had died in the hundreds. The Moscow Soviet reported the deaths of 228 Red Guards. The deaths of those who had fought against them went unnoticed by Russia's new authorities.[76]

For the first time in six days, Paustovskii ventured beyond the walls of his refuge in an alley near the Nikitskaia Gate. "Across the trolley tracks lay a dead horse, its yellow teeth bared," he remembered. "Frozen blood lay like ribbons upon the stones around our gate. Houses, riddled by machine-gun bullets, were dropping sharp bits of glass from their frames, and you could hear it tinkling all around us." No one wanted to break the stillness that followed the battle. Paustovskii saw a detachment of Red Guards marching along Bolshaia Nikitskaia and thought that "the red ends of their cigarettes, flickering in the darkness, looked like a soundless skirmish." Then, the sound of singing broke the night's cold silence: "Nobody gives us our salvation / Not God, the tsar, nor heroes. / We shall strike the final blow for liberation / With our hands alone."[77]

While the Muscovites buried their dead, Lenin and the Soviet of People's Commissars struggled to put their oft-stated promises into practice. Not merely in decreeing that the estates of Russia's noble lords would become the property of the people but in every other possible area of Russian life, they emphasized equality. Urged on by Aleksandra Kollontai, the champion of down-trodden women all across Russia, they legislated absolute equality between men and women and set out to create the legal and social framework to ensure that "spouses do not lose or lessen their own identity as free and equal and independent personalities" in marriage.[78] This meant substituting civil ceremony for religious sacrament, liberalizing grounds for divorce, and providing state facilities to care for offspring, whether the product of marriage or some more transient encounter. All this marked the beginning of the sexual revolution (and its corresponding sexual excesses) that characterized Russian life during the 1920s, and produced the phenomenon one commentator called "life without control,"[79] which, in turn, led Lenin to denounce the so-called "glass-of-water theory" of sex. "You must be aware of the famous theory that, in communist society, the satisfaction of sexual desires, of love, will be as simple and unimportant as drinking a glass of water," he once told a visitor. "This glass-of-water theory has made our young people mad, quite mad," he went on. "Of course, thirst must be satisfied," he concluded. "But will the normal man in normal circumstances lie down in the gutter and drink out of a puddle, or out of a glass with a rim greasy from many lips?"[80]

Other early decrees of the Soviet of People's Commissars—already widely known by its Russian acronym Sovnarkom—deprived Russia's more well-to-do citizens of real estate, stocks, bonds, gold, and pensions, while local soviets levied extortionist taxes, demanded forced loans, and imposed prison sentences that could be commuted for high ransoms against industrialists, merchants, and aristocrats.[81] Quickly and brutally, Sovnarkom's agents seized Russians' wealth and Russia's banks, proclaiming "in the name of the Revolution, [that] the criminal speculation of the banking bourgeoisie . . . has brought the army to starvation, created chaos in all areas of economic life, and threatens to lead the people of the capitals and other large cities to the brink of famine."[82] So thoroughly did Sovnarkom and its local agencies take control of Russians' wealth that the first months of 1918 found former aristocrats and army officers hawking newspapers and cigarettes on street corners, while any number of their wives and daughters were driven into prostitution. A rare sociological study of Moscow's prostitutes done in the mid-1920s found that some thirty-five hundred came from merchant, professional, or other "bourgeois" origins, and that nearly a thousand more had been born into some level of the former tsarist aristocracy.[83]

But the flood of decrees that made life more miserable for the upper classes by confiscating their property and wealth could not relieve the hunger and cold of the urban masses. Nor could repeated accusations that "bourgeois sabotage" lay at the root of Russia's desperate problems afford much comfort to city dwellers that winter. Now that the Sovnarkom sat in the Provisional Government's place, it had to make good on Lenin's promise of bread to Petrograd's proletarians, and it would take more than the reassurances of Marxist theory mixed with outcries against the wrong-doings of the bourgeoisie to make food appear. Stated most simply, Petrograd's bakers needed at least 725 tons of flour a day to bake enough bread to feed the city, but in mid-October they were receiving a scant 215.[84] The city's food supply had hung by the slenderest of threads that had stretched to the breaking point on the day after the Winter Palace fell, when officials in the fallen government's Ministry of Food called for a general strike to protest the Bolsheviks' seizure of power. That same day, officials in the not yet nationalized State Bank refused to disburse any cash for food supplies. Everywhere the Black Market flourished, as a pound of stringy beef rose beyond an average worker's entire daily wage.[85] Only the most desperate measures could rescue Petrograd and the soldiers who stood between its gates and the armies of Germany from near-certain starvation.[86]

All too aware of the curse of Russian inefficiency, Lenin sent workers and soldiers "to search for provisions, first of all, right here in Piter," and quickly uncovered nearly fifty thousand tons of food in freight cars that had been shunted onto railway sidings and forgotten. Insisting that "it is intolerable to allow poor people to starve while the rich hold back large reserves of food for the purpose of speculation," the Military Revolutionary Committee urged all proletarians "who know the whereabouts of hidden reserves of food" to report them to the authorities at Smolnyi. Within a few days, this effort uncovered hundreds of tons of sugar, tea, and coffee secreted in warehouses in and around the capital, not to mention over a thousand tons of grain that some enterprising investigators discovered riding peacefully at anchor on four huge Neva River barges that had been mixed in with dozens of others loaded with firewood.[87]

Such extraordinary finds allowed the proud Bolsheviks to increase by half the bread ration that the Provisional Government had lowered to a half-pound during its last week in office.[88] But even Lenin's hopeful commissars knew that they could not expect to ward off starvation in Petrograd much beyond the first of the year. "For God's sake, use the *most* energetic and *revolutionary* measures in order to send *grain, grain,* and *more grain!!!*" Lenin telegraphed to several special envoys in the provinces. "Otherwise, Piter may perish."[89] Like Moscow and the other large Russian industrial cities that depended upon food shipped from distant regions to feed their legions of factory workers, Petrograd demanded ongoing and reliable sources of grain to end its food crisis. Yet, as Lenin and the Sovnarkom knew all too well, the grain they needed belonged to peasants deeply suspicious of any authority. As they had done for the better part of three years, these cautious peasants continued to hoard their harvests against that long-hoped for moment when much-needed manufactured goods would become available.[90]

As the Provisional Government already had done on a smaller scale, Lenin sent armed expeditions into the countryside "to obtain grain through voluntary sales at firm prices" if possible, but to seize it by force if all else failed. On Lenin's orders, Petrograd's Military Revolutionary Committee therefore sent five hundred Kronstadt sailors, chosen for their dedication to the revolution, into Russia's grain-growing regions at the end of October with orders "to put a stop to plundering shipments of food, to regulate the whole matter of obtaining provisions, and to oversee on the spot the dispatch of grain shipments" to Petrograd. Yet Lenin never hesitated to mix carrots with the Sovnarkom's sticks whenever it seemed worthwhile. Therefore, apart from groups of armed sailors, the Petrograd authorities during the first ten days of November sent something over seven thousand workers and sailors bearing cloth, thread, and tools into the countryside with orders to exchange their goods for grain at the best possible rates.[91]

In early 1918, grain finally began to flow toward Russia's cities, particularly from the more temperate parts of Siberia, where Tobolsk province alone held over a million tons of vital foodstuffs in storage.[92] Between January and March, Siberia's peasants exchanged nearly a hundred thousand tons of grain for over sixty million rubles' worth of manufactured goods.[93] Yet the supply of manufactured goods that Bolshevik trading expeditions had at their disposal proved far too limited to continue the exchange for long. Persuasion and barter soon had to give way to force and armed Bolshevik contingents seized the grain that Russia's cities needed so desperately. "It may seem that this is a struggle only for bread," Lenin wrote as the crisis threatened to overwhelm them. "In fact, the very future of socialism is at stake."[94]

To convince a reluctant peasantry and a hostile bourgeoisie of the virtues of socialism required persuasion buttressed by coercion, but as opposition to the Bolsheviks became better organized, the balance shifted sharply toward coercion early in 1918. "The bourgeoisie, the landowners, and all the rich classes are making desperate efforts to undermine the revolution," Lenin wrote in a brief note to Feliks Dzerzhinskii at the beginning of December. "The bourgeoisie are prepared to commit the most heinous crimes . . . [and] have even gone so far as to sabotage food distribution, thereby menacing millions of people with famine."[95] Dzerzhinskii urged that the Sovnarkom organize a special bureau to combat counterrevolution, and on December 7 the Sovnarkom named him to head an All-Russian Extraordinary Commission to Combat Counterrevolution and Sabotage, known immediately by its Russian acronym, Cheka. Established "to wipe out all acts of counterrevolution and sabotage anywhere in Russia,"[96] the Cheka began with a modest staff and limited resources, and quickly grew into one of the most fearsome security police agencies in the history of the modern world.

Known for the brutal, sadistic men who served in its ranks, and credited with killing tens of thousands of Russians, the Cheka was headed by one of the most austere and self-denying men ever to enter the ranks of Russia's revolutionaries. Feliks Dzerzhinskii, son of one of Poland's legion of impoverished aristocratic families, had entered the Russian revolutionary movement at the age of seventeen, and had fought against the tsarist government in Vilno, Kovno, Warsaw, St. Petersburg, and Moscow. Tall and slender, with aquiline features and a sharply sculpted Vandyke beard that gave him an air disconcertingly reminiscent of those fanatic Spanish grandees who once had served the Inquisition, Dzerzhinskii was known for his fearless dedication to the Bolsheviks' cause. Perpetually raised eyebrows gave him a look of permanent disbelief, while widely spaced, heavily-lidded almond eyes, set above extremely high cheekbones, projected an ascetic image that men often mistook for cruelty. Nearly a third of an adult life spent inside the prisons of Nicholas II or in Siberia had molded in Dzerzhinskii that flint-hard revolutionary spirit that never broke in the face of adversity. To his new duties as the Cheka's head, he brought a prisoner's unruffled patience and a professional revolutionary's single-minded dedication, combining legendary austerity with an incredible capacity for work. Dzerzhinskii was known to sleep in the Cheka's offices when the burden of work demanded it, and chastised his deputies when they tried to ease his burdens by bringing him better food than that available to Petrograd's hungry workers during the austere years of the civil war.[97] As head of the Cheka, Dzerzhinskii dealt with sedition, espionage, and death by meting out much larger doses of the same. "It is necessary to show the greatest strictness, pitilessness, [and] directness in the very beginning," one of his deputies once explained. "Deserved punishment must follow the crime; then many fewer victims fall on both sides."[98]

As the devout moral guardian of the revolution, Dzerzhinskii's "Red Terror" drew its authority from the self-righteousness of its master and from the polemical justifications of Lenin. "The bourgeoisie of international imperialism killed ten million and mutilated twenty million human beings in 'its' war to decide whether British or German robbers should rule the whole world," Lenin explained to American workers in 1918. "If our war, the war of the oppressed and exploited against the oppressors and exploiters, will cost half a million or a million victims in all countries, the bourgeoisie will say that the former sacrifices were justified, [but that] the latter [was] criminal."[99] For Lenin and Dzerzhinskii, terror therefore could have a moral justification if practiced in the name of social justice and the people's welfare, just as Europe's statesmen had justified the Great War's slaughter in the name of independence and national defense.

The Bolsheviks used the same appeals to the ever-present danger of "bourgeois counterrevolution and sabotage" with which they defended the Cheka's reign of terror to justify their suppression of the long-promised Constituent Assembly, whose meeting Russians of many political persuasions had awaited since Nicholas's abdication. From the first days of the February Revolution, Russian politicians had looked to the Constituent Assembly to legitimize tsarism's overthrow and establish the governmental framework for a democratic Russia, and the Provisional Government often had postponed critical decisions until the Assembly met for that reason. Even the Bolsheviks had cheered the Assembly on occasion as a forum from which the masses would speak their will, and less than a week before the armed uprising broke out in Petrograd, Trotskii had criticized Kerenskii's government for its failure to summon the Constituent Assembly to settle the burning questions of peace, land, and democracy. "We have nothing in common with this government of national treason," Trotskii had exclaimed on October 20. "We [Bolsheviks] therefore turn to the people: Long live an immediate, honest, democratic peace! All power to the Soviets! All land to the People! Long live the Constituent Assembly!"[100]

Trotskii had struck a nerve left too long exposed, for while Russia's politicians continually deferred critical issues until the Constituent Assembly met, none who tried to govern between February and October had dared to summon it into session. The Constituent Assembly had become an ideal, and it had become difficult for men who had anticipated it for so long to see it in terms of real politics. Its constant postponement eventually gave the Bolsheviks, the one party in Russia whose leaders raised considerations of *realpolitik* above all others, a deadly weapon against their idealistic opponents, and they wielded it with consummate skill. "The constant postponement of the Constituent Assembly capped the edifice of tardinesses," Viktor Chernov later wrote as he reflected upon 1917's many lost opportunities. "It gave the Bolsheviks one of their strongest trumps."[101]

Perhaps none placed more blind faith in the mere fact of the Assembly's eventual convocation than Chernov's Socialist Revolutionaries. For them, it was a hallowed dream, the culmination of many decades' quest, and the perfect embodiment of popular sovereignty in Russia. "Democracy was inconceivable without the Constituent Assembly," one of their leading spokesmen explained some years later. "Even the revolution itself, as the ultimate expression of the popular will, ought, it seemed, to find its highest expression through the Constituent Assembly."[102] Such blind faith reaped generous rewards, for no political party in Russia won more support from the masses when elections for the Assembly took place in mid-November than the Socialist Revolutionaries. Although workers in industrial centers voted solidly for the Bolsheviks, the sheer masses of peasants who supported Socialist Revolutionary candidates outvoted them by more than two to one. With just more than half of the Assembly's seats at their command, Chernov's party of the peasants stood ready to control its direction and its policies if Lenin and the Sovnarkom allowed it to meet.[103]

Although always ready to play upon Russians' hopes for the Constituent Assembly whenever it suited their purposes, the Bolsheviks had never placed any faith in parliamentary democracy. "Life and the revolution push the Constituent Assembly into the background," Lenin explained in April. "The question of the Constituent Assembly is subordinate to the course and outcome of the class struggle between the bourgeoisie and the proletariat," he added at the end of July, while Stalin insisted that the ultimate source of power lay not in the Constituent Assembly but in the "workers and peasants who will fight for a new [form of] revolutionary justice." As a popular Bolshevik pamphlet insisted some weeks later, "the Constituent Assembly must meet in Petrograd, so that the revolutionary people, and especially the revolutionary garrison, can watch it and direct it."[104]

Almost fifteen years before, when his unyielding demand for rigid party discipline had split the ranks of Russian social democracy at its Second Party Congress, Lenin had insisted that democratic principles must always yield to party interests. Once the Bolsheviks had seized power, he urged that elections to the Constituent Assembly be postponed, but had to give way when the majority of the Central Committee refused to support such an extreme course.[105] "It's clearly a mistake which can prove very costly," he argued, when his comrades remained unmoved by his plea for resolute action. "Let us hope that the revolution will not have to pay for it with its life."[106] The Socialist Revolutionaries' stunning victory in the voting a week later made it clear that the Bolsheviks would need stern measures indeed to remain in power. "We may have to dissolve the Constituent Assembly with bayonets," Petrograd's commissar for press, propaganda, and agitation remarked.[107] "It is pretty difficult to argue on the essence of any matter with public men of the [Bolsheviks'] type," the newspaper *Novaia zhizn* replied. "The only argument to convince them is the bayonet."[108] The Bolsheviks, not their opponents, used force. Ten days later they arrested the entire All-Russian Electoral Commission on the grounds that they had conspired to deny the right to vote to two million Russians languishing in German and Austrian prisoner of war camps, detained them for nearly a week at Smolnyi, and then announced their replacement by a commission of Bolsheviks headed by the cruel and doctrinaire Mikhail Uritskii.[109] Still critical of the Bolsheviks' decision to exclude moderate socialists from the Sovnarkom, and, more generally, dissatisfied with the direction the revolution seemed to be taking, Gorkii asked, "What will the revolution offer that is new . . . ? How will it change the bestial Russian way of life?"[110]

Before the middle of December, Lenin provided an answer of sorts. "A republic of Soviets is a higher form of democracy than the usual bourgeois republic with a Constituent Assembly," he explained in his famous "Theses on the Constituent Assembly." "The only chance of securing a painless solution to the crisis that has arisen owing to the divergence between the elections of the Constituent Assembly . . . and the interests of the working and exploited classes," he continued, "[is] for the Constituent Assembly . . . to proclaim that it unreservedly recognizes Soviet power, the Soviet revolution, and its policy on the questions of peace, the land, and workers' control [of industry]. . . . Any attempt to tie the hands of Soviet power in this struggle," Lenin concluded ominously, "would be tantamount to aiding counter-revolution."[111]

Lenin had the weapons to deal with that type of "counterrevolution" even though the Socialist Revolutionaries had won the majority of Russians' ballots because the Bolsheviks' strength centered in Petrograd, Moscow, the Baltic Fleet, and among the soldiers on the northern and northwest fronts, from which force could best be applied against the Constituent Assembly when it assembled. Scattered across some eight million square miles of territory, Russia's peasants had no immediate practical way to express their displeasure if the Bolsheviks set their preferences aside, while loyal workers, soldiers, and sailors could impose the Bolsheviks' will with ease in key urban strongholds. Such efforts as the Committee for the Defense of the Constituent Assembly, which fearful and desperate opposition parties formed at the end of November, simply could not stand against the force of the Bolsheviks' urban allies. Nor could the Committee's pitiful effort to hold a symbolic opening of the Constituent Assembly a month in advance of its announced convocation produce more than failure. When a handful of Socialist Revolutionary and Kadet delegates marched to the Taurida Palace for a symbolic meeting on November 28, Lenin and the Sovnarkom simply responded by ordering their arrest as "enemies of the people."[112]

For Russia's idealistic liberals and dreamy socialists, who loved noble words and symbolic gestures so much, nothing proved more difficult to confront than the Bolsheviks' readiness to meet propaganda with censorship and peaceful demonstrations with brute force applied in the name of defending the interests of their nation's toiling masses. No matter how lofty its delegates' ideals, or how hopeful their dreams, there could be no future for the Constituent Assembly other than the one Lenin had envisioned when he warned that it must support Soviet power without reservation or be crushed as a retrograde and counterrevolutionary menace to the "will and interests of the working and exploited classes."[113] The Assembly's first official session on January 5, 1918, therefore became little more than a ritual to be got through for the sake of appearances, "a lost day," Lenin said, spent among "people from another world."[114]

The impending confrontation between brokers of ideals and brokers of power generated tension as thick as any Petrograd fog when the All-Russian Constituent Assembly gathered on the afternoon of January 5. Just four days before, while Lenin was being driven home from a meeting—"deep in thought, meditating upon the construction of proletarian happiness" according to the Bolsheviks' official account—four shots had been fired at him from a crowd. Thanks to the quick reflexes of a companion, who received a bullet wound on his hand as he pulled Lenin's head out of the line of fire, the chairman of the Sovnarkom escaped harm, but the incident had made the Bolsheviks doubly vigilant as the Constituent Assembly prepared to assemble. Some of the Socialist Revolutionaries thought Petrograd looked more like an armed camp than a capital city as Bolshevik loyalists, armed to the teeth, filled the city's streets, ready to fall upon their adversaries at a moment's notice. Machine guns guarded Smolnyi and the Taurida Palace, while pistols and bayonets gleamed evilly in the spectator's gallery.[115] With "bandoleers of cartridges draped coquettishly across their shoulders, and grenades hanging obtrusively from their belts,"[116] sailors from the Baltic Fleet stood guard outside the Palace that day, scrutinizing the delegates' credentials and refusing entry to any they mistrusted. "How dare you exclude us! Don't you know who we are?" one outraged deputy exclaimed. "I know," a young sailor replied sagely. "You are the servants of the People. We will give you orders and you shall obey us!"[117]

In an amazing juxtaposition of opposites worthy of any ever set down by Dostoevskii, the Bolsheviks arranged for the Constituent Assembly to meet in the Taurida Palace, which had been restored to its pre-revolutionary splendor in the few short weeks since the October Revolution, and then filled its galleries with the dregs of Petrograd's proletarian society. At least half of them were drunk, the Socialist Revolutionary delegate Boris Sokolov estimated, laughing, whistling, and cursing as they looked on. Like a migrating flock, they moved from one buffet to another, gorging themselves as they went and vomiting where they stood.[118]

On the Assembly floor itself, the Socialist Revolutionary delegates insisted upon playing out their legislative charade, going through the motions of electing as president Viktor Chernov, who treated them to a long and rambling speech that many found disappointing. A better, more contentious speech by Tseretelli followed a shorter speech by the Bolsheviks' Bukharin, although neither spoke to much purpose. At one point, the Assembly threatened to explode in a bloodbath when a dedicated Bolshevik sailor denounced "the counter-revolutionary majority . . . [for] trying to block the progress of the workers' and peasants' movement."[119] Unwilling to face the dissolution they knew must come, the Socialist Revolutionary majority kept the Assembly in session until 4:40 on the morning of January 6. When the delegates departed, vowing to return to continue their deliberations later in the day, the Bolsheviks announced that the Assembly had reached its end because it served "only as a cover for the struggle of bourgeois counter-revolution for the overthrow of the power of the Soviets."[120] To emphasize that point, the Sovnarkom placed guards at all entrances to the Taurida Palace.

"It was a hard, boring and irksome day in the elegant rooms of the Taurida Palace," Lenin wrote. "It was as though history had accidentally, or by mistake, turned its clock back."[121] Boris Sokolov penned a different epitaph. "On our side stood legality, high ideals, and faith in the triumph of democracy," he wrote from exile. "On their side stood a readiness for action, machine guns, rifles, and pistols. And the mob stood behind them."[122] Intoxicated by their first heady draughts of power, the armed defenders of the "people's democracy" preferred bullets to ballots. "We will not," one of them announced as the Constituent Assembly dispersed, "exchange our rifles for a voting paper."[123]

CHAPTER XVIII

Peace and War

More than any other nation in Europe, Russia had borne the terrible ravages of war, and now, more than any other, she needed peace. As the war's fourth winter began, Russians could bear no more. In less than forty months of fighting, they had lost more men in battle than any of their allies would lose in fifty. Five times as many Russian soldiers died from disease as did their allies, and the numbers of civilians killed, wounded, infected by disease, or driven from their homes had soared beyond ten million.[1] Russians must have peace, "a just, democratic peace," but, most of all, "an immediate peace," Lenin had told the Second All-Russian Congress of Soviets on October 26, 1917,[2] and the tumultuous ovation that greeted his words spoke to the depths of Russians' exhaustion. "War," concluded a senior member of America's recently-arrived Red Cross Commission, "is dead in the hearts of men."[3]

Certainly, it seemed so in Petrograd. In the heady atmosphere of the Second Congress of Soviets, when Lenin read his Decree on Peace, Russia's soldiers, sailors, and working men and women began to believe that the war's end had really come. From their vantage point in the center of the hall, John Reed and Albert Rhys Williams saw grizzled soldiers weeping at the thought of peace, while younger men cheered it and threw their caps into the air. *"Pust budet konets voine!"*—"Let it mean an end to the war!"—Williams heard a worker mutter in a reverent half-daze.[4] "They actually were beginning to believe in the nearness of peace," Sukhanov remembered as he recalled the "excited, animated faces" of men and women "permeated by a faith that all would now go well."[5]

More thoughtful Russians feared otherwise, however. Worried about the fate that awaited Russia at the hands of Lenin's "pack of scoundrels," Zinaida

Gippius felt none of the joy that swept over the crowd at Smolnyi. "Oh God, we haven't seen the end of it yet!" she exclaimed.[6] "O poor, O sinful land!" she wrote in a poem she bitterly entitled "Merriment." "Behaving like madmen, the people have murdered their own freedom."[7] Locked away in the apartment where she had entertained Kerenskii and Savinkov in better days, Gippius peered into "a nocturnal void filled with black, clotting blood" as she tried to bring the future into focus.[8] In a black mood, she thought black, terrible thoughts, as she wrote an awful prophecy: "CIVIL WAR WITHOUT END AND WITHOUT LIMIT."[9]

Gippius's black pessimism could not darken what she called "October's merriment," as Petrograd's masses celebrated Lenin's promise of peace. But it was one thing to call for peace, to proclaim that "our government's first duty [is] to offer an immediate peace," and quite another to end the killing along Russia's western front. None could doubt that Russia's Western allies would oppose such a peace, especially since the Bolsheviks promised to publish the secret treaties that their governments had concluded with the ministers of Nicholas II and the Provisional Government. Nor was it by any means certain what the German reaction would be to a proposal of peace from men who referred to the kaiser as "Wilhelm the Hangman" and summoned Germany's workers to rise against their government as a means for securing "a just, democratic peace."[10]

The road to peace therefore promised to be difficult. "We never promised that the war could be ended at one stroke, by driving bayonets into the ground," Lenin cautioned a group of workers and soldiers from the front soon after they had read his Peace Decree.[11] Russia's workers, peasants, and soldiers therefore must face the hard fact that, as Lenin warned his listeners at the Second Congress of Soviets, "the governments [of Europe] and the bourgeoisie will make every effort to unite their forces and drown the workers' and peasants' revolution in blood."[12] Allied officers had been seen fighting on the side of the Committee of Salvation when Krasnov's cavalry had advanced against Petrograd,[13] and none of the Allied embassies made any effort to conceal their deep dislike for Russia's new rulers.[14] All refused to recognize the Sovnarkom as Russia's legitimate government, and none would communicate with it officially.[15] As anti-Bolshevik forces began to gravitate toward *Stavka* after the collapse of their efforts to seize control of Moscow and Petrograd, the Allies began to communicate directly with General Nikolai Dukhonin, who had begun to act as Russia's supreme commander after Kerenskii's flight.[16]

For a moment, the Allies seemed about to recognize *Stavka* and the deposed, discredited statesmen who gathered there as Russia's government. Unable to communicate officially with Russia's allies, Trotskii, as people's commissar for foreign affairs, had to settle for intermediaries: Captain Jacques Sadoul of the French Military Mission, Bruce Lockhart of the British Embassy, and Colonel Richard Robins of the American Red Cross. Each of these men quickly came to respect Trotskii, and each incurred the disdain of his government as a result. Trotskii, Robins once said, was "a four kind son of a bitch, but the greatest Jew since Christ." Robins dismissed out of hand the charges that Trotskii had long been a German agent, scheming to hand Russia and her precious natural resources over to the Germans. "If the German General Staff bought Trotskii," he told his British colleagues bluntly, "they bought a lemon."[17] Stubbornly, the governments of Russia's allies persisted, as Sadoul once wrote, "in denying that the earth turns . . . [and] claiming that the Bolshevik government does not exist."[18]

Yet the Allies' stubborn hatred was not foremost in the minds of Lenin and Trotskii as they launched the Bolsheviks' peace offensive in November 1917. Their most pressing problem was to negotiate an armistice and begin serious peace discussions with the Germans, and they allowed a crude array of ideological baggage to complicate their effort far more than was necessary or sensible. "To arouse the masses of Germany, of Austria-Hungary, as well as of the Entente—this was what we hoped to achieve by entering into peace negotiations," Trotskii later explained.[19] "We all stood in favor of agitation, for revolutionizing the working class of Germany, Austria-Hungary, and of all Europe."[20] Certain that Germany stood at the brink of class war, Lenin and Trotskii insisted that Russia's best route to peace lay in urging enemy soldiers and workers to throw off the imperialist yoke and make peace with their Russian breathren. "Only the revolutionary struggle of the toiling masses against the existing governments can bring Europe nearer to a [democratic] peace," Trotskii announced. "Its full realization can only be guaranteed by the victorious proletarian revolution in all capitalist countries."[21]

Fearful that their revolution would perish if revolutions in more advanced nations did not follow, the Bolsheviks urged socialists in Germany, Austria-Hungary, and the lands of their allies "to oppose to the program of imperialism brought forward by their ruling classes their own revolutionary program,"[22] and they laid down a barrage of propaganda all along Europe's eastern front to bolster that effort among the armies of their enemies. Still seeking revolution with a dedication exceeded only by his passionate belief that "out of democracy will be born the new world—richer, braver, freer, more beautiful," John Reed joined his fellow journalist Albert Rhys Williams in the Bureau of International Propaganda that Trotskii's Commissariat of Foreign Affairs organized at the beginning of November.[23] "We immediately began the publication of a series of daily propaganda newspapers," he explained later. "The first of these was in German, *Die Fackel (The Torch)*, issued in editions of half a million a day . . . [which] a regularly-organized system of couriers brought to the front trenches for distribution." With two million rubles of Soviet government funds to support their effort, Reed and Williams had antiwar newspapers that summoned soldiers to revolution in German, Hungarian, Czech, Rumanian, and Croatian pouring off their presses within a fortnight. They even produced an illustrated German weekly called *The Russian Revolution in Pictures.* [24]

Propaganda combined with fraternization could stir discontent among the rank-and-file of Russia's enemies, but it would take time for it to have its hoped-for impact. In the meantime, Russia's new leaders realized that the jack-booted, straight-backed officers of Wilhelmian Germany must be brought to the negotiating table if only to gain a respite for their war-weary soldiers. Late on November 7, Sovnarkom therefore ordered General Dukhonin to propose "an immediate cessation of hostilities for the purpose of opening peace negotiations." [25] In the midst of exploring the possibilities for establishing a rival government with Allied support, Dukhonin at first gave no indication that the order had reached his headquarters. To gain time, he telegraphed the War Ministry asking that it confirm the telegram's authenticity, but sent no word to Sovnarkom at any time during the next twenty hours. [26] Finally, just after midnight on November 9, Lenin, Stalin, and Ensign Nikolai Krylenko, the thirty-two-year-old people's commissar for war, contacted Dukhonin on the Hughes apparatus.

"The People's Commissars are here and are awaiting your reply. . . ," they began.

"Is it proposed to enter into negotiations for a separate armistice, and with whom?" Dukhonin asked in reply. "[Am I to negotiate] only with the Germans, or also with the Turks? Or are the negotiations to lead us to a general armistice?"

"The text of the telegram sent to you was absolutely precise and clear," the angry Sovnarkom representatives replied. "We insist that envoys be dispatched immediately and that we receive hourly progress reports about the negotiations. . . ."

"Only a central government, supported by the army and the nation at large, can command sufficient authority and prestige with the enemy to give these negotiations the necessary authoritativeness to obtain results," Dukhonin replied.

"Are you categorically refusing to give us a direct answer and carry out our instructions?"

"I have given you a precise answer about the reasons why it is impossible for me to carry out your orders," Dukhonin concluded.

"Then, you are relieved of your duties for your refusal to obey government orders," the people's commissars announced. "Ensign Krylenko is now appointed supreme commander." [27]

Ensign Nikolai Vasilevich Krylenko, a graduate in history and law from St. Petersburg University who had first gone to war in April 1916 as punishment for repeated illegal political activities, now commanded Russia's armies. A Bolshevik since 1904, and a member of the Party's Central Committee since 1915, he had never held a position of military command until he helped to lead the Bolsheviks' armed uprising in Petrograd. Now he was supreme commander of Russia's land forces, direct heir to the tradition established by Grand Duke Nikolai Nikolaevich, and carried on by Nicholas II, Alekseev, Brusilov, Kornilov, Kerenskii, and Dukhonin.[28] "Ensign Krylenko has been named the new supreme commander," Lenin announced in a radio message to all units of the army. "Soldiers! The business of peace is in your hands. Do not allow counter-revolutionary generals to subvert the great cause of peace. Place them under guard so as to avoid lynchings unworthy of a revolutionary army and prevent these generals from escaping the trial that awaits them. . . . Maintain the strictest revolutionary and military order."[29] While the Allied military attachés at *Stavka* continued to communicate with Dukhonin as the head of Russia's government until his arrest by *Stavka*'s garrison ten days later, Krylenko and a retinue of Red Guards and Baltic sailors began the journey from Petrograd to take command.[30]

Krylenko arrived at *Stavka* on the morning of November 20 to find Dukhonin under arrest and the garrison in a sullen mood after learning that Kornilov, along with Denikin, Markov, Romanovskii, and Lukomskii all had escaped from Bykhov jail.[31] "Where is Kerenskii? Where is Kornilov?" one of the soldiers shouted. "Dukhonin will also escape!" That moment, the mob's mood turned to rage. "I gazed into their faces and did not recognize them," one war correspondent wrote. "The faces were darkened and distorted. The men looked like hungry wolves."[32] In fury, the crowd threw themselves on the defenseless Dukhonin, stabbing with bayonets and clubbing with rifle butts. "He received sixteen bayonet thrusts from the front and twenty-one from behind," France's General Niessel reported. "Krylenko showed himself incapable of controlling the detachment who escorted him."[33] There was little Krylenko could have done against the crowd's fury. Although strongly built, he was not a large man and the mob merely thrust him aside when he tried to bar their way. "The hatred of the people has boiled over," he explained sadly. "Be worthy of your newly won liberty," he commanded Russia's soldiers. "The revolutionary people are terrible in war, but should be gentle after victory."[34] For Dukhonin, such high-minded exhortations had come too late.

Dukhonin had been left to face Russia's angry soldiers at Mogilёv for ten days after Lenin had relieved him as acting supreme commander partly because Krylenko had traveled to *Stavka* by way of Dvinsk, certainly not on the direct route from Petrograd, but a key point near the front, and from there he dispatched three emissaries to the German lines just after noon on November 13. Four hours later, these three men—a lieutenant from the Kiev Hussars, a military surgeon, and a volunteer—crossed into no-man's-land, a trumpeter announcing their advance under the protection of a white flag. Three hundred meters ahead they entered the enemy lines. Quickly, German officers blindfolded them and took them to their division headquarters. By the next afternoon, they were en route to Petrograd with word that negotiations could begin within the week in Brest-Litovsk at the headquarters of General von Hoffmann, who for some months had commanded Germany's armies in the east, even though Prince Leopold of Bavaria actually held the title of commander in chief.[35]

A key figure in Germany's intelligence service on the eastern front since the first days of the Great War, Max von Hoffmann's abilities were not reflected in his looks. The symmetry of his closely-cropped round head scarcely disturbed by small ears that pressed flat against his skull, he stared at the world through small, closely-set eyes which endowed him with a mildly porcine appearance that belied the brilliant mind that lay within. Once described as "composed of equal parts of steel and whalebone," he had the courage of his convictions and the patience to wait for time to prove him right. Hoffmann had followed Russia's military collapse with the closest attention. Deeply confident that the revolutionary virus would sap his enemy's vital forces, he felt certain that he could win on the eastern front without committing any of the reserves of men or materiel so desperately needed by Hindenburg and Ludendorff, who now had taken Falkenhayn's place in the west. Patience and steady nerves were the best weapons with which to fight Russia's crumbling armies, and Nature had bestowed both upon Hoffmann in generous measure. Now, in the citadel which rose above the blackened ruins of Brest-Litovsk, he waited to see what proposals the Russians would bring.[36] With him waited fourteen other representatives: five from Germany, four from Austria-Hungary, three from Turkey, and two from Bulgaria. The Turks had sent a general to lead their representatives. The Bulgarians had sent a colonel, and the Austrians only a lieutenant-colonel.[37]

Although Hoffmann had come to Brest-Litovsk with his allies, the Russians did not. From the moment they had received copies of Lenin's Peace Decree, Russia's allies had maintained an angry, stony silence.[38] Unperturbed, Trotskii tried just after Krylenko's emissaries returned to Petrograd to get the Allies to take part in the peace talks, and even offered to delay the Russians' arrival at Brest-Litovsk for a few days to give their allies a chance to join them. "If the allied peoples do not send their representatives," he warned, "we shall begin negotiations with the Germans ourselves." Trotskii still hoped to conclude a universal peace. If that proved impossible, Russia must conclude one on her own, for, as he informed the Allies, "the Russian army and the Russian people cannot, and will not, wait any longer [for peace]."[39] Still, the Allies remained silent. "The Soviet of People's Commissars was, and still is, of the opinion that it is necessary to begin simultaneous negotiations with all the Allies for the purpose of achieving the most rapid armistice on all fronts," Trotskii announced two days later. "The Allied governments and their diplomatic representatives in Russia are kindly asked to reply as to whether or not they wish to take part in the negotiations opening on November 19/December 2 at five o'clock in the afternoon."[40] Again, the Allies sent no reply.

The peace delegation Trotskii assembled in Dvinsk on November 19 therefore included only Russians. Carefully chosen to represent the social forces of the Bolshevik revolution, its twenty-eight members included workers, peasants, soldiers, sailors, women, and, above all, loyal Bolsheviks. Adolf Ioffe, destined to become one of the leading diplomats of the early Bolshevik era, led the delegation. Lieutenant Colonel Fokke, a tsarist officer who served as one of the delegation's military experts, thought him "unpleasant and smugly contemptuous," and remembered that his overlong hair, badly-trimmed beard, worn bowler hat, and too-large black overcoat contrasted sharply with the precise manners and immaculate appearance of his German counterpart. The delegation's other leading figures included what Fokke called its two "lions," a play on the first names of Kamenev and Karakhan, Lev, which meant lion in Russian. Still only partly reconciled to Lenin's and Trotskii's arrogant seizure of power, Kamenev had continued to be an uneasy ally on the Bolsheviks' right wing, while Karakhan, in Fokke's words, remained "a typical Armenian, almost a caricature of that 'Eastern type' who is able to shift from a sort of dreamy inertia to fast-moving agitation in a single moment." Anastasiia Bitsenko, a taciturn woman of peasant origin, added yet another dimension to the delegation. Present as a symbol of the Bolsheviks' proclaimed dedication to female equality, Bitsenko had served seventeen years of penal servitude in Siberia for assassinating a Russian general. A woman who preferred to remain apart, Bitsenko's only passion was the revolution. "She reminds one," Austria's foreign minister Count Czernin once remarked, "of a beast of prey seeing its victim at hand and preparing to fall upon it and rend it."[41]

If Germany's senior generals in the east were unused to negotiating with a female terrorist and a handful of dedicated revolutionaries, they were even less accustomed to dealing with the assortment of proletarians that rounded out Ioffe's delegation. Setting aside his fellow technical military advisers, Fokke described the remainder of the delegation as a "menangerie." Fëdor Olich, a tall, muscular sailor from the Baltic Fleet, was the "seawolf" in this group. Nikolai Beliakov, a short, stocky, middle-aged soldier reminded Fokke of a "sullen badger," and Pavel Obukhov, a factory worker, seemed most like an "insolent ferret." Each of these three unique figures had been chosen to represent one of the revolutionary groups that formed the base of the Bolsheviks' power, and there was a fourth, who had come to the delegation in a more unusual manner. En route to Petrograd's Warsaw Station on the evening of their departure, Ioffe and Kamenev had suddenly remembered that they had forgotten to include a peasant in their delegation.

"We've forgotten the Russian peasantry! We have no one among us to represent the many millions of rural toilers!" one of them suddenly exclaimed. At that very moment, Fokke recalled, Russia's would-be diplomats spied a bent figure in a typical peasant's black homespun coat and knapsack trudging down a side street. They swerved in his direction.

"Where are you heading?" Russia's peace delegation asked.

"To the station, comrades."

"Jump in. We'll give you a lift!. . . ."

"Wait! This isn't where I'm going, comrades!" the peasant protested as they drew up to the Warsaw Station. "It's the Nikolaevskii Station I need. I'm heading for Moscow. . . ."

"What party do you belong to?" they asked, fearful that the peasant's political loyalties might disqualify him from taking part in their venture.

"I'm a Socialist Revolutionary, comrades. We're all Socialist Revolutionaries."

"Left- or right-wing?" they queried further.

"Left, comrades. As far left as you can get."

"Well, you don't need to go to the country right this moment," the delegation insisted, now certain that they had found their delegate. "Come with us to meet with the enemy at Brest. We're going to make peace with the Germans."

Now, the Russian delegation was complete. Roman Stashkov, a simple villager, would be the "plenipotentiary representative of the Russian peasantry" at Brest-Litovsk.[42] "A typical Russian figure with long grey curls and an enormous untrimmed beard," according to Hoffmann, Stashkov won the hearts of the Germans the very first night when, in response to an orderly's question about whether he preferred claret or hock with dinner, said that he would drink whichever one was stronger.[43]

At 5:30 on the afternoon of November 19/December 2, just hours after the Mogilëv garrison arrested Dukhonin, and the day before Krylenko reached *Stavka*, Russia's official peace delegation crossed the German lines. Three days of negotiations followed in the Brest Citadel. Hoffmann proposed a cessation of hostilities, with each side holding its present positions during the negotiations, while Ioffe called for a six months' armistice, the German evacuation of the islands they had seized in Moon Sound during their recently-ended Riga offensive, and a promise that the German High Command would move no troops from the eastern to the western front against Russia's still-silent Allies.[44] Although he "refused curtly and energetically" even to consider the Russian demand for Germany's evacuation of any territory, and insisted that the armistice could not be for more than twenty-eight days, Hoffmann agreed that no German troops "except those who were already being moved, or had already received orders to go" would be moved from the eastern front during the armistice. Ioffe responded by asking that the Russians be given a free hand to distribute antiwar propaganda among the German armies during the armistice, to which Hoffmann replied that such a request seemed "superfluous in view of the fact that we are ready to conclude peace, but that the distribution of propagandistic brochures in other [Entente] countries would seem to be entirely appropriate."[45] Further, he agreed, "as it seemed quite impossible to prevent all intercourse," that it should be limited to certain points. "In this way," he later explained, "it would be possible to exercise some control, and to intercept the greater part of the propaganda literature that might be expected."[46] He could not, he insisted, agree to the Russians' proposal for a general armistice on all fronts because Ioffe's delegation clearly did not have the authority to negotiate for their Western allies.[47] This raised a hurdle that Ioffe could not surmount. His masters at Smolnyi clearly had in mind a general armistice, not a separate one, and they had answered his urgent request for new instructions with orders that it was impossible to depart from the stated points of his original proposals.[48] Anxious to help save face for a man whom he respected for being "extraordinarily intelligent," Hoffmann agreed to a week's recess so that Ioffe could return to Petrograd on November 24 for new instructions.[49]

Even before Ioffe returned, Trotskii informed Russia's allies that the meetings at Brest had been halted for a week "in order to make it possible to inform the people and governments of the Allied nations about the negotiations and the direction they are taking." Again he invited the Allies to take part when the negotiations were reopened. Again, they remained silent. On November 29/December 12, the Soviet delegation returned to Brest-Litovsk, this time without its symbolic sailor, soldier, worker, and peasant,[50] and the negotiations moved quickly. On December 2/15, the two delegations signed a formal armistice until January 1/14, 1918, with a provision for automatic thirty-day renewals unless either side gave a week's notice.[51] Peace had not yet come, but, at no time since July 1914 had it been closer. "A separate armistice is not yet a separate peace," Trotskii told the Allies a few days later, "but it means a danger of a separate peace."[52] For the Allies, a separate peace posed a significant danger. If peace came in the east, Germany could shift her eastern front armies to the west, and offset the advantage that the Allies expected to reap from the arrival of fresh divisions of young, vigorous, well-equipped Americans. "If all went smoothly at Brest-Litovsk, if the people there worked with real energy," Ludendorff later wrote, "we could expect to have our forces ready for a successful attack in the West [by about the middle of March]."[53]

New American armies might help the Allies in the west, but they could do nothing in the east, where the Russians had no defenses left against the Germans. Bravely exclaiming that "we have not overthrown the Tsar and the bourgeoisie in order to kneel down before the German Kaiser," Trotskii had spoken of creating "a powerful army of soldiers and Red Guards, strong with revolutionary enthusiasm," to carry on the fight and "summon all peoples to a holy war against the militarists of all countries" if the Germans tried to impose "an unjust and undemocratic peace."[54] Yet he knew only too well that Russia could not hope to hold the front if the Germans reopened hostilities. Her economy ravaged by war, her networks of transportation, supply, and finance shattered by the revolutionary events of the past ten months, Russia lay in ruins.

Russian soldiers had walked away from the war by the hundreds of thousands during the past few months, and the huge wave of men welling up toward the rear now rose higher still. "The numbers of infantry do not in any way correspond to the magnitude of the front we occupy," one commander reported. "Many sections of the front have been completely abandoned by the units assigned to defend them and are utterly unprotected. Under such conditions, the front has become nothing more than a geographical designation."⁵⁵ For the moment, the Bolsheviks could hope that Russia's turmoil had become so chaotic that the Germans' usual information networks had not yet been able to sort it out and prepare the analytical appraisals to which the German General Staff had grown accustomed. But, even if true, that situation could not be expected to last for very long.

Yet, Fate had not stacked all the cards against the Russians. As they prepared to return to Brest-Litovsk, the Germans knew only too well that their armies were strong only in comparison to those of Russia and that their resources had been stretched disastrously thin by war on Europe's two major fronts. In the west, her submarine fleet had not proved the great equalizer that Hindenburg and Ludendorff had envisioned. England's sea lanes had not been cut, and it now seemed doubtful that U-boats could do more than delay only briefly the arrival of American divisions in France. Convinced that Germany must launch an offensive in the west if she hoped for victory, Ludendorff had urged a massive attack. "That the attack in the West would be one of the most difficult operations in history I was perfectly sure," he later wrote, but he saw no other way for Germany to survive the war. "Attack is the strongest form of combat," he insisted. "It alone is decisive. Military history proves it on every page." Yet the resources for any attack could come only from the east, and they must come quickly. "Delay," Ludendorff knew all too well, "could only serve the enemy, since he was expecting reinforcements."⁵⁶

Therefore, when Germany sent her negotiators back to Brest-Litovsk early in December, they went with a sense of urgency only slightly less pressing than the Russians'. Many influential people hoped, as the newspaper *Vörwarts* wrote, that "no Russian will have to regard as a misfortune for his country the peace which is now being concluded," and that peace therefore would come quickly.⁵⁷ As for the Austrians, their foreign minister Count Ottokar Czernin confided to a friend that the most they could hope for was "to settle with Russia as speedily as possible, then break through the determination of the Entente to exterminate us, and then make peace, even at a loss." Czernin knew that the Habsburg Monarchy would never again stand among Europe's Great Powers as it had of old. At best, he hoped to preserve a vestige of its former splendor. To him, Brest-Litovsk offered a hope to "come out of [the war] . . . at last, albeit rather mauled."⁵⁸

Now led by Germany's secretary of state for foreign affairs Baron Rikhard von Kühlmann, the German delegation boasted some of the nation's most able diplomats. The Austrians sent Czernin as their chief spokesman, the Bulgarians sent their minister of justice, and the Turks dispatched both their grand vizir and their foreign minister to speak for their interests. For the moment, Lenin and Trotskii left Russia's fate in the hands of Ioffe, but strengthened his resources by adding to his suite the dean of Russia's early Marxist historians, Mikhail Pokrovskii, and General Aleksandr Samoilo, a tsarist general more ready to serve the Bolsheviks than most of his colleagues. Assembling in Brest-Litovsk on December 7/20, these awkward apostles of the new order and the elegant defenders of the old prepared for the first direct clash between Bolshevism and the West.

If Hoffmann had been impressed by Ioffe's intelligence at their earlier meetings, certainly Czernin had a far lower opinion of his trustworthiness. "They are strange creatures, these Bolsheviks," he wrote in his diary after their first dinner together. "They talk of freedom and reconciliation of the peoples of the world, of peace and unity, and withal they are said to be the most cruel tyrants history has ever known. They are simply exterminating the bourgeoisie, and their arguments are machine-guns and the gallows. My talk today with Ioffe has shown me," he concluded, "that these people are not honest . . . for, to oppress decent citizens . . . and then talk at the same time of the universal blessing of freedom—it is sheer lying."[59] Thus, Czernin's finely-honed diplomatic instincts—a sixth sense, perhaps—warned him of the barrier destined to rise again and again between West and East because the Western powers and the Bolsheviks lived by very different values and applied them in very different ways. One side thought mainly in terms of long-held traditions, of projecting the past into the future. The other thought in terms of the "impending collapse of capitalism," the "inevitable world revolution," and the "dawning dictatorship of the proletariat." Inevitably, the two had to clash.

At Kühlmann's invitation, Ioffe presented Russia's conditions for peace when the conference convened late on the afternoon of December 9/22. Based upon the program Lenin had drawn up a fortnight earlier, Ioffe's Six Points all emphasized the Bolsheviks' commitment to their long-used (and not infrequently-abused) slogan of "no annexations or indemnities."[60] Czernin and Kühlmann replied that they shared "the opinion that the principles laid down by the Russian delegation form the basis for the discussion of . . . [a just and general] peace," but warned that they could accept them "only in case all belligerents [including the nations of the Entente] without exception pledge themselves" to do the same. In any case, they announced that the Central Powers had "no intention to annex by force the territories seized during the war."[61] With unconcealed delight, Ioffe telegraphed to Petrograd that the Central Powers had accepted Russia's principles. "Even our enemies, who only recently predicted that the Germans would not even talk to us . . . must now admit that our diplomacy has met with great success," Trotskii exulted from his office in Russia's Commissariat of Foreign Affairs. "Germany," he continued, "gives in not merely to the force of truth but to the fear of revolution which menaces the very existence of the bourgeois regime."[62]

However, as Trotskii would learn all too soon, what Fokke once called Ioffe's "naive optimism" had failed to perceive the finer distinctions that the Germans drew to clarify their view.[63] In this case, Hoffmann undertook the less than pleasant task of setting the record straight at lunch on December 14/27. The Central Powers took the position, he told Ioffe, "that it was not forcible annexation if portions of the former Russian Empire [he had in mind Courland, Lithuania, and Poland] decided, of their own free will, and by determination of their existing political representatives, on a separation from the Union of Russian States, and on being united to the German Empire or any other State."[64] Ioffe, Hoffmann remembered, "looked as if he had received a blow on the head,"[65] and Fokke thought that Russia's chief spokesman looked suddenly "dismayed, exhausted, and crushed." Ever-conscious of Russia's past greatness, Pokrovskii wept openly behind his small, round, wire-rimmed glasses as he asked Hoffmann how he could speak of "peace without annexations when Germany was tearing eighteen provinces away from the Russian state."[66] Fearful that the negotiations would collapse, Czernin seemed "beside himself," and spoke openly of making a separate peace if the Russians should leave Brest-Litovsk in protest against Germany's demands.[67] More coldly calculating than any of his Austrian and Russian counterparts, Hoffmann insisted that "there was no question of the negotiations being broken off by the Russians." By then, he knew how desperately Lenin needed peace. "The only possibility for the Bolsheviks to remain in power was by signing a peace," he later explained. "They were obliged to accept the conditions of the Central Powers, however hard they might be."[68]

On the Russian side, Fokke sadly concluded that "any agreement about 'peace without annexations' was a mere specter."[69] As the delegations parted on December 15/28 for a twelve-day recess, Trotskii and Lenin understood only too well how far their premature optimism had led them astray. Russia's only hope to avoid what Hoffmann privately referred to as the "hard" conditions of the Central Powers[70] lay in either broadening the Brest-Litovsk discussions into a general peace conference by convincing her Allies to join them, or prodding the proletariat of Germany to overthrow the kaiser as Russia's working classes had toppled his cousin. Therefore, on December 17/30, the day after Ioffe returned to Petrograd, Trotskii once again urged the leaders of Great Britain, France, and the United States to reconsider their earlier refusal to take part in the Brest-Litovsk meetings.

Not yet having learned to separate the quiet workings of international diplomacy from the noisier rantings of international revolutionary propaganda, Trotskii addressed the people of the Allied nations as well as their leaders, and did so in terms that no self-respecting government could tolerate. It was not enough to speak only of the self-determination of the peoples of Alsace-Lorraine, Galicia, Poznań, and Bohemia, he insisted. Allied governments also must admit the rights of independence and self-determination of the peoples in the British and French colonial territories of Ireland, Egypt, India, Madagascar, and Indochina. "The Russian revolution has opened the door to an immediate general peace on the basis of compromise," Trotskii went on, as he reminded the Allied statesmen that only ten days remained—"ten days that will decide the fate of hundreds of thousands and millions of human lives"—in which they could come to the negotiating table. "If the Allied governments, in the blind stubbornness that characterizes the declining and dying classes, again refuses to take part in negotiations," Trotskii warned, as he shifted his words from Europe's statesmen to the people they governed, "then the working class will be confronted by the iron necessity of tearing power from those hands who cannot, or will not, give the people peace."[71]

Aside from the fact that their war aims now clashed bitterly, no Allied government would consent to being lectured to in such a fashion by the people's commissar of foreign affairs of a government whose messianic revolutionary preachings they thought repugnant. Although neither communicated directly with the Russians, French and English statesmen quickly let it be known that they rejected any thought of joining such discussions, while the Bolsheviks themselves had only insulting words for President Woodrow Wilson's moderately sympathetic replies. "Mr. Wilson serves American war industry just as Kaiser Wilhelm serves the iron and steel industry of Germany," *Pravda* announced scornfully. "One gives his speeches in the style of a Quaker Republican —the other wraps himself in the mists of Prussian-Protestant-Absolutist phraseology, but at bottom it is all the same."[72]

Now certain that he must make peace without the Allies' support, Lenin decided to postpone the final reckoning at Brest-Litovsk as long as possible in the vain hope that Central Europe's proletarians might yet be stirred to revolution. He therefore sent Trotskii to Brest-Litovsk in place of Ioffe because, as he explained, "to delay the negotiations, we need someone who knows how to do the delaying." Trotskii's was to be a thankless, arduous task. "I felt as if I were going to a torture chamber," he later wrote, as he recalled his first journey to Brest-Litovsk.[73] Although he greeted an honor guard at Dvinsk with the bold promise that "we did not overthrow the bourgeoisie in order to bow our heads before foreign imperialists and their government," and vowed that the Russian people would "accept only an honest, democratic peace,"[74] Trotskii later admitted that, at that point, "it was absolutely evident to me that we were not in a position to go on fighting." As he passed beyond Dvinsk, he found many of the trenches nearly empty. "Peace," he muttered to himself as his train approached the crossing point. "Peace, whatever happens!"[75]

If Trotskii and Lenin realized that they could not renew the war, they failed to perceive the full cost of delaying its final moment of reckoning, for the German position had begun to harden more than even Kühlmann or Hoffmann could have predicted. Far from being the covetous imperialist that the Russians thought him to be, Hoffmann had sought only minimal territorial gains in the east that would bring very precise strategic improvements in Germany's eastern frontier. Certain that Germany would reap no benefit from acquiring millions of new Polish subjects, Hoffmann had told the kaiser emphatically on New Year's Day that "the new Polish frontier ought to be drawn in such a way that it should bring to the Empire the smallest possible number of Polish subjects and that there should be only a few unimportant corrections of the frontier." In sum, he explained, "I was an enemy of any settlement of the Polish question which would increase in Germany the number of subjects of Polish nationality."[76]

Hoffmann's wise opinion had made sense to Kühlmann and the kaiser, but it had not found favor with the German High Command, where Hindenburg and Ludendorff demanded broad new eastern territories for Germany. Outraged that their nation's statesmen seemed ready to fritter away the fruits of their armies' hard-won eastern victories, Germany's senior generals now insisted that "the result of the peace [must] correspond to the sacrifices and achievements of the German nation and army, and that the terms of peace [must] increase our material power."[77] In an instant, both Left and Right took up their cause, and focused their anger especially upon Germany's chief negotiator. "If a prize had been offered for showing how a brilliant military position may be utterly ruined, Baron von Kühlmann would have won it," one editorial exploded. "The German people have now to choose between Hindenburg and Ludendorff on the one hand and Kühlmann and [his ally Chancellor Count von] Hertling on the other." Confident of his ground, the editorial's author concluded that, when faced with the choice, the German people would "rally in unanimous love round their two heroes."[78] If that proved true, "peace without annexations" would no longer be even a specter; it would become an impossibility.

With powerful forces in Kühlmann's rear demanding more territory in the east, with Trotskii trying to postpone a final reckoning in the desperate hope that revolution might yet break out in Germany, and with Czernin nearly as desperate to "accelerate the signing of some kind of a treaty at all costs" to give Austria the *Brotfrieden*—Bread Peace—with Russia or the Ukraine needed to rescue her starving people, the Brest-Litovsk negotiations resumed on December 26/January 8, the day after the Russians celebrated Christmas, and a week after the New Year's coming in the West.[79] Trotskii made it clear from the first that the Central Powers' representatives faced a far different adversary than they had in Ioffe. Scornful, even contemptuous of his enemies, he rejected all friendly contact with them and steadfastly refused to act in any way like the representative of an exhausted nation obliged to sue for peace. Mephistophelian in appearance, and equally brilliant as polemicist, orator, historian, diplomat, revolutionary tactician, and military commander, Trotskii proved to be the Bolsheviks' man for all seasons. Second only to Lenin in his uncanny ability to turn difficult situations to advantage, he proved second to none in his genius for probing men's hearts and bending them to his will. "Diabolically intelligent, diabolically scornful," one commentator wrote, "he was destined to be both the Michael and the Lucifer of the Revolution."[80]

Trotskii arrived at Brest-Litovsk to find that others who had fought the Central Powers had come before him to treat with the enemy. Throughout 1917, Ukrainian nationalists had been a festering thorn in the southwest flank of Russia's troubled body politic as they moved ever closer to declaring full independence. Conflicts with the Ukrainian Central Rada, or national assembly, late in June had been a major factor in shattering Russia's first coalition government on the eve of the July Days, and almost immediately after the Bolsheviks seized power in Petrograd the Rada had declared full independence. Not all Ukrainians had supported that decision, and a small group sympathetic to the Bolsheviks had established a rival government in Kharkov and appealed to Petrograd for support. Nonetheless, several days before Trotskii arrived, Ukrainians representing the Central Rada in Kiev and calling themselves representatives of the Ukrainian People's Republic, had appeared at Brest-Litovsk. Claiming that no peace could be binding upon the Ukraine, the breadbasket of Eastern Europe, unless "accepted and ratified by the government of the Ukrainian People's Republic," they demanded to negotiate a separate peace with the Central Powers.[81]

To make light of a bad situation, Trotskii insisted that because the Ukrainians considered themselves independent and because they had entered the negotiations apart from the Russians, the Central Powers must deal with them separately.[82] In truth, both Russia and Austria desperately needed the Ukraine's grain. From Russia, Trotskii received reports that some regiments had no food, and he must have remembered only too well how he had used the terrible food shortages in Petrograd to the Bolsheviks' advantage in October. At the same time, Czernin received daily reports from Vienna that forecasted famine in a matter of days. "We shall have thousands perishing in a few weeks," he confided to his diary on January 7/20. "I cannot, and dare not, look on and see hundreds of thousands starve." Czernin therefore was ready to consider paying the Ukrainians' price: cession of the Polish city of Cholm and the lands around it to the Ukrainian People's Republic, even though that would earn the Austrians the Poles' deepest hatred.[83] The issue of the Ukraine thus hung like the sword of Damocles over both Czernin and Trotskii. Without its grain, both of their nations faced starvation.

While Czernin faced the near-certainty of famine in Austria, Trotskii crossed swords with Kühlmann. From the outset, he understood that the gap between their aspirations could never be bridged, but Lenin had said to delay, and Trotskii would prove a master at doing so. Day after day, he engaged Kühlmann in debate, rising to subtle discussions of first principles that ranged far beyond the concrete territorial issues that divided them. As their debate became more theoretical and turned increasingly upon abstractions, Trotskii switched from Russian to German to argue its finer points in Kühlmann's own language. Conflicting visions of the destinies of nations rose to a resounding clash of opposing world views between them, as Kühlmann laid out with elegant grace a defense of traditional European diplomacy in which statesmen respected history's legal continuity and strived to maintain the intricately balanced cantilever into which the fate of nations had become so inextricably fitted.[84]

Trotskii responded as an unabashed apostle of revolution, the spokesman for a brave new world ungoverned and ungovernable by the traditional formulas which Russia's revolution had overthrown. "We are revolutionaries, but we are also realists," he insisted when Kühlmann claimed that the peoples in Russia's former western lands had chosen to link their destinies with Germany in freely-stated self-determination. "We prefer to speak directly about annexations and not clothe their real name in a pseudonym."[85] "I would take the liberty to propose," he said after he had read the phrase about living in peace and friendship that the Central Powers proposed to include in the draft treaty's preamble, "that the second phrase [about friendship] be deleted. . . . Such declarations," he went on, "have never yet characterized the real relations between states."[86] At one point setting aside the debate on higher principles, Trotskii took time to transform the conference "for a few minutes" into what he called "a class in Marxist instruction for beginners," in which he pointed out to General von Hoffmann that, "in a class society, every government rests upon force. . . . What startles and antagonizes the governments of other countries about our actions," he went on, "is that we arrest not the strikers but the capitalists who subject the workers to lockouts, and that we do not shoot peasants who demand land but arrest those landlords and officers who try to shoot the peasants."[87]

For a month, Trotskii had his delay, with Kühlmann rising to cross swords with him at every opportunity until the limits on Ludendorff's patience at General Headquarters wore dangerously thin and only a few grains remained in the hourglass of Austria's food supply. Yet Czernin had little recourse but to watch the days pass, hoping that time would not run out, but fearful that it would. From Vienna, he received reports of a "serious strike movement." At the same time, Germany insisted that she could send no help because she too faced serious food shortages.[88] For a moment, it seemed that the Bolsheviks' tactics of delay might actually have postponed the day of reckoning long enough for mass hunger to fuel a mass revolution that would free Russia from the Germans' imposed peace. However, pressed by Ludendorff, who wanted to throw divisions from the eastern front into his upcoming western offensive, Hoffmann finally broke the impasse by signing a peace with the Ukrainians. Then, on January 5/18, he laid before Trotskii a detailed map of the former Russian territories that must come under German hegemony. "As far as the line drawn on this map is concerned," Trotskii remarked sarcastically, "we would be extremely grateful if you would explain to us the principles and considerations that guided you in its preparation." Ruthlessly, Hoffmann took up his challenge. "The line in question," he stated bluntly, "was dictated by military considerations. It guarantees to the people living on this side of it an orderly state structure and the fulfillment of self-determination."[89]

The negotiations had reached an impasse, for Trotskii had no recourse against German annexation of former Russian territories. "The position of the opposing side on the political-territorial issue has now been fully clarified and it can be summed up in the following way," he concluded, emphasizing every one of the words in his brief speech. "Germany and Austria-Hungary are cutting off from the domains of the former Russian Empire territories more than 150,000 square kilometers in size."[90] Such a decision, he insisted, could not be made without consulting Lenin and his colleagues on the Sovnarkom. He therefore asked for a nine-day recess in order for his government to reach a final decision.

Trotskii's measured tones revealed nothing of his inner turmoil. "It is impossible to sign their peace," he insisted to Lenin. Confident that German public opinion would not permit a renewal of hostilities in the east, Trotskii urged Lenin to "announce the termination of the war and demobilization without signing any peace."[91] Russia's new formula thus would be "No War, No Peace." However, willing to sign even a ruinous peace to extricate Russia from the Great War, Lenin had serious doubts about the more subtle formula that Trotskii proposed. "It's risky, very risky," he warned. Convinced that "at the moment there is nothing more important in the world than our revolution," he dared not face the danger of a new German offensive. "We cannot take the risk," he explained,[92] and went on to insist that "it would be absolutely impermissible tactics to stake the fate of the socialist revolution, which has already begun in Russia, merely on the chance that the German revolution may begin in the immediate future." Russia would be wiser to sign the peace now offered rather than have a more ruinous one imposed upon her a few weeks hence. "There can be not the slightest doubt," he warned, "that our army is absolutely in no condition at the present moment . . . to beat back a [new] German offensive." If the army would not fight and, in many places, had simply melted away, and if its peasant majority would "unreservedly declare in favor of a peace with annexations," then Russia must make peace. Without a respite at home, Lenin concluded, Russia's socialist revolution faced almost certain collapse.[93]

Along with Nikolai Bukharin's entire Moscow group, Trotskii thought Lenin mistaken, although he did not share Bukharin's wildly unrealistic conviction that the Russian proletariat could escape the Germans' heavy demands by declaring a revolutionary war against Germany and Austria-Hungary.[94] Supported by Stalin, Trotskii continued to emphasize the moral force of his "no war, no peace" program as a way "to give clear and incontestable proof of the deadly hatred that exists between us and the rulers of Germany to the workers of Europe." At the very least, he insisted, that would "strike a decisive blow against the legend about secret ties with the Hohenzollerns" which had clouded the

Bolsheviks' relations with workers at home and abroad ever since Lenin's "sealed train" journey at the end of March.[95]

For once, Lenin's awesome powers of persuasion could not sway comrades outraged by his insistence upon peace at any price. "If there were five hundred courageous men in Petrograd, we would put you in prison!" one of them exclaimed, the fires of revolutionary war burning hotly in his breast. "If you will calculate the probabilities," Lenin replied coldly, "you will see that it is much more likely that I will send you [to jail] than you [will send] me."[96] At a special meeting on January 8, thirty-two leading party members voted for the revolutionary war that the hot-headed Bukharin now demanded, while sixteen supported Trotskii and only fifteen stood with Lenin.[97] Angrily, Lenin rebuked his opponents: "Now you wish that we [should] perish and that the capitalistic military governments should emerge victorious in the name of our [old] revolutionary formula [for a people's peace]."[98] This time, he spoke to no avail. "[They] want to stick to the old *tactical* position, and stubbornly refuse to see the *change* that has taken place, the new *objective* situation that has arisen," he wrote in his notes after the meeting. "In their zealous repetition of old slogans, [they] have not even taken into consideration the fact that we Bolsheviks have now all become defensists."[99] To Lenin, Bukharin's position seemed nothing less than a policy of unmitigated insanity. "The position of the Germans," he warned as Bukharin's supporters continued to insist upon launching a revolutionary war against German imperialism, "is such that in an offensive they could take Reval and Petrograd with [their] bare hands."[100]

"Around Ilich there emerged a sort of void," Krupskaia remembered as she recalled those tense days when Russia's infant workers' state was not yet three months old. "The voracious beast of German imperialism had seized us by the throat."[101] The consequences were frightening to contemplate. As Lenin warned again and again during those days, "German militarism is a beast that moves swiftly."[102] Lenin now turned to Trotskii to fill the political void that had formed around him and to strike a compromise that could free Russia's struggling Soviet government from the "beast's" inexorably closing jaws. Certain that Bukharin's revolutionary war promised disaster, Lenin agreed for the moment to support Trotskii's policy of "no war, no peace," which, if it failed, would result in only a harsher peace. "Let's suppose that we refuse to sign a peace and the Germans move to the attack. What do you plan to do then?" Trotskii remembered him asking. "We'll sign the peace at the point of a bayonet," Trotskii replied. "The situation then will be clear to all the world." Trotskii promised that under no conditions would he support Bukharin's call for a revolutionary war if his policy of "no war, no peace" failed.[103] Fearful that it would cost Russia dearly, Lenin nonetheless agreed to put Trotskii's plan to the test.[104] Trotskii now would return to Brest-Litovsk, prolong the negotiations for as long as possible, and then, at what he considered the appropriate moment, present his new formula to the Central Powers.

Trotskii's long-awaited moment came on January 28/February 10, two days after the Central Powers had formally signed their peace treaty with representatives of the Ukrainian Rada. "We declare to all peoples and governments that we are getting out of this war," he began. He paused. Then he continued, his voice clear and metallic, emphasizing each word, "At the same time, we announce that the conditions of peace set before us by Germany and Austria-Hungary fundamentally work against the interests of all peoples [and] . . . we cannot inscribe the signature of the Russian Revolution beneath conditions that bring with them oppression, grief, and misfortune to millions of human beings." Every ear in the room strained not to miss a word. "In refusing to sign an annexationist peace," Trotskii concluded, "Russia announces that the state of war with Germany, Austria-Hungary, Bulgaria, and Turkey is at an end. Orders for general demobilization have already gone out to the Russian armed forces."[105]

Tension, thick and heavy-charged, blanketed the room as Trotskii finished. "Unerhört!" Hoffmann gasped. "Shocking!" "Unheard of!" Further down the table, it seemed to Fokke that Trotskii's declaration had "burst like thunder from a clear sky." Russia would no longer fight, but neither would she sign, Trotskii had just said with unconcealed contempt, "conditions which German and Austro-Hungarian imperialism would inscribe with the sword upon the body of the living peoples involved." "Well, what will happen now?" a German major asked Fokke as the Russians boarded their train. "Is it really possible that we shall again go to war?" That question lay at the very center of everyone's thoughts. As a huge wave of industrial unrest welled up across Central Europe, would the Germans dare to launch a new offensive? If so, how far—and how fast—were they prepared to march? For the moment, Fokke could only shrug his shoulders,[106] while Trotskii hurried back to Petrograd to bask for a brief moment in the comforting misapprehension that perhaps "no war" meant peace after all. As he prepared to leave Brest-Litovsk, one of the German diplomats had told Ioffe as much, and Germany's representative in Petrograd held the same opinion. "They cannot continue their aggression without revealing their cannibal teeth, dripping with human blood," *Pravda* assured its readers.[107] Lenin's intuition told him that the chances were very good that Trotskii's formula would fail. Still, for a brief moment, even he allowed his hopes to rise above his better judgment. "*Les apparences*—appearances—are saved," he remarked at one point, "and here we are, out of the war."[108]

But the Bolsheviks had taken their readings of German intentions from the remarks of civilian statesmen and had failed to reckon with Germany's High Command. "With our rifles idle in our hands, we shall watch the whole situation being transformed to our own disadvantage [if we do not act]," Ludendorff raged. Renewed hostilities could only strengthen Germany's position in the east. Even more, he went on, "we may, perhaps, give Bolshevism its death blow, thereby ameliorating our internal situation and helping our relations with the best elements in Russia."[109] In any case, Ludendorff and his allies in the High Command demanded action. "We could not possibly leave matters in this condition," he later wrote. "At any moment fresh dangers might arise while we were fighting for our lives in the West. . . . It was a military absurdity, or worse, to sit still."[110]

For the better part of six days, the Bolsheviks lived with their illusions of peace. Then their dream broke apart with all the suddenness of shattered crystal. "Today, at 7:30 p.m.," General Samoilo telegraphed on February 16* from Brest-Litovsk, "I received official notification from General von Hoffmann that, at noon on February 18, the armistice concluded with the Russian Republic will expire and a state of war will begin again."[111] Nothing could have suited Hoffmann better. "We are going to start hostilities against the Bolsheviks," he wrote with the self-satisfied confidence of one whose mission fully coincided with his moral principles. "Otherwise, these brutes will . . . quietly get together a new revolutionary army and turn the whole of Europe into a pig-sty . . . The whole of Russia," he concluded, "is no more than a vast heap of maggots—a squalid, swarming mass."[112] On the appointed day, Hoffmann unleashed fifty-three divisions against the near-empty Russian trenches along his front, concentrating sixteen of them against Petrograd.[113] "This beast leaps fast," Lenin exclaimed.[114] Without a moment's hesitation, Hoffmann advanced toward Reval, Pskov, and Petrograd in the northwest, and into the Ukraine in the south to protect the Central Powers' newly-won grain supplies. Dvinsk, from which Krylenko had sent Russia's first peace emissaries and from which Ioffe and Trotskii had begun their ill-fated journeys to Brest-Litovsk, fell on the first day, yielding up vast quantities of artillery to the Germans.[115] "No peace, no war," Woodrow Wilson's special envoy to Russia remarked, "is war!"[116]

War it might be, but not for long. The moment he received Samoilo's telegram, Lenin launched a campaign for peace even more intense than the one he had waged to win approval for the Bolsheviks' armed uprising in October. Then he had fought to make the revolution; now he threw every resource into a struggle to preserve it. The revolutionary war that Bukharin's allies demanded, Lenin insisted at a Central Committee meeting on the evening of February 18, could not, and must not, be fought. "The peasant will not fight a revolutionary war," he warned, "and will surely turn against anyone who openly supports one." No matter how much the workers' districts of Russia's cities might blaze with indignation at the new German attack, the hard truth was that the experiment with "no war, no peace" had failed, and the price must be paid. "We cannot play around with war," he concluded. "[Even] if we give them Finland, Latvia, and Estonia . . . it will not ruin the revolution." Finally, after several narrow defeats, Lenin's motion to notify the Germans that Russia would sign their

*This date is according to the Gregorian calendar, to which Lenin's government switched on February 1/14, 1918.

terms passed the Central Committee by the narrow margin of seven to five, Trotskii having kept his promise that if the "no war, no peace" formula failed, he would support an immediate peace.[117]

Fearing to waste even a minute, Lenin insisted upon dispatching a telegram to Berlin just after midnight on February 19. "The Council of People's Commissars protests against the German Government moving troops against the Russian Soviet Republic," it began sternly, only to shift quickly to the language of capitulation. "Under the circumstances," it went on, "the Soviet of People's Commissars finds itself forced to announce its agreement to sign those conditions which were proposed by the delegations of the Quadruple Alliance at Brest-Litovsk."[118] Now it was Hoffmann's turn to delay. While his divisions advanced unopposed in what he called "the most comical war I have ever known,"[119] Lenin and Trotskii paced the halls of Smolnyi awaiting the German reply. "[Trotskii] seems to be in a devil of a hurry," Hoffmann confided to his diary. "We are not." A day passed, and his only response was that the Russian offer to capitulate must be put in writing and signed. In almost unseemly haste, Lenin dispatched a courier with the proper papers, only to be met by more silence. Now, the Russians would have to wait until the German armies had reached the objectives that would cut Estonia and Latvia off from Russia completely.

To the men at Smolnyi, the fate of their revolution seemed to hang in the balance, and that balance seemed to be tipping precariously in the wrong direction. "Yesterday, we still sat firmly in the saddle," Trotskii remembered Lenin saying after another day of silence had passed. "Today we are only holding fast to the mane."[120] In dark moments, both men feared that there would be no peace until they and their government had been obliterated. Had the foes of their revolution—the imperialists of all nations—struck a secret pact for their destruction? Certainly, Lenin and Trotskii had good reason to think so, if they remembered that they had just annulled almost fifteen billion rubles' (roughly seven billion dollars) worth of Allied war loans.[121] Now, dark specters of the world's imperialists united against them loomed before the Central Committee. Hastily, the Sovnarkom warned Russia's soldiers, workers, and peasants that, in "fulfilling the task with which it has been charged by the capitalists of all countries, German militarism wants to strangle the Russian and Ukrainian workers and peasants, to return the land to the landowners, the mills and factories to the bankers, and power to the monarchy." Summoning Russia's proletarians "to defend the Republic of Soviets against the hordes of bourgeois-imperialist Germany," Lenin and his comrades proclaimed a "revolutionary defense" of Russia "under the supervision of Red Guards." All must do their duty. All who did not, the proclamation announced in ruthless tones, "are to be shot."

For forty-eight hours, the Bolsheviks threw themselves into preparations for a last-ditch defense of Russia and their revolution. "All Soviets and revolutionary

organizations are ordered to defend every position to the last drop of blood,"
the Sovnarkom proclaimed, as its agents marshaled every able-bodied man and
woman to defend "Petrograd, Kiev, and all towns, townships, villages and
hamlets along the line of the new front."[122] "All of us, including Lenin,"
Trotskii confessed in his autobiography, "had the impression that evidently the
Germans had reached an agreement with the Entente about the destruction of
the Soviets, and that a peace built upon the bones of the Russian Revolution
was being prepared on the western front."[123] The next day, hasty offers of aid
from France, England, and the United States eased those fears, but many
Bolsheviks remained deeply suspicious. Even after Trotskii's most urgent pleadings, the Central Committee agreed to accept Allied aid by a margin of one vote,
and only with the iron-clad proviso that Russia retain absolute independence in
her foreign policy. Unable to attend the meeting, Lenin sent his vote by messenger a few minutes later. "I request," he scribbled in a brief note, "that my vote
be recorded *in favor* of taking potatoes and weapons from the bandits of Anglo-
French imperialism."[124]

Reality proved less terrifying than the Bolsheviks' imaginings, for the dilemmas posed by the Allies' promises of aid never had to be resolved. On the
morning of February 23, a courier brought Germany's new conditions, and
although they proved more brutal than those proposed at Brest Litovsk, they
nonetheless offered peace. In addition to the losses demanded at Brest-Litovsk,
Russia now must surrender all of Courland, Latvia, and Estonia, withdraw all
troops and Red Guards from Finland and the Ukraine, and recognize the
Central Rada as the legitimate government of the Ukraine. Russia had forty-
eight hours to decide.[125] By the time the courier reached Smolnyi, half that time
had already passed.

Lenin did not remind his opponents that the worst of his oft-stated fears had
come to pass, but he warned that the Germans' conditions must be accepted
without question. If his comrades refused, he insisted, he would resign from the
government and the Central Committee. Always the supreme realist, he now
painted the situation in brutally stark terms: "For a revolutionary war, we need
an army. We don't have one. That means we sign the conditions." Still, Bukharin resisted. There remained no room for theoretical meanderings, Lenin insisted bluntly. "These conditions have to be signed," he concluded. "If you don't
sign them, then you will be signing the death sentence of the Soviet government
three weeks from now. . . . Of that, I haven't even the slightest shadow of a
doubt." Stubbornly, Bukharin and three of his allies voted against peace. Four
more, including Trotskii, abstained. Six—enough to make the vote binding on
the Sovnarkom—voted with Lenin to end the war.[126]

Many more than Bukharin and his allies stood against Lenin that night. At
a joint meeting of the Petrograd Soviet and the Central Executive Committee
of the Congress of Soviets, Lenin faced jeers and curses, even from such long
and true friends as Aleksandra Kollontai, who berated him for "compromising

with imperialism." This time, Lenin spoke coldly, every word a lash biting into the hearts of the working men and women who sat before him. "You must sign this shameful peace in order to save the world revolution, in order to hold fast to . . . its only foothold—the Soviet Republic," he stated flatly. "You think that the path of the proletarian revolution is strewn with roses?" he went on. "The revolution is not a pleasure trip!" True revolutionaries must be prepared to follow the hardest roads, crawling, if need be, "through dirt and dung to communism."[127] Although some never retracted their bitter criticisms of Lenin's course that night, Trotskii was not among them. "I consider it my duty to say to this authoritative assembly," he told a joint meeting of the supreme organs of the Soviet government some seven months later, "that, in that hour when many of us, myself included, wavered about whether or not to sign the Peace of Brest-Litovsk, only Comrade Lenin, with stubbornness and incomparable intuition, stood against many of us. . . . And now, we must admit that we were wrong."[128]

There remained only to sign the treaty. Ironically, none of the leading diplomats who had traded verbal thrusts during December and January reappeared at Brest-Litovsk for this final act. Hoping to convince the Germans of a dramatic shift in Russian policy, Trotskii had resigned as commissar of foreign affairs,[129] leaving Grigorii Sokolnikov, future people's commissar for finance and Stalin's ambassador to Great Britain, to lead the Russian delegation. Count Czernin and Kühlmann had gone to Bucharest to negotiate a separate peace with Russia's former Rumanian ally, leaving men of lesser rank to complete what they had begun. Over them all loomed the shadow of General von Hoffmann, voice of the High Command and apostle of a peace negotiated at gunpoint.

To Sokolnikov's request that their arrival on the afternoon of February 28 mark the end of hostilities, Hoffmann firmly refused. Only when the treaty had been signed would he halt his divisions. Then the Germans proposed that they discuss the treaty's provisions. Still a few months away from his thirtieth birthday, but in no way intimidated by the situation in which he found himself, Sokolnikov coldly insisted that discussions were not necessary. Russia, he explained, had come to sign, not negotiate. Therefore, he would agree to no discussions and to hear no explanations. Announcing that this was "a peace which Russia, grinding its teeth, is forced to accept," he declared that "we are going to sign immediately the treaty presented to us as an ultimatum but at the same time we refuse to enter into any discussion of its terms."[130] No effort on the part of Germany's representatives could sway him from his insistence that this was not a negotiated peace and that, as such, its signing bore no discussion. At 5:50 p.m. on March 3, 1918, Sokolnikov had his way.[131] Peace between Russia, Germany, Austria-Hungary, Bulgaria, and Turkey had come.

The Germans had been every bit as brutal as Lenin had feared, but he now had the peace the Bolsheviks needed to survive. Russians "must measure to the very bottom that abyss of defeat, dismemberment, enslavement, and humiliation into which we have now been pushed," he insisted, but he nonetheless saw more cause for hope than for despair. "Our natural wealth, our manpower and the splendid impetus which the great revolution has given to the creative powers of the people are ample material to build a truly mighty and abundant Russia," he promised. To do so, Russians must borrow from the Germans, just as they had done in the long-ago days of Peter the Great. "Yes, learn from the Germans!" Lenin exclaimed. "That is just what the Russian Soviet Socialist Republic requires in order to cease being wretched and impotent and become mighty and abundant for all time."[132] Others shared his view. "Keep in step with the revolution's pace! / We've an indefatigable foe to face!" Aleksandr Blok wrote in "The Twelve," the most famous poem to come out of the Russian—or any other—Revolution, which appeared on the day Sokolnikov's delegation returned. "Forward, forward, always forward," Blok urged. "Working people ever onward!"[133]

EPILOGUE

Burned when the retreating Russian armies had evacuated it in July 1916, the blackened timbers of Brest-Litovsk's ravaged buildings stood in stark contrast to the whiteness of the late winter snow as Russia's peace delegation prepared to return to Petrograd. As Sokolnikov and his comrades made their way northeastward that March, they passed the newly-fallen city of Dvinsk, Russian just a fortnight ago but now held by Germany's occupying armies. Beyond Dvinsk, the war's devastation faded quickly, for the Germans had moved too swiftly and the Russians had offered too little resistance for much destruction to have been accomplished during the twelve-day February war. Further to the northeast, the white limestone walls of Pskov, medieval Russia's gateway to the West, rose above the confluence of the rivers Pskova and Velikaia. Just a year and a day before, Pskov had witnessed the tsar's final defeat. Now it marked the point of the Germans' farthest advance in the northeast. No German conqueror had come to Pskov since Aleksandr Nevskii, Prince of Novgorod the Great and Grand Prince of Russia, had driven the Teutonic knights back in defeat across the lake's crumbling spring ice on April 5, 1242. Now, for the first time in history, German lords trod proudly upon the eastern shore of Pskov's Lake Peipus, able to look westward across more than a thousand miles of waters and lands that lay in German hands.

In the region Sokolnikov's party entered beyond Pskov, the ruin was of a different kind, less immediately obvious but no less complete. Here Russia's revolutionary canker had consumed both moral fiber and surface flesh, leaving behind devastation that could not be erased by new buildings and more material comforts. Russia's new revolutionary order had yet to find the institutions and values needed to anchor itself firmly in time and space. War had rendered life cheap; the revolution now made it transient. Everyone in Russia seemed to be on the move across the vast Eurasian land mass in those days, crammed into railway cars that moved for hours and halted for days. Some were forced to cling to any part of the train that offered a handhold, "hanging on like bunches of grapes," one observer remarked, and often dying from exposure in the subzero weather, but desperate to move nonetheless.[1] Generally, the wave flowed eastward, its major direction given by soldiers leaving the front, anxious to resume the life they had left behind, searching for that narrow path which, the peasant proverb had promised, would lead them home once again.

Perhaps the compulsion to move served as an antidote to the need to stand, for waiting had become a way of life in revolutionary Russia. Although the revolution had brought shortages of everything else, it seemed to have created a surplus of time. Long, dark, serpentine queues stretched along the white snow of revolutionary Russia's city streets. As winter deepened, the queues lengthened, a sure sign that shortages had grown even more desperate. "For queue work only" notices appeared in the want ads of city newspapers, as people with nothing to sell but their time tried to reap some profit from the long hours spent in waiting. Then, as now, Russian queue etiquette dictated that women with infants receive preferential treatment. Soon, women began to rent infants and small children to take with them so that they could be served ahead of the others.[2] More and more, people simply stood, without complaint, often without conversation or comment. The future had slipped entirely out of focus, perhaps because it seemed so ominous as the Bolsheviks began to hunt down their enemies. "All roads now lead to prison," one of their intended victims wrote that winter. "I am tired, exhausted, partly with work and excitement, partly with hunger."[3]

In those days, no place displayed the anomalies of revolutionary life more dramatically than Petrograd. Russia's beautiful capital—the monument to the Romanov master builders that once had stood, in a poet's words, "steadfast like Russia herself"[4]—had gone to ruin. To some, it seemed that the city of Peter and Catherine the Great was dying.[5] Gippius had called it a "frightful, never-before-seen Petersburg" that fall.[6] Rarely was the snow cleared from the city's streets now, although the Bolsheviks made an effort to marshal the former propertied classes to do the work. "The Princess Obolenskaia has been ordered to go and clear snow off the Fontanka Quay," a dismayed Comte de Robien reported at the beginning of February 1918. "Others have to sweep the tram lines at night."[7] Everywhere, the contradictions that always had made Russia such an enigma to western observers grew more prevalent and more complex. "Careless, charming, infuriating Russia, where women who have nothing to eat wear necklaces of priceless pearls," de Robien exclaimed. "This is Russia in a nutshell, where the cold reaches twenty degrees [below zero] and there are no covered cabs!"[8]

But there was no contradiction in the way that the law of the jungle ruled Russia's cities at night, as bandits roamed darkened streets armed with rifles, hand grenades, and even machine guns. In Petrograd alone, on any given night in January 1918, according to one report, an average of five hundred private homes and apartments and another three hundred shops fell victim to their attacks.[9] Even high rank among the Bolsheviks offered little protection. One night, bandits stopped the sleigh of the chief of Petrograd's Cheka, and allowed him to go on his way only after they had stripped him of everything including his clothing.[10] Not even Lenin escaped. Once on his way to an orphans' New Year's party at the forest school that Krupskaia had organized in Sokolniki, he was waylaid by bandits who seized his car and left him, his sister, his driver, and his bodyguard to make their way as best they could on foot.[11]

A way had to be found to bring Russia out of the chaos into which the untried road to proletarian revolution had taken her. For her masses, freedom at first had meant emancipation from the constraints of authority, whether in the army, the factory, on the farm, or in society at large, and for the better part of a year, they had reveled in their new-found liberation. Now, if Russia and the Russians hoped to survive, food, fuel, and raw materials had to be produced, transportation had to be restored, factories had to be run, cities made to function, and an army rebuilt. Discipline and order must return if Russia were not to collapse into chaos. "Learn discipline from the Germans," Lenin insisted at the Extraordinary Seventh Congress of the Russian Communist Party at the beginning of March. "If we do not, we, as a people, are doomed [and] we shall live in eternal slavery. . . . We should have but one slogan," he concluded. "Learn the art of war properly and put the railways in order. We must produce order."[12]

Perhaps the greatest truth Lenin had spoken to Russia's toilers during the hard days of February's twelve-day war had been his warning that revolution's path was not "strewn with roses," but led instead "over thorns and briars." Lenin had returned to Russia in April 1917 triumphantly bearing the slogan "Peace, land, and bread," to launch the Bolsheviks upon the course that had brought them to power in October. But their victory did not immediately transform the lives of Russia's workers and peasants into the better world that Lenin had seemed to promise during the previous spring and summer. That world, he now explained, lay in the future, perhaps far in the future, and much labor and great sacrifice would be needed before the promise could become a reality. World revolution was "for the time being . . . a very beautiful fairy tale." Sternly, Lenin now asked the party faithful at the Extraordinary Seventh Congress: "Is it proper for a serious revolutionary to believe in fairy tales?" Somewhat softening his tone, he added, "one may dream about . . . revolution on a world-wide scale, for it will come," but one must never again confuse the dream with the reality. "An epoch of most grievous defeats is ahead of us," he concluded. "It is with us now."[13]

Certainly it was. During 1917, the value of industrial goods produced in Russia had fallen by one-half, while the productivity of her factory workers declined by two-thirds. Coal production fell by almost a third in 1917, and half of the output of her mines could not be moved because her rail network had crumbled into a tangle of slovenly-maintained rolling stock and ill-kept track. Production of iron and steel dropped, as did Russia's output of oil. According to one report at the beginning of 1918, the price of gasoline in Petrograd (now available for nongovernment cars only on the black market) had risen by 25,000 percent since 1913. Shortages of fuel and raw materials closed two out of every five Moscow mills that worked with iron and steel and idled a quarter of Russia's blast furnaces before the end of 1917.[14]

To repair the damage of 1917's revolutionary chaos required a great deal more than setting ruined machinery to rights, restoring neglected railroads, getting workers back on the job, and increasing production, for the Brest-Litovsk Treaty had deepened dramatically the economic damage wrought by Russia's forty-four months of war and revolution. Brest-Litovsk cost Russia almost two million square kilometers of territory, over sixty-two million people, nine thousand factories, mills, and refineries, a third of her crops, 80 percent of her sugar refineries, 73 percent of her iron mines, and 75 percent of her coal fields. Put another way, the state that the Bolsheviks claimed to rule when they transferred Russia's capital to Moscow in mid-March included only two-thirds of the former Russian Empire's arable land, a seventh of its sugar beet fields, two-thirds of its industrial enterprises, a mere quarter of its coal mines, just more than a quarter of its iron foundries, and less than three-fifths of its former population.[15]

Moreover, the "grievous defeats" of which Lenin had spoken at the Seventh Congress were not to be confined to the terrible losses imposed by the Brest-Litovsk Treaty. For the next three years the Bolsheviks' seizure of power would be challenged from many sides within Russia herself, in bloody uprisings all over her outlying regions, away from the cities where the Bolsheviks had their greatest strength. A major center of protest developed in the south, in the region that had been for centuries the homeland of Russia's Cossack legions. The Territory of the Don, which centered upon those fertile lands where the Don River flowed to the Sea of Azov and into the Black Sea beyond, had little room for Bolsheviks in the late fall of 1917 except in the large industrial city of Rostov, where urban proletarians supported Lenin's program. For hundreds of miles in every direction around Rostov, General Aleksei Kaledin, the seemingly shy, mild-mannered hetman of the Don Cossacks who had spoken out so fiercely on behalf of order and strong government at the Moscow State Conference in August 1917, held sway. At the beginning of November, the men who had sided with him in Moscow began to make their way south to his headquarters in Novocherkassk, a city famous for its sparkling wines, to offer their support against the Bolsheviks. One of the first to arrive was General Mikhail Alekseev, "retired" after serving for eleven days as *Stavka*'s chief of staff in the wake of the Kornilov "rebellion."

Within days after the fall of the Winter Palace, Alekseev, now suffering the ravages of the intestinal cancer that would take his life in less than a year, quietly had slipped away from Petrograd to Novocherkassk, where he intended to build an army to march against the Bolsheviks.

Within a month, the Kadets' steely and professorial Pavel Miliukov also arrived, still rigidly insistent that any discussion of the social and economic reforms that Russia's masses demanded must wait until political order had been established. "Fat Rodzianko" came also—Rodzianko, whose pleadings for a cabinet made up of reliable men responsible to the Duma Nicholas II had dismissed with such ill-advised scorn. Then, men and women all across Russia had known him as the president of the Duma, who summoned Russians to take up the burdens of civic responsibility and to fight on for final victory over the armies of Germany and Austria. Now, his influence in Russia's capitals had vanished, and Rodzianko, the man with the huge body and stentorian voice, had been obliged to make the long and arduous journey to Novocherkassk disguised as a cripple confined to a wheel chair.

The February Revolution had raised a barrier that separated men like Miliukov and Rodzianko from those who remained dedicated monarchists. That barrier evaporated at the beginning of November as monarchists and moderate republicans set their differences aside to stand together against their common Bolshevik enemies. Sometimes singly, sometimes in groups, they came to Novocherkassk—politicians, noblemen, officers, cadets, and a few old soldiers—all drawn to the Territory of the Don, where, General Denikin later wrote, "the names of leaders whom popular legend linked with the Don shone as a beacon amid the surrounding gloom."[16] At Alekseev's urging, they began to form the Volunteer Army, the first of the "White" armies that took shape along Russia's borders to challenge the Bolsheviks' hold upon the Russian land. Always their numbers were small, and they often fought with weapons they begged, borrowed, or stole, but their quality was extremely high. After facing them in battle, Bolshevik commanders often overestimated their numbers by several times their real strength.

None among these first White army volunteers proved to be more dedicated in their hatred of the Bolsheviks than the men from Bykhov prison. For almost two months after they had surrendered to Kerenskii's government, General Kornilov and the men who had stood with him in his "rebellion" had been incarcerated at Bykhov, where they had lived in relative comfort protected by Kornilov's loyal *Tekintsy.* Escape would have been a simple matter in those days, but, as one of them explained, "we wished to have our day in court."[17] Thus, they had waited through September and October, confident that the commission of inquiry appointed to look into their "rebellion" would exonerate them. The Bolsheviks' victories had shattered their hopes for justice. Certain that they could not remain at Bykhov, Kornilov decided to march south with his *Tekintsy,* while he urged Denikin, Lukomskii, Romanovskii, and Markov—the four senior generals with whom he had shared his prison—to travel separately. They would meet at Novocherkassk. From there they would join the campaign against the Bolsheviks.

With the elegant, aristocratic, and fastidious Markov brashly swaggering in the arrogant manner of a private soldier whom the revolution had liberated from all constraints of military discipline, the more nondescript Romanovskii masquerading as an ensign in the engineers, the unfailingly military Lukomskii posing as a civilian, and Denikin, the son of a Russian serf, posing as a Polish official, the four generals began their separate journeys, each arriving safely in Novocherkassk without incident. Kornilov arrived last, dirty and dressed in rags, on December 6, the six-week anniversary of the Bolsheviks' victory in Petrograd. He had traveled part of the way with his *Tekintsy* and had fought through several Bolshevik ambushes. After that, he had decided to go on alone in the disguise of a poor peasant and had traveled for the better part of three weeks through hostile territories before he reached Kaledin's headquarters. With Kornilov's arrival, the formation of the Volunteer Army began in earnest.[18]

During the next three years, a half-dozen more White armies arose in Russia's borderlands to challenge the republic of workers and peasants that Lenin and the Bolsheviks had created. Intervention by fourteen nations added to the conflict's bitterness and complexity, as statesmen fearful that Lenin's promised world revolution might come to pass sent troops or supported rival governments that the Whites had established in parts of Siberia, the Far North, the Crimea, the Kuban, and the Ukraine. In the spring of 1918, the Germans marched triumphantly into the Ukraine, while the Turks moved further into the Transcaucasus. England sent forty thousand troops to the White Sea lands around Murmansk and Arkhangelsk, while another forty thousand former Czech prisoners of war from the tsarist era formed the famous Czech legion that stormed the Siberian center of Ekaterinburg a few days after the Bolsheviks there executed Nicholas, Aleksandra, and their children. Ten thousand Americans and a handful of Italians followed sixty thousand Japanese into Siberia, while the French, the Greeks, and the English sent military units to aid the anti-Bolshevik forces in South Russia. Movements for national independence in the Ukraine, Georgia, Armenia, and the Baltic provinces added to the fighting, and the Poles rounded off the conflicts of those years by attempting to wrest parts of the Ukraine and White Russia from their battered Bolshevik adversaries.

All across the Russian land killing raged, as brothers fought brothers, and sons fought fathers, leaving regions twice and thrice devastated. Never had the Great War seen such raw cruelty, as Reds and Whites tortured and killed each other with unmatched callousness. "A few days ago, three sealed wagons with the inscription 'fresh meat, destination Petrograd' arrived at one of the Petrograd stations," de Robien wrote in his diary on January 26, 1918. "When the wagons were opened," he continued, "they were found to be filled with piles of stiffened corpses of Red Guards, covered with frozen blood, with grimacing faces, placed in obscene positions."[19] De Robien's was but one of many accounts of brutality to emerge from Russia's terrible civil war. On another occasion, General Wrangel ordered 370 Bolshevik prisoners to be shot on the spot just to show others what fate men could expect if they fought against his armies.[20] When Kornilov was killed by an enemy shell in April, Bolsheviks unearthed his body and spent two days dragging it through the streets of Ekaterinodar, mangling it and hanging it from tree limbs before they burned the broken corpse at a local slaughter house.[21] Another group of White officers met their deaths at the bottom of the harbor at Novorossiisk. Some months later, a diver found them, according to General Denikin's account, "livid, greenish, swollen, mangled corpses kept upright owing to the weights tied to their legs [so that they] stood in serried ranks, swaying to and fro, as if talking to one another."[22]

Their passage through the Armageddon of the Great War ended, Russians now faced a new struggle, not of great set-piece battles between opposing armies that counted their numbers in the hundreds of thousands, but of smaller more brutal encounters between desperate men who fought not for land, or raw materials, or foreign markets, or even for national pride. Men now killed for principles that transcended the frontiers of nations. Bolsheviks fought to bring into being a new order of freedom and equality that they conceived of as being founded on forms of social and economic justice that capitalist societies could not allow to exist. Against them stood men who thought that freedom and prosperity could come only from the free reign of precisely those principles that the Bolsheviks opposed. No longer could the laws that had governed the conduct of men and nations at war restrain them, for they struggled not for territory, but for a vision of the world to come that could exist only at the expense of the other's annihilation.

Thus, as the Bolshevik enemies of old Russia fought to free themselves from its grip, the defenders of capitalistic, propertied Russia struggled to eradicate the revolutionary ideals that had taken root in their midst. "A revolutionary class . . . must crush [the propertied classes]," Lenin once wrote. "We shall suppress the resistance of the propertied [classes] with the same means by which the propertied [classes] suppressed the proletariat. Other means have not been invented."[23] The reply of the propertied classes had been penned nearly three-quarters of a century before, as the revolutionary tide of 1848 had swept eastward across Europe, and its truth applied with even greater force now, as civil war cast its terrible shadow across the Russian land. "For a long time, only two real forces have existed in Europe—Revolution and Russia," the poet-diplomat Fëdor Tiutchev had written then. "No treaties are possible between them. The existence of one means the death of the other."[24]

NOTES

Prologue

1. Quoted in Oron J. Hale, *The Great Illusion, 1900–1914*, p. 3.
2. Winston S. Churchill, *The World Crisis, 1911–1914*, p. 188.
3. Captain B. H. Liddell Hart, *The Real War, 1914–1918*, p. 35.
4. I. S. Bloch [Bliokh], *The Future of War in Its Technical, Economic, and Political Relations*, p. 6. Parts of this chapter contain, in an abbreviated form, source materials first used in my book *In War's Dark Shadow: The Russians Before the Great War.*
5. Valerii Briusov, "Griadushchie gunny," in his *Sobranie sochinenii v semi tomakh*, I, p. 433.
6. Aleksandr Blok, introduction to "Vozmezdie," in his *Sobranie sochinenii v shesti tomakh*, III, p. 188.
7. A. A. Brusilov, *Moi vospominaniia*, p. 62; *Letters of Tsar Nicholas and Empress Marie*, p. 139.
8. K. Sidorov (ed.), "Bor'ba so stachechnym dvizheniem nakanune mirovoi voiny," pp. 96–97.
9. N. V. Gogol, "Mertvye dushi," in his *Sobranie sochinenii*, V, p. 259.
10. N. V. Riasanovsky, *Nicholas I and Official Nationality in Russia, 1825–1855*, p. 85.
11. M. P. Pogodin, *Istoriko-politicheskie pis'ma i zapiski v prodolzhenii Krymskoi Voiny, 1853–1856gg*, pp. 3, 254.
12. Major-General Sir Alfred Knox, *With the Russian Army, 1914–1917*, I, p. xxi.
13. M. T. Florinsky, *Russia*, II, p. 1230. This assumes a European population of 443,500,000 (see Karl Baedeker, *Russia, 1914*, p. xxxiii).
14. Florinsky, *Russia*, p. 1231.
15. General Basil Gourko [Vasilii Gurko], *War and Revolution in Russia, 1914–1917*, p. 120.
16. "Commerce entre l'Allemagne et la Russie depuis le debut de la guerre." Rapport

de Heilmann, Consul de France à Riga au Ministère des Affaires Etrangères, in "Guerre 1914–1918. Russie," pp. 294–305.

17. General Iu. N. Danilov, *Rossiia v mirovoi voine, 1914–1915gg.*, pp. 256–258; N. N. Golovine, *The Russian Army in the World War*, pp. 32, 126–127; N. Iakovlev, *1 Avgusta 1914*, pp. 33–37; Norman Stone, *The Eastern Front, 1914–1917*, pp. 30–32; E. Barsukov, *Podgotovka russkoi armii k voine v artilleriiskom otnoshenii*, pp. 56–57.

18. Brusilov, *Moi vospominaniia*, p. 63.

19. A. M. Zaionchkovskii, *Podgotovka Rossii k imperialisticheskoi voine*, pp. 123–124.

20. Knox, *With the Russian Army*, I, p. xxxiii.

21. *Ibid.*, pp. 124–125; General N. N. Golovin, *Voennye usiliia Rossii v mirovoi voine* I, pp. 56, 61.

22. S. K. Dobrorolskii, *Die Mobilmachung der russischen Armee 1914*, p. 28; V. S. Diakin *et al.*, eds., *Istoriia rabochikh Leningrada*, I, pp. 184–185.

23. Field Marshal Paul von Hindenburg, *Out of My Life*, p. 48.

24. General Friedrich von Bernhardi, *Germany and the Next War*, pp. 11, 14, 37, 27, 258, 105–106, 287, 114, 154.

25. Quoted in Barbara Tuchman, *The Proud Tower*, pp. 379–380.

26. Quoted in Prince Bernhard von Bülow, *Memoirs*, I, p. 418.

27. See, for example, Bülow, *Memoirs*, II, pp. 355.

28. Quoted in Barbara Tuchman, *The Guns of August*, p. 23.

29. Lord Thomas Newton, *Lord Lansdowne*, p. 199.

30. S. Iu. Vitte [Witte], *Vospominaniia*, II, p. 292; S. Iu. Witte, *The Memoirs of Count Witte*, p. 189.

31. Quoted in *Nikolai II: Materialy dlia kharakteristiki lichnosti i tsarstvovaniia*, p. 23.

32. Alexander, Grand Duke of Russia, *Once a Grand Duke*, p. 61.

33. Quoted in S. G. Sviatikov, *Obshchestvennoe dvizhenie v Rossii, 1700–1895*, II, p. 197.

34. Florinsky, *Russia*, II, pp. 1141–1142.

35. Quoted in Alexander, *Once a Grand Duke*, pp. 168–169.

36. Aleksandra's jottings in Nicholas's diary are quoted in Sir Bernard Pares, *The Fall of the Russian Monarchy*, p. 36. See also Aleksandra's letter to Nicholas, 12 November 1916, in *Letters of the Tsaritsa to the Tsar, 1914–1917*, p. 441.

37. These quotations are taken from Aleksandra's letters to Nicholas, dated 20 September 1914, 22 August 1915, 28 August 1915, 9 September 1915, 4 December 1916, and 13–14 December 1916, in *Letters of the Tsaritsa to the Tsar*, pp. 3, 114, 127, 152, 442, and 454–455, and from Nicholas's letter to Aleksandra of 14 December 1916, in *Letters of the Tsar to the Tsaritsa*, pp. 307–308.

38. *Nikolai II: Materialy*, pp. 60–62; Harrison Salisbury, *Black Night, White Snow*, pp. 205, 657; Alexander, *Once a Grand Duke*, pp. 181–182.

39. Quoted in Pares, *Fall of the Russian Monarchy*, p. 129.

40. Pierre Gilliard, *Thirteen Years at the Russian Court*, p. 83; Anna Viroubova [Vyrubova], *Memories of the Russian Court*, p. 23.

41. The quote about the first meeting of Nicholas and Aleksandra with Rasputin in the previous paragraph comes from Nicholas's diary and is quoted in Salisbury, *Black Night, White Snow*, p. 176. The information about hemophilia comes from Robert K. Massie, *Nicholas and Alexandra*, p. 190.

42. Quoted in Pierre Gilliard, *Le destin tragique de Nicholas II et de sa famille*, p. 93.

43. A. N. Kuropatkin, "Dnevnik A. N. Kuropatkina," II, p. 106.

44. Quoted in Florinsky, *Russia*, II, p. 1285.

45. Bülow, *Memoirs*, II, p. 153.

46. *Ibid.*, p. 153.

47. Vitte, *Vospominaniia*, II, p. 477.

48. This theme runs through all three volumes of Maurice Paléologue, *La Russie des Tsars Pendant la Grande Guerre*, and begins to appear in his official reports in the spring of 1915; see "Télégramme de M. Paléologue au Ministère des Affaires Etrangères," (Secret), Petrograd, 15 mai 1915, and "Télégramme secret de M. Paléologue au Ministère des Affaires Etrangères," Petrograd, le 26 juillet 1915.

49. Quoted in Churchill, *World Crisis*, p. 195.

50. G. P. Gooch, *Before the War*, II, p. 345.

51. Quoted in Sidney B. Fay, *The Origins of the World War*, I, pp. 445–446.

52. Luigi Albertini, *The Origins of the War of 1914*, II, pp. 74–88; R. W. Seton-Watson, *Sarajevo*, pp. 70–75; and Lincoln, *In War's Dark Shadow*, pp. 422–424.

53. See Lincoln, *In War's Dark Shadow*, pp. 421–426, for a brief discussion of these events, which are covered in exhaustive detail in Seton-Watson, *Sarajevo*, and Vladimir Dedijer, *The Road to Sarajevo*.

54. Sir Maurice de Bunsen to Sir Edward Grey, Vienna, 5 July 1914, in G. P. Gooch and Harold Temperley, eds., *British Documents on the Origins of the War, 1898–1914*, XI, p. 32.

55. Viscount Edward Grey, *Twenty-Five Years, 1892–1916*, I, p. 209.

56. Quoted in Albertini, *Origins of the War*, II, p. 123.

57. Quoted in Dwight F. Lee, *Europe's Crucial Years*, pp. 378–379, and Albertini, *Origins of the War*, II, p. 138.

58. Quoted in N. P. Poletika, *Vozniknovenie pervoi mirovoi voiny (iiul'skii krizis 1914g.)*, p. 49.

59. Quoted in Lee, *Europe's Crucial Years*, p. 395.

60. Quoted in Sidney B. Fay, *The Origins of the World War*, II, p. 337.

61. Danilov, *Rossiia v mirovoi voine*, p. 14.

62. Quoted in Albertini, *Origins of the War*, II, p. 558.

63. Quoted in *ibid.*

64. S. D. Sazonov, *Vospominaniia*, pp. 242–243; M. V. Rodzianko, *Khrushenie imperii*, p. 94.

65. Sazonov, *Vospominaniia*, pp. 248–249.

66. *Ibid.* See also "Zapiski Sverbeeva," in "Nachalo voiny 1914g. Podennaia zapis' b. Ministerstva Inostrannykh Del," pp. 28–30.

67. Quoted in "Zapiski Sverbeeva," in "Nachalo voiny 1914g.," p. 31.

68. Quoted in Pierre Renouvin, *The Immediate Origins of the War*, p. 245.

69. For a summary of Schlieffen's plan, see Tuchman, *Guns of August*, pp. 33–44. The details of the plan, its evolution, and the changes inserted during the two decades before the Great War put it to the final test are covered in Gerhard Ritter, *The Schlieffen Plan*.

70. Sazonov, *Vospominaniia*, p. 151.

71. *Ibid.*, pp. 257–260.

72. V. A. Sukhomlinov, "Rossiia khochet mira, a gotova k voine," pp. 4, 6.

73. Paléologue, *La Russie des Tsars*, I, p. 44.

Chapter I: "It Is a Wide Road That Leads to the War"

1. Grand Duchess Maria Pavlovna, *Education of a Princess*, p. 162.

2. Viroubova [Vyrubova], *Memories*, p. 104.

3. Paléologue, *La Russie des Tsars*, I, pp. 45–46.

4. Quoted in Mikhail Lemke, *250 dnei v tsarskoi stavke (25 sentiabria 1915–2 iiulia 1916)*, p. 7.

5. Knox, *With the Russian Army*, I, p. 39.

6. Maria Pavlovna, *Education of a Princess*, p. 163.
7. Paléologue, *La Russie des Tsars*, I, pp. 46–47.
8. Pares, *Fall of the Russian Monarchy*, pp. 187–188.
9. Maria Pavlovna, *Education of a Princess*, p. 163.
10. General A. I. Spiridonovich, *Velikaia voina i fevral'skaia revoliutsiia, 1914–1917gg.*, I, pp. 13–14; Paléologue, *La Russie des Tsars*, I, pp. 51–52.
11. Quoted in K. Mochul'skii, *Andrei Belyi*, pp. 42, 191.
12. Quoted in Wiktor Woroszylski, *The Life of Mayakovsky*, p. 132.
13. Zinaida Gippius, *Siniaia kniga*, pp. 9, 12; Andrei Belyi, *Nachalo veka*, pp. 173–174.
14. Paléologue, *La Russie des Tsars*, I, p. 74.
15. Quoted in P. N. Miliukov, *Vospominaniia, 1859–1917*, II, p. 190.
16. Danilov, *Rossiia v mirovoi voine*, pp. 110–111.
17. Miliukov, *Vospominaniia*, II, p. 190.
18. *Ibid.*, pp. 183–184.
19. *Ibid.*, p. 184; Viktor Chernov, *Rozhdenie revoliutsionnoi Rossii (fevral'skaia revoliutsiia)*, p. 75.
20. Quoted in Pierre Gilliard, *Le destin tragique*, p. 91.
21. Quoted in Miliukov, *Vospominaniia*, II, p. 184.
22. Quoted in *ibid.*
23. Pares, *Fall of the Russian Monarchy*, p. 219.
24. Paléologue, *La Russie des Tsars*, I, p. 106.
25. Brusilov, *Moi vospominaniia*, pp. 82–83.
26. Quoted in Florinsky, *Russia*, II, p. 1378.
27. Quoted in Tuchman, *Guns of August*, p. 349.
28. *Niva*, XLV, No. 32 (9 avgusta 1914), pp. 639–640g; Paléologue, *La Russie des Tsars*, I, pp. 63–65; Miliukov, *Vospominaniia*, II, pp. 189–191; Florinsky, *Russia*, II, pp. 1368–1369.
29. Gilliard, *Le destin tragique*, p. 91.
30. "O nezhelatel'nosti dal'neishago rosta promyshlennosti v samom gorode S.-Peterburge, v sviazi s politicheskami manifestatsiami i stachkami rabochikh," iiul' 1914g., sheets 2–3; O. Chaadaeva, ed., "Iiulskie stachki i demonstratsii 1914g.," pp. 139, 147, 149–155; "Bor'ba so stachechnym dvizheniem nakanune mirovoi voiny," p. 96.
31. N. N. Golovin, *Iz istorii kampanii 1914 goda na russkom fronte: Nachalo voiny*, pp. 85–86; N. Iakovlev, *1 avgust 1914*, p. 28.
32. Paléologue, *La Russie des Tsars*, I, p. 58.
33. Sir George Buchanan, *My Mission to Russia*, I, p. 215.
34. General N. N. Golovin, *Voennye usiliia Rossii v mirovoi voine*, I, pp. 86–87.
35. V. A. Sukhomlinov, "Dnevnik Generala Sukhomlinova," entries for 21 and 23 July (o.s.), p. 221.
36. Buchanan, *My Mission*, I, p. 214.
37. Quoted in Paléologue, *La Russie des Tsars*, I, p. 85.
38. *Ibid.*, p. 87.
39. "Posledniaia voina v istorii Evropy," p. 620v.
40. Paléologue, *La Russie des Tsars*, I, pp. 85, 89, 87.
41. *Ibid.*, pp. 84–88; Buchanan, *My Mission*, I, pp. 214–215; Major General Dubenskii (ed.), *Ego Imperatorskoe Velichestvo Gosudar' Imperator Nikolai Aleksandrovich v deistvuiushchei armii*, I, pp. xiii–xvi.
42. Sir Bernard Pares, *Day by Day with the Russian Army, 1914–1915*, p. 10.
43. Quoted in Tuchman, *Guns of August*, p. 348.
44. Aleksandra to Nicholas, 24 September 1914, *Letters of the Tsaritsa to the Tsar*, p. 9.

45. Brand Whitlock, *Belgium*, I, pp. 151–219; Hugh Gibson, *A Journal from Our Legation in Belgium*, pp. 198–270.

46. Paléologue, *La Russie des Tsars*, I, p. 55.

47. *Materialy po istorii franko-russkikh otnoshenii za 1910–1914gg.*, pp. 698–703. See also *Manevrennyi period 1914 goda: Vostochno-Prusskaia operatsiia*, p. 9; V. A. Emets, "O roli russkoi armii v pervyi period mirovoi voiny 1914–1918 gg.," pp. 61–62; N. N. Golovin, "The Russian War Plan of 1914," p. 571.

48. Quoted in Major-General Sir Edmund Ironside, *Tannenberg*, p. 43.

49. Quoted in Paléologue, *La Russie des Tsars*, I, p. 56.

50. Sukhomlinov, "Dnevnik," p. 221.

51. Aleksandra to Nicholas, 16 June 1915, in *Letters of the Tsaritsa to the Tsar*, p. 97. For the grand duke's comments to Rasputin (his actual words were: "Set one foot in here and I'll hang you!"), see A. Bubnov, *V tsarskoi stavke*, p. 37.

52. General A. Mosolov, *Pri dvore Imperatora*, p. 72.

53. Quoted in Pares, *Day by Day with the Russian Army*, pp. 17–18.

54. Golovin, *Iz istorii kampanii 1914 goda: Nachalo voiny*, p. 231; Stone, *The Eastern Front*, pp. 20–25; A. Kersnovskii, *Istoriia russkoi armii*, III, pp. 637–638.

55. See Grand Duke Andrei Vladimirovich, *Dnevnik za 1915g.*, pp. 49–50.

56. Quoted in Golovin, *Iz istorii kampanii 1914 goda: Nachalo voiny*, I, p. 231.

57. Brusilov, *Moi vospominaniia*, p. 76; Knox, *With the Russian Army*, I, p. 42.

58. Gourko, *War and Revolution in Russia*, p. 8; Knox, *With the Russian Army*, I, p. 42.

59. Zaionchkovskii, *Podgotovka*, p. 200.

60. Quoted in I. I. Rostunov, *Russkii front pervoi mirovoi voiny*, p. 91.

61. *Ibid.*, pp. 200–230; A. L. Sidorov, *Finansovoe polozhenie Rossii v gody pervoi mirovoi voiny (1914–1917)*, pp. 53–57; Stone, *The Eastern Front*, pp. 32–34.

62. Zaionchkovskii, *Podgotovka*, pp. 256–280; A. L. Sidorov, "Iz istorii podgotovki tsarizma k voine," pp. 120–150; Sergei Dobrorol'skii, "Nashi strategicheskie shansy v 1914 godu," pp. 20–38; Stone, *The Eastern Front*, pp. 34–36.

63. P. A. Zaionchkovskii, *Samoderzhavie i russkaia armiia na rubezhe XIX–XX stoletii*, pp. 168–248, 333; H. P. Stein, "Der Offizier des russischen Heeres im Zeitabschnitt zwischen Reform und Revolution, 1861–1905," pp. 380–410.

64. Stone, *The Eastern Front*, pp. 25–32. For Sukhomlinov's view of this conflict, see V. Sukhomlinov, *Vospominaniia*, pp. 308–312.

65. Stone, *The Eastern Front*, p. 58.

66. Golovin, "Russian War Plan of 1914," p. 573.

67. *Ibid.*, pp. 566–568.

68. Iu. N. Danilov, *Velikii kniaz' Nikolai Nikolaevich*, pp. 125–129; Golovin, *Iz istorii kampanii 1914 goda: Nachalo voiny*, pp. 73–77; Rostunov, *Russkii front*, pp. 109–110.

69. Rostunov, *Russkii front*, pp. 109–112; Stone, *The Eastern Front*, p. 48.

70. A. A. Manikovskii, *Boevoe snabzhenie russkoi armii v mirovuiu voinu*, I, pp. 119–121, 147–150, 152–157.

71. Golovin, "War Plan of 1914," p. 573.

72. Rostunov, *Russkii front*, pp. 97–98.

73. *Ibid.*, pp. 99–100.

74. Danilov, *Rossiia v mirovoi voine*, p. 153.

75. Gourko, *War and Revolution in Russia*, p. 35.

76. Danilov, *Rossiia v mirovoi voine*, pp. 191–192; Rostunov, *Russkii front*, pp. 100–101.

77. N. Zhitkov, "Prodfurazhnoe snabzhenie russkikh armii v mirovuiu voinu," pp. 65–81; Rostunov, *Russkii front*, pp. 100–101; Golovin, *Voennye usiliia Rossii*, II, pp. 69–71.

78. Stone, *The Eastern Front*, p. 51; Knox, *With the Russian Army*, I, p. xxv.
79. Quoted in Golovin, *Iz istorii kampanii 1914 goda: Nachalo voiny*, p. 111.
80. Gourko, *War and Revolution in Russia*, p. 37.
81. *Ibid.*, p. 27.
82. L. N. Tolstoi, *Voina i Mir*, in his *Sobranie sochinenii v dvadtsati tomakh*, VI, p. 7.
83. Quoted in Knox, *With the Russian Army*, I, p. 50.

Chapter II: The Fall Campaigns

1. Knox, *With the Russian Army*, I, pp. xxi–xxiii; *The Times History of the War*, I, pp. 508–510; Stone, *The Eastern Front*, p. 170.
2. Quotes from *The Times History of the War*, I, p. 498, 488.
3. Manikovskii, *Boevoe snabzhenie russkoi armii v mirovuiu voinu*, pp. 153–155.
4. *Ibid.*, 149–151, 154–164; and the very important materials in Archives de la Guerre, Service historique de l'armée de la terre, Château de Vincennes [abbreviated hereafter as AG-CV], file 10N90 (especially "Russian Munitions Tables," August 1916 and December 1917, and "Situation de l'armament de l'armée Russe, février 1917" in file 10N73.
5. See especially the reports of Major C. Dunlop to Major-General C. E. Caldwell from Vladivostok during 1916 and 1917, AG-CV, file 10N93.
6. See the materials from AG-CV referred to in note 4, the biweekly reports of the French military attaché in Petrograd about Allied arms shipments to Russia that are in the same files, and Manikovskii, *Boevoe snabzhenie russkoi armii*, pp. 298, 364–365, 398.
7. Danilov, *Rossiia v mirovoi voine*, pp. 124–125; A. Bubnov, *V tsarskoi stavke*, pp. 27–28; P. K. Kondzerovskii, *V Stavke Verkhovnogo, 1914–1917*, pp. 19–30.
8. Knox, *With the Russian Army*, I, p. 46; Pares, *The Fall of The Russian Monarchy*, pp. 20–21.
9. A. Samoilo, *Dve zhizni*, pp. 142–145; Bubnov, *V tsarskoi stavke*, pp. 29–32; Danilov, *Rossiia v mirovoi voine*, pp. 125–127.
10. Bubnov, *V tsarskoi stavke*, pp. 31–32.
11. Ironside, *Tannenberg*, p. 25.
12. Gurko, *War and Revolution in Russia*, p. 11.
13. Général Marquis de Laguiche, "Rapport secret du Général de Laguiche, Attaché Militaire à Monsieur le Ministre de la Guerre (Etat-Major de l'Armée 2e Bureau)," le 13/26 octobre 1914.
14. General Max von Hoffmann, *The War of Lost Opportunities*, pp. 12–13.
15. Gurko, *War and Revolution in Russia*, pp. 59–70; Golovin, *Iz istorii kampanii 1914 goda: Nacholo voiny*, pp. 109–118, 176–192; Rostunov, *Russkii front*, pp. 117–118; Ironside, *Tannenberg*, pp. 42–43; Stone, *The Eastern Front*, pp. 55–56.
16. Ironside, *Tannenberg*, p. 68.
17. Knox, *With the Russian Army*, I, p. 205. See also A. G. Kavtaradze, "Pavel Karlovich Rennenkampf," cols. 1019–1020.
18. Quoted in *Manevrennyi period 1914 goda: Vostochno-Prusskaia operatsiia*, p. 12.
19. Hoffmann, *War of Lost Opportunities*, p. 16.
20. Quoted in Tuchman, *Guns of August*, p. 307.
21. Golovin, *Iz istorii kampanii 1914 goda: Nachalo voiny*, pp. 109–131; Hoffmann, *War of Lost Opportunities*, pp. 12–18; Ironside, *Tannenberg*, pp. 60–81.
22. Quoted in Golovin, *Iz istorii kampanii 1914 goda: Nachalo voiny*, p. 134.
23. Quoted in *ibid.*
24. Radus-Zenkovich, *Ocherk vstrechennogo boia po opytu Gumbinnenskoi operatsii v avguste 1914g.*, pp. 50–51.

25. *Ibid.,* p. 53.
26. Quoted in *ibid.,* p. 76. See also pp. 66–75.
27. Quoted in Golovin, *Iz istorii kampanii 1914 goda: Nachalo voiny,* p. 155.
28. *Ibid.,* pp. 139–152; Kersnovskii, *Istoriia russkoi armii,* III, pp. 640–642.
29. "Raport Generala Rennenkampfa k Generalu Zhilinskomu," 7/20 avgusta 1914g., in *Manevrennyi period 1914 goda. Vostochno-Prusskaia Operatsiia,* p. 205.
30. "Raport Generala Rennenkampfa k Generalu Zhilinskomu," 9/22 avgusta 1914g., in *ibid.,* p. 202.
31. Hoffmann, *War of Lost Opportunities,* pp. 20–23; Ironside, *Tannenberg,* pp. 105–107.
32. General Erich Ludendorff, *My War Memoirs, 1914–1918,* I, pp. 41–42.
33. *Ibid.,* p. 45.
34. Hindenburg, *Out of My Life,* p. 84.
35. *Ibid.,* p. 80.
36. Quoted in Danilov, *Velikii kniaz' Nikolai Nikolaevich,* p. 132.
37. Hoffmann, *War of Lost Opportunities,* pp. 26–28.
38. Ludendorff, *My War Memoirs,* p. 49.
39. "Telegramma Generala Zhilinskogo k Generalu Samsonovu, 10/23 avgusta 1914g.," in *Manevrennyi period 1914 goda: Vostochno-Prusskaia Operatsiia,* p. 263.
40. "Direktiva Nachal'nika shtaba armii general-leitenanta Mileanta, 13/26 avgusta 1914g.," and "Prikaz komanduiushchago armiei general-ad"iutanta generala ot kavalerii Rennenkampfa, 13/26 avgusta 1914g.," in *ibid,* pp. 228–229.
41. Stone, *The Eastern Front,* p. 62.
42. "Telegramma Generala Zhilinskago k Generalu Samsonovu, 10/23 avgusta 1914g.," and "Direktiva voiskam 2-oi armii ot komanduiushchago 2-oi armiei generala to kavalerii Samsonova, 10/23 avgusta 1914g.," in *Manevrennyi period 1914 goda: Vostochno-Prusskaia Operatsiia,* pp. 263–264; A. M. Zaionchkovskii, *Mirovaia voina 1914–1918gg.,* I, pp. 138–140.
43. Stone, *The Eastern Front,* p. 63.
44. Ironside, *Tannenberg,* pp. 145–146, 151.
45. "Direktiva komandiram 1-go, 6-go, 13-go, 16-go korpusam, No. 6308, ot Generala Postovskago, 10/23 avgusta 1914g.," in *Manevrennyi period 1914 goda: Vostochno-Prusskaia Operatsiia,* p. 264.
46. General Hermann von François, "Kriticheskoe issledovanie srazheniia na Mazurskikh ozerakh v sentiabre 1914g.," pp. 36–37; "Aleksandr Vasil'evich Samsonov"; cols. 520–521; Ironside, *Tannenberg,* p. 26; Gurko, *War and Revolution in Russia,* p. 13; Knox, *With the Russian Army,* I, pp. 60–62.
47. "Raport polkovnika Krymova k Generalu Samsonovu, 10/23 avgusta 1914g.," in *Manevrennyi period 1914 goda: Vostochno-Prusskaia Operatsiia,* pp. 260–261.
48. P. N. Bogdanovich, *Vtorzhenie v vostochnuiu prussiiu v avguste 1914 goda,* pp. 59–60.
49. Winston S. Churchill, *The Unknown War,* p. 195.
50. A. Kolenkovskii, *Manevrennyi period pervoi mirovoi imperialisticheskoi voiny 1914g.,* pp. 191–194.
51. Hoffmann, *War of Lost Opportunities,* p. 27
52. Bogdanovich, *Vtorzhenie, v* p. 101; Ludendorff, *My War Memoirs,* pp. 49–50; Zaionchkovskii, *Mirovaia voina,* I, pp. 141–142.
53. Quoted in Golovin, *Iz istorii kampanii 1914 goda: Nachalo voiny,* p. 224.
54. *Ibid.,* p. 225.
55. Danilov, *Rossiia v mirovoi voine,,* pp. 149–150.
56. Hindenburg, *Out of My Life,* p. 95.

57. *Ibid.*, p. 96.
58. Quoted in Bogdanovich, *Vtorzhenie, v* pp. 143–144.
59. Hoffmann, *War of Lost Opportunities*, p. 29.
60. Quoted in Bogdanovich, *Vtorzhenie, v* p. 158.
61. "Pis'mo Generala Oranovskago k Generalu Mileantu, 13/26 avgusta 1914g.," in *Manevrennyi period 1914 goda: Vostochno-Prusskaia Operatsiia*, p. 220.
62. Ironside, *Tannenberg*, p. 206.
63. *Ibid*, pp. 208–209.
64. Quoted in Golovin, *Iz istorii kampanii 1914 goda: Nachalo voiny*, p. 339.
65. *Ibid.*, pp. 312–314; Bogdanovich, *Vtorzhenie*, pp. 236–238.
66. Quoted in Bogdanovich, *Vtorzhenie* p. 238. See also Golovin, *Iz istorii kampanii 1914 goda: Nachalo voiny*, pp. 336–340.
67. Golovin, *Iz istorii kampanii 1914 goda: Nachalo voiny*, pp. 362–369; Stone, *The Eastern Front*, p. 311, note 33.
68. Hindenburg, *Out of My Life*, p. 102.
69. Churchill, *The Unknown War*, pp. 218–220; Golovin, *Iz istorii kampanii 1914 goda: Nachalo voiny*, pp. 370–379.
70. "Telegramma Generala Rennenkampfa k Generalu Zhilinskomu, 27 avgusta/9 sentiabria 1914," in *Manevrennyi period 1914 goda: Vostochno-Prusskaia Operatsiia*, pp. 379–380.
71. Ironside, *Tannenberg*, p. 236. See also pp. 234–239.
72. Hindenburg, *Out of My Life*, p. 106.
73. Quoted in Ironside, *Tannenberg*, p. 245, and O. O. Gruzenberg, *Yesterday*, p. 145. See also pp. 246–256, and the copies of telegrams and army directives in *Manevrennyi period 1914 goda: Vostochno-Prusskaia Operatsiia*, pp. 386–434.
74. Knox, *With the Russian Army*, I, p. 49.
75. Brusilov, *Moi vospominaniia*, pp. 76–77.
76. Beloi, A. *Galitsiiskaia bitva*, pp. 55–57; Kersnovskii, *Istoriia russkoi armii*, III, p. 654. Quotes are from Churchill, *The Unknown War*, p. 46.
77. Golovin, *Iz istorii kampanii 1914 goda: Galitsiiskaia bitva*, pp. 81–83; Zaionchkovskii, *Mirovaia voina 1914–1918gg.*, I, pp. 149–151; Rostunov, *Russkii front pervoi mirovoi voiny*, pp. 131–132.
78. Quoted in Churchill, *The Unknown War*, p. 151.
79. Beloi, *Galitsiiskaia bitva*, pp. 105–106.
80. Knox, *With the Russian Army*, I, p. 205.
81. *Ibid.*, pp. 104–137; Rostunov, *Russkii front pervoi mirovoi voiny*, pp. 131–140.
82. Quoted in Stone, *The Eastern Front*, p. 88.
83. Beloi, *Galitsiiskaia bitva*, pp. 152–154.
84. *Ibid.*, p. 189.
85. Zaionchkovskii, *Mirovaia voina 1914–1918gg.*, I, pp. 160–161.
86. Quoted in Golovin, *Iz istorii kampanii 1914 goda: Galitsiiskaia bitva*, p. 480.
87. Quoted in I. I. Rostunov, *General Brusilov*, p. 72.
88. Churchill, *The Unknown War*, p. 161.
89. *Ibid.*, Golovin, *Iz istorii kampanii 1914 goda: Galitsiiskaia bitva*, pp. 486–487; Beloi, *Galitsiiskaia bitva*, pp. 184–209; Brusilov, *Moi vospominaniia*, pp. 105–108.
90. Josef Białynia Chołodecki, *Lwów w czasie okupacji rosyjskiej (3 wrzesnia 1914–22 czerwca 1915)*, pp. 35–67.
91. Kolenkovskii, *Manevrennyi period*, pp. 242–264; Rostunov, *Russkii front pervoi mirovoi voiny*, pp. 140–153; N. N. Golovin, *Iz istorii kampanii 1914 goda: Dni pereloma Galitsiiskoi bitvy (1-3 sentiabria novago stilia)*, pp. 163–181.
92. Ludendorff, *My War Memories, 1914–1918*, p. 71.

93. See the telegrams that *Stavka* exchanged with Ivanov, and Ruzskii between September 7/20 and September 16/29, 1914, in *Manevrennyi period 1914g: Varshavsko-Ivangorodskaia Operatsiia*, pp. 24–36; and Rostunov, *Russkii front*, pp. 162–168.

94. *Manevrennyi period 1914 g: Varshavsko-Ivangorodskaia Operatsiia*, p. 13.

95. "Telegramma Generalam Ruzskomu i Ivanovu ot Generala Ianushkevicha, 15/28 sentiabria 1914g.," in *ibid.*, pp. 33–34.

96. *Manevrennyi period 1914g: Varshavsko-Ivangorodskaia Operatsiia*, p. 13.

97. Hindenburg, *Out of My Life*, pp. 118, 117.

98. Churchill, *The Unknown War*, p. 243.

99. Hindenburg, *Out of My Life*, pp. 117–118.

100. Hoffmann, *War of Lost Opportunities*, pp. 50–51.

101. "Razgovor generala Danilova s generalom Bonch-Bruevichem, 19 sentiabria/2 oktiabria 1914," in *Manevrennyi period 1914g: Varshavsko-Ivangorodskaia Operatsiia*, p. 40.

102. I have drawn the account of the Warsaw-Ivangorod campaign that follows from the following: Kolenkovskii, *Manevrennyi period*, pp. 265–283; Rostunov, *Russkii front*, pp. 154–175; Stone, *The Eastern Front*, pp. 96–99; and the relevant documents in *Manevrennyi period 1914g: Varshavsko-Ivangorodskaia Operatsiia*.

103. Hoffmann, *War of Lost Opportunities*, p. 50.

104. "Raport Generala Millera Cencralu Alekseevu, 28 sentiabria/11 oktiabria 1914g.," in *Manevrennyi period 1914g: Varshavsko-Ivangorodskaia Operatsiia*, p. 224; "Komanduiushchim 2, 4, 5, i 9 armiiami ot Generala Ivanova, 26 sentiabria/9 oktiabria 1914g.," in *ibid.*, pp. 150–151.

105. K. Popov, *Vospominaniia kavkazskago grenadera 1914–1920, gg.* p. 27.

106. Hindenburg, *Out of My Life*, pp. 121–122.

107. *Ibid.*, p. 124.

108. *Ibid.*, p. 123.

109. *Manevrennyi period 1914 goda: Lodzinskaia operatsiia*, p. 101.

110. *Ibid.*, pp. 150–151. See also G. K. Korol'kov, *Lodzinskaia operatsiia 2 noiabria–19 dekabria 1914g.*, pp. 4–42.

111. Korol'kov, *Lodzinskaia operatsiia*, pp. 56–74.

112. Knox, *With the Russian Army*, I, p. 204.

113. Kolenkovskii, *Manevrennyi period*, pp. 296–299.

114. Churchill, *The Unknown War*, p. 262.

115. Quoted in Golovin, *The Russian Army in the World War*, p. 216.

116. Danilov, *Rossiia v mirovoi voine*, pp. 238–241.

117. Quoted in Paléologue, *La Russie des Tsars*, I, pp. 207–208.

118. Buchanan, *My Mission to Russia*, I, p. 218.

119. Danilov, *Rossiia v mirovoi voine*, pp. 241–242.

120. Kolenkovskii, *Manevrennyi period*, pp. 297–303; Stone, *The Eastern Front*, pp. 106–107; Hoffmann, *War of Lost Opportunities*, pp. 72–75.

121. Lincoln, *The Romanovs*, p. 732.

122. Quoted in Paléologue, *La Russie des Tsars*, I, pp. 222–223.

123. Golovin, *Voennye usiliia Rossii v mirovoi voine*, I, pp. 153–159; II, p. 130; *Rossiia v mirovoi voine 1914–1918 v tsifrakh*, p. 30.

124. Nicholas to Aleksandra, 19 November 1914, in *Letters of the Tsar to the Tsaritsa*, p. 14.

125. Danilov, *Rossiia v mirovoi voine*, pp. 246–249.

126. *Ibid.*, p. 252.

127. Knox, *With the Russian Army*, I, p. 217.

128. Nicholas to Aleksandra, 19 November 1914, in *Letters of the Tsar to the Tsaritsa*, p. 14.

129. Commandant Langlois, "Rapport du Commandant Langlois sur sa Seconde Mission en Russie, 10 avril 1915," Section I, Chapter 2, p. 5; Knox, *With the Russian Army*, I, p. 217.

130. Manikovskii, *Boevoe snabzhenie russkoi armii*, I, p. 294; Radus-Zenkovich, *Ocherk vstrechennago boia*, pp. 66–75.

131. Brusilov, *Moi vospominaniia*, p. 105.

132. Knox, *With the Russian Army*, I, p. 219.

133. "Lettre (Secrète) de M. Paléologue à Ministère des Affaires Etrangères," No. 677, Petrograd, le 28 septembre 1914"; Paléologue, *La Russie de Tsars*, I, pp. 231–232; Buchanan, *My Mission to Russia*, I, pp. 219–221.

134. Gippius, *Siniaia kniga*, p. 14.

135. Buchanan, *My Mission to Russia*, I, p. 219.

Chapter III: The War's First Winter

1. Konstantin Paustovskii, *Sobranie sochinenii v shesti tomakh* (Moscow, 1957), III, p. 348.

2. L. N. Tolstoi, "Strashnyi vopros."

3. O. Chaadaeva (ed.), "Soldatskie pis'ma v gody mirovoi voiny," pp. 123–125.

4. "Zhurnal obshchago prisutstviia Akmolinskago oblastnago pravleniia," sheets 38–39; *Rossiia v mirovoi voine, 1914–1918 goda (v tsifrakh)*, pp. 17–18.

5. "Poiasnitel'nyi tekst k svodnomu otchetu po vsem fabrichnym uchastkam petrogradskoi fabrichnyi inspektsii za 1915 god," sheets 8–14; "Chislo fabrichno-zavodskikh rabochikh bastovavshikh v 1912, 1913, 1914, 1915, i 1916gg," sheet 185.

6. "Kopiia sekretnoi zapiski Ego Prevoskhoditel'stvu Gospodinu Direktoru Departamenta Politsii," 25 ianvaria 1915g. sheets 2–3.

7. *Trudy komissii po obsledovaniiu sanitarnykh posledstvii voiny 1914–1920 godov*, pp. 160–165; *Izvestiia glavnago komiteta Vserossiiskago Zemskago Soiuza*, No. 11, pp. 56–58.

8. Vsevolod Vishnevskii, "Moi vospominaniia," p. 653; Pares, *Day by Day with the Russian Army*, pp. 16, 65.

9. *Izvestiia glavnago komiteta Vserossiiskago Zemskago Soiuza*, No. 34, pp. 47–60.

10. Stanley Washburn, *Field Notes from the Russian Front*, pp. 52, 55–57.

11. Maria Pavlovna, *Education of a Princess*, p. 196.

12. Viroubova, *Memories*, p. 109.

13. *Reports of Military Observers Attached to the Armies in Manchuria during the Russo-Japanese War*, IV, pp. 192–193, 196.

14. K. P. Serapin, "O ranakh ot razryvnoi avstriiskoi puli na iugo-zapadnom fronte," pp. 435–449; Commandant Langlois, "Rapport du Commandant Langlois sur sa Troisième Mission en Russie," 20 juin 1915, Part I, Chapter 4, p. 1.

15. Miliukov, *Vospominaniia*, II, p. 199.

16. Paustovskii, *Sobranie sochinenii*, III, p. 339.

17. M. V. Rodzianko, *The Reign of Rasputin*, pp. 115–116.

18. Viroubova, *Memories*, p. 110; Letters of Aleksandra to Nicholas, Nos. 83 and 84, June 12, 1915, in *Letters of the Tsaritsa to the Tsar*, pp. 88, 91. My italics.

19. Maria Pavlovna, *Education of a Princess*, p. 196.

20. Paléologue, *La Russie des Tsars*, I, p. 235.

21. Meriel Buchanan, *Dissolution of an Empire*, p. 36.

22. Maria Pavlovna, *Education of a Princess*, pp. 196–197.

23. Viktor Chernov, *Rozhdenie revoliutsionnoi Rossii*, p. 122; Lincoln, *The Romanovs*, pp. 630–631; Pares, *Fall of the Russian Monarchy*, p. 133.

24. V. Semennikov, *Romanovy i Germanskie vliianiia vo vremia mirovoi voiny*, pp. 76–118; S. P. Mel'gunov, *Vospominaniia i dnevniki*, II, p. 186, entry for October 4, 1914.

25. Paléologue, *La Russie des Tsars*, I, p. 263.

26. F. M. Dostoevskii, *Brat'ia Karamazovykh*, in *Polnoe sobranie sochineniia* XV, p. 175.

27. *Ibid.*, XIV, p. 26.

28. Quoted in René Fülöp-Miller, *Rasputin*, p. 215.

29. Lili Dehn, *The Real Tsaritsa*, pp. 100–101.

30. Paléologue, *La Russie des Tsars*, I, p. 308.

31. Kokovtsev, *Iz moego proshlago*, II, 40.

32. Quoted in Salisbury, *Black Night, White Snow*, p. 169.

33. Robert Massie, *Nicholas and Alexandra*, p. 218.

34. Quoted in Chernov, *Rozhdenie revoliutsionnoi Rossii*, p. 114.

35. Maria Pavlovna, *Education of a Princess,*, p. 179.

36. Miliukov, *Vospominaniia*, II, p. 199.

37. V. I. Gurko, *Features and Figures of the Past*, p. 543, 521.

38. "Dopros N. A. Maklakova, 14 iiunia 1917 goda," pp. 202–204.

39. "Dopros M. V. Rodzianko, 4 sentiabria 1917 goda," p. 124. See also Gurko, *Features and Figures of the Past*, p. 521.

40. A. P. Pogrebinskii, "K istorii soiuzov zemstv i gorodov v gody imperialisticheskoi voiny," pp. 40–42; N. A. Ivanov, "Zemskii i gorodskoi soiuzy (Vserossiiskii zemskii soiuz pomoshchi bol'nym i ranenym voinam i Vserossiiskii soiuz gorodov)," col. 675.

41. P. E. Shchegolev (ed.), *Padenie tsarskogo rezhima*, VII, pp. 370–371; Miliukov, *Vospominaniia*, II, pp. 200–201.

42. T. J. Polner, Prince Vladimir A. Obolensky, and Sergei P. Turin, *Russian Local Government During the War and the Union of Zemstvos*, pp. 188–190; William Gleason, "The All-Russian Union of Zemstvos and World War I," p. 370.

43. Gurko, *Features and Figures of the Past*, p. 539.

44. Pogrebinskii, "K istorii soiuzov zemstv i gorodov v gody imperialisticheskoi voiny," p. 44; Polner, Obolensky, and Turin, *Russian Local Government During the War*, pp. 60–63.

45. Polner, Obolensky, and Turin, *Russian Local Government During the War*, pp. 89–91, 101.

46. Miliukov, *Vospominaniia*, II, p. 202.

47. "Dopros M. V. Rodzianko, 4 sentiabria 1917 goda," p. 123.

48. *Ibid.*, pp. 123–126.

49. Florinsky, *Russia*, II, pp. 1363, 1368–1369.

50. S. O. Zagorsky, *State Control of Industry in Russia During the War*, p. 46; Polner, Obolensky, and Turin, *Russian Local Government During the War*, pp. 246–247.

51. Zagorsky, *State Control of Industry*, p. 46.

52. Polner, Obolensky, and Turin, *Russian Local Government During the War*, pp. 62, 246–250; Zagorsky, *State Control of Industry*, pp. 139–141.

53. P. A. Khromov, *Ekonomicheskoe razvitie Rossii v XIX–XX vekakh, 1800–1917gg.*, pp. 198–199; 456–462.

54. "Liste des usines de la région de Petrograd, possédant un outillage susceptible d'être utilisé à la fabrication des projectiles," A. L. Sidorov, "K istorii toplivnogo krizisa v Rossii v gody pervoi mirovoi voiny (1914–1917)," pp. 46, 70.

55. "O nezhelatel'nosti dal'neishago rosta promyshlennosti v samom gorode S.-Peterburge, v sviazi s politicheskimi manifestatsiami i stachkami rabochikh," iiul' 1914.

56. "Vsepoddanneishii doklad ministra torgovli i promyshlennosti, Kn. V. Shakhovskoi, 20 fevralia 1917g," sheet 50; Iu. I. Kir'ianov, "Vliianie pervoi mirovoi voiny na izmenenie chislennosti i sostava rabochikh Rossii," pp. 91–92; Zagorsky, *State Control of Industry*, p.

36; B. M. Kochakov, "Petrograd v gody pervoi mirovoi voiny i Fevral'skoi burzhuazno-demokraticheskoi revoliutsii," p. 940.

57. "Pis'ma I. Bazili i kniazia N. Kudasheva Ministru Inostrannykh Del S. D. Sazonovu," 14/27 dekabria 1914–26 dekabria 1914/8 ianvaria 1915, in M. N. Pokrovskii, ed., "Stavka i ministerstvo inostrannykh del," pp. 29–37.

58. Manikovskii, *Boevoe snabzhenie russkoi armii*, pp. 119–129, 147–150, 152–157; Stone, *The Eastern Front*, pp. 144–145.

59. Zhitkov, "Prodfurazhnoe snabzhenie russkikh armii," pp. 74–75; A. N. Antsiferov, A. D. Bilimovich, M. O. Batshev, and D. N. Ivantsov, *Russian Agriculture During the War*, pp. 139–140, 212–214.

60. "Raport Vserossiiskago Soiuza Torgovlia i Promyshlennosti, 14 avgusta 1917g.," sheet 27; Sidorov, "K istorii toplivnogo krizisa v Rossii," pp. 27–36, 68; Gurko, *Features and Figures of the Past*, p. 545.

61. "Poiasnitel'nyi tekst k svodnomu otchetu za 1915 god po Vladimirskoi gubernii," 20 aprelia 1916, sheets 64–74; M. F. Fedorov, *Voina i dorogovizna zhizni 1916g. Doklad predsedatelia Petrogradskoi gorodskoi finansovoi komisii M. P. Fedorova*, pp. 44–54, 66–69; Kochakov, "Petrograd v gody pervoi mirovoi voiny," p. 938; Antsiferov, Bilimovich, Batshev, and Ivantsov, *Russian Agriculture During the War*, pp. 219–221; I. P. Leiberov, "Petrogradskii proletariat v gody pervoi mirovoi voiny," pp. 465–467.

62. "O dopushchenii nekotorykh otstuplenii ot pravil o rabote zhenshchin, podrostkov i maloletnikh," 3 marta 1915g., sheet 4.

63. "Delo o dopushchenii maloletnykh i zhenshchin k nochnym i podzemnym rabotam v kamennougol'nykh kopiakh Evropeiskoi Rossii, 3 marta 1915g.," sheets 376–383; "Poiasnitel'nyi tekst k svodnomu otchetu po vsem fabrichnym uchastkam Petrogradskoi fabrichnoi inspektsii za 1915 god," sheet 12; "Spravka mezhduvedomstvennoi komissii pri Ministerstve zemledeliia, 16 fevralia 1916g.," sheet 50; Zagorsky, *State Control of Industry*, pp. 55–56; P. M. Ekzempliarskii, *Istoriia goroda Ivanova* I, pp. 310–313; N. V. Kuznetsov, "V bor'be protiv imperialisticheskoi voiny i samoderzhaviia (1914-fevral' 1917g.)," pp. 242–243.

64. "Polozhenie rabochego klassa," in Iu. I. Korablev (ed.), *Rabochee dvizhenie v Petrograde v 1912–1917gg.*," pp. 267–268; Kochakov, "Petrograd v gody pervoi mirovoi voiny," p. 950.

65. Quoted in Leiberov, "Petrogradskii proletariat v gody pervoi mirovoi voiny," pp. 475–476.

66. "Kopiia sekretnoi zapiski Ego Prevoskhoditel'stvu Gospodinu Direktoru Departamenta Politsii." 25 ianvaria 1915g., sheet 2; Chernov, *Rozhdenie revoliutsionnoi Rossii*, p. 290.

67. A. Badaev, *The Bolsheviks in the Tsarist Duma*, pp. 209–219.

68. Quoted in *ibid.*, p. 219.

69. "Listovka Peterburgskogo komiteta RSDRP po povodu aresta bol'shevistskoi fraktsii IV Gosudarstvennoi dumy, 11 noiabria 1914g.," in Korablev (ed.), *Rabochee dvizhenie v Petrograde v 1912–1917gg.*, pp. 261–262.

70. "Chislo fabrichno-zavodskikh rabochikh bastovavshikh v 1912, 1913, 1914, 1915, i 1916gg.," sheet 185; M. G. Fleer (ed.), *Rabochee dvizhenie v gody voiny*, pp. 28–29; F. A. Romanov, *Rabochee i professional'noe dvizhenie, v* pp. 48–49.

71. Badaev, *Bolsheviks in the Tsarist Duma*, p. 224.

72. Quoted in Kochakov, "Petrograd v gody pervoi mirovoi voiny," p. 961; Kuznetsov, "V bor'be protiv imperialisticheskoi voiny i samoderzhaviia," pp. 241–242; "Doklad petrogradskogo gradonachal'nika glavnomu nachal'niku Petrogradskogo voennogo okruga," 12 noiabria 1914g., and "Iz doklada nachal'nika Petrogradskogo okhrannogo otdeleniia ministru vnutrennikh del," 12 noiabria 1914g.," pp. 263–266.

73. Quoted in V. Ia. Laverychev, *Tsarizm i rabochii vopros v Rossii (1861–1917gg.)*, p. 275.

74. "Chislo fabrichno-zavodskikh rabochikh bastovavshikh v 1912, 1913, 1914, 1915, i 1916gg.," sheet 185.

75. "Proklamatsiia kostromskikh zhenshchin-rabotnits k soldatam," iiun' 1915g., and "Iz protokola pokazanii pomoshchnika kostromskogo politseimeistera," 6 iiunia 1915g.," in M. Inozemtsev (ed.), "Iz istorii rabochego dvizheniia vo vremia mirovoi voiny (stachechnoe dvizhenie v Kostromskoi gubernii)," pp. 12, 10; Fleer (ed.), *Rabochee dvizhenie v gody voiny*, pp. 211–214.

76. Nicholas II, *Journal intime de Nicolas II (juillet 1914–juillet 1918)*, pp. 52–71; Major General Dubenskii (ed.), *Ego Imperatorskoe Velichestvo Gosudar Imperator Nikolai Aleksandrovich v deistvuiushchei armii*, vols II–III.

77. Paléologue, *La Russie des Tsars*, I, pp. 267–270.

78. This is the account of Staff Surgeon Fedorov, who reported it to General A. I. Spiridonovich; *Velikaia voina i fevral'skaia revoliutsiia 1914–1917gg.*, I, pp. 85–86. See also Vyrubova's version in Viroubova, *Memories*, pp. 118–122.

79. Paléologue, *La Russie des Tsars*, I, pp. 350–352; "Dopros V. F. Dzhunkovskogo, 7 iiunia 1917 goda," pp. 101–105.

80. Letter of Aleksandra to Nicholas, 22 June 1915, *Letters of the Tsaritsa to the Tsar*, pp. 105–106.

81. "Dopros Dzhunkovskogo, 7 iiunia 1917 goda," p. 103.

82. Quoted in Paléologue, *La Russie des Tsars*, I, p. 289.

83. Miliukov, *Vospominaniia*, II, pp. 195–196.

84. Brusilov, *Moi vospominaniia*, pp. 153–154.

85. K. F. Shatsillo, "Delo polkovnika Miasoedova," pp. 103–116; Gurko, *Features and Figures of the Past*, p. 551.

86. Paléologue, *La Russie des Tsars*, I, pp. 267–268; Gurko, *Features and Figures of the Past*, p. 682.

87. Quoted in Paléologue, *La Russie des Tsars*, I, p. 371.

88. Paustovskii, *Sobranie sochinenii*, III, p. 355.

Chapter IV: The Road to Disaster

1. Quoted in Fedor [Theodore] Dan, *The Origins of Bolshevism*, p. 101.

2. Richard G. Robbins, Jr., *Famine in Russia, 1891–1892*, pp. 171–187.

3. Paléologue, *La Russie des Tsars*, I, p. 304.

4. Nicholas V. Riasanovsky, *Nicholas I and Official Nationality*, p. 85.

5. Quoted in S. S. Ol'denburg, *Tsarstvovanie Imperatora Nikolaia II*, I, p. 343.

6. Nicholas I to Prince Ivan Paskevich, January 17, 1854, in Theodor Schiemann, *Geschichte Russlands unter Kaiser Nikolaus I*, IV, p. 305.

7. Quoted in Paléologue, *La Russie des Tsars*, I, p. 304.

8. Ludendorff, *My War Memories*, p. 118. The description of the Russian defeat at the forest of Augustowo that follows is taken from the following: Zaionchkovskii, *Mirovaia voina, 1914–1918gg.*, I, pp. 269–276; Rostunov, *Russkoi front*, pp. 208–223; Kersnovskii, *Istoriia russkoi armii*, III, pp. 696–706; Ludendorff, *My War Memories*, pp. 113–137; Hoffmann, *War of Lost Opportunities*, pp. 79–91; Hindenburg, *Out of My Life*, pp. 129–139; Churchill, *The Unknown War*, pp. 289–300; Knox, *With the Russian Army*, I, pp. 235–241; Stone, *The Eastern Front*, pp. 115–119.

9. Hoffmann, *War of Lost Opportunities*, p. 86.

10. Ludendorff, *My War Memories*, pp. 123–124.

11. Hindenburg, *Out of My Life*, p. 137.

12. Ludendorff, *My War Memories*, p. 125.
13. Quoted in Paléologue, *La Russie des Tsars*, I, p. 305.
14. Hindenburg, *Out of My Life*, p. 137.
15. Gurko, *Features and Figures of the Past*, p. 549.
16. Quoted in Knox, *With the Russian Army*, I, p. 249.
17. Ludendorff, *My War Memories*, p. 132.
18. Quoted in A. Borisov, "Prasnyshskaia operatsiia," p. 32.
19. Ludendorff, *My War Memories*, p. 132.
20. See Zaionchkovskii, *Mirovaia voina*, I, pp. 276–278; *Strategicheskii ocherk voiny 1914–1918gg.*, III, pp. 811–83; and, especially, Borisov, "Prasnyshskaia operatsiia," pp. 27–35.
21. Paléologue, *La Russie des Tsars*, I, pp. 307–308.
22. Stone, *The Eastern Front*, pp. 113–116, 314.
23. Quoted in *ibid.*, p. 114.
24. V. Iakovlev, "Peremyshl'," pp. 306–308; Danilov, *Rossiia v mirovoi voine*, pp. 313–314.
25. Brusilov, *Moi vospominaniia*, pp. 150, 148.
26. Quoted in Rostunov, *General Brusilov*, p. 95.
27. Danilov, *Rossiia v mirovoi voine*, pp. 292, 280.
28. *Ibid.*, p. 250.
29. Quoted in Paléologue, *La Russie des Tsars*, I, p. 305.
30. Quoted in *Ibid.*, I, p. 268.
31. Stone, *The Eastern Front*, pp. 122–126.
32. Quoted in M. Bonch-Bruevich, *Poteria nami Galitsii*, I, pp. 78–80. See also Danilov, *Rossiia v mirovoi voine*, pp. 314–316.
33. Danilov, *Velikii kniaz' Nikolai Nikolaevich*, pp. 169–170.
34. Stone, *The Eastern Front*, pp. 120–121.
35. General Erich von Falkenhayn, *General Headquarters 1914–1916 and Its Critical Decisions*, p. 76.
36. Quoted in Bonch-Bruevich, *Poteria nami Galitsii*, I, pp. 90, 98, 106.
37. Knox, *With the Russian Army*, I, p. 289; Stone, *The Eastern Front*, p. 130.
38. Falkenhayn *General Headquarters*, p. 80.
39. *Ibid.*, pp. 81–82; *Sbornik dokumentov mirovoi imperialisticheskoi voiny na russkom fronte (1914–1917gg.): Gorlitskaia operatsiia*, pp. 12–13.
40. *Sbornik dokumentov: Gorlitskaia operatsiia*, p. 13.
41. "Dopolnenii k svodke No. 124 o svedenii o protivnike, poluchennykh v shtabe glavnokomanduiushchego armiiami iugo-zapadnogo fronta, 31 marta/13 aprelia i 5/18 aprelia 1915g.," in *Sbornik dokumentov: Gorlitskaia operatsiia*, pp. 48–50.
42. "Prikaz nachal'nikam shtabov 3 i 4-i armii, ot Generala Dragomirova," 16/29 aprelia 1915g., in *ibid.*, pp. 91–92. See also Danilov, *Rossiia v mirovoi voine*, pp. 335–336.
43. Knox, *With the Russian Army*, I, p. 282.
44. *Sbornik dokumentov: Gorlitskaia operatsiia*, p. 12. In addition to the crushing quantities of shell that they had by their guns when the attack began, Mackensen's gunners had almost two million shells (1,200 per gun) in reserve.
45. Popov, *Vospominaniia kavkazskago grenadera 1914–1920gg.*, p. 89.
46. Knox, *With the Russian Army*, I, p. 282.
47. *Ibid;* Churchill, *The Unknown War*, pp. 314–315.
48. "Razgovor po apparatu Generala Danilova s Generalom Dragomirovym," in *Sbornik dokumentov: Gorlitskaia operatsiia*, p. 129; Strokov, *Vooruzhennye sily*, pp. 296–297.
49. Langlois, "Rapport du Commandant Langlois sur sa Troisième Mission en Russie," 20 juin 1915, Part I, Chapter 2, p. 6.

50. "Telegramma to Generala Dragomirova k Nachal'niku Shtaba Verkhovnogo Glav-nokomanduiushchego ot 27-go aprelia/ 10 maia 1915g.," in *Sbornik dokumentov: Gorlitskaia operatsiia,* pp. 323–324.

51. Langlois, "Rapport du Commandant Langlois sur sa Troisième Mission en Russie," 20 juin 1915, Part I, Chapter 2, p. 4.

52. Quoted in Stone, *The Eastern Front,* p. 139. See also pp. 136–141.

53. Quoted in *ibid.,* p. 140.

54. *Sbornik dokumentov: Gorlitskaia operatsiia,* p. 375. See also pp. 376–384 for more than two dozen desperate requests for shells and cartridges from Ivanov, Radko-Dimitriev, and Brusilov.

55. Quoted in Anton I. Denikin, *The Career of a Tsarist Officer,* p. 253.

56. "Pis'mo Ianushkevicha Sukhomlinovu, 21 maia 1915g.," and "Pis'mo Sukhom-linova Ianushkevichu, 23 maia 1915g.," in I. A. Blinov (ed.), "Perepiska V. A. Sukhomlinova s N. N. Ianushkevichem," pp. 63–65.

57. *Sbornik dokumentov: Gorlitskaia operatsiia,* pp. 140–143; Rostunov, *Russkii front pervoi mirovoi voiny,* pp. 243–245.

58. "Dokladnaia zapiska velikago kniazia Nikolaia Nikolaevicha tsariu, 24 maia 1915g.," quoted in V. A. Emets, *Ocherki vneshnei politiki Rossii, 1914–1917gg.,* p. 173.

59. Falkenhayn, *General Headquarters,* p. 104.

60. *Rossiia v mirovoi voine 1914–1918 goda (v tsifrakh),* p. 30.

61. Knox, *With the Russian Army,* I, pp. 290–292; Stone, *The Eastern Front,* p. 172. Zaionchkovskii, *Mirovaia voina, 1914–1918gg.* I, pp. 315–317.

62. Ludendorff, *My War Memories,* pp. 147–148.

63. *Ibid.,* pp. 148–149; Hoffmann, *War of Lost Opportunities,* pp. 105–108; Falken-hayn, *General Headquarters,* pp. 112–120.

64. Stone, *The Eastern Front,* pp. 174–175.

65. *Ibid.,* p. 131.

66. *Rossiia v mirovoi voine 1914–1918 goda (v tsifrakh),* p. 30.

67. Knox, *With the Russian Army,* I, p. xxviii.

68. Stone, *The Eastern Front,* p. 163.

69. Quoted in Knox, *With the Russian Army,* I, pp. xxviii–xxix.

70. *Ibid.,* p. xxi; Stone, *The Eastern Front,* p. 217; *Rossiia v mirovoi voine 1914–1918 goda (v tsifrakh),* pp. 17–20.

71. A. L. Sidorov, *Ekonomicheskoe polozhenie Rossii v gody pervoi mirovoi voiny,* pp. 336–339; Zagorsky, *State Control of Industry,* pp. 69–71.

72. Langlois, "Rapport du Commandant Langlois sur sa Seconde Mission en Russie," 10 avril 1915, Part I, Chapter 2, pp. 8–10; "Rapport du Commandant Langlois sur sa Troisième Mission en Russie, Part I, Chapter 5, p. 14.

73. Langlois, "Rapport du Commandant Langlois sur sa Troisième Mission en Russie," Part I, Chapter 5, p. 19; Manikovskii, *Boevoe snabzhenie russkoi armii v mirovuiu voinu,* I, pp. 77–82; I. V. Maevskii, *Ekonomika russkoi promyshlennosti v usloviiakh pervoi mirovoi voiny,* pp. 50–53; Sidorov, *Ekonomicheskoe polozhenie Rossii,* pp. 22–31.

74. "Report from Major Dunlop at Vladivostok," p. 1.

75. Quoted by Paléologue, *La Russie des Tsars,* II, pp. 11–12.

76. Quoted in *ibid.,* I, p. 244.

77. Churchill, *The Unknown War,* p. 319.

Chapter V: Russia's Great Retreat

1. "Zasedanie 24 iiulia 1915 goda," in A. N. Iakhontov, "Tiazhelye dni (sekretnye zasedaniia Soveta Ministrov 16 iiulia–2 sentiabria 1915 goda)," p. 24; Knox, *With the Russian*

Army, I, p. 312. For casualty figures see Golovin, *The Russian Army in the World War*, p. 222.

2. Gruzenberg, *Yesterday*, p. 140; Brusilov, *Moi vospominaniia*, pp. 158–159; Ol'denburg, *Tsarstvovanie Imperatora Nikolaia II*, II, p. 167n.

3. Quotes from Allan K. Wildman, *The End of the Russian Imperial Army*, p. 92, and Letter of Aleksandra to Nicholas, June 14, 1915, *Letters of the Tsaritsa to the Tsar*, pp. 93–94.

4. See letters of Aleksandra to Nicholas of April 25, 1916, and June 14, 1916, in *Letters of the Tsaritsa to the Tsar*, pp. 327, 354, and Pares, *Fall of the Russian Monarchy*, p. 342.

5. Alexander, *Once a Grand Duke*, p. 139; Paléologue, *La Russie des Tsars*, II, p. 41. Other quotes are in Wildman, *End of the Russian Imperial Army*, p. 92.

6. Ol'denburg, *Tsarstvovanie Imperatora Nikolaia II*, II, p. 168.

7. Wildman, *End of the Russian Imperial Army*, p. 92.

8. Lincoln, *In War's Dark Shadow*, pp. 211–222.

9. Letter of Aleksandra to Nicholas, September 17, 1915, in *Letters of the Tsaritsa to the Tsar*, p. 171. Nicholas's remarks to his mother are quoted in Howard D. Mehlinger and John M. Thompson, *Count Witte and the Tsarist Government in the 1905 Revolution*, p. 62.

10. Quoted in Gruzenberg, *Yesterday*, p. 123. See also pp. 104–124; Hans Rogger, "The Beilis Case" pp. 615–629; A. S. Tager, *The Decay of Czarism*, pp. 60–82; and A. S. Tager (ed.), "Protsess Beilisa v otsenke departamenta politsii," pp. 85–125.

11. Gruzenberg, *Yesterday*, pp. 157–158; Louis Greenberg, *The Jews in Russia*, II, p. 94. See also pp. 95–98. Quotes in Paléologue, *La Russie des Tsars*, II, p. 16.

12. "Dokumenty o presledovanii evreev," Nos. 30, 32, pp. 260–262; "Zasedaniia 4, 6 i 9 avgusta 1915 goda," in Iakhontov, "Tiazhelye dni," pp. 42–43, 48, 57; Gurko, *Figures and Figures of the Past*, pp. 42–43; Paléologue, *La Russie des Tsars*, II, p. 37.

13. "Dokumenty o presledovanii evreev," Nos. 9, 21, 22, pp. 250–251, 256–257.

14. Gruzenberg, *Yesterday*, pp. 145–146; L. Andreev, M. Gor'kii, F. Sologub (eds.), *Shchit*, pp. 59, 61, 64. See also quotes in Greenberg, *Jews in Russia*, II, pp. 100–101.

15. Andreev, Gor'kii, and Sologub (eds.), *Shchit*, pp. 61, 31, 167; Greenberg, *Jews in Russia*, II, p. 119.

16. "Zasedaniia 4, 6, i 9 avgusta 1915 goda," in Iakhontov, "Tiazhelye dni," pp. 48, 47, 57, 46.

17. Gippius, *Siniaia kniga*, pp. 21, 17, 22; Paléologue, *La Russie des Tsars*, II, pp. 41–42.

18. Gippius, *Siniaia kniga*, p. 30; Knox, *With the Russian Army*, I, p. 309.

19. Gippius, *Siniaia kniga*, p. 22; Paléologue, *La Russie des Tsars*, II, p. 31.

20. Letters of Aleksandra to Nicholas, June 10, 11, and 12, 1915, in *Letters of the Tsaritsa to the Tsar*, pp. 86–91.

21. Quoted in Michael Cherniavsky (ed.), *Prologue to Revolution*, pp. 13–14.

22. Golovin, *The Russian Army in the World War*, p. 108; A. A. Polivanov, *Iz dnevnikov i vospominanii po dolzhnosti voennogo ministra i ego pomoshchnika, 1907–1916gg.*, edited by A. M. Zaionchkovskii (Moscow, 1924), pp. 184–186.

23. Polivanov, *Iz dnevnikov*, p. 184.

24. Paléologue, *La Russie des Tsars*, II, p. 28.

25. Danilov, *Rossiia v mirovoi voine*, pp. 352–353.

26. Brusilov, *Moi vospominaniia*, p. 167.

27. *Rossiia v mirovoi voine, 1914–1917 (v tsifrakh)*, p. 32; Nicholas's speech quoted in Golovin, *The Russian Army in the World War*, p. 66. See also p. 99, and Stone, *The Eastern Front*, p. 166.

28. Quoted in Polivanov, *Iz dnevnikov*, p. 186.

29. Danilov, *Rossiia v mirovoi voine*, p. 112; Wildman, *End of the Russian Imperial Army*, pp. 102.

30. Langlois, "Rapport du Commandant Langlois sur sa Troisième Mission en Russie," 20 juin 1915, Chapter IV, pp. 2–3; Peter Kenez, "A Profile of the Pre-Revolutionary Officer Corps," pp. 147–149; M. N. Gerasimov, *Probuzhdenie*, p. 51.

31. Quotes from Ianushkevich's letter are taken from Polivanov, *Iz dnevnikov*, p. 185, and "Zasedanie 24 iiulia 1915 goda, in Iakhontov, "Tiazhelye dni," p. 24. See also Danilov, *Rossiia v mirovoi voine*, p. 112.

32. Sazonov's remarks were recorded by Grand Duke Andrei Vladimirovich in his diary, *Dnevnik*, p. 69. See also Stone, *The Eastern Front*, p. 169.

33. Knox, *With the Russian Army*, I, p. 255.

34. Quoted in *ibid.*, p. 287.

35. Falkenhayn, *General Headquarters*, pp. 114–116; *Der Weltkrieg 1914 bis 1918*, VII, pp. 341–342.

36. Falkenhayn, *General Headquarters*, p. 108.

37. Knox, *With the Russian Army*, I, p. 290.

38. Golovin, *The Russian Army in the World War*, p. 221.

39. Knox, *With the Russian Army*, I, p. 319.

40. Rostunov, *Russkii front*, pp. 255–257.

41. G. K. Korol'kov, *Prashnyshskoe srazhenie, iiul' 1915g.* pp. 58–61; Rostunov, *Russkii front pervoi mirovoi voiny*, pp. 257–258.

42. Quoted in Polivanov, *Iz dnevnikov*, p. 186.

43. Quoted in Golovin, *The Russian Army in the World War*, p. 67.

44. This brief account of the first days of Gallwitz's offensive is drawn from the following sources. Korol'kov, *Prasnyshskoe srazhenie*, pp. 60–80; *Der Weltkrieg 1914 bis 1918*, VIII, pp. 275–285; Rostunov, *Russkii front*, pp. 258–259; Knox, *With the Russian Army*, I, pp. 310–315.

45. Paléologue, *La Russie des Tsars*, II, p. 36. Lloyd George's remarks are quoted in brief by Sir Bernard Pares, *My Russian Memoirs*, p. 339.

46. Rostunov, *Russkii front*, pp. 260–262; Knox, *With the Russian Army*, I, p. 325; G. Korol'kov, *Srazhenie pod Shavli, passim*.

47. Lemke, *250 dnei*, pp. 223–255; Knox, *With the Russian Army*, I, pp. 324–328; Stone, *With the Russian Army*, p. 187.

48. "Zasedanie 4 avgusta 1915 goda," in Iakhontov, "Tiazhelye dni," p. 37.

49. Hoffmann, *War of Lost Opportunities*, p. 111.

50. Quoted in "Les télégrammes secrets de M. Paléologue au Ministère des Affaires Etrangères." Petrograd, le 26 juillet et 2 septembre 1915, pp. 169, 206.

51. Grand Duke Andrei Vladimirovich, *Dnevnik*, pp. 49–50.

52. Quoted in Paléologue, *La Russie des Tsars*, II, p. 45. See also *Der Weltkrieg 1914 bis 1918*, VIII, pp. 340–480.

53. Lieutenant-General Schwarz, *La Defense d'Ivangorod en 1914–1915*, pp. vi–viii, 131, 143–148.

54. Quoted in Andrei Vladimirovich, *Dnevnik*, p. 66.

55. Lemke, *250 dnei*, pp. 214–217. See also Knox, *With the Russian Army*, pp. 320–321; Stone, *The Eastern Front*, pp. 181–182; Churchill, *The Unknown War*, pp. 327–328; Hindenburg, *Out of My Life*, p. 142.

56. Knox, *With the Russian Army*, pp. 321–322.

57. Ludendorff, *My War Memories*, p. 162; Hoffmann, *War of Lost Opportunities*, p. 116.

58. Hindenburg, *Out of My Life*, p. 142; Knox, *With the Russian Army*, I, p. 305; Hoffmann, *War of Lost Opportunities*, p. 111.

59. "Zasedanie 30 iiulia 1915 goda," in Iakhontov, "Tiazhelye dni," pp. 32–33; Gurko, *Features and Figures of the Past,* p. 558.

60. Knox, *With the Russian Army,* I, pp. 305, 322–323; Gurko, *War and Revolution in Russia,* p. 151.

61. "Zasedanie 4 avgusta 1915 goda," in Iakhontov, "Tiazhelye dni," p. 37.

62. Paustovskii, *Sobranie sochinenii,* III, p. 435.

63. "Zasedanie 30 iiulia 1915 goda," in Iakhontov, "Tiazhelye dni," p. 33.

64. "Zasedanie 16 avgusta 1915 goda," in *ibid.,* p. 74.

65. Paustovskii, *Sobranie sochinenii,* III, pp. 467–468.

66. Quoted in Stone, *The Eastern Front,* p. 183.

67. "Zasedanie 6 avgusta 1915 goda," in Iakhontov, "Tizhelye dni," p. 52.

68. Quoted in Golovin, *Russian Army in the World War,* pp. 228–229.

69. "Zasedanie 6 avgusta 1915 goda," in Iakhontov, "Tiazhelye dni," pp. 52–53.

70. *Ibid.,* p. 54.

71. "Zasedaniia 6 i 10 avgusta 1915 goda," in *ibid.,* pp. 55, 60.

72. "Zasedaniia 10, 6, i 21 avgusta 1915 goda," in *ibid.,* pp. 61, 55, 98.

73. Quoted by Paléologue, *La Russie des Tsars,* II, p. 62.

74. "Zasedanie 6 avgusta 1915 goda," in Iakhontov, "Tiazhelye dni," p. 54.

75. Quoted in Ol'denburg, *Tsarstvovanie Imperatora Nikolaia II,* II, p. 176.

76. Letter of Nicholas to Aleksandra, August 25, 1915, in *Letters of the Tsar to the Tsaritsa,* p. 71; Letter of Aleksandra to Nicholas, August 22, 1915, in *Letters of the Tsaritsa to the Tsar,* p. 114.

77. Letter of Nicholas to Aleksandra, August 27, 1915, in *Letters of the Tsar to the Tsaritsa,* p. 73.

78. Letter of Nicholas to Aleksandra, August 25, 1915, in *Letters of the Tsar to the Tsaritsa,* pp. 71–72; Letter of Aleksandra to Nicholas, August 22, 1915, in *Letters of the Tsaritsa to the Tsar,* pp. 114, 116.

79. Letters of Aleksandra to Nicholas, September 9, 1915, November 4, 1916, November 12, 1916, and December 16, 1916, in *Letters of the Tsaritsa to the Tsar,* pp. 152, 433, 441, 454.

80. Letters of Aleksandra to Nicholas, August 22, 25, and 29, 1915, in *ibid.,* pp. 115, 121, 129.

81. Letter of Aleksandra to Nicholas, September 12, 1915, in *ibid.,* p. 159.

82. Letter of Nicholas to Aleksandra, September 23, 1916, in *Letters of the Tsar to the Tsaritsa,* p. 269; Letters of Aleksandra to Nicholas, September 27, 1916, August 23, September 15, 1915, December 5, 1916, and December 13, 1916, and June 22, 1915, in *Letters of the Tsaritsa to the Tsar,* pp. 416, 417, 117, 166, 444, 454, 106.

83. Letter of Nicholas to Aleksandra, December 31, 1915, in *Letters of the Tsar to the Tsaritsa,* p. 122.

84. Denikin, *Career of a Tsarist Officer,* p. 273. See also Commandant Langlois, "Rapport du Commandant Langlois sur sa Quatrième Mission en Russie," Part II, Chapter 3, p. 4.

85. Quoted in Gilliard, *Le destin tragique,* p. 35.

86. Brusilov, *Moi vospominaniia,* p. 159.

87. Quoted in Knox, *With the Russian Army,* II, p. 333.

88. Paustovskii, *Sobranie sochinenii,* III, p. 476.

Chapter VI: The Tsar Takes Command

1. Dubenskii, (ed.), *Ego Imperatorskoe Velichestvo Gosudar Imperator Nikolai Aleksandrovich v deistvuiushchei armii,* IV, pp. 22–27.

2. Letter of Nicholas to Aleksandra, August 27, 1915, in *Letters of the Tsar to the Tsaritsa*, p. 74.

3. Brusilov, *Moi vospominaniia*, p. 180.

4. Sidorov, *Ekonomicheskoe polozhenie Rossii*, pp. 213–220; P. I. Liashchenko, *Istoriia narodnogo khoziaistva SSSR*, II, p. 632–633.

5. *Rossiia v mirovoi voine 1914–1918 goda (v tsifrakh)*, p. 30; Manikovskii, *Boevoe snabzhenie russkoi armii*, I, pp. 123–128, 289–298; Stone, *The Eastern Front*, p. 212.

6. The exact estimates of weapons needed during the fall of 1915 were: 1,479 76-mm light field guns, 1,056 114-mm and 122-mm howitzers, 710 107-mm field guns, and 125 152-mm heavy guns. Between the beginning of August and the end of October, the army received from all domestic and foreign sources: 606 75-mm guns, 140 114-mm and 122-mm howitzers, and 28 152-mm heavy guns. Alekseev's staff made the following estimates of the army's monthly ammunition needs: 2.5 million rounds of 76-mm ammunition, 50,000 107-mm shells, 200,000 122-mm shells, and 110,000 152-mm shells, while Russia's factories produced the following quantities of shell each month: 735,000 (76-mm), 20,250 (107-mm), 30,908 (122-mm), and 10,983 (152-mm). Manikovskii, *Boevoe snabzhenie russkoi armii*, I, pp. 178–187, 344–361, 379–386.

7. Major General Dubenskii, *Ego Imperatorskoe Velichestvo Gosudar' Imperator Nikolai Aleksandrovich v deistvuiushchei armii*, IV, pp. 17–18.

8. Nikolai Nikolaevich's response is quoted in Danilov, *Velikii kniaz' Nikolai Nikolaevich*, p. 269; Letters of Aleksandra to Nicholas, June 25 and August 22, 1915, in *Letters of the Tsaritsa to the Tsar*, pp. 110–111, 115; Letter of Nicholas to Aleksandra, August 25, 1915, in *Letters of the Tsar to the Tsaritsa*, pp. 70–71.

9. Brusilov, *Moi vospominaniia*, p. 248.

10. Andrei Vladimirovich, *Dnevnik*, p. 78.

11. Quoted in Rodzianko, *Reign of Rasputin*, p. 151.

12. The conversations between the Empresses Maria Feodorovna and Aleksandra and others of the imperial family were recorded by Andrei Vladimirovich in his diary the next day. See Andrei Vladimirovich, *Dnevnik*, pp. 76–78.

13. A. O. Arutiunian, *Kavkazskii front 1914–1917gg.*, pp. 120–121.

14. Quoted in Paléologue, *La Russie des tsars*, II, pp. 282, 281.

15. N. G. Korsun, *Pervaia mirovaia voina na Kavkazskom fronte*, pp. 20–21.

16. Arutiunian, *Kavkazskii front*, pp. 139–144; Zaionchkovskii, *Mirovaia voina*, I, pp. 287–290.

17. N. G. Korsun, *Sarykamyshskaia operatsiia*, pp. 20–21.

18. Arutiunian, *Kavkazskii front*, pp. 145–151.

19. *Ibid.*, pp. 174–202, 227–249; N. G. Korsun, *Alashkertskaia i Khamadanskaia operatsii na kavkazskom fronte mirovoi voiny v 1915 godu*, pp. 50–151.

20. Letter of Nicholas to Aleksandra, August 25, 1915, in *Letters of the Tsar to the Tsaritsa*, p. 70.

21. Brusilov, *Moi vospominaniia*, pp. 207–208, 154–155, 191.

22. Letter of Nicholas to Aleksandra, February 24, 1917, in *Letters of the Tsar to the Tsaritsa*, p. 315, and Miliukov, *Vospominaniia*, II, p. 222.

23. Letters of Nicholas to Aleksandra, August 31 and September 9, 1915, December 17, 1916, in *Letters of the Tsar to the Tsaritsa*, pp. 78, 85, 311; Quoted in Paléologue, *An Ambassador's Memoirs*, II, p. 88.

24. Letters of Nicholas to Aleksandra, September 7, October 6, October 31, November 12, and November 26, 1915, in *ibid.*, pp. 108–109, 83, 113, 102, 95–96, and letter of Aleksandra to Nicholas, August 22, 1915, in *Letters of the Tsaritsa to the Tsar*, p. 114.

25. Bubnov, *V tsarskoi stavke*, pp. 179–181; Kondzerovskii, *V Stavke verkhovnogo*, pp. 75–80, 90–95; Otets Georgii Shavel'skii, *Vospominaniia*, I, pp. 323–366; Letters of Nicholas to Aleksandra, August 31, 1915, December 6, 1916, and February 23, 1917, in *Letters of the Tsar to the Tsaritsa*, pp. 78, 300, 313.

26. Letters of Nicholas to Aleksandra, September 7, October 9, November 30, and December 31, 1915, in *Letters of the Tsar to the Tsaritsa*, pp. 98, 115, 123, 84.

27. Quoted in *ibid.*, p. xi.

28. "Mikhail Vasil'evich Alekseev", cols. 379–380; Lemke, *250 dnei*, pp. 139–141.

29. Gurko, *War and Revolution in Russia*, pp. 10–11.

30. Lemke, *250 dnei*, pp. 149, 142.

31. *Ibid.*, pp. 142–143, 148; Bubnov, *V tsarskoi stavke*, p. 168; Letter of Aleksandra to Nicholas, August 29, 1915, in *Letters of the Tsaritsa to the Tsar*, p. 128.

32. Letters of Aleksandra to Nicholas, November 5, December 4, November 11, 1916, and June 22, 1915, *Letters of the Tsaritsa to the Tsar*, pp. 434, 443, 440, 106. Lincoln, *The Romanovs*, p. 696; Kondzerovskii, *V Stavke verkhovnogo*, pp. 79–80; and Katkov, *Russia 1917*, .p. 40.

33. Lemke, *250 dnei*, p. 143.

34. Brusilov, *Moi vospominaniia*, p. 254.

35. Commandant Langlois, "Rapport du Commandant Langlois sur sa Sixième Mission en Russie," 12 juin 1916, Part I, Chapter 1, p. 1.

36. *Ibid.*, pp. 1–2; Knox, *With the Russian Army*, I, p. 249; Stone, *The Eastern Front*, pp. 225–226.

37. Langlois, "Rapport du Commandant Langlois sur sa Sixième Mission en Russie," Part I, Chapter 1, pp. 1–2; Stone, *The Eastern Front*, pp. 225–229.

38. Letters of Aleksandra to Nicholas, November 11, 12, and 13, 1915, in *Letters of the Tsaritsa to the Tsar*, pp. 216, 217, 220; Langlois, "Rapport du Commandant Langlois sur sa Sixième Mission en Russie," Part I, Chapter 1, p. 2.

39. Ol'denburg, *Tsarstvovanie Imperatora Nikolaia II*, II, pp. 188–189.

40. Paléologue, *La Russie des Tsars*, II, p. 127.

41. Quoted in *ibid.*, p. 83. See also, Knox, *With the Russian Army*, I, p. 352.

42. Knox, *With the Russian Army*, pp. 352–353.

43. Letter of Nicholas to Aleksandra, August 27, 1915, in *Letters of the Tsar to the Tsaritsa*, p. 73.

44. Quoted in Golovin, *The Russian Army in the World War*, p. 240.

45. Katkov, *Russia 1917*, pp. 67–68.

46. "Télégramme secret de M. Paléologue au Ministère des Affaires Etrangères," Petrograd, le 26 juillet 1915.

47. Quoted in Florinsky, *Russia*, II, p. 1336.

48. Ol'denburg, *Tsarstvovanie Imperatora Nikolaia II*, II, p. 187.

49. Quoted in Paléologue, *La Russie des Tsars*, II, p. 140.

50. Quoted in *ibid.*, I, pp. 314–315.

51. Quoted in *ibid.*, pp. 314, 315.

52. Buchanan, *My Mission to Russia*, I, p. 237; "Télégramme secret de M. Paléologue au Ministère des Affaires Etrangères," Petrograd, le 27 août 1915 p. 199; Paléologue, *La Russie des Tsars*, II, pp. 56–60.

53. Langlois, "Rapport du Commandant Langlois sur sa Quatrième Mission en Russie," 16 septembre, 1915, Part II, Chapter 3, pp. 2–3; Knox, *With the Russian Army*, I, p. 331.

54. Lemke, *250 dnei*, p. 644.

55. *Rossiia v mirovoi voine (v tsifrakh)*, table 9.

56. Wildman, *End of the Russian Imperial Army*, p. 99.

57. Quoted in Knox, *With the Russian Army*, II, p. 412.

58. P. A. Zaionchkovskii, *Samoderzhavie i russkaia armiia*, pp. 117–118; Golovin, *The Russian Army in the World War*, pp. 21–22.

59. Quoted in Golovin, *The Russian Army in the World War*, p. 110.

60. Quoted in "Prikaz glavnokomanduiushchego armiiami Iugo-Zapadnogo fronta N. I. Ivanova," 22 sentiabria 1915g., in A. L. Sidorov (ed.), *Revoliutsionnoe dvizhenie*, p. 128.

61. Golovin, *Russian Army in the World War*, pp. 101–102.

62. Quoted in M. I. Akhun and V. A. Petrov (eds.), *Bol'sheviki i armiia v 1905–1917gg.*, p. 168. For Krivoshein's comments see: "Zasedanie 4 avgusta 1915 goda," in Iakhontov, "Tiazhelye dni," p. 38. See also Golovin, *Russian Army in the World War*, p. 123.

63. O. A. Chaadaeva (ed.), "Soldatskie pis'ma," No. 43, p. 132.

64. "Prikaz glavnokomanduiushchego armiiami Iugo-Zapadnogo fronta N. I. Ivanova," 22 sentiabria 1915g., in Sidorov (ed.), *Revoliutsionnoe dvizhenie*, pp. 128–129.

65. *Ibid.* See also "Prikaz No. 150 po 19-i pekhotnoi zapasnoi brigade," 15 noiabria 1915g., in Akhun and Petrov (eds.), *Bol'sheviki i armiia v 1905–1917gg.*, p. 166.

66. "Zasedanie 4 avgusta 1915 goda," in Iakhontov, "Tiazhelye dni," p. 38.

67. Golovin, *Russian Army in the World War*, pp. 56–58.

68. "Donesenie Astrakhanskogo gubernatora I. N. Sokolovskogo," 11 sentiabria 1915g., in Sidorov (ed.), *Revoliutsionnoe dvizhenie*, p. 122.

69. See a variety of documents published in *ibid.*, pp. 115–128, especially "Donesenie nachal'nika Petrogradskogo okhrannogo otdeleniia K. I. Globacheva," 11 sentiabria 1915g., and "Raport ispolniaiushchego dolzhnost' iaroslavskogo politseimeistera R. G. Dolivo-Dobrovol'skogo," 12 sentiabria 1915g.

70. Chaadaeva (ed.), "Soldatskie pis'ma," No. 44, p. 132.

71. *Ibid.*, Nos. 97 and 91, pp. 141–142.

72. Iu. A. Pisarev, *Serbiia i Chernogoriia v pervoi mirovoi voine*, pp. 130–155; Zaionchkovskii, *Mirovaia voina*, I, pp. 347–351; Churchill, *The Unknown War*, pp. 331–348.

73. Falkenhayn, *General Headquarters*, pp. 179–186.

74. *Ibid.*, p. 209.

75. Brusilov, *Moi vospominaniia*, pp. 199–200.

76. Lemke, *250 dnei*, p. 545.

77. Ol'denburg, *Tsarstvovanie Imperatora Nikolaia II*, II, p. 197.

78. Alekseev made these remarks in a letter he wrote to Zhilinskii on 18/31 January 1916, reprinted in *Nastuplenie Iugo-Zapadnogo fronta v mae–iiune 1915g. Sbornik dokumentov*, p. 36.

79. *Les Armées françaises dans la Grande Guerre*, IV, pt. 1, annexes, pp. 87–88.

80. *Nastuplenie Iugo-Zapadnogo fronta*, p. 39.

81. R. R. Palmer and Joel Colton, *A History of the Modern World*, p. 680.

82. Falkenhayn's Christmas 1915 report, from which the quotes in this paragraph are taken, is published in Falkenhayn, *General Headquarters*, pp. 209–218. The passages quoted here are from pp. 216–217.

83. *Ibid.*, p. 224.

84. Palmer and Colton, *History of the Modern World*, p. 680.

85. Liddell Hart, *The Real War*, p. 216.

86. Churchill, *The Unknown War*, pp. 354–355.

87. Quoted in Liddell Hart, *The Real War*, p. 218.

88. Quoted in *ibid.*, p. 221.

89. *Mezhdunarodnye otnosheniia v epokhu imperializma*, IX pp. 573–574.

90. *Nastuplenie Iugo-Zapadnogo fronta*, pp. 56–57.

91. M. D. Bonch-Bruevich, *Vsia vlast' sovetam*, pp. 102–103.

92. Knox, *With the Russian Army*, II, p. 410; Stone, *The Eastern Front*, pp. 227–230.

93. Lemke, *250 dnei*, p. 685. See also pp. 616–627, 686–697.
94. Quoted in Liddell Hart, *The Real War*, p. 224.
95. Letter of Nicholas to Aleksandra, June 23, 1916, in *Letters of the Tsar to the Tsaritsa*, p. 217. See also A. I. Denikin, *The Russian Turmoil*, pp. 19–21.
96. Quoted in Paléologue, *La Russie des Tsars*, II, p. 43.
97. Quoted in Miliukov, *Vospominaniia*, II, p. 208.
98. Quoted in Paléologue, *La Russie des Tsars*, II, p. 43.
99. Gippius, *Siniaia kniga*, p. 29.
100. Quoted in Paléologue, *La Russie des Tsars*, II, p. 44.
101. Quoted in Cherniavsky (ed.), *Prologue to Revolution*, p. 16.
102. Quoted in Paléologue, *La Russie des Tsars*, II, p. 43.
103. Quoted in Cherniavsky (ed.), *Prologue to Revolution*, p. 17.

Chapter VII: New Ways of Politics

1. Chaadaeva (ed.), "Soldatskie pis'ma," No. 29, p. 129; E. Glukhovtseva, "Bezhentsy vo vremia voiny 1914–1917gg.," No. 10, sheet 124.
2. Chaadaeva (ed.), "Soldatskie pis'ma," No. 10, p. 125.
3. *Ibid.*, No. 38, pp. 130–131.
4. "Otnoshenie tovarishcha ministra vnutrennikh del V. F. Dzhunkovskogo," 12 ianvaria 1915g., in A. M. Anfimov (ed.), *Krest'ianskoe dvizhenie*, No. 130, p. 231.
5. "Predstavlenie i. d. prokurora Permskogo okruzhnogo suda P. B. Orlovskogo," 16 aprelia 1915g., in *ibid.*, No. 120, p. 220.
6. "Iz doneseniia saratovskogo gubernatora A. A. Shirinskogo-Shikhmatova," 16 iiulia 1915g., in *ibid.*, No. 134, p. 234.
7. See, for example, the documents numbered 78, 87, 115, 116, 126, 128, 129, 139 in *ibid.*, pp. 159, 172–173, 212–216, 226–230, 238–239. The report of "mute dissatisfaction" comes from document 87, pp. 172–173.
8. I. S. Kliuzhev, "Dnevnik," No. 17/148.
9. Fleer, *Rabochee dvizhenie v gody voiny*, pp. 51–102, 211–214.
10. "Raport o rasstrele Ivanovo-Voznesenskikh rabochikh," sheets 9–12; "Spravka departamenta politsii za avgust 1915," "Pis'mo upravliaiushchego ministerstvom vnutrennikh del kn. Shcherbatova voennomu ministru Polivanovu ot 13 avgusta 1915g.," "Proklamatsiia Komiteta Tverskoi gruppy RSDRP, avgust 1915g.," "Ob"iasneniia voennogo ministerstva po zaprosu Gosudarstvennoi Dumy ot 14 avgusta 1915g.," and "Obvinitel'nyi akt po delu o krest'ianakh Petre Arkad'eve i drugikh v chisle 30 chelovek, obviniaemykh po 1 ch. 123 ugol. ulozh. ot 15 iiulia 1916g.," in M. Inozemtsev (ed.), "Rasstrel Ivanovo-Voznesenskikh rabo-chikh v 1915g.," pp. 98, 104, 110–113, 116.
11. "Chislo fabrichno-zavodskikh rabochikh bastovavshikh v 1912, 1913, 1914, 1915, i 1916gg.," p. 185; Leiberov, "Petrogradskii proletariat v gody pervoi mirovoi voiny," pp. 483–484.
12. "Pokazaniia P. N. Miliukova, 4 avgusta 1917 goda," p. 307.
13. Quoted in B. Dvinov, *Pervaia mirovaia voina i rossiiskaia sotsial-demokratiia*, p. 49.
14. "Dopros A. I. Guchkova, 2 avgusta 1917 goda," p. 294.
15. Letters of Aleksandra to Nicholas, September 2 and 11, 1915, in *Letters from the Tsaritsa to the Tsar*, pp. 135, 156.
16. V. I. Startsev, *Russkaia burzhuaziia i samoderzhavie v 1905–1917*, pp. 132–136.
17. Quoted T. I. Polner, *Zhiznennyi put' kniazia Georgiia Evgenievicha L'vova*, p. 211.
18. Quoted in N. Lapin (ed.), "Progressivnyi blok v 1915–1917gg.," *Krasnyi arkhiv*, L-LI, pp. 119–120.
19. M. V. Rodzianko, *Krushenie imperii*, pp. 114, 113.

20. *Utro Rossii,* May 23, 1915, quoted in B. S. Seiranian, *Bor'ba bol'shevikov protiv voenno-promyshlennykh komitetov,* p. 115.

21. Quoted in Zagorsky, *State Control of Industry,* p. 84.

22. Polivanov, *Iz dnevnikov i vospominanii,* p. 205; A. P. Pogrebinskii, "Voenno-promyshlennye komitety," p. 160; *Deiatel'nost' moskovskago oblastnago voenno-promyshlennago komiteta,* p. iii; Jo Ann Ruckman, *The Moscow Business Elite,* pp. 50–53, 190–195.

23. For quotes, see Ol'denburg, *Tsarstvovanie Imperatora Nikolaia II,* II, p. 168n.

24. Quoted in Diakin, *Russkaia burzhuaziia i tsarizm,* p. 74.

25. Pogrebinskii, "Voenno-promyshlennye komitety," pp. 161–183; Sidorov, *Ekonomicheskoe polozhenie Rossii,* pp. 191–212.

26. Kliuzhev, "Dnevnik," p. 168.

27. Quoted in Diakin, *Russkaia burzhuaziia i tsarism,* p. 80.

28. "Pokazaniia P. N. Miliukova, 4 avgusta 1917 goda," p. 313.

29. Quoted in Pares, *Fall of the Russian Monarchy,* p. 155.

30. *Ibid.,* p. 143.

31. Quoted in Diakin, *Russkaia burzhuaziia i tsarizm,* p. 89.

32. Letter of Aleksandra to Nicholas, June 16 and 17, 1915, in *Letters of the Tsaritsa to the Tsar,* pp. 97, 100.

33. Quoted in Diakin, *Russkaia burzhuaziia i tsarizm,* p. 81.

34. Ol'denburg, *Tsarstvovanie Imperatora Nikolaia II,* II, p. 172.

35. Quoted in Diakin, *Russkaia burzhuaziia i tsarizm,* p. 86.

36. Quoted in Pogrebinskii, "Voenno-promyshlennye komitety," p. 178.

37. Rodzianko, *Krushenie imperii,* p. 126.

38. Paléologue, *La Russie des Tsars,* II, p. 34. Excerpts from the Duma speeches are taken from: Dubenskii (ed.), *Ego Imperatorskoe Velichestvo Gosudar' Imperator Nikolai Aleksandrovich v deistvuiushchei armii,* IV, pp. 4–6, and Startsev, *Russkaia burzhuaziia i samoderzhavie,* pp. 142–143.

39. "Télégramme secret de M. Paléologue au Ministère des Affaires Etrangères," Petrograd, le 4 juin 1915, p. 125.

40. Quoted in Startsev, *Russkaia burzhuaziia i samoderzhavie,* pp. 144–146.

41. Paléologue, *La Russie des Tsars,* II, p. 35.

42. Quoted in Diakin, *Russkaia burzhuaziia i tsarizm,* p. 96.

43. "Donesenie nachal'nika moskovskogo okhrannogo otdeleniia direktoru departamenta politsii o soveshchanii u P. P. Riabushinskogo," 14 avgusta 1915g., pp. 20–22; A. Grunt, "Progressivnyi blok," pp. 110–111.

44. Quoted in Ol'denburg, *Tsarstvovanie Imperatora Nikolaia II,* II, p. 181.

45. Miliukov, *Vospominaniia,* II, p. 218.

46. Lapin (ed.), "Progressivnyi blok" pp. 118–121; A. Grunt, "Progressivnyi blok," pp. 109–110; Gurko, *Features and Figures of the Past,* pp. 445–446.

47. Lapin, "Progressivnyi blok," pp. 122–126.

48. *Ibid.,* pp. 137, 159; "Donesenie nachal'nika petrogradskogo okhrannogo otdeleniia v departamente politsii ob obrazovanii 'progressivnogo bloka,' " pp. 26–29.

49. Polivanov, *Iz dnevnikov,* pp. 224–225.

50. Letter of Aleksandra to Nicholas, August 28, 1915, in *Letters of the Tsaritsa to the Tsar,* p. 127.

51. Letter of Aleksandra to Nicholas, August 24, 1915, in *ibid.,* p. 120; Letter of Nicholas to Aleksandra, August 24, 1915, in *Letters of the Tsar to the Tsaritsa,* p. 71.

52. "Zasedanie 26 avgusta 1915 goda," in Iakhontov, "Tiazhelye dni," pp. 107–108.

53. *Ibid.,* pp. 109–110.

54. "Zasedanie 28 avgusta 1915 goda," in Iakhontov, "Tiazhelye dni," p. 120.

55. *Ibid.,* pp. 124, 127–128.

56. Letter of Aleksandra to Nicholas, August 29, 1915, in *Letters of the Tsaritsa to the Tsar*, p. 130.

57. "Zasedanie 2 sentiabria 1915 goda," in Iakhontov, "Tiazhelye dni," pp. 128, 132–133.

58. *Ibid.*, pp. 128, 136.

59. Polivanov, *Iz dnevnikov*, p. 236.

60. Letters of Aleksandra to Nicholas, September 11, 12, 15, 1915, in *Letters of the Tsaritsa to the Tsar*, pp. 156–157, 159–60, 166.

61. Letter of Nicholas to Aleksandra, September 15, 1915, in *Letters of the Tsar to the Tsaritsa*, p. 90; Gurko, *Features and Figures from the Past*, p. 580.

62. Gurko, *Features and Figures from the Past*, pp. 580–581; Letter of Nicholas to Aleksandra, September 16, 1915, in *Letters of the Tsar to the Tsaritsa*, p. 90; "Zasedanie 2 sentiabria 1915 goda," in Iakhontov, "Tiazhelye dni," p. 136.

63. Paléologue, *La Russie des Tsars*, II, p. 86; Gurko, *Features and Figures from the Past*, p. 581.

64. Rodzianko, *Krushenie imperii*, p. 133.

65. Quoted in Startsev, *Russkaia burzhuaziia i samoderzhaviia*, p. 176.

66. Rodzianko, *Krushenie imperii*, p. 133.

67. Sidorov, *Ekonomicheskoe polozhenie Rossii*, pp. 58–59, note 16; Rodzianko, *Krushenie imperii*, pp. 113–117.

68. Miliukov, *Vospominaniia*, II, p. 204.

69. Quoted in Sidorov, *Ekonomicheskoe polozhenie Rossii*, p. 74. See also pp. 69–84; A. S. Lukomskii, *Vospominaniia*, I, p. 65; T. D. Krupina, "Politicheskii krizis 1915g. i sozdanie osobogo soveshchaniia po oborone," *Istoricheskie zapiski*, LXXXIII (1969), pp. 66–67.

70. Krupina, "Politicheskii krizis 1915g.," pp. 68–73, and Sidorov, *Ekonomicheskoe polozhenie Rossii*, pp. 102–103. See also pp. 86–94.

71. Sidorov, *Ekonomicheskoe polozhenie Rossii*, pp. 109–172; Stone, *The Eastern Front*, pp. 210–211.

72. Stone, *The Eastern Front*, p. 201.

73. Diakin, *Russkaia burzhuaziia i tsarizm*, pp. 123–124.

74. V. P. Semennikov, *Politika Romanovykh nakanune revoliutsii*, p. 98.

75. "Donesenie nachal'nika moskovskogo okhrannogo otdeleniia direktoru departamenta politsii o gorodskom s"ezde," p. 50.

76. *Ibid.*, p. 51.

77. "Rezoliutsiia s"ezda predstavitelei gorodov v Moskve," p. 57.

78. Diakin, *Russkaia burzhuaziia i tsarizm*, p. 122.

79. "Obrashchenie kniazia L'vova k tsariu," pp. 59–60.

80. "Iz doklada chlenov deputatsii, izbrannoi sobraniem upolnomochennykh gubernskikh zemstv 9-go sentiabria 1915g.," p. 58.

81. Startsev, *Russkaia burzhuaziia i samoderzhavie*, p. 179.

82. Polivanov, *Iz dnevnikov*, p. 141, note 1; "Dopros A. N. Khvostova, 18 marta 1917g.," pp. 3–5; Diakin, *Russkaia burzhuaziia i tsarizm*, p. 128.

83. Letters of Aleksandra to Nicholas, September 11, and two letters of September 17, 1915, in *Letters of the Tsaritsa to the Tsar*, pp. 157, 170, 175.

84. Letter of Nicholas to Aleksandra, September 18, 1915, in *Letters of the Tsar to the Tsaritsa*, p. 92.

85. Quoted in Pares, *The Fall of the Russian Monarchy*, p. 286.

86. *Ibid.*, pp. 286–287.

87. Quoted in Diakin, *Russkaia burzhuaziia i tsarizm*, pp. 130.

88. Semennikov, *Romanovy i germanskie vliianiia*, pp. 76–118.

89. "Doklad Otdeleniia po okhraneniiu obshchestvennoi bezopasnosti i poriadka v stolitse" (Sovershenno sekretno), 17 dekabria 1915g., sheets 118–125; "Otchet o prodovol'stvii Petrograda merami pravitel'stva upolnomochennago po prodovol'stviiu Petrograda za pervoe polugodie (s oktiabria 1915g. po marta 1916g. vkliuchitel'no)," sheets 16–67; Diakin, *Russkaia burzhuaziia i tsarizm*, pp. 132–133.

90. "Pokazaniia S. P. Beletskogo," pp. 122–125.

91. "Dopros K. D. Kafafova," p. 135.

92. Diakin, *Russkaia burzhuaziia i tsarizm*, pp. 135–140.

93. M. P. Fedorov, *Voina i dorogovizna zhizni 1916g.*, pp. 34–35.

94. Lapin, "Progressivnyi blok," pp. 145–146, 148, 153.

95. Quoted in Diakin, *Russkaia burzhuaziia i tsarizm*, p. 141.

96. Lapin (ed.), "Progressivnyi blok," p. 151.

97. The foregoing quotes are from Ol'denburg, *Tsarstvovanie Imperatora Nikolaia II*, II, p. 186.

98. Florinsky, *Russia*, II, p. 1367.

99. Quoted in Ol'denburg, *Tsarstvovanie Imperatora Nikolaia II*, II, p. 189.

100. Lapin (ed.), "Progressivnyi blok," p. 153.

101. Letter of Aleksandra to Nicholas, January 9, 1916, in *Letters of the Tsaritsa to the Tsar*, p. 261.

102. Rodzianko, *Krushenie imperii*, pp. 142–143.

103. Paléologue, *La Russie des Tsars*, II, p. 168.

104. Lapin (ed.), "Progressivnyi blok," p. 184.

105. Letter of Nicholas to Aleksandra, January 9, 1916, in *Letters of the Tsar to the Tsaritsa*, p. 133.

106. Quoted in *ibid.*, p. 129 note.

107. Paléologue, *La Russie des Tsars*, II, p. 170.

108. Letters of Aleksandra to Nicholas, January 4, 7, and 8, 1916, in *Letters of the Tsaritsa to the Tsar*, pp. 251, 256, 259.

109. K. Betskii and P. Pavlov, *Russkii rokambol' (prikliucheniia I. F. Manasevicha-Manuilova)*, pp. 161–166.

110. Paléologue, *La Russie des Tsars*, II, p. 172.

111. Quoted in Florinsky, *Russia*, II, p. 1364.

112. Letter of Aleksandra to Nicholas, August 22, 1915, in *Letters of the Tsaritsa to the Tsar*, p. 114.

113. Letter of Aleksandra to Nicholas, June 10, 1915, in *ibid.*, p. 86.

Chapter VIII: The Roots of Upheaval

1. Lapin (ed.), "Progressivnyi blok," p. 153.

2. Quoted in Pares, *Fall of the Russian Monarchy*, p. 302.

3. Quoted in *ibid.*, p. 301.

4. V. O. Kliuchevskii, *Sochineniia*, p. 313. Some of the following descriptions of peasant and proletarian living and working conditions include, in abbreviated form, quotes from source materials first used in chapters II and IV of my book *In War's Dark Shadow*.

5. For a dispassionate contemporary report on peasant life in Russia, see A. I. Shingarev, *Vymiraiushchaia derevnia*.

6. A. M. Anfimov, *Rossiiskaia derevnia v gody pervoi mirovoi voiny*, pp. 188–192, 196–197; S. N. Prokopovich, *Voina i narodnoe khoziaistvo*, pp. 229–231; "Predstavlenie Volynskoi gubernskoi zemskoi upravy," 11 dekabria 1915g., pp. 93–98; "Ofitsial'noe pis'mo Podol'skago Gubernatora Glavnokomanduiushchemu Armiiami Iugo-Zapadnago fronta," 6 fevralia 1916g., sheet 32.

7. Anfimov, *Rossiiskaia derevnia*, p. 94.

8. *Ibid.*

9. "O meropriiatiiakh Ministerstva Zemledeliia," 10 oktiabria 1916g., sheets 31–32.

10. "Predstavlenie ministra zemledeliia A. A. Bobrinskogo v Sovet Ministrov," 10 oktiabria 1916g., p. 16.

11. Anfimov, *Rossiiskaia derevnia*, pp. 94–98.

12. "Spravka. V tseliakh vospolneniia ubyli rabochikh rukh v sel'skom khoziaistve priniaty nizhesleduiushchaia mery," sheets 198–199; "Spravka mezhduvedomstvennoi komissii pri ministerstve zemledeliia," 16 fevralia 1916g., sheet 127.

13. "Ofitsial'noe pis'mo ot Shtaba Omskago Voennago Okruga," 21 marta 1916g., sheet 164; "Vypiska iz soobshcheniia Volynskago Gubernatora ot 4 marta 1916 goda, za No. 1615," sheet 167.

14. Anfimov, *Rossiiskaia derevnia*, pp. 276–294; Sidorov, *Ekonomicheskoe polozhenie*, pp. 465–472; George Yaney, *The Urge to Mobilize*, p. 409.

15. Yaney, *Urge to Mobilize*, p. 410.

16. *Ibid.*, pp. 411–413.

17. *Ibid.*; Anfimov, *Rossiiskaia derevnia*, pp. 308–312.

18. Quoted in Anfimov, *Rossiiskaia derevnia*, p. 313.

19. "Khronika krest'iansksogo dvizheniia 1916g.," pp. 501–505.

20. M. E. Saltykov-Shchedrin, "Dnevnik provintsiala v Peterburge," p. 271.

21. Quoted in Diane Koenker, *Moscow Workers and the 1917 Revolution*, p. 52.

22. *Ibid.*, p. 80; Leiberov, "Petrogradskii proletariat v gody pervoi mirovoi voiny," p. 466.

23. P. I. Liashchenko, *Ocherki agrarnoi evoliutsii Rossii*, pp. 143–159; Petr Okhtin, *Bezzemel'nyi proletariat v Rossii*, pp. 28–36.

24. Koenker, *Moscow Workers*, pp. 50–52.

25. P. A. Moiseenko, *Vospominaniia*, p. 17.

26. Robert E. Johnson, *Peasant and Proletarian*, pp. 68–72.

27. Quotes from S. Lapitskaia, *Byt rabochikh trekhgornoi manufaktury*, pp. 62–63. See also S. Gvozdev, *Zapiski fabrichnago inspektora*, pp. 147–148; N. K. Druzhinin (ed.), *Usloviia byta rabochiklh v dorevoliutsionnoi Rossii*, pp. 13–14.

28. Ivan Belousov, *Ushedshaia Moskva*, p. 43.

29. Gvozdev, *Zapiski fabrichnago inspektora*, pp. 155–156.

30. Quotes from Lapitskaia, *Byt rabochikh trekhgornoi manufaktury*, p. 54.

31. N. Krupskaia, *Iz dalekykh vremen*, pp. 9–10.

32. V. Iu. Krupianskaia, "Evoliutsiia semeino-bytovogo uklada rabochikh," pp. 274–275.

33. Quoted in Lapitskaia, *Byt rabochikh trekhgornoi manufaktury*, pp. 67–68.

34. Quoted in I. A. Baklanova, "Formirovanie i polozhenie promyshlennogo proletariata," p. 143.

35. D. G. Kutsentov, "Naselenie Peterburga. Polozhenie peterburgskikh rabochikh," pp. 204–206.

36. Koenker, *Moscow Workers*, p. 55; J. H. Bater, *St. Petersburg*, p. 352; Z. V. Stepanov, *Rabochie Petrograda v period podgotovki i provedeniia oktiabr'skogo vooruzhennogo vosstaniia*, pp. 58–60.

37. Koenker, *Moscow Workers*, p. 57.

38. Quoted in Iu. I. Kir'ianov, *Rabochie iuga Rossii, 1914–fevral' 1917g.*, pp. 103–104.

39. "Spisok gubernii i oblastei," 1916g., sheet 20; "Spravka mezhduvedomstvennoi komissii pri Ministerstve zemledeliia," 16 fevralia 1916g., sheet 50; Sidorov, *Ekonomicheskoe polozhenie Rossii*, p. 416.

40. Quoted in Lincoln, *In War's Dark Shadow*, pp. 125–126. See also pp. 122–128 for

a description of Russia's worst slums—particularly Moscow's Khitrovka—that is especially indebted to V. A. Giliarovskii's brilliant accounts in *Moskva i Moskvichi*, pp. 18–38, 43–60, 275–304, and *Trushchobnye liudi*, II, pp. 74–80, 167–175.

41. N. K. Druzhinin (ed.), *Usloviia byta rabochikh*, pp. 52–53.

42. Quotes from Koenker, *Moscow Workers*, p. 63.

43. Kir'ianov, *Rabochie iuga Rossii*, p. 104.

44. Koenker, *Moscow Workers*, p. 63, note 84.

45. *Ibid.*, pp. 63–64.

46. I. I. Ianzhul, *Iz vospominanii i perepiski*, pp. 126–133; P. A. Peskov, *Fabrichnyi byt vladimirskoi gubernii.*, pp. 125–130; O. A. Parasun'ko, *Polozhenie i bor'ba rabochego klassa Ukrainy*, pp. 184–185; E. E. Kruze, "Rabochie Peterburga v gody novogo revoliutsionnogo pod"ema," p. 406.

47. Koenker, *Moscow Workers*, pp. 83–87; Zagorsky, *State Control of Industry*, Appendix XXV, p. 341; K. A. Pazhitnov, *Polozhenie rabochego klassa v Rossii*, III, pp. 47–54; A. I. Utkin, "Ekonomicheskoe polozhenie moskovskikh rabochikh," pp. 41–53; L. S. Gaponenko, "Polozhenie rabochego klassa Rossii nakanune oktiabr'skoi revoliutsii," pp. 7–10; Iu. N. Kir'ianov, "Ekonomicheskoe polozhenie rabochikh iuga Rossii v gody pervoi mirovoi voiny," pp. 34–35, 40.

48. Sidorov, *Ekonomicheskoe polozhenie Rossii*, p. 412; L. S. Gaponenko, *Rabochii klass Rossii v 1917 godu*, pp. 102–106.

49. Sidorov, *Ekonomicheskoe polozhenie Rossii*, p. 414; Leiberov, "Petrogradskii proletariat," pp. 464–466; Koenker, *Moscow Workers*, p. 80.

50. Koenker, *Moscow Workers*, pp. 82–83; Sidorov, *Ekonomicheskoe polozhenie*, pp. 416–417; Stepanov, *Rabochie Petrograda*, p. 34; S. A. Smith, *Red Petrograd*, pp. 23–26.

51. Quoted in Koenker, *Moscow Workers*, p. 67.

52. Zagorsky, *State Control of Industry*, pp. 53–56; "Dobycha i razrabotka topliva v Rossii v 1917 godu," 1 iiulia 1917g., pp. 20–32; "Kopiia sekretnoi zapiski Ego Prevoskhoditel'stvu Gospodinu Direktoru Departamenta Politsii," sheets 2–6, 27–32.

53. Fedorov, *Voina i dorogovizna zhizni*, p. 42; Stepanov, *Rabochie Petrograda*, pp. 36, 42–50, 59–65; "Predvaritel'nye svedeniia, 1–11 ianvaria 1916 goda," sheets 45–73; "Doklad Ministru Vnutrennikh Del Otdeleniia po okhraneniiu obshchestvennoi bezopasnosti i poriadka v stolitse," (sovershenno sekretno), 17 dekabria 1915g., sheets 119–125.

54. A. G. Rashin, *Formirovanie rabochego klassa Rossii*, p. 105.

55. Smith, *Red Petrograd*, p. 8; Koenker, *Moscow Workers*, pp. 57–58.

56. Fleer, *Rabochee dvizhenie v gody voiny*, pp. 211–214; M. Inozemtsev (ed.), "Rasstrel Ivanovo-Voznesenskikh rabochikh v 1915g.," pp. 97–118; P. M. Ekzempliarskii, *Istoriia goroda Ivanova*, I, pp. 321–326.

57. "Chislo fabrichno-zavodskikh rabochikh bastovavshikh v 1912, 1913, 1914, 1915, i 1916gg.," p. 185; Fleer, *Rabochee dvizhenie v gody voiny*, pp. 102–181; Smith, *Red Petrograd*, p. 50; "Doklad nachal'nika Petrogradskogo okhrannogo otdeleniia," 9 ianvaria 1916g., pp. 398–401; Leiberov, "Petrogradskii proletariat v gody pervoi mirovoi voiny," pp. 500–503.

58. Koenker, *Moscow Workers*, p. 60; Smith, *Red Petrograd*, p. 34.

59. S. S. Volk, "Prosveshchenie i shkola v Peterburge," in Kochakov (ed.), *Ocherki istorii Leningrada*, III, pp. 578–582.

60. Koenker, *Moscow Workers*, p. 60.

61. Quoted in Ekzempliarskii, *Istoriia goroda Ivanova*, I, p. 184.

62. Quoted in *Gosudarstvennaia deiatel'nost' predsedatelia Soveta Ministrov Stats-Sekretaria Petra Arkad'evicha Stolypina*, I, pp. 2–4, 8.

63. "Rech' P. A. Stolypina 5-go dekabria 1908g.," in M. P. Bok, *Vospominaniia o moem ottse P. A. Stolypina*, pp. 290–292; "Rech' P. A. Stolypina, 10 maia 1907 goda," in *ibid.*, p. 248.

64. Nicholas to the Dowager Empress Maria Feodorovna, September 10, 1911, *Letters of Nicholas and Marie*, pp. 265–266; Kokovtsev, *Iz moego proshlago*, I, pp. 475–477; P. G. Kurler, *Gibel' imperatorskoi Rossii* (Berlin, 1923), pp. 121–134; A. Girs, "Smert' Stolypina. Iz vospominanii byvshago Kievskago Gubernatora," in A. Stolypin, *P.A. Stolypin, 1862–1911* (Paris, 1927), pp. 86–102.

65. Quoted in Bertrand D. Wolfe, *Three Who Made a Revolution*, p. 361. See also Edward Crankshaw, *The Shadow of the Winter Palace*, p. 369.

66. Quoted in Hugh Seton-Watson, *The Russian Empire, 1801–1917*, p. 635.

67. I. V. Babushkin, *Vospominaniia*, p. 50.

68. Iu. Martov, *Zapiski sotsialdemokrata* pp. 19–40; Israel Getzler, *Martov*, pp. 1–26; Lincoln, *In War's Dark Shadow*, pp. 178–180.

69. Adam B. Ulam, *The Bolsheviks*, pp. 96–110; Wolfe, *Three Who Made a Revolution*, pp. 65–85; Iu. Z. Polevoi, *Zarozhdenie marksizma v Rossii*, pp. 407–422; Lincoln, *In War's Dark Shadow*, pp. 170–174, 180–184.

70. Quoted in Leopold H. Haimson, *Russian Marxists*, p. 89. See also Allan K. Wildman, *The Making of a Workers' Revolution*, pp. 174–188; V. Levitskii, *Za chetvert' veka*, I, pt. 1, pp. 84–86.

71. Haimson, *Russian Marxists*, p. 84. See also pp. 75–91, and John L. H. Keep, *The Rise of Social Democracy in Russia*, pp. 49–66.

72. V. I. Lenin, "Nasushchennye zadachi nashego dvizheniia," in *Sochineniia*, 4th edn., IV, pp. 342, 344–345; Keep, *The Rise of Social Democracy in Russia*, p. 68.

73. V. I. Lenin, "Chto delat'?" in *Sochineniia*, V, pp. 356, 478, 438.

74. Quoted in Wolfe, *Three Who Made a Revolution*, pp. 235–237. See also Dmitrii Kuz'min, "Kazanskaia demonstratsiia 1876g. i G. V. Plekhanov," pp. 7–40.

75. Quoted in Abraham Ascher, *Pavel Akselrod*, p. 188.

76. Quoted in Getzler, *Martov*, p. 87.

77. Quoted in Haimson, *Russian Marxists*, p. 185.

78. Quoted in Getzler, *Martov*, p. 89.

79. Dan, *Origins of Bolshevism*, p. 402.

80. Getzler, *Martov*, p. 146.

81. Quoted in *ibid.*, pp. 146, 144.

82. Quoted in *ibid.*, p. 146.

83. Quoted in Dan, *Origins of Bolshevism*, p. 402.

84. *Ibid.*, p. 402.

85. *Ibid.*, p. 403.

86. Quoted in Wolfe, *Three Who Made a Revolution*, p. 361.

87. Quoted in Getzler, *Martov*, p. 144.

88. Brusilov, *Moi vospominaniia*, p. 210.

89. *Ibid.*, p. 212. Brusilov's description of his fellow generals' reaction is quoted in Stone, *The Eastern Front*, p. 234.

Chapter IX: The Brusilov Offensive

1. Ludendorff, *My War Memories*, p. 171.

2. Hindenburg, *Out of My Life*, p. 149.

3. *Ibid.*, pp. 148–149.

4. Langlois, "Rapport du Commandant Langlois sur sa Quatrième Mission en Russie," 16 septembre 1915, Part I, Chapter 1, p. 1.

5. Langlois, "Rapport du Commandant Langlois sur sa Sixième Mission en Russie," 12 juin 1916, Chapter V, p. 1.

6. *Ibid.*

7. Général Marquis de Laguiche, "Rapport secret du Général de Laguiche, Attaché Militaire à l'Ambassade de France, à Monsieur le Ministre de la Guerre (Etat-Major de l'Armée, 2e Bureau)"; Letter of Nicholas to Aleksandra, February 6, 1916, in *Letters of the Tsar to the Tsaritsa*, p. 145.

8. Brusilov, *Moi vospominaniia*, p. 210. See also remarks by Kuropatkin quoted in Stone, *The Eastern Front*, p. 234.

9. Quoted in Langlois, "Rapport du Commandant Langlois sur sa Quatrième Mission en Russie," Part I, Chapter 1, p. 6.

10. Brusilov, *Moi vospominaniia*, p. 230.

11. Rostunov, *General Brusilov, passim;* "Viacheslav Napoleonovich Klembovskii," col. 419; "Dmitrii Mikhailovich Karbyshev, cols. 24–25."

12. Brusilov, *Moi vospominaniia*, p. 199.

13. Quoted in Stone, *The Eastern Front*, p. 232.

14. Manikovskii, *Boevoe snabzhenie russkoi armii*, pp. 129, 157, 267–269, 298, 361, 370.

15. "Munitions expediées par les départements techniques et de l'artillerie vers l'interieur de la Russie durant les mois anterieurs à juillet 1916" (secret).

16. Knox, *With the Russian Army*, II, p. 422; "Russian Munitions Tables" (secret), pp. 1–44.

17. Langlois, "Rapport du Commandant Langlois sur sa Quatrième Mission en Russie," Part I, Chapter 1, pp. 1–4.

18. Polivanov, *Iz dnevnikov i vospominanii*, p. 229.

19. Pares, *My Russian Memoirs*, p. 393.

20. Hindenburg, *Out of My Life*, p. 150.

21. V. N. Klembovskii (ed.), *Strategicheskii ocherk*, V, pp. 26–27; Rostunov, *General Brusilov*, pp. 112–113.

22. Brusilov, *Moi vospominaniia*, p. 200.

23. Letters of Aleksandra to Nicholas, January 9, March 6, and March 12, 1916, in *Letters of the Tsaritsa to the Tsar*, pp. 260, 290, 297; Letter of Nicholas to Aleksandra, March 10, 1916, in *Letters of the Tsar to the Tsaritsa*, p. 155.

24. Knox, *With the Russian Army*, I, p. 356; II, pp. 510–511.

25. *Sbornik dokumentov: Nastuplenie Iugo-Zapadnogo fronta v mae–iiune 1916 goda*, pp. 68–74.

26. Brusilov, *Moi vospominaniia*, pp. 211–212.

27. Nicholas to Aleksandra, March 31, 1916, in *Letters of the Tsar to the Tsaritsa*, p. 162.

28. Brusilov, *Moi vospominaniia*, pp. 208–211.

29. Stone, *The Eastern Front*, p. 235.

30. Golovin, *The Russian Army in the World War*, pp. 220–221.

31. Quoted in G. Belov, "Russkii polkovodets A. A. Brusilov," p. 43.

32. Quoted in *ibid.*, p. 44.

33. Quoted in Golovin, *The Russian Army in the World War*, p. 95.

34. Pares, *Fall of the Russian Monarchy*, pp. 359–362; Stone, *The Eastern Front*, pp. 235–238.

35. Knox, *With the Russian Army*, I, p. 287; Pares, *My Russian Memoirs*, p. 393.

36. Stone, *The Eastern Front*, p. 242.

37. *Ibid.*, p. 243.

38. *Sbornik dokumentov: Nastuplenie Iugo-Zapadnogo fronta v mae–iiune 1916g.* pp. 174–189.

39. Brusilov, *Moi vospominaniia*, p. 218.
40. L. V. Vetoshnikov, *Brusilovskii proryv*, p. 56.
41. Quoted in Stone, *The Eastern Front*, p. 234.
42. Brusilov, *Moi vospominaniia*, pp. 231–232.
43. Brusilov, *Moi vospominaniia*, p. 230.
44. Quoted in Stone, *The Eastern Front*, p. 249.
45. Brusilov, *Moi vospominaniia*, pp. 230–231; Knox, *With the Russian Army*, II, pp. 440–442.
46. Belov, "Russkii polkovodets A. A. Brusilov," p. 44; Rostunov, *General Brusilov*, pp. 140–142; Letters of Nicholas to Aleksandra, May 27 and 29, 1916, in *Letters of the Tsar to the Tsaritsa*, pp. 191, 193; Paléologue, *La Russie des tsars*, II, p. 284.
47. Sbornik dokumentov: *Nastuplenie Iugo-Zapadnogo fronta v mae–iiune 1916 goda*, p. 290.
48. Quoted in Stone, *The Eastern Front*, p. 250.
49. *Ibid.*, pp. 250–254; Rostunov, *General Brusilov*, pp. 141–142; Rostunov, *Russkii front*, pp. 311–318; Sbornik dokumentov: *Nastuplenie Iugo-Zapadnogo fronta v mae–iiune 1916 goda*, pp. 269–276, 520; Klembovskii (ed.), *Strategicheskii ocherk*, V, pp. 36–46. Zaionchkovskii, *Pervaia mirovaia voina*, II, pp. 49–54.
50. Quoted in Stone, *The Eastern Front*, p. 243.
51. Ludendorff, *My War Memories*, p. 220; Hindenburg, *Out of My Life*, pp. 164, 158.
52. Ludendorff, *My War Memories*, pp. 220–221, 225; Hindenburg, *Out of My Life*, p. 158; Hoffmann, *War of Lost Opportunities*, p. 149.
53. Stone, *The Eastern Front*, pp. 255–256.
54. Quoted in Stone, *The Eastern Front*, p. 257. See also pp. 255–257, and Brusilov, *Moi vospominaniia*, pp. 231–235.
55. Quoted in Stone, *The Eastern Front*, p. 257.
56. Sbornik dokumentov: *Nastuplenie Iugo-Zapadnogo fronta v mae–iiune 1916 goda*, p. 389.
57. Klembovskii (ed.), *Strategicheskii ocherk*, V, pp. 62–66; Rostunov, *Russkii front*, pp. 321–323; Stone, *The Eastern Front*, pp. 260–261; Knox, *With the Russian Army*, II, pp. 450–451.
58. Brusilov, *Moi vospominaniia*, p. 237.
59. *Ibid.*, p. 241.
60. Letter of Nicholas to Aleksandra, June 22, 1916, in *Letters of the Tsar to the Tsaritsa*, p. 216.
61. Rostunov, *General Brusilov*, pp. 154–155; Kersnovskii, *Istoriia russkoi armii*, IV, pp. 805–807.
62. Letter of Nicholas to Aleksandra, October 9, 1915, in *Letters of the Tsar to the Tsaritsa*, p. 99.
63. Knox, *With The Russian Army*, II, p. 472.
64. *Ibid.*, p. 463; Brusilov, *Moi vospominaniia*, p. 242.
65. Brusilov, *Moi vospominaniia*, p. 242; Rodzianko, *Krushenie imperii*, p. 170.
66. Knox, *With the Russian Army*, II, pp. 466–467, 470; Rodzianko, *Krushenie imperii*, p. 173.
67. Letter of Nicholas to Aleksandra, July 17, 1916, in *Letters of the Tsar to the Tsaritsa*, p. 231.
68. Rodzianko, *Krushenie imperii*, pp. 173–174.
69. Knox, *With the Russian Army*, II, p. 473.
70. Letter of Nicholas to Aleksandra, August 18, 1916, in *Letters of the Tsar to the Tsaritsa*, p. 250.
71. Stone, *The Eastern Front*, p. 225.

72. Zaionchkovskii, *Mirovaia voina*, II, pp. 63–65.

73. Stone, *The Eastern Front*, p. 270.

74. *Sbornik dokumentov: Nastuplenie Iugo-Zapadnogo fronta v mae–iiune 1916g.*, p. 291.

75. Klembovskii (ed.), *Strategicheskii ocherk*, V, appendix 15, p. 120.

76. *Ibid.*, pp. 73, 108–109.

77. Buchanan, *My Mission to Russia*, II, p. 2.

78. Hindenburg, *Out of My Life*, p. 199.

79. Quoted in Paléologue, *La Russie des Tsars*, II, p. 330.

80. Quoted in Buchanan, *My Mission to Russia*, II, p. 2.

81. Quoted in Stone, *The Eastern Front*, p. 265. See also pp. 264–266.

82. Quoted in Buchanan, *My Mission to Russia*, II, p. 2.

83. Pares, *My Russian Memoirs*, p. 407; A. M. Zaionchkovskii, "Vsepoddanneishaia zapiska," pp. 29–31.

84. Stone, *The Eastern Front*, p. 265.

85. Knox, *With the Russian Army*, II, pp. 473, 484.

86. Stone, *The Eastern Front*, p. 272.

87. Hindenburg, *Out of My Life*, p. 273.

88. "Doklad petrogradskogo okhrannogo otdeleniia, oktiabr' 1916g.," p. 12.

89. Wildman, *End of the Russian Imperial Army*, p. 115.

90. Quoted in *ibid.*

91. "Zakliuchenie voennogo prokurora ob"edinennogo suda 12-i armii D. M. Matiasa," dekabr' 1916g.," p. 272.

92. "Vyborka iz pis'ma za 28 oktiabria 406 p. Shchigrovskii polk. D. Nikolaeva, g. Kishinev, Skulianskaia rogata," in Chaadaeva (ed.), "Soldatskie pis'ma," p. 144.

93. Quoted in Pares, *Fall of the Russian Monarchy*, p. 375.

94. Quoted in Paléologue, *La Russie des Tsars*, III, p. 24.

95. Paustovskii, *Sobranie sochinenii*, III, pp. 546–548.

Chapter X: "Dancing a Last Tango"

1. Paustovskii, *Sobranie sochinenii*, III, p. 548.

2. V. V. Shul'gin, *Dni*, p. 76.

3. Marc Slonim, *From Chekhov to the Revolution*, p. 104.

4. See the appropriate portions of Solovev's famous *Three Conversations* in S. Frank's *A Solovyov Anthology;* K. Mochul'skii, *Vladimir Solov'ev*, pp. 253–261; Samuel D. Cioran, *Vladimir Solov'ev*, pp. 62–67; and Lincoln, *In War's Dark Shadow*, pp. 351–352.

5. Quoted in Sergei Hackel, *The Poet and the Revolution*, p. 2. See also Bernice Glatser Rosenthal, *Dmitri Sergeevich Merezhkovsky and the Silver Age*, pp. 106–108, 128–129, and Slonin, *From Chekhov to the Revolution*, pp. 111–112.

6. Valerii Briusov, "Griadushchie gunny," in his *Sobranie sochinenii v semi tomakh*, I, p. 433. The following discussion of the ideas of Briusov, Blok, Belyi, and their Symbolist contemporaries contains some quotations from their writings which appeared, in a less abbreviated form, in Chapter X of my book *In War's Dark Shadow*.

7. Billington, *The Icon and the Axe*, p. 507.

8. Valerii Briusov, "Kon' bled," in his *Sobranie sochinenii*, I, pp. 442–444.

9. Quoted in Avril Pyman, *The Life of Aleksandr Blok: The Distant Thunder*, p. 229.

10. Andrei Belyi, *Nachalo veka*, pp. 321–322, 328.

11. Quoted in Pyman, *Distant Thunder*, p. 241.

12. Oleg Maslenikov, *The Frenzied Poets*, pp. 206–207.

13. Viacheslav Ivanov, *Po zvedzam: stat'i i aforizmy*, p. 372.

14. Renato Poggioli, *The Poets of Russia, 1890–1930*, p. 76.
15. Quotes from Slonim, *From Chekhov to the Revolution*, pp. 108, 110; and Billington, *The Icon and the Axe*, p. 508.
16. Aleksandr Blok, "Kometa," in his *Sobranie sochinenii v shesti tomakh*, III, pp. 87–88.
17. Avril Pyman, *The Life of Aleksandr Blok: The Release of Harmony*, pp. 91–92.
18. Quotes from Marc Slonim, *Russian Theatre*, p. 193; and Pyman, *Release of Harmony*, p. 86.
19. "Pis'mo Aleksandra Bloka k materi, 4 aprelia 1910g.," in Aleksandr Blok, *Pis'ma Aleksandra Bloka k rodnym*, II, p. 68.
20. Quoted in Pyman, *Release of Harmony*, p. 120.
21. Blok's introduction to "Vozmezdie," in his *Sobranie sochinenii v shesti tomakh*, III, pp. 187–188.
22. A. S. Tager (ed.), "Protsess Beilisa," pp. 85–125; A. S. Tager, *The Decay of Czarism*, pp. 18–82.
23. Quoted in Pyman, *Release of Harmony*, p. 119.
24. Paustovskii, *Sobranie sochinenii*, p. 311; Gippius, *Siniaia kniga*, pp. 13–14.
25. Quoted in Pyman, *Release of Harmony.*, p. 119.
26. Quoted in *Ibid.*, p. 135.
27. "Letter of Blok to Belyi, 11 March 1911," in V. I. Orlov (ed.), *Aleksandr Blok i Andrei Belyi*, p. 249.
28. Quoted in Zinaida Gippius, *Dmitrii Merezhkovskii*, p. 208.
29. "Pis'mo Aleksandra Bloka k materi, 8 marta 1911g.," in Blok, *Pis'ma Aleksandra Bloka k rodnym*, II, p. 132.
30. Quoted in Andrei Bely [Belyi], *Petersburg*, p. xxiv.
31. *Ibid.*
32. *Ibid.*, p. 168.
33. *Ibid.*, p. 63.
34. *Ibid.*, pp. 11, 50, 170, 11, 63, 9, 11, 209, 186, 207, 65.
35. *Ibid.*, p. 64.
36. Quoted in Pyman, *Release of Harmony*, p. 119.
37. Bely [Belyi], *Petersburg*, p. 65.
38. Quoted in Pyman, *Release of Harmony*, p. 140.
39. Quoted in Gippius, *Dmitrii Merezhkovskii*, p. 208.
40. A. G. Rashin, "Gramotnost' i narodnoe obrazovanie," Jeffrey Brooks, "The Zemstvo and the Education of the People," pp. 269–274.
41. Ilya Ehrenburg, *People and Life, Memoirs of 1891–1917*, p. 173.
42. Quoted in Edward J. Brown, *Mayakovsky*, pp. 43–45.
43. Quoted in Vladimir Mayakovsky [Maiakovskii], *The Bedbug and Selected Poetry*, p. 18.
44. Quoted in *ibid.*, p. 19.
45. V. Pertsov, *Maiakovskii: Zhizn' i tvorchestvo do velikoi oktiabr'skoi sotsialisticheskoi revoliutsii, 1893–1917gg.*, pp. 100–258; E. Usievich, *Vladimir Maiakovskii*, pp. 5–37; Brown, *Mayakovsky*, pp. 40–46.
46. Quoted in Usievich, *Vladimir Maiakovskii*, p. 37.
47. An excerpt of this account by Maiakovskii's fellow futurist Benedikt Livshits is in Woroszylski, *Life of Mayakovsky*, p. 138. See also Viktor Shklovsky, *Mayakovsky and His Circle*, pp. 66–67.
48. This is Ivan Bunin's recollection, excerpted in Woroszylski, *Life of Mayakovsky*, p. 139.
49. Bunin's and Livshits's recollections excerpted in *ibid.*, pp. 138–139.

50. *Ibid.*

51. Vladimir Maiakovskii, "Vam!", in his *Sobranie sochinenii v vos'mi tomakh,* I, p. 73.

52. "Birzhevye vedomosti," February 13, 1915, excerpted in Woroszylski, *Life of Mayakovsky,* p. 140.

53. This remark is taken from the diary of Boris Iurkovskii, excerpted in Woroszylski, *Life of Mayakovsky,* p. 144.

54. Pertsov, *Maiakovskii, 1893–1917gg.,* pp. 357–363; Shklovsky, *Mayakovsky and His Circle,* p. 88.

55. Vladimir Maiakovskii, "Voina i Mir," in his *Sobranie sochinenii,* I, pp. 160, 164, 170, 173–175.

56. *Ibid.,* pp. 187, 189.

57. Quoted in Brown, *Mayakovsky,* p. 115.

58. Shklovsky, *Mayakovsky and His Circle,* p. 87.

59. Mikhail Artsybashev, *Sanine. A Russian Love Novel,* trans. Percy Pinkerton, p. 122.

60. *Ibid.,* pp. 122–123.

61. Quoted in Richard Stites, *The Women's Liberation Movement in Russia,* p. 187.

62. Quoted in *ibid.,* p. 266.

63. Quoted in *ibid.,* p. 251.

64. Beatrice Farnsworth, *Aleksandra Kollontai,* p. 3.

65. *Ibid.,* pp. 29–67; A. M. Itkina, *Revoliutsioner, tribun, diplomat,* pp. 11–38; Stites, *Women's Liberation Movement in Russia,* pp. 249–252.

66. Aleksandra Kollontai, *Love and the New Morality,* pp. 22–25.

67. Louise Bryant, *Mirrors of Moscow,* p. 113.

68. Quoted in Stites, *Women's Liberation Movement in Russia,* p. 355.

69. Kollontai, *Love and the New Morality,* p. 25.

70. Quoted in Stites, *Women's Liberation Movement in Russia,* p. 355.

71. Quoted in *ibid.,* p. 267.

72. Quoted in Prince Felix Youssoupoff [Iusupov], *Avant l'exil,* p. 154; Salisbury, *Black Night, White Snow,* p. 294.

73. Letter of Aleksandra to Nicholas, October 30, 1916, in *Letters of the Tsaritsa to the Tsar,* p. 429; Billington, *The Icon and the Axe,* p. 500.

74. Quoted in Salisbury, *Black Night, White Snow,* p. 302. See also Aron Simanovich, *Raspoutine,* pp. 36–40.

75. Fülöp-Miller, *Holy Devil,* pp. 181–182.

76. Prince Felix Youssoupoff [Iusupov], *Rasputin,* pp. 90–91.

77. Alex de Jonge, *The Life and Times of Grigorii Rasputin,* pp. 223–225. Quote from p. 224.

78. Quoted in Fülöp-Miller, *Holy Devil,* p. 271.

79. Colin Wilson, *Rasputin and the Fall of the Romanovs,* p. 181.

80. Quoted in Youssoupoff, *Avant l'exil,* p. 199.

81. "Letter of Aleksandra to Nicholas, December 5, 1916," in *Letters of the Tsaritsa to the Tsar,* p. 444.

82. Shul'gin, *Dni,* p. 115.

83. N. V. Gogol', *Polnoe sobranie sochinenii,* VIII, p. 41.

84. Quoted in de Jonge, *Life and Times of Grigorii Rasputin,* p. 219.

85. *Ibid.*

86. "Pokazaniia S. P. Beletskogo, 24 iiunia 1917 goda," pp. 171–174, 332–336; Lincoln, *The Romanovs,* pp. 700–701.

87. Viroubova [Vyrubova], *Memories,* p. 162.

88. Khvostov's testimony before the Supreme Investigating Commission, "Dopros A. N. Khvostova, 18 marta 1917g.," p. 48.

89. *Ibid.*, pp. 48–49; Fülöp-Miller, *Holy Devil*, pp. 108–109.

90. "Letter of Aleksandra to Nicholas, September 10, 1915," in *Letters of the Tsaritsa to the Tsar*, p. 240; V. P. Semennikov, *Politika Romanovykh nakanune revoliutsii*, pp. 136–137; Pares, *Fall of the Russian Monarchy*, p. 264.

91. Testimony of Khvostov and Protopopov before the Supreme Investigating Commission, "Dopros A. N. Khvostova, 18 marta 1917g." pp. 17–20; and "Dopros A. D. Protopopova, 21 marta 1917g.," pp. 177–180; Fülöp-Miller, *Holy Devil*, pp. 106–107.

92. Semennikov, *Politika Romanovykh nakanune revoliutsii*, pp. 130–140.

93. Paléologue, *La Russie des Tsars*, III, p. 111; "Dopros A. N. Khvostova, 18 marta 1917g.," pp. 16–21.

94. Pares, *Fall of the Russian Monarchy*, p. 284.

95. "Dopros kniazia M. M. Andronikova, 8 aprelia 1917g.," p. 11.

96. Fülöp-Miller, *Holy Devil*, p. 104; Pares, *Fall of the Russian Monarchy*, p. 283.

97. "Dopros Grafa Frederiksa," p. 46.

98. Letter of Aleksandra to Nicholas, December 1, 1915, in *Letters of the Tsaritsa to the Tsar*, p. 231.

99. "Dopros A. N. Khvostova, 18 marta 1917"; "Dopros kniazia M. M. Andronikova, 6 aprelia 1917, pp. 9–23; "Dopros A. D. Protopopova, 8 aprelia 1917"; "Pokazanie A. D. Protopopova," pp. 20–22, 361–402; II, pp. 9–23; IV, pp. 157–187.

100. Bunin's account quoted in Woroszylski, *Life of Mayakovsky*, p. 139.

101. Aleksei Tolstoi, *Khozhdenie po mukam*, in *Izbrannye sochineniia v shesti tomakh* III, pp. 8–10.

102. "Pokazanie A. D. Protopopova," IV, pp. 22–23.

103. Shul'gin, *Dni*, pp. 100–101.

104. Quoted in Florinsky, *Russia*, II, p. 1365.

105. Letters of Aleksandra to Nicholas, December 13 and December 14, 1916, in *Letters of the Tsaritsa to the Tsar*, pp. 454–455.

Chapter XI: "Is This Stupidity, or Is This Treason?"

1. Quoted in Raymond Pearson, *The Russian Moderates*, p. 78.

2. Rodzianko, *Krushenie imperii*, p. 149; Gurko, *Features and Figures of the Past*, p. 428.

3. Rodzianko, *Krushenie imperii*, p. 149.

4. *Ibid.*, 150; Paléologue, *La Russie des Tsars*, II, p. 196.

5. Published in Dubenskii (ed.), *Ego Imperatorskoe Velichestvo Gosudar' Imperator Nikolai Aleksandrovich v deistvuiushchei armii*, IV, p. 221.

6. Miliukov, *Vospominaniia*, II, p. 226.

7. Paléologue, *La Russie des Tsars*, II, p. 196; Dubenskii (ed.), *Nikolai Aleksandrovich v deistvuiushchei armii,*, IV, p. 221.

8. Rodzianko, *Krushenie imperii*, pp. 149–150.

9. *Ibid.*, p. 151.

10. Quoted in Diakin, *Russkaia burzhuaziia*, p. 170.

11. *Ibid.*

12. Quoted in Tsuyoshi Hasegawa, *The February Revolution*, p. 46.

13. "Donesenie nachal'nika moskovskogo okhrannogo otdeleniia . . . o VI s"ezde partii k.-d.," 25 fevralia, 1916g., p. 73.

14. "Svodka moskovskogo okhrannogo otdeleniia o 'nastroenii obshchestva' na 29 fevralia 1916g.," in *ibid.*, p. 76.

15. "Donesenie nachal'nika moskovskogo okhrannogo otdeleniia . . . VI s "ezde partii k.-d., 25 fevralia 1916g., p. 73.

16. Letters of Aleksandra to Nicholas, September 11, 1915, and March 2, 1916, in *Letters of the Tsaritsa to the Tsar*, pp. 157, 283.

17. B. V. Anan'ich, R. Sh. Ganelin, B. B. Dubentsov, V. S. Diakin, and S. I. Potolov, *Krizis samoderzhaviia v Rossii, 1895–1917gg.*, p. 583.

18. *Ibid.*, pp. 582–584; testimony of Khvostov, Beletskii, and Polivanov in Shchegolev (ed.), *Padenie*, I, pp. 7–110; IV, pp. 335–415; VII, pp. 79–80; Pares, *Fall of the Russian Monarchy*, pp. 310–317.

19. Ananich *et. al.*, *Krizis samoderzhaviia*, p. 583; Pares, *Fall of the Russian Monarchy*, pp. 312–314.

20. "Dopros A. A. Khvostova, 12 iiulia 1917 goda," p. 458; Shulgin, *Dni*, p. 79; "Pokazaniia Kniazia V. M. Volkonskogo, 19 iiulia 1917 goda," p. 138; "Pokazaniia A. N. Naumova, 4 aprelia 1917, goda," p. 345.

21. "Dopros gr. P. N. Ignat'eva," pp. 17–18; "Dopros Generala A. A. Polivanova, 25 avgusta 1917 goda," p. 78.

22. Letter of Nicholas to Aleksandra, January 28, 1916, in *Letters of the Tsar to the Tsaritsa*, p. 140; Letters of Aleksandra to Nicholas, January 29, and April 5, 1916, in *Letters of the Tsaritsa to the Tsar*, pp. 268, 317.

23. "Dopros Generala A. A. Polivanova," 25 avgusta 1917g.," pp. 81–82; Rodzianko, *Krushenie imperii*, p. 155.

24. Quoted in Knox, *With the Russian Army*, II, p. 414.

25. Letter of Aleksandra to Nicholas, January 9, 1916, in *Letters of the Tsaritsa to the Tsar*, p. 260.

26. Letter of Aleksandra to Nicholas, March 12, 1916, in *ibid.*, p. 297.

27. Letter of Nicholas to Aleksandra, March 10, 1916, in *Letters of the Tsar to the Tsaritsa*, p. 155.

28. Knox, *With the Russian Army*, II, p. 412; Rodzianko, *Krushenie imperii*, p. 156.

29. Paléologue, *La Russie des Tsars*, II, p. 236.

30. "Svodka moskovskogo okhrannogo otdeleniia o 'nastroenii obshchestva' na 29 fevralia 1916g.," p. 75.

31. Quoted in Ananich *et al.*, *Krizis samoderzhaviia*, p. 585.

32. "Kopiia sekretnoi zapiski ego prevoskhoditel'stvu gospodinu direktoru departamenta politsii," 20 iiulia 1915g; "Sekretnyi doklad Ego prevoskhoditel'stvu gospodinu tovarishchu ministra vnutrennikh del," 13 avgusta 1915g; and "Sekretnaia zapiska otdeleniia po okhraneniiu obshchestvennoi bezopasnosti i poriadke v stolitse," 2 sentiabria 1915g., sheets 34–35, 40–42, 70–71.

33. Pearson, *The Russian Moderates*, p. 89; Pares, *Fall of the Russian Monarchy*, p. 105.

34. See especially the letters of Aleksandra to Nicholas, September 2 and 11, 1915, in *Letters of the Tsaritsa to the Tsar*, pp. 135, 156.

35. Lemke, *250 dnei*, p. 341.

36. Quoted in Thomas Riha, *A Russian European*, p. 244.

37. Pearson, *The Russian Moderates*, p. 81.

38. Quoted in Riha, *A Russian European*, p. 245.

39. Quoted in Pearson, *The Russian Moderates*, p. 83

40. Quoted in *ibid.*, p. 82.

41. Quoted in Riha, *A Russian European*, p. 246.

42. Pearson, *The Russian Moderates*, pp. 83–88.

43. Quotes from Riha, *A Russian European*, pp. 247–248.

44. "Donesenie nachal'nika moskovskogo okhrannogo otdeleniia direktoru departamenta politsii o vystupleniiakh A. I. Konovalova sredi moskovskikh promyshlennikov," p. 140.

45. Quoted in Rodzianko, *Krushenie imperii,* p. 159.
46. Quoted in Diakin, *Russkaia burzhuaziia,* p. 202.
47. Pearson, *The Russian Moderates,* pp. 92–96.
48. Rodzianko, *Krushenie imperii,* p. 162.
49. "Doklad Generala Alekseeva Nikolaiu II," p. 264.
50. Letter of Nicholas to Aleksandra, June 11, 1916, in *Letters of the Tsar to the Tsaritsa,* p. 207.
51. Rodzianko, *Krushenie imperii,* pp. 166–167.
52. Letter of Nicholas to Aleksandra, June 25, 1916, in *Letters of the Tsar to the Tsaritsa,* p. 219.
53. "Dopros B. V. Sturmera, 22 marta 1917g.," pp. 224–225.
54. Letter of Aleksandra to Nicholas, March 17, 1916, in *Letters of the Tsaritsa to the Tsar,* p. 305.
55. *Ibid.,* pp. 240–246.
56. Paléologue, *La Russie des Tsars,* II, p. 325; Buchanan, *My Mission to Russia,* II, pp. 18–20.
57. Rodzianko, *Krushenie imperii,* p. 169.
58. Miliukov, *Vospominaniia,* II, p. 217.
59. Quoted in *ibid.,* p. 259.
60. Sidorov, *Ekonomicheskoe polozhenie Rossii,* pp. 479–499; Diakin, *Russkaia burzhuaziia,* pp. 207–216; P. B. Volobuev, *Ekonomicheskaia politika vremennogo pravitel'stva,* pp. 385–444; A. M. Anfimov, *Rossiiskaia derevnia v gody pervoi mirovoi voiny,* pp. 305–317; A. G. Shliapnikov (ed.), *Kanun semnadtsatogo goda,* II, pp. 63–72.
61. Thomas Fallows, "Politics and the War Effort in Russia," pp. 70–90; Yaney, *The Urge to Mobilize,* pp. 427–442.
62. Betskii and Pavlov, *Russkii rokambol',* p. 191.
63. Quoted in Rodzianko, *Krushenie imperii,* pp. 168–169.
64. Paléologue, *La Russie des Tsars,* II, p. 172.
65. Letters of Aleksandra to Nicholas, September 7 and 22, 1916, in *Letters of the Tsaritsa to the Tsar,* pp. 394, 408–409; Pares, *Fall of the Russian Monarchy,* p. 378.
66. Letters of Aleksandra to Nicholas, September 7 and November 11, 1916, in *Letters of the Tsaritsa to the Tsar,* pp. 408–409, 440.
67. Letters of Nicholas to Aleksandra, September 9 and 10, 1916, in *Letters of the Tsar to the Tsaritsa,* pp. 256–258.
68. Quoted in Pares, *Fall of the Russian Monarchy,* pp. 376–378.
69. "Pokazaniia kn. V. M. Volkonskogo," 19 iiulia 1917 goda, p. 138; Purishkevich, *Dnevnik,* ·p. 48.
70. Quoted in Diakin, *Russkaia burzhuaziia i tsarizm,* p. 231.
71. "Donesenie nachal'nika moskovskogo okhrannogo otdeleniia . . . o soveshchaniiakh u A. I. Konovalova," 12 oktiabria 1916g., p. 141.
72. Letters of Aleksandra to Nicholas, September 7 and 9, 1916, in *Letters of the Tsaritsa to the Tsar,* pp. 394–395.
73. "Zapiska Badmaeva o lechenii naslednika," and "Telegrammy A. D. Protopopova Badmaevu," in V. P. Semennikov (ed.), *Za kulisami tsarizma,* pp. 17–18; "Dopros A. D. Protopopova, 21 marta 1917 goda.," pp. 114–116. Rasputin quoted in V. M. Purishkevich, *Dnevnik,* p. 77. For an account of Protopopov's general symptoms, see "Dopros M. V. Rodzianko," pp. 141–144, and Paléologue, *La Russie des Tsars,* III, pp. 38–39.
74. "Dopros M. V. Rodzianko," p. 145; Buchanan, *My Mission to Russia,* II, p. 51; Palélogue, *La Russie des Tsars,* III, p. 39.
75. See Shingarev's comments in "Soveshchanie s A. D. Protopopovym u M. V. Rodzianko, 19 oktiabria 1916 goda," II, p. 100.

76. *Ibid.*, p. 99. See also Florinsky, *The End of the Russian Empire*, pp. 89–93; E. D. Chermenskii, *IV Gosudarstvennaia duma*, pp. 198–202.

77. Quoted in Riha, *A Russian European*, p. 265.

78. F. A. Golder, *Documents of Russian History*, p. 165.

79. Miliukov, *Vospominaniia*, II, p. 277.

80. Quoted in Pearson, *The Russian Moderates*, p. 116.

81. Diakin, *Russkaia burzhuaziia*, p. 244.

82. Letter of Nicholas to Aleksandra, November 8, 1916, in *Letters of the Tsar to the Tsaritsa*, p. 295; Letters of Aleksandra to Nicholas, November 9, 1916, in *Letters of the Tsaritsa to the Tsar*, p.437.

83. Miliukov, *Vospominaniia*, II, p. 277.

84. Shul'gin, *Dni*, p. 86.

85. Letters of Aleksandra to Nicholas, November 10, 11, and 12, 1916, in *Letters of the Tsaritsa to the Tsar*, pp. 438–442.

86. Letter of Nicholas to Aleksandra, December 4, 1916, in *Letters of the Tsar to the Tsaritsa*, p. 299, and Letter of Aleksandra to Nicholas, December 4, 1916, in *Letters of the Tsaritsa to the Tsar*, pp. 442–443.

87. Diakin, *Russkaia burzhuaziia*, pp. 242–256; Chermenskii, *IV gosudarstvennaia duma*, pp. 222–229; Pearson, *The Russian Moderates*, pp. 119–123; Hasegawa, *The February Revolution*, pp. 57–58.

88. Purishkevich, *Dnevnik*, pp. 7–8.

89. *Stenograficheskii otchet. Gosudarstvennaia Duma*, Chetvertyi sozyv, Sessiia V, zasedanie shestoe, 19 noiabria 1916g., (Petrograd, 1916), cols. 260–288; Purishkevich, *Dnevnik*, pp. 5–6. See also Buchanan's "Report No. 261, to Lord Grey" of December 5, 1916, pp. 163–170; and S. B. Liubosh, *Russkii fashist Vladimir Purishkevich*, passim.

90. Purishkevich, *Dnevnik*, p. 6.

91. Shul'gin, *Dni*, p. 86.

92. Purishkevich, *Dnevnik*, p. 11.

93. Youssoupoff, *Avant l'exil*, p. 229.

94. Purishkevich, *Dnevnik*, pp. 11–13.

95. *Ibid.*, pp. 17–23; Youssoupoff, *Avant l'exil*, pp. 211–212, 229–231.

96. Purishkevich, *Dnevnik*, p. 48.

97. *Ibid.*, p. 62; Youssoupoff, *Avant l'exil*, pp. 235–236.

98. Purishkevich, *Dnevnik*, pp. 63–64.

99. Youssoupoff, *Avant l'exil*, p. 238.

100. This account of Rasputin's last hours is taken from Purishkevich's diary and Iusupov's memoirs. Admittedly, such testimony poses problems for the historian, because the absence of other firsthand accounts makes it difficult to substantiate their claims. However, the two accounts agree on many details, and even if they may not be fully accurate on all lesser counts, they remain the only firsthand evidence available. Purishkevich, *Dnevnik*, pp. 64–81; Youssoupoff, *Avant l'exil*, pp. 236–251.

101. Grand Duke Andrei Vladimirovich, "Iz dnevnika A. V. Romanova za 1916–1917gg.," p. 187; Youssoupoff, *Avant l'exil*, pp. 260–262; "Télégramme de M. Paléologue au Ministère des Affaires Etrangères." Petrograd, 1 janvier 1917.

102. Buchanan, *My Mission to Russia*, p. 51.

103. Letter of Nicholas to Aleksandra, December 14, 1916, in *Letters of the Tsar to the Tsaritsa*, p. 307.

104. "Dopros kniazia N. D. Golitsyna," p. 251.

105. Quoted in Ol'denburg, *Tsarstvovanie Imperatora Nikolaia II*, II, p. 231.

106. Andrei Vladimirovich, "Iz dnevnika," p. 188.

107. "Ubiistvo Rasputina. Ofitsial'noe doznanie," p. 82.

108. Andrei Vladimirovich, "Iz dnevnika," pp. 191–192.
109. Nicholas II, *Journal intime de Nicholas II,* p. 73.
110. Buchanan, *My Mission to Russia,* II, p. 46.
111. Letters of Grand Duke Aleksandr Mikhailovich to Nicholas II, December 25, 1916, and January 1, 25, and February 4, 1917, in *Lettres des Grands-Ducs à Nicholas II,* pp. 212, 214, 217, 207; Alexander, *Once a Grand Duke* pp. 283–284.
112. The materials in the Minister of Internal Affairs' private chancery archive (TsGIAL, fond 1282), especially the files numbered 1165 and 1167 (both in opis' 1), provide overwhelming evidence of Protopopov's appalling concern with trivia and petty influence-peddling.
113. "Télégramme secret de M. Paléologue au Ministère des Affaires Etrangères," Petrograd, 14 janvier 1917, pp. 78–79.
114. Paustovskii, *Sobranie sochinenii,* III, p. 546.

Chapter XII: "Down with the Tsar!"

1. "Doklad petrogradskogo okhrannogo otdeleniia osobomu otdelu departamenta politsii, oktiabr' 1916g.," pp. 10–12.
2. Hasegawa, *The February Revolution,* p. 200.
3. Quoted in *ibid.,* p. 201.
4. I. A. Baklanova, *Rabochie Petrograda,* pp. 23–27, 31–35; S. G. Strumilin, *Zarabotnaia plata i proizvoditel'nost',* pp. 11–13.
5. A. N. Rodzianko, "Pis'mo kn. Zin. N. Iusupovoi grafine Sumarokovoi-El'ston," 12 fevralia 1917g., p. 246.
6. Diakin, *Russkaia burzhuaziia,* pp. 313–315.
7. "Pokazaniia A. D. Protopopova," p. 21.
8. M. N. Pokrovskii (ed.), "Ekonomicheskoe polozhenie Rossii pered revoliutsiei," p. 67.
9. Quoted in Sidorov, *Ekonomicheskoe polozhenie Rossii,* p. 626; A. N. Rodzianko, "Pis'mo kn. Zin. N. Iusupovoi grafine Sumarokovoi-El'ston, 12 fevralia 1917g., p. 246.
10. M. N. Pokrovskii (ed.), "Politicheskoe polozhenie Rossii nakanune fevral'skoi revoliutsii v zhandarmskom osveshchenii," p. 15.
11. Quoted in Diakin, *Russkaia burzhuaziia i tsarism,* p. 314.
12. M. V. Rodzianko, "Zapiska M. V. Rodzianko," p. 88.
13. A. N. Rodzianko, "Pis'mo kn. Zin. N. Iusupovoi, 12 fevralia 1917g., p. 245.
14. Letter of Aleksandra to Nicholas, December 14, 1916, in *Letters of the Tsaritsa to the Tsar,* pp. 454–456.
15. Quoted in Chermenskii, *IV gosudarstvennaia duma,* p. 277.
16. Shul'gin, *Dni,* p. 142.
17. Quoted in Hasegawa, *The February Revolution,* pp. 183–184.
18. Paléologue, *La Russie des Tsars,* III, p. 92.
19. Shul'gin, *Dni,* p. 139.
20. Quoted in Hasegawa, *The February Revolution,* p. 183.
21. Quoted in *ibid.*
22. Quoted in Pearson, *The Russian Moderates,* p. 137.
23. Quoted in *ibid.,* pp. 132–133.
24. M. E. Solov'ev, "Partiia bol'shevikov—organizator pobedy fevral'skoi burzhuazno-demokraticheskoi revoliutsii v 1917g. v Rossii," pp. 199–200; M. Mitel'man, B. Glebov, and A. Ul'ianskii, *Istoriia Putilovskogo zavoda,* pp. 457–458; "Vedomost' promyshlennykh predpriiatii Petrograda, bastovavshikh 9 ianvaria 1917g.," pp. 523–526; M. I. Akhun and V. A. Petrov, *1917 god v Petrograde: Khronika sobytii i bibliografiia,* pp. 2–3.

25. E. N. Burdzhalov, *Vtoraia russkaia revoliutsiia.* I, pp. 85–93; I. I. Mints, *Istoriia velikogo oktiabria* I, pp. 470–472; I. P. Leiberov, *Na shturm samoderzhaviia.* pp. 113–115; Romanov, *Rabochee i professional'noe dvizhenie,* pp. 194–195.

26. "Zapiska otdeleniia po okhraneniiu obshchestvennoi bezopastnosti i poriadka v stolitse," 14 fevralia 1917g.; "Vedomost' predpriniatiiam goroda Petrograda, rabotaiushchim na Gosudarstvennuiu oboronu, masterovye koikh prekratili raboty 14-go fevralia 1917 goda."

27. Quoted in Hasegawa, *The February Revolution,* p. 210.

28. Mints, *Istoriia velikogo oktiabria,* I, pp. 474–476; Burdzhalov, *Vtoraia russkaia revoliutsiia,* I, pp. 108–109.

29. Buchanan, Sir George, "Report No. 65 to the Right Honourable Arthur J. Balfour," Petrograd, March 20, 1917, p. 543. The lines from *Masquerade* are quoted from V. A. Manuilov, "Lermontov," p. 307. The censor's opinion of it is quoted in Marc Slonim, *The Epic of Russian Literature,* p. 114.

30. "Dopros Generala S. S. Khabalova, 22 marta 1917g.," pp. 184–185; "Pokazaniia A. D. Protopopova," p. 96.

31. Buchanan, "Report No. 65 to Balfour," p. 544; "Télégramme secret de M. Paléologue au Ministère des Affaires Estrangères," Petrograd, 14 janvier 1917, p. 78; Paléologue, *La Russie des Tsars,* III, p. 213; Claude Anet, *La révolution russe,* I, p. 9.

32. Hasegawa, *The February Revolution,* pp. 215–216.

33. Quoted in I. P. Leiberov, "Petrogradskii proletariat v fevral'skoi revoliutsii 1917g.," p. 512.

34. Quoted in Hasegawa, *The February Revolution,* p. 217.

35. I. I. Mil'chik, *Rabochii Fevral',* pp. 59–60.

36. Mints, *Istoriia velikogo oktiabria,* I, pp. 488–492; E. P. Onufriev, *Za nevskoi zastavoi,* pp. 126–127; Leiberov, "Petrogradskii proletariat v fevral'skoi revoliutsii 1917g.," pp. 512–515.

37. Burdzhalov, *Vtoraia russkaia revoliutsiia,* I, pp. 119–129; Mitel'man, Glebov, and Ul'ianskii, *Istoriia putilovskogo zavoda,* I, pp. 461–463.

38. A. G. Shliapnikov, *Semnadtsatyi god,* I, pp. 79–87.

39. N. N. Sukhanov, *The Russian Revolution, 1917,* p. 5.

40. Hasegawa, *The February Revolution,* pp. 228–230; "Dopros Generala S. S. Khabalova," pp. 184–186.

41. Letter of Nicholas to Aleksandra, February 24, 1917, in *Letters of the Tsar to the Tsaritsa,* p. 315.

42. A. D. Protopopov, "Iz dnevnika A. D. Protopopova," p. 177; "Pokazaniia A. D. Protopopova," p. 96; E. I. Martynov, *Tsarskaia armiia v fevral'skom perevorote,* p. 74.

43. Gippius, *Siniaia kniga,* p. 72.

44. Comte Louis de Robien, p. 7.

45. *Ibid;* Paléologue, *La Russie des Tsars,* III, p. 214; Leon Trotsky, *History of the Russian Revolution,* I, p. 102.

46. Quoted in Hasegawa, *The February Revolution,* p. 233.

47. Mints, *Istoriia velikogo oktiabria,* I, pp. 502–503.

48. I. Gordienko, *Iz boevogo proshlogo,* p. 57.

49. V. Kaiurov, "Shest dnei fevral'skoi revoliutsii," p. 159.

50. Trotsky, *History of the Russian Revolution,* I, p. 105.

51. Hasegawa, *The February Revolution,* pp. 233–238. Igor Leiberov, a Soviet historian who has studied the question with considerable care, estimates the total at the much higher figure of 214,111, and the venerable I. I. Mints accepts his estimate. Leiberov's inflated figure, however, seems to include the *putilovtsy,* who, in fact, made no move to join the surging strike movement on either February 23 or 24. See Leiberov, *Na shturm samoderzhaviia,* pp. 136–144, and Mints, *Istoriia velikogo oktiabria,* I, pp. 500–501. Hasegawa's estimates, based

on a fresh reading of the archival materials and Okhrana reports, are, in the final analysis, more precise.

52. Leiberov, *Na shturm samoderzhaviia*, pp. 145–146.

53. A. Blok, *Poslednie dni imperatorskoi vlasti*, p. 53; A. Tarasov-Rodionov, *Fevral'*, p. 85.

54. Quoted in Hasegawa, *The February Revolution*, p. 239.

55. Martynov, *Tsarskaia armiia*, p. 75; Rodzianko, *Krushenie imperii*, p. 203; "Dopros M. V. Rodzianko," pp. 145–146; Buchanan, *My Mission to Russia*, II, p. 51.

56. Martynov, *Tsarskaia armiia*, pp. 76–77.

57. A. Kondrat'ev, "Vospominaniia o podpol'noi rabote v Petrograde," p. 64.

58. Hasegawa, *The February Revolution*, p. 247.

59. Trotsky, *History of the Russian Revolution*, I, pp. 108–109.

60. Leiberov, *Na shturm samoderzhaviia*, p. 161; Hasegawa, *The February Revolution*, p. 248.

61. Mints, *Istoriia velikogo oktiabria*, I, p. 512; Martynov, *Tsarskaia armiia*, pp. 78–79; Hasegawa, *The February Revolution*, pp. 248–249.

62. Quoted in Mitel'man, Glebov, and Ul'ianskii, *Istoriia Putilovskogo zavoda*, I, pp. 468–469.

63. Quoted in Trotsky, *History of the Russian Revolution*, I, p. 108.

64. Quoted in Leiberov, *Na shturm samoderzhaviia*, p. 178. See also: P. P. Aleksandrov, "Ot fevralia k oktiabriu," pp. 52–53; Martynov, *Tsarskaia armiia*, pp. 79–80; Hasegawa, *The February Revolution*, pp. 253–254; Burdzhalov, *Vtoraia russkaia revoliutsiia*, I, pp. 146–147.

65. Quoted in Hasegawa, *The February Revolution*, p. 254.

66. Quoted in *ibid.*, p. 258. See also pp. 254–259.

67. Letters of Nicholas to Aleksandra, February 25 and 26, 1917, in *Letters of the Tsar to the Tsaritsa*, pp. 315–316; Nicholas II, *Journal intime*, p. 92; D. N. Dubenskii, "Kak proizoshel perevorot v Rossii," pp. 36–39.

68. "Dopros Generala S. S. Khabalova," p. 190.

69. *Ibid.*

70. *Ibid.*, p. 191.

71. *Ibid.*; Martynov, *Tsarskaia armiia*, pp. 81–82; Burdzhalov, *Vtoraia russkaia revoliutsiia*, I, pp. 151–153.

72. "Dopros Generala S. S. Khabalova," pp. 191–192; Hasegawa, *The February Revolution*, p. 264.

73. Paléologue, *La Russie des Tsars*, I, p. 170; III, p. 217.

74. Quoted in Burdzhalov, *Vtoraia russkaia revoliutsiia*, I, p. 151.

75. Gippius, *Siniaia kniga*, p. 76.

76. *Ibid.*, pp. 76–77.

77. Quoted in Hasegawa, *The February Revolution*, p. 264.

78. Mints, *Istoriia velikogo oktiabria*, I, pp. 522–523.

79. "Telegramma Generala Khabalova Generalu Alekseevu 26 fevralia 1917g., No. 3703," p. 5.

80. Quoted in Marc Ferro, *La Révolution de 1917*, I, p. 69.

81. Quoted in Burdzhalov, *Vtoraia russkaia revoliutsiia*, I, p. 180. See also Hasegawa, *The February Revolution*, pp. 267–272; Mints, *Istoriia velikov oktiabria*, I, pp. 522–526; Martynov, *Tsarskaia armiia*, pp. 85–88.

82. Burdzhalov, *Vtoraia russkaia revoliutsiia*, I, pp. 177–178.

83. Sukhanov, *The Russian Revolution*, p. 29.

84. Burdzhalov, *Vtoraia russkaia revoliutsiia*, I, pp. 177–179.

85. Quoted in *ibid.*, p. 182.

86. Paléologue, *La Russie des Tsars*, III, p. 218.

87. Gippius, *Siniaia knigna*, pp. 78–79.

88. Quoted in Mints, *Istoriia velikogo oktiabria*, I, p. 534.

89. Quoted in Hasegawa, *The February Revolution*, pp. 279–280.

90. Leiberov, "Petrogradskii proletariat v fevral'skoi revoliutsii 1917g.," pp. 528–529.

91. Leiberov, *Na shturm samoderzhaviia*, pp. 220–223; Burdzhalov, *Vtoraia russkaia revoliutsiia*, I, p. 193; Hasegawa, *The February Revolution*, pp. 282–283.

92. Leiberov, *Na shturm samoderzhaviia*, pp. 224–225.

93. Viktor Shklovsky, *A Sentimental Journey*, pp. 11–13; I. M. Gordienko, *Iz boevogo proshlogo, 1914–1918gg.*, p. 63.

94. Shklovsky, *Sentimental Journey*, p. 11; Burdzhalov, *Vtoraia russkaia revoliutsiia*, I, p. 193.

95. Quoted in Wildman, *End of the Russian Imperial Army*, p. 149. See also pp. 148–151.

96. "Telegramma Generala S. S. Khabalova tsariu, 27 fevralia 1917g., sheet 7."

97. Quoted in Hasegawa, *The February Revolution*, p. 275.

98. "Dopros grafa Frederiksa, 2 iiunia 1917g.," p. 38.

99. "Telegramma Rodzianko tsariu 27 fevralia 1917g.," pp. 6–7.

100. Quoted in Nicolas de Basily, *Memoirs* p. 106.

101. Pares, *Fall of the Russian Monarchy*, pp. 436, 470.

102. "Razgovor po priamomu provodu gen. Alekseeva s vel. kniazem Mikhailom Aleksandrovichem 27 fevralia 1917g.," pp. 11–12.

103. Dubenskii, "Kak proizoshel perevorot v Rossii," pp. 43–45; Hasegawa, *The February Revolution*, pp. 461–463.

104. Ivanov's abortive counterrevolutionary effort has long been mishandled by Soviet and Western historians alike. Tsuyoshi Hasegawa has at last dealt with it in an even-handed manner, putting the whole campaign in its proper perspective, and, for the first time, clearly explaining what Ivanov, *Stavka*, and Nicholas all expected from the expedition, and why it failed. See Hasegawa, *The February Revolution*, pp. 460–486.

105. Hasegawa, *The February Revolution*, pp. 303–309.

106. Sukhanov, *The Russian Revolution, 1917*, p. 57.

107. *Ibid.*, pp. 55–58.

108. Shul'gin, *Dni*, pp. 178–179.

109. Miliukov, *Vospominaniia*, II, p. 293.

110. Alexander Kerensky [Kerenskii], *The Catastrophe*, p. 21.

111. Shul'gin, *Dni*, p. 163.

112. Quoted in Wildman, *End of the Russian Imperial Army*, p. 112.

113. Dubenskii, "Kak proizoshel perevorot v Rossii," pp. 42–52; Katkov, *Russia 1917*, pp. 310–313; A. S. Lukomskii, *Vospominaniia*, pp. 125–131; Hasegawa, *The February Revolution*, pp. 437–441.

114. Dubenskii, "Kak proizshel perevorot v Rossii," p. 53.

115. N. V. Ruzskii, "Prebyvanie Nikolaia II v Pskove," pp. 145–146.

116. Andrei Vladimirovich, "Iz dnevnika," p. 208.

117. Ruszkii, "Prebyvanie Nikolaia II v Pskove," pp. 146–147.

118. "Telegramma gen. Alekseeva tsariu 1 marta 1917g.," pp. 53–54.

119. Ruzskii, "Prebyvanie Nikolaia II v Pskove," p. 148.

120. "Razgovor po priamomu provodu gen. Ruzskogo s Rodzianko 1 marta 1917g.," p. 56.

121. *Ibid.*, pp. 56–59.

122. "Telegramma generala Danilova generalu Alekseevu 2 marta 1917g.," p. 63.

123. Alekseev's dispatch, as included in "Razgovor po priamomu provodu generala Everta s generalem Klembovskim, 2 marta 1917g.," p. 67.

124. *Ibid.*, pp. 72–73.

125. Ruzskii, "Prebyvanie Nikolaia II v Pskove," pp. 154–155.

126. "Telegramma generala Alekseeva tsariu 2 marta 1917g.," p. 73.

127. Ruzskii, "Prebyvanie Nikolaia II v Pskove," p. 155.

128. General S. S. Savvich, "Priniatie Nikolaem II resheniia ob otrechenii ot prestola," in Shchegolev (ed.), *Otrechenie Nikolaia II*, pp. 176–177.

129. "Telegramma Nikolaia II nachal'niku shtaba verkhovnogo glavkomanduiushchego generalu Alekseevu," in Shchegolev (ed.), *Otrechenie Nikolaia II*, p. 205.

130. "Telegramma Nikolaia II predsedateliu Gosudarstvennoi Dumy Rodzianko," in *ibid.*, p. 205.

131. Ruzskii, "Prebyvanie Nikolaia II v Pskove," p. 155.

132. Witold S. Sworakowskii, "The Authorship of the Abdication Documents of Nicholas II," p. 282.

133. Quoted in Hasegawa, *The February Revolution*, p. 511. See also pp. 509–512; Katkov, *Russia 1917*, pp. 335–340; S. P. Mel'gunov, *Martovskie dni 1917 goda*, pp. 190–193.

134. Shul'gin, *Dni*, pp. 266–268; Andrei Vladimirovich, "Iz dnevnika," p. 207.

135. Shul'gin, *Dni*, p. 268.

136. Quoted in Hasegawa, *The February Revolution*, p. 512.

137. Quoted in Andrei Vladimirovich, "Iz dnevnika," p. 207.

138. Quoted in Hasegawa, *The February Revolution*, p. 512.

139. Shul'gin, *Dni*, pp. 268–271; Andrei Vladimirovich, "Iz dnevnika," p. 207.

140. Quoted from Lincoln, *The Romanovs*, p. 725.

141. Shul'gin, *Dni*, p. 275.

142. "Dopros A. I. Guchkova, 2 avgusta 1917g.," pp. 268–269.

143. "Dopros D. N. Dubenskogo, 9 avgusta 1917g.," p. 393.

144. Nicholas II, *Journal intime*, p. 93.

145. For the Provisional Government's decision on March 7 (at 9:45 p.m.) to arrest Nicholas, Aleksandra, and their children, and to assemble them all at Tsarskoe Selo's Alexander Palace, see "Delo Vremennogo Komiteta po organizatsii novago Pravitel'stva, 1–27 marta 1917g.," sheet 70. See also Baroness Sophie Buxhoeveden, *The Life and Tragedy of Alexandra Feodorovna*, p. 271.

146. For a brief account of Nicholas's and Aleksandra's imprisonment and their last days, see Lincoln, *The Romanovs*, pp. 727–746.

147. de Robien, *Diary of a Diplomat*, p. 21.

148. Pitirim Sorokin, *Leaves from a Russian Diary*, p. 13.

149. V. V. Maiakovskii, "Revoliutsiia. Poetokhronika," in his *Sobranie sochinenii*, I, pp. 224–225.

150. This version of Kerenskii's remarks is reconstructed from Kerenskii's own reminiscences and those of another eyewitness who reported them to Ambassador Paléologue. Kerensky, *The Catastrophe*, p. 70; Paléologue, *La Russie des Tsars*, III, pp. 240–247.

151. Shul'gin, *Dni*, p. 307.

152. "Minutes of the Petrograd Soviet, March [3] 16, 1917," in Golder, *Documents*, p. 299.

Chapter XIII: New Men and Old Policies

1. Shul'gin, *Dni*, pp. 297–298.

2. *Ibid.*, pp. 299–300, 310; Miliukov, *Vospominaniia*, II, pp. 316–319.

3. Shul'gin, *Dni*, p. 308.

4. Nikolai Sukhanov, *Zapiskii o revoliutsii*, I, p. 231.

5. S. D. Mstislavskii [Maslovskii], *Piat' dnei*, p. 56.

6. Sukhanov, *Zapiski*, I, p. 201.

7. *Ibid*, p. 229.

8. *Ibid*., p. 232.

9. Quoted in Burdzhalov, *Vtoraia russkaia revoliutsiia*, I, p. 275.

10. Quotes from Wildman, *End of the Russian Imperial Army*, p. 184, and note 46, p. 185.

11. Quoted in V. I. Miller, "Nachalo demokratizatsii staroi armii v dni fevral'skoi revoliutsii (zasedanie Petrogradskogo Soveta 1 marta 1917g. i Prikaz No. 1)," pp. 33–35.

12. Quoted in *ibid.*, p. 37.

13. "Prikaz No. 1 Petrogradskogo Soveta rabochikh i soldatskikh deputatov po voiskam Petrogradskogo voennogo okruga," 1 marta 1917g., in Gavrilov (ed.), *Voiskovye Komitety*, pp. 17–18. See also Iu. S. Tokarev, *Petrogradskii Sovet Rabochikh i Soldatskikh Deputatov v marte–aprele 1917g.*, pp. 56–65.

14. Quoted in Wildman, *End of the Russian Imperial Army*, p. 186.

15. Shul'gin, *Dni*, p. 223.

16. *Ibid.*, p. 157.

17. Hasegawa, *The February Revolution*, pp. 522–524, and especially Table 7 on p. 523.

18. Miliukov, *Vospominaniia*, II, pp. 310–312; Sukhanov, *Zapiski*, I, pp. 321–328; Polner, *Zhiznennyi put' kniazia Georgiia Evgenievicha L'vova*, pp. 232–234.

19. Shul'gin, *Dni*, p. 168.

20. Kerensky, *The Catastrophe*, pp. 29, 59; Miliukov, *Vospominaniia*, II, p. 304.

21. Sukhanov, *Zapiski*, I, pp. 314–315.

22. In the absence of a complete copy of the March 3, 1917 issue of *Izvestiia Petrogradskogo Soveta rabochikh i soldatskikh deputatov*, which contains the fullest version of Kerenskii's speech, I have been obliged to assemble the foregoing compilation of Kerenskii's remarks from excerpts of his speech included in the following: V. I. Startsev, *Vnutrenniaia politika Vremennogo Pravitel'stva pervogo sostava*, pp. 82–83; Burdzhalov, *Vtoraia Russkaia Revoliutsiia*, I, pp. 320–321; Sukhanov, *Zapiski*, I, pp. 314–317.

23. Quoted in Startsev, *Vnutrenniaia politika*, p. 83.

24. Sukhanov, *Zapiski*, I, p. 340.

25. Quoted in Florinsky, *Russia*, II, p. 1391. On the final negotiations between the Provisional Government and the Soviet's Executive Committee, see Tokarev, *Petrogradskii Sovet*, pp. 92–103.

26. Sukhanov, *Zapiski*, I, pp. 340, 346.

27. Gippius, *Siniaia kniga*, p. 96.

28. Quoted in William Henry Chamberlin, *The Russian Revolution*, I, p. 101.

29. Sukhanov, *Zapiski*, I, p. 346.

30. Quoted in Trotsky, *History of the Russian Revolution*, I, p. 199.

31. Paustovskii, *Sobranie sochinenii*, III, pp. 569, 574.

32. Shklovsky, *Sentimental Journey*, p. 16.

33. Quoted in Trotsky, *History of the Russian Revolution*, I, p. 196.

34. Paustovskii, *Sobranie sochinenii*, III, p. 573.

35. Kerensky, *The Catastrophe*, p. 75.

36. Quoted in Salisbury, *Black Night, White Snow*, p. 373.

37. Quoted in Florinsky, *Russia*, II, p. 1378.

38. "Télégramme secret de M. Paléologue au Ministère des Affaires Etrangères," Petrograd, 15 novembre 1916, p. 61, and "Télégramme très confidentiel de M. Paléologue au Ministère des Affaires Etrangères, novembre 1916. pp. 65–66.

39. Paléologue, *La Russie des Tsars*, III, p. 176.

40. *Ibid.*, p. 187.

41. *Ibid.*, pp. 244, 246, 249–250.

42. "Miliukov's Explanation of His Policies and of His Resignation from the Government," in Robert Paul Browder and Alexander F. Kerensky (eds.), *The Russian Provisional Government, 1917* III, p. 1272.

43. Miliukov, *Vospominaniia*, II, p. 336.

44. "Miliukov's Note on Policy of Provisional Government," 5 (18) March 1917, in Golder, *Documents*, p. 324.

45. "First Declaration of the Provisional Government, March 7, 1917," in Browder and Kerensky (eds.), *Russian Provisional Government*, I, p. 157–158.

46. Paléologue, *La Russie des Tsars*, III, p. 256.

47. "Zasedanie voennogo soveshchaniia 1 fevralia 1917g.," p. 50.

48. "Notes and Tables on the Present and Prospective Munition Output of Russia" (Secret), pp. 8–17; "Ob"iasnitel'naia zapiska k programme snabzheniia armii glavneishimi predmetami artilleriiskago dovol'stviia na period do 1 ianvaria 1918g." (Sekretno), pp. 1–27.

49. Buchanan, *Mission to Russia*, II, p. 90.

50. Paléologue, *La Russie des Tsars*, III, p. 255.

51. Published in David R. Francis, *Russia from the American Embassy*, pp. 50–51.

52. A. V. Ignat'ev, *Vneshniaia politika vremennogo pravitel'stva*, pp. 126–128.

53. Knox, *With the Russian Army*, II, p. 585.

54. Paléologue, *La Russie des Tsars*, III, 265.

55. Quoted in Francis, *From the American Embassy*, pp. 97–98, and in Paléologue, *La Russie des Tsars*, III, p. 257 note.

56. "K Narodam vsego mira," published in Sukhanov, *Zapiski*, II, pp. 234–235.

57. Sukhanov, *Zapiski*, II, p. 204.

58. *Ibid.*, p. 352.

59. "Two Positions," in Golder, *Documents*, p. 329.

60. "Press Interview with Miliukov," March 23, 1917, in Browder and Kerensky (eds.), *Russian Provisional Government*, II, p. 1045.

61. "K narodam vsego mira," in Sukhanov, *Zapiski*, II, p. 235.

62. Quoted in Ol'denburg, *Tsarstvovanie Nikolaia II*, II, p. 188.

63. Quoted in Wolfe, *Three Who Made a Revolution*, p. 429.

64. V. B. Stankevich, "Vospominaniia," p. 415; Miliukov, *Vospominaniia*, II, pp. 373–374; Trotsky, *History of the Russian Revolution*, I, pp. 228–229.

65. Anet, *La révolution russe à Petrograd et aux armées*, I, p. 211.

66. Stankevich, "Vospominaniia," p. 414.

67. Quoted in Rex A. Wade, *The Russian Search for Peace*, p. 18.

68. *Ibid.*, pp. 19–25.

69. Quoted in G. I. Zlokazov, *Petrogradskii Sovet Rabochikh i Soldatskikh Deputatov*, pp. 112–113.

70. Stankevich, "Vospominaniia," p. 414.

71. "Press Interview with Miliukov," in Browder and Kerensky (eds.), *Russian Provisional Government*, II, p. 1044.

72. "The Provisional Government and War Aims," in Golder, *Documents*, pp. 329–330. See also C. Jay Smith, Jr., *The Russian Struggle for Power*, pp. 471–472.

73. Miliukov, *Vospominaniia*, II, p. 346.

74. "Miliukov's Denial That the Government Has Renounced the Agreement on Constantinople and the Straits," April 1 (14), 1917, in Browder and Kerensky (eds.), *Russian Provisional Government*, II, p. 1058.

75. Quoted in Wade, *Russian Search for Peace*, p. 31.

76. Quoted in Oliver H. Radkey, *The Agrarian Foes of Bolshevism*, p. 158.

77. S. I. Shidlovskii, "Vospominaniia," p. 304. See also Miliukov, *Vospominaniia*, II, pp. 367–368; and Virgil D. Medlin and Steven L. Parsons (eds.), *V. D. Nabokov and the Russian Provisional Government, 1917,* p. 116.

78. Radkey, *Agrarian Foes of Bolshevism,* p. 158.

79. Ignat'ev, *Vneshniaia politika vremennogo pravitel'stva,* p. 188.

80. "The Note of April 18," in Browder and Kerensky (eds.), *Russian Provisional Government,* II, p. 1098.

81. Trotsky, *History of the Russian Revolution,* I, p. 335.

82. Irakli Tsereteli, "Reminiscences of the February Revolution: The April Crisis," I, pp. 105–107.

83. Quoted in Ferro, *La Révolution de 1917,* I, p. 320.

84. V. I. Lenin, "The War and Russian Social Democracy," in *Collected Works,* XXI, p. 28.

85. *Ibid.,* p. 34. See also V. A. Lavrin, *Bol'shevistskaia partiia v nachale pervoi mirovoi imperialisticheskoi voiny,* pp. 47–61.

86. V. I. Lenin, "The Tasks of Revolutionary Social-Democracy in the European War," p. 18.

87. "Lenin's Theses on the War," in Olga Hess Gankin and H. H. Fisher (eds.), *The Bolsheviks and the World War,* pp. 140–142. For the reaction of other Bolsheviks to Lenin's "Theses," see Lavrin, *Bol'shevistskaia partiia* pp. 22–31.

88. V. I. Lenin, "The State of Affairs in Russian Social Democracy," p. 285.

89. V. I. Lenin, "The Collapse of the Second International," pp. 215–216. For information about those who supported (or, at least, sympathized with) Lenin's position, see S. V. Tiutiukin, *Voina, mir, revoliutsiia,* pp. 226–246.

90. V. I. Lenin, "Lecture on the 1905 Revolution," p. 253.

91. N. K. Krupskaia, *Vospominaniia,* p. 286.

92. V. I. Lenin, "Draft Theses, March 4 (17), 1917," pp. 287–289.

93. V. I. Lenin, "Letters from Afar," p. 297.

94. Krupskaia, *Vospominaniia,* p. 289.

95. *Ibid.,* 293–296; Lenin, *Collected Works,* XXIII, pp. 416–418, note 158.

96. V. I. Lenin, "Farewell Letter to the Swiss Workers," p. 371–372.

97. Krupskaia, *Vospominaniia,* p. 297.

98. F. F. Raskol'nikov, *Na boevykh postakh,* pp. 62–68; *Petrogradskie bol'sheviki v oktiabr'skoi revoliutsii,* pp. 64–68; A. L. Fraiman (ed.), *Oktiabr'skoe vooruzhennoe vosstanie* I, pp. 178–181.

99. Quoted in V. D. Bonch-Bruevich, *Na boevykh postakh fevral'skoi i oktiabr'skoi revoliutsii,* p. 15.

100. Quoted in Alexander Rabinowitch, *Prelude to Revolution,* p. 36.

101. "Aprel'skie tezisy V. I. Lenina," 4 aprelia 1917g., No. 1, pp. 3–4.

102. Alexander Rabinowitch, *The Bolsheviks Come to Power,* p. 162.

103. Quotes from Rabinowitch, *Prelude to Revolution,* pp. 36–41.

104. Quoted in *ibid.,* p. 41. See also p. 40.

105. Sukhanov, *Russian Revolution, 1917,* p. 290.

106. "Rezoliutsiia TsK RSDRP(b) 20 aprelia (3 maia) 1917 goda," No. 739, and "Rezoliutsiia Tsentral'nogo Komiteta RSDRP(b), priniataia 21 aprelia (4 maia) 1917g.," No. 757, pp. 726, 737.

107. V. Rakhmetov (ed.), "Aprel'skie dni 1917 goda v Petrograde"; Trotsky, *History of the Russian Revolution,* I, p. 340; Irakli Tsereteli, "Reminiscences of the February Revolution: The April Crisis—II," p. 184.

108. Quoted in Trotsky, *History of the Russian Revolution,* I, p. 345. See also the detailed account of the engineering student Georgii Tolstoi, who observed several shooting

incidents as he marched along the Nevskii that day, in Rakhmetov (ed.), "Aprel'skie dni 1917 goda v Petrograde," pp. 36–48.

109. Miliukov, *Vospominaniia*, II, p. 363; Kerensky, *The Catastrophe*, p. 136.

110. Miliukov, *Vospominaniia*, II, p. 363.

111. Quoted in Ferro, *La Révolution de 1917*, I. p. 320.

112. Quoted in Rex A. Wade, *Red Guards and Workers' Militias*, pp. 89–90.

113. Quoted in Wade, *Russian Search for Peace*, p. 45.

114. Quoted in Trotsky, *History of the Russian Revolution*, I, p. 344.

115. "Declaration of the Provisional Government Reviewing Its Accomplishments and Calling for the Support and Cooperation of All the Vital Forces in the Nation," in Browder and Kerenskii (eds), *Russian Provisional Government*, III, pp. 1249–1251.

116. Quoted in Wade, *Russian Search for Peace*, p. 42; "The Government's Explanatory Note of April 22," in Browder and Kerensky (eds), *Russian Provisional Government*, II, p. 1100.

117. "*Novoe Vremia* on the Government Declaration," in Browder and Kerenskii (eds.), *Russian Provisional Government*, II, p. 1253.

118. "Letter of Kerenskii Advocating the Addition of Other Democratic Representatives to the Government," and "Prince L'vov Asks Chkheidze to Bring the Question of Coalition Before the Soviet Executive Committee," in *ibid.*, pp. 1252–1253.

119. "Soviet Acceptance of the Government's Supplementary Explanatory Note," in *ibid.*, p. 1242.

120. "The Meeting of the Members of the Four State Dumas: The Speech of Guchkov," in *ibid.*, pp. 1259–1260.

121. Miliukov, *Vospominaniia*, II, p. 369.

122. Paléologue, *La Russie des Tsars*, III, p. 341.

123. "Caractère et Développement de la Révolution Russe," 8 juin 1917. p. 111.

124. Paléologue, *La Russie des Tsars*, III, p. 336.

125. Excerpts published in Buchanan, *My Mission to Russia*, pp. 114.

126. "The Meeting of the Members of the Four State Dumas: The Speech of Prince L'vov," in Browder and Kerensky (eds.), *Russian Provisional Government*, III, pp. 1258–1259.

127. "Guchkov's Letter of Resignation to Prince L'vov, May 1, 1917," in *ibid.*, pp. 1267–1268; Miliukov, *Vospominaniia*, II, p. 369.

128. "The Soviet Decisison to Enter the Government," in Browder and Kerensky (eds.), *Russian Provisional Government*, p. 1269.

129. Stankevich, "Vospominaniia," pp. 427–429. See also Zlokazov, *Petrogradskii Sovet*, pp. 202–205; Mints, *Istoriia Velikogo Oktiabria*, II, pp. 330–334.

130. "Miliukov's Explanation of His Policies and of His Resignation from the Government," in Browder and Kerensky (eds.), *Russian Provisional Government*, III, p. 1271.

131. Chernov, *Great Russian Revolution*, p. 289.

132. Paustovskii, *Sobranie sochinenii*, III, pp. 576–577.

133. "The Declaration of May 5 of the New Coalition Government," in Browder and Kerensky (eds), *Russian Provisional Government*, p. 1277.

134. Quotes from Florinsky, *Russia*, II, p. 1387.

135. Paléologue, *La Russie des Tsars*, III, pp. 339–340.

136. Lockhart, *British Agent*, pp. 176–177.

137. Kerensky, *The Catastrophe*, p. 195.

138. *Ibid.*, p. 185.

139. Quoted in Chamberlin, *The Russian Revolution*, I, p. 151.

Chapter XIV: "To Keep the Swing from Going Over the Top"

1. Quoted in Knox, *With the Russian Army*, II, p. 581.
2. Paustovskii, *Sobranie sochinenii*, III, p. 569.
3. Paléologue, *La Russie des Tsars*, III, p. 258.
4. Knox, *With the Russian Army*, II, p. 575.
5. Paustovskii, *Sobranie sochinenii*, III, p. 580.
6. A. N. Voznesenskii, *Moskva v 1917 godu*, pp. 33–34.
7. Paustovskii, *Sobranie sochinenii*, III, p. 580.
8. Pis'ma Aleksandra Bloka k materi, 23 marta 1917g., in Aleksandr Blok, *Pis'ma Aleksandra Bloka k rodnym*, p. 339.
9. Pis'mo Aleksandra Bloka k materi, 2 aprelia 1917g., in *ibid.*, pp. 343–344.
10. Paustovskii, *Sobranie sochinenii*, III, p. 577.
11. Knox, *With the Russian Army*, II, pp. 581–582.
12. *Ibid.*, p. 573.
13. Quoted in Pyman, *Release of Harmony*, II, p. 255.
14. L. S. Gaponenko, *Rabochii klass Rossii v 1917 godu*, p. 195.
15. *Ibid.*, p. 188.
16. I. A. Baklanova, *Rabochie Petrograda v periode mirnogo razvitiia revoliutsii*, p. 38.
17. Quoted in *ibid.*, p. 39.
18. Quoted in B. M. Freidlin, *Ocherki istorii rabochego dvizheniia v Rossii v 1917g.*, p. 206.
19. Baklanova, *Rabochie Petrograda*, p. 41.
20. Quoted in *ibid.*, p. 52.
21. Gordienko, *Iz boevogo proshlogo*, p. 103.
22. "Izvestiia komiteta Eia Imperatorskoi Vysochestva Velikoi Kniazhny Tatiany Nikolaevny," 15 fevralia 1917g., sheet no. 118.
23. Quoted in Baklanova, *Rabochie Petrograda*, p. 53.
24. Z. V. Stepanov, "Ekonomicheskoe polozhenie rabochikh. Bor'ba za rabochii kontrol' nad proizvodstvom i raspredeleniem," pp. 40–41; Baklanova, *Rabochie Petrograda*, p. 53.
25. Z. V. Stepanov, *Rabochie Petrograda v period podgotovki i provedeniia oktiabr'skogo vooruzhennogo vosstaniia*, p. 51.
26. Quoted in Freidlin, *Ocherki istorii rabochego dvizheniia*, p. 77.
27. "Pis'mo rabochikh zavodov Sysertskogo gornogo okruga," 19 marta 1917g., No. 465, p. 521.
28. Ferro, *La Révolution de 1917*, p. 174.
29. "Pervomaiskaia listovka Moskovskogo oblastnogo biuro TsK i Moskovskogo komiteta RSDRP(b)," (ne pozdnee 18 aprelia 1917g.), No. 64, p. 95.
30. Keep, *The Russian Revolution*, p. 69.
31. Quoted in Freidlin, *Ocherki istorii rabochego dvizheniia*, p. 80; "The Agreement on Working Conditions in Petrograd," March 10, 1917, in Browder and Kerensky (eds.), *Russian Provisional Government*, II, p. 712.
32. Quoted in P. B. Volobuev, *Proletariat i burzhuaziia Rossii v 1917g.*, p. 153.
33. "Rezoliutsiia sobraniia zhenshchin-rabotnits Moskovskogo raiona Petrograda," 7 marta 1917g., No. 385, p. 470.
34. *Ibid.* See also quote in S. A. Smith, *Red Petrograd*, p. 67.
35. Quoted in Smith, *Red Petrograd*, pp. 66–67.
36. "Statement on the Eight-Hour Day by the Chairman of the Council of the Congress of Representatives of Industry and Trade (N. N. Kutler)," in Browder and Kerensky (eds.), *Russian Provisional Government*, II, p. 717.

37. Quoted in Gaponenko, *Rabochii klass Rossii*, pp. 348–349.

38. "The First of May," in Browder and Kerensky (eds.), *Russian Provisional Government*, II, p. 718.

39. Quoted in Volobuev, *Proletariat i burzhuaziia Rossii*, p. 121. See also pp. 116–124; Gaponenko, *Rabochii klass Rossii*, pp. 349–351; Freidlin, *Ocherki istorii rabochego dvizheniia*, pp. 82–93.

40. Volobuev, *Proletariat i burzhuaziia Rossii*, pp. 124–125.

41. Quoted in *ibid.*, p. 125.

42. *Ibid.*, pp. 124–138; Gaponenko, *Rabochii klass Rossii*, pp. 188–190, 352–363.

43. Keep, *The Russian Revolution*, pp. 72–73.

44. Quoted in Chamberlin, *The Russian Revolution*, I, pp. 267–268.

45. Ferro, *La Révolution de 1917*, I, p. 175.

46. Quoted in Keep, *The Russian Revolution*, p. 84.

47. "Svodka svedenii o sostave delegatov I Petrogradskoi konferentsii fabrichno-zavodskikh komitetov"; Oskar Anweiler, *The Soviets*, pp. 126–127.

48. Quoted in Smith, *Red Petrograd*, p. 93. See also pp. 80–88, and Freidlin, *Ocherki istorii rabochego dvizheniia*, pp. 128–129.

49. "Account of the Activities of the Factory Committees and Their Relation to the Trade Union Movement," Browder and Kerensky (eds.), *Russian Provisional Government*, II, p. 725.

50. "Rezoliutsiia ob"edinennogo zasedaniia prezidiumov zavodskogo komiteta rabochikh i komiteta sluzhashchikh moskovskogo zavoda 'Dinamo'," 19 maia 1917g., No. 229, p. 280.

51. "Rezoliutsiia sobraniia rabochikh Sestroetskogo oruzheinogo zavoda o razgruzke Petrograda," 20 maia 1917g.

52. "Rezoliutsiia I Petrogradskoi konferentsii fabrichno-zavodskikh komitetov," 3 iiunia 1917g., No. 242, p. 291. See also Baklanova, *Rabochie Petrograda v period mirnogo razvitiia revoliutsii*, pp. 131–137; Gaponenko, *Rabochii klass Rossii*, pp. 373–376.

53. Tsuyoshi Hasegawa, "The Formation of the Militia in the February Revolution"; Wade, *Red Guards and Workers' Militias*, pp. 51–74.

54. Quoted in Wade, *Red Guards and Workers' Militias*, pp. 75–76.

55. Quoted in *ibid.*, p. 76.

56. Quoted in *ibid.*, pp. 89–90.

57. Quoted in Keep, *The Russian Revolution*, p. 93.

58. Gaponenko, *Rabochii klass Rossii*, pp. 308–310. Dan's remarks are quoted in Wade, *Red Guards and Workers' Militias*, p. 91.

59. Quoted in Wade, *Red Guards and Workers' Militias*, p. 100.

60. Quoted in Gaponenko, *Rabochii klass Rossii*, p. 311.

61. Quoted in Keep, *The Russian Revolution*, p. 76. See also pp. 74–75; Freidlin, *Ocherki istorii rabochego dvizheniia*, pp. 115–118; Robert Wilton, *Russia's Agony*, p. 180.

62. Freidlin, *Ocherki istorii rabochego dvizheniia*, p. 117.

63. Quoted in Gaponenko, *Rabochii klass Rossii*, p. 372.

64. "Vedomost' predpriiatiiam goroda Petrograda, rabotaiushchim na gosudarstvennuiu oboronu, masterovye koikh prekratili raboty 14-go fevralia 1917 goda," sheets 40–41.

65. Trotsky, *History of the Russian Revolution*, I, p. 417.

66. See especially Koenker, *Moscow Workers*, pp. 299–301, 312, and Appendix I, p. 384.

67. Gaponenko, *Rabochii klass Rossii*, pp. 378–386.

68. Trotsky, *History of the Russian Revolution*, I, p. 421.

69. Sukhanov, *Zapiski*, IV, p. 136.

70. Quotes from Chernov, *Great Russian Revolution*, p. 230, and Chamberlin, *Russian Revolution*, I, p. 153.

71. Chernov, *Great Russian Revolution*, p. 231.

72. Quoted in *ibid.*, p. 232.

73. See A. A. Novosel'skii and N. S. Chaev, "Krest'ianskaia voina pod predvoditel'st-vom S. T. Razina," pp. 277–311; Paul Avrich, *Russian Rebels 1600–1800*, pp. 10–122.

74. Quoted in Florinsky, *Russia*, I, p. 507.

75. Quoted in Lincoln, *The Romanovs*, p. 226.

76. N. F. Dubrovin, *Pugachev i ego soobshchenniki* and John T. Alexander, *Autocratic Politics in a National Crisis, passim;* and Avrich, *Russian Rebels*, pp. 180–254.

77. "Obzor polozheniia Rossii za tri mesiatsa revoliutsii po dannym otdela snoshenii s provintsiei Vremennago Komiteta Gosudarstvennoi Dumy," sheet 241.

78. "Pis'mo pomeshchitsy N. A. Trusovoi," 9 marta 1917g., No. 635, p. 672.

79. "Iz svodki svedenii glavnogo upravleniia po delam militsii MVD," No. 679, pp. 704–705.

80. "Telegramma pomeshchika Shmidta," No. 653, p. 683.

81. Quoted in Chamberlin, *The Russian Revolution*, I, p. 246.

82. Ia. Iakovlev (ed.), "Mart–Mai 1917g.," pp. 55, 43.

83. Richard Pipes, *Russia Under the Old Regime*, p. 157.

84. "Obzor polozheniia Rossii za tri mesiatsa revoliutsii," sheet 244; "Doklad instruktora provintsial'nogo otdela Moskovskogo Sovета Rabochikh Deputatov Shevtsova," 12 aprelia 1917g., No. 580, p. 593.

85. "Rezoliutsiia skhoda krest'ian Kul'pinskoi volosti Volokolamskogo uczda Moskov-skoi gubernii," 1 aprelia 1917g., No. 682, p. 707.

86. "Obzor polozheniia Rossii za tri mesiatsa revoliutsii," sheet 244.

87. Ferro, *La Révolution de 1917*, p. 187.

88. Quoted in N. A. Kravchuk, *Massovoe krest'ianskoe dvizhenie v Rossii nakanune oktiabria*, pp. 103–104.

89. Paustovskii, *Sobranie sochinenii*, III, p. 581.

90. "Pis'mo soldata 430-go Valkskogo polka P. Sokkonena," No. 664, p. 659.

91. "Obzor polozheniia Rossii za tri mesiatsa revoliutsii," sheets 251–252.

92. "Doklad komissara Ranenburgskogo uczda Riazanskoi gubernii, ot 28 iiulia 1917g., No. 2613," pp. 187–191; P. N. Pershin, *Agrarnaia revoliutsiia v Rossii*, I, pp. 287–293; Graeme J. Gill, *Peasants and Government in the Russian Revolution* pp. 40–46; Keep, *The Russian Revolution*, pp. 186–216.

93. Quoted in Launcelot A. Owen, *The Russian Peasant Movement, 1906–1917*, pp. 180–181.

94. Quoted in Keep, p. 213. See also pp. 210–212.

95. "Obzor polozheniia Rossii za tri mesiatsa revoliutsii," sheet 260.

96. Kravchuk, *Massovoe krest'ianskoe dvizhenie*, pp. 21–23.

97. "The First Declaration of the Provisional Government," and "Prince L'vov States the Aims and Hopes of the New Government," in Browder and Kerensky (eds.), *Russian Provisional Government*, I, pp. 157–159.

98. Keep, *The Russian Revolution*, p. 162.

99. "Pis'mo ministra zemledeliia A. I. Shingareva," 30 aprelia 1917g., No. 274, p. 327.

100. *Ibid.*, p. 326; A. I. Shingarev, *Vymiraiushchaia derevnia*, pp. 239–269.

101. "Postanovlenie Vremennogo pravitel'stva po zemel'nomu voprosu," 19 marta 1917g., No. 345, p. 439.

102. Quoted in Chamberlin, *The Russian Revolution*, I, p. 245. See also, "Postanovlenie Vremennogo pravitel'stva po zemel'nomu voprosu," 19 marta 1917g., No. 345, p. 439.

103. Startsev, *Vnutrenniaia politika vremennogo pravitel'stva*, p. 214.

104. Quoted in Kravchuk, *Massovoe krest'ianskoe dvizhenie v Rossii*, p. 23.
105. Quoted in Pershin, *Agrarnaia revoliutsiia v Rossii*, I, p. 293.
106. Quoted in Keep, *The Russian Revolution*, p. 235. See also pp. 236–237, and Pershin, *Agrarnaia revoliutsiia v Rossii*, I, pp. 341–344.
107. Quoted in Keep, *The Russian Revolution*, p. 235.
108. Quoted in Chamberlin, *The Russian Revolution*, I, p. 248.
109. Quoted in William G. Rosenberg, *Liberals in the Russian Revolution*, p. 152.
110. Quoted in *ibid.*
111. Kerensky, *The Catastrophe*, p. 245.
112. Alexander Kerensky, *Russia and History's Turning Point*, p. 324.
113. Quoted in Rosenberg, *Liberals in the Russian Revolution*, p. 169.
114. Quoted in Kerensky, *Russia and History's Turning Point*, p. 303.
115. Rosenberg, *Liberals in the Russian Revolution*, pp. 157–169.
116. Quoted in *ibid.*, p. 173.
117. "Raport glavnokomanduiushchego voiskami Petrogradskogo voennogo okruga general-maiora P. A. Polovtseva, 17 iiunia 1917g., No. 473, p. 525.
118. Quoted in Trotsky, *History of the Russian Revolution*, II, p. 8; I, p. 457.
119. Quoted in Alexander Rabinowitch, *Prelude to Revolution*, p. 97.
120. Sukhanov, *Zapiski*, IV, pp. 330–331.
121. *Ibid.*, pp. 339–340.
122. Quoted in Rabinowitch, *Prelude to Revolution*, pp. 114–115.
123. Trotsky, *History of the Russian Revolution*, I, p. 301.
124. Quoted in Rabinowitch, *Prelude to Revolution*, pp. 112, 124, 117–118.
125. Quoted in *ibid.*, pp. 121–122, 133.
126. Rabinowitch, *The Bolsheviks Come to Power*, p. xxxiii.
127. Quoted in Rabinowitch, *Prelude to Revolution*, p. 134.
128. Sukhanov, *Zapiski*, IV, p. 361.
129. Trotsky, *History of the Russian Revolution*, II, p. 12.
130. Sukhanov, *Zapiski*, IV, p. 361.
131. Quoted in Rosenberg, *Liberals in the Russian Revolution*, p. 174. See also pp. 170–175, and Richard Pipes, *The Formation of the Soviet Union*, pp. 50–60.
132. This description of Bleichman and his remarks is compiled from "Testimony of the Origins and Progress of the Armed Uprising" (compiled by a Special Commission of Inquiry in late July), in Browder and Kerensky (eds.), *Russian Provisional Government*, III, p. 1338, and Trotsky, *History of the Russian Revolution*, II, pp. 13–14. See also O. N. Znamenskii, *Iiul'skii krizis 1917 goda*, pp. 45–46, 57–60, 63–64.
133. Wade, *Red Guards and Workers' Militias*, pp. 117–118; Mints, *Istoriia Velikogo Oktiabria*, II, pp. 598–599.
134. Mitel'man, *Istoriia putilovskogo zavoda*, I, p. 535.
135. *Ibid.*, pp. 533–537; Znamenskii, *Iiul'skii krizis*, pp. 61–63.
136. Quoted in Mitel'man, *Istoriia putilovskogo zavoda*, I, p. 535.
137. "The Events of July 3 in Petrograd," (from *Russkiia Vedomosti*), in Browder and Kerensky (eds.), *Russian Provisional Government*, III, p. 1335.
138. Quoted in Rabinowitch, *Prelude to Revolution*, pp. 163, 184.
139. Quoted in *ibid.*, p. 182.
140. Sukhanov, *Zapiski*, IV, pp. 424–425. For details on the role of the Kronstadt sailors in the July crisis, see Israel Getzler, *Kronstadt 1917–1921*, pp. 111–152, and A. V. Bogdanov, *Moriaki-Baltiitsy v 1917g.*, pp. 112–128.
141. Quoted in Trotsky, *History of the Russian Revolution*, II, p. 40. See also Znamenskii, *Iiul'skii krizis*, pp. 77–106, and O. A. Lidak, "Iiul'skie sobytie 1917 goda," in M. N. Pokrovskii (ed.), *Ocherki po istorii oktiabr'skoi revoliutsii*, II, pp. 281–299.

142. Sukhanov, *Zapiski*, IV, pp. 440–441.
143. Pares, *My Russian Memoirs*, p. 465.
144. "Report of the Public Prosecutor on the Investigation of the Charges Against the Bolsheviks," in Browder and Kerensky (eds.), *Russian Provisional Government*, III, p. 1376.
145. "Lenin on 'The Question of the Bolshevik Leaders Appearing Before the Courts,' " in *ibid.*, p. 1370.
146. Quoted in Rosenberg, *Liberals in the Russian Revolution*, p. 179.
147. "The Declaration of the Provisional Government," July 8, 1917, in Browder and Kerensky (eds.), *Russian Provisional Government*, III, pp. 1386–1387.
148. "The Debate and Resolution of VTsIK and the Executive Committee of the Soviets of Peasants' Deputies," in *ibid.*, p. 1394.
149. "An Appeal to All the Population," in *ibid.*, p. 1394.
150. Maxim Gorky, *Untimely Thoughts*, pp. 75, 72.
151. "Pis'mo Aleksandra Bloka k materi," 30 iiunia 1917g., in Blok, *Pis'ma Aleksandra Bloka k rodnym*, II, p. 381.

Chapter XV: Kerenskii Takes Charge

1. Gorky, *Untimely Thoughts*, pp. 39–40, 47, 72, 15–16, 55.
2. M. D. Bonch-Bruevich, *From Tsarist General to Red Army Commander*, pp. 123–124.
3. "Soveshchanie v *Stavke* 18 marta 1917 goda," p. 11.
4. "Vyderzhka iz doklada o poezdke na front chlenov Gosudarstvennoi Dumy Maslennikova i P. M. Shmakova," p. 57.
5. Quoted in Knox, *With the Russian Army*, II, p. 633. For a similar account see Anet, *La révolution russe*, I, p. 223.
6. Quoted in Wildman, *End of the Russian Imperial Army*, p. 222.
7. Quoted in *ibid.*
8. "The Government's Initial Statement on Foreign Policy," March 4, 1917, and "Press Interview with Miliukov," March 22, 1917, in Browder and Kerensky (eds.), *Russian Provisional Government*, II, pp. 1042–1045.
9. Quoted in Akhun and Petrov, *Bol'sheviki i armiia*, p. 168.
10. Quoted in A. E. Ioffe, *Russko-frantsuzskie otnosheniia v 1917g.*, p. 181.
11. Denikin, *Russian Turmoil*, p. 70.
12. Chernov, *Great Russian Revolution*, pp. 316, 314.
13. Denikin, *Russian Turmoil*, p. 68.
14. Knox, *With the Russian Army*, II, pp. 580–581.
15. "Prikaz No. 1 Petrogradskogo Soveta rabochikh i soldatskikh deputatov po voiskam Petrogradskogo voennogo okruga," 1 marta 1917g., and "Prikaz No. 2 ispolnitel'nogo komiteta Petrogradskogo Soveta rabochikh i soldatskikh deputatov po voiskam Petrogradskogo voennogo okruga," 5 marta 1917g., pp. 17–21.
16. Knox, *With the Russian Army*, II, p. 603.
17. V. I. Miller, *Soldatskie komitety russkoi armii v 1917g*, p. 70.
18. "Pis'mo voennogo ministra A. I. Guchkova," 9 marta 1917g., No. 333, pp. 429–430.
19. "Predpisanie vremenno ispolniaiushchego obiazannosti verkhovnogo glavnokomanduiushchego generala M. V. Alekseeva," 11 marta 1917g.
20. "Soobshchenie vremeno-ispolniaiushchego dolzhnost' verkhovnogo glavnokomansuiushchego generala M. V. Alekseeva," No. 24, p. 40.
21. "Pis'mo v Gosudarstvennuiu Dumu," 23 aprelia 1917g., sheet 26.
22. Sukhanov, *Zapiski*, II, p. 297.

23. "Kerensky's Address to the Delegation from the 7th Army," in Browder and Kerensky (eds.), *Russian Provisional Government*, II, p. 909.

24. Quoted in Wildman, *End of the Russian Imperial Army*, p. 299.

25. Sukhanov, *Zapiski*, II, p. 314.

26. W. S. Woytinsky, *Stormy Passage*, p. 260.

27. Sukhanov, *Zapiski*, II, p. 307.

28. *Ibid.*, p. 308.

29. Quoted in Wildman, *End of the Russian Imperial Army*, p. 309.

30. Knox, *With the Russian Army*, II, p. 598.

31. "Pis'mo komanduiushchego 5-i armiei A. M. Dragomirova," 29 marta 1917g., pp. 43–45.

32. "Pis'mo verkhovnogo glavnokomanduiushchego M. V. Alekseeva voennomu ministru A. I. Guchkovu," 16 aprelia 1917g.

33. Quoted in Gorky, *Untimely Thoughts*, p. 11.

34. "Vyderzhka iz doklada o poezdke na front chlenov Gosudarstvennoi Dumy Maslennikova i P. M. Shmakova," p. 55.

35. "Vozzvanie Petrogradskogo Soveta rabochikh i soldatskikh deputatov o prekrashchenii brataniia s protivnikom," 30 aprelia 1917g., p. 77.

36. "Iz sostavlennogo Moskovskoi voennoi tsenzuroi obzora frontovykh pisem," p. 91.

37. Gorky, *Untimely Thoughts*, p. 11.

38. These excerpts from remarks made at the Petrograd Conference are taken from General Denikin's account, which quotes his colleagues at considerable length. Denikin, *The Russian Turmoil*, pp. 177–186.

39. "Draft of a Resolution of the Petrograd Soviet on Soldiers' Rights," March 9, 1917, in Browder and Kerensky (eds.), *Russian Provisional Government*, II, pp. 878–879.

40. "Guchkov's Letter of Resignation to Prince L'vov," May 1, 1917, in *ibid.*, III, pp. 1267–1268.

41. "Order No. 8 on the Rights of Servicemen (Declaration of Soldiers' Rights)," May 11, 1917, in *ibid.*, II, pp. 881–883.

42. Quoted in Denikin, *Russian Turmoil*, p. 175.

43. Quotes from Kerensky, *The Catastrophe*, pp. 193–195.

44. Denikin, *The Russian Turmoil*, p. 257.

45. Gorky, *Untimely Thoughts*, pp. 32, 34, 38, 39.

46. Kerensky, *The Catastrophe*, pp. 193, 195.

47. Quoted in Sukhanov, *Zapiski*, IV, pp. 135–136.

48. *Ibid.*, p. 153.

49. Brusilov, *Moi vospominaniia*, p. 275; Maria Botchkareva, *Yashka*: p. 205.

50. Quoted in Gorky, *Untimely Thoughts*, p. 198.

51. Sukhanov, *Zapiski*, IV, p. 137.

52. "Telegramma gen. Brusilova verkh. glavnokomanduiushchemu," 16 maia 1917g.

53. "Telegramma gen. Alekseeva Glavnokomanduiushchemu Iugo-Zapadnym frontom," "Telegramma gen. Brusilova voen. ministru i verkh. glavnokomanduiushchemu," 20 maia 1917g., and "Telegramma gen. Alekseeva Glavnokomanduiushchemu Iugo-Zapadnym frontom," 21 maia 1917g.

54. Pares, *Russian Memories*, p. 452.

55. "Prisiaga revoliutsionera-volontera."

56. "Vozzvanie," p. 69.

57. Pares, *Russian Memories*, p. 453.

58. The following account of Bochkareva's career is taken mainly from Botchkareva, *Yashka*. A colorful tale, and highly exaggerated at times ("Iashka" portrays herself as being on close terms with Rodzianko, Kerenskii, Polivanov, and Brusilov, all of whom, she insists,

sought her advice on more than one occasion, although none mentions her in their accounts of the period), Bochkareva's account remains the only firsthand account by a member of these unique military organizations that we have.

59. Botchkareva, *Yashka*, pp. 73–74.

60. *Ibid.*, p. 157.

61. *Ibid.*

62. "Pis'mo zhenshchiny-dobrovol'tsa voennomu ministru"; Botchkareva, *Yashka*, p. 160.

63. "The Revel Battalion of Death," in Golder (ed.), *Documents*, p. 432.

64. Quoted in Knox, *With the Russian Army*, II, p. 649.

65. *Rossiia v mirovoi voine (v tsifrakh)*, p. 26.

66. "Kratkii otchet vazhneishikh sobytii kampanii 1917g. na russkom fronte," p. 181.

67. For the Russian position at these meetings, which clearly conflicted with the French and British one, see V. A. Emets, "Petrogradskaia konferentsiia 1917g. i Frantsiia"; D. S. Babichev, "Rossiia na parizhskoi soiuznicheskoi konferentsii 1916g. po ekonomicheskim voprosam"; A. V. Ignat'ev, *Russko-angliiskie otnosheniia nakanune oktiabr'skoi revoliutsii*, pp. 102–122; and M. M. Karliner, "Angliia i Petrogradskaia konferentsiia Antanty 1917g." Summary minutes of the military sessions are in E. Martynov (ed.), "Konferentsiia soiuznikov v Petrograde v 1917 godu."

68. Kerensky, *Russia and History's Turning Point*, p. 271.

69. "Telegramma generala Gutora verkhovnomu glavnokomanduiushchemu," 30 maia 1917g.

70. Knox, *With the Russian Army*, II, pp. 639–640.

71. Brusilov, *Moi vospominaniia*, p. 264.

72. "Kerensky's Order to the Army and the Fleet," June 16, 1917, in Golder (ed.), *Documents*, pp. 426–427.

73. "Appeal from the Provisional Government," June 20, 1917, in Browder and Kerensky (eds.), *Russian Provisional Government*, II, p. 943.

74. Quoted in Kerensky, *Russia and History's Turning Point*, p. 282.

75. Brusilov, *Moi vospominaniia*, p. 275.

76. Knox, *With the Russian Army*, II, pp. 639, 636.

77. Kerensky, *Russia and History's Turning Point*, p. 285.

78. "Kerensky Requests Honors for the Regiments Leading the Offensive," in Browder and Kerensky (eds.), *Russian Provisional Government*, II, p. 943.

79. Denikin, *Russian Turmoil*, p. 273; "Excerpts from the Protocols of the Conference of Government and Military Leaders at *Stavka* at Mogilev on July 16," in Browder and Kerensky (eds.), *Russian Provisional Government*, II, p. 993.

80. A. M. Zaionchkovskii, *Kampaniia 1917 goda* pp. 58–75; Rostunov, *Russkii front*, pp. 355–359; Kerensky, *Russia and History's Turning Point*, pp. 284–287; Knox, *With the Russian Army*, II, pp. 640–646.

81. "Kerensky Requests Honors for the Regiments Leading the Offensive," July 18, 1917, in Browder and Kerensky (eds.), *Russian Provisional Government*, II, p. 943.

82. Quoted in Kerensky, *Russia and History's Turning Point*, p. 286.

83. Knox, *With the Russian Army*, II, pp. 645–646.

84. "The Attack on the Southwestern Front: the 11th Army," in Browder and Kerensky (eds.), *Russian Provisional Government*, II, p. 967.

85. "Protokol Soveshchaniia, byvshego 16-go iiulia 1917 goda v Stavke," p. 23.

86. *Ibid.*, p. 22.

87. "Kratkii ocherk vazhneishikh sobytii kampanii 1917g. na russkom fronte," (ed.), p. 181.

88. M. V. Alekseev, "Iz dnevnika Generala M. V. Alekseeva," p. 21.

89. "The Attack on the Southwestern Front: the 11th Army," in Browder and Kerensky (eds.), *Russian Provisional Government,* II, p. 967.

90. "Report of the Commander of the 11th Army to *Stavka,*" July 12, 1917, in *ibid.,* p. 968.

91. "Attack on Members of the Executive Committee of the Soviet," July 3, 1917, in Golder (ed.), *Documents,* p. 433.

92. "*Izvestiia* on the Sokolov Incident," in Browder and Kerensky (eds.), *Russian Provisional Government,* II, p. 955.

93. "Pis'mo verkhovnogo glavnokomanduiushchego A. F. Kerenskomu, 11 iiulia 1917g."

94. "Kerensky's Explanation of the Restoration of the Death Penalty," 13 July, 1917, in Browder and Kerensky (eds.), *Russian Provisional Government,* II, p. 982.

95. "Protokol Soveshchaniia, byvshego 16-go iiulia 1917 goda v Stavke," pp. 21, 30, 26, 49.

96. "Pis'mo M. V. Alekseeva ministru-predsedateliu A. F. Kerenskomu ot 20 iiulia 1917g.," p. 29.

97. Kerensky, *The Catastrophe,* p. 290.

98. I have taken the materials for the following biographical summary of Kornilov from the following: E. I. Martynov, *Kornilov,* pp. 11–22; George Katkov, *The Kornilov Affair,* pp. 39–45; Rabinowitch, *Bolsheviks Come to Power,* pp. 96–109; A. S. Lukomskii, *Vospominaniia,* pp. 217–252.

99. General Knox heard this account from Kornilov himself in October 1916. Knox, *With the Russian Army,* II, pp. 488–490.

100. Browder and Kerensky (eds.), *Russian Provisional Government,* II, p. 1023.

101. Quoted in Rabinowitch, *Bolsheviks Come to Power,* p. 97; Denikin, *Russian Turmoil,* p. 298.

102. "Telegramma predsedatelia vremennogo komiteta gosudarstvennoi dumy M. V. Rodzianko," 2 marta 1917g.

103. Knox, *With the Russian Army,* II, p. 608.

104. Mints, *Istoriia Velikogo Oktiabria,* II, pp. 317–318.

105. Quoted in Katkov, *Kornilov Affair,* p. 31.

106. N. Ia. Ivanov, *Kornilovshchina i ee razgrom,* pp. 40–41; Kerensky, *The Catastrophe,* pp. 292–295.

107. Quoted in Salisbury, *Black Night, White Snow,* p. 218.

108. "Protokol soveshchaniia byvshego 16-go iiulia 1917 goda v Stavke," pp. 33–35.

109. Brusilov, *Moi vospominaniia,* p. 276.

110. Ivanov, *Kornilovshchina i ee razgrom,* p. 41.

111. Kerensky, *The Catastrophe,* pp. 306–307.

112. "Privetstvennaia telegramma Moskovskogo soveshchaniia obshchestvennykh deiatelei generalu L. G. Kornilovu," 10 avgusta 1917g., No. 361, p. 360.

113. V. Vladimirova, *Kontr-revoliutsiia v 1917g. (Kornilovshchina),* p. 62.

114. Kerensky, *The Catastrophe,* p. 309.

115. Lukomskii, *Vospominaniia,* I, pp. 227–228; Martynov, *Kornilov,* p. 48.

116. Kerensky, *The Catastrophe,* p. 280; "The Kornilov Deposition," in Katkov, *The Kornilov Affair,* pp. 172–173.

117. Lukomskii, *Vospominaniia,* I, p. 227.

118. Rabinowitch, *Bolsheviks Come to Power,* p. 110.

119. M. N. Pokrovskii and Ia. A. Iakovlev (eds.), *Gosudarstvennoe soveshchanie,* p. 105.

120. *Ibid.,* pp. 60, 101, 106, 109–110, 117.

121. The following account of Kornilov's visit to Moscow is drawn from the following: Rabinowitch, *Bolsheviks Come to Power,* pp. 113–115; Katkov, *The Kornilov Affair,* pp. 59–63; Vladimirova, *Kontr-revoliutsiia,* pp. 70–92; Martynov, *Kornilov,* pp. 62–73.

122. Vladimirova, *Kontr-revoliutsiia*, p. 84.

123. Quoted in Martynov, *Kornilov*, p. 67.

124. Quoted in *ibid.*, p. 69.

125. "The Kornilov Deposition," in Katkov, *The Kornilov Affair*, p. 173.

126. These excerpts from Kornilov's speech are taken from Pokrovskii and Iakovlev (eds.), *Gosudarstvennoe soveshchanie*, pp. 61–66.

127. Quotes from the speeches given at the Moscow Conference by Kaledin and Maklakov on August 14, and Kerenskii on the evening of August 15, are from *ibid.*, pp. 75, 116, and 307. The Kadet N. I. Astrov's remark "It's not politics, it's hysterics," as he listened to Kerenskii's concluding speech, is quoted in Katkov, *The Kornilov Affair*, p. 62.

128. "*Novoe Vremia* on Kornilov's Speech," in Browder and Kerensky (eds.), *Russian Provisional Government*, III, pp. 1515–1516.

129. "*Izvestiia* on the Conference," in *ibid.*, p. 1520.

130. "Half a Year of the Revolution," in *ibid.*, p. 1521.

131. Quoted in Gippius, *Siniaia kniga*, p. 174.

132. "The Recollections of A. I. Putilov," in Browder and Kerensky (eds.), *Russian Provisional Government*, III, pp. 1528–1530; James D. White, "The Kornilov Affair: A Study in Counter-Revolution," pp. 187–190.

133. Hoffmann, *War of Lost Opportunities*, p. 189; Zaionchkovskii, *Kampaniia 1917 goda*, pp. 95–96.

134. "The Kornilov Deposition," in Katkov, *The Kornilov Affair*, p. 174; Rostunov, *Russkii front*, pp. 366–371. For Voitinskii's version, see Woytinsky, *Stormy Passage*, pp. 341–342, and "The Report of Assistant Commissar Voitinskii on the Breakthrough of August 19," in Browder and Kerensky (eds.), *Russian Provisional Government*, II, pp. 1033–1034.

135. Hoffmann, *War of Lost Opportunities*, p. 189.

136. "The Kornilov Deposition," in Katkov, *The Kornilov Affair*, pp. 174–175.

137. "*Izvestiia* Urges Unity in the Democracy to Save Russia and Revolution," in Browder and Kerensky (eds.), *Russian Provisional Government*, II, p. 1033.

138. "But Where Is the Truth?", in *ibid.*, p. 1037.

139. Pokrovskii and Iakovlev (eds.), *Gosudarstvennoe soveshchanie*, p. 63.

140. "But Where Is the Truth?", in Browder and Kerensky (eds.), *Russian Provisional Government*, II, p. 1037.

141. Ivanov, *Kornilovshchina i ee razgrom*, pp. 79–81.

142. Numerous efforts have been made to unravel the incredible tangle of events that took place between the fall of Riga on August 19 and Kornilov's arrest on Kerenskii's orders on September 1. By far the best—and the one to which my own account is much indebted —is the masterly study by Alexander Rabinowitch in *The Bolsheviks Come to Power*, pp. 116–150.

143. "The Kornilov Deposition," in Katkov, *The Kornilov Affair*, p. 186. See also pp. 184–187.

144. "Zapis' razgovora, sostoiavshegosia 23 avgusta 1917g. v Stavke mezhdu upravliaiushchim Voennym i Morskim ministerstvami B. V. Savinkovym i verkhovnym glavnokomanduiushchim generalom L. G. Kornilovym," No. 419; "A More Detailed Account of Savinkov's Conversation with Kornilov at Stavka," in Browder and Kerensky (eds.), *Russian Provisional Government*, III, pp. 1557–1558. See also Lukomskii, *Vospominaniia*, I, pp. 234–237; "The Kornilov Deposition," in Katkov, *The Kornilov Affair*, p. 175; and B. V. Savinkov, "General Kornilov," p. 192.

145. Lukomskii, *Vospominaniia*, p. 239.

146. Kerensky, *Russia and History's Turning Point*, p. 343; "From the Memoirs of V. N. L'vov," in Browder and Kerensky (eds.), *Russian Provisional Government*, III, p. 1562; A. F. Kerensky, *The Prelude to Bolshevism*, p. 158.

147. Kerensky, *Prelude to Bolshevism*, p. 161; Kerensky, *Russia and History's Turning Point*, p. 342.

148. Lukomskii, *Vospominaniia*, I, p. 238.

149. "Kornilov Deposition," in Katkov, *The Kornilov Affair*, p. 179.

150. Lukomskii, *Vospominaniia*, I, pp. 239–240.

151. The foregoing quotes are taken from Kerensky, *Prelude to Bolshevism*, pp. 162–168, and Kerensky, *Russia and History's Turning Point*, pp. 344–346.

152. "Zapis' razgovora po priamomu provodu ministra-predsedatelia A. F. Kerenskogo s verkhovnym glavnokomanduiushchim generalom L. G. Kornilovym," 26 avgusta 1917g., No. 443.

153. P. N. Miliukov, *Istoriia vtoroi russkoi revoliutsii*, I, pt. 2, p. 216.

154. "Radio-Telegram from Kerensky to All the Country, No. 4163," in Browder and Kerensky (eds.), *Russian Provisional Government*, III, p. 1572.

155. Quoted in Martynov, *Kornilov*, p. 110.

156. "Telegram of the Minister of Foreign Affairs from the Representative of the Ministry at Stavka," in Browder and Kerensky (eds.), *Russian Provisional Government*, III, p. 1574.

157. "Order to the Forces of Petrograd," in *ibid.*, p. 1581. Krymov's remark is quoted in General Denikin, *Russian Turmoil*, p. 68.

158. "The Arming of the Workers," in Browder and Kerenskii (eds.), *Russian Provisional Government*, III, p. 1590.

159. "Members of the Native Division Deny Knowingly Participating in the Revolt and Pledge Support to the Government and the Soviet," in *ibid.*, p. 1607.

160. "General Samarin's Account of the Role and Suicide of Krymov," in *ibid.*, p. 1589; Martynov, *Kornilov*, pp. 150–151; Kerensky, *Prelude to Bolshevism*, p. 188.

161. Florinsky, *Russia*, II, p. 1440.

162. Martynov, *Kornilov*, pp. 162–165; A. L. Fraiman (ed.), *Oktiabr'skoe vooruzhennoe vosstanie*, II, pp. 169–170.

163. Rabinowitch, *Bolsheviks Come to Power*, pp. 165–167.

164. Quoted in Pyman, *Release of Harmony*, pp. 268–269.

165. Gorky, *Untimely Thoughts*, p. 83.

166. Wade, *Red Guards and Workers' Militias*, p. 161.

167. Gorky, *Untimely Thoughts*, p. 83.

168. Fraiman (ed.), *Oktiabr'skoe vooruzhennoe vosstanie*, II, pp. 171–193.

169. Quoted in Rabinowitch, *Bolsheviks Come to Power*, p. 180.

Chapter XVI: Lenin Seizes Power

1. Nikolai Valentinov, *Encounters with Lenin*, p. 67.

2. Lenin, "Chto delat'?", p. 478.

3. V. I. Lenin, "On Compromises," p. 306.

4. L. F. Karamysheva, *Bor'ba bol'shevikov za petrogradskii sovet, mart–oktiabr' 1917g.*, pp. 157–170; Trotsky, *History of the Russian Revolution*, II, pp. 299–300; Robert V. Daniels, *Red October*, pp. 48–50.

5. Lenin, "On Compromises," p. 310.

6. V. I. Lenin, "The Bolsheviks Must Assume Power," p. 20; V. I. Lenin, "Marxism and Insurrection," pp. 23, 25, 27.

7. Stephen F. Cohen, *Bukharin and the Bolshevik Revolution*, pp. 47–49.

8. Quoted in Trotsky, *History of the Russian Revolution*, III, p. 133.

9. V. I. Lenin, "The Crisis Has Matured," pp. 77, 82.

10. V. I. Lenin, "Can the Bolsheviks Retain State Power?", pp. 94, 130.

11. "Rezoliutsiia sovmestnogo soveshchaniia Tsentral'nogo i Peterburgskogo komitetov RSDRP(b) i bol'shevistskoi fraktsii demokraticheskogo soveshchaniia 'Tekushchii moment i zadachi proletariata,'" No. 59, p. 75.

12. P. N. Mikhrin, "Vystupleniia," p. 122; M. V. Fofanova, "Poslednee podpol'e," pp. 344–347; Krupskaia, Vospominaniia o Lenine, pp. 334–335.

13. V. I. Lenin, "To Workers, Peasants, and Soldiers!", p. 138.

14. V. I. Lenin, "Advice of an Onlooker," pp. 179, 181.

15. V. I. Lenin, "Letter to the Central Committee, the Moscow and Petrograd Committees, and the Bolshevik Members of the Petrograd and Moscow Soviets," and "Letter to the Bolshevik Comrades Attending the Congress of Soviets of the Northern Region," pp. 141, 183, 187.

16. Lenin, "Advice of an Onlooker," pp. 180–181.

17. Mikhrin, "Vystupleniia," pp. 119–124; Fraiman (ed.), Oktiabr'skoe vooruzhennoe vosstanie, II, pp. 230–231, including note 223.

18. Sukhanov, Russian Revolution 1917, p. 556.

19. Quoted in Rabinowitch, Bolsheviks Come to Power, p. 202.

20. Iu. N. Flakserman, "10 oktiabria 1917 goda," pp. 266–267, 269; Petrogradskie bol'sheviki v oktiabr'skoi revoliutsii, pp. 338–339; Sukhanov, Zapiski, VII, pp. 32–33.

21. "Iz protokola zasedaniia Tsentral'nogo Komiteta RSDRP(b)," 10 oktiabria 1917g., No. 14, pp. 48–49.

22. Trotsky, History of the Russian Revolution, III, p. 148.

23. Flakserman, "10 oktiabria 1917 goda," pp. 267–268.

24. "Rezoliutsiia Tsentral'nogo Komiteta RSDRP(b)," 10 oktiabria 1917g., No. 15.

25. Quoted in Ulam, The Bolsheviks, p. 365; Rabinowitch, Bolsheviks Come to Power, pp. 204–205.

26. Trotsky, History of the Russian Revolution, III, p. 155.

27. Quoted in ibid., p. 151.

28. Quoted in Wolfe, Three Who Made a Revolution, p. 323.

29. Trotsky, History of the Russian Revolution, III, p. 75.

30. Lenin, "Can the Bolsheviks Retain State Power?", p. 111.

31. Quoted in Rabinowitch, Bolsheviks Come to Power, p. 226. See also pp. 225–228.

32. A. L. Popov (ed.), Oktiabr'skii perevorot, p. 155.

33. "Rezoliutsiia Petrogradskogo Soveta rabochikh i soldatskikh deputatov o neobkhodimosti peredachi vlasti Sovetam," 9 oktiabria 1917g., No. 103, p. 140; S. S. Tarasova and M. B. Keirim-Markus (comps.), Velikaia oktiabr'skaia sotsialisticheskaia revoliutsiia, pp. 370–371; "Postanovlenie kollegii voennogo otdela ispolnitel'nogo komiteta Petrogradskogo Soveta rabochikh i soldatskikh deputatov," 11 oktiabria 1917g., No. 9, pp. 38–39.

34. Sukhanov, Zapiski, VII, p. 76.

35. Quoted in S. Mel'gunov, Kak bol'sheviki zakhvatili vlast', p. 35.

36. Isaac Deutscher, The Prophet Armed, pp. 300–302; Ulam, The Bolsheviks, pp. 366–367.

37. "Protokol zasedaniia Tsentral'nogo Komiteta RSDRP(b)," pp. 93, 91.

38. Ibid., p. 93. Riazanov's remarks are quoted in Mel'gunov, Kak bol'sheviki zakhvatili vlast', p. 36.

39. Quoted in Daniels, Red October, pp. 97–98.

40. V. I. Lenin, "Letter to the Central Committee of the R.S.D.L.P (B.)"; V. I. Lenin, "Letter to Bolshevik Party Members," p. 217.

41. Trotsky, History of the Russian Revolution, III, p. 165.

42. John Reed, Ten Days That Shook the World, p. 49.

43. Quoted in ibid., p. xix.

44. Quoted in ibid., p. xxiii.

45. This summary of Reed's biography and views is much indebted to Bertram Wolfe's striking introductory essay in *ibid.*, pp. vii–xlv.

46. *Ibid.*, p. 49.

47. Ilya Ehrenburg, *First Years of Revolution, 1918–1921*, pp. 10, 30.

48. "Iz protokola Petrogradskoi konferentsii po voprosu o bezrabotnitse s izlozheniem doklada M. K. Vladimirova," oktiabria 1917g., pp. 351–352; V. I. Startsev, *Krakh kerenshchiny*, pp. 176–178; Reed, *Ten Days*, pp. 12–13, 18–19.

49. de Robien, *Diary*, p. 123.

50. Reed, *Ten Days*, pp. 36, 49–50.

51. *Ibid.*, p. 41.

52. *Ibid.*, pp. 75–76.

53. Quoted in Mel'gunov, *Kak bol'sheviki zakhvatili vlast'*, p. 36.

54. Trotsky, *History of the Russian Revolution*, III, pp. 167–173. In 1917, Marxists generally attributed Engels's *Revolution and Counter-Revolution in Germany* to Marx because it had first appeared in *The New York Daily News* under Marx's signature. Only the later publication of the correspondence between Marx and Engels revealed Engels's authorship. See Lenin, *Collected Works*, XXVI, p. 531, note 6.

55. Quoted in Rabinowitch, *Bolsheviks Come to Power*, p. 241.

56. "Obrashchenie k soldatam Petrogradskogo garnizona," No. 30, p. 63.

57. Sukhanov, *Zapiski*, VII, pp. 91–92.

58. Quoted in Trotsky, *History of the Russian Revolution*, III, p. 179.

59. Popov (ed.), *Oktiabr'skii perevorot*, p. 161.

60. Quoted in Rabinowitch, *Bolsheviks Come to Power*, p. 245. See also pp. 244–246.

61. Trotsky, *History of the Russian Revolution*, III, p. 182; B. M. Kochakov, "Bol'-shevizatsiia Petrogradskogo garnizona v 1917g.," pp. 178–182.

62. *Ibid.*, pp. 182–183.

63. Quoted in Wade, *Red Guards and Workers' Militias*, p. 139.

64. *Ibid.*, pp. 147, 158–164; Mitel'man *et al.*, *Istoriia putilovskogo zavoda*, I, pp. 569–573; V. V. Petrash, *Moriaki baltiiskogo flota v bor'be za pobedu oktiabria*, pp. 217–248; V. I. Startsev, "Voenno-revoliutsionnyi komitet i krasnaia gvardiia v oktiabr'skom vooruzhennom vosstanii," pp. 112–141.

65. Mitel'man, *et al.*, *Istoriia putilovskogo zavoda*, I, p. 569. The recollection of the Franko-Russkii Factory worker is quoted in Wade, *Red Guards and Workers' Militias*, p. 167.

66. Quoted in V. I. Startsev, "O vybore momenta dlia Oktiabr'skogo vooruzhennogo vosstaniia," p. 74.

67. Reed, *Ten Days*, p. 41; Trotsky, *History of the Russian Revolution*, p. 202.

68. Quoted in Reed, *Ten Days*, p. 89.

69. Fofanova, "Poslednee podpol'e," p. 348.

70. Lenin, "Letter to Central Committee Members, October 24, 1917," pp. 234–235.

71. *Ibid.*

72. L. N. Tolstoi, *Voina i Mir*, in his *Sobranie sochinenii* (Moscow, 1962), VI, p. 220.

73. Reed, *Ten Days*, p. 96.

74. *Ibid.*, p. 113.

75. de Robien, *Diary*, pp. 130–131.

76. *Ibid.*, p. 134.

77. *Ibid.*, p. xxv.

78. *Ibid.*, p. 132; Anet, *La révolution russe*, II, p. 229; Albert Rhys Williams, *Journey into Revolution*, pp. 100–102; Reed, *Ten Days*, pp. 132–134.

79. Williams, *Journey into Revolution*, pp. 103, 117–118; Trotsky, *History of the Russian Revolution*, III, p. 232.

80. V. I. Lenin, "Letter to Comrades," p. 195.

81. Tarasova and Keirit-Markus (comps.), *Velikaia oktiabr'skaia sotsialisticheskaia revoliutsiia,* IV, p. 564.

82. de Robien, *Diary,* p. 134.

83. Quoted in Rabinowitch, *Bolsheviks Come to Power,* p. 256.

84. "Kerensky's Speech Before the Council of the Republic, October 24, 1917," in Browder and Kerensky (eds.), *Russian Provisional Government,* III, pp. 1772–1778.

85. "Dan's Speech Before the Council," and "Martov's Speech and the Resolutions Offered to the Council," in *ibid.,* pp. 1778–1780.

86. N. F. Slavin, "Oktiabr'skoe vooruzhennoe vosstanie i Predparlament," pp. 226–228.

87. Kerensky, *The Catastrophe,* pp. 327–328, 330; F. Dan, "K istorii poslednikh dnei vremennogo pravitel'stva," pp. 170–175.

88. Quoted in Mel'gunov, *Kak bol'sheviki zakhvatili vlast',* p. 77.

89. Fraiman (ed.) *Oktiabr'skoe vooruzhennoe vosstanie,* II, pp. 307–308.

90. Sukhanov, *Zapiski,* VII, p. 160.

91. Quoted in Mel'gunov, *Kak bol'sheviki zakhvatili vlast',* p. 108.

92. Gippius, *Siniaia kniga,* pp. 210–212.

93. Reed, *Ten Days,* p. 102.

94. Krupskaia, *Vospominaniia,* p. 332.

95. A. V. Lunacharskii, "Smol'nyi v velikuiu noch'," p. 411.

96. Fofanova, "Poslednee podpol'e," p. 349.

97. Quoted in Daniels, *Red October,* p. 160; E. A. Rakh'ia, "Moi vospominaniia o Vladimire Il'iche," pp. 432–433; E. A. Rakh'ia, "Edem v Smol'nyi," pp. 368–371.

98. V. A. Antonov-Ovseenko, *V semnadtsatom godu,* p. 310.

99. E. Iiunga, *Kreiser "Avrora,"* pp. 75–81; Mints (ed.), *Oktiabr'skoe vooruzhennoe vosstanie,* II, pp. 317–325; Rabinowitch, *Bolsheviks Come to Power,* pp. 269–270; A. V. Belyshev, "Avrora v dni oktiabria," pp. 370–371; Ia. G. Temkin and S. V. Shestakov, *Nezabyvaemye dni,* pp. 137–142.

100. Quoted in Rabinowitch, *Bolsheviks Come to Power,* p. 269.

101. A. V. Liverovskii, "Poslednie chasy Vremennogo pravitel'stva," pp. 40–41.

102. "Zapis' razgovora po priamomu provodu general-kvartirmeistera Shtaba verkhovnogo glavnokomanduiushchego generala M. K. Diterikhsa s generalom dlia poruchenii B. A. Levitskim," 25 oktiabria 1917g., p. 340.

103. "Pokazaniia praporshchika Knirshi"; Francis, *Russia from the American Embassy,* pp. 179–180; V. I. Startsev, "Begstvo Kerenskogo," pp. 204–205.

104. Quoted in Temkin and Shestakov, *Nezabyvaemye dni,* p. 164.

105. P. N. Maliantovich, "V Zimnem Dvortse 25–26 oktiabria 1917 goda. Iz vospominanii," p. 118.

106. Quoted in Mel'gunov, *Kak bol'sheviki zakhvatili vlast',* p. 109.

107. Reed, *Ten Days,* p. 118.

108. Williams, *Journey into Revolution,* p. 111.

109. Maliantovich, "V Zimnem Dvortse," p. 120; A. S. Izgoev, "Piat' let v Sovetskoi Rossii (Otryvki vospominanii i zametki)," p. 20.

110. N. Podvoiskii, "Vziatie Zimnego," p. 147.

111. A. Sinegub, "Zashchita Zimnego Dvortsa, 25 oktiabria 1917g.," p. 165.

112. Maliantovich, "V Zimnem Dvortse," pp. 123–124.

113. Fraiman (ed.), *Oktiabr'skoe vooruzhennoe vosstanie,* II, pp. 346–347.

114. Quoted in Rabinowitch, *Bolsheviks Come to Power,* pp. 285–286.

115. Maliantovich, "V Zimnem Dvortse," p. 120.

116. Reed, *Ten Days,* p. 119.

117. Quoted in Mel'gunov, *Kak bol'sheviki zakhvatili vlast',* p. 127. See also, Iiunga, *Kreiser "Avrora,"* pp. 81–96.

118. Buchnanan, *My Mission to Russia,* II, p. 208.

119. Mel'gunov, *Kak bol'sheviki zakhvatili vlast',* pp. 128–129; Reed, *Ten Days,* p. 136; Williams, *Journey into Revolution,* p. 118; Fraiman (ed.), *Oktiabr'skoe vooruzhennoe vosstanie,* II, p. 351; Trotsky, *History of the Russian Revolution,* III, p. 274.

120. Knox, *With the Russian Army,* II, p. 709, quoting the account of General Polovtsev's aide-de-camp Captain Ragosin.

121. Sinegub, "Zashchita Zimnego Dvortsa," p. 163.

122. *Ibid.,* p. 187.

123. Maliantovich, "V Zimnem Dvortse," p. 129.

124. *Ibid.,* p. 129.

125. *Ibid.,* p. 130.

126. *Ibid.*

127. *Ibid.*

128. Quoted in Fraiman (ed.), *Oktiabr'skoe vooruzhennoe vosstanie,* II, p. 358. For Antonov-Ovseenko's version, see Antonov-Ovseenko, *V semnadtsatom godu,* p. 319.

129. Liverovskii, "Poslednye chasy," p. 47.

130. V. A. Antonov-Ovseenko, *Zapiski o grazhdanskoi voine* (Moscow, 1924), I, pp. 19–20.

131. The translation of this order is from a Russian poster reprinted in Reed, *Ten Days,* p. 267.

132. de Robien, *Diary,* p. 164.

133. Quoted in L. Trotskii, *Moia zhizn'. Opyt avtobiografii,* II, p. 59.

Chapter XVII: Birth Pangs of a New Order

1. Sukhanov, *Zapiski,* VII, p. 203. See also "Declaration of the Socialist Revolutionary Faction," in Browder and Kerensky (eds.), *Russian Provisional Government,* III, p. 1796.

2. Quoted in Getzler, *Martov,* p. 163. See also pp. 37–39 about Martov's Siberian exile and illness.

3. Sukhanov, *Zapiski,* pp. 203–204.

4. V. D. Bonch-Bruevich, "Kak pisal Vladimir Il'ich dekret o zemle," p. 435.

5. Reed, *Ten Days,* p. 36.

6. V. I. Lenin, "To Workers, Soldiers, and Peasants!" pp. 247–248.

7. Krupskaia, *Vospominaniia,* pp. 334–335; E. A. Lutskii, "Podgotovka proekta dekreta o zemle," pp. 233–248; V. N. Ginev, *Agrarnyi vopros i melkoburzhuaznye partii v Rossii v 1917g.,* pp. 219–224.

8. Bonch-Bruevich, "Kak pisal Vladimir Il'ich dekret o zemle," p. 437.

9. Trotsky, *History of the Russian Revolution,* III, p. 334.

10. Bonch-Bruevich, "Kak pisal Vladimir Il'ich dekret o zemle," pp. 438–440.

11. Mints, *Istoriia velikogo oktiabria,* II, p. 1111.

12. Quoted in *ibid.,* p. 1112.

13. Reed, *Ten Days,* pp. 170–172.

14. Sukhanov, *Zapiski,* VII, p. 253.

15. Williams, *Journey into Revolution,* p. 125.

16. V. I. Lenin, "Report on Peace, October 26," pp. 249–253.

17. Reed, *Ten Days,* pp. 177–178.

18. Quoted in Michael Kettle, *The Allies and the Russian Collapse, March 1917–March 1918,* p. 109.

19. George F. Kennan, *Russia Leaves the War,* p. 75.

20. Krupskaia, *Vospominaniia,* p. 337.

21. Bonch-Bruevich, "Kak pisal Vladimir Il'ich dekret o zemle," p. 437.

22. V. I. Lenin, "Report on Land," pp. 257–261.

23. Krupskaia, *Vospominaniia*, p. 338.

24. Bonch-Bruevich, "Kak pisal Vladimir Il'ich dekret o zemle," p. 437

25. Trotskii, *Moia zhizn'*, II, pp. 59–60.

26. Krupskaia, *Vospominaniia*, p. 338.

27. Quoted in Mel'nikov, *Kak bol'sheviki zakhvatili vlast'*. p, 180.

28. O. F. Solov'ev, *Velikii Oktiabr' i ego protivniki*, pp. 76–77; Mints, *Istoriia velikogo oktiabria*, III, pp. 108–109.

29. A. L. Popov (ed.), *Oktiabr'skii perevorot*, p. 339.

30. Mel'gunov, *Kak bol'sheviki zakhvatili vlast'*, pp. 184–185.

31. "Pokazanie praporshchika E. V. Zelenskogo," p. 183.

32. V. S. Vasiukov, *Predystoriia interventsii, fevral' 1917–mart 1918* (Moscow, 1968), pp. 210–211.

33. Mel'gunov, *Kak bol'sheviki zakhvatili vlast'*, p. 181.

34. Vasiukov, *Predystoriia interventsii*, pp. 207–209.

35. Quoted in Mel'gunov, *Kak bol'sheviki zakhvatili vlast'*, pp. 187–188.

36. A. Shliapnikov (ed.), "Oktiabr'skii perevorot i stavka," *Krasnyi arkhiv*, IX (1925), p. 161.

37. Buchanan, *My Mission to Russia*, II, p. 211.

38. I. S. Lutovinov, *Likvidatsiia miatezha Kerenskogo-Krasnova*, pp. 7–8; A. L. Fraiman, *Forpost sotsialisticheskoi revoliutsii*, pp. 18–20.

39. "Zapis' razgovora po priamomu provodu nachal'nika Shtaba verkhovnogo glavnokomanduiushchego generala N. N. Dukhonina," 25 oktiabria 1917g., No. 591, p. 412.

40. "Zapis' razgovora po priamomu provodu nachal'nika Shtaba verkhovnogo glavnokomanduiushchego generala N. N. Dukhonina s general-kvartirmeisterom shtaba Severnogo fronta generalom V. L. Baranovskim i komissarom Vremennogo pravitel'stva na Severnom fronte V. S. Voitinskim o polozhenii na fronte," 26–27 oktiabria 1917g., p. 605.

41. "Iz zapisi razgovora po priamomu provodu nachal'nika Shtaba verkhovnogo glavnokomanduiushchego generala N. N. Dukhonina s nachal'nikom shtaba fronta generalom S. G. Lukirskim o prodvizhenii voisk k g. Petrogradu na podavlenie vosstaniia," 27 oktiabria 1917g., No. 828; Fraiman (ed.), *Oktiabr'skoe vooruzhennoe vosstanie*, II, pp. 376–377; Lutovinov, *Likvidatsiia miatezha Kerenskogo-Krasnova*, pp. 16–17.

42. "Prikaz A. F. Kerenskogo o naznachenii generala P. N. Krasnova komanduiushchim vooruzhennymi silami Petrogradskogo voennogo okruga," 27 oktiabria 1917g., No. 822.

43. Quoted in Mints, *Istoriia velikogo oktiabria*, III, p. 106.

44. Fraiman (ed.), *Oktiabr'skoe vooruzhennoe vosstanie*, II, p. 377.

45. Quoted in Mel'gunov, *Kak bol'sheviki zakhvatili vlast'*, pp. 161, 163, 165. See also pp. 162–164.

46. Raskol'nikov, *Na boevykh postakh*, p. 212.

47. Quoted in Reed, *Ten Days*, p. 235.

48. "Direct-Line Conversations [of V. I. Lenin] with Helsingfors, October 27, 1917," in Lenin, *Collected Works*, XXVI, pp. 266–268; Raskol'nikov, *Na boevykh postakh*, p. 211.

49. Mitel'man, *et al.*, *Istoriia putilovskogo zavoda*, pp. 595–597.

50. Louise Bryant, *Six Red Months in Russia*, p. 178.

51. A. M. Konev, *Krasnaia gvardiia na zashchite Oktiabria*, p. 67.

52. "Obrashchenie k naseleniiu Petrograda o podavlenii iunkerskogo miatezha," No. 669, p. 280; Fraiman, *Forpost sotsialisticheskoi revoliutsii*, p. 55.

53. Z. N. Shul'man, "Vziatie Vladimirskogo uchilishcha," in Mushtukov (ed.), *Petrograd v dni velikogo oktiabria*, pp. 476–485; Fraiman, *Forpost sotsialisticheskoi revoliutsii*, pp. 59–61; V. I. Startsev, *Ocherki po istorii Petrogradskoi Krasnoi Gvardii i rabochei militsii* pp. 198–199.

54. Quoted in Reed, *Ten Days,* p. 235.

55. Lutovinov, *Likvidatsiia miatezha Kerenskogo-Krasnova,* pp. 32–33.

56. Fraiman, *Forpost sotsialisticheskoi revoliutsii,* pp. 62–99; Mints, *Istoriia velikogo oktiabria,* III, pp. 156–188; Fraiman (ed.), *Oktiabr'skoe vooruzhennoe vosstanie,* II, pp. 390–418.

57. Kerensky, *Russia and History's Turning Point,* p. 446.

58. Buchanan, *My Mission to Russia,* II, p. 216; Chernov, *Great Russian Revolution,* p. 409.

59. Cohen, *Bukharin and the Bolshevik Revolution,* p. 49.

60. Mel'gunov, *Kak bol'sheviki zakhvatili vlast',* p. 277.

61. "Telefonogramma predsedatelia Moskovskogo Soveta rabochikh deputatov V. P. Nogina i chlena TsIK Sovetov rabochikh i soldatskikh deputatov V. P. Miliutina v Moskovskii Sovet rabochikh deputatov o pobede vooruzhennogo vosstaniia v Petrograde," 25 oktiabria 1917g., No. 267, p. 251.

62. "Protokol zasedaniia Moskovskogo komiteta RSDRP(b)," and "Protokol dnevnogo zasedaniia uzkogo sostava Moskovskogo oblastnogo biuro RSDRP(b)," 25 oktiabria 1917g., Nos. 259–269.

63. Koenker, *Moscow Workers and the 1917 Revolution,* p. 335.

64. Quoted in Mints, *Istoriia velikogo oktiabria,* III, p. 195.

65. A. N. Voznesenskii, *Moskva v 1917 godu,* p. 170; Mints, *Istoriia velikogo oktiabria,* III, p. 244.

66. *Oktiabr' v Moskve,* p. 355; G. A. Trukan, *Oktiabr' v Tsentral'noi Rossii,* pp. 269–260.

67. "Obrashchenie moskovskogo soveta professional'nykh soiuzov 'ko vsem rabochim i rabotnitsam' s prizyvom splotit'sia vokrug voenno-revoliutsionnogo komiteta," pp. 399–400.

68. A. Ia. Grunt, *Pobeda oktiabr'skoi revoliutsii v Moskve,* pp. 164–165; *Oktiabr' v Moskve,* pp. 361–362; Mints, *Istoriia velikogo oktiabria,* III, pp. 221–222, especially notes 105 and 106.

69. Quoted in *Oktiabr' v Moskve,* p. 370.

70. Koenker, *Moscow Workers and the 1917 Revolution,* p. 333.

71. Paustovskii, *Sobranie sochinenii,* III, pp. 585–597.

72. Quoted in *Oktiabr' v Moskve,* p. 378.

73. *Ibid.,* pp. 379–386.

74. Grunt, *Pobeda oktiabr'skoi revoliutsii v Moskve,* pp. 191–193; Mints, *Istoriia velikogo oktiabria,* III, p. 245.

75. *Oktiabr' v Moskve,* pp. 388–389.

76. Mints, *Istoriia velikogo oktiabria,* III, p. 291, note 342.

77. Paustovskii, *Sobranie sochinenii,* III, pp. 598–599.

78. Quoted in Stites, *Women's Liberation Movement in Russia,* p. 363.

79. *Ibid.,* p. 358.

80. Quoted in *ibid.,* p. 377.

81. Chamberlin, *Russian Revolution,* I, pp. 359–362.

82. Quoted in A. Gindin, *Kak bol'sheviki natsionalizirovali chastnye banki,* p. 45.

83. Stites, *Women's Liberation Movement in Russia,* p. 372; Ariadna Tyrkova-Williams, *From Liberty to Brest-Litovsk,* pp. 433–434; de Robien, *Diary,* p. 222.

84. Fraiman, *Forpost sotsialisticheskoi revoliutsii,* p. 289; M. I. Davydov, *Bor'ba za khleb,* pp. 20–21.

85. Fraiman (ed.), *Oktiabr'skoe vooruzhennoe vosstanie,* II, pp. 518–519.

86. Davydov, *Bor'ba za khleb,* p. 22; "Svedeniia o nalichii na frontakh prodovol'stviia, furazha i obespechennosti v dniakh po sostoianiiu na 20 oktiabria 1917g., po dannym glavnogo polevogo intendanta," No. 470.

87. Fraiman, *Forpost sotsialisticheskoi revoliutsii*, 292–293; Davydov, *Bor'ba za khleb*, pp. 23–24; Konev, *Krasnaia gvardiia na zashchite oktiabria*, pp. 114–115.

88. Mints, *Istoriia velikogo oktiabria*, III, p. 861.

89. Quoted in I. A. Gladkov, *Ocherki stroitel'stva sovetskogo planovogo khoziaistva v 1917–1918gg.*, p. 266.

90. Yaney, *Urge to Mobilize*, pp. 455–458, 485–489.

91. Davydov, *Bor'ba za khleb*, pp. 26–27.

92. Konev, *Krasnaia gvardiia na zashchite oktiabria*, p. 110.

93. Gladkov, *Ocherki stroitel'stva sovetskogo planovogo khoziaistva*, pp. 263–267.

94. Quoted in Konev, *Krasnaia gvardiia na zashchite oktiabria*, p. 108.

95. V. I. Lenin, "Note to F. E. Dzerzhinsky with a Draft of a Decree on Fighting Counter-Revolutionaries and Saboteurs," p. 374.

96. *Iz istorii Vserossiiskoi Chrezvychainoi Komissii, 1917–1921gg.*, p. 79.

97. A. F. Khatskevich, *Soldat velikikh boev*, pp. 44–252; N. Zubov, *F. E. Dzerzhinskii*, pp. 206–343; Chamberlin, *Russian Revolution*, II, pp. 76–77.

98. Quoted in Chamberlin, *Russian Revolution*, II, p. 78.

99. Quoted in *ibid.*, II, p. 77.

100. Quoted in Sukhanov, *Zapiski*, VI, pp. 250–251.

101. Chernov, *Great Russian Revolution*, p. 410.

102. Boris Sokolov, "Zashchita Vserossiiskago Uchreditel'nago Sobraniia," p. 15.

103. Oliver H. Radkey, *The Sickle Under the Hammer*, pp. 281–282; N. Rubinshtein, *Bol'sheviki i uchreditel'noe sobranie*, pp. 53–61; I. S. Malchevskii (ed.), *Vserossiiskoe uchreditel'noe sobranie*, pp. 114–115.

104. Quoted in Anweiler, *The Soviets*, p. 211.

105. *Ibid.*, p. 213.

106. Quoted in Leon Trotsky, *Lenin: Notes for a Biographer*, p. 111.

107. Quoted in Chamberlin, *Russian Revolution*, I, p. 365.

108. Tyrkova-Williams, *From Liberty to Brest-Litovsk*, p. 335.

109. Rubinshtein, *Bol'sheviki i uchreditel'noe sobranie*, pp. 28–30; Keep, *The Russian Revolution*, p. 326; Oliver H. Radkey, *Election to the Russian Constituent Assembly of 1917*, pp. 49–50.

110. Gorky, *Untimely Thoughts*, p. 99.

111. V. I. Lenin, "Theses on the Constituent Assembly," pp. 379, 383.

112. Rubinshtein, *Bol'sheviki i uchreditel'noe sobranie*, pp. 62–63; "Decree on the Arrest of the Leaders of the Civil War Against the Revolution," November 28, 1917, in Lenin, *Collected Works*, XXVI, p. 351.

113. Lenin, "Theses on the Constituent Assembly," p. 382.

114. V. I. Lenin, "People from Another World," p. 431.

115. Radkey, *Sickle Under the Hammer*, pp. 333–335, 386; Chamberlin, *Russian Revolution*, I, pp. 368–369; Sokolov, "Zashchita Vserossiiskago Uchreditel'nago Sobraniia," pp. 46–48.

116. Quoted in Keep, *The Russian Revolution*, p. 329.

117. Tyrkova-Williams, *From Liberty to Brest-Litovsk*, p. 353.

118. Sokolov, "Zashchita Vserossiiskago Uchreditel'nago Sobraniia," p. 67.

119. Quoted in Keep, *The Russian Revolution*, p. 332.

120. Quoted in Chamberlin, *Russian Revolution*, I, p. 370.

121. Lenin, "People from Another World," pp. 431–432.

122. Sokolov, "Zashchita Vserossiiskago Uchreditel'nago Sobraniia," p. 69.

123. Quoted in Tyrkova-Williams, *From Liberty to Brest-Litovsk*, p. 369.

Chapter XVIII: Peace and War

1. Ts. Urlanis, *Voiny i narodonaselenie Evropy*, pp. 297–300, 465, and *passim*.
2. Lenin, "Report on Peace, October 26, 1917," pp. 249–250.
3. Quoted in Kennan, *Russia Leaves the War*, p. 85.
4. Reed, *Ten Days*, p. 178; Williams, *Journey into Revolution*, p. 132.
5. Sukhanov, *Zapiski*, VII, p. 253.
6. Gippius, *Siniaia kniga*, VII, pp. 219, 221.
7. Zinaida Gippius, "Vesel'e," p. 84.
8. Published in Temira Pachmuss, *Zinaida Hippius*, p. 199.
9. Gippius, *Siniaia kniga*, p. 221.
10. Lenin, "Report on Peace, October 26," pp. 253, 249.
11. V. I. Lenin, "Speech at a Joint Meeting of the Petrograd Soviet of Workers' and Soldiers' Deputies and Delegates from the Fronts," November 4, 1917, p. 293.
12. Lenin, "Report on Peace, October 26," p. 253.
13. Williams, *Journey into Revolution*, p. 144.
14. Kennan, *Russia Leaves the War*, pp. 80–84; Kettle, *The Allies and the Russian Collapse*, pp. 106–111; John W. Wheeler-Bennett, *The Forgotten Peace*, pp. 70–71; Judah L. Magnes, *Russia and Germany at Brest-Litovsk*, pp. 14–16; Buchanan, *My Mission to Russia*, II, pp. 215–224.
15. Robert D. Warth, *The Allies and the Russian Revolution from the Fall of the Monarchy to the Peace of Brest-Litovsk*, p. 169.
16. A. O. Chubar'ian, *Brestskii mir*, pp. 54–56; I. I. Mints, *God 1918-i*, pp. 10–11.
17. Quoted in Lockhart, *British Agent*, p. 222.
18. Quoted in Deutscher, *The Prophet Armed*, p. 351. See also William Hard, *Raymond Robins' Own Story*, pp. 68–70.
19. Trotsky, *Lenin*, p. 99.
20. Trotskii, *Moia zhizn'*, II, p. 119.
21. Quoted in Merle Fainsod, *International Socialism and the World War*, p. 168.
22. Quoted in *ibid.*
23. Quoted in Reed, *Ten Days*, p. xxix.
24. Quoted in Fainsod, *International Socialism and the World War*, p. 170. See also pp. 169, 171.
25. "Pravitel'stvennoe predpisanie Verkhovnomu Glavnokomanduiushchemu Dukhoninu," 7/20 noiabria 1917g., No. 4, pp. 15–16.
26. "Razgovor po priamomu provodu gen. Dukhonina s voennym ministrom gen. Manikovskim i nach. gen. shtaba gen. Marushevskim," 9 noiabria 1917g. p. 198; Kennan, *Russia Leaves the War*, pp. 89–90.
27. "Razgovor Pravitel'stva so Stavkoi po priamomu provodu," 9/22 noiabria 1917g., No. 6.
28. "Nikolai Vasil'evich Krylenko, cols. 194–195."
29. "Obrashchenie Soveta Narodnykh Komissarov k armii i flotu po voprosu o mire," 9/22 noiabria 1917g., No. 7, p. 20.
30. See, for example, "Obrashchenie nachal'nikov voennykh missii pri shtabe verkh. glavnokomanduiushchego na imia gen. Dukhonina 11 noiabria 1917g.," and "Zaiavlenie nachal'nika frantsuzskoi voennoi missii pri shtabe verkh. glavnokomanduiushchego," 12 noiabria 1917g., pp. 211–212, 214–215.
31. James Bunyan and H. H. Fisher, *The Bolshevik Revolution, 1917–1918*, p. 406.
32. Quoted in Tyrkova-Williams, *From Liberty to Brest-Litovsk*, p. 311.
33. General A. Niessel, *Le Triomphe des Bolcheviks et la Paix de Brest-Litovsk*, p. 110.

34. Quoted in Tyrkova-Williams, *From Liberty to Brest-Litovsk,* p. 310.

35. "Soobshchenie Russkoi delegatsii o peregovorakh s Germanskim komandovaniem o peremirii," 14/27 noiabria 1917g., No. 12, pp. 26–27.

36. Wheeler-Bennett, *The Forgotten Peace,* pp. 77–79.

37. D. G. Fokke, "Na stsene i za kulisami Brestskoi tragikomedii (memuary uchastnika Brest-Litovskikh mirnykh peregorov)," p. 29.

38. "Obrashchenie Sovetskogo Pravitel'stva k poslam soiuznykh derzhav," 8/21 noiabria 1917g.," No. 5.

39. "Obrashchenie Sovetskogo Pravitel'stvo k pravitel'stvam i narodam voiuiushchikh stran," 15/28 noiabria 1917g.," No. 13.

40. "Soobshchenie Narodnogo Komissara Inostrannykh Del o predstoiashchem otkrytii peregovorov o peremirii, 17/30 noiabria 1917g.," No. 15.

41. Fokke, "Na stsene i za kulisami Brestskoi tragikomedii," pp. 15–16; Count Ottokar Czernin, *In the World War,* p. 245.

42. Fokke, "Na stsene i za kulisami Brestskoi tragikomedii," pp. 16–17.

43. Hoffmann, *War of Lost Opportunities,* p. 201.

44. Chubar'ian, *Brestskii mir,* pp. 88–89.

45. Quoted in Chubar'ian, *Brestskii mir,* p. 90.

46. Hoffmann, *War of Lost Opportunities,* p. 199.

47. *Ibid.,* p. 198; Wheeler-Bennett, *Forgotten Peace,* pp. 89–91; Chubar'ian, *Brestskii mir,* pp. 87–91; S. I. Blinov, *Vneshniaia politika Sovetskoi Rossii,* pp. 102–104; Niessel, *Le triomphe des Bolcheviks et la Paix de Brest-Litovsk,* pp. 165–171.

48. Chubar'ian, *Brestskii mir,* p. 89.

49. Hoffmann, *War of Lost Opportunities,* p. 200–201.

50. Fokke, "Na stsene i za kulisami Brestskoi tragikomedii," pp. 68–69.

51. Bunyan and Fisher, *The Bolshevik Revolution,* pp. 273–274.

52. Quoted in Wheeler-Bennett, *Forgotten Peace,* p. 92.

53. Erich Ludendorff, *Ludendorff's Own Story, August 1914–November 1918,* II, pp. 166–167.

54. Quoted in Deutscher, *The Prophet Armed,* p. 359.

55. M. D. Bonch-Bruevich, *Vsia vlast' sovetam,* p. 231.

56. Ludendorff, *Ludendorff's Own Story,* II, p. 165.

57. Quoted in Wheeler-Bennett, *Forgotten Peace,* p. 110.

58. Czernin, *In the World War,* p. 242.

59. *Ibid.,* p. 246; Fokke, "Na stsene i za kulisami Brestskoi tragikomedii," pp. 95–98.

60. V. I. Lenin, "Outline Programme for Peace Negotiations"; "The Russian Conditions of Peace," in Bunyan and Fisher, *Bolshevik Revolution,* pp. 477–478.

61. "Czernin's Reply for the Quadruple Alliance," December 25, 1918, in Bunyan and Fisher, *Bolshevik Revolution,* pp. 479–481.

62. Quoted in *ibid.,* pp. 484–485.

63. Fokke, "Na stsene i za kulisami Brestskoi tragikomedii," p. 118.

64. Hoffmann, *War of Lost Opportunities,* p. 209.

65. *Ibid.,* p. 209.

66. Fokke, "Na stsene i za kulisami Brestskoi tragikomedii," p. 130.

67. Hoffmann, *War of Lost Opportunities,* pp. 209–210; Czernin, *In the World War,* pp. 253–254; Bunyan and Fisher, *Bolshevik Revolution,* pp. 482–483.

68. Hoffmann, *War of Lost Opportunities,* p. 210.

69. Fokke, "Na stsene i za kulisami Brestskoi tragikomedii," p. 118.

70. Hoffmann, *War of Lost Opportunities,* p. 210.

71. "Obrashchenie Narodnogo Komissariata Inostrannykh Del k narodam i pravitel'stvam soiuznykh stran," 17/30 dekabria 1917g.

72. Quoted in Kennan, *Russia Leaves the War*, p. 263. This editorial appeared on December 22, 1917/January 4, 1918, just four days before Wilson delivered his famous "Fourteen Points" speech to Congress. However, as Kennan notes, *Pravda*'s commentary was "so sweeping in its qualifications of the President's utterances generally as to render further comment on this particular speech almost superfluous." (*Ibid.*, p. 262.)

73. Trotskii, *Moia zhizn'*, II, p. 87.

74. Fokke, "Na stsene i za kulisami Brestskoi tragikomedii," p. 127.

75. Quoted in Trotsky, *Lenin*, p. 99.

76. Hoffmann, *War of Lost Opportunities*, pp. 212–213.

77. Ludendorff, *The General Staff*, II, p. 524.

78. Quoted in Wheeler-Bennett, *Forgotten Peace*, p. 135.

79. *Ibid.*, p. 151; Czernin, *In the World War*, pp. 258–261.

80. Wheeler-Bennett, *Forgotten Peace*, p. 152.

81. "Declaration of the Ukrainian Delegation," December 29, 1917/January 10, 1918, in Bunyan and Fisher, *Bolshevik Revolution*, p. 492.

82. Bunyan and Fisher, *Bolshevik Revolution*, pp. 492–493.

83. Czernin, *In the World War*, pp. 267–268.

84. The full debates of the Brest-Litovsk conference are in *Mirnye peregovory v Brest-Litovske* (Moscow, 1926). Useful summaries (often with excerpts) are to be found in Wheeler-Bennett, *Forgotten Peace*, 157–175; Bunyan and Fisher, *Bolshevik Revolution*, pp. 489–498; Fokke, Na stsene i za kulisami Brestskoi tragikomedii," pp. 134–172; Deutscher, *The Prophet Armed*, pp. 361–372.

85. Trotskii, *Moia zhizn'*, II, p. 100.

86. Quoted in Deutscher, *The Prophet Armed*, p. 366.

87. Trotskii, *Moia zhizn'*, II, p. 99.

88. Czernin, *In the World War*, p. 266.

89. Quoted in Fokke, "Na stsene i za kulisami Brestskoi tragikomedii," p. 167.

90. "Vechernee zasedanie 5/18 ianvaria 1918g.," in L. Trotskii, *Sochineniia* XVII, pt. 1, p. 51.

91. Quoted in Wheeler-Bennett, *Forgotten Peace*, pp. 185–186.

92. Quoted in Trotsky, *Lenin*, p. 100.

93. V. I. Lenin, "Theses on the Question of the Immediate Conclusion of a Separate and Annexationist Peace."

94. Cohen, *Bukharin and the Bolshevik Revolution*, pp. 62–63; Chubar'ian, *Brestskii mir*, pp. 130–131.

95. Trotskii, *Moia zhizn'*, II, pp. 108–109.

96. Quoted in Wheeler-Bennett, *Forgotten Peace*, p. 188.

97. Chubar'ian, *Brestskii mir*, p. 134.

98. Quoted in David Shub, *Lenin*, p. 297.

99. V. I. Lenin, "Afterword to the Theses on the Question of the Immediate Conclusion of a Separate and Annexationist Peace," p. 452.

100. Quoted in Wheeler-Bennett, *Forgotten Peace*, p. 192.

101. Krupskaia, *Vospominaniia o Lenine*, pp. 380, 379.

102. Quoted in Trotsky, *Lenin*, p. 104.

103. Trotskii, *Moia zhizn'*, II, p. 111.

104. Chubar'ian, *Brestskii mir*, p. 135.

105. Fokke, "Na stsene i za kulisami Brestskoi tragikomedii," pp. 205–206.

106. *Ibid.*, pp. 206–207.

107. Quoted in Wheeler-Bennett, *Forgotten Treaty*, p. 237. See also pp. 236, 238.

108. Quoted in Trotsky, *Lenin*, p. 105.

109. Ludendorff, *The General Staff*, II, p. 550.

110. Ludendorff, *Ludendorff's Own Story*, II, pp. 183–184.
111. "Soobshchenie generala Samoilo Narodnomu Komissaru Inostrannykh Del," 16 fevralia 1918g.
112. Quoted in Wheeler-Bennett, *Forgotten Peace*, pp. 243–244.
113. Mints, *God 1918-i*, p. 85; Blinov, *Vneshniaia politika Sovetskoi Rossii*, p. 137.
114. Quoted in Trotsky, *Lenin*, p. 106.
115. Mints, *God 1918-i*, pp. 86–87.
116. Edgar Sisson, *One Hundred Red Days*, p. 326.
117. "Zasedanie TsK RSDRP, 18 fevralia 1918g. (vecherom)."
118. "Radiogramma Soveta Narodnykh Komissarov Pravitel'stvu Germanii," 19 fevralia 1918g., No. 68.
119. Quoted in Wheeler-Bennett, *Forgotten Peace*, p. 245.
120. Quoted in *ibid.*, p. 252–253.
121. Bunyan and Fisher, *Bolshevik Revolution*, pp. 602–603.
122. V. I. Lenin, "The Socialist Fatherland Is in Danger!", February 21, 1918, in *Collected Works*, XXVII, pp. 30–33.
123. Trotskii, *Moia zhizn'*, II, pp. 116–117.
124. "Zasedanie TsK RSDRP, 22 fevralia 1918g.," p. 208.
125. "Postanovlenie Soveta Narodnykh Komissarov o priniatii germanskikh uslovii mira." 23 fevralia 1918g., No. 73.
126. "Zasedanie TsK RSDRP, 23 fevralia 1918g.," pp. 211–213.
127. Quoted in Wheeler-Bennett, *Forgotten Peace*, p. 260.
128. Quoted in Trotskii, *Moia zhizn'*, II, p. 123.
129. Trotsky, *Lenin*, p. 106.
130. Quoted in Wheeler-Bennett, *Forgotten Peace*, pp. 268–269.
131. Chubar'ian, *Brestskii mir*, p. 135.
132. V. I. Lenin, "The Chief Task of Our Day."
133. Aleksandr Blok, "Dvenadtsat'," pp. 497–498.

Epilogue

1. de Robien, *Diary*, p. 204.
2. Beatty, *Red Heart of Russia*, p. 314.
3. Sorokin, *Leaves from a Russian Diary* p. 108.
4. A. S. Pushkin, "Mednyi vsadnik," p. 382.
5. Sorokin, *Leaves from a Russian Diary*, p. 133.
6. Gippius, *Dmitrii Merezhkovskii*, p. 224.
7. de Robien, *Diary*, p. 218.
8. *ibid.*, p. 223.
9. Tyrkova-Williams, *From Liberty to Brest-Litovsk*, p. 395.
10. Lockhart, *British Agent*, p. 329.
11. Krupskaia, *Vospominaniia o Lenine*, p. 419.
12. V. I. Lenin, "Political Report of the Central Committee to the Extraordinary Seventh Congress of the R.C.P.(b)," March 7, 1918, pp. 106, 102, 108.
13. *Ibid.*, pp. 102, 104, 109.
14. Chubar'ian, *Brestskii mir*, pp. 98–100; Liashchenko, *Istoriia narodnogo khoziaistva SSSR*, II, pp. 669–674; de Robien, *Diary*, pp. 202–203.
15. Bunyan and Fisher, *Bolshevik Revolution*, pp. 523–524.
16. Anton Denikin, *The White Army*, p. 28.
17. Lukomskii, *Vospominaniia*, I, p. 266.
18. *Ibid.*, pp. 269–274, and note on pp. 278–279.

19. de Robien, *Diary,* p. 206.

20. P. N. Wrangel, *Memoirs,* p. 59.

21. Denikin, *White Army,* p. 30.

22. *Ibid.,* pp. 96–98.

23. Quoted in Chamberlin, *Russian Revolution,* II, p. 77.

24. F. I. Tiutchev, "La Russie et la Révolution," in *Polnoe sobranie sochinenii F. I. Tiutcheva,* p. 344.

WORKS AND
SOURCES CITED

The following is not intended to be an exhaustive compilation of all materials available about Russia during the Great War and the revolutions of 1917, nor is it even a comprehensive listing of all the materials consulted in the research for this book. It is not in any sense, then, a bibliography. As its title states, it is a list of the works and sources cited in the notes to this book.

KEY TO ABBREVIATIONS:

AdAE: Archives des Affaires Etrangères, Paris.
AG-CV: Archives de la Guerre. Service historique de l'armée de la terre. Château de Vincennes, Vincennes.
PRO: Public Records Office, London.
TsGIAL: Tsentral'nyi gosudarstvennyi istoricheskii arkhiv SSSR, Leningrad.

Akhun, M. I., and Petrov, V. A. *1917 god v Petrograde: Khronika sobytii i bibliografiia.* Leningrad, 1933.
——— (eds.). *Bol'sheviki i armiia v 1905–1917gg.* Leningrad, 1929.
Albertini, Luigi. *The Origins of the War of 1914.* 3 vols. Oxford, 1965.
"Aleksandr Vasil'evich Samsonov," *Sovetskaia istoricheskaia entsiklopediia,* XII, cols. 520–521.
Aleksandra Feodorovna, Empress. *Letters of the Tsaritsa to the Tsar, 1914–1916.* Edited by Sir Bernard Pares. London, 1923.
Aleksandrov, P. P. "Ot fevralia k oktiabriu," in Petrov *et al.* (eds.), *V Ogne revoliutsionnykh boev,* II, pp. 51–61.
Alekseev, M. V. "Iz dnevnika Generala M. V. Alekseeva," *Rusky Historicky Archiv,* Prague, 1929, I, pp. 15–56.
Alekseev, S. A., ed. *Fevral'skaia revoliutsiia.* 2 vols. Moscow-Leningrad, 1925.
Alexander, Grand Duke of Russia [Aleksandr Mikhailovich]. *Once a Grand Duke.* Translated. New York, 1932.

Alexander, John T. *Autocratic Politics in a National Crisis: The Imperial Russian Government and Pugachev's Revolt, 1773–1775.* Bloomington, 1969.

Ananich, B. V., Ganelin, R. Sh., Dubentsov, B. B., Diakin, V. S., and Potolov, S. I. *Krizis samoderzhaviia v Rossii, 1895–1917gg.* Leningrad, 1984.

Andreev, L., Gor'kii, M., and Sologub, F. (eds.). *Shchit: Literaturnyi sbornik.* Moscow, 1916.

Andrei Vladimirovich, Grand Duke. *Dnevnik byvshego Velikogo Kniazia Andreia Vladimirovicha* za 1915g. Edited by V. P. Semennikov. Leningrad-Moscow, 1925.

————. "Iz dnevnika A. V. Romanova za 1916–1917gg.," *Krasnyi arkhiv,* XXVI (1928), pp. 185–210.

Anet, Claude. *La révolution russe à Petrograd et aux armées.* 4 vols. Paris, 1919.

Anfimov, A. M. *Rossiiskaia derevnia v gody pervoi mirovoi voiny (1914–fevral' 1917g).* Moscow, 1962.

———— (ed.). *Krest'ianskoe dvizhenie v Rossii v gody pervoi mirovoi voiny, iiul' 1914g.–fevral' 1917g. Sbornik dokumentov.* Moscow-Leningrad, 1965.

————*et al.* (eds.). *Ekonomicheskoe polozhenie Rossii nakanune velikoi oktiabr'skoi sotsialisticheskoi revoliutsii. Dokumenty i materialy.* 3 vols. Leningrad, 1967.

Antonov-Ovseenko, V. A. *V semnadtsdatom godu.* Moscow, 1933.

————. *Zapiski o grazhdanskoi voine.* Vol. I. Moscow, 1924.

Antsiferov, A. N., Bilomovich, A. D., Batshev, M. O., and Ivantsov, D. N. *Russian Agriculture During the War.* New Haven, 1930.

Anweiler, Oskar. *The Soviets: The Russian Workers', Peasants', and Soldiers' Councils, 1905–1921.* Translated from the German by Ruth Hein. New York, 1974.

"Aprel'skie tezisy V. I. Lenina," 4 aprelia 1917g., in Gaponenko (ed.), *Revoliutsionnoe dvizhenie v Rossii v aprele 1917g.,* No. 1, pp. 3–6.

Les Armées francaises dans la Grande Guerre. 11 vols. Paris, 1922–1939.

Arutiunian, A. O. *Kavkazskii front 1914–1917gg.* Erevan, 1971.

Ascher, Abraham. *Pavel Akselrod and the Development of Menshevism.* Cambridge, Mass., 1972.

Avdeev, N., (ed.). "Vokrug Gatchiny," *Krasnyi arkhiv,* IX (1925), pp. 171–194.

Avrich, Paul. *Russian Rebels, 1600–1800.* New York, 1972.

Avvakumov, S. I., *et al.* (eds.). *Oktiabr'skoe vooruzhennoe vosstanie v Petrograde. Sbornik statei.* Moscow-Leningrad, 1957.

Babichev, D. S. "Rossiia na parizhskoi soiuznicheskoi konferentsii 1916g. po ekonomicheskim voprosam," *Istoricheskie zapiski,* LXXXIII (1969), pp. 38–57.

Babushkin, I. V. *Vospominaniia Ivana Vasil'evicha Babushkina, 1893–1900.* Moscow, 1951.

Badaev, A. *The Bolsheviks in the Tsarist Duma.* Translated. New York, n.d.

Badmaev, Zhamsaran. "Zapiska Badmaeva o lechenii naslednika," in Semennikov (ed.), *Za kulisami tsarizma,* p. 17.

Baedeker, Karl. *Russia, 1914.* London, 1971.

Baklanova, I. A. "Formirovanie i polozhenie promyshlennogo proletariata. Rabochee dvizhenie 60-e gody—nachalo 90-kh godov," in Diakin *et al.* (eds.). *Istoriia rabochikh Leningrada,* I, pp. 124–178.

————. *Rabochie Petrograda v period mirnogo razvitiia revoliutsii, mart–iiun' 1917g.* Leningrad, 1978.

Bank, N., Zakharenko, N., and Eventov, I. (eds.), *Nemerknushchie gody: Ocherki i vospominaniia o krasnom Petrograde, 1917–1918.* Leningrad, 1957.

Barsukov, E. *Podgotovka russkoi armii k voine v artilleriiskom otnoshenii.* Moscow, 1926.

Basily, Nicolas de. *Diplomat of Imperial Russia, 1903–1917: Memoirs.* Stanford, 1973.

Bater, J. H. *St. Petersburg: Industrialization and Change.* London, 1976.

Beatty, Bessie. *The Red Heart of Russia.* New York, 1918.

Beloi, A. *Galitsiiskaia bitva.* Moscow-Leningrad, 1929.

Bclousov, Ivan. *Ushedshaia Moskva: Zapiski po lichnym vospominaniiam s nachala 1870-kh godov*. Moscow, 1927.

Belov, G. "Russkii polkovodets A. A. Brusilov," *Voenno-istoricheskii zhurnal*, No. 10 (1962), pp. 41–55.

Belyi, Andrei [Boris Bugaev]. *Nachalo veka*. Leningrad, 1933.

——— [Bely]. *Petersburg*. Translated, annotated, and introduced by Robert A. Maguire and John F. Malmstadt. Bloomington, Ind., and London, 1978.

Belyshev, A. V. *"Avrora* v dni oktiabria," in Mushkutov (ed.), *Petrograd v dni velikogo oktiabria*, pp. 367–373.

Bernhardi, General Friedrich von. *Germany and the Next War*. Translated by Allen H. Powles. New York and London, 1914.

Betskii, K., and Pavlov, P. *Russkii rokambol' (prikliucheniia I. F. Manasevicha–Manuilova)*. Leningrad, 1925.

Billington, James. *The Icon and the Axe: An Interpretive History of Russian Culture*. London, 1966.

Bing, Edward J. (ed.). *Letters of Tsar Nicholas and Empress Marie*. London, Translated from the French and Russian. 1937.

Blinov, I. A. (ed.). "Perepiska V. A. Sukhomlinova s N. N. Ianushkevichem," *Krasnyi Arkhiv*, III (1923), pp. 29–74.

Blinov, S. A. *Vneshniaia politika Sovetskoi Rossii: Pervyi god proletarskoi diktatury*. Moscow, 1973.

Bliokh [Bloch], I. S. *The Future of War in Its Technical, Economic, and Political Relations*. Boston, 1902.

Blok, Aleksandr. "Dvenadtsat'," in *Izbrannie proizvedeniia*, pp. 489–499.

———. *Izbrannye proizvedeniia*. Leningrad, 1970.

———. *Pis'ma Aleksandra Bloka k rodnym*. Edited by M. A. Beketova. 2 vols. Moscow-Leningrad, 1932.

———. *Poslednie dni imperatorskoi vlasti*. Petrograd, 1921.

———. *Sobranie sochinenii v shesti tomakh*. 6 vols. Moscow, 1971.

Bogdanov, A. V. *Moriaki-Baltiitsy v 1917g*. Moscow, 1955.

Bogdanovich, P. N. *Vtorzhenie v vostochnuiu Prussiiu v avguste 1914 goda. Vospominaniia ofitsera general'nogo shtaba armii generala Samsonova*. Buenos Aires, 1964.

Bok, M. P. *Vospominaniia o moem ottse P. A. Stolypina*. New York, 1953.

Bonch-Bruevich, M. D. *From Tsarist General to Red Army Commander*. Translated from the Russian by Vladimir Vezy. Moscow, 1966.

———. *Poteria nami Galitsii*. 2 vols. Moscow, 1920–1926.

———. *Vsia vlast' sovetam*. Moscow, 1964.

Bonch-Bruevich, V. D. "Kak napisal Vladimir Il'ich dekret o zemle," in Mushtukov (ed.), *Petrograd v dni velikogo oktiabria*, pp. 435–440.

———. *Na boevikh postakh fevral'skoi i oktiabr'skoi revoliutsii*. Moscow, 1931.

"Bor'ba so stachechnym dvizheniem nakanune mirovoi voiny," *Krasnyi arkhiv*, XXXIV (1929), pp. 95–125.

Borisov, A. "Prasnyshskaia operatsiia," *Voenno-istoricheskii zhurnal*, No. 3 (March 1941), pp. 27–35.

Botchkareva [Bochkareva], Maria. *Yashka: My Life as Peasant, Officer, and Exile*, as set down by Isaac Don Levine. New York, 1919.

Briusov, Valerii. *Sobranie sochinenii v semi tomakh*. 7 vols. Moscow, 1973.

Brooks, Jeffrey. "The Zemstvo and the Education of the People," in Emmons and Vucinich (eds.), *The Zemstvo in Russia*, pp. 243–278.

Browder, Robert Paul, and Kerensky, Alexander F. (eds.). *The Russian Provisional Government, 1917*. Stanford, 1961.

Brown, Edward J. *Mayakovsky: A Poet in Revolution.* Princeton, 1973.

Brusilov, A. A. *Moi vospominaniia.* Moscow, 1963.

Bryant, Louise. *Mirrors of Moscow.* New York, 1923. Reprinted 1973.

――――. *Six Red Months in Russia: An Observer's Account of Russia Before and During the Proletarian Dictatorship.* New York, 1918.

Bubnov, A. *V tsarskoi stavke: Vospominaniia admirala Bubnova.* New York, 1955.

Buchanan, Sir George. *My Mission to Russia.* 2 vols. New York, 1970.

――――. "Report No. 261 to Lord Grey," December 5, 1916. PRO, FO371/2752.

――――. "Report No. 65 to the Right Honourable Arthur J. Balfour," Petrograd, March 20, 1917. PRO, FO371/2995.

Budkevich, S. (ed.). "Vsepoddanneishaia zapiska Generala A. M. Zaionchkovskogo 20 oktiabria 1916g," *Krasnyi arkhiv,* LVIII (1933), pp. 24–25.

Bukhbinder, N. (ed.). "Na fronte v predoktiabr'skie dni. Po sekretnym materialam Stavki," *Krasnaia Letopis',* No. 6 (1923), pp. 9–64.

Bülow, Prince Bernhard von. *Memoirs of Prince von Bülow.* Translated by F. A. Voigt. 4 vols. Boston, 1931–1932.

Bunyan, James, and Fisher, H. H. *The Bolshevik Revolution, 1917–1918: Documents and Materials.* Stanford, 1934.

Burdzhalov, E. N. *Vtoraia russkaia revoliutsiia.* 2 vols. Moscow, 1967.

Buxhoeveden, Baroness Sophie. *The Life and Tragedy of Alexandra Feodorovna, Empress of Russia.* London, 1928.

"Caractère et Développement de la Révolution Russe," Groupe de l'Avant. 2e et 3e Bureaux. 8 juin 1917, (Secret Report prepared by the General Staff of the French Army), AdAE, Guerre 1914–1918, Russie, Dossier Général No. 653/100–124.

Chaadaeva, O. A. (ed.). "Iiulskie stachki i demonstratsii 1914g," *Krasnyi arkhiv,* XCV (1939), pp. 137–155.

――――. "Soldatskie pis'ma v gody mirovoi voiny (1915–1917gg.)," *Krasnyi arkhiv,* LXV–LXVI (1934), pp. 118–163.

Chamberlin, William Henry. *The Russian Revolution.* 2 vols. New York, 1965.

Cherepakhov, M. S., and Fingerit, E. M. (eds.). *Russkaia periodicheskaia pechat', 1895–oktiabr' 1917gg.* Moscow, 1957.

Chermenskii, E. D. *IV Gosudarstvennaia duma i sverzhenie tsarizma v Rossii.* Moscow, 1976.

Cherniavsky, Michael (ed.). *Prologue to Revolution: The Notes of A. N. Iakhontov on the Secret Meetings of the Council of Ministers, 1915.* Englewood Cliffs, N.J., 1967.

Chernov, Viktor. *Rozhdenie revoliutsionnoi Rossii (fevral'skaia revoliutsiia).* Paris, Prague, New York, 1934.

"Chislo fabrichno-zavodskikh rabochikh bastovavshikh v 1912, 1913, 1914, 1915, i 1916 gg.," TsGIAL, fond 1276, opis' 9, delo No. 137b.

Chołodecki, Josef Białynia. *Lwów w czasie okupacji rosyjskiej (3 wrzesnia 1914 - 22 czerwca 1915).* Lwów, 1930.

Chubar'ian, A. O. *Brestskii mir.* Moscow, 1964.

Chugaev, D. A. (ed.). *Petrogradskii voenno-revoliutsionnyi komitet. Dokumenty i materialy,* 3 vols. Moscow, 1966.

――――. *Revoliutsionnoe dvizhenie v Rossii nakanune oktiabr'skogo vooruzhennogo vosstaniia, 1–24 oktiabria 1917g.* Moscow, 1962.

――――. *Revoliutsionnoe dvizhenie v Rossii v avguste 1917g. Razgrom kornilovskogo miatezha.* Moscow, 1959.

――――. *Revoliutsionnoe dvizhenie v Rossii v mae-iiune 1917g. Iiunskaia demonstratsiia.* Moscow, 1959.

――――. *Revoliutsionnoe dvizhenie v Rossii v sentiabre 1917g. Obshchenatsional'nyi krizis.* Moscow, 1961.

————. *Triumfal'noe shestvie sovetskoi vlasti.* 2 vols. Moscow, 1963.

Churchill, Winston S. *The World Crisis, 1911–1914.* London, 1923

Cioran, Samuel D. *Vladimir Solov'ev and the Knighthood of the Divine Sophia.* Waterloo, 1977.

Cohen, Stephen F. *Bukharin and the Bolshevik Revolution: A Political Biography, 1888–1938.* Oxford and New York, 1980.

Crankshaw, Edward. *The Shadow of the Winter Palace: Russia's Drift to Revolution, 1825–1917.* New York, 1976.

Czernin, Count Ottokar. *In the World War.* New York and London, 1920.

Dan, Fedor. "K istorii poslednikh dnei vremennogo pravitel'stva," *Letopis' revoliutsii,* I (1923), pp. 163–175.

————. [Theodore]. *The Origins of Bolshevism.* Edited and translated by Joel Carmichael. New York, 1964.

Daniels, Robert V. *Red October: The Bolshevik Revolution of 1917.* New York, 1967.

Danilov, General Iu. N. *Rossiia v mirovoi voine, 1914–1915gg.* Berlin, 1924.

————. *Velikii kniaz' Nikolai Nikolaevich.* Paris, 1930.

Davydov, M. I. *Bor'ba za khleb: Prodovol'stvennaia politika kommunisticheskoi partii i sovetskogo gosudarstva v gody grazhdanskoi voiny, 1917–1920gg.* Moscow, 1971.

"Decree on the Arrest of the Leaders of the Civil War against the Revolution," November 28, 1917, in Lenin, *Collected Works,* XXVI, p. 351.

Dedijer, Vladimir. *The Road to Sarajevo.* London and New York, 1966.

Dehn, Lili. *The Real Tsaritsa.* Boston, 1922.

Deiatel'nost' moskovskago oblastnago voenno-promyshlennago komiteta i ego otdelov po 31 ianvaria 1916g. Moscow, 1916.

"Delo o dopushchenii maloletnikh i zhenshchin k nochnym i podzemnym rabotam v kamennougol'nykh kopiakh Evropeiskoi Rossii, 3 marta 1915g," TsGIAL, fond 23, opis' 27, delo No. 803.

"Delo Vremennago Komiteta po organizatsii novago pravitel'stva, 1–27 marta 1917g," TsGIAL, fond 1278, opis' 10, delo No. 9.

Denikin, Anton I. *The Career of a Tsarist Officer: Memoirs, 1872–1916.* Translated by Margaret Patoski. Minneapolis, 1975.

————. *The Russian Turmoil. Memoirs: Military, Social, Political.* Translated. London, 1922.

————. *The White Army.* Translated from the Russian by Catherine Zvegintzov. London, 1930.

Deutscher, Isaac. *The Prophet Armed. Trotsky: 1879–1921.* New York, 1954.

————. *Stalin: A Political Biography.* New York, 1967.

Diakin, V. S. *Russkaia burzhuaziia i tsarizm v gody pervoi mirovoi voiny, 1914–1917.* Leningrad, 1967.

———— et al. (eds.). *Istoriia rabochikh Leningrada.* 2 vols. Leningrad, 1972.

Dmitrienko, S. L., and Keirit-Markus, M. B. (comps). *Velikaia oktiabr'skaia sotsialisticheskaia revoliutsiia: Khronika sobytii, 26 iiulia-11 sentiabria 1917 goda.* Moscow, 1960.

"Dmitrii Mikhailovich Karbyshev," *Sovetskaia istoricheskaia entsiklopediia,* VII, cols. 24–25.

Dobrorolskii, S. K. *Die Mobilmachung der russischen Armee 1914.* Berlin, 1922.

————. "Nashi strategicheskie shansy v 1914 godu," *Voina i Mir,* No. 16 (1924), pp. 20–38.

"Dobycha i razrabotka topliva v Rossii v 1917 godu," 1 iiulia 1917g., TsGIAL, fond 23, opis' 15, delo No. 305.

"Doklad Generala Alekseeva Nikolaiu II," in Semennikov (ed.), *Monarkhiia pered krusheniem,* pp. 259–266.

"Doklad instruktora provintsial'nogo otdela Moskovskogo Soveta Rabochikh Deputatov Shevtsova Moskovskomu Sovetu o polozhenii krest'ian v Mikhailovskom uezde Riazanskoi gubernii i Venevskom uezde Tul'skoi gubernii," 12 aprelia 1917g., in Gaponenko (ed.), *Revoliutsionnoe dvizhenie v aprele 1917g.*, No. 580, pp. 592–594.

"Doklad komissara Ranenburgskogo uezda Riazanskoi gubernii, ot 28 iiulia 1917g., No. 2613," in Martynov (ed.), "Agrarnoe dvizhenie v 1917 godu," pp. 187–191.

"Doklad Ministru Vnutrennikh Del otdeleniia po okhraneniiu obshchestvennoi bezopasnosti i poriadka v stolitse," (sovershenno sekretno), 17 dekabria 1915g., TsGIAL, fond 1276, opis' 11, delo No. 167.

"Doklad nachal'nika Petrogradskogo okhrannogo otdeleniia ministru vnutrennikh del o zabastovkakh, demonstratsiiakh, i stolknoveniiakh rabochikh i soldat s politsiei," (sekretno), 9 ianvaria 1916g., in Korablev (ed.), *Rabochee dvizhenie v Petrograde*, pp. 398–401.

"Doklad otdeleniia po okhraneniiu obshchestvennoi bezopasnosti i poriadka v stolitse," (sovershenno sekretno), 17 dekabria 1915, TsGIAL, fond 1276, opis' 11, delo no 167.

"Doklad petrogradskogo gradonachal'nika glavnomu nachal'niku Petrogradskogo voennogo okruga ob uchastii zapasnykh nizhnikh chinov i ratnikov opolcheniia v zabastovke rabochikh zavoda Russkogo obshchestva dlia izgotovleniia snariadov i voennikh pripasov 12 noiabria 1914g.," in Korablev (ed.), *Rabochee dvizhenie v Petrograde*, pp. 264–266.

"Doklad petrogradskogo okhrannogo otdeleniia osobomu otdelu departamenta politsii, oktiabr' 1916g," in Pokrovskii (ed.), "Politicheskoe polozhenie Rossii," pp. 1–35.

"Dokumenty o presledovanii evreev," *Arkhiv russkoi revoliutsii*, XIX, pp. 245–284.

Dokumenty vneshnei politiki SSSR. vol. 1. Moscow, 1957.

"Donesenie Astrakhanskogo gubernatora I. N. Sokolovskogo upravliaiushchemu Ministerstva Vnutrennikh del N. B. Shcherbatova o volnenii naseleniia i ratnikov v Astrakhani," 11 sentiabria 1915g., in Sidorov (ed.), *Revoliutsionnoe dvizhenie*, pp. 122–123.

"Donesenie nachal'nika moskovskogo okhrannogo otdeleniia direktoru departmenta politsii o gorodskoi s"ezde," 8 sentiabria 1915g., in Grave (ed.), *Burzhuaziia*, pp. 49–52.

"Donesenie nachal'nika petrogradskogo okhrannogo otdeleniia v departamente politsii ob obrazovanii 'progressivnogo bloka,' " in Grave (ed.), *Burzhuaziia*, pp. 26–29.

"Donesenie nachal'nika moskovskogo okhrannogo otdeleniia direktoru departamenta politsii o VI s"ezde partii k.-d.," 25 fevralia 1916g., in Grave (ed.), *Burzhuaziia*, pp. 73–75.

"Donesenie nachal'nika moskovskogo okhrannogo otdeleniia direktoru departamenta politsii o soveshchanii u P. P. Riabushinskogo," 14 avgusta 1915g., in Grave (ed.), *Burzhuaziia*, pp. 20–22.

"Donesenie nachal'nika moskovskogo okhrannogo otdeleniia direktoru departamenta politsii o soveshchaniiakh u A. I. Konovalova," 12 oktiabria 1916g., in Grave (ed.), *Burzhuaziia*, pp. 141–144.

"Donesenie nachal'nika moskovskogo okhrannogo otdeleniia direktoru departamenta politsii o vystupleniiakh A. I. Konovalova sredi moskovskikh promyshlennikov," in Grave (ed.), *Burzhuaziia*, pp. 139–140.

"Donesenie nachal'nika Petrogradskogo okhrannogo otdeleniia K. I. Globacheva upravliaiushchemu Ministerstrom vnutrennich del N. B. Shcherbatovu o volnenii ratnikov opolcheniia na Nikolaevskom vokzale v Petrograde," 11 sentiabria 1915g., in Sidorov (ed.), *Revoliutsionnoe dvizhenie*, pp. 115–119.

"Dopros kniazia M. M. Andronikova, 8 aprelia, 1917 goda," in Shchegolev (ed.), *Padenie tsarskogo rezhima*, II, pp. 9–23.

"Dopros D. N. Dubenskogo, 9 avgusta 1917g.," in Shchegolev (ed.), *Padenie tsarskogo rezhima*, VI, pp. 373–415.

"Dopros V. F. Dzhunkovskogo, 7 iiunia 1917 goda," in Shchegolev (ed.), *Padenie tsarskogo rezhima*, V, pp. 100–122.

"Dopros Grafa Frederiksa, 2 iiunia 1917 goda," in Shchegolev (ed.), *Padenie tsarskogo rezhima*, V, pp. 32–49.

"Dopros kniazia N. D. Golitsyna," in Shchegolev (ed.), *Padenie tsarskogo rezhima*, II, pp. 249–272.

"Dopros A. I. Guchkova, 2 avgusta 1917 goda," in Shchegolev (ed.), *Padenie tsarskogo rezhima*, VI, pp. 248–294.

"Dopros gr. P. N. Ignat'eva," in Shchegolev (ed.), *Padenie tsarskogo rezhima*, VI, pp. 1–26.

"Dopros K. D. Kafafova, 14 aprelia 1917g.," in Shchegolev (ed.), *Padenie tsarskogo rezhima*, II, pp. 134–145.

"Dopros Generala S. S. Khabalova, 22 marta 1917g.," in Shchegolev (ed.), *Padenie tsarskogo rezhima*, I, pp. 1182–219.

"Dopros A. N. Khvostova, 18 marta 1917g.," in Shchegolev (ed.), *Padenie tsarskogo rezhima*, I, pp. 1–53.

"Dopros A. N. Khvostova, 12 iiulia 1917 goda," in Shchegolev (ed.), *Padenie tsarskogo rezhima*, V, pp. 444–473.

"Dopros N. A. Maklakova, 14 iiunia 1917 goda," in Shchegolev (ed.), *Padenie tsarskogo rezhima*, V, pp. 190–211.

"Dopros Generala A. A. Polivanova, 25 avgusta 1917 goda," in Shchegolev (ed.), *Padenie tsarskogo rezhima*, VII, pp. 54–87.

"Dopros A. D. Protopopova, 21 marta 1917 goda," in Shchegolev (ed.), *Padenie tsarskogo rezhima*, I, pp. 111–181.

"Dopros A. D. Protopopova, 8 aprelia 1917 goda," in Shchegolev (ed.), *Padenie tsarskogo rezhima*, II, 9–23.

"Dopros M. V. Rodzianko, 4 sentiabria 1917 goda," in Shchegolev (ed.), *Padenie tsarskogo rezhima*, VII, pp. 116–175.

"Dopros B. V. Stiurmera, 22 marta 1917 goda," in Shchegolev (ed.), *Padenie tsarskogo rezhima*, I, pp. 221–230.

Dostoevskii, F. M. *Brat'ia Karamazovykh*, in *Polnoe sobranie sochineniia*, vols. XIV–XV. Leningrad, 1976.

Druzhinin, N. K. (ed.). *Usloviia byta rabochikh v dorevoliutsionnoi Rossii*. Moscow, 1958.

Dubenskii, Major General (ed.). *Ego Imperatorskoe Velichestvo Gosudar' Imperator Nikolai Aleksandrovich v deistvuiushchei armii*. 4 vols. Petrograd, 1915–1916.

———. "Kak proizoshel perevorot v Rossii," in Shchegolev (ed.), *Otrechenie Nikolaia II*, pp. 31–78.

Dubrovin, N. F. *Pugachev i ego soobshchenniki*. 2 vols. St. Petersburg, 1884.

Dunlop, Major C. "Reports to Major-General C. E. Caldwell, 1916–1917," AG-CV, file 10N93.

Dvinov, V. *Pervaia mirovaia voina i rossiiskaia sotsial-demokratiia*. Inter-University Project on the History of the Menshevik Movement. Paper No. 10. New York, 1962.

Ehrenburg, Ilya. *People and Life. Memoirs of 1891–1917*. Translated by Anna Bostock and Yvonne Kapp. London, 1961.

———. *First Years of Revolution, 1918–1921*. Translated by Anna Bostock, London, 1962.

Ekzempliarskii, P. M. *Istoriia goroda Ivanova*. 2 vols. Ivanovo, 1958.

Emets, V. A. *Ocherki vneshnei politiki Rossii, 1914–1917gg*. Moscow, 1977.

———. "O roli russkoi armii v pervyi period mirovoi voiny 1914–1918gg.," *Istoricheskie zapiski*, LXXVII (1967), pp. 57–84.

———. "Petrogradskaia konferentsiia 1917g. i Frantsiia," *Istoricheskie zapiski*, LXXXIII (1969), pp. 23–37.

Emmons, Terence, and Vucinich, Wayne S. (eds.). *The Zemstvo in Russia: An Experiment in Local Self-Government*. London and New York, 1982.

Ermanskii, O. A. *Iz perezhitogo, 1887–1921*. Moscow, 1927.

Fainsod, Merle. *International Socialism and the World War*. New York, 1966.

Falkenhayn, General Erich von. *General Headquarters 1914–1916 and Its Critical Decisions.* Translated from the German. London, 1919.

Fallows, Thomas. "Politics and the War Effort in Russia: The Union of Zemstvos and Organization of the Food Supply, 1914–1916," *Slavic Review,* XXXVII (1978), pp. 70–90.

Farnsworth, Beatrice. *Aleksandra Kollontai: Socialism, Feminism, and the Bolshevik Revolution.* Stanford, 1980.

Fay, Sidney B. *The Origins of the World War.* 2 vols. New York, 1930.

Fedorov, M. F. *Voina i dorogovizna zhizni 1916g. Doklad predsedatelia Petrogradskoi gorodskoi finansovoi komisii M. P. Fedorova.* Petrograd, 1916.

Ferro, Marc. *La Révolution de 1917.* 2 vols. Paris, 1967, 1976.

"Fevral'skaia revoliutsiia 1917 goda," *Krasnyi arkhiv,* XXI (1927), pp. 3–78.

Flakserman, Iu. N. "10 oktiabria 1917 goda," in Mushtukov (ed.), *Petrograd v dni velikogo oktiabria,* pp. 264–269.

Fleer, M. G. (ed.). *Rabochee dvizhenie v gody voiny.* Moscow, 1925.

Florinsky, M. T. *The End of the Russian Empire.* New Haven, 1931.

————. *Russia: A History and an Interpretation.* 2 vols. New York, 1968.

Fofanova, M. V. "Poslednee podpol'e," in *Ob Il'iche,* pp. 344–350.

Fokke, D. G. "Na stsene i za kulisami Brestskoi tragikomedii (memuary uchastnika Brest-Litovskikh mirnykh peregovorov)," *Arkhiv russkoi revoliutsii,* XX (1930), pp. 5–207.

Fraiman, A. L. (ed.). *Istoriia rabochikh Leningrada.* Vol. II. Leningrad, 1972.

————. *Forpost sotsialisticheskoi revoliutsii: Petrograd v pervye mesiatsy Sovetskoi vlasti.* Leningrad, 1969.

———— (ed.). *Oktiabr'skoe vooruzhennoe vosstanie.* 2 vols. Leningrad, 1967.

Francis, David R. *Russia from the American Embassy, April 1916–November 1918.* New York, 1921.

François, General Hermann von. "Kriticheskoe issledovanie srazheniia na Mazurskikh ozerakh v sentiabre 1914g.," *Voina i Mir: Vestnik voennoi nauki i tekhniki,* No. 12 (1924), pp. 36–59.

Frank, S. *A Solovyov Anthology.* London, 1950.

Freidlin, B. M. *Ocherki istorii rabochego dvizheniia v Rossii v 1917g.* Moscow, 1967.

Frenkin, M. *Russkaia armiia i revoliutsiia, 1917–1918.* Munich, 1978.

Fülöp-Miller, René. *Rasputin: The Holy Devil.* New York, 1928.

Gankin, Olga Hess, and Fisher, H. H. (eds.). *The Bolsheviks and the World War.* Stanford, 1960.

Gaponenko, L. S. "Polozhenie rabochego klassa Rossii nakanune oktiabr'skoi revoliutsii," *Istoricheskie zapiski,* LXXXIII (1969), pp. 3–22.

————. *Rabòchii klass Rossii v 1917 godu.* Moscow, 1970.

———— (ed.). *Revoliutsionnoe dvizhenie v Rossii posle sverzheniia samoderzhaviia.* Moscow, 1957.

———— (ed.). *Revoliutsionnoe dvizhenie v Rossii v aprele 1917g. Aprel'skii krizis.* Moscow, 1958.

———— (ed.). *Revoliutsionnoe dvizhenie v russkoi armii, 27 fevralia–24 oktiabria 1917 goda. Sbornik dokumentov.* Moscow, 1968.

Gavrilov, L. M. (ed.). *Voiskovye komitety deistvuiushchei armii, mart 1917g.–mart 1918g.* Moscow, 1982.

Gerasimov, M. N. *Probuzhdenie.* Moscow, 1965.

Getzler, Israel. *Kronstadt 1917–1921: The Fate of a Soviet Democracy.* Cambridge, 1983.

————. *Martov: A Political Biography of a Russian Social Democrat.* Cambridge and Melbourne, 1967.

Gibson, Hugh. *A Journal from Our Legation in Belgium.* Garden City and New York, 1917.

Giliarovskii, V. A. *Moskva i Moskvichi*. Moscow, 1979.
———. *Trushchobnye liudi*. In *Sochineniia V. A. Giliarovskogo*. 4 vols. Moscow, 1967.
Gill, Graeme J. *Peasants and Government in the Russian Revolution*. London, 1979.
Gilliard, Pierre. *Thirteen Years at the Russian Court*. Translated by F. Appleby Holt. New York, 1970.
———. *Le destin tragique de Nicolas II et de sa famille*. Paris, 1938.
Gindin, A. *Kak bol'sheviki natsionalizirovali chastnye banki (Fakti i dokumenty posleoktiabr'-skikh dnei v Petrograde)*. Moscow, 1962.
Ginev, V. N. *Agrarnyi vopros i melkoburzhuaznye partii v Rossii v 1917g*. Leningrad, 1977.
Gippius, Zinaida. *Dmitrii Merezhkovskii*. Paris, 1951.
———. *Siniaia kniga: Peterburgskii dnevnik, 1914–1918gg*. Belgrad, 1929.
———. [Hippius]. "Vesel'e," in Markov and Sparks (eds.), *Modern Russian Poetry*, p. 84.
Gladkov, I. A. *Ocherki stroitel'stva sovetskogo planovogo khoziaistva v 1917–1918gg*. Moscow, 1950.
Gleason, William. *Alexander Guchkov and the End of the Russian Empire*. (Transactions of the American Philosophical Society), LXXIII, pt. 3 (1983).
———. "The All-Russian Union of Zemstvos and World War I," in Emmons and Vucinich (eds.), *The Zemstvo in Russia*, pp. 365–382.
Glukhovtseva, E. "Bezhentsy vo vremia voiny 1914–1917gg," TsGIAL, fond 1322, opis' 1, delo No. 10.
Gogol', N. V. *Polnoe sobranie sochinenii N. V. Gogolia*. 10 vols. Moscow, 1913.
———. *Sobranie sochinenii*. 6 vols. Moscow, 1959.
Golder, F. A. *Documents of Russian History, 1914–1917*. Translated by Emanuel Aronsberg. New York, 1927.
Golikov, G. N. (ed.). *Oktiabr'skoe vooruzhennoe vosstanie v Petrograde*. Moscow, 1957.
——— et al. (eds.). *Vospominaniia o Vladimire Il'iche Lenine*. 5 vols. Moscow, 1969.
Golovin [Golovine], N. N. *Iz istorii kampanii 1914 goda. Dni pereloma Galitsiiskoi bitvy (1–3 sentiabria novago stilia)*. Paris, 1940.
———. *Iz istorii kampanii 1914 goda na russkom fronte. Galitsiiskaia bitva: pervyi period do 1 sentiabria novago stilia*. Paris, 1930.
———. *Iz istorii kampanii 1914 goda na russkom fronte. Nachalo voiny i operatsii v vostochnoi Prussii*. Prague, 1926.
———. *Iz istorii kampanii 1914 goda na russkom fronte. Plan voiny*. Paris, 1936.
———. *The Russian Army in the World War*. New Haven, 1931.
———. "The Russian War Plan of 1914," *The Slavonic and East European Review*, XIV, No. 42 (April 1936), pp. 564–584.
———. *Voennye usiliia Rossii v mirovoi voine*. 2 vols. Paris, 1939.
Gooch, G. P. *Before the War: Studies in Diplomacy*. 2 vols. New York, 1967.
———, and Temperley, Harold (eds.). *British Documents on the Origins of the War, 1898–1914*. London, 1933.
Gordienko, I. M. *Iz boevogo proshlogo, 1914–1918gg*. Moscow, 1957.
Gorky, Maksim. *Untimely Thoughts: Essays on Revolution, Culture, and the Bolsheviks, 1917–1918*. Translated from the Russian with an Introduction and Notes by Herman Ermolaev. New York, 1968.
Gorodetskii, V. P. (ed.). *Istoriia russkoi literatury: Literatura 1840-kh godov*. Moscow-Leningrad, 1955.
Gosudarstvennaia deiatel'nost' predsedatelia Soveta Ministrov Stats-Sekretaria Petra Arkad'evicha Stolypina. 2 vols., St. Petersburg, 1911.
Gourko [Gurko], General Basil [Vasilii]. *War and Revolution in Russia, 1914–1917*. Translated from the Russian. New York, 1919.
Grave, B. B. (ed.). *Burzhuaziia nakanune fevral'skoi revoliutsii*. Moscow-Leningrad, 1927.

Greenberg, Louis. *The Jews in Russia.* 2 vols. New Haven, 1951.

Grey, Viscount Edward. *Twenty-Five Years, 1892–1916.* 2 vols. London, 1925.

Grunt, A. Ia. *Pobeda oktiabr'skoi revoliutsii v Moskve, fevral'-oktiabr' 1917g.* Moscow, 1961.

———. "Progressivnyi blok," *Voprosy istorii,* No. 3–4 (1945), pp. 108–117.

Gruzenberg, O. O. *Yesterday: Memoirs of a Russian-Jewish Lawyer.* Edited and with an Introduction by Don C. Rawson. Translated by Don C. Rawson and Tatiana Tipton. Berkeley and Los Angeles, 1981.

"Guerre 1914–1918. Russie," AdAE, Dossier Général No. 641 (juillet 1914–décembre 1915).

"Guerre 1914–1918. Russie," AdAE, Dossier Général No. 642 (janvier–31 mars 1916).

"Guerre 1914–1918. Russie," AdAE, Dossier Général No. 644 (1 août–31 octobre 1916).

"Guerre 1914–1918. Russie," AdAE, Dossier Général No. 645 (1 novembre–31 décembre 1916).

"Guerre 1914–1918. Russie," AdAE, Dossier Général No. 646 (1 janvier–28 février 1917).

"Guerre 1914–1918. Russie," AdAE, Dossier Général No. 647 (1 mars 1917–19 mars 1917).

"Guerre 1914–1918. Russie," AdAE, Dossier Général No. 653 (juin 1917).

Gurko, Vladimir I. *Features and Figures of the Past: Government and Opinion in the Reign of Nicholas II.* Stanford, 1939.

Gvozdev, S. *Zapiski fabrichnago inspektora. Iz nabliudenii i praktiki v periode 1894–1908gg.* Moscow, 1911.

Hackel, Sergei. *The Poet and the Revolution: Aleksandr Blok's 'The Twelve.'* Oxford, 1976.

Haimson, Leopold. *The Russian Marxists and the Origins of Bolshevism.* Cambridge, Mass., 1955.

Hale, Oron J. *The Great Illusion, 1900–1914.* New York, 1971.

Hamilton, General Sir Ian. *A Staff Officer's Scrap-Book during the Russo-Japanese War.* London, 1912.

Hard, William. *Raymond Robins' Own Story.* New York, 1971.

Harding, Neil. *Lenin's Political Thought.* Vol. II. New York, 1981.

Hasegawa, Tsuyoshi. *The February Revolution: Petrograd, 1917.* Seattle and London, 1981.

———. "The Formation of the Militia in the February Revolution: An Aspect of the Origins of Dual Power," *Slavic Review,* XXXII, No. 2 (June 1973), pp. 303–322.

Hindenburg, Field Marshal Paul von. *Out of My Life.* Translated by F. A. Holt. London and New York, 1920.

Hoffmann, General Max von. *The War of Lost Opportunities.* Translated from the German. New York, 1925.

Iakhontov, A. N. "Tiazhelye dni (sekretnye zasedaniia Soveta Ministrov 16 iiulia–2 sentiabria 1915 goda)," *Arkhiv russkoi revoliutsii,* XVII (1926), pp. 5–136.

Iakovlev, Ia. (ed.). "Mart–Mai 1917g.," *Krasnyi arkhiv,* XV (1926), pp. 30–60.

Iakovlev, N. *1 Avgusta 1914.* Moscow, 1974.

Iakovlev, V. "Peremyshl'," *Inzhenernyi zhurnal,* No. 4 (April 1915), pp. 273–334.

Ianushkevich, N. N. "Pis'ma k V. A. Sukhomlinovu," in Blinov (ed.), "Perepiska," pp. 29–74.

Ianzhul, I. I. *Iz vospominanii i perepiski fabrichnago inspektora pervago prizyva. Materialy dlia istorii russkago rabochego voprosa i fabrichnago zakonodatel'stva.* St. Petersburg, 1907.

Ignat'ev, A. V. *Russko-angliiskie otnoshenie nakanune oktiabr'skoe revoliutsii.* Moscow, 1966.

———. *Vneshniaia politika vremennogo pravitel'stva.* Moscow, 1974.

Iiunga, E. *Kreiser "Avrora."* Moscow, 1949.

Inozemtsev, M. (ed.). "Iz istorii rabochego dvizheniia vo vremia mirovoi voiny (stachechnoe dvizhenie v Kostromskoi gubernii)," *Krasnyi arkhiv,* LXVII (1934), pp. 5–27.

——— (ed.). "Rasstrel Ivanovo-Voznesenskikh rabochikh v 1915g.," *Krasnyi arkhiv,* LXVIII (1935), pp. 97–118.

Ioffre, A. E. *Russko-frantsuzskie otnosheniia v 1917g. (fevral'-oktiabr').* Moscow, 1958.

Ironside, Major-General Sir Edmund. *Tannenberg: The First Thirty Days in East Prussia.* Edinburgh and London, 1925.

Itkina, A. M. *Revoliutsioner, tribun, diplomat: Ocherk zhizni Aleksandry Mikhailovny Kollontai.* Moscow, 1964.

Ivanov, L. M. (ed.). *Rossiiskii proletariat: Oblik, bor'ba, gegemoniia.* Moscow, 1970.

Ivanov, N. A. "Zemskii i gorodskoi soiuzy (Vserossiiskii zemiskii soiuz pomoshchi bol'nym i ranennym voinam i Vserossiiskii soiuz gorodov," *Sovetskaia Istoricheskaia Entsiklopediia,* V, cols. 675–676.

Ivanov, N. Ia. *Kornilovshchina i ee razgrom: Iz istorii bor'by s kontrrevoliutsiei v 1917g.* Leningrad, 1965.

Ivanov, Viacheslav. *Po zvezdam: stat'i i aforizmy.* St. Petersburg, 1909.

"Iz doklada chlenov deputatsii, izbrannoi sobraniem upolnomochennykh gubernskikh zemstv 9-go sentiabria 1915g., in Grave (ed.), *Burzhuaziia,* pp. 57–58.

"Iz doklada nachal'nika Petrogradskogo okhrannogo otdeleniia ministru vnutrennikh del o zabastovkakh rabochikh na zavodakh Petrograda, 12 noiabria 1914g.," in Korablev (ed.), *Rabochee dvizhenie v Petrograde,* pp. 263–264.

"Iz doneseniia saratovskogo gubernatora A. A. Shirinskogo-Shikhmatova v Departament politsii o soprotivlenii krestianok-soldatok provedeniiu zemleustroitel'nykh rabot po vliianiem agitatsii i pisem s fronta," 16 iiulia 1915g., in Anfimov (ed.), *Krest'ianskoe dvizhenie,* No. 134, p. 234.

Iz istorii Vserossiiskoi Chrezvychainoi Komissii, 1917–1921gg. Sbornik dokumentov. Moscow, 1958.

"Iz protokola Petrogradskoi konferentsii po voprosu o bezrabotnitse s izlozheniem doklada M. K. Vladimirova o prodovol'stvennom polozhenii v Petrograde, vystuplenii i rezoliutsii po etomu voprosu," 15 oktiabria 1917g., in Sidorov (ed.), *Ekonomicheskoe polozhenii Rossii nakanune Velikoi Oktiabr'skoi Sotsialisticheskoi Revoliutsii,* II, pp. 351–353.

"Iz protokola zasedaniia Tsentral'nogo Komiteta RSDRP (b) o podgotovke vooruzhennogo vosstaniia i sozdanii Politicheskogo Biuro TsK," 10 oktiabria 1917g., in Golikov (ed.), *Oktiabr'skoe vooruzhennoe vosstanie v Petrograde,* No. 14, pp. 48–49.

"Iz sostavlennogo Moskovskoi voennoi tsenzuroi obzora frontovykh pisem," in Gaponenko (ed.), *Revoliutsionnoe dvizhenie v russkoi armii,* pp. 90–94.

"Iz svodki svedenii glavnogo upravleniia po delam militsii MVD o krest'ianskom dvizhenii v Volynskoi, Voronezhskoi, Kurskoi, Minskoi, Mogilëvskoi, Novgorodskoi, Penzenskoi, Pskovskoi, Riazanskoi, Tambovskoi, Tul'skoi, Ufimskoi, i Iaroslavskoi guberniiakh," in Gaponenko (ed.), *Revoliutsionnoe dvizhenie v Rossii posle sverzheniia samoderzhaviia,* No. 679, pp. 704–705.

"Iz zapisi razgovora po priamomu provodu nachal'nika shtaba verkhovnogo glavnokomanduiushchego generala N. N. Dukhonina s nachal'nikom shtaba fronta generalom S. G. Lukirskim o prodvizhenii voisk k g. Petrogradu na podavlenie vosstanie," 27 oktiabria 1917g., in Golikov (ed.), *Oktiabr'skoe vooruzhennoe vosstanie v Petrograde,* No. 828, pp. 614–615.

Izgoev, A. S. "Piat' let v Sovetskoi Rossii (Otryvki vospominanii i zametki)," *Arkhiv russkoi revoliutsii,* X (1923), pp. 5–55.

Izvestiia glavnago komiteta Vserossiiskago Zemskago Soiuza. Nos. 1–65. Moscow, 1914–1917.

"Izvestiia komiteta Eia Imperatorskoi Vysochestva Velikoi Kniazhny Tatiany Nikolaevny," 15 fevralia 1917g., TsGIAL, fond 1322, opis' 1, delo No. 10.

Johnson, Robert E. *Peasant and Proletarian: The Working Class of Moscow in the Late Nineteenth Century.* New Brunswick, 1979.

Jonge, Alex de. *The Life and Times of Grigorii Rasputin.* New York, 1982.

Kabanov, P. I. (ed.). *Proletariat vo glave osvoboditel'nogo dvizheniia v Rossii.* Moscow, 1971.

Kaiurov, V. "Shest dnei fevral'skoi revoliutsii," *Proletarskaia revoliutsiia*, No. 1, XIII (1923).

Kakurin, N. E. (ed.), *Razlozhenie armii v 1917 godu*. Moscow-Leningrad, 1925.

Karamysheva, L. F. *Bor'ba bol'shevikov za petrogradskii sovet, mart–oktiabr' 1917g*. Leningrad, 1964.

Karliner, M. M. "Angliia i Petrogradskaia konferentsiia Antanty 1917g., in Trukhanovskii (ed.), *Mezhdunarodnye otnosheniia*, pp. 322–358.

Katkov, George. *The Kornilov Affair: Kerenskii and the Break-Up of the Russian Army*. London and New York, 1980.

———. *Russia 1917: The February Revolution*. New York, 1967.

Kavtaradze, A. G. "Pavel Karlovich Rennenkampf," *Sovetskaia istoricheskaia entsiklopediia*. XI, cols. 1019–1020.

Keep, John L. H. *The Russian Revolution: A Study in Mass Mobilization*. New York, 1976.

Kenez, Peter A. "A Profile of the Pre-Revolutionary Officer Corps," *California Slavic Studies*, VII, pp. 121–158.

Kennan, George D. *Russia Leaves the War*. Princeton, 1956.

Kerensky [Kerenskii], Alexander. *The Catastrophe: Kerensky's Own Story of the Russian Revolution*. New York, 1927.

———. *The Prelude to Bolshevism: The Kornilov Rebellion*. London, 1919.

———. *Russia and History's Turning Point*. New York, 1965.

Kersnovskii, A. *Istoriia russkoi armii*. 4 vols. Belgrad, 1933–1938.

Kettle, Michael. *The Allies and the Russian Collapse, March 1917–March 1918*. London, 1981.

Khatsevich, A. F. *Soldat velikikh boev. Zhizn' i deiatel'nost' F. E. Dzerzhinskogo*. Minsk, 1961.

Khromov, P. A. *Ekonomicheskoe razvitie Rossii v XIX–XX vekakh, 1800–1917*, Moscow, 1950.

"Khronika krest'ianskogo dvizheniia 1916g.," in Anfimov (ed.). *Krest'ianskoe dvizhenie v Rossii*, pp. 501–505.

Kir'ianov, Iu. I. "Ekonomicheskoe polozhenie rabochikh iuga Rossii v gody pervoi mirovoi voiny," *Istoricheskie zapiski*, LXXII (1962), pp. 30–62.

———. *Rabochie Iuga Rossii, 1914–fevral' 1917g*. Moscow, 1971.

———. "Vliianie pervoi mirovoi voiny na izmenenie chislennosti i sostava rabochikh Rossii," *Voprosy istorii*, No. 10 (October 1960), pp. 89–101.

Klembovskii, V. N. (ed.). *Stratigecheskii ocherk voiny 1914–1918gg. Positsionnaia voina i proryv avstriitsev iugo-zapadnym frontom*. 7 vols. Moscow, 1920.

Kliuchevskii, V. O. *Sochineniia*. 8 vols. Moscow, 1956.

Kliuzhev, I. S. "Dnevnik," TsGIAL, fond 669, opis' 1, dela Nos. 16 and 17.

Knox, Major-General Sir Alfred. *With the Russian Army, 1914–1917. Being Chiefly Extracts from the Diary of a Military Attaché*. London, 1921.

Kochakov, B. M. "Bol'shevizatsiia Petrogradskogo garnizona v 1917g.," in Avvakumov et al. (eds.), *Oktiabr'skoe vooruzhennoe vosstanie v Petrograde*, pp. 142–183.

——— (ed.). *Ocherki istorii Leningrada. Period imperializma i burzhuazno–demokraticheskikh revoliutsii, 1895–1917gg*. Moscow–Leningrad, 1956.

———. "Petrograd v gody pervoi mirovoi voiny i Fevral'skoi burzhuazno-demokraticheskoi revoliutsii," in Kochakov (ed.), *Ocherki istorii Leningrada*, III, pp. 932–1000.

Koenker, Diane. *Moscow Workers and the 1917 Revolution*. Princeton, 1981.

Kokovtsev, Graf V. N. *Iz moego proshlago. Vospominaniia, 1903–1919*. 2 vols. Paris, 1933.

Kolenkovskii, A. *Manevrennyi period pervoi mirovoi imperialistielieskoi voiny 1914 g*. Moscow, 1940.

Kollontai, Aleksandra. *Love and the New Morality*. Edited and translated by Alix Holt. Bristol, 1972.

Kondrat'ev, A. "Vospominaniia o pod'pol'noi rabote v Petrograde," *Krasnaia letopis'*, No. 7 (1923).

Kondzerovskii, General P. K. *V Stavke Verkhovnogo, 1914–1917. Vospominaniia Dezhurnogo Generala pri Verkhovnom Glavnokomanduiushchem.* Paris, 1967.

Konev, A. M. *Krasnaia gvardiia na zashchite oktiabria.* Moscow, 1978.

"Kopiia sekretnoi zapiski Ego Prevoskhoditel'stvu gospodinu Direktoru Departamenta Politsii. Otdeleniia po okhraneniiu obshchestvennoi bezopastnosti i poriadka v stolitse." 25 ianvaria 1915g., TsGIAL, fond 1405, opis' 530, delo No. 1058/2–6.

"Kopiia sekretnoi zapiski ego prevoskhoditel'stvu gospodinu direktoru departamenta politsii. Otdeleniia po okhraneniiu obshchestvennoi bezopastnosti i poriadka v stolitse," 20 iiulia 1915g., TsGIAL, fond 1405, opis' 530, delo No. 1058/34–35.

Korablev, Iu. I. (ed.). *Rabochee dvizhenie v Petrograde v 1912–1917gg. Dokumenty i materialy.* Leningrad, 1958.

Korol'kov, G. K. *Lodzinskaia operatsiia 2 noiabria–19 dekabria 1914g.* Moscow, 1934.

———. *Prasnyshskoe srazhenie, iiul' 1915g.* Moscow, 1928.

———. *Srazhenie pod Shavli.* Moscow, 1926.

Korsun, N. G. *Alashkertskaia i Khamadanskaia operatsiia na kavkazskom fronte mirovoi voiny v 1915 godu.* Moscow, 1940.

———. *Pervaia mirovaia voina na Kavkazskom fronte.* Moscow, 1946.

———. *Sarykamyshskaia operatsiia.* Moscow, 1937.

"Kratkii ocherk vazhneishikh sobytii kampanii 1917g. na russkom fronte," in Kakurin (ed.), *Razlozhenie armii*, pp. 179–184.

Kravchuk, N. A. *Massovoe krest'ianskoe dvizhenie v Rossii nakanune oktiabria (mart-oktiabr' 1917g., po materialam velikorusskikh gubernii Evropeiskoi Rossii).* Moscow, 1971.

Krupina, T. D. "Politicheskii krizis 1915g.i sozdanie osobogo soveshchaniia po oborone," *Istoricheskie zapiski*, LXXXIII (1969), pp. 58–75.

Krupianskaia, V. Iu. "Evoliutsiia semeino-bytovogo uklada rabochikh," in Ivanov (ed.), *Rossiiskii proletariat*, pp. 271–289.

Krupskaia, N. K. *Iz dalekykh vremen.* Moscow-Leningrad, 1930.

———. *Vospominaniia o Lenine.* Moscow, 1968.

Kruze, E. E. "Rabochie Peterburga v gody novogo revoliutsionnogo pod"ema," in Diakin et al. (eds.), *Istoriia rabochikh Leningrada*, I, pp. 385–460.

Kuropatkin, General A. N. "Dnevnik A. N. Kuropatkina." *Krasnyi arkhiv*, II (1922), pp. 5–112; V (1924), pp. 82–101; VII (1924), pp. 55–69; VIII (1925), pp. 70–100; LXVIII (1935), pp. 65–96; LXIX–LXX (1935), pp. 101–127.

Kutsentov, D. G. "Naselenie Peterburga. Polozhenie peterburgskikh rabochikh," in Kochakov (ed.), *Ocherki istorii Leningrada*, II, pp. 170–230.

Kuz'min, Dmitrii. "Kazanskaia demonstratsiia 1876g. i G. V. Plekhanov," *Katorga i ssylka*, 44 (1928), pp. 7–40.

Kuznetsov, N. V. "V bor'be protiv imperialisticheskoi voiny i samoderzhaviia (1914–fevral' 1917gg.)," in Kabanov (ed.), *Proletariat vo glave osvoboditel'nogo dvizheniia v Rossii*, pp. 239–271.

Laguiche, Général Marquis de. "Rapport secret du Général de Laguiche, Attaché Militaire, à Monsieur le Ministre de la Guerre (Etat-Major de l'Armée, 2e Bureau)," le 13/26 octobre 1914, AG-CV, file 7N757.

———. "Rapport secret du Général de Laguiche, attaché militaire à l'Ambassade de France, à Monsieur le Ministre de la Guerre (Etat-Major de l'Armée, 2e Bureau)," 6/19 avril 1916, AG-CV, file 7N757.

Langlois, Commandant. "Rapport du Commandant Langlois sur sa Mission en Russie," 5 février 1915, AG-CV, file 7N1547.

————. "Rapport du Commandant Langlois sur sa Seconde Mission en Russie," 10 avril 1915, AG-CV, file 7N1547.

————. "Rapport du Commandant Langlois sur sa Troisième Mission en Russie," 20 juin 1915, AG-CV, file 7N1547.

————. "Rapport du Commandant Langlois sur sa Quatrième Mission en Russie," 16 septembre 1915, AG-CV, file 7N1547.

————. "Rapport du Commandant Langlois sur sa Sixième Mission en Russie," 12 juin 1916, AG-CVm file 7N1547.

Lapin, N. (ed.). "Progressivnyi blok v 1915–1917gg.," *Krasnyi arkhiv*, L–LI (1932), pp. 117–150; LII (1932), pp. 143–196.

Lapitskaia, S. *Byt rabochikh trekhgornoi manufaktury.* Moscow, 1935.

Laverychev, V. Ia. *Tsarizm i rabochii vopros v Rossii (1861–1917gg.)* Moscow, 1972.

Lavrin, V. A. *Bol'shevistskaia partiia v nachale pervoi mirovoi imperialisticheskoi voiny.* Moscow, 1972.

Lee, Dwight E. *Europe's Crucial Years: The Diplomatic Background of World War I, 1902–1914.* Hanover, 1974.

Leiberov, I. P. *Na shturm samoderzhaviia. Petrogradskii proletariat v gody pervoi mirovoi voiny i fevral'skoi revoliutsii.* Moscow, 1979.

————. "Petrogradskii proletariat v fevral'skoi revoliutsii 1917g.," in Diakin *et al.* (eds.), *Istoriia rabochikh Leningrada,* I, pp. 512–538.

————. "Petrogradskii proletariat v gody pervoi mirovoi voiny," in Diakin *et al.* (eds.), *Istoriia rabochikh Leningrada,* I, pp. 461–511.

Lemke, Mikhail. *250 dnei v tsarskoi stavke (25 sentiabria 1915–2 iiulia 1916).* Petersburg, 1920.

Lenin, V. I. "Advice of an Onlooker," in *Collected Works,* XXVI, pp. 179–181.

————. "Afterword to the Theses on the Question of the Immediate Conclusion of a Separate and Annexationist Peace," in *Collected Works,* XXVI, pp. 451–452.

————. "The Bolsheviks Must Assume Power," in *Collected Works,* XXVI, pp. 19–21.

————. "Can the Bolsheviks Retain State Power?" in *Collected Works,* XXVI, pp. 87–136.

————. "The Chief Task of Our Day," in *Collected Works,* XXVII, pp. 159–163.

————. "Chto delat'?", in *Sochineniia,* V, pp. 319–494.

————. "The Collapse of the Second International," in *Collected Works,* XXI, pp. 207–259.

————. *Collected Works.* Various translators, 45 vols. Moscow, 1960–1970.

————. "The Crisis Has Matured," in *Collected Works,* XXVI, pp. 74–86.

————. "Direct-Line Conversations [of V.I. Lenin] with Helsingfors, October 27, 1917," in Lenin, Collected Works, XXVI, pp. 266–269.

————. "Draft Theses," March 4 (17), 1917, in *Collected Works,* XXIII, pp. 287–291.

————. "Farewell Letter to the Swiss Workers," in *Collected Works,* XXIII, pp. 367–373.

————. "Lecture on the 1905 Revolution," in *Collected Works,* XXIII, pp. 236–253.

————. "Letter to the Bolshevik Comrades Attending the Congress of Soviets of the Northern Region," in *Collected Works,* XXVI, 182–187.

————. "Letter to Bolshevik Party Members," in *Collected Works,* XXVI, pp. 216–219.

————. "Letter to the Central Committee, the Moscow and Petrograd Committees, and the Bolshevik Members of the Petrograd and Moscow Soviets," in *Collected Works,* XXVI, pp. 140–141.

————. "Letter to Central Committee Members, October 24, 1914," in *Collected Works,* XXVI, pp. 234–235.

————. "Letter to the Central Committee of the R.S.D.L.P. (B.)," in *Collected Works,* XXVI, pp. 223–227.

————. "Letter to Comrades," in *Collected Works,* XXVI, pp. 195–215.

————. "Letters from Afar," in *Collected Works,* XXIII, pp. 297–342.

————. "Marxism and Insurrection," in *Collected Works*, XXVI, pp. 22–27.

————. "Note to F. E. Dzerzhinsky with a Draft of a Decree on Fighting Counter-Revolutionariesand Saboteurs," in*Collected Works*, XXVI, pp. 374–376.

————. "On Compromises," in *Collected Works*, XXV, pp. 305–310.

————. "Outline Programme for Peace Negotiations," in *Collected Works*, XXVI, pp. 349–350.

————. "People from Another World," in *Collected Works*, XXVI, pp. 431–433.

————. "Political Report of the Central Committee to the Extraordinary Seventh Congress of the R.C.P.(b)," March 7, 1918, in *Collected Works*, XXVII, pp. 87–109.

————. "The Position and Tasks of the Socialist International," in *Collected Works*, XXI, pp. 35–41.

————. "Report on Land," in *Collected Works*, XXVI, pp. 257–261.

————. "Report on Peace, October 26," in *Collected Works*, XXVI, pp. 249–253.

————. *Sochineniia*, 40 vols. 4th ed. Moscow, 1952–58.

————. "The Socialist Fatherland Is in Danger!" February 21, 1918, in *Collected Works*, XXVII, pp. 30–33.

————. "Speech at a Joint Meeting of the Petrograd Soviet of Workers' and Soldiers' Deputies and Delegates from the Fronts," November 4, 1917, in *Collected Works*, XXVI, pp. 293–295.

————. "The State of Affairs in Russian Social Democracy," in *Collected Works*, XXI, pp. 281–286.

————. "The Tasks of Revolutionary Social-Democracy in the European War," in *Collected Works*, XXI, pp. 15–19.

————. "Theses on the Constituent Assembly," in *Collected Works*, XXVI, pp. 379–383.

————. "Theses on the Question of the Immediate Conclusion of a Separate and Annexationist Peace," in *Collected Works*, XXVI, pp. 442–450.

————. "To Workers, Peasants, and Soldiers!" in *Collected Works*, XXVI, pp. 137–139.

————. "To Workers, Soldiers, and Peasants!" in *Collected Works*, XXVI, pp. 247–248.

————. "The Voice of an Honest French Socialist," in *Collected Works*, XXI, pp. 349–356.

————. "The War and Russian Social Democracy," in *Collected Works*, XXI, pp. 27–34.

Lettres des Grands-Ducs à Nicolas II. Traduit du Russe par M. Lichnevsky. Paris, 1926.

Levitskii, V. [V. O. Tsederbaum]. *Za chetvert' veka. Revoliutsionnye vospominaniia*. vol. I. Moscow, 1926.

Liashchenko, P. I. *Istoriia narodnogo khoziaistva SSSR*. 2 vols. Moscow, 1956.

Liddell Hart, B. H. *The Real War, 1914–1918*. Boston, 1930.

————. *Ocherki agrarnoi evoliutsii Rossii*. Moscow-Petrograd, 1923.

Lincoln, W. Bruce. *In War's Dark Shadow: The Russians Before the Great War*. New York, 1983.

————. *The Romanovs: Autocrats of All the Russias*. New York, 1981.

"Liste des usines de la région de Petrograd, possédant un outillage susceptible d'être utilisé à la fabrication des Projectiles," février 1915, AG-CV, file 10N69.

Liverovskii, A. V. "Poslednie chasy Vremennogo pravitel'stva. Dnevnik ministra Liverovskogo," *Istoricheskii arkhiv*, No. 6 (1960), pp. 40–48.

Liubosh, S. B. *Russkii fashist Vladimir Purishkevich*. Leningrad, 1925.

Lockhart, R. H. Bruce. *British Agent*. Garden City, 1933.

Ludendorff, Erich. *The General Staff and Its Problems: The History of the Relations between the High Command and the German Imperial Government as Revealed by Official Documents*. Translated by F. A. Holt. New York, n.d.

————. *Ludendorff's Own Story, August 1914–November 1918: The Great War from the Siege of Liège to the Signing of the Armistice as Viewed from the Grand Headquarters of the German Army*. Translated from the German. 2 vols. New York and London, 1919.

────. *My War Memoirs, 1914–1918.* Translated from the German. London, 1919.

Lukomskii, A. S. *Vospominaniia generala A. S. Lukomskago.* 2 vols. Berlin, 1922.

Lunacharskii, A. V. "Smol'nyi v velikuiu noch'," in Mushtukov (ed.), *Petrograd v dni velikogo oktiabria,* pp. 410–412.

Lutovinov, I. S. *Likvidatsiia miatezha Kerenskogo-Krasnova.* Moscow-Leningrad, 1965.

Lutskii, E. A. "Podgotovka proekta dekreta o zemle," *Oktiabr' i grazhdanskaia voina v SSSR,* pp. 233–248.

Maevskii, I. V. *Ekonomika russkoi promyshlennosti v usloviiakh pervoi mirovoi voiny.* Moscow, 1937.

Magnes, Judah L. *Russia and Germany at Brest-Litovsk.* New York, 1919.

Malchevskii, I. S. (ed.). *Vserossiiskoe uchreditel'noe sobranie.* Moscow-Leningrad, 1930.

Maliantovich, P. N. "V Zimnem Dvortse 25–26 oktiabria 1917 goda. Iz vospominanii," *Byloe,* XII, No. 6 (Iiun' 1918), pp. 111–141.

Manevrennyi period 1914 goda. Lodzinskaia operatsiia: Sbornik dokumentov. Moscow-Leningrad, 1936.

Manevrennyi period 1914 goda. Varshavsko-Ivangorodskaia Operatsiia: Sbornik dokumentov. Moscow, 1938.

Manevrennyi period 1914 goda. Vostochno-Prusskaia Operatsiia: Sbornik dokumentov. Moscow, 1939.

Manikovskii, A. A. *Boevoe snabzhenie russkoi armii v mirovuiu voinu.* 2nd ed. Vol. 1. Moscow-Leningrad, 1930.

Manuilov, V. A. "Lermontov," in Gorodetskii (ed.), *Istoriia russkoi literatury,* VII, pp. 263–378.

Maria Feodorovna, Empress. *Letters of Tsar Nicholas and Empress Marie.* Translated from the French and Russian. Edited by Edward J. Bing. London, 1937.

Maria Pavlovna, Grand Duchess. *Education of a Princess: A Memoir.* Translated from the French and Russian. New York, 1934.

Markov, Vladimir, and Sparks, Merrill (eds.). *Modern Russian Poetry.* Indianapolis, Kansas City, and New York, 1967.

Martov, Iu. *Zapiski sotsialdemokrata.* Berlin, Petersburg, Moscow, 1922.

Martynov, E. I. "Konferentsiia soiuznikov v Petrograde v 1917 godu," *Krasnyi arkhiv,* XX (1927), pp. 39–55.

────. *Kornilov: Popytka voennogo perevorota.* Leningrad, 1927.

────. *Tsarskaia armiia v fevral'skom perevorote.* Leningrad, 1927.

Martynov, M. (ed.). "Agrarnoe dvizhenie v 1917 godu po dokumentam glavnogo zemel'nogo komiteta," *Krasnyi arkhiv,* XIV (1926), pp. 182–226.

Maslenikov, Oleg. *The Frenzied Poets: Andrei Biely and the Russian Symbolists.* Berkeley, 1952.

Massie, Robert, K. *Nicholas and Alexandra.* New York, 1978.

"Materialy iz kantseliarii Ministra Vnutrennikh Del," TsGIAL, fond 1282, opis' 1, dela 1165 and 1167.

Materialy po istorii franko-russkikh otnoshenii za 1910–1914gg. Moscow, 1922.

Mayakovsky, Vladimir. *The Bedbug and Selected Poetry.* Edited by Patricia Blake, translated by Max Hayward and George Reavey. New York and Cleveland, 1970.

──── [Maiakovskii]. *Sobranie sochinenii v vos'mi tomakh.* Moscow, 1968.

Medlin, Virgil D., and Parsons, Steven L. (eds.). *V. D. Nabokov and the Russian Provisional Government, 1917.* With an Introduction by Robert P. Browder. New Haven and London, 1976.

Medvedev, Roy A. *Let History Judge: The Origins and Consequences of Stalinism.* Translated by Colleen Taylor. Edited by David Joravsky and Georges Haupt. New York, 1971.

Mehlinger, Howard D., and Thompson, John M. *Count Witte and the Tsarist Government in the 1905 Revolution.* Bloomington, 1972.

Mel'gunov, S. P. *Kak bol'sheviki zakhvatili vlast'.* Paris, 1953.

———. *Martovskie dni 1917 goda.* Paris, 1961.

———. *Vospominaniia i dnevniki.* 3 vols. Paris, 1964.

Mezhdunarodnye otnosheniia v epokhu imperializma. Dokumenty iz arkhivov tsarskogo i vremennogo pravitel'stva, 1878–1917gg. Seriia III, 1914–1917. 10 vols. Moscow-Leningrad, 1931–1938.

"Mikhail Vasil'evich Alekseev," *Sovetskaia istoricheskaia entsiklopediia,* I, cols. 379–380.

Mikhrin, P. N. "Vystupleniia," in Mints (ed.), *Lenin i oktiabr'skoe vooruzhennoe vosstanie v Petrograde,* pp. 119–124.

Mil'chik, I. I. *Rabochii Fevral'.* Moscow-Leningrad, 1931.

Miliukov, P. N. *Istoriia vtoroi russkoi revoliutsii.* Sofia, 1921.

———. *Vospominaniia, 1859–1917.* New York, 1955.

Miller, V. I. "Nachalo demokratizatsii staroi armii v dni fevral'skoi revoliutsii (zasedanie Petrogradskogo Soveta, 1 marta 1917g. i Prikaz No. 1)," *Istoriia SSSR,* No. 6 (noiabr'–dekabr', 1966), pp. 26–43.

———. *Soldatskie komitety russkoi armii v 1917g: vozniknovenie i nachal'nyi period deiatel'nosti.* Moscow, 1974.

Ministerstvo inostrannykh del SSSR. *Dokumenty vneshnei politiki SSSR.* Vol. 1. Moscow, 1957.

Mints, I. I. *God 1918-i.* Moscow, 1982.

———. *Istoriia velikogo oktiabria.* 3 vols. Moscow, 1967.

——— (ed.). *Lenin i oktiabr'skoe vooruzhennoe vosstanie v Petrograde. materialy Vsesoiuznoi nauchnoi sessii, sostoiavsheisia 13–16 noiabria 1962g. v Leningrade.* Moscow, 1964.

Mirnye peregovory v Brest-Litovske. Moscow, 1926.

Mitel'man, M., Glebov, B., and Ul'ianskii, A. *Istoriia putilovskogo zavoda, 1789–1917gg.* Vol. I. Moscow-Leningrad, 1941.

Mochul'skii, K. *Andrei Belyi.* Paris, 1955.

———. *Vladimir Solov'ev: Zhizn' i uchenie.* 2nd ed. Paris, 1948.

Moiseenko, P. A. *Vospominaniia starogo revoliutsionera.* Moscow, 1966.

Mosolov, General A. *Pri dvore Imperatora.* Riga, n.d.

Mstislavskii [Maslovskii], S. D. *Piat' dnei: Nachalo i konets Fevral'skoi revoliutsii.* Moscow, 1922.

"Munitions expediées par les départements techniques et de l'artillerie vers l'interieur de la Russie durant les mois anterieurs à juillet 1916," (Secret), AG-CV, file No. 7N758.

Murashov, S. I. (ed.). *Partiia bol'shevikov v gody pervoi mirovoi voiny. Sverzhenie monarkhii v Rossii.* Moscow, 1963.

Mushtukov, V. E. (ed.). *Petrograd v dni velikogo oktiabria. Vospominaniia uchastnikov revoliutsionnykh sobytii v Petrograde v 1917 godu.* Leningrad, 1967.

"Nachalo voiny 1914g. Podennaia zapis' b. Ministerstva Inostrannykh Del," *Krasnyi arkhiv,* IV (1923), pp. 6–62.

"Nakanune peremiriia," *Krasnyi arkhiv,* XXIII (1927), pp. 195–249.

Nastuplenie Iugo-Zapadnogo fronta v mae-iiune 1915g. Sbornik dokumentov. Moscow, 1950.

Newton, Lord Thomas. *Lord Lansdowne: A Biography.* London, 1929.

Nicholas II. *Journal intime de Nicolas II (juillet 1914–juillet 1918).* Paris, 1934.

———. *Letters of Tsar Nicholas and Empress Marie.* Edited by Edward J. Bing. Translated from the French and the Russian. London, 1937.

———. *Letters of the Tsar to the Tsaritsa, 1914–1917.* Translated by A. L. Hynes. London, 1929.

Niessel, General A. *Le Triomphe des Bolcheviks et la Paix de Brest-Litovsk: Souvenirs, 1917–1918*. Paris, 1940.

Nikolai II: Materialy dlia kharakteristiki lichnosti i tsarstvovaniia. Moscow, 1917.

"Nikolai Vasil'evich Krylenko," *Sovetskaia istoricheskaia entsiklopediia*, VIII, cols. 194–195.

Niva: Illiustrirovannyi zhurnal literatury, politiki, i sovremennoi zhizni. Volumes for 1914–1917.

"Notes and Tables on the Present and Prospective Munition Output of Russia" (Secret). January-February 1917. AG-CV, File 10N91, pp. 8–17.

Novosel'skii, A. A. (ed.). *Ocherki istorii SSSR*. Moscow, 1955.

————, and Chaev, N. S. "Krest'ianskaia voina pod predvoditel'stvom S. T. Razina," in Novosel'skii (ed.), *Ocherkii*, pp. 277–311.

"O dopushchenii nekotorykh otstuplenii ot pravil o rabote zhenshchin, podrostkov i maloletnikh," 3 marta 1915g., TsGIAL, fond 1276, opis' 11, delo No. 100.

"O meropriiatiiakh Ministerstva Zemledeliia, vyzvannykh obstoiatel'stvami voennago vremeni, i potrebnikh dlia osushchestvleniia etikh meropriiatii kreditakh," 10 oktiabria 1916g., TsGIAL, fond 1276, opis' 12, delo No. 1062.

"O nezhelatel'nosti dal'neishago rosta promyshlennosti v samom gorode S.-Peterburge, v sviazi s politicheskami manifestatsiami i stachkami rabochikh," iiul' 1914g. TsGIAL. fond 1276, opis' 10, delo No. 127.

Ob Il'iche. Leningrad, 1970.

"Ob"iasnitel'naia zapiska k programme snabzhjeniia armii glavneishimi predmetami artilleriiskago dovol'stviia na period do 1 ianvaria 1918g.," (sekretno). Petrograd, 1917. AG-CV, File 10N69, pp. 1–27.

"Obrashchenie k naseleniiu Petrograda o podavlenii iunkerskogo miatezha," in Chugaev (ed.), *Petrogradskii voenno-revoliutsionnyi komitet*, I, No. 669, pp. 280–281.

"Obrashchenie k soldatam Petrogradskogo garnizona ob ispolnenii prikazov i rasporiazhenii po garnizonu tol'ko za podpis'iu voenno-revoliutsionnogo komiteta," in Chugaev (ed.), *Petrogradskii voenno-revoliutsionnyi komitet*, I, No. 30, p. 63.

"Obrashchenie kniazia L'vova k tsariu," in Grave (ed.), *Burzhuaziia*, pp. 59–60.

"Obrashchenie moskovskogo soveta professional'nykh soiuzov 'ko vsem rabochim i rabotnitsam' s prizyvom splotit'sia vokrug voenno-revoliutsionnogo komiteta," in *Podgotovka i pobeda oktiabr'skoi revoliutsii v Moskve*, pp. 399–400.

"Obrashchenie nachal'nikov voennykh missii pri shtabe verkh. glavnokomanduiushchego na imia gen. Dukhonina 11 noiabria 1917g. ob ukreplenii poriadka i distsipliny v russkoi armii," in "Nakanune peremiriia," pp. 211–212.

"Obraschenie Narodnogo Komissariata Inostrannykh Del k narodam i pravitel'stvam soiuznykh stran," 17/30 dekabria 1917g., in *Dokumenty vneshnei politiki SSSR*, I, pp. 69–70.

"Obrashchenie Soveta Narodnykh Komissarov k armii i flotu po voprosu o mire. Radio vsem. polkovym, divizionnym, korpusnym, armeiskim i drugim komitetam, vsem soldatam Revoliutsionnoi Armii i matrosam Revoliutsionnogo Flota," 9/22 noiabria 1917g., *Dokumenty vneshnei politiki SSSR*, I, No. 7, pp. 19–20.

"Obrashchenie Sovetskogo Pravitel'stva k poslam soiuznykh derzhav s predlozheniem nemedlennogo peremiriia i otkrytiia mirnykh peregovorov," 8/21 noiabria 1917g.," in *Dokumenty vneshnei politiki SSSR*, No. 5, I, pp. 16–17.

"Obrashchenie Sovetskogo Pravitel'stvo k pravitel'stvam i narodam voiuiushchikh stran s predlozheniem prisoedinit'sia k peregovoram o peremirii," 15/28 noiabria 1917g., *Dokumenty vneshnei politiki SSSR*, I, No. 13, pp. 29–30.

"Obzor polozheniia Rossii za tri mesiatsa revoliutsii po dannym otdela snozhenii s provintsiei Vremennago Komiteta Gosudarstvennoi Dumy," TsGIAL, fond 1278, opis' 10, delo No. 4.

"Ofitsial'noe pis'mo ot Shtaba Omskago Voennago Okruga," 21 marta 1916g., TsGIAL, fond 1291, opis' 132, delo No. 397–1916.

"Ofitsial'noe pis'mo Podol'skago Gubernatora Glavnokomanduiushchemu Armiiami Iugo-Zapadnago fronta," 6 fevralia 1916g., TsGIAL, fond 1291, opis' 132, delo No. 397–1916.

Okhtin, Petr. *Bezzemel'nyi proletariat v Rossi.* Moscow, 1905.

Oktiabr' i grazhdanskaia voina v SSSR: Sbornik statei k 70-letiiu akademika I. I. Mintsa. Moscow, 1966.

Oktiabr' v Moskve. Moscow, 1967.

Ol'denburg, S. S. *Tsarstvovanie Imperatora Nikolaia II.* 2 vols. Belgrad, 1939.

Onufriev, E. P. *Za nevskoi zastavoi.* Moscow, 1968.

Orlov, V. I. (ed.). *Aleksandr Blok i Andrei Belyi. Perepiska.* Moscow, 1940.

"Otchet o prodovol'stvii Petrograda merami pravitel'stva upolnomochennago po prodovol'st-viiu Petrograda za pervoe polugodie (s oktiabria 1915g. po marta 1916g. vkliuchitel'no)," TsGIAL, fond 1276, opis' 15, delo No. 48.

"Otnoshenie tovarishcha ministra vnutrennikh del V. F. Dzhunkovskogo nachal'niku shtaba verkhovnogo glavnokomanduiushchego N. N. Ianaushkevichu ob otkaze krest'ian Sara-tovskoi gubernii ot platezhe povinnostei i arendnykh deneg za zemliu pod vliianiem pisem s fronta," 12 ianvaria 1915g., in Anfimov (ed.), *Krest'ianskoe dvozhenie,* No. 130, pp. 231–232.

Owen, Launcelot A. *The Russian Peasant Movement, 1906–1917.* New York, 1963.

Pachmuss, Temira. *Zinaida Hippius: An Intellectual Profile.* Carbondale, 1971.

Paléologue, Maurice. *La Russie des Tsars pendant la Grande Guerre.* 3 vols. Paris, 1921.

Lettre (Secrète) de M. Paléologue au Ministère des Affaires Etrangères, No. 677, Petrograd, le 28 septembre 1914, AdAE, Guerre 1914–1918, Russie, Dossier Général No. 641/25.

————. "Télégramme de M. Paléologue au Ministerè des Affaires Etrangères, Petrograd, le 15 mai, 1915, (Secret), Dechiffrement, No. 647, AdAE Guerre 1914–1918, Russie, Dossier Général, No. 641/110–111.

————. "Télégramme secret de M. Paléologue au Ministère des Affaires Etrangères," Petrograd, le 4 juin 1915, AdAE, Guerre 1914–1918, Russie, Dossier Général No. 641/125.

————. "Télégramme secret de M. Paléologue au Ministère des Affaires Etrangères," Petrograd, le 26 juillet 1915, AdAE, Guerre 1914–1918, Russie, Dossier Général No. 641/169.

————. "Télégramme secret de M. Paléologue au Ministère des Affaires Etrangères," Petrograd, le 27 août 1915, AdAE, Guerre 1914–1918, Russie, Dossier Général No. 641/199.

————. "Télégramme secret de M. Paléologue au Ministère des Affaires Etrangères," Petrograd, le 2 septembre 1915, AdAE, Guerre 1914–1918, Russie, Dossier Général No. 641/205–208.

————. "Télégramme secret de M. Paléologue au Ministère des Affaires Etrangères," Petrograd, le 15 novembre 1916, AdAE, Guerre 1914–1918, Russie, Dossier Général No. 645/60–62.

————. "Télégramme très confidentiel de M. Paléologue au Ministère des Affaires Etran-gères," Petrograd, novembre 1916, AdAE, Guerre 1914–1918, Russie, Dossier Général No. 645/65–66.

————. "Télégramme de M. Paléologue au Ministère des Affaires Etrangères," Petrograd, le 1 janvier 1917, AdAE, Guerre 1914–1918, Russie, Dossier Général No. 646/1.

————. "Télégramme secret de M. Paléologue au Ministère des Affaires Etrangères," Petrograd, le 14 janvier 1917, AdAE, Guerre 1914–1918, Russie, Dossier Général No. 646/78–79.

Palmer, R. R., and Colton, Joel. *A History of the Modern World.* New York, 1957.

Parasun'ko, O. A. *Polozhenie i bor'ba rabochego klasssa Ukrainy.* Kiev, 1963.

Pares, Sir Bernard. *Day by Day with the Russian Army, 1914–1915.* Boston and New York, 1915.

————. *The Fall of the Russian Monarchy: A Study of the Evidence.* New York, 1961 (originally published 1939).

————— (ed.). *Letters of the Tsaritsa to the Tsar, 1914–1916.* London, 1923.

—————. *My Russian Memoirs.* London, 1931.

Paustovsky, Konstantin. *Sobranie sochinenii v shesti tomakh.* Moscow, 1956. 6 vols.

Pazhitnov, K. A. *Polozhenie rabochego klassa v Rossii.* 3 vols. Leningrad, 1924.

Pearson, Raymond. *The Russian Moderates and the Crisis of Tsarism, 1914–1917.* London, 1977.

Pershin, P. N. *Agrarnaia revoliutsiia v Rossii: Istoriko-ekonomicheskoi issledovanie.* 2 vols. Moscow, 1966.

Pertsov, V. *Maiakovskii: Zhizn' i tvorchestvo do velikoi oktiabr'skoi sotsialisticheskoi revoliutsii, 1893–1917gg.* Moscow-Leningrad, 1950.

"Pervomaiskaia listovka moskovskogo oblastnogo biuro TsK i Moskovskogo komiteta RSDRP(b)," in Gaponenko (ed.), *Revoliutsionnoe dvizhenie v Rossii v aprele 1917g.,* No. 64, pp. 94–95.

Peskov, P. A. *Fabrichnyi byt vladimirskoi gubernii. Otchet za 1882–1883gg.* St. Petersburg, 1884.

Petrash, V. V. *Moriaki baltiiskogo flota v bor'be za pobedu oktiabria.* Moscow-Leningrad, 1966.

Petrogradskie bol'sheviki v oktiabr'skoi revoliutsii. Leningrad, 1957.

Petrov, F. N. *et al.* (eds.) *V Ogne revoliutsionnykh boev. Sbornik vospominanii starykh bol'shevikov-pitertsev.* 2 vols. Moscow, 1971.

Pipes, Richard. *The Formation of the Soviet Union.* Cambridge, Mass., 1964.

—————. *Russia Under the Old Regime.* New York, 1974.

Pisarev, Iu. A. *Serbiia i Chernogoriia v pervoi mirovoi voine.* Moscow, 1968.

"Pis'mo M. V. Alekseeva ministru-predsedateliu A. F. Kerenskomu ot 20 iiulia 1917g.," in Alekseev, "Iz dnevnika Generala M. V. Alekseeva," p. 29.

"Pis'mo komanduiushchego 5-i armiei A. M. Dragomirova glavnokomanduiushchemu armiiami Severnogo fronta N. V. Russkomu o vystuplenii soldat protiv voiny i nepovinovenii ikh nachal'stvu," 29 marta 1917g., in Gaponenko (ed.), *Revoliutsionnoe dvizhenie v russkoi armii,* pp. 43–45.

"Pis'mo ministra zemledeliia A. I. Shingareva chlenu Vremennogo komiteta Gosudarstvennoi dumy S. I. Shidlovskomu o merakh bor'by s krest'ianskim dvizheniem," 30 aprelia 1917g., in Gaponenko (ed.), *Revoliutsionnoe dvizhenie v Rossii v aprele 1917g.,* No. 274, pp. 326–328.

"Pis'mo pomeshchitsy N. A. Trusovoi predsedateliu Vremennogo komiteta Gosudarstvennoi dumy M. V. Rodzianko o krest'ianskom dvizhenii v Ustiuzhenskom uezde Novogorodskoi gubernii," 9 marta 1917g., in Gaponenko (ed.), *Revoliutsionnoe dvizhenie v Rossii posle sverzheniia samoderzhaviia,* No. 635, pp. 672–673.

"Pis'mo rabochikh zavodov Sysertskogo gornogo okruga ministru iustitsii A. F. Kerenskomu s trebovaniem priniat' mery dlia uluchsheniia ikh tiazhelogo polozheniia," 19 marta 1917g., in Gaponenko (ed.), *Revoliutsionnoe dvizhenie v Rossii posle sverzheniia samoderzhaviia,* p. 521.

"Pis'mo soldata 430-go Valkskogo polka P. Sokkonena v gazetu 'Izvestiia Petrogradskogo Soveta rabochikh i soldatskikh deputatov' o podderzhke pomeshchikov kulakami dereven' Zhernovka i Malinovka Petrogradskogo uezda Petrogradskoi gubernii," in Gaponenko (ed.), *Revoliutsionnoe dvizhenie v Rossii v aprele 1917g.,* No. 664, pp. 658–659.

"Pis'mo verkhovnogo glavnokomanduiushchego M. V. Alekseeva voennomu ministru A. I. Guchkovu o moral'nom sostoianii armii," 16 aprelia 1917g., in Gaponenko (ed.), *Revoliutsionnoe dvizhenie v russkoi armii,* p. 61.

"Pis'mo verkhovnogo glavnokomanduiushchego A. F. Kerenskomu, 11 iiulia 1917g.," in Bukhbinder (ed.), "Na fronte v predoktiabr'skie dni," pp. 16–17.

"Pis'mo v Gosudarstvennuiu Dumu," 23 aprelia 1917g., TsGIAL, fond 1278, opis' 5, delo No. 1219/26.

"Pis'mo voennogo ministra A. I. Guchkova nachal'niku shtaba verkhovnogo glavnokomandu-iuschchego generalu M. V. Alekseevu ob otsutstvii real'noi vlasti i Vremennogo pravi-tel'stva i nevozmozhnosti prisylki popolnenii na front v sviazi s revoliutsionnym nastro-eniem soldat," 9 marta 1917g., No. 333, in Gaponenko (ed.), *Revoliutsionnoe dvizhenie v Rossii posle sverzheniia samoderzhaviia*, pp. 429–430.

"Pis'mo zhenshchiny-dobrovol'tsa voennomu ministru," in Kakurin (ed.), *Razlozhenie armii*, p. 70.

Podgotovka i pobeda oktiabr'skoi revoliutsii v Moskve. Dokumenty i materialy. Moscow, 1955.

Podvoiskii, N. "Vziatie Zimnego," in Bank, Zakharenko, and Eventov (eds.), *Nemerknush-chie gody*, pp. 142–151.

Poggioli, Renato. *The Poets of Russia, 1890–1930.* Cambridge, Mass., 1960.

Pogodin, M. P. *Istoriko-politicheskie pis'ma i zapiski v prodolzhenii Krymskoi Voiny, 1853–1856gg.* Moscow, 1874.

Pogrebinskii, A. P. "K istorii soiuzov zemstv i gorodov v gody imperialisticheskoi voiny," *Istoricheskii zapiski*, XII (1941), pp. 39–60.

———. "Voenno-promyshlennye komitety," *Istoricheskie zapiski*, XI (1941), pp. 160–200.

"Poiasnitel'nyi tekst k svodnomu otchetu po vsem fabrichnym uchastkam Petrogradskoi fabrichnoi inspektsii za 1915 god," TsGIAL, fond 23, opis' 19, delo No. 38.

"Poiasnitel'nyi tekst k svodnomu otchetu za 1915 god po Vladimirskoi gubernii, 20 aprelia 1916 g.," Ministerstvo torgovlia i promyshlennosti. TsGIAL, fond 23, opis' 19, delo No. 38.

"Pokazaniia S. P. Beletskogo, 17 maia–20 iiulia 1917g.," in Shchegolev (ed.), *Padenie tsar-skogo rezhima*, IV, pp. 119–533.

"Pokazaniia S. P. Beletskogo, 24 iiunia 1917 goda," in Shchegolev (ed.), *Padenie tsarskogo rezhima*, IV, pp. 117–533.

"Pokazaniia P. N. Miliukova, 4 avgusta 1917 goda," in Shchegolev (ed.), *Padenie tsarskogo rezhima*, VI, pp. 294–319.

"Pokazaniia A. N. Naumova, 4 aprelia 1917 goda," in Shchegolev (ed.), *Padenie tsarskogo rezhima*, I, pp. 329–360.

"Pokazanie praporshchika Knirshi," in Avdeev (ed.), "Vokrug Gatchiny," pp. 179–182.

"Pokazanie praporshchika E. V. Zelenskogo," in Tobolin (ed.), "Zagovor monarkhisticheskoi organizatsii V. M. Purishkevicha," pp. 182–184.

"Pokazaniia A. D. Protopopova, 20 aprelia–18 sentiabria 1917 goda," in Shchegolev (ed.), *Padenie tsarskogo rezhima*, I, pp. 10–26, 361–402; II, pp. 9–23; IV, 1–116, 157–187.

"Pokazaniia kniazia V. M. Volkonskogo, 19 iiulia 1917 goda," in Shchegolev (ed.), *Padenie tsarskogo rezhima*, VI, pp. 128–142.

Pokrovskii, M. N., (ed.), "Ekonomicheskoe polozhenie Rossii pered revoliutsiei," *Krasnyi arkhiv*, X (1925), pp. 67–94.

———. *Ocherki po istorii oktiabr'skoi revoliutsii.* 2 Vols. Moscow-Leningrad, 1927.

———. "Politicheskoe polozhenie Rossii nakanune fevral'skoi revoliutsii v zhandarmskom osveshchenii," *Krasnyi arkhiv*, XVII (1926), pp. 1–35.

———. "Stavka i ministerstvo inostrannykh del," *Krasnyi arkhiv*, XXVI (1928), pp. 1–50.

———. and Iakovlev, Ia. A. (eds.). *Gosudarstvennoe soveshchanie: Stenograficheskii otchet.* Moscow-Leningrad, 1930.

Poletika, N. P. *Vozniknovenie pervoi mirovoi voiny (iiul'skii krizis 1914g.).* Moscow, 1964.

Polevoi, Iu. Z. *Zarozhdenie marksizma v Rossii, 1883–1894gg.* Moscow, 1959.

Polivanov, A. A. *Iz dnevnikov i vospominanii po dolzhnosti voennogo ministra i ego pomoshch-nika, 1907–1916gg.* Edited by A. M. Zaionchkovskii. Moscow, 1924.

Polner, T. I. *Zhizennyi put' kniazia Georgiia Evgenievicha L'vova.* Paris, 1932.

——, Obolenskii, Prince Vladimir A., and Turin, Sergei P. *Russian Local Government During the War and the Union of Zemstovos.* New Haven, 1930.

Popov, A. L. *Oktiabr'skii perevorot: fakty i dokumenty.* Petrograd, 1918.

Popov, K. *Vospominaniia kavkazskago grenadera, 1914–1920gg.* Belgrad, 1925.

"Posledniaia voina v istorii Evropy," *Niva*, XLV, No. 31 (2 avgusta 1914g.), pp. 620b–620v.

"Postanovlenie kollegii voennogo otdela ispolnitel'nogo komiteta Petrogradskogo Soveta rabochikh i soldatskikh deputatov o garnizonnom soveshchanii i revoliutsionnom shtabe po oborone Petrograda," 11 oktiabria 1917g., in Chugaev (ed.), *Petrogradskii voenno-revoliutsionnyi komitet,* I, No. 9, pp. 38–39.

"Postanovlenie Soveta Narodnykh Komissarov o priniatii germanskikh uslovii mira," 23 fevralia 1918g., in *Dokumenty vneshnei politiki SSSR,* I, No. 73, pp. 112–114.

"Postanovlenie Vremennogo pravitel'stva po zemel'nomu voprosu," 19 marta 1917g., in Gaponenko (ed.), *Revoliutsionnoe dvizhenie v Rossii posle sverzheniia samoderzhaviia,* No. 345, p. 439.

"Pravitel'stvennoe predpisanie Verkhovnomu Glavnokomanduiushchemu Dukhoninu," 7/20 noiabria 1917g., in *Dokumenty vneshnei politiki SSSR,* I, No. 4, pp. 15–16.

"Predpisanie vremenno ispolniaiushchego obiazannosti verkhovnogo glavnokomanduiushchego generala M. V. Alekseeva komanduiushchim frontami i armiami o priniatii mer protiv rasprostraneniia revoliutsionnoi propagandy v deistvuiushchei armii," 11 marta 1917g., in Gaponenko (ed.), *Revoliutsionnoe dvizhenie v Rossii posle sverzheniia samoderzhaviia,* pp. 627–628.

"Predstavlenie i. d. prokurora Permskogo okruzhnogo suda P. B. Orlovskogo prokuroru Kazanskoi sudebnoi palaty V. A. Val'tsu ob antivoennykh vyskazyvanniiakh krest'ianina Staroputinskoi vol. Okhanskogo u. S. P. Zav'ialova," 16 aprelia 1915g., in Anfimov (ed.), *Krest'ianskoe dvizhenie,* No. 120, pp. 219–220.

"Predstavlenie ministra zemledeliia A. A. Bobrinskogo v Sovet Ministrov o vliianii voiny na sostoianie sel'skago khoziaistva i neobkhodimykh meropriiatiiakh dlia ego pod"ema," 10 oktiabria 1916g., in Anfimov et al. (eds.). *Ekonomicheskoe polozhenie Rossii,* III, pp. 16–32.

"Predstavlenie Volynskoi gubernskoi zemskoi upravy ministru zemledeliia A. N. Naumovu ob ushcherbe sel'skomu khoziaistvu gubernii, nanesennom voennymi deistviiami, i planakh vosstanovleniia pogolov'ia skota," 11 dekabria 1915g., in Anfimov et al. (eds.). *Ekonomicheskoe polozhenie Rossii,* III, pp. 93–98.

"Predvaritel'nye svedeniia o probytii prodovol'stvennykh gruzov pervoi neobkhodimosti po zheleznym dorogam v Petrograd; otpravke v Finliandiiu i ostal'nye mestnosti Imperii i ostatkakh dlia Petrograda s 1 po 11 ianvaria 1916 goda," TsGIAL, fond 457, opis' 1, delo No. 1017.

"Prikaz A. F. Kerenskogo o naznachenii generala P. N. Krasnova komanduiushchim vooruzhennymi silami Petrogradskogo voennogo okruga," 27 oktiabria 1917g., in Golikov (ed.), *Oktiabr'skoe vooruzhennoe vosstanie v Petrograde,* No. 822, p. 609.

"Prikaz glavnokomanduiushchego armiiami Iugo-Zapadnogo fronta N. I. Ivanova o merakh po ukrepleniiu distsipliny v armii," No. 1240, 22 sentiabria 1915g., in Sidorov (ed.), *Revoliutsionnoe dvizhenie,* pp. 128–129.

"Prikaz No. 1 Petrogradskogo Soveta rabochikh i soldatskikh deputatov po voiskam Petrogradskogo voennogo okruga," 1 marta 1917g., in Gavrilov (ed.), *Voiskovye komitety,* pp. 17–18.

"Prikaz No. 2 ispolnitel'nogo komiteta Petrogradskogo Soveta rabochikh i soldatskikh deputatov po voiskam Petrogradskogo voennogo okruga," in Gavrilov (ed.), *Voiskovye komitety,* pp. 19–21.

"Prikaz No. 51 i. d. glavkoverkha M. V. Alekseeva o vvedenii v deistvie Vremennogo polozheniia ob organizatsii chinov deistvuiushchei armii," in Gavrilov (ed.), *Voiskovye komitety,* pp. 58–65.

"Prisiaga revoliutsionera-volontera," in Kakurin (ed.), *Razlozhenie armii,* p. 69.

"Privetstvennasia telegramma Moskovskogo soveshchaniia obshchestvennykh deiatelei generalu L. G. Kornilovu," 10 avgusta 1917g., in Chugaev (ed.), *Revoliutsionnoe dvizhenie v Rossii v avguste 1917g.,* No. 361, p. 360.

Prokopovich, S. N. *Voina i narodnoe khoziaistvo,* Moscow, 1918.

"Protokol dnevnogo zasedaniia uzkogo sostava Moskovskogo oblastnogo biuro RSDRP(b)," 25 oktiabria 1917g., in Chugaev (ed.), *Triumfal'noe shestvie sovetskoi vlasti,* No. 269, p. 253.

"Protokol soveshchaniia, byvshego 16-go iiulia 1917g. v Stavke," in Bukhbinder (ed.), "Na fronte v predoktiabr'skie dni," pp. 19–51.

"Protokol zasedaniia Moskovskogo komiteta RSDRP(b)," 25 oktiabria 1917g., in Chugaev (ed.), *Triumfal'noe shestvie sovetskikh vlasti,* I, No. 268, pp. 251–252.

"Protokol zasedaniia Tsentral'nogo Komiteta RSDRP(b)," 16 oktiabria 1917g., in Chugaev (ed.), *Revoliutsionnoe dvizhenie v Rossii nakanune oktiabr'skogo vooruzhennogo vosstaniia,* pp. 87–95.

Protokoly tsentral'nogo komiteta RSDRP (b), avgust 1917-fevral' 1918. Moscow, 1958.

Protopopov, A. D. "Iz dnevnika A. D. Protopopova," *Krasnyi arkhiv,* X (1925), pp. 175–183.

———. "Telegrammy A. D. Protopopova Badmaevu," in Semennikov (ed.), *Za kulisami tsarizma,* pp. 17–18.

Purishkevich, V. M. *Dnevnik chlena Gosudarstvennoi Dumy Vladimira Mitrofanovicha Purishkevicha.* Riga, n.d.

Pushkin, A. S. "Mednyi Vsadnik," in *Polnoe sobranie sochinenii A. S. Pushkina.* vol IV. Moscow, 1963, pp. 380–397.

Pyman, Avril. *The Life of Aleksandr Blok: The Distant Thunder, 1880–1908.* Oxford, 1979.

———. *The Life of Aleksandr Blok: The Release of Harmony, 1908–1921.* Oxford, 1980.

Rabinowitch, Alexander. *The Bolsheviks Come to Power: The Revolution of 1917 in Petrograd.* New York, 1976.

———. *Prelude to Revolution: The Petrograd Bolsheviks and the July 1917 Uprising.* Bloomington, 1968.

"Radiogramma Soveta Narodnykh Komissarov Pravitel'stvu Germanii," 19 fevralia 1918g., in *Dokumenty vneshnei politiki SSSR,* I, No. 68, p. 106.

Radkey, Oliver H. *The Agrarian Foes of Bolshevism: Promise and Default of the Russian Socialist Revolutionaries, February to October 1917.* New York, 1958.

———. *Election to the Russian Constituent Assembly of 1917.* Cambridge, Mass., 1950.

———. *The Sickle Under the Hammer: The Russian Socialist Revolutionaries in the Early Months of Soviet Rule.* New York and London, 1963.

Radus-Zenkovich, V. A. *Ocherk vstrechennogo boia po opytu Gumbinenskoi operatsii v avguste 1914g. Kritiko-istoricheskoe issledovanie.* Moscow, 1921.

Rakh'ia, E. A. "Edem v Smol'nyi," in *Ob Il'iche,* pp. 366–371.

———. "Moi vospominaniia o Vladimire Il'iche," in Golikov et. al. (eds.), *Vospominaniia o Vladimire Il'iche Lenine,* II, pp. 430–434.

Rakhmetov, V. (ed.). "Aprel'skie dni 1917 goda v Petrograde," *Krasnyi arkhiv,* XXXIII (1929), pp. 36–48.

"Raport glavnokomanduiushchego voiskami Petrogradskogo voennogo okruga general-maiora P. A. Polovtseva voennomu i morskomu ministru A. F. Kerenskomu o narastanii nedovol'stva Vremennym pravitel'stvom v chastiakh Petrogradskogo garnizona," 17 iiunia 1917g., in Chugaev (ed.), *Revoliutsionnoe dvizhenie v Rossii v mae–iiune 1917g.,* No. 473, p. 525.

"Raport ispolnaiushchego dolzhnost' iaroslavskogo politseimeistera R. G. Dolivo-Dobrovol'-skogo gubernatoru S.D. Evreinovu o revoliutsionnoi agitatsii fel'dfebelia A. Zaderenko sredi rabochikh Iaroslavskoi manufaktury," 12 sentiabria 1915g., in Sidorov (ed.), *Revoliutsionnoe dvizhenie*, pp. 125–127.

"Raport komandira 47-go pekhotnogo Ukrainskogo polka A. N. Tushina nachal'niku 12-i pekhotnoi divizii M. S. Pustovoitenko ob otkaze soldat idti v nastuplenie i o nedoverii ikh k ofitseram," 28 marta 1917g., in Gaponenko (ed.), *Revoliutsionnoe dvizhenie v russkoi armii*, pp. 42–43.

"Raport o rasstrele Ivanovo-Voznesenskikh rabochikh," TsGIAL, fond 1276, opis' 11, delo No. 172.

"Raport Vserossiiskago Soiuza Torgovlia i Promyshlennosti," 14 avgusta 1917g., TsGIAL, fond 23, opis' 9, delo No. 351.

Rashin, A. G. *Formirovanie rabochego klassa Rossii*. Moscow, 1958.

———. "Gramotnost' i narodnoe obrazovanie v Rossii v XIX i nachale XX vv.," *Istoricheskie zapiski*, XXXVII (1951), pp. 44–50.

Raskol'nikov, F. F. *Na boevykh postakh*. Moscow, 1964.

"Razgovor po priamomu provodu gen. Alekseeva s vel. kniazem Mikhailom Alek sandrovichem 27 fevralia 1917g.," in "Fevral'skaia revoliutsiia 1917 goda," pp. 11–12.

"Razgovor po priamomu provodu gen. Dukhonina s voennym ministrom gen. Manikovskim i nach. gen. shtaba gen. Marushevskim, 9 noiabria 1917g. ob uvol'nenii Dukhonina ot dolzhnosti glavkoverka," in "Nakanune peremiriia," pp. 197–200.

"Razgovor po priamomu provodu generala Everta s generalem Klembovskim." 2 marta 1917g., 2 marta 1917g., in "Fevral'skaia revoliutsiia 1917 goda," p. 67.

"Razgovor po priamomu provodu gen. Ruzskago s Rodzianko 1 marta 1917g.," in "Fevral'-skaia revoliutsiia 1917 goda," p. 56.

"Razgovor Pravitel'stva so Stavkoi po priamomu provodu," 9/22 noiabria 1917g., *Dokumenty vneshnei politiki SSSR*, I, No. 6, pp. 17–19.

Reed, John. *Ten Days That Shook the World*. New York, 1960 (originally published 1919).

Renouvin, Pierre. *The Immediate Origins of the War (28th June–4th August 1914)*. Translated by T. C. Hume. New York, 1969.

"Report from Major C. Dunlop at Vladivostok," February 5, 1917, AG-CV, file 10N93.

"Reports of the French Military Attaché in Petrograd, 1914–1918," AG-CV, file 10N73.

Reports of Military Observers Attached to the Armies in Manchuria During the Russo-Japanese War. 5 vols. Washington, D. C., 1907.

"Rezoliutsiia ob"edinennogo zasedaniia prezidiumov zavodskogo komiteta rabochikh i komiteta sluzhashchikh moskovskogo zavoda 'Dinamo' o neobkhodimosti vvedeniia rabochego kontrolia nad proizvodstvom," 19 maia 1917g., in Chugaev (ed.), *Revoliutsionnoe dvizhenie v Rossii v mae–iiune 1917g.*, No. 229, p. 280.

"Rezoliutsiia I Petrogradskoi konferentsii fabrichno-zavodskikh komitetov ob ekonomicheskikh merakh bor'by s razrukhoi," 3 iiunia 1917g., in Chugaev (ed.), *Revoliutsionnoe dvizhenie v Rossii v mae–iiune 1917g.*, No. 242, pp. 290–291.

"Rezoliutsiia Petrogradskogo Soveta rabochikh i soldatskikh deputatov o neobkhodimosti peredachi vlasti Sovetam," 9 oktiabria 1917g., in Chugaev (ed.), *Revoliutsionnoe dvizhenie v Rossii nakanune oktiabr'skogo vooruzhennogo vosstaniia*, No. 103, p. 140.

"Rezoliutsiia s "ezda predstavitelei gorodov v Moskve," in Grave (ed.), *Burzhuaziia*, pp. 57–58.

"Rezoliutsiia skhoda krest'ian Kul'pinskoi volosti Volokolamskogo uezda Moskovskogo gubernii s privetstviem Moskovskomu Sovetu rabochikh deputatov i trebovaniem natsionalizatsii zemli i otmeny otrubnogo zemlevladeniia," 1 aprelia 1917g., in Gaponenko (ed.), *Revoliutsionnoe dvizhenie v Rossii posle sverzheniia samoderzhaviia*, No. 682, p. 707.

"Rezoliutsiia sobraniia rabochikh Sestroetskogo oruzheinogo zavoda o razgruske Petrograda," 20 maia 1917g., in Chugaev (ed.), *Revoliutsionnoe dvizhenie v Rossii v mae-iiune 1917g.*, No. 230, pp. 280–281.

"Rezoliutsiia sobraniia zhenshchin-rabotnits Moskovskogo raiona Petrograda s trebovaniem demokraticheskoi respubliki, ustanovleniia vos'michasovogo rabochego dnia, ravnopraviia zhenshchin, sotsial'nogo strakhovaniia i okhrany truda," 7 marta 1917g., in Gaponenko (ed.), *Revoliutsionnoe dvizhenie v Rossii posle sverzheniia samoderzhaviia*, No. 384, p. 470.

"Rezoliutsiia sovmestnogo soveshchaniia tsentral'nogo i Peterburgskogo komitetov RSDRP(b) i bol'shevistskoi fraktsii demokraticheskogo soveshchaniia 'Tekushchii moment i zadachi proletariata,' " 24 sentiabria 1917g., in Chugaev (ed.), *Revoliutsionnoe dvizhenie v Rossii v sentiabre 1917g.*, No. 59 pp. 74–75.

"Rezoliutsiia Tsentral'nogo Komiteta RSDRP(b) o podgotovke vooruzhennogo vosstaniia, predlozhennaia V. I. Leninym," 10 oktiabria 1917g., in Golikov (ed.), *Oktiabr'skoe vooruzhennoe vosstanie v Petrograde*, No. 15, p. 49.

"Rezoliutsiia TsK RSDRP(b) 20 aprelia (3 maia) 1917 goda o krizise v sviazi s notoi Vremennogo pravitel'stva ot 18 aprelia (1 maia) 1917g.," document No. 739, in Gaponenko (ed.), *Revoliutsionnoe dvizhenie v Rossii v aprele 1917g. Aprel'skii krizis*, pp. 726–727.

"Rezoliutsiia Tsentral'nogo Komiteta RSDRP(b), priniataia 21 aprelia (4 maia) 1917g.," document No. 757, in Gaponenko (ed.), *Revoliutsionnoe dvizhenie v Rossii v aprele 1917g. Aprel'skii krizis*, pp. 737–738.

Riasanovsky, N. V. *A History of Russia.* New York, 1977.

———. *Nicholas I and Official Nationality in Russia, 1825–1855.* Berkeley and Los Angeles, 1959.

Riha, Thomas. *A Russian European: Paul Miliukov in Russian Politics.* Notre Dame and London, 1969.

Ritter, Gerhard. *The Schlieffen Plan: Critique of a Myth.* Translated by Andrew and Eva Wilson. London, 1958.

Robbins, Richard G., Jr. *Famine in Russia, 1891–1892.* New York and London, 1975.

Robien, Comte Louis de. *The Diary of a Diplomat in Russia, 1917–1918.* Translated from the French by Camilla Sykes. London, 1969.

Rodzianko, A. N. "Pis'ma Anny Nikolaevny Rodzianko kn. Zin. N. Iusupovoi grafine Sumarokovoi-El'ston," in Sadikov (ed.), "K istorii poslednykh dnei tsarskogo rezhima," pp. 236–239.

Rodzianko, M. V. *Khrushenie imperii.* Leningrad, 1929.

———. *The Reign of Rasputin: An Empire's Collapse. The Memoirs of M. V. Rodzianko.* Translated by Catherine Zvegintzoff, with an Introduction by Sir Bernard Pares and a new Introduction by David R. Jones. Gulf Breeze, Fla., 1973.

———. "Zapiska M. V. Rodzianko," in Pokrovskii (ed.), "Ekonomicheskoe polozhenie Rossii," pp. 69–86.

Rogger, Hans. "The Beilis Case: Anti-Semitism and Politics in the Reign of Nicholas II," *Slavic Review*, XXV, No. 4 (1966), pp. 615–629.

Romanov, F. A. *Rabochee i professional'noe dvizhenie v gody pervoi mirovoi voiny i vtoroi russkoi revoliutsii (1914–fevral' 1917 goda). Istoricheskii ocherk.* Moscow, 1949.

Rosenberg, William G. *Liberals in the Russian Revolution: The Constitutional Democratic Party, 1917–1921.* Princeton, 1974.

Rosenthal, Bernice Glatser. *Dmitrii Sergeevich Merezhkovsky and the Silver Age: The Development of a Revolutionary Mentality.* The Hague, 1975.

Rossiia v mirovoi voine 1914–1918 goda (v tsifrakh). Moscow, 1925.

Rostunov, I. I. *General Brusilov.* Moscow, 1964.

———. *Russkii front pervoi mirovoi voiny.* Moscow, 1976.

Rubinshtein, N. *Bol'sheviki i uchreditel'noe sobranie.* Moscow, 1938.

Ruckman, Jo Ann. *The Moscow Business Elite: A Social and Cultural Portrait of Two Generations, 1840–1905.* DeKalb, Ill., 1984.

"Russian Munitions Tables, (secret) August 1916," AG-CV, 10N90.

"Russian Munitions Tables, (secret) December 1917," AG-CV, 10N90.

Ruzskii, N. V. "Prebyvanie Nikolaia II v Pskove 1 i 2 marta 1917g. (Beseda s generala S. N. Vil'chkovskim)," in Shchegolev (ed.), *Otrechenie Nikolaia II,* pp. 145–155.

Sadikov, P. (ed.). "K istorii poslednykh dnei tsarskogo rezhima (1916–1917gg.)," *Krasnyi arkhiv,* XIV (1926), pp. 227–249.

Salisbury, Harrison. *Black Night, White Snow: Russia's Revolutions, 1905–1917.* New York, 1978.

Saltykov-Shchedrin, M. E. "Dnevnik provintsiala v Peterburge," in his *Sobranie sochinenii,* vol. X. Moscow, 1970.

———. *Sobranie Sochineniia v desiati tomakh.* Moscow, 1968–1970.

Samoilo, A. *Dve zhizni.* Moscow, 1958.

Savinkov, B. V. "General Kornilov," *Byloe,* XXXI, No. 3 (1925), pp. 186–197.

Sazonov, S. D. *Vospominaniia.* Paris, 1927.

Sbornik dokumentov mirovoi imperialisticheskoi voiny na russkom fronte (1914–1917gg.): Gorlitskaia operatsiia. Moscow, 1941.

Sbornik dokumentov mirovoi imperialisticheskoi voiny na russkom fronte. Nastuplenie Iugo-Zapadnogo Fronta v mae–iiune 1916 goda. Moscow, 1940.

Schiemann, Theodor. *Geschichte Russlands unter Kaiser Nikolaus I.* 4 vols. Berlin, 1908–1919.

Schwarz, Lieutenant-General. *La Défense d'Ivangorod en 1914–1915: Extraits des mémoires du Lieutenant-General Schwarz, Gouverneur de la Place.* Translated by Th. Gutchkoff. Nancy-Paris-Strasbourg, 1922.

Seiranian, B. S. *Bor'ba bol'shevikov protiv voenno-promyshlennykh komitetov.* Erevan, 1961.

"Sekretnaia zapiska otdeleniia po okhraneniiu obshchestvennoi bezopasnosti i poriadke v stolitse," 2 sentiabria 1915g., TsGIAL, fond 1405, opis' 530, delo No. 1058/69–73.

"Sekretnyi doklad Ego prevoskhoditel'stvu gospodinu tovarishchu ministra vnutrennikh del. Otdeleniia po okhraneniiu obshchestvennoi bezopasnosti i poriadka v stolitse," 13 avgusta 1915g., TsGIAL, fond 1405, opis' 530, delo No. 1058/34–75.

Semennikov, V. *Politika Romanovykh nakanune revoliutsii (ot antanty - k Germanii) po novym dokumentam.* Moscow-Leningrad, 1926.

———. *Romanovy i germanskie vliianiia 1914–1917gg.* Leningrad, 1929.

———. *Za kulisami tsarizma (arkhiv tibetskogo vracha Badmaeva).* Leningrad, 1925.

——— (ed.). *Monarkhiia pered krusheniem 1914–1917gg. Bumagi Nikolaia II i drugie dokumenty.* Moscow-Leningrad, 1927.

Serapin, K. P. "O ranakh ot razryvnoi avstriiskoi puli na iugo-zapadnom fronte," *Voenno-meditsinskii zhurnal,* CCXLII (April 1915), pp. 434–449.

Seton-Watson, Hugh. *The Russian Empire, 1801–1917.* Oxford, 1967.

Seton-Watson, R. W. *Sarajevo: A Study in the Origins of the Great War.* London, 1926.

Shatsillo, K. F. "Delo polkovnika Miasoedova," *Voprosy istorii,* No. 4 (1967), pp. 103–116.

Shavel'skii, Otets Georgii. *Vospominaniia poslednego protopresvitera russkoi armii i flota.* 2 vols. New York, 1954.

Shchegolev, P. E. (ed.). *Otrechenie Nikolaia II: Vospominaniia ochevidtsev, dokumenty.* Leningrad, 1927.

———. *Padenie tsarskogo rezhima.* 7 vols. Moscow-Leningrad, 1924–1927.

Shidlovskii, S. I. "Vospominaniia," in Alekseev (ed.), *Fevral'skaia revoliutsiia,* II, pp. 282–315.

Shingarev, A. I. *Vymiraiushchaia derevnia* St. Petersburg, 1907. Republished as an appendix to K. M. Shuvaev, *Staraia i novaia derevnia.* Moscow, 1937.

Shklovsky, Viktor. *Mayakovsky and His Circle.* Edited and translated by Lily Feiler. New York, 1972.

————. *A Sentimental Journey: Memoirs, 1917–1922.* Translated by Richard Sheldon. Ithaca, 1970.

Shliapnikov, A. G. (ed.). *Kanun semnadtsatogo goda.* 2 vols. Moscow, 1922.

———— (ed.). "Oktiabr'skii perevorot i stavka," *Krasnyi arkhiv,* VIII (1925), pp. 153–175; IX (1925), pp. 156–170.

————. *Semnadtsatyi god.* 4 vols. Moscow-Leningrad, 1923–1931.

Shub, David. *Lenin: A Biography.* Garden City, 1948.

Shul'gin, V. V. *Dni.* Belgrade, 1925.

Shul'man, Z. N. "Vziatie Vladimirskogo uchilishcha," in Mustukov (ed.), *Petrograd v dni velikogo oktiabria,* pp. 476–485.

Sidorov, A. L. *Ekonomicheskoe polozhenie Rossii nakanune velikoi oktiabr'skoi sotsialisticheskoi revoliutsii. Dokumenty i materialy.* 3 vols. Moscow-Leningrad, 1957.

————. *Ekonomicheskoe polozhenie Rossii v gody pervoi mirovoi voiny.* Moscow, 1973.

————. *Finansovoe polozhenie Rossii v gody pervoi mirovoi voiny (1914–1917),.* Moscow, 1960.

————. "Iz istorii podgotovki tsarizma k voine," *Istoricheskii arkhiv* No. 2 (1962), pp. 120–150.

————. "K istorii toplivnogo krizisa v Rossii v gody pervoi mirovoi voiny (1914–1917)," *Istoricheskie zapiski,* LIX (1957), pp. 26–83.

———— (ed.). *Revoliutsionnoe dvizhenie v armii i na flote v gody pervoi mirovoi voiny, 1914–fevral' 1917.* Moscow, 1966.

Sidorov, K. (ed.). "Bor'ba so stachechnym dvizheniem nakanune mirovoi voiny," *Krasnyi arkhiv,* XXXIV (1929), pp. 95–125.

Simanovich, Aron. *Raspoutine.* Paris, 1930.

Sinegub, A. "Zashchita Zimnego Dvortsa, 25 oktiabria 1917g.," *Arkhiv russkoi revoliutsii,* IV (1922), pp. 121–197.

Sisson, Edgar. *One Hundred Red Days: A Personal Chronicle of the Bolshevik Revolution.* New Haven, 1931.

"Situation de l'armament de l'armée Russe, février 1917," AG-CV, 10N73.

Slavin, N. F. "Oktiabr'skoe vooruzhennoe vosstanie i predparlament," in Mints (ed.), *Lenin i oktiabr'skoe vooruzhennoe vosstanie v Petrograde,* pp. 222–231.

Slonim, Marc. *The Epic of Russian Literature from Its Origins Through Tolstoi.* New York, 1964.

————. *From Chekhov to the Revolution: Russian Literature, 1900–1917.* New York, 1962.

————. *Russian Theatre: From the Empire to the Soviets.* Cleveland and New York, 1961.

Smith, C. Jay. *The Russian Struggle for Power, 1914–1917: A Study of Russian Foreign Policy During the First World War.* New York, 1956.

Smith, S. A. *Red Petrograd: Revolution in the Factories, 1917–1918.* Cambridge, England, and New York, 1983.

Sokolov, Boris, "Zashchita Vserossiiskago Uchreditel'nago Sobraniia," *Arkhiv russkoi revoliutsii,* XIII (1924), pp. 5–70.

Solov'ev, M. E. "Partiia bol'shevikov—organizator pobedy fevral'skoi burzhuazno-demokraticheskoi revoliutsii v 1917g. v Rossii," in Murashov (ed.), *Partiia bol'shevikov v gody pervoi mirovoi voiny,* pp. 186–255.

Solov'ev, O. F. *Velikii Oktiabr' i ego protivniki: O roli soiuza Antanty s vnutrennei kontrrevoliutsiei v razviazyvanii interventsii i grazhdanskoi voiny, oktiabr' 1917–iiul' 1918.* Moscow, 1968.

"Soobshchenie generala Samoilo Narodnomu Komissaru Inostrannykh Del," 16 fevralia 1918g., in *Dokumenty vneshnei politiki SSSR*, I, p. 105.

"Soobshchenie Narodnogo Komissara Inostrannykh Del o predstoiashchem otkrytii peregovorov o peremirii, 17/30 noiabria 1917g.," in *Dokumenty vneshnei politiki SSSR*, I, No. 15, pp. 31–32.

"Soobshchenie Russkoi delegatsii o peregovorakh s Germanskim komandovaniem o peremirii, 14/27 noiabria 1917g.," in *Dokumenty vneshnei politiki SSSR*, No. 12, pp. 26–28.

"Soobshchenie vremenno-ispolniaiushchego dolzhnost' verkhovnogo glavnokomanduiushchego generala M. V. Alekseeva pomoshchniku glavnokomanduiushchego armiiami Rumynskogo fronta V. V. Sakharovu o neobkhodimosti shirokogo rasprostraneniia komitetov v voiskakh," in Gaponenko (ed.), *Revoliutsionnoe dvizhenie v russkoi armii*, No. 24, p. 40.

Sorokin, Pitirim. *Leaves from a Russian Diary*. New York, 1924.

"Soveshchanie s A. D. Protopopovym u M. V. Rodzianko, 19 oktiabria 1916 goda," in Shliapnikov (ed.), *Kanun semnadtsatogo goda*, II, pp. 99–107.

"Soveshchanie v *Stavke* 18 marta 1917 goda," in Kakurin (ed.), *Razlozhenie armii*, pp. 10–11.

Spiridonovich, General A. I. *Velikaia voina i fevral'skaia revoliutsiia, 1914–1917gg*. New York, 1960.

"Spisok gubernii i oblastei s pokazaniem: (1) Chisla imeiushchikhsia u khoziaev voennoplennykh, (2) Chisla voennoplennykh, podlezhashchikh dostavke iz za Urala, (3) Obshchego chisla voennoplennykh, kotoroe predlozheno imet' na rabotakh v 1916g.," TsGIAL, fond 1291, opis' 132, delo No. 397/20.

"Spravka mezhduvedomstvennoi komisssii pri ministerstve zemledeliia, 16 fevralia 1916g.," TsGIAL, fond 1291, opis' 132, delo No. 397–1916.

"Spravka. V tseliakh vospolneniia ubyli rabochikh rukh v sel'skom khoziaistve priniaty nizhesleduiushchaia mery," mart' 1916g, TsGIAL, fond 1291, opis' 132, delo No. 397–1916.

Stankevich, V. B. "Vospominaniia," in Alekseev (ed.), *Fevral'skaia revoliutsiia*, II, pp. 399–429.

Startsev, V. I. "Begstvo Kerenskogo," *Voprosy istorii*, XLI, No. 11 (1966), pp. 204–206.

———. *Krakh kerenshchiny*. Leningrad, 1982.

———. "O vybore momenta dlia oktiabr'skogo vooruzhennogo vosstaniia," in Mints (ed.), *Lenin i oktiabr'skoe vooruzhennoe vosstanie v Petrograde*, pp. 68–81.

———. *Ocherki po istorii petrogradskoi krasnoi gvardii i rabochei militsii*. Moscow-Leningrad, 1965.

———. *Russkaia burzhuaziia i samoderzhavie v 1905–1917: Bor'ba vokrug "otvetstvennogo ministerstva" i "pravitel'stva doveriia."* Leningrad, 1977.

———. *Vnutrenniaia politika Vremennogo Pravitel'stva pervogo sostava*. Leningrad, 1980.

———. "Voenno-revoliutsionnyi komitet i krasnaia gvardiia v oktiabr'skom vooruzhennom vosstanii," in Avvakumov *et al.* (eds.), *Oktiabr'skoe vooruzhennoe vosstanie v Petrograde*, pp. 1112–141.

Stein, H. P. "Der Offizier des russischen Heeres im Zeitabschnitt zwischen Reform und Revolution, 1861–1905," *Forschungen zur osteuropäischen Geschichte*, XIII (1967), pp. 346–507.

Stenograficheskii otchet. Gosudarstvennaia Duma. Chetvertyi sozyv, Sessiia V. Petrograd, 1916.

Stepanov, Z. V. "Ekonomicheskoe polozhenie rabochikh. Bor'ba za rabochii kontrol' nad proizvodstvom i raspredeleniem," in Fraiman (ed.), *Istoriia rabochikh Leningrada*, II, pp. 40–51.

———. *Rabochie Petrograda v period podgotovki i provedeniia oktiabr'skogo vooruzhennogo vosstaniia, avgust–oktiabr' 1917g*. Moscow-Leningrad, 1965.

Stites, Richard. *The Women's Liberation Movement in Russia: Feminism, Nihilism, and Bolshevism, 1860–1930.* Princeton, 1978.

Stone, Norman. *The Eastern Front, 1914–1917.* London, 1975.

Strategicheskii ocherk voiny 1914–1918gg. 7 vols. Moscow, 1920–1923.

Strumilin, S. G. *Zarabotnaia plata i proizvoditel'nost' truda v russkoi promyshlennosti, 1913–1922gg.* Moscow, 1923.

Sukhanov, N. N. *The Russian Revolution, 1917.* Edited, abridged, and translated by Joel Carmichael. Oxford, 1955.

———. *Zapiski o revoliutsii.* 7 vols. Berlin-Petersburg-Moscow, 1922.

Sukhomlinov, V. A. "Dnevnik Generala Sukhomlinova," *Dela i dni: Istoricheskii zhurnal,* I (1920), pp. 219–239.

———. "Pis'ma k N. N. Ianushkevichu," in Blinov (ed.), "Perepiska," pp. 29–74.

———. "Rossiia khochet mira, a gotova k voine," *Birzhevye Vedomosti* (27 fevralia 1914 goda), reprinted in M. Lemke, *250 dnei v tsarskoi stavke,* pp. 4–6.

———. *Vospominaniia.* Berlin, 1924.

"Svedeniia o nalichii na frontakh prodovol'stviia, furazha i obespechennosti v dniakh po sostoianiiu na 20 oktiabria 1917g., do dannym glavnogo polevogo intendanta," in Sidorov (ed.), *Ekonomicheskoe polozhenie Rossii nakanune velikoi oktiabr'skoi sotsialisticheskoi revoliutsii,* II, No. 470, p. 291.

Sverbeev, S. "Zapiski Sverbeeva," in "Nachalo voiny 1914g. Podennaia zapis' b. Ministerstva Inostrannykh Del," pp. 6–62.

Sviatikov, S. G. *Obshchestvennoe dvizhenie v Rossii, 1700–1895.* 2 vols. Rostov-on-the-Don, 1905.

"Svodka moskovskogo okhrannogo otdeleniia o 'nastroenii obshchestva' na 29 fevralia 1916g.," in Grave (ed.), *Burzhuaziia,* pp. 75–81.

"Svodka svedenii o sostave delegatov I Petrogradskoi konferentsii fabrichno-zavodskikh komitetov," in Chugaev (ed.), *Revoliutsionnoe dvizhenie v Rossii v mae—iiune 1917g.,* No. 244, p. 293.

Sworakowski, Witold. "The Authorship of the Abdication Documents of Nicholas II," *Russian Review,* XXX, No. 3 (July 1971), pp. 277–286.

Tager, A. S. *The Decay of Czarism.* Philadelphia, 1935.

———. (ed.). "Protsess Beilisa v otsenke departamenta politsii," *Krasnyi arkhiv,* XLIV (1931), pp. 85–125.

Tarasov-Rodionov, A. *Fevral'.* Moscow, 1931.

Tarasova, S. S., and Keirit-Markus, M. B. (comps.). *Velikaia oktiabr'skaia sotsialisticheskaia revoliutsiia: Khronika sobytii, 12 sentiabria–25 oktiabria 1917 goda.* Moscow, 1961.

"Telefonogramma predsedatelia Moskovskogo Soveta rabochikh, deputatov V. P. Nogina i chlena TsIK Sovetov rabochikh i soldatskikh deputatov V. P. Miliutina v Moskovskii Sovet rabochikh deputatov o pobede vooruzhennogo vosstaniia v Petrograde," 25 oktiabria 1917g., in Chugaev (ed.), *Triumfal'noe shestvie sovetskoi vlasti,* I, No. 267, p. 251.

"Telegramma gen. Alekseeva Glavnokomanduiushchemu Iugo-Zapadnym frontom," 18 maia 1917g., in Kakurin (ed.), *Razlozhenie armii,* p. 66.

"Telegramma gen. Alekseeva Glavnokomanduiushchemu Iugo-Zapadnym frontom," 21 maia 1917g., in Kakurin (ed.), *Razlozhenie armii,* p. 67.

"Telegramma gen. Alekseeva tsariu 1 marta 1917g.," in "Fevral'skaia revoliutsiia 1917 goda," pp. 53–54.

"Telegramma generala Alekseeva tsariu 2 marta 1917g.," in "Fevral'skaia revoliutsiia 1917 goda," p. 73.

"Telegramma gen. Brusilova verkh. glavnokomanduiushchemu," 16 maia 1917g., in Kakurin (ed.), *Razlozhenie armii,* p. 65.

"Telegramma gen. Brusilova voennomu ministru i verkh. glavnokomanduiushchemu," 20 maia 1917g., in Kakurin (ed.), *Razlozhenie armii,* pp. 66–67.

"Telegramma generala Danilova generalu Alekseevu 2 marta 1917g.," in "Fevral'skaia revoliutsiia 1917 goda," p. 63.

"Telegramma generala Gutora verkhovnomu glavnokomanduiushchemu," 30 maia 1917g., in Kakurin (ed.), *Razlozhenie armii,* p. 86.

"Telegramma generala Khabalova generalu Alekseevu 26 fevralia 1917g., No. 3703," in "Fevral'skaia revoliutsiia 1917 goda," p. 5.

"Telegramma generala S. S. Khabalova Tsariu, 27 fevralia 1917g.," TsGIAL, fond 1282, opis' 1, delo No. 737/70.

"Telegramma pomeshchika Shmidta ministru-predsedateliu G. E. L'vovu o razgrome krest'ianami ego imeniia v Menzelinskom uezde Ufimskoi gubernii," in Gaponenko (ed.), *Revoliutsionnoe dvizhenie v Rossii posle sverzheniia samoderzhaviia,* No. 653, p. 683.

"Telegramma predsedatelia Vremennogo komiteta Gosudarstvennoi dumy M. V. Rodzianko nachal'niku shtaba verkhovnogo glavnokomanduiushchego generalu M. V. Alekseevu o naznachenii generala L. G. Kornilova glavnokomanduiushchim Petrogradskim voennym okrugom," 2 marta 1917g., in Gaponenko (ed.), *Revoliutsionnoe dvizhenie v Rossii posle sverzheniia samoderzhaviia,* pp. 409–410.

"Telegramma Rodzianko tsariu 27 fevralia 1917g.," in "Fevral'skaia revoliutsiia 1917 goda," *Krasny Arkhiv,* XXI, 1927, pp. 6–7.

Temkin, Ia. G. and Shestakov, S. V. *Nezabyvaemye deni.* Moscow, 1959.

The Times History of the War. 21 vols. London, 1914–1921.

Tiutchev, F. I. *Polnoe sobranie sochinenii F. I. Tiutcheva.* St. Petersburg, 1913.

Tiutiukin, S. V. *Voina, mir, revoliutsiia. Ideinaia bor'ba v rabochem dvizhenii Rossii 1914–1917gg.* Moscow, 1972.

Tobolin, I. (ed.). "Zagovor monarkhisticheskoi organizatsii V. M. Purishkevicha," *Krasnyi arkhiv,* XXVI (1928), pp. 169–185.

Tokarev, Iu. S. *Petrogradskii Sovet Rabochikh i Soldatskikh Deputatov v marte–aprele 1917g.* Leningrad, 1976.

Tolstoi, A. K. *Izbrannye sochineniia v shesti tomakh.* Moscow, 1951.

Tolstoi, L. N. *Sobranie sochinenii.* 20 vols. Moscow, 1960–1966.

———. "Strashnyi vopros," *Russkie vedomosti,* No. 306 (November 6, 1891).

Trotsky, [Trotskii], L. *The History of the Russian Revolution.* Translated by Max Eastman. 3 vols. in 1. Ann Arbor, 1960 (originally published 1932).

———. *Lenin: Notes for a Biographer.* Translated by Tamara Deutscher, with an Introduction by Bertram D. Wolfe. New York, 1971.

———. *Moia zhizn', Opyt avtobiografii.* Berlin, 1930.

———. *Sochineniia.* Vol. XVII. Moscow-Leningrad, 1926.

Trudy komissii po obsledovaniiu sanitarnykh posledstvii voiny 1914–1917 godov. Moscow-Leningrad, 1923.

Trukan, G. A. *Oktiabr' v Tsentral'noi Rossii.* Moscow, 1967.

Trukhanovskii, V. G. (ed.). *Mezhdunarodnye otnosheniia: Politika, diplomatiia XVI–XX veka. Sbornik statei k 80-letiiu akademika I. M. Maiskogo.* Moscow, 1964.

Tseretelli, Irakli. "Reminiscences of the February Revolution: The April Crisis," *The Russian Review,* XIV, No. 2 (April 1955), pp. 91–108; XIV, No. 3 (July 1955), pp. 184–200; XIV, No. 4 (October 1955), pp. 301–321; XV, No. 1 (January 1956), pp. 37–48.

Tuchman, Barbara. *The Guns of August.* New York, 1962.

———. *The Proud Tower: A Portrait of the World Before the War, 1890–1914.* New York, 1966.

Tyrkova-Williams, Ariadne. *From Liberty to Brest-Litovsk: The First Year of the Russian Revolution.* London, 1919.

"Ubiistvo Rasputina. Ofitsial'noe doznanie." *Byloe*, No. 1 (July 1917), pp. 64–83.

Ulam, Adam B. *The Bolsheviks: The Intellectual and Political History of the Triumph of Communism in Russia*. New York, 1965.

——. *Stalin: The Man and His Era*. New York, 1973.

Urlanis, B. Ts. *Voiny i narodonaselenie Evropy: Liudskie poteri vooruzhennykh sil evropeiskikh stranv voinakh XVII–XXvv. istoriko-statisticheskoe issledovanie*. Moscow, 1960.

Usievich, E. *Vladimir Maiakovskii: Ocherk zhizni i tvorchestva*. Moscow, 1950.

Utkin, A. I. "Ekonomicheskaia polozhenie moskovskikh rabochikh posle pervoi russkoi revoliutsii," *Vestnik Moskovskogo gosudarstvennogo universiteta*, No. 1 (1974), pp. 41–53.

Valentinov, Nikolai [N. V. Volskii]. *Encounters with Lenin*. London, 1968.

Vasiukov, V. S. *Predystoriia interventsii, fevral' 1917–mart 1918*. Moscow, 1968.

"Vechernee zasedanie 5/18 ianvaria 1918g.," in Trotskii, *Sochineniia*, XVII, pt. 1, pp. 51–65.

"Vedomost' predpriiatiiam goroda Petrograda, rabotaiushchim na gosudarstvennuiu oboronu, masterovye koikh prekratili raboty 14-go fevralia 1917 goda," TsGIAL, fond 1405, opis' 530, delo No. 953/40–41.

"Vedomost' promyshlennykh predpriiatii Petrograda, bastovavshikh 9 ianvaria 1917g., sostavlennaia Petrogradskim okhrannym otdeleniem," in Korablev (ed.), *Rabochee dvizhenie v Petrograde v 1912–1917gg.*, pp. 523–526.

Vetoshnikov, L. V. *Brusilovskii proryv. Operativno-stratigecheskii ocherk*. Moscow, 1940.

"Viacheslav Napoleonovich Klembovskii," *Sovetskaia istoricheskaia entsiklopediia*, VII. col. 419.

Vishnevskii, Vsevolod. "Moi vospominaniia (1914–1921gg.)," in *Sobranie sochinenii*, vol. 2.

——. *Sobranie sochinenii v piati tomalch*. Moscow, 1954.

Vitte, S. Iu. *Vospominaniia*. 3 vols., Moscow, 1960.

—— [Witte]. *The Memoirs of Count Witte*. Translated by Abraham Yarmolinsky. New York, 1967.

Vladimirova, V. *Kontr-revoliutsiia v 1917g. (Kornilovshchina)*. Moscow, 1924.

Volobuev, P. B. *Ekonomicheskaia politika vremennogo pravitel'stva*. Moscow, 1962.

——. *Proletariat: burzhuaziia Rossii v 1917g.* Moscow, 1964.

Voznesenskii, A. N. *Moskva v 1917 godu*. Moscow-Leningrad, 1928.

"Vozzvanie," in Kakurin (ed.), *Razlozhenie armii*, pp. 68–69.

"Vozzvanie Petrogradskogo Soveta rabochikh i soldatskikh deputatov o prekrashchenii brataniia s protivnikom," 30 aprelia 1917g., in Gaponenko (ed.), *Revoliutsionnoe dvizhenie v russkoi armii*, pp. 76–79.

"Vsepoddanneishii doklad ministra torgovli i promyshlennosti, Kn. V. Shakhovskoi, 20 fevralia 1917g," TsGIAL, fond 40, opis' 1, delo No. 73.

"Vyderzhka iz doklada o poezdke na front chlenov Gosudarstvennoi Dumy Maslennikova i P. M. Shmakova," 11–19 aprelia 1917g., in Kakurin (ed.), *Razlozhenie armii*, pp. 50–57.

"Vypiska iz soobshchenii Volynskago Gubernatora ot 4 marta 1916 goda, za No. 1615," TsGIAL, fond 1291, opis' 132, delo No. 397–1916.

Viroubova [Vyrubova]. *Memories of the Russian Court*. New York, 1923.

Wade, Rex. A. *Red Guards and Workers Militias in the Russian Revolution*. Stanford, 1984.

——. *The Russian Search for Peace, February–October 1917*. Stanford, 1969.

Walling, William E. *The Socialists and the War*. New York, 1915.

Warth, Robert D. *The Allies and the Russian Revolution from the Fall of the Monarchy to the Peace of Brest-Litovsk*. Durham, N.C., 1954.

Washburn, Stanley. *Field Notes from the Russian Front*. London, 1915.

Der Weltkrieg 1914 bis 1918. Die militärischen operationen zu Lande. Bearbeitet im Reichsarchiv. 14 vols. Berlin, 1925–1944.

Wheeler-Bennett, John W. *The Forgotten Peace: Brest-Litovsk, March 1918.* New York, 1939.

White, James D. "The Kornilov Affair: A Study in Counter-Revolution," *Soviet Studies,* XX, No. 2 (October 1968), pp. 187–205.

Whitlock, Brand. *Belgium: A Personal Narrative.* 2 vols. New York, 1919.

Wildman, Allan K. *The End of the Russian Imperial Army: The Old Army and the Soldiers' Revolt, March–April 1917.* Princeton, 1980.

———. *The Making of a Workers' Revolution: Russian Social Democracy, 1891–1903.* Chicago, 1967.

Williams, Albert Rhys. *Journey into Revolution: Petrograd, 1917–1918.* Edited by Lucita Williams. Chicago, 1969.

Wilson, Colin. *Rasputin and the Fall of the Romanovs.* New York, 1964.

Wilton, Robert. *Russia's Agony.* London, 1918.

Wolfe, Bertram D. *Three Who Made a Revolution: A Biographical History.* New York, 1948.

Woroszylski, Wiktor. *The Life of Mayakovsky.* Translated from the Polish by Boleslaw Taborski. New York, 1970.

Woytinsky [Voitinskii], W. S. *Stormy Passage: A Personal History Through Two Russian Revolutions to Democracy and Freedom: 1905–1960,* with an Introduction by Adolf A. Berle. New York, 1961.

Wrangel, P. N. *The Memoirs of General Wrangel.* Translated by Sophie Goulston. London, 1930.

Yaney, George. *The Urge to Mobilize: Agrarian Reform in Russia, 1861–1930.* Urbana, Ill., 1982.

Youssoupoff [Iusupov], Prince Felix. *Rasputin.* Translated from the Russian by Oswald Rayner. New York, 1928.

———. *Avant l'exil, 1887–1919.* Paris, 1952.

Zagorsky, S. O. *State Control of Industry in Russia During the War.* New Haven, 1928.

"Zaiavlenie nachal'nika frantsuzskoi voennoi missii pri shtabe verkh. glavnokomanduiushchego ot 12 noiabria 1917g. na imia gen. Dukhonina o nepriznanii Frantsiei vlasti Soveta Narodnykh Komissarov," in "Nakanune peremiriia," pp. 214–215.

Zaionchkovskii, A. M. *Kampaniia 1917 goda* [volume VII of *Stratigecheskii ocherk voiny 1914–1918gg.*]. Moscow, 1923.

———. *Mirovaia voina 1914–1918gg.* 2 vols. Moscow, 1938.

———. *Podgotovka Rossii k imperialisticheskoi voine. Ocherki voennoi podgotovki i pervonachal'nykh planov.* Moscow, 1926.

———. "Vsepoddanneishaia zapiska *Generala A. M. Zaionchkovskogo* 20 oktiabria 1916g." Edited by S. Budkevich. *Krasnyi arkhiv.* LVIII (1933), pp. 24–45.

Zaionchkovskii, P. A. *Samoderzhavie i russkaia armiia na rubezhe XIX–XX stoletii.* Moscow, 1973.

"Zakliuchenie voennogo prokurora ob "edinennogo suda 12-i armii D. M. Matiasa po delu ob otkaze soldat 1-go batal'ona 17-go Sibirskogo strelkovogo polka idti v nastuplenie v dekabre 1916g.," in Sidorov (ed.), *Revoliutsionnoe dvizhenie,* pp. 270–275.

"Zapis' razgovora po priamomu provodu general-kvartirmeistera Shtaba verkhovnogo glavnokomanduiushchego generala M. K. Diterikhsa s generalom dlia poruchenii B. A. Levitskim o khode vooruzhennogo vosstaniia v g. Petrograde," 25 oktiabria 1917g., in Golikov (ed.), *Oktiabr'skoe vooruzhennoe vosstanie v Petrograde,* p. 340.

"Zapis' razgovora po primomu provodu ministra-predsedatelia A. F. Kerenskogo s verkhovnym glavnokomanduiushchim generalom L. G. Kornilovym o podtverzhdenii polnomochii V. N. L'vova na vedenie peregovorov o gosudarstvennom perevorote," 26 avgusta 1917g., in Chugaev (ed.), *Revoliutsionnoe dvizhenie v Rossii v avguste 1917g.* Moscow, 1959, No. 443, pp. 443–444.

"Zapis' razgovora po priamomu provodu nachal'nika Shtaba verkhovnogo glavnokomandu-iushchego generala N. N. Dukhonina s general kvartirmeistrom shtaba Severnogo fronta generalom V. L. Baranovskim i komissarom Vremennogo pravitel'stva na Severnom fronte V. S. Voitinskim o polozhenii na fronte," 26–27 oktiabria 1917g., in Golikov (ed.), *Oktiabr'skoe vooruzhennoe vosstanie v Petrograde*, No. 821, pp. 603–609.

"Zapis' razgovora po priamomu provodu nachal'nika Shtaba verkhovnogo glavnokomandu-iushchego generala N. N. Dukhonina s glavnokomanduiushchim Severnym frontom generalom V. A. Cheremisovym i nachal'nikom shtaba Severnogo fronta generalom S. G. Lukirskim o vooruzhennom vosstanii v g. Petrograde i otmene prikaza o dvizhenii voisk na g. Petrograd," 25 oktiabria 1917g., in Golikov (ed.), *Oktiabr'skoe vooruzhennoe vosstanie v Petrograde*, No. 591, pp. 411–413.

"Zapisrazgovora po priamomu provodu nachal'nika shtaba verkhovnogo glavnokomandu-iushchego generala N. N. Dukhonina s komissarom Vremennogo pravitel'stva pri Stavke V. B. Stankevichem o khode vooruzhennogo vosstanie v g. Petrograde," 25 oktiabria 1917g., in Golikov (ed.), *Oktiabr'skoe vooruzhennoe vosstanie v Petrograde*, pp. 409–410.

"Zapis' razgovora, sostoiavshegosia 23 avgusta 1917g. v Stavke mezhdu upravliaiushchim Voennym i Morskim ministerstvami B. V. Savinkovym i verkhovnym glavnokomandu-iushchim generalom L. G. Kornilovym, o napravlenii v Petrograde konnogo korpusa i izmenenii sostava Vremennogo pravitel'stva za shchet usileniia v nem kontrrevoliutsion-nykh elementov," in Chugaev (ed.), *Revoliutsionnoe dvizhenie v Rossii v avguste 1917g.*, No. 419, pp. 421–423.

"Zapiska otdeleniia po okhraneniiu obshchestvennoi bezopastnosti i poriadka v stolitse," 14 fevralia 1917g., TsGIAL, fond 1405, opis' 530, delo No. 953.

"Zasedanie TsK RSDRP, 18 fevralia 1918g. (vecherom)," in *Protokoly tsentral'nogo komiteta*, pp. 200–205.

"Zasedanaie TsK RSDRP, 22 fevralia 1918g.," in *Protokoly tsentral'nogo komiteta*, pp. 206–208.

"Zasedanie TsK RSDRP, 23 fevralia 1918g.," in *Protokoly tsentral'nogo komiteta*, pp. 211–215.

"Zasedanie voennogo soveshchaniia 1 fevralia 1917g.," in "Konferentsiia soiuznikov v Petro-grade v 1917 godu," *Krasnyi arkhiv*, XX (1927), pp. 42–55.

Zhitkov, N. "Prodfurazhnoe snabzhenie russkikh armii v mirovuyu voinu," *Voenno-istori-cheskii zhurnal*, No. 12 (December 1940), pp. 65–81.

"Zhurnal obshchago prisutstviia Akmolinskago oblastnago pravleniia," 21 ianvaria 1916g, TsGIAL, fond 1291, opis' 32, delo No. 397–1916.

Zlokazov, G. I. *Petrogradskii Sovet Rabochikh i Soldatskikh Deputatov v period mirnogo razvitiia revoliutsii: fevral'–iiun' 1917g.* Moscow, 1969.

Znamenskii, O. N. *Iiul'skii krizis 1917 goda.* Moscow-Leningrad, 1964.

Zubov, N. *F. E. Dzerzhinskii. Biografiia.* Moscow, 1963.

INDEX